CONTROL OF COMPLEX SYSTEMS

To my parents, George and Evgenia. Your love and support was and will continue to be my biggest inspiration and source of strength

—Kyriakos G. Vamvoudakis

'To my family'

—Sarangapani Jagannathan

CONTROL OF COMPLEX SYSTEMS

Theory and Applications

Edited by

KYRIAKOS G. VAMVOUDAKIS
Blacksburg, VA, United States

SARANGAPANI JAGANNATHAN
Rolla, MO, United States

ELSEVIER

AMSTERDAM • BOSTON • HEIDELBERG • LONDON
NEW YORK • OXFORD • PARIS • SAN DIEGO
SAN FRANCISCO • SINGAPORE • SYDNEY • TOKYO
Butterworth-Heinemann is an imprint of Elsevier

B
H

Butterworth-Heinemann is an imprint of Elsevier
The Boulevard, Langford Lane, Kidlington, Oxford OX5 1GB, UK
50 Hampshire Street, 5th Floor, Cambridge, MA 02139, USA

Library of Congress Cataloging-in-Publication Data
A catalog record for this book is available from the Library of Congress

British Library Cataloguing-in-Publication Data
A catalogue record for this book is available from the British Library

ISBN: 978-0-12-805246-4

For information on all Butterworth Heinemann publications visit our website at https://www.elsevier.com/

Publisher: Joe Hayton
Acquisition Editor: Sonnini R. Yura
Editorial Project Manager: Mariana Kühl Leme
Editorial Project Manager Intern: Ana Claudia Abad Garcia.
Production Project Manager: Sruthi Satheesh
Cover Designer: Matthew Limbert

Typeset by SPi Global, India

CONTENTS

CONTRIBUTORS

C.T. Abdallah
The University of New Mexico, Albuquerque, NM, United States

P.J. Antsaklis
University of Notre Dame, Notre Dame, IN, United States

J.M. Bailey
Northeast Georgia Medical Center, Gainesville, GA, United States

T. Bian
New York University, Brooklyn, NY, United States

S. Bogdan
University of Zagreb, Zagreb, Croatia

M. Car
University of Zagreb, Zagreb, Croatia

S. Commuri
University of Oklahoma, Norman, OK, United States

D.A. Copp
University of California, Santa Barbara, CA, United States

P.J. Cruz
The University of New Mexico, Albuquerque, NM, United States

W.E. Dixon
University of Florida, Gainesville, FL, United States

R. Fierro
The University of New Mexico, Albuquerque, NM, United States

E. Garcia
Infoscitex Corporation, Wright-Patterson AFB, OH, United States

W.M. Haddad
Georgia Institute of Technology, Atlanta, GA, United States

R. Harvey
University of Central Florida, Orlando, FL, United States

G. Heredia
University of Seville, Seville, Spain

G. Herrmann
University of Bristol, Bristol, United Kingdom

J.P. Hespanha
University of California, Santa Barbara, CA, United States

S.P. Hou
Singapore Institute of Manufacturing Technology, A*STAR, Singapore, Singapore

H. Hu
University of Virginia, Charlottesville, VA, United States

J. Huang
The Chinese University of Hong Kong, Shatin, Hong Kong

A. Ivanović
University of Zagreb, Zagreb, Croatia

S. Jagannathan
Missouri University of Science and Technology, Rolla, MO, United States

Z.-P. Jiang
New York University, Brooklyn, NY, United States

R. Kamalapurkar
University of Florida, Gainesville, FL, United States

B. Kiumarsi
University of Texas at Arlington Research Institute, Fort Worth, TX, United States

A.C. Kizilkale
Polytechnique Montreal; Research Group in Decision Analysis (GERAD), Montreal, QC, Canada

Z. Kovacic
University of Zagreb, Zagreb, Croatia

F.L. Lewis
University of Texas at Arlington Research Institute, Fort Worth, TX, United States

Z. Lin
University of Virginia, Charlottesville, VA, United States

D. Liu
School of Automation and Electrical Engineering, University of Science and Technology Beijing, Beijing, China

B. Luo
The State Key Laboratory of Management and Control for Complex Systems, Institute of Automation, Chinese Academy of Sciences, Beijing, China

A. Mai
University of Oklahoma, Norman, OK, United States

R.P. Malhamé
Polytechnique Montreal; Research Group in Decision Analysis (GERAD), Montreal, QC, Canada

N. Meskin
Qatar University, Doha, Qatar

H. Modares
University of Texas at Arlington Research Institute, Fort Worth, TX, United States

K.H. Movric
Czech Technical University in Prague (CVUT), Czech Republic

J. Na
Kunming University of Science and Technology, Kunming, China

K.S. Narendra
Yale University, New Haven, CT, United States

M. Nasiruddin Mahyuddin
University of Science, Malaysia, Pulau Pinang, Malaysia

R.W. Newcomb
University of Maryland, College Park, MD, United States

A. Ollero
University of Seville, Seville, Spain

I. Palunko
University of Dubrovnik, Dubrovnik, Croatia

F. Petric
University of Zagreb, Zagreb, Croatia

Z. Qu
University of Central Florida, Orlando, FL, United States

S. Ramazani
Louisiana Tech University, Ruston, LA, United States

A. Sahoo
The DEI Group, Millersville, MD, United States

M. Sebek
Czech Technical University in Prague (CVUT), Czech Republic

R.R. Selmic
Louisiana Tech University, Ruston, LA, United States

R. Song
School of Automation and Electrical Engineering, University of Science and Technology Beijing, Beijing, China

D. Tolić
University of Zagreb, Zagreb, Croatia

K.G. Vamvoudakis
Virginia Tech, VA, United States

S. Vera
University of Seville, Seville, Spain

P. Walters
University of Florida, Gainesville, FL, United States

Q. Wei
The State Key Laboratory of Management and Control for Complex Systems, Institute of Automation, Chinese Academy of Sciences, Beijing, China

S.Y. Yoon
University of New Hampshire, Durham, NH, United States

M. de Queiroz
Louisiana State University, Baton Rouge, LA, United States

ABOUT THE EDITORS

Kyriakos G. Vamvoudakis was born in Athens, Greece. He received his diploma (a 5-year degree, equivalent to an MS degree) in electronic and computer engineering from the Technical University of Crete, Greece, in 2006 with the highest honors. After moving to the United States, he studied at the University of Texas with Frank L. Lewis as his advisor, and he received his MS and PhD degrees in electrical engineering in 2008 and 2011 respectively. From May 2011 to January 2012 he worked as an adjunct professor and faculty research associate at the University of Texas at Arlington and at the Automation and Robotics Research Institute. During the period from 2012 to 2016 he was a project research scientist at the Center for Control, Dynamical Systems and Computation at the University of California, Santa Barbara. He is now an assistant professor in the Department of Aerospace and Ocean Engineering at Virginia Polytechnic Institute and State University. His research interests include approximate dynamic programming, game theory, nonlinear control, and optimal control. Recently, his research has focused on network security, smart grid, and multiagent optimization. He is the recipient of several international awards, including the 2016 International Neural Network Society Young Investigator Award, the Best Paper Award for Autonomous/Unmanned Vehicles at the 27th Army Science Conference in 2010, the Best Presentation Award at the World Congress of Computational Intelligence in 2010, and the Best Researcher Award from the Automation and Robotics Research Institute in 2011. He is a member of the honor societies Tau Beta Pi, Eta Kappa Nu, and Golden Key, and is listed in *Who's Who in the World*, *Who's Who in Science and Engineering*, and *Who's Who in America*. He has served on various international program committees and has organized special sessions for several international conferences. He is currently a member of the Technical Committee on Intelligent Control of the IEEE Control Systems Society, a member of the Technical Committee on Adaptive Dynamic Programming and Reinforcement Learning of the IEEE Computational Intelligence Society, an associate editor of the *Journal of Optimization Theory and Applications*, an associate editor on the IEEE Control Systems Society Conference Editorial Board, Editor in Chief of *Communications in Control Science and Engineering*, a registered electrical/computer engineer (professional engineer), and a member of the Technical Chamber of Greece. He is a Senior Member of the IEEE.

Sarangapani Jagannathan is at the Missouri University of Science and Technology (formerly University of Missouri-Rolla), where he is Rutledge-Emerson Endowed Chair Professor of Electrical and Computer Engineering and Site Director for the NSF Industry/University Cooperative Research Center on Intelligent Maintenance Systems. His research interests include neural network control, adaptive event-triggered control, secure networked control systems, prognostics, and autonomous systems/robotics. He has coauthored 138 peer-reviewed journal articles, with many in various IEEE transactions, 245 refereed IEEE conference articles, and several book chapters, has authored/coedited 5 books, has received 19 US patents, 1 patent defense publication and several pending. He has supervised to graduation 23 doctoral and 30 MS-level students, and his total funding is in excess of $15.9 million from various US federal and industrial members, with more than $9.8 million to his shared credit. He was a coeditor of the Institution of Engineering and Technology book series on control from 2010 through 2013, and now serves on many editorial boards. He has received many awards, including a 2000 NSF CAREER Award, and has been on organizing committees for several IEEE conferences. He has served as Chair of the IEEE Control Systems Society Technical Committee on Intelligent Control. He is a Fellow of the IEEE, a Fellow of the Institute of Measurement and Control, United Kingdom, and a Fellow of the Institution of Engineering and Technology, United Kingdom.

PREFACE

The book presents a variety of state-of-the-art methods for the control of complex systems. These methods cover the theory and applications of complex dynamical systems. Given the presence of modeling uncertainties, the unavailability of the model, and the possibility of cooperative/non-cooperative goals and malicious attacks compromising the security of networked systems, there is a need for novel approaches to respond to situations not programmed or anticipated in the design. This book has five sections. The first section provides an introduction to complex systems and background on the control mechanisms that will be used in the subsequent sections of the book to solve several problems. The second section provides techniques from adaptive control and neuroscience. Several adaptive learning algorithms from different perspectives are presented in the third section. Cyber-physical systems are an example of complex dynamical systems, and cooperative control for such systems is included as a collection of chapters in the fourth section. Finally, the fifth section provides real-world examples of complex systems.

In the first section of the book, we summarize in this preface the main contributions from each chapter to put them in context. In Chapter 1 we provide background on complex systems and several feedback control techniques that will be used in the book. In the second section of the book, Chapter 2 is by Narendra, who along with his colleagues during the last 3 decades has attempted to use a general method based on multiple models to control rapidly time-varying plants. More specifically, they have proposed four distinct methods based on (1) switching, (2) switching and tuning, (3) interactive/evolutionary adaptation, and (4) second level adaptation. In this chapter, these four methods are critically examined. Since many of the problems considered were formulated only recently, they are open-ended and hopefully will be of interest to a wide audience, including both beginners and experts. Chapter 3, by Garcia and Antsaklis, considers the problem of stabilization of uncertain systems where feedback measurements are transmitted through a limited-bandwidth communication network. An adaptive model-based event-triggered control framework is presented where a model of the plant dynamics is implemented in the controller node which provides estimates of the plant states between event-based measurement updates. Further, a sensor collocated Kalman filter is employed to estimate the plant parameters, which are also transmitted to the controller node to update the plant model. The adaptive model-based event-triggered control framework provides a significant improvement in terms of network utilization by allowing longer time intervals between updates. This chapter focuses on several applications of adaptive model-based event-triggered control framework to stabilize uncertain dynamical systems by estimating unknown system parameters and disturbances while minimizing network communication. The last chapter in the second section (Chapter 4) is by Haddad, Hou, Bailey, and Meskin and uses dynamical system theory to develop a mechanistic mean field model for neural activity to study the anesthetic cascade. The proposed synaptic drive firing rate model predicts the conscious-to-unconscious transition as the implied anesthetic concentration increases, where excitatory neural activity is characterized by a Poincaré-Andronov-Hopf bifurcation with the awake-state transitioning to a stable limit cycle and then subsequently to an asymptotically stable unconscious equilibrium state. This chapter concludes with several numerical experiments to illustrate the different aspects of the proposed theory.

In the third section of the book, Kiumarsi, Modares, and Lewis present adaptive solutions to the optimal tracking problem of nonlinear discrete-time and continuous-time systems. These methods find the optimal feedback and feedforward parts of the control input simultaneously, without requiring complete knowledge of the system dynamics. First, an augmented system composed of the error system dynamics and the reference trajectory dynamics is formed to introduce a new nonquadratic discounted performance function for the optimal tracking control problem. This encodes the input constrains caused by the actuator saturation into the optimization problem. Then the tracking Bellman equation and the tracking Hamilton-Jacobi-Bellman equation for both discrete-time and continuous-time systems are derived. Finally, to obviate the requirement of complete knowledge of the system dynamics in finding the Hamilton-Jacobi-Bellman solution, integral reinforcement learning and off-policy reinforcement learning algorithms are developed for continuous-time systems, and a reinforcement learning algorithm on an actor-critic structure is developed for discrete-time systems. The chapter that follows (Chapter 6) is by Copp and Hespanha and considers the estimation and control of systems with parametric uncertainty by combining moving-horizon estimation and model predictive control into a single min-max optimization. This is done by inclusion of the state, inputs, and uncertain parameters as optimization variables. Learning the true values of the uncertain parameters requires a sufficiently large number of past measurements and that the system be persistently exciting. The true values of the uncertain parameters may change over time, and the optimization computes future control inputs that adapt to changing estimates of the uncertain parameters in order to better control the uncertain system. The chapter concludes with several linear and nonlinear examples with parametric uncertainties. In Chapter 7, Bian and Jiang present preliminary results on the development of a stochastic adaptive dynamic programming theory for the data-driven, model free-based design of robust optimal controllers for continuous-time stochastic systems. Two types of optimal control problems—the discounted problem and the biased problem—are investigated. Reinforcement learning and adaptive dynamic programming techniques are employed to design stochastic adaptive optimal controllers through online successive approximations of optimal solutions, and three practical examples arising from biological motor control and vehicle suspension control are illustrated. In Chapter 8, in a similar vein to Chapter 7, Kamalapurkar, Walters, and Dixon develop a model-based reinforcement learning algorithm to solve approximate optimal regulation problems online under a persistent excitation-like rank condition. The development is based on the observation that, given a model of the system, model-based reinforcement learning can be implemented by evaluation of the Bellman error at any number of desired points in the state space. Chapter 9 is by Wei, Song, Liu, and Luo, who propose a heterogeneous policy iteration-based adaptive dynamic programming algorithm to solve the optimal distributed coordination control problems for multiagent differential graphical games. Convergence analysis is performed to show that the iterative value functions of heterogeneous multiagent differential graphical games can converge to the Nash equilibrium. The final chapter in this section (Chapter 10) is by Vamvoudakis. This chapter is motivated by emerging networking applications characterized by measurement attacks, time-varying connectivity, jamming attacks, and lack of centralized control and access to information. Although the agents in the team can be endowed with behaviors designed a priori, they often need to learn new behaviors, such as how the agent, or the entire team, will gradually improve its performance. The algorithm is robust against changes in measurements and topology, and has to guarantee optimal performance at all times. For this reason the cost functions are defined for the case of measurement attacks as a zero-sum game problem for each agent and a non-zero-sum game for all the agents in the team, and for the jamming attacks as a max-min time problem, where there is a team of agents who want to maximize the time they are connected and a jammer who wants to minimize this time.

In the fourth section of the book, which is titled "Networked Systems and Cooperative Control," there are eight chapters. The first chapter in this section (Chapter 11) is by Sahoo and Jagannathan, who propose a novel approximation-based event-triggered control of multiinput-multioutput uncertain nonlinear continuous-time systems in affine form. The controller is approximated by use of a linearly parameterized neural network in the context of event-based sampling. After the neural network approximation property has been revisited in the context of event-based sampling, a stabilizing control scheme is introduced first and subsequently an optimal regulator is designed by use of neural networks. A suite of novel weight update laws for tuning the neural network weights at the aperiodic event-trigger or sampling instants is proposed to relax the requirement of knowledge of the complete system dynamics and reduce the computation. In Chapter 12, Tolić, Palunko, Ivanović, Car, and Bogdan compute transmission rates and delays that provably stabilize multiagent systems in the presence of disturbances and noise. Specifically, to consider multiagent systems characterized by sets of equilibrium points, the notions of Lp-stability and Lp-detectability with respect to a set are employed. The computed transmission rates are experimentally verified for a group of of-the-shelf quadcopters. Chapter 13, by Ramazani, Selmic, and de Queiroz, considers the problem of formation acquisition for a group of n agents that belong to a plane Q and r coordinators that move in a plane W outside the plane Q. Graph coning and triangulation concepts are used to create a three-dimensional framework that retains the infinitesimal and minimal rigidity characteristics of the framework of n planar agents. The chapter that follows (Chapter 14) is by Huang, and presents a more general formulation of the cooperative output regulation problem for linear multiagent systems that includes certain existing versions of the cooperative output regulation problem as special cases. In Chapter 15, Cruz, Fierro, and Abdallah focus on a multirobotic network made of aerial relays and ground sensors that are deployed to sense areas of interest in an environment populated with obstacles. They use potential field methods to guide the ground mobile sensors to target regions. On the other hand, antenna diversity and reinforcement learning are combined to control the flying relays such that they increase the received signal strength among themselves and the mobile sensors. The authors conclude by demonstrating the effectiveness of their control strategy through simulation tests contrasting independent and cooperative learning. In Chapter 16, Hu, Yoon, and Lin study the flocking control problem for a group of nonholonomic vehicles with discrete-time dynamics and in the presence of an arbitrarily large communication delay. A potential function-based low-gain control law is designed for each agent. The effectiveness of the controller is demonstrated by numerical examples. On a similar note, Chapter 17, by Harvey and Qu, investigates input-output stability analysis tools and output-feedback networked control design for linear and nonlinear dynamical systems. Specifically, the chapter focuses on two classes of passivity-short systems: input feedforward passive systems and output feedback passive systems. The last chapter in this section (Chapter 18), by Hengster Movric and Sebek, presents sufficient conditions for cooperative stability analysis and design schemes for various identical agent multiagent system state synchronization distributed controls. Cooperative state feedback, observer design, static output feedback, and delayed state feedback are addressed.

The final section of the book contains five chapters that are more related to applications. Namely, in Chapter 19, Newcomb extends the stereographic circle operation to complex-valued functions and show that $z = y_1 \circ y_2$ is positive real when y_1 and y_2 are. Several physical interpretations are given, including z being the input impedance when an admittance-based Richards section of parameter y_1 is loaded in the admittance y_2. Kizilkale and Malhamé in Chapter 20 propose a mean field game theory-based control architecture as a load control mechanism to limit the required flows of information, and produce local-constraints conscious decentralized individual controls which aggregate to a desired mean behavior. The corresponding system of mean field Nash equilibrium-inducing equations is developed, and existence and uniqueness properties and numerical simulation results are presented. In Chapter 21, Vera, Petric, Heredia, and Kovacic present a new optimal trajectory planning method based on pseudospectral collocation methods which are especially suitable for multivehicle problems. They consider several kinds of applications and scenarios with multiple aerial and ground vehicles. Experimental results with teams of unmanned aerial vehicles and unmanned ground vehicles are provided to validate the multivehicle trajectories obtained by the methods. In Chapter 22, Mai and Commuri propose a direct neural dynamic programming-based learning scheme to dynamically adapt and improve the performance of a prosthetic ankle in real time. In this approach the measured gait data from the subject are used to recognize the amputee's desired gait and thereby specify a displacement profile for the prosthetic joint. The direct neural dynamic programming controller then adaptively compensates for the unmodeled dynamics and disturbances to achieve closed-loop stability with guaranteed tracking and optimized performance. The performance of the controller is studied through numerical simulations in the presence of noises, uncertainties in terrain interaction, disturbance torques, variations in gait parameters, and changes in gait. The last chapter in this section and of the book (Chapter 23) is by Herrmann, Na, and Nasiruddin Mahyuddin. This chapter provides a novel continuous-time adaptive parameter estimation algorithm which is robust to disturbances. This is shown in theoretical and practical analysis. The novel algorithm has been proven to work as a parameter estimation algorithm in connection with an observer in a vehicular context and together with a demand-tracking controller for a robotic arm. In both cases the novel algorithm has provided superior results in terms of parameter estimation and control performance in comparison with the traditional least squares method or the gradient descent algorithm. Moreover, it is shown that the algorithm lends itself easily to adaptive optimal control.

In summary, this book presents a variety of challenging theoretical problems coupled with real practical applications for complex dynamical systems. The future directions are also covered in individual chapters.

As a final note, we dedicate this volume to Frank L. Lewis, on the occasion of his 65th birthday, for his seminal contributions to the field of control. Frank is the author of 6 US patents, 327 journal articles, 411 conference papers, 15 books, including *Optimal Control*, *Optimal Estimation*, *Applied Optimal Control and Estimation*, *Aircraft Control and Simulation*, *Control of Robot Manipulators*, *Neural Network Control*, *High-Level Feedback Control With Neural Networks*, and the IEEE reprint volume *Robot Control*, 44 chapters, and edited 11 journal special issues. Frank is a fellow of the IEEE, the International Federation on Automatic Control, the UK Institute of Measurement and Control, and the National Academy of Inventors, he is a registered professional engineer (Texas), a chartered engineer (United Kingdom), and is a Distinguished Scholar Professor, a Distinguished Teaching Professor, and Moncrief-O'Donnell Chair at the University of Texas at Arlington Research Institute. Frank has inspired each of us to work on challenging control problems that can impact several areas in engineering. We hope you enjoy reading this volume as much as we appreciated the opportunity to edit it.

Kyriakos G. Vamvoudakis
Blacksburg, VA, United States

Sarangapani Jagannathan
Rolla, MO, United States

SECTION 1

Introduction and Preface

CHAPTER ONE

Introduction to Complex Systems and Feedback Control

K.G. Vamvoudakis*, S. Jagannathan†
*Virginia Tech, VA, United States
†Missouri University of Science and Technology, Rolla, MO, United States

1 INTRODUCTION

The meaning of the word *complex* is different from that of the word *complicated*. One should think of the most elaborate mechanical watches—for example, the *Henry Graves Supercomplication*, which has 920 individual parts. The pieces of such complicated systems can be well understood only in isolation, and the whole can be reassembled from its parts. In such a system the components work in unison to accomplish a function. But one key defect may bring the entire system to an halt. Redundancy needs to be built in when system failure is not an option.

The term *complex systems* [1], or *complexity science*, is often used to describe the loosely organized academic field that has grown up around the study of such systems. What distinguishes a complex system from a pure multiagent system is the focus on top-level properties and features such as self-similarity, complexity, and self-organization. A multiagent system is defined as a system composed of multiple interacting agents, whereas in complex systems the agents and the system exhibit complicated behavior and self-similar collectivity of interacting agents. Such systems can have feedback loops, spontaneity, and emergent and hierarchical organization, and sometimes they are characterized by a high degree of adaptive capacity, giving them resilience in the face of perturbation. The world we see around us is a world of forms at all levels from the microscopic to the macroscopic—for example, elementary particles, atoms, molecules, macromolecules, cells, plants, animals, ecosystems composed of plant and animal organisms, planets, stars, and galaxies. Cosmic forms are dynamically evolving with

Control of Complex Systems
http://dx.doi.org/10.1016/B978-0-12-805246-4.00001-X

time, while all the forms of things contained in physical reality constitute a whole, the universe, which is the most complex system.

The universe is a dynamic reality of complex grid interconnections and relationships at all levels. The individual components of a complex system interact, so the state of the original system can be determined at any time by the continuous interaction of its parts. The reductive science predominated previously argued that any complex system eventually breaks down into key ingredients and the fundamental interactions between fundamental elements described mathematically through fundamental equations.

Today science of complexity has been removed from this simple reductive perception. The old reductionism of the complex to the simple has been replaced by the fundamental principle of self-organization. According to the principle that a set can be coordinated and self-organized, this can develop multiple types of relationships and resonances. Also, a level of self-organization cannot be reduced to a lower level. This means that there is a complete overthrow of the reducing mode of interpretation of physical reality.

Other important properties of complex systems are homeostasis, communication, cooperation, spatial and temporal organization, and reproduction. Communication and cooperation occur at all levels, from the agent to the system level. The forces driving cooperation between agents in such a system, in some cases, can be analyzed with game theory [2].

We can thus conclude that complexity science is not a single theory—it encompasses more than one theoretical framework and is highly interdisciplinary, seeking the answers to some fundamental questions about living, adaptable, and evolving systems. Complexity theory is the third revolution in the field of modern science (ie, in addition to the theory of relativity and quantum theory) with major consequences for modern technology. Human technology originally evolved as material technology, while in the 21st century we have rapid evolution in information technology. In physics, a state of matter is one of the distinct forms that matter takes on. Four states of matter are observable in everyday life: solid, liquid, gas, and plasma. The first three conditions are observed at relatively low temperatures, while a plasma is observed at very high temperatures. Even material is divided into inanimate or living. In inanimate matter there is a lack of life functions such as reproduction, adaptation, consumption of matter and energy, subsistence, information processing, and consciousness/self-consciousness. Many other states are known to exist only in extreme situations, such as Bose-Einstein condensates, neutron-degenerate matter and quark-gluon plasma, which occur in situations of extreme cold, extreme density and extremely high-energy color-charged matter respectively. Some other states are believed to be possible but remain theoretical for now. Energy was initially divided into mechanical energy, electrical energy, chemical energy, etc., while in a more detailed description it is divided into motor, electromagnetic, and nuclear energy because every other form of energy can be reduced to them.

The physical theory is the basis on which the technology is developed and the technical problems are solved. Complexity theory is the basis of many important technological applications in all areas of modern technology (matter, energy, information) particularly where the modeling is difficult or impossible through prior theory. Mathematical models of complex systems are of three types: black-box (phenomenological), white-box (mechanistic, based on the first principles) and grey-box (mixtures of phenomenological and mechanistic models). In black-box models, the individual-based (mechanistic) mechanisms of a complex dynamic system remain hidden. Black-box models are completely non-mechanistic. They are phenomenological and ignore a composition and internal structure of a complex system. We cannot investigate interactions of subsystems of such a non-transparent model. A white-box model of complex dynamic system has 'transparent walls' and directly shows underlying mechanisms. All events at micro-, meso- and macro-levels of a dynamic system are directly visible at all stages of its white-box model evolution. In most cases mathematical modelers use the heavy black-box mathematical methods, which cannot produce mechanistic models of complex dynamic systems. Grey-box models are intermediate and combine black-box and white-box approaches.

2 TYPES OF COMPLEXITY IN NATURE

The theories of complexity appeared in the mid-20th century in the scientific fields of physics, biology, chemistry, engineering, and mathematics. But the novel science of complexity is interdisciplinary. To that extent the purpose of studying complex phenomena and their mechanisms tends to be to deal with societal problems as mentioned in [3]. Some key concepts of the theory of complexity are, "chaos," "attractors," "nonlinearities," "self-organization," and "feedback mechanisms." All these behaviors can be seen today in a wide range of areas, which extend from cosmology to ecology and the social sciences.

The science of complexity marks a significant difference in approach against the Newtonian scientific model, which focused only on binary oppositions, reductionism, linear causality, and identification of only stable equilibria. But dynamic behavior dominates the world, through complexity, randomness, and disorder, which cease to constitute marginal phenomena of science and are revealed as fundamental manifestations of nature.

2.1 Cosmology and Microphysics

Today we understand that the entire known universe [4] is a complex system that includes an endless sequence of complex phenomena from the microscopic to the macroscopic level. Initially we felt that any complex system can be broken down into simple particles and simple forms of force (eg, gravity) and that every phenomenon in the microcosm or macrocosm is the result of an interaction of elementary particles through the fundamental forces. The key feature of complexity is the fact that a complex system appears to consist of several independent subsystems while maintaining a holistic entity and revealing an unbroken unity of the parties. The modern physical space is not empty but is a dynamic complex reality, capable of expanding and producing or absorbing material particles. The elementary particles are not at the level of quarks and below, but are revealed as complex filaments with internal structure and movement. In accordance with the principle of dualism, the elementary particles are not simple even though sometimes they show local-particle and sometimes nonlocal wave disposition. Finally, other key trends are the one-sided flow of time from past to future as well as the unilateral increase of entropy or one-sided flow of heat from high to low temperatures.

2.2 Fluids

A fluid away from thermodynamic equilibrium with marked differences in temperature and speed displays an impressive amount of complexity and self-organization, such as turbulence or macroscopic harmonic oscillations/periodicity [5, 6]. It has been observed that when the temperature exceeds a threshold, then the microscopic chaos and random movement of the particles is transformed into a uniform visual circular motion so that the particles (on the order of trillions) begin to rotate uniformly with the same frequency as an invisible "conductor" to coordinate simultaneously throughout the space. Other similar phenomena have been observed in liquids in turbulence conditions or in cases of differential rotation of a fluid.

2.3 Chemistry

In chemical reactors we can observe complex chemical reactions which are self-organizing, and they are similar to those of fluids (see Section 2.2) Oscillating reaction is a complex mixture of reacting chemical compounds in which the concentration of one or more components exhibits periodic changes, or where sudden property changes occur after a predictable induction time. This phenomenon is known as a chemical clock. To understand the importance of this phenomenon, it suffices to think that the reaction of chemical species at a point in the reactor is normally independent of the way a chemical reaction happens in other parts of the reactor. This is because the chemical reaction of two molecules to synthesize new molecules is a random phenomenon of several microscopic conflicts. Again here we have an incomprehensible agent that coordinates all these chemical reactions within the entire volume of the reactor. If the system is quite far from thermodynamic equilibrium, then several random chemical reactions are happening that produce different products each time in the chemical reactor. The chemical self-organization of complex systems produces chemical structures in space and time again as if an invisible conductor coordinates every moment of the infinite number of chemical molecules within the chemical reactor.

2.4 Ecosystems

Earth's ecosystem [7] includes the physiochemical environment (ie, land, sea, air, and the plants and animals that livc and grow in it). If we recall the historical sequence that within 4 billion years led the amorphous warm land mass to today's complex form of Earth's physicochemical and biological environment, then we will understand the function of self-organization on a planetary level. In this planetary complex system, there are many complex physiochemical and biological events happening and all the environment resembles a planetary living organism.

2.5 Social Systems

While social systems are highly complex, there are still relatively simple collective behaviors that are not well understood, as stated in [8]. These include commercial fads, market cycles, and social bubbles and busts. Understanding the fluctuating dynamics and predictability of markets will continue to be a major challenge. It is important to emphasize that studies of complex systems are not necessarily about predicting the market, but are about understanding its predictability or lack thereof. More generally, there are many complex social challenges associated with complex social systems, ranging from military challenges to school/education system failures and health care errors and problems with quality of service. Moreover, other major challenges remain in our inability to address fundamental social ills such as poverty, drug abuse, and crime. It is helpful to focus on one of these, and the current military context is a convenient one to clarify some aspects of social systems from a complex systems perspective.

2.6 Biology

The chemical clocks and chemical waves produced by chemical self-organization are the beginning of our understanding the biological self-organization seen from simple protozoa up to developed complex biological organisms. The biological world consists of numerous living organisms at the level of simple unicellular organisms to the level of plants and animals that are complex multicellular organisms. The simplest living organism is the cell. Within the cell, thousands or millions of chemical or physical functions occur in a completely coordinated manner with each other so that the cell may feed and reproduce. The genetic code of life is in the heart of each cell, and is inscribed in DNA, which is a set of rules by which information is encoded within genetic material. Each different kind of living organism contains a different form of DNA. The DNA can copy itself, and thus living organisms reproduce by copying themselves. The course of the conversion of simple chemical molecules into complex macromolecules and from there into complex cells and complex organizations is a miracle of self-organization.

2.7 Human and Cognitive Systems

As human physical complexity and self-organization operation tends to another depth and this leads to the appearance of all human functions (ie, linguistic, logic, consciousness, and self-consciousness). The human brain is the most complex system in the universe [9]. It can get information from the environment and generate new information to be applied to the world. At the human level, this shows all those psychosomatic functions that seem to defy a simple scientific explanation. Human thought can describe almost all other phenomena but is unable to understand itself. Humans live in a world, but also look beyond this world, seeking the meaning of it. Finally, humans look for the causes of order and self-organization of the world at all levels.

The problem of understanding the brain and mind can be understood quite generally through the role of relationships between patterns in the world and patterns of neuronal activity and synaptic change. While the physical/biological structure of the system is the brain, the properties of the patterns identify the psychofunctioning of the mind. The relationship between external and internal patterns are further augmented by relationships between patterns within the brain. The functional role of patterns is achieved through the ability of internal patterns to represent both concrete and abstract entities and processes, ranging from the process of sensory-motor response to internal dialog. This complex nonlinear dynamic system has a great richness of valid statements that can be made about it, but an integrated understanding of the brain-mind system cannot be captured by perspectives that are limited in their approach through the particular methods of the researchers involved. Indeed, the potential contributions of the diverse approaches to studies of the brain and mind have been limited by the internal dynamics of the scientific and engineering approaches.

The study of complex systems aspects of cognitive systems, including the description of patterns in the world and patterns in mind, the description of complex systems, and the limitations on information processing that are possible for complex systems, is relevant to the application of cognitive studies to the understanding of human factors in man-machine systems, and more generally the design of systems that include both humans and computer-based information systems as functional systems. Such hybrid systems, reflect the importance of the converging technology approach.

2.8 Human Engineered Systems

All of the aforementioned systems in nature can provide insight into how to design engineered systems. Large-scale engineered systems typically consist of many concurrently operating and interacting elements—human and nonhuman—which as an ensemble either produce a result or provide a service or a set of services. Such systems exhibit similar emergent and multiscalar properties. It has become increasingly clear that systems engineering is the engineering of complex systems. This realization is now fairly widely shared, at least implicitly, and complex systems ideas, tools, and techniques have been applied to systems engineering problems for some time.

3 CHARACTERISTICS OF COMPLEX SYSTEMS

According to [10], some of the most important characteristics of complex systems are:

- The number of elements is sufficiently large that conventional descriptions (eg, a system of differential equations) are not only impractical but cease to assist in our understanding the system. Moreover, the elements interact dynamically, and the interactions can be physical or involve the exchange of information. Such interactions are rich—that is, any element or subsystem in the system is affected by and affects several other elements or subsystems.
- The interactions are nonlinear: small changes in inputs, physical interactions, or stimuli can cause large effects or very significant changes in outputs.
- Interactions are primarily but not exclusively with immediate neighbors, and the nature of the influence is modulated.
- Any interaction can feed back onto itself directly or after a number of intervening stages. Such feedback can vary in quality. This is known as *recurrency*.
- Such systems may be open, and it may be difficult or impossible to define system boundaries.
- Complex systems operate under far from equilibrium conditions. There has to be a constant flow of energy to maintain the organization of the system.
- Complex systems have a history. They evolve, and their past is co-responsible for their present behavior.
- Elements in the system may be ignorant of the behavior of the system as a whole, responding only to the information or physical stimuli available to them locally.

The common goals of the algorithms that will be developed for complex systems are given below—in the chapters that will follow, the reader will see several of these goals:

- *Performance.* The decision maker is generally trying to solve one of two problems: (1) maximize the system-of-systems performance subject to a cost constraint or (2) minimize additional cost under performance constraints.
- *Stability.* Stability theory has a strong mathematical tradition mainly through Lyapunov theory and its many extensions to allow for continuous, discrete, impulsive, hybrid, and time-delay systems. An interesting view is taken in engineering science, where the emphasis is on responses to the influence of many external impacts, such as inputs, disturbances, and interconnections.
- *Robustness.* Robustness is the ability of a system to continue to operate correctly across a wide range of operational conditions, and to fail

gracefully outside that range. In this chapter we argue against a seemingly common design paradigm that attempts to achieve robustness by predicting the conditions in which a system will operate and then carefully designing the system to operate well in those conditions.

- *Flexibility and self-healing properties.* By *flexibility* we mean the ease with which a system can change its configuration in response to a change in the environment.

4 FEEDBACK CONTROL TECHNIQUES IN COMPLEX SYSTEMS

Feedback control [11–16] is the basic mechanism by which systems, whether mechanical, electrical, or biological, maintain their equilibrium or homeostasis. For the higher life forms the conditions under which life can exist are quite narrow. Feedback control may be defined as the use of difference signals, determined by a comparison of the actual values of system variables with their desired values (because the system is always making small corrections to get to the right place), as a means of controlling a system. An everyday example of a feedback control system is automobile speed control, which uses the difference between the actual and the desired speed to vary the fuel flow rate. Since the system output is used to regulate its input, such a device is said to be a *closed-loop control system*. An *open-loop control system*, on the other hand, uses a controller or control actuator to obtain the desired response but without knowledge of the output.

Feedback is a universal concept that appears in natural systems, interaction of species, and biological systems, including the basic cell and muscle control systems in the body. Darwin [17] showed that feedback over long periods is responsible for natural selection of species. Volterra [18] showed that feedback is responsible for the interactions of predator-prey species in closed ecosystems. Adam Smith [19] showed that feedback occurs in the balance of the international economies of nations. Feedback in human engineered systems has many benefits. Using feedback, one can move the closed-loop poles so that the system performance improves. Performance can be improved in terms of fast transient response, percent overshoot, and oscillation frequency. One can obtain improved performance in the presence of disturbances, such as wind gusts in aircraft control systems. One can also select a suitable compensator to obtain good steady-state tracking by studying the steady-state error.

When the true plant parameters are unknown, the controller parameters are either estimated directly (direct schemes) or computed by the solving of the same design equations with use of plant parameter estimates (indirect schemes). Figs. 1–3 show three of the existing controller topologies. The use of feedback in control systems has a lot of advantages, including increased accuracy (eg, gets to the desired final position more accurately because small errors will be corrected in subsequent measurement cycles), less sensitivity to nonlinearities, and reduced sensitivity to noise in the input signal. But if the designer selects a high control gain, this can cause instability, it can couple noise from sensors into the dynamics of a system, and it can increase the overall complexity of a system.

Understanding complex systems does not mean that we can predict their behavior exactly, it is not just about massive databases, or massive simulations, even though these are important tools of research in complex systems. The main role of research in the study of complex systems is recognizing what we can and cannot say about complex systems given a certain level (or scale) of description, and how we can generalize across diverse types of complex systems. It is just as important to know what we can know, as to know.

In most cases, one cannot provide an accurate model of a complex system. A model that gives an exact representation of the behavior will be too complex to design a control algorithm for it.

Hence to provide tools for complex systems, an important feature of control theory is versatile modeling. One typically starts by considering a finite-dimensional linear system even if the system is nonlinear and has uncertainties. The algorithms developed need to be robust with respect to modeling and disturbances.

4.1 Continuous-Time and Discrete-Time Control Systems

Dynamical systems can be classified as discrete-time and continuous-time dynamical systems.

For a discrete-time nonlinear dynamical system [20], we denote time by $k \in \mathbb{N}_{>0}$, and the system is specified by the equations

$$x(k+1) = f(x(k)) + g(x(k))u(k), \quad x(0) = x_0,$$

where $x \in \mathbb{R}^n$ is the state of the system, $u \in \mathbb{R}^m$ is the control input, and $f : \mathbb{R}^n \to \mathbb{R}^n$ and $g : \mathbb{R}^n \to \mathbb{R}^{n \times m}$ are continuous functions of x.

For a continuous-time nonlinear dynamical system, we denote time by $k \in \mathbb{R}_{>0}$, and the system is specified by the equations

$$\dot{x} = f(x(t)) + g(x(t))u(t), \quad x(0) = x_0,$$

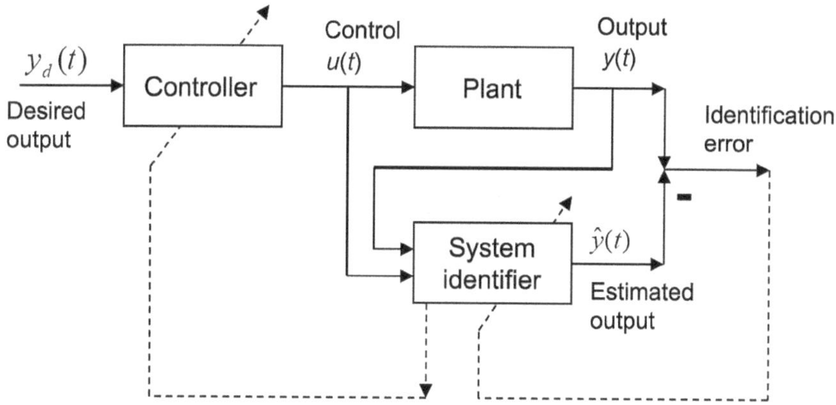

Fig. 1 Indirect control scheme.

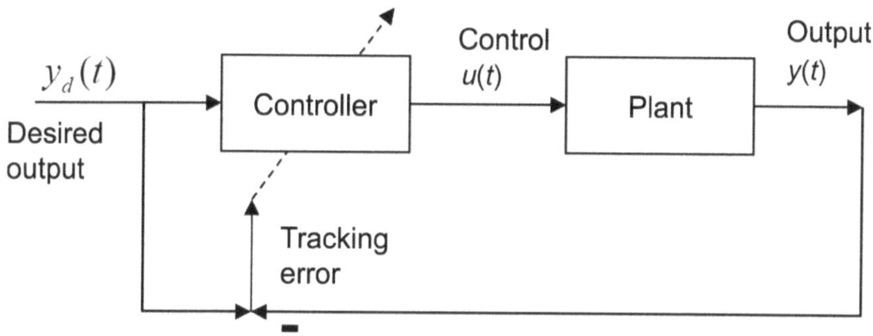

Fig. 2 Direct control scheme.

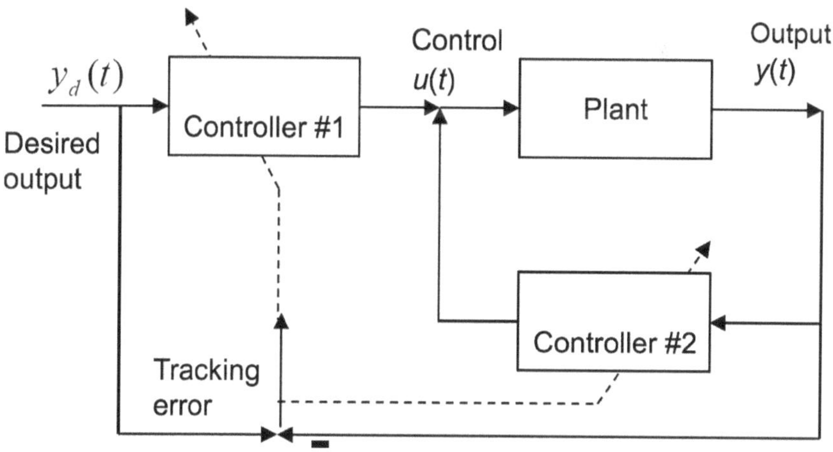

Fig. 3 Feedback-feedforward control scheme.

where $x \in \mathbb{R}^n$ is the state of the system, $u \in \mathbb{R}^m$ is the control input, and $f : \mathbb{R}^n \to \mathbb{R}^n$ and $g : \mathbb{R}^n \to \mathbb{R}^{n \times m}$ are continuous functions of x.

In linear systems [21] we make the assignment $A := f(\cdot)$ and $B := g(\cdot)$, where A and B are matrices of appropriate dimensions.

A control systems engineer needs to pick appropriately the control input u such that the system is operating in the desired way in terms of stability and performance.

4.2 Time-Invariant/Linear Time-Varying Control Systems and the Notions of Controllability and Observability

We define as linear time-varying systems, systems that evolve in continuous time $t \in \mathbb{R}_{>0}$ through linear state-space equations as

$$\dot{x}(t) = A(t)x(t) + B(t)u(t),$$
$$y(t) = C(t)x(t) + D(t)u(t),$$

where $x(t) \in \mathbb{R}^n$ denotes the system state, $u(t) \in \mathbb{R}^m$ denotes the system inputs, $y(t) \in \mathbb{R}^p$ denotes the system outputs, and $A(t) \in \mathbb{R}^{n \times n}$, $B(t) \in \mathbb{R}^{n \times m}$, $C(t) \in \mathbb{R}^{p \times n}$, and $D(t) \in \mathbb{R}^{p \times m}$ are matrices of appropriate dimensions.

Linear time-varying systems are useful in many application areas. They frequently arise as models of mechanical or electrical systems whose parameters (eg, the stiffness of a spring or the inductance of a coil) change in time. As we will see, linear time-varying systems also arise when one linearizes a nonlinear system around a trajectory, which is a very common approach in practice. Faced with a nonlinear system, one often uses the full nonlinear dynamics to design an optimal trajectory to guide the system from its initial state to a desired final state. However, ensuring that the system will actually track this trajectory in the presence of disturbances is not easy. One solution is to linearize the nonlinear system (ie, approximate it by a linear system) around the optimal trajectory; the approximation is accurate as long as the nonlinear system does not drift too far from the optimal trajectory. The result of the linearization is a linear time-varying system, which can be controlled by several methods. If the control design is done well, the state of the nonlinear system will always stay close to the optimal trajectory, hence ensuring that the linear approximation remains valid.

Linear time-invariant (LTI) systems are a special class of linear time-varying systems. These systems are described by state equations of the form

$$\dot{x}(t) = Ax(t) + Bu(t),$$
$$y(t) = Cx(t) + Du(t),$$

where the matrices $A \in \mathbb{R}^{n\times n}$, $B \in \mathbb{R}^{n\times m}$, $C \in \mathbb{R}^{p\times n}$, and $D \in \mathbb{R}^{p\times m}$ are constant matrices $\forall t \in \mathbb{R}_{>0}$ of appropriate dimensions. LTI systems are somewhat easier to deal with and will be treated as a special case of the more general linear time-varying systems. Linear systems are ideally suited for research on control systems theory because of their simplicity to derive clean formulas, although most systems in nature are nonlinear. The main point of linear systems theory is to exploit the algebraic structure to develop tractable "algorithms" that allow us to answer analysis questions which appear intractable by themselves.

For example, let us consider the LTI system

$$\dot{x}(t) = Ax(t) + Bu(t). \tag{1}$$

Given $x_0 \in \mathbb{R}^n$, $T > 0$, and a continuous function (input trajectory) $u(\cdot) : [0, T] \to \mathbb{R}^m$, we can show that there exists a unique function $x(\cdot) : [0, T] \to \mathbb{R}^n$ such that

$$x(0) = x_0, \quad \text{and} \quad \dot{x}(t) = Ax(t) + Bu(t), \quad \forall t \in [0, T].$$

This function is called the *state trajectory* with initial condition x_0 under the input $u(\cdot)$. It is obvious that $u(\cdot)$ does not even need to be continuous, provided one appropriately qualifies the statement $\forall t \in [0, T]$. The LTI system is called *controllable* if and only if $x_0 \in \mathbb{R}^n$, $\forall \hat{x} \in \mathbb{R}^n$, and $\forall T > 0$, and there exists $u(\cdot) : [0, T] \to \mathbb{R}^m$ such that the solution of the system with initial condition x_0 under the input $u(\cdot)$ is such that $x(T) = \hat{x}$. Controllability is clearly an interesting property for a system to have. If the system is controllable, then we can guide it from any initial state to any final state by selecting an appropriate input. If it is not, there may be some desirable parts of the state space that we cannot reach from some initial states. Unfortunately, determining whether a system is controllable directly from the definition is impossible. This would require our calculating all trajectories that start at all initial conditions. Except for trivial cases (such as the linear system $\dot{x}(t) = u(t)$, $t \geq 0$) this calculation is intractable, since the initial states, x_0, the times T of interest, and the possible input trajectories, $u(\cdot) : [0, T] \to \mathbb{R}^m$, are all infinite.

On the other hand, observability (mathematical dual to controllability) is a measure of how well internal states of a system can be inferred from knowledge of its external outputs. The concept of observability was firstly introduced by Kalman [22, 23] for linear dynamic systems.

The controllability property is summarized in Theorem 1:

Definition 1. Consider a system of the form

$$\dot{x}(t) = Ax(t) + Bu(t).$$

The pair (A, B) is controllable if, given a duration $T > 0$ and two arbitrary points x_0 and $x_T \in \mathbb{R}^n$, there exists a piecewise continuous function $t \to \bar{u}(t)$ from $[0, T]$ to $\mathbb{R}^m$ such that the integral curve $\bar{x}(t)$ generated by $\bar{u}$ with $\bar{x}(0) = x_0$ satisfies $\bar{x}(T) = x_T$.

Theorem 1. *A system of the form*

$$\dot{x}(t) = Ax(t) + Bu(t)$$

is controllable if and only if the matrix $\begin{bmatrix} B & AB & \cdots & A^{n-1}B \end{bmatrix} \in \mathbb{R}^{n \times nm}$ *has rank* n.

Similarly, we can define the notion of observability.

Definition 2. A system of the form

$$\dot{x}(t) = Ax(t) + Bu(t),$$
$$y = Cx(t)$$

is said to be observable if, for any possible sequence of state and control vectors, the current state can be determined in finite time with use of only the outputs. Less formally, this means that from the system's outputs it is possible to determine the behavior of the entire system. If a system is not observable, this means the current values of some of its states cannot be determined through output sensors. This implies that their value is unknown to the controller.

Now we can state Theorem 2:

Theorem 2. *A system of the form*

$$\dot{x}(t) = Ax(t) + Bu(t),$$
$$y = Cx(t)$$

is observable if and only if the matrix $\begin{bmatrix} C \\ CA \\ \vdots \\ CA^{n-1} \end{bmatrix}$ *has rank* n.

Note that the previous theorems of controllability and observability were defined for LTI systems. One has to be very careful when extending those theorems to linear time-varying systems.

4.3 Basic Concepts of a Control System

There are seven basic concepts of a control system (see Fig. 4):

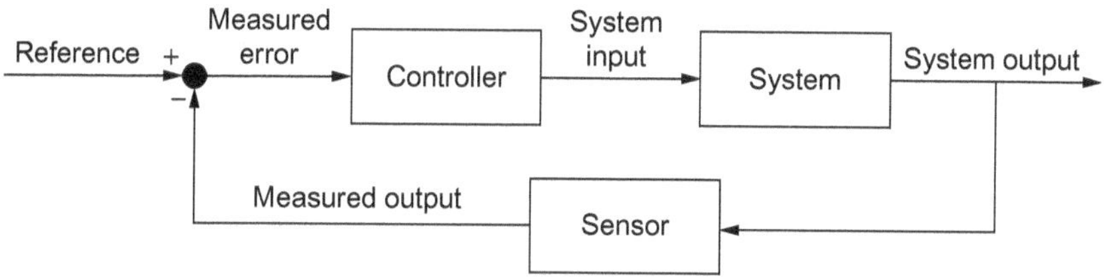

Fig. 4 Basic components of a control system.

1. **Plant:** a physical object to be controlled such as a mechanical device, a heating furnace, a chemical reactor, or a spacecraft.
2. **Controlled variable:** the variable controlled, generally refers to the system output.
3. **Expected value:** the desired value of controlled variable based on the requirement; often it is used as the reference input.
4. **Controller:** an agent that can calculate the required control signal.
5. **Actuator:** a mechanical device that takes energy, usually created by air, electricity, or a liquid, and converts it into some kind of motion.
6. **Sensor:** a device that measures a physical quantity and converts it into a signal which can be read by an observer or an instrument.
7. **Disturbance:** the unexpected factors disturbing the normal functional relationship between the controlling and controlled parameter variations.

5 CONTROL TECHNIQUES

Autonomous systems have been studied for many years with the hope of achieving human-like (or more) performance in solving certain problems [24]. There has been a recent resurgence in the field of machine intelligence and autonomy owing to the introduction of new topologies, training algorithms, and VLSI implementation techniques. The potential benefits of intelligent systems such as parallel distributed processing, high computation rates, fault tolerance, and adaptive capability have lured researchers from different fields to seek solutions to their complicated problems. Autonomy is a capability that enables a particular action of a system to be automatic or, within programmed boundaries, "self-governing," whereas in automation the system functions with no/little human operator involvement for well-defined tasks that have a predetermined response. Decentralization, uncertainty, and complexity are several issues that cannot be handled with classical automatic control methods. The power of adaptation and machine intelligence is the underlying foundation of autonomous technology that enables manpower efficiencies, rapid response in harsh environments, and capabilities beyond human limits and across operational domains.

There is a need to integrate human cognitive models to improve human-agent feedback loops, optimize trust/transparency, and improve data decision models. Furthermore, networks of autonomous complex systems must have secure communication protocols while their operators expand shared perception and problem solving across multiple agents and improve guidance and control. Complex systems consist of cooperating/noncooperating humans and manned/unmanned vehicles, and their interactions and structure of group communication protocols can yield unexpected behaviors. Cyber-physical systems have received much attention because of their potential application in a variety of systems, such as biological models (eg, infectious diseases) and social networks, human-robot interaction systems, the Internet, transportation systems, cyber security (eg, malware spreading), unmanned aerial/underwater vehicles, sensor networks, power networks, and mobile robotics. Those systems are large, complex, dynamic, and highly nonlinear in their global behavior.

The help of machine intelligence will result in the development of technologies that ensure robot system and self-protection capabilities in unanticipated and dynamic environments. In the following sections we provide a brief background of basic control techniques that will be used in several contexts in the chapters that follow.

5.1 Optimal Control

Optimal control [25–30] has emerged as one of the fundamental design philosophies of modern control systems design. Optimal control policies satisfy the specified system performance while minimizing a structured cost index which describes the balance between desired performance and available control resources. From a mathematical point of view the solution of the optimal control problem is based on the solution of the underlying Hamilton-Jacobi-Bellman equation. Until recently, because of the intractability of this nonlinear differential equation for continuous-time systems, which form the object of interest in this work, only particular solutions were available (eg, for the LTI case, the Hamilton-Jacobi-Bellman becomes the Riccati equation).

Optimal control design algorithms using the Riccati equation are well known. Good design algorithms for the linear-quadratic regulator (LQR) [25–28] are available in MATLAB Control System Toolbox and elsewhere. The structure and design procedure for the LQR is shown in Fig. 5 for

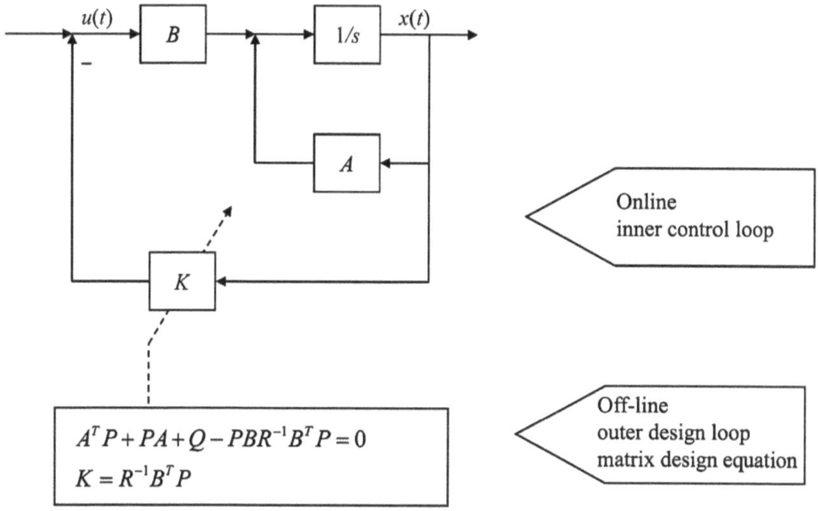

Fig. 5 Linear quadratic regulator.

system (1) with the performance to be minimized given as

$$J_{\mathrm{LQR}} = \frac{1}{2}\int_0^{\infty}(x^TQx + u^TRu)dt, \tag{2}$$

where the user-defined matrices are $Q \succeq 0$ and $R \succ 0$, of appropriate dimensions.

Depending on how these design parameters are selected, the closed-loop system will exhibit a different response. Generally speaking, selecting Q to be large means that, to keep J_{LQR} small, the state $x(t)$ must be smaller. On the other hand, selecting R to be large means that the control input $u(t)$ must be smaller to keep J_{LQR} small. This means that larger values of Q generally result in the poles of the closed-loop system matrix being further left in the frequency domain plane, so the state decays faster to zero. On the other hand, larger R means that less control effort is used, so the poles generally move more slowly, resulting in larger values of the state $x(t)$. In terms of eigenvalues, the eigenvalues of Q should be nonnegative, while those of R should be positive. If both matrices are selected to be diagonal, this means that all the entries of R must be positive, while those of Q should be positive, with possibly some zeros on its diagonal. Note that R is then invertible. Since the plant is linear and the performance index is quadratic, the problem of determining the state variable feedback control to minimize J_{LQR} is called the LQR problem. The word *regulator* refers to the fact that the function of this feedback is to regulate the states to zero. This is in contrast to tracker problems, where the objective is to make the output follow a prescribed (usually nonzero) reference command.

To find the optimal feedback control K we proceed as follows. Suppose there exists a constant matrix P and $u = -Kx$ such that

$$\frac{d}{dt}(x^T Px) = -x^T(Q + K^T RK)x. \tag{3}$$

Then substituting this in Eq. (2) yields

$$J_{\text{LQR}} = -\frac{1}{2}\int_0^\infty \frac{d}{dt}(x^T Px)dt = \frac{1}{2}x^T(0)Px(0), \tag{4}$$

where we assumed that the closed-loop system is stable so that $x(t)$ goes to zero as time t goes to infinity. Eq. (4) means that J is now independent of K. It is a constant that depends only on the auxiliary matrix P and the initial conditions.

Now we can find K so that assumption (3) does indeed hold. To accomplish this, we differentiate Eq. (3) and after using Eq. (1) with $u = -Kx$, the Eq. (3) becomes

$$x^T[(A - BK)^T P + P(A - BK) + Q + K^T RK]x = 0.$$

Note that this equation has to hold for every $x(t)$. Therefore the term in brackets must be identically equal to zero. Thus proceeding one sees that

$$(A - BK)^T P + P(A - BK) + Q + K^T RK = 0.$$

This is a matrix quadratic equation. Exactly as for the scalar case, one may complete the squares, though this procedure is a bit complicated for matrices. Suppose we select $K = R^{-1}B^T P$, then we have,

$$\begin{aligned}&(A - BR^{-1}B^T P)^T P + P(A - BR^{-1}B^T P) + Q + (R^{-1}B^T P)^T\\&R(R^{-1}B^T P) = 0.\end{aligned} \tag{5}$$

This result is of extreme importance in modern control theory. Eq. (5) is known as the algebraic Riccati equation. It is named after Riccati, an Italian who lived in the 19th century and used a similar equation in the study of heat flow. It is a matrix quadratic equation that can be solved for the auxiliary matrix P given (A, B, Q, R). Then the optimal gain K is given by $K = R^{-1}B^T P$. The minimal value of the performance index obtained with this gain is given by $\frac{1}{2}x^T(0)Px(0)$, which depends only on the initial condition. This mean that the cost of our using the selected K can be computed from the initial conditions before the control is ever applied to the system.

The LQR has many good properties, but these limitations:

- It requires full state feedback from all the states. These can be hard to measure.
- Its function is regulation—that is, to guarantee stability (ie, to regulate all the states to zero).
- It has no structure. All the inputs are linear combinations of all the states.
- It relies on off-line design methods that require full knowledge of the systems dynamics. However, finding the dynamics model requires great effort in system identification techniques. Also, system models are not usually completely accurate
- It minimizes only quadratic performance indices (eg, energies). What about minimum fuel, minimum time, etc.?
- It is an off-line design procedure, and so does not allow real-time changes in system performance index requirements.

5.2 Adaptive Control

The history of adaptive control began in the early 1950s. With the passing of the years, many articles and books have been published. These research activities have proposed solutions for basic problems and for broader classes of systems. Especially interest in nonlinear adaptive control began in the mid-1980s. Adaptive control [31–37] is a powerful tool that deals with modeling uncertainties in nonlinear (and linear) systems by online tuning of parameters. Very important research activities include online identification and pattern recognition inside the feedback control loop. It is well known that global stability properties of model reference adaptive systems are guaranteed under the "matching assumption" that the model order is not lower than that of the unknown plant. This restrictive assumption is likely to be violated in applications. Hence it is important to determine stability and robustness properties of adaptive schemes with respect to modeling errors.

To see how adaptive control works, we will define the control error $r(t)$ with dynamics given by

$$\dot{r} = f(x) - \tau,$$

where τ is a control input and $f(x)$ is some unknown nonlinearity.

Now we assume that $f(x)$ is known to be of structure

$$f(x) = W^T\phi(x),$$

where W is an unknown parameter and $\phi(x)$ is a known basis set or regression vector. Note that this structure is linear in the parameter W. Also the regression matrix $\phi(x)$ depends on the system and changes for different systems.

Then the error dynamics are

$$\dot{r} = W^T\phi(x) - \tau.$$

A scheme of an adaptive controller is shown in Fig. 6.

We define the estimate for the nonlinear function as $\hat{f}(x) = \hat{W}^T\phi(x)$. We select the controller as

$$\tau = \hat{f}(x) + K_v r = \hat{W}^T\phi(x) + K_v r,$$

where $\hat{W}(t)$ a time-varying estimate of the unknown parameter W and control gain $K_v \succ 0$ is any positive definite symmetric matrix.

Then the closed-loop system becomes

$$\dot{r} = W^T\phi(x) - \tau = W^T\phi(x) - \hat{W}^T\phi(x) - K_v r := \tilde{W}^T\phi(x) - K_v r,$$

where $\tilde{W} := W - \hat{W}(t)$.

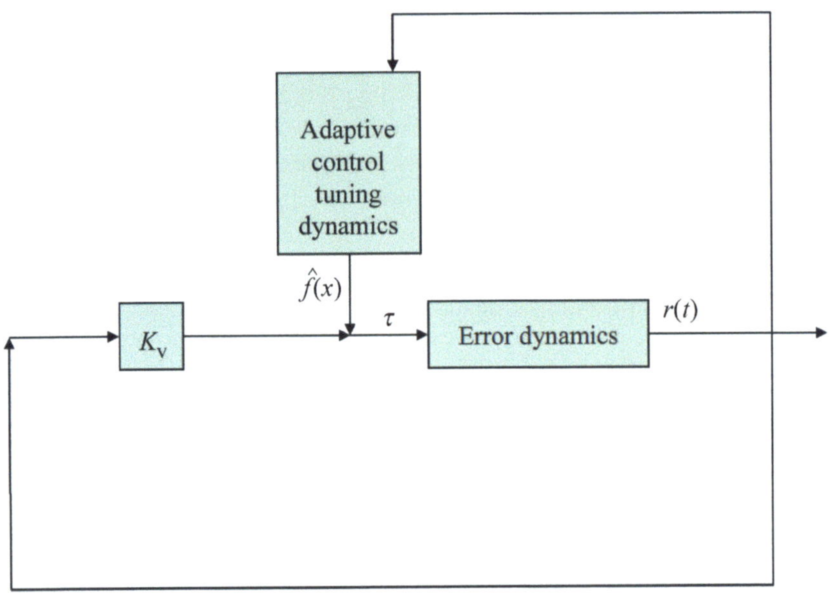

Fig. 6 Adaptive controller.

Let the parameter estimate be updated with

$$\dot{\hat{W}} = F\phi(x)r^T,$$

where the tuning gain $F \succ 0$ is any positive definite symmetric matrix. This is the internal (tuning) dynamics of the adaptive controller. The state of the adaptive controller is given by the parameter estimates. With use of the adaptive controller, the closed-loop system is asymptotically stable (the equilibrium point of the system rests at the origin)—that is, the control error $r(t)$ goes to zero. If an additional persistence of excitation condition holds, the parameter estimates converge to the actual unknown parameters. The typical behavior of adaptive controllers is to guarantee that the errors go to zero and the parameter estimates converge. This assumes that the parameterization $f(x) = W^T\phi(x)$ holds exactly and that there are no disturbances in the system.

5.3 Robust Control

Robust control [38–43] deals explicitly with uncertainty in its approach to controller design. Controllers designed by robust control methods tend to be able to cope with small differences between the true system and the nominal model used for the design. The early methods of Bode [67] and others were fairly robust; the state-space methods invented in the 1960s and 1970s were sometimes found to lack robustness. Examples of modern robust control techniques include H-infinity loop shaping developed by McFarlane and Glover [44] and sliding mode control (SMC) developed by Utkin et al. [45]. Robust methods aim to achieve robust performance and/or stability in the presence of small modeling errors.

To see how robust control works, we will define in a similar manner as in the adaptive control technique the control error $r(t)$. Hence the dynamics are given by $\dot{r} = f(x) - \tau$. We also know a fixed nominal value or estimate $\hat{f}(x)$ for unknown $f(x)$, and that the estimation error $\tilde{f} := f(x) - \hat{f}(x)$ is bounded like $\left\|\tilde{f}(x)\right\| \leq F(x)$, with $F(x)$ a known upper bound function, possibly nonlinear. Note that robust control assumes we have less information than with the adaptive controller, which needs the linear in the parameters structure $f(x) = W^T\phi(x)$, with W an unknown parameter vector and $\phi(x)$ a known basis set or regression vector. A scheme for robust control is shown in Fig. 7.

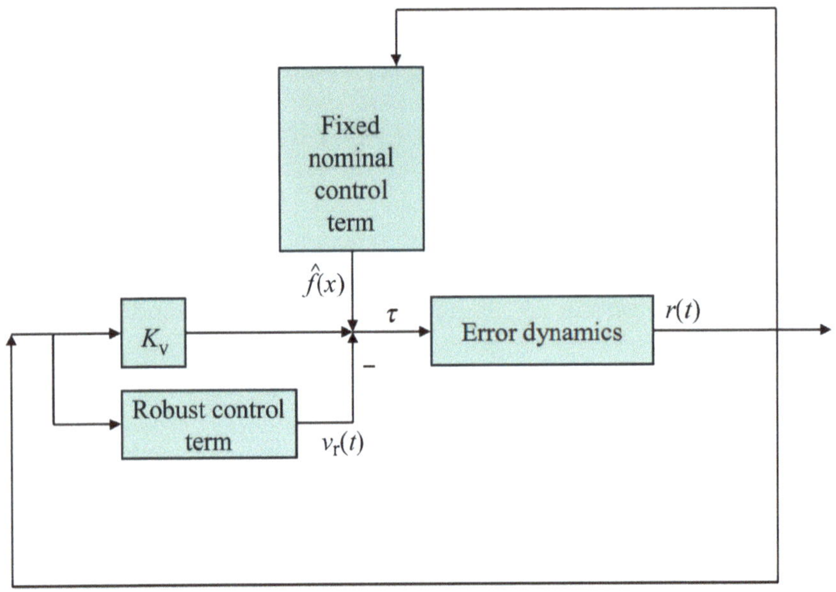

Fig. 7 Robust controller.

For robust control we need to select the controller

$$\tau = \hat{f}(x) + K_v r - \nu_r,$$

where ν_r a robust control term given by,

$$\nu_r = \begin{cases} -r\frac{F(x)}{\|r\|}, & \|r\| \geq \epsilon, \\ \\ -r\frac{F(x)}{\epsilon}, & \|r\| < \epsilon, \end{cases}$$

where $\epsilon \in \mathbb{R}^+$ is a small design parameter. Then the closed-loop dynamics become

$$\dot{r} = f(x) - \tau = f(x) - \left(\hat{f}(x) + K_v r - \nu_r\right) := \tilde{F}(x) - K_v r + \nu_r.$$

This robust controller is easier to implement than the adaptive controller because it does not have any dynamics. As long as the performance of the aforementioned robust controller, the closed-loop system is bounded stable with $\|r\|$ bounded with a magnitude near ϵ. The error does not go to zero but does indeed stay small.

5.4 Intelligent Control

Intelligent control, the discipline where control algorithms are developed by emulation of certain characteristics of biological systems, is being fueled by recent advancements in computing technology and is emerging as a technology that may open avenues for significant technological advances. The characteristics include adaptation and learning, planning under large uncertainty, and coping with large amounts of data. Today, the area of intelligent control tends to encompass everything that is not characterized as conventional control; it has, however, shifting boundaries and what is called "intelligent control" today will probably be called "control" tomorrow.

Intelligent control [46–56] uses various artificial intelligence computing approaches, such as neural networks, Bayesian probability, fuzzy logic, machine learning, and self-organizing systems: neural networks and fuzzy logic systems for feedback control; discrete event systems, Dempster-Shafer theory, Petri nets, and decision-making supervisory control systems, evolutionary computation, and genetic algorithms to control a dynamic system.

Intelligent control can also be used in intelligent machinery monitoring, and in repair in advanced sensor processing, including Kalman filtering and sensor fusion. An intelligent control structure that uses reinforcement learning and especially an actor-critic structure is shown in Fig. 8

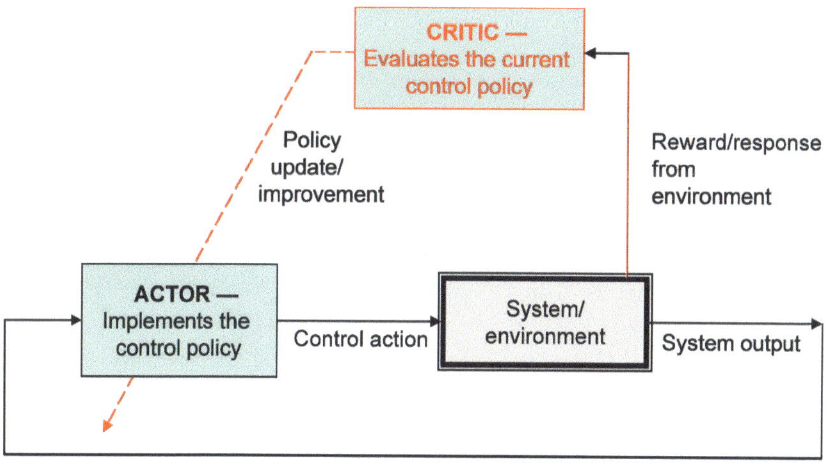

Fig. 8 Actor-critic structure for online adaptive optimal control.

5.5 Game-Theoretic Control

One can model strategic interactions in a complex system (networked society) as a multiplayer simultaneous-move game [2, 57, 58]. The agents involved have conflicting objectives and it is natural to consider optimization based on individual payoffs or costs. A game theory approach may be able to induce the world of complex networks to more readily surrender their secrets. To characterize a game one needs to specify several items—namely, the players that make the decisions, the rules that define the actions allowed by the players and their effects, the information structure which specifies what each player knows before making each decision, and the objective which specifies the goal of each player. Also, there are several concepts including, zero-sum versus non-zero-sum games, noncooperative versus cooperative games, complete versus incomplete information, deterministic versus stochastic, static versus dynamic games.

5.6 Cooperative and Distributed Control

In recent years there has been an increasing need of advanced techniques for control and optimization of complex large-scale dynamical systems. Fortunately, along with the increase of complexity there have been great technological advances in the fields of communication networks and embedded computer technology. The natural approach to control such complex systems is therefore to distribute the computational burden among different agents that, through appropriate communication, cooperatively achieve system-wide goals. Synchronization of all the agents interacting through the network to a leader behavior has been the main subject of consensus and distributed control algorithms [59–62] since the work of Tsitsiklis [63].

5.7 Bandwidth Effective Control

Shared congestion- and energy-saving objectives demand that it should be rigorously decided when all information through a network should be transmitted. For that reason one needs to design "bandwidth"-efficient controllers that can function in event-driven environments and update their values only when needed. Bandwidth-efficient policy design is a newly developed framework that can potentially enhance the performance of applications that have limited resources and bandwidth and offers a new point of view, with respect to conventional time-driven strategies, on

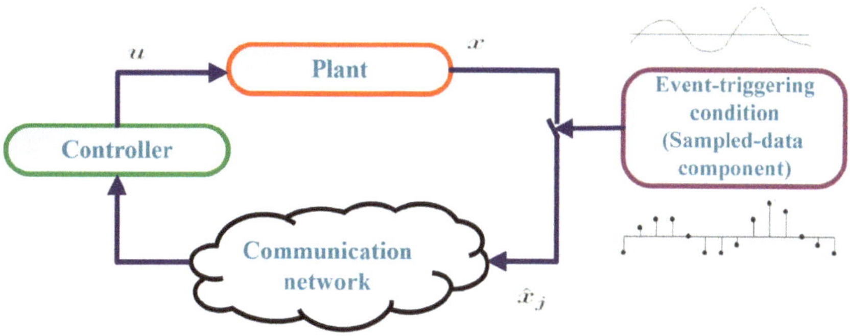

Fig. 9 Event-triggered controller.

how information could be sampled for control purposes. The bandwidth-efficient control algorithms [64–66] are composed of a feedback controller updated on the basis of the sampled state, and the event-triggering mechanism, which determines when the controller has to be transmitted from a stability and performance point of view. This can reduce the computation and communication resources significantly (Fig. 9).

6 CONCLUSION

This introductory chapter has summarized the fundamental notions and concepts of complex systems along with the basic feedback control techniques that are used extensively to force a desired behavior. The goal was to offer basic and key content of interest. Currently, a number of outstanding books and journals are available to help researchers, scientists, and engineers master the use of complex control theory and the contributions it brings to develop new theoretical approaches and new applications. The references in this and the following chapters include some of them.

The remaining chapters in this book provide readers with a clear picture of the current state of the art in complex systems through novel theoretical results and applications.

REFERENCES

[1] Y. Bar-Yam, Dynamics of Complex Systems, vol. 213, Addison-Wesley, Reading, MA, 1997.

[2] T. Basar, G.J. Olsder, G.J. Clsder, T. Basar, T. Baser, G.J. Olsder, Dynamic Noncooperative Game Theory, vol. 200, SIAM, Philadelphia, PA, 1995.

[3] G.W. Flake, The Computational Beauty of Nature: Computer Explorations of Fractals, Chaos, Complex Systems, and Adaptation, MIT Press, Cambridge, MA, 1998.

[4] E. Laszlo, Evolution: The Grand Synthesis, New Science Library/Shambhala Publications, Boston, 1987.
[5] H. Haken, Information and Self-Organization: A Macroscopic Approach to Complex Systems, Springer Science & Business Media, Berlin, 2006.
[6] W.M. Haddad, V. Chellaboina, S.G. Nersesov, Thermodynamics: A Dynamical Systems Approach: A Dynamical Systems Approach, Princeton University Press, Princeton, NJ, 2009.
[7] D. Larsen-Freeman, L. Cameron, Complex Systems and Applied Linguistics, Oxford University Press, Oxford, 2008.
[8] R.K. Sawyer, Social Emergence: Societies as Complex Systems, Cambridge University Press, Cambridge, 2005.
[9] R. Serra, G. Zanarini, Complex Systems and Cognitive Processes, Springer Science & Business Media, Berlin, 2013.
[10] P. Cilliers, D. Spurrett, Complexity and Post-Modernism: Understanding Complex Systems, S. Afr. J. Philos. 18 (2) (1999) 258–274.
[11] R.C. Dorf, Modern Control Systems, Addison-Wesley Longman Publishing Co., Inc., Reading, MA, 1995.
[12] G.F. Franklin, J.D. Powell, A. Emami-Naeini, J.D. Powell, Feedback Control of Dynamic Systems, vol. 3, Addison-Wesley, Reading, MA, 1994.
[13] J.P. Hespanha, Linear Systems Theory, Princeton University Press, Princeton, NJ, 2009.
[14] W.S. Levine, The Control Handbook, CRC Press, Boca Raton, 1996.
[15] E.D. Sontag, Mathematical Control Theory: Deterministic Finite Dimensional Systems, vol. 6, Springer, Berlin, 1998.
[16] G.C. Goodwin, S.F. Graebe, M.E. Salgado, Control System Design, vol. 240, Prentice Hall, Englewood Cliffs, NJ, 2001.
[17] C. Darwin, W.F. Bynum, The Origin of Species by Means of Natural Selection: Or, the Preservation of Favored Races in the Struggle for Life, AL Burt, New York, 2009.
[18] V. Volterra, Theory of Functionals and of Integral and Integro-Differential Equations, Courier Dover Publications, Mineola, NY, 2005.
[19] M. Dobb, M.H. Dobb, Theories of Value and Distribution Since Adam Smith: Ideology and Economic Theory, Cambridge University Press, Cambridge, 1975.
[20] H.K. Khalil, J. Grizzle, Nonlinear Systems, vol. 3, Prentice Hall, Upper Saddle River, NJ, 2002.
[21] P.J. Antsaklis, A.N. Michel, Linear Systems, Springer, Berlin, 2006.
[22] R.E. Kalman, Mathematical Description of Linear Dynamical Systems, J. Soc. Ind. Appl. Math. A Control 1 (2) (1963) 152–192.
[23] R. Kalman, On the general theory of control systems, IRE Trans. Autom. Control 4 (3) (1959) 110.
[24] K.G. Vamvoudakis, P.J. Antsaklis, W.E. Dixon, J.P. Hespanha, F.L. Lewis, H. Modares, B. Kiumarsi, Autonomy and machine intelligence in complex systems: a tutorial, in: American Control Conference (ACC), July 2015, 2015, pp. 5062–5079.
[25] A.E. Bryson, Applied Optimal Control: Optimization, Estimation and Control, Halsted Press Book', Taylor & Francis, New York, 1975.
[26] D.E. Kirk, Optimal Control Theory: An Introduction, Dover Books on Electrical Engineering, Dover Publications, Mineola, NY, 2012.
[27] F.L. Lewis, D. Vrabie, V.L. Syrmos, Optimal Control, EngineeringPro Collection, Wiley, New York, 2012.
[28] M. Athans, P.L. Falb, Optimal Control: An Introduction to the Theory and Its Applications, Dover Books on Engineering Series, Dover Publications, Mineola, NY, 2006.

[29] R.F. Stengel, Optimal Control and Estimation, Dover Books on Advanced Mathematics, Dover Publications, Mineola, NY, 1986.
[30] D. Liberzon, Calculus of Variations and Optimal Control Theory: A Concise Introduction, Princeton University Press, Princeton, NJ, 2011.
[31] P. Ioannou, B. Fidan, Adaptive Control Tutorial, Advances in Design and Control, Society for Industrial and Applied Mathematics (SIAM), Philadelphia, PA, 2006.
[32] G. Tao, Adaptive Control Design and Analysis, Adaptive and Cognitive Dynamic Systems: Signal Processing, Learning, Communications and Control, Wiley, New York, 2003.
[33] S. Sastry, M. Bodson, Adaptive Control: Stability, Convergence and Robustness, Dover Books on Electrical Engineering Series, Dover Publications, Mineola, NY, 2011.
[34] K.J. Åström, B. Wittenmark, Adaptive Control, Dover Books on Electrical Engineering, second ed., Dover Publications, Mineola, NY, 2013.
[35] N. Hovakimyan, C. Cao, L1 Adaptive Control Theory: Guaranteed Robustness With Fast Adaptation, Advances in Design and Control, Society for Industrial and Applied Mathematics (SIAM), Philadelphia, PA, 2010.
[36] P. Ioannou, J. Sun, Robust Adaptive Control, Dover Publications, Mineola, NY, 2013.
[37] M. Krstić, I. Kanellakopoulos, P.V. Kokotović, Nonlinear and Adaptive Control Design, Adaptive and Learning Systems for Signal Processing, Communications, and Control, Wiley, New York, 1995.
[38] M. Green, D.J.N. Limebeer, Linear Robust Control, Dover Books on Electrical Engineering Series, Dover Publications, Incorporated, Mineola, NY, 2012.
[39] B.A. Francis, P.P. Khargonekar, Robust Control Theory, The IMA Volumes in Mathematics and its Applications, Springer, New York, 2011.
[40] P.D. Christofides, Nonlinear and Robust Control of PDE Systems: Methods and Applications to Transport-Reaction Processes, Springer Science & Business Media, 2012.
[41] S.O.R. Moheimani, Perspectives in Robust Control, Advances in Pattern Recognition, Springer, Berlin, 2001.
[42] T.E. Djaferis, Robust Control Design: A Polynomial Approach: A Polynomial Approach, Springer US, New York, 1995.
[43] I. Petersen, V.A. Ugrinovskii, A.V. Savkin, Robust Control Design Using H-∞ Methods, Communications and Control Engineering, Springer, London, 2000.
[44] D.C. McFarlane, K. Glover, Robust Controller Design Using Normalized Coprime Factor Plant Descriptions, Lecture Notes in Control and Information Sciences, Springer, Berlin, 1990.
[45] V. Utkin, J. Guldner, M. Shijun, Sliding Mode Control in Electro-Mechanical Systems, Automation and Control Engineering, Taylor & Francis, New York, 1999.
[46] P.J. Antsaklis, K.M. Passino, Introduction to Intelligent Control Systems With High Degrees of Autonomy, Kluwer Academic Publishers, Norwell, MA, 1993.
[47] K.P. Valavanis, Applications of Intelligent Control to Engineering Systems: In Honour of Dr. G. J. Vachtsevanos, Intelligent Systems, Control and Automation: Science and Engineering, Springer, Berlin, 2009.
[48] F.W. Lewis, S. Jagannathan, A. Yesildirak, Neural Network Control of Robot Manipulators and Non-Linear Systems, Series in Systems and Control, Taylor & Francis, New York, 1998.
[49] J. Nie, D.A. Linkens, Fuzzy-Neural Control: Principles, Algorithms, and Applications, Prentice Hall, Upper Saddle River, NJ, 1995.
[50] S.J. Schiff, Neural Control Engineering: The Emerging Intersection Between Control Theory and Neuroscience, Computational Neuroscience, MIT Press, Cambridge, MA, 2012.
[51] O. Omidvar, D.L. Elliott, Neural Systems for Control, Elsevier Science, Amsterdam, 1997.

[52] E.N. Sanchez, A.Y. Alanis, A.Y.A. García, A.G. Loukianov, Discrete-Time High Order Neural Control: Trained With Kalman Filtering, Studies in Computational Intelligence, Springer, Berlin, 2008.
[53] F.L. Lewis, J. Campos, R. Selmic, Neuro-Fuzzy Control of Industrial Systems With Actuator Nonlinearities, Frontiers in Applied Mathematics, Society for Industrial and Applied Mathematics (SIAM), Philadelphia, PA, 2002.
[54] F.L. Lewis, D. Liu, Reinforcement Learning and Approximate Dynamic Programming for Feedback Control, IEEE Press Series on Computational Intelligence, Wiley, New York, 2013.
[55] D. Vrabie, K.G. Vamvoudakis, F.L. Lewis, Institution of Engineering and Technology, Optimal Adaptive Control and Differential Games by Reinforcement Learning Principles, IET Control Engineering Series, Institution of Engineering and Technology, United Kingdom, 2013.
[56] A.M. Meystel, J.S. Albus, Intelligent Systems: Architecture, Design, and Control, John Wiley & Sons, Inc., New York, 2000.
[57] W.B. Arthur, S.N. Durlauf, D.A. Lane, The Economy as an Evolving Complex System II, vol. 28, Addison-Wesley, Reading, MA, 1997.
[58] C. Camerer, Behavioral Game Theory: Experiments in Strategic Interaction, Princeton University Press, Princeton, NJ, 2003.
[59] Y. Cao, W. Yu, W. Ren, G. Chen, An overview of recent progress in the study of distributed multi-agent coordination, IEEE Trans. Ind. Inf. 9 (1) (2013) 427–438.
[60] A. Jadbabaie, J. Lin, et al., Coordination of groups of mobile autonomous agents using nearest neighbor rules, IEEE Trans. Autom. Control 48 (6) (2003) 988–1001.
[61] R. Olfati-Saber, A. Fax, R.M. Murray, Consensus and cooperation in networked multi-agent systems, Proc. IEEE 95 (1) (2007) 215–233.
[62] W. Ren, R.W. Beard, et al., Consensus seeking in multiagent systems under dynamically changing interaction topologies, IEEE Trans. Autom. Control 50 (5) (2005) 655–661.
[63] J.N. Tsitsiklis, Problems in Decentralized Decision Making and Computation, Tech. Rep., DTIC Document, 1984.
[64] W. Heemels, J. Sandee, P. Van Den Bosch, Analysis of event-driven controllers for linear systems, Int. J. Control 81 (4) (2008) 571–590.
[65] X. Wang, M.D. Lemmon, Event-triggering in distributed networked control systems, IEEE Trans. Autom. Control 56 (3) (2011) 586–601.
[66] P. Tabuada, Event-triggered real-time scheduling of stabilizing control tasks, IEEE Trans. Autom. Control 52 (9) (2007) 1680–1685.
[67] H.W. Bode, Network analysis and feedback amplifier design, Van Nostrand, New York, USA, 1945.

SECTION 2

Adaptive Control and Neuroscience

CHAPTER TWO

Hierarchical Adaptive Control of Rapidly Time-Varying Systems Using Multiple Models

K.S. Narendra
Yale University, New Haven, CT, United States

1 INTRODUCTION

Adaptive control was born in the 1950s to deal with the control of linear dynamical systems with constant unknown parameters. In the following 5 decades the field made major advances, and stability and robustness issues were thoroughly investigated. The cardinal requirements for good controller design have always been accuracy, stability, and robustness. Achieving these in the presence of uncertainty, complexity, nonlinearity, and time variations has been the challenge for the control theorist. From the very beginning the general philosophy of adaptive control has been to determine adaptive laws which ensure the stability of the overall nonlinear system, and subsequently to optimize its performance within the general framework of stability.

Even though the raison d'être of adaptive control is to cope with time-varying environments, for mathematical tractability, researchers confined their attention to time-invariant systems. The accepted philosophy was, and continues to be, that if an adaptive system is fast and accurate when the parameters are constant, then it would also prove to be satisfactory even when the parameters vary with time, provided their rate of variation is less than the speed of adaptation of the system.

The limitations of classical adaptive control became evident in applications in which the controllers were called on to respond rapidly and accurately to sudden and large changes in plant parameters. Classical adaptive control, which performs satisfactorily when the parameter errors are small (or vary slowly with time), was found to exhibit large transients

Control of Complex Systems
http://dx.doi.org/10.1016/B978-0-12-805246-4.00002-1

when the parameters had large amplitudes or changed rapidly with time. Problems in which the controlled processes exhibit such characteristics arise in both nontechnical and technical applications. These include medical emergencies, trading on the stock market, and conflict management using counterterror measures, as well as engineering problems such as energy management, aircraft and automobile control, manufacturing, and security. Even under such adverse conditions, the system is often expected to perform satisfactorily.

When the environments in which the decisions have to be made are relatively constant, information can be collected slowly and used to take appropriate action. However, the situation is quite different when critical decisions have to be made rapidly as the environment changes. In such cases there is incomplete, uncertain and unreliable information and there are limited resources. The aim in such cases is to use the available resources as effectively as possible to find adequate responses.

One of the main difficulties when one is dealing with systems with varying parameters is in characterizing them. System parameters can vary in arbitrarily many ways, some of which are shown in Fig. 1. They may be small and rapid as in Fig. 1b, or slowly varying with large excursions as in Fig. 1c. Such variations have been addressed in the adaptive literature. However, when the parameter variations are large, rapidly varying (ie, contain high frequencies), even discontinuous, or a combination of these, new methods are needed.

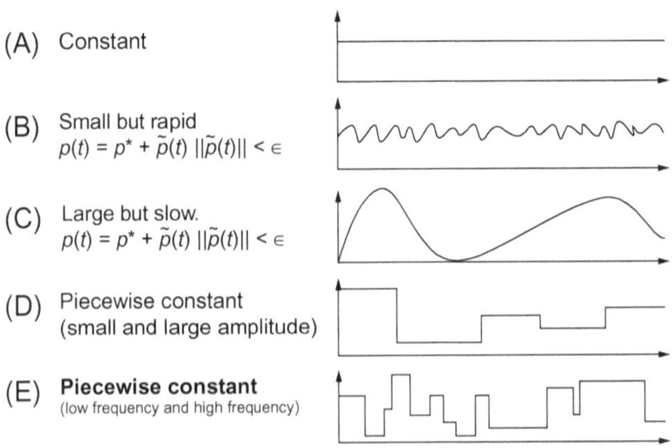

Fig. 1 Time-varying parameters.

It is the latter class of problems that is of interest in this chapter. While adaptive controller design has traditionally been based on a single model of the plant, our interest will be in the use of multiple identification models to improve the speed and accuracy of the response.

2 MATHEMATICAL PRELIMINARIES

While this chapter is primarily concerned with multiple model-based methods for adaptive control, for comparison purposes we first treat the classical indirect approach to adaptive control using a single model.

2.1 The Classical Adaptive Control Problem

Direct and indirect adaptive control are two distinct well-known approaches that have been used since the beginning for controlling a dynamical system with unknown parameters. While no identification model is used in the former, the latter relies on such a model for estimating the parameters of the unknown system, even as it is in operation. These estimates, in turn, are used to determine the controller parameters (based on the certainty equivalence principle) (see Fig. 2).

2.1.1 Adaptive Control Using a Single Model

A linear time-invariant system (the plant) Σ_p is described by the state equation

$$\Sigma_p : \dot{x}_p(t) = A_p x_p(t) + b_p u(t), \tag{1}$$

where $x_p(t), b_p(t) \in R^n$, $A_p \in R^{n \times n}$, $u(t) : R^+ \rightarrow R$, and A_p and b_p are in companion form. The last row of A_p is the vector θ_p^T, whose elements are $a_{p1}, a_{p2}, \ldots, a_{pn}$.

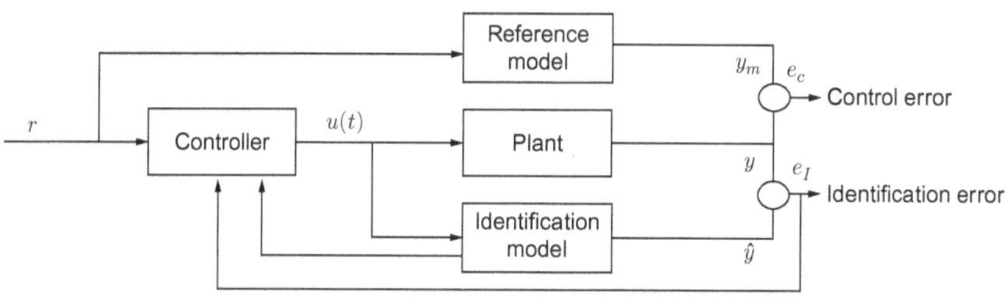

Fig. 2 Adaptive control using a single identification model.

A stable reference model $\Sigma_{\rm m}$ is described by the differential equation

$$\Sigma_{\rm m} : \dot{x}_{\rm m}(t) = A_{\rm m}x_{\rm m}(t) + b_{\rm m}r(t), \tag{2}$$

where $r(t) : R^+ \to R$ is a known bounded, piecewise continuous reference signal. The matrix $A_{\rm m}$ is also in companion form, and its last row is $\theta_{\rm m}^T$. The response $x_{\rm m}(t)$ of $\Sigma_{\rm m}$ to the reference input $r(t)$ is assumed to be the desired output, and the objective of adaptation is to determine the input $u(t)$ to the plant $\Sigma_{\rm p}$ which results in the bounded output $x_{\rm p}(t)$ tracking $x_{\rm m}(t)$ asymptotically with zero error—that is, $\lim_{t\to\infty} ||x_{\rm m}(t) - x_{\rm p}(t)|| = 0$.

In the classical indirect adaptive approach, an identification model Σ_1 is used to identify $\Sigma_{\rm p}$. It is described by the differential equation

$$\Sigma_1 : \dot{x}_1(t) = A_{\rm m}x_1(t) + [A_1(t) - A_{\rm m}(t)]x_{\rm p}(t) + bu(t), \tag{3}$$

where $A_{\rm m}$ is a stable matrix and $A_1(t)$ is the estimate at time t of the unknown matrix $A_{\rm p}$. More specifically, it is assumed that A_1 is in companion form and that the last row of $A_1(t)$, denoted by $\theta_1^T(t) = [a_1(t), \ldots, a_{\rm m}(t)]$, is the estimate of the vector $\theta_{\rm p}^T$. $\theta_1(t)$ is adjusted adaptively using the identification error from the unknown plant (see Fig. 2).

Comment 1. The series parallel structure in Eq. (3) is assumed to ensure the stability of the identification model, even when the plant is unstable. Also, while $A_{\rm m}$ in Eq. (3) can be any stable matrix, it is assumed to be identical to that of the reference model to simplify the control problem.

Defining the output error as $e_1(t) = x_1(t) - x_{\rm p}(t)$ and the parameter error as $\tilde{\theta}_1(t) = \theta_1(t) - \theta_{\rm p}$, we can write the error equation as

$$\dot{e}_1(t) = A_{\rm m}e_1(t) + b\tilde{\theta}_1^T x_{\rm p}(t). \tag{4}$$

If the Lyapunov function candidate $V(e_1, \tilde{\theta}_1)$ is chosen as

$$V(e_1, \tilde{\theta}_1) = e_1^T Pe_1 + \tilde{\theta}_1{}^T \tilde{\theta}_1, \tag{5}$$

where P is the positive definite matrix which is the unique solution of the Lyapunov equation $A_{\rm m}^T P + PA_{\rm m} = -Q$, $Q = Q^T > 0$, it follows that the adaptive law

$$\dot{\theta}_1 = \dot{\tilde{\theta}}_1 = -e_1^T Pbx_{\rm p} \tag{6}$$

results in the time derivative of $V(e_1, \tilde{\theta}_1)$ along any trajectory being

$$\dot{V}(e_1, \tilde{\theta}_1) = -e_1^T Qe_1 \leq 0. \tag{7}$$

This ensures the stability of the identification model and the boundedness of the state error $e_1(t)$ and the parameter error $\tilde{\theta}_1(t)$.

To ensure the stability of the controlled plant and the boundedness of $x_p(t)$, a feedback signal (the control input) is chosen as $u(t) = r(t) - k^T(t)x_p(t)$, where $k(t) = \theta_i(t) - \theta_m$. With use of the well-known certainty equivalence principle, this yields the error equation

$$\dot{e_c}(t) = \dot{x}_p - \dot{x}_m = A_m e_c(t) + b\tilde{k}^T(t)x_p(t), \tag{8}$$

where $\tilde{k}(t) = k(t) - k^\star$ and $k^\star = \theta_p - \theta_m$. Following standard arguments, one can show that $e_c \in \mathcal{L}^2 \cap \mathcal{L}^\infty$. Further, by use of Barbalat's lemma, it follows that $\lim_{t\to\infty} e_c(t) = 0$ and hence $\lim_{t\to\infty} ||x_m(t) - x_p(t)|| = 0$.

Comment 2. In the approach discussed above, the identification error $e_1(t)$ was used to determine the estimate of the plant parameter, and the controller parameter $k(t)$ was computed from this estimate. In the direct approach the control parameter $k(t)$ is adaptively adjusted directly with use of the control error $e_c(t)$.

2.1.2 Multiple Models

The concept of the use of multiple models is not new in control theory. Multiple Kalman filters were studied in the 1970s to improve the accuracy of the state estimate in control problems [1–3]. This was followed by several applications [4–7]. In all cases no switching was involved and no stability results were reported. In the late 1980s, switching was introduced in the context of adaptive control by Martensson [8], and results obtained by Fu and Barmish [9], Poolla and Cusumano [10], Miller and Davison [11], and Miller [12] followed. In all cases, the emphasis was on when to switch to the next controller in a predetermined sequence, and it was soon realized that such methods had little practical utility. In contrast to these direct switching methods, a second class proposed by Middleton et al. [13], Narendra and Balakrishnan [14], and Morse [15] depended on indirect switching, where multiple identification models were used.

The use of switching and tuning was proposed by Narendra and Balakrishnan [14] and was later studied extensively by them [16, 17]. The ideas and proofs of stability in these articles, as well as those of [15, 18], have been responsible for the great interest in this area and the wide application of the results in different fields.

In the following, we will use "identification models," "environments," and "unknown parameter vectors" synonymously. This is because the different models have the same structure and differ only in the parameter vectors defining them. If these parameters are constants, they may correspond to different environments in which the plants are called upon to

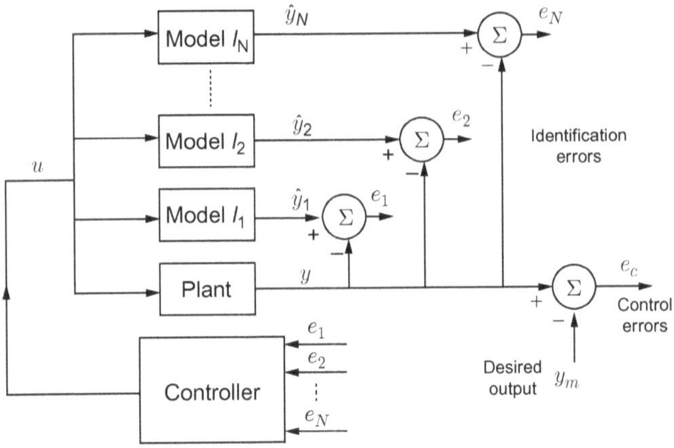

Fig. 3 Multiple models for identification and control.

operate. Since the input to all the models is the same as the input to the plant, the outputs of the different identification models can be considered as representing the outputs of the plant in different "environments." The central idea of the multiple model-based approach is to exploit the information obtained from the different models.

The proposed architecture for multiple model-based control is shown in Fig. 3. An unknown plant P has an input u and an output y. Multiple stable identification models $I_1, I_2, \ldots, I_N$ are set up to operate in parallel with the unknown plant. If in some cases the plant is unstable, the identification models are chosen to be of the series-parallel type to ensure stability (see Comment 1).

Obviously, the determination of the "best" model at any instant and the corresponding control action chosen depend on the prior information provided to the designer. Broadly speaking, identification is based on a performance function depending on the identification error (ie, $e_i = \hat{y}_i - y$), and control is performed to reduce some criterion based on the control error $e_c(t) = y(t) - y_m(t)$.

Comment 3. When an identification model is adaptive, the corresponding control is also adaptive. Adaptive identifiers and controllers can be initialized and terminated as required, depending on the strategy used.

Comment 4. It is worth noting that while there are multiple identifiers, only one control input can be applied to the plant at every instant. In general applications, this corresponds to having multiple advisors to make a single decision. This is shown in Fig. 3.

To determine the accuracy of any identification model (fixed or adaptive), different performance criteria have been proposed. If $e_j(t)$ is the identification error for a continuous-time plant (or $e_j(k)$ for a discrete-time plant), these are listed below:

Continuous time	Discrete time
$J_1(t) = e_i^2(t)$	$J_1(k) = e_i^2(k)$
$J_2(t) = \int_0^t e_i^2(\tau)d\tau$	$J_2(k) = \sum_{k=0}^{N} e_i^2(k)$
$J_3(t) = \alpha e_i^2(t) + \beta \int_0^t e_i^2(\tau)d\tau$	$J_3(k) = \alpha e_i^2(k) + \beta \sum_{k=0}^{N} e_i^2(k)$
$J_4(t) = \alpha e_i^2(t) + \beta \int_0^t \mathrm{e}^{-\gamma(t-\tau)} e_j^2(\tau)d\tau$	$J_4(k) = \alpha e_i^2(k) + \beta \sum_{k=0}^{N} \gamma^k e_i^2(k)$,
$\gamma > 0$	$0 < \gamma < 1$.

J_1 represents the instantaneous value of $e_i^2(t)$ ($e_i^2(k)$), while J_2 represents an integral (sum), and J_4 includes J_1 and a weighted value of J_2 (to ensure boundedness). Depending on the method used, one of the above performance criteria can be chosen and at every instant the model that best approximates the plant ($J_{\min} = \mathrm{Min}_i J_i(t)$) can be determined.

3 MULTIPLE MODEL-BASED ADAPTIVE CONTROL

The methods based on multiple models for the adaptive control of time-varying plants use the information provided by the different identification models described in the previous section to determine the control input at any instant.

Comment 5. The models chosen to identify the plant in Fig. 3 may be discrete or continuous, fixed or adaptive, linear or nonlinear, deterministic or stochastic.

From Fig. 3 it is clear that the designer has ample freedom at every stage in determining the control input. This includes the choice of the number and type of models to be used, their location in parameter space, their structure, the criterion for choosing an identification model at any instant, and the choice of the control input based on the identification model. In this section we describe four methods which enjoy advantages which make them preferable in different situations. These are (chronologically) (1) switching, (2) switching and tuning, (3) interactive/evolutionary adaptation, and (4) second-level adaptation. These methods can be considered as modifications and extensions of conventional adaptive control treated in the previous section.

3.1 Switching

3.1.1 *Multiple Fixed Models*

As described in Section 2, one of a number of criteria can be used to determine which of the models best approximates the plant at any instant. If the model parameters are constant, they are referred to as *fixed models* and if the parameters are updated adaptively as the plant is in operation they are referred to as *adaptive models*. We first consider in this subsection only fixed models (or models with constant parameters). Corresponding to each of them is a fixed controller so that the model and the controller together match the reference model. During the operation of the system, one merely has to identify the specific environment selected at any instant and apply the corresponding controller. The specific structure used for the controllers is shown in Fig. 4.

In practice, the fixed models must be distributed throughout the region of uncertainty in parameter space. If a model is sufficiently close to the plant and the controller corresponding to it stabilizes the plant, the overall

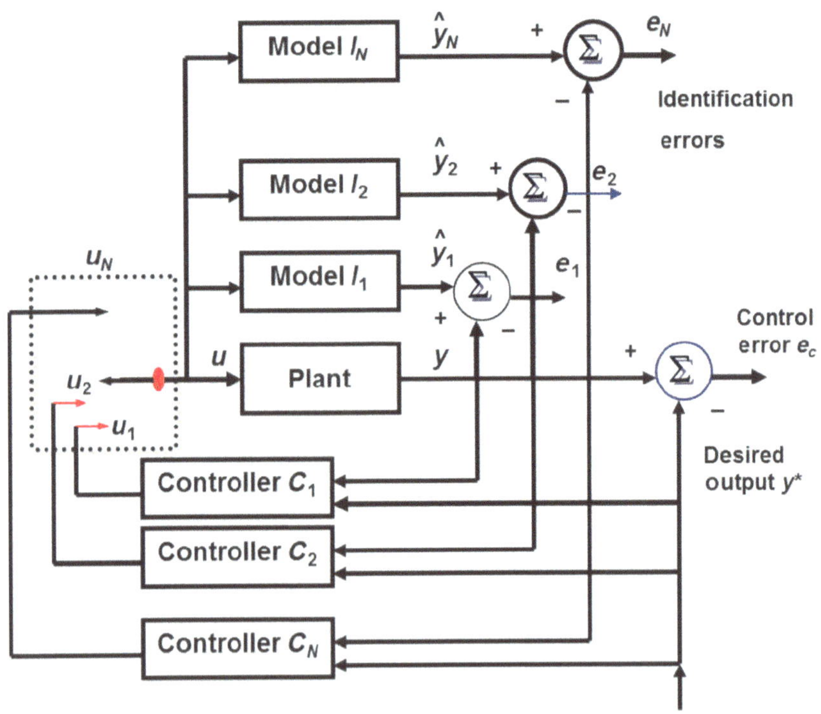

Fig. 4 Multiple models for identification and control (switching).

adaptive process is stable. At the same time, since both the plant and the model are constant but different, there is invariably a residual output error between the two for large values of t. This is shown in Fig. 5A and B.

When the conditions specified above are not satisfied (ie, model separation is large), instability might result, and this is indicated in Fig. 5C and D.

From the preceding discussion it is seen that, in general, a large number of models may be needed in applications. Design of adaptive systems using the above approach has been proposed in aircraft systems using thousands of models.

Assuming that a specific identification model and a corresponding controller have been chosen in the limit, it is seen that all other models are no longer functional. To have bounded outputs, they need to be stable.

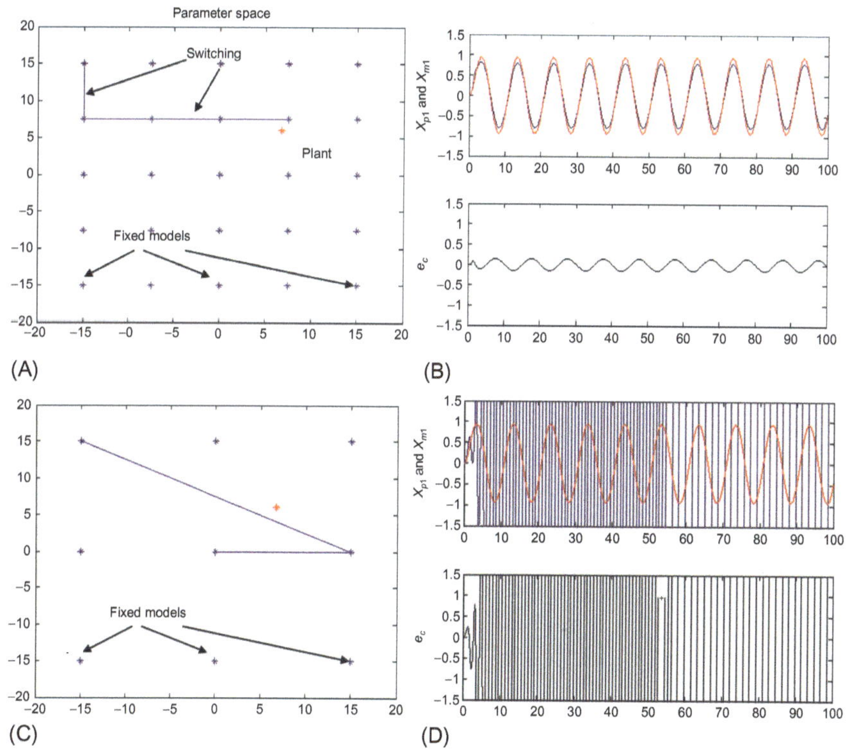

Fig. 5 Stable and unstable adaptive switching: (A) stable switching (25 models); (B) x_p and x_m *(top)*, e_c *(bottom)*; (C) unstable switching (nine models); (D) x_p and x_m *(top)*, e_c *(bottom)*.

If the plant is known to be stable a priori in the entire region of uncertainty, the models will also be stable and the above problem does not arise. However, when the plant (and consequently the models) can be unstable, suitable precautions have to be taken (see Comment 1).

In view of the above considerations, when multiple models are used, the following questions have to be addressed:

(a) When should switching occur?

(b) Are there precautions to be taken while one is switching between models?

(c) Will switching stop after a finite time?

(d) Is the switching scheme stable?

(e) When and why should model performance be terminated?

Comment 6 (Stability and robustness). As stated in Section 1, the cardinal requirements for good controller design have always been accuracy, stability, and robustness. From the beginning, methods proposed for adaptively controlling a plant have been accepted by the control community only when their stability has been rigorously proved. Since this chapter is primarily devoted to concepts and ideas, proofs are not included here. The reader is referred to [17–19] for proofs of the three methods included in this subsection.

Comment 7. In some situations, depending upon the criterion chosen, arbitrarily fast switching between models may occur. To prevent such arbitrarily fast switching, a "dwell time" is included between two switches.

3.1.2 Multiple Adaptive Models

In place of the single adaptive identification model used in Section 2, N identification models $\Sigma_1, \Sigma_2, \ldots, \Sigma_N$ can be set up to obtain N estimates at every instant of the unknown parameter vector θ_p of the plant. These are described by the differential equation

$$\Sigma_i : \dot{x}_i(t) = A_{\rm m} x_i(t) + [A_i(t) - A_{\rm m}] x_{\rm p}(t) + bu(t) \quad i \in \Omega = \{1, 2, \ldots, N\}, \tag{9}$$

with $x_i(t_0) = x_{\rm p}(t_0)$. Σ_i is the ith identification model, whose output is $x_i(t)$ and $i \in \{1, 2, \ldots, N\}$.

Comment 8. The N identification models are described by identical differential equations and differ only in the initial estimates of the parameter vector $\theta_{\rm p}$, defined as $\theta_i(t)$. $\theta_i^T(t)$ $(i \in \Omega)$ are the last rows of the N matrices A_i. Depending on the prior information available, the initial values of the parameter estimates θ_i are chosen to cover the region in parameter space where the unknown plant parameter vector is known to lie.

From the above assumptions, similar error equations and adaptive laws are obtained for the N models and have the forms

$$\dot{e}_i(t) = A_{\mathrm{m}} e_i(t) + b\tilde{\theta}_i^T(t) x_{\mathrm{p}}(t) \quad e_i(0) = 0 \quad i \in \Omega \tag{10}$$

and

$$\dot{\theta}_i(t) = \dot{\tilde{\theta}}_i(t) = -e_i^T P b x_{\mathrm{p}}(t) \quad \theta_i(t_0) = \theta_{i0}, \tag{11}$$

where $x_i(t) - x_{\mathrm{p}}(t) = e_i(t)$ and $\theta_i(t) - \theta_{\mathrm{p}} = \tilde{\theta}_i(t)$.

As described in Section 2, all the adaptive models converge to the plant with errors $e_i(t)$ tending to zero. At every instant the system with minimum identification error (ie, $J_1(t)$) is chosen.

Comment 9. It was stated earlier that conventional adaptive control results in a satisfactory response when the uncertainties are small. When the number N of adaptive models used is large, at least one of the models is close to the plant, and hence the responses observed are satisfactory (see Fig. 6). It is also worth mentioning that $e_i(t) \in \mathcal{L}^2 \cap \mathcal{L}^\infty$ $(i \in \Omega)$, so both criterion J_1 and criterion J_2 are bounded for all the models, and hence no special efforts are needed to terminate the adaptive process.

Comment 10. Obviously, the above procedure cannot be used in time-varying contexts. For the procedure to be applicable in such cases, all the models have to be reinitialized periodically (see Section 3.5).

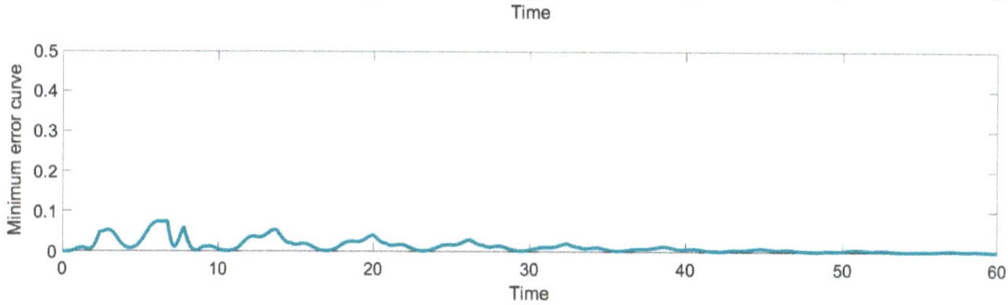

Fig. 6 Multiple adaptive models—all models converge with $e_i \to 0$.

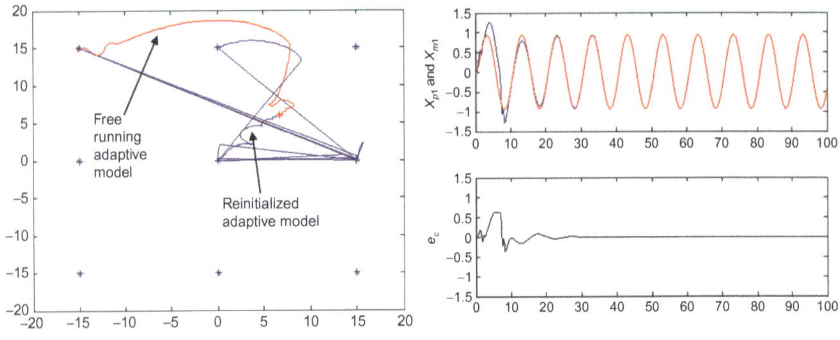

Fig. 7 Switching and tuning: (A) parameter adaptation; (B) x_p, x_m, and e_c.

In Fig. 6 the identification of a time-invariant plant by means of multiple adaptive models is shown. In Fig. 6A the responses of three models are indicated. It is seen that some of the models have large output errors initially. Since at every instant $\inf_i ||e_i(t)||$ determines the choice of the controller, the output error is seen to be small.

3.2 Switching and Tuning [14, 16, 17, 20]

In many cases, switching between fixed models results in an unsatisfactory response. More specifically, when the plant parameter varies with time (see Section 5), chattering of the parameter values is observed. To avoid such responses, an approach based on switching and tuning was proposed by Narendra and Balakrishnan [16]. Switching resulted in fast but coarse adaptation to avoid catastrophes, and tuning (ie, adaptation) was used to improve performance on a slower timescale. In [17], the approach was demonstrated to be stable.

While the approach described in Section 3.1 based on multiple adaptive models is satisfactory, it is computationally intensive as compared with the method based on fixed models. It was therefore proposed in [17] that $N-2$ fixed models and two adaptive models should be used to adaptively control systems of higher order with many unknown parameters. The fixed models, as described earlier, are located in the region of uncertainty, and on the basis of their responses and the performance criterion used, one of them is chosen at every instant. At the same time, an adaptive model is initiated at the latter value and is included in the set of identification models to be evaluated. If, at a later stage, a different *fixed model* yields the best response, the first adaptive model is terminated, and the second one is initiated at the new value.

Comment 11. While the stability of the approach has been demonstrated, as mentioned in Comment 7, rapid switching between two models has been observed occasionally in simulation studies. To avoid such behavior in practical applications, a second free-running adaptive model was also used. Thus at any time, only N fixed models and two adaptive models are involved in the adaptive process.

Over a period of roughly 2 decades, this method has been found to work satisfactorily in industrial applications.

3.3 Second-Level Adaptation [19, 21, 22]

While interactive or mutual adaptation is the next approach that was proposed chronologically, we now consider the fourth approach, which uses second-level adaptation, since it is closer in spirit to the methods discussed thus far.

Despite the great success of adaptation based on switching and tuning, it (along with the method based on switching) suffers from several shortcomings. The number of models needed in both cases is generally large since at least one model has parameters sufficiently close to that of the plant, if the response is to be satisfactory. A second drawback of this method is that very little information contained in all the models is used explicitly in the decision-making process.

Second-level adaptation, which was introduced by Han and Narendra [19], is a radical departure from either of the methods considered so far. In this approach, the identification of the unknown plant parameters is based on collective information provided by all the models during the adaptive process. While switching is fast but coarse, and switching and tuning is piecewise continuous, second-level adaptation results in a smooth control input (see Fig. 7).

3.3.1 *Adaptive Models*

As described earlier, if the plant and reference model are described by the equations

$$\dot{x}_p = A_p x_p + bu(t), \tag{12}$$

$$\dot{x_m} = A_m x_m + br(t), \tag{13}$$

the identification models are described by the equation

$$\dot{x}_i = A_m x_i + [A_i - A_m]x_p + bu. \tag{14}$$

It is assumed that A_p lies in the convex hull of the matrices A_i (or equivalently, θ_p lies in the convex hull of the vectors θ_i). If the initial condition $x_p(0) = x_{p0}$ of the plant is known, all the models are started with the same initial conditions. In such a case, the error equations of all the models have the form

$$\dot{e}_i = A_m e_i + b\tilde{\theta}_i^T(t)x_p(t) \quad e_i(0) = 0. \tag{15}$$

It has been shown that if all the models are adaptive, and θ_p lies in the convex hull $\theta_i(0)$, then it lies in the convex hull of $\theta_i(t)$ as adaptation proceeds. In particular, if

$$\theta_p = \sum_{i=1}^{N} \alpha_i \theta_i(t_0); \quad \alpha_i > 0 \quad \sum_{i=1}^{N} \alpha_i = 1, \tag{16}$$

then it also satisfies the equation

$$\theta_p = \sum_{i=1}^{N} \alpha_i \theta_i(t_t) \tag{17}$$

as adaptation proceeds. This is depicted pictorially in parameter space (for $n = 2$) in Fig. 8. Hence α_i can be considered as an alternate parameterization of the plant, and identification of the system reduces to the estimation of the constant vector α.

If the dimension of the unknown parameter vector θ_p is n, a minimum of $(n+1)$ models are needed to contain θ_p in their convex hull. Eq. (17) can also be used to express the output of the plant $x_p(t)$ in terms of the outputs $x_i(t)$ of the identification models. Because of linearity we have, (for $N = n+1$)

$$\sum_{i=1}^{n+1} \alpha_i e_i(t) = 0, \tag{18}$$

where $e_i(t)$ are the identification errors. Since these errors are continuously monitored, Eq. (18) can be used to determine α_i and hence θ_p, and, in turn, the feedback control vector.

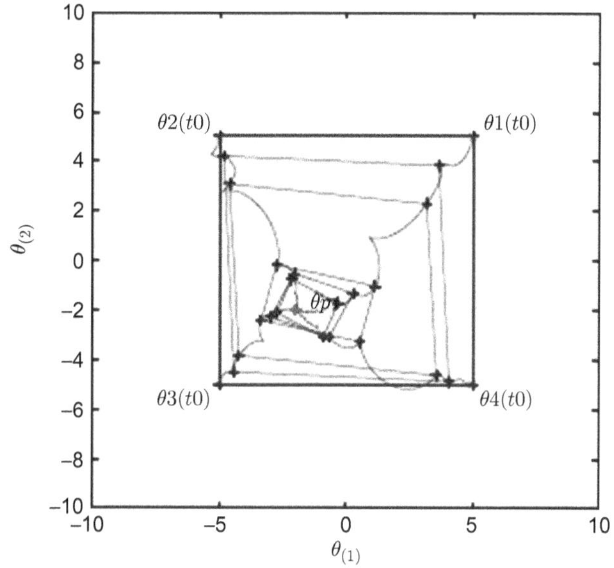

Fig. 8 Convex hull property in second-level adaptation.

3.3.2 Algebraic Computation of α

Let $E(t) \in R^{n \times n+1}$ be a matrix whose ith column is $e_i(t)$. Eq. (18) can then be expressed as

$$E(t)\alpha = 0. \tag{19}$$

Since $\sum_{i=1}^{n+1} \alpha_i = 1$, $\alpha_{n+1} = 1 - \sum_{i=1}^{n} \alpha_i$. Eq. (19) then reduces to

$$\bar{E}(t)\bar{\alpha} = -\alpha_{n+1}e_{n+1}(t), \tag{20}$$

where the columns of $\bar{E}$ are $e_i(t) - e_{n+1}(t)$ and $\bar{\alpha} = [\alpha_1, \alpha_2, \ldots, \alpha_n]^T$. Denoting $-\alpha_{n+1}e_{n+1}(t) = \bar{l}(t)$, we can solve Eq. (20) for $\bar{\alpha}$, algebraically (by inverting the matrix $\bar{E}$) as $\bar{\alpha} = E^{-1}(t)\bar{l}$.

3.3.3 Dynamic Estimation of $\bar{\alpha}$

A robust method for estimating $\bar{\alpha}$ while the system is in operation is by the use of a differential equation. Premultiplying both sides of Eq. (20) by $\bar{E}^T(t)$, we have the algebraic equation

$$\bar{E}^T(t)\bar{E}(t)\bar{\alpha} = \bar{E}^T(t)\bar{l}(t). \tag{21}$$

Let the estimate $\hat{\alpha}(t)$ be an estimate of $\bar{\alpha}$ at time t, which satisfies the differential equation

$$\dot{\hat{\alpha}}(t) = -\bar{E}^T(t)\bar{E}(t)\hat{\alpha}(t) + \bar{E}^T\bar{l}(t). \tag{22}$$

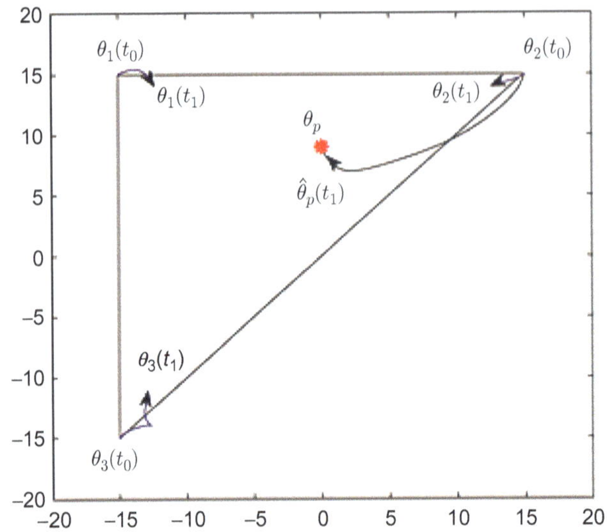

Fig. 9 Second-level and conventional adaptation.

If we define the error vector $\tilde{\alpha}(t) = \hat{\alpha}(t) - \bar{\alpha}$, it follows that $\tilde{\alpha}(t)$ satisfies the differential equation

$$\dot{\tilde{\alpha}}(t) = -\bar{E}^T(t)\bar{E}(t)\tilde{\alpha}(t), \tag{23}$$

which can be shown to be asymptotically stable if the input is persistently exciting. This implies that $\hat{\alpha}(t)$ tends to the constant vector $\bar{\alpha}$. In simulation studies it is found that second-level adaptation is an order of magnitude faster than the other methods, making it the prime candidate for adaptively controlling a rapidly time-varying system at the present time (see Fig. 9 and Section 5).

3.3.4 Multiple Fixed Models

While second-level adaptation with multiple adaptive models is very fast, it suffers from the same disadvantages mentioned in Section 3.1.2 in the discussion of switching (with multiple adaptive models). It cannot be directly extended to time-varying systems without reinitialization.

After adaptive methods had been described in [19] for estimation of plant parameters, Han and Narendra realized that fixed models could also be used for the same purpose (ie, θ_i in the previous discussion can be kept constant) while α and θ_p are being estimated. With very few changes, either algebraic computation or dynamic estimation described earlier can be used to determine $\hat{\alpha}(t)$. The errors $e_i(t)$ in this case are the errors between the outputs of the fixed models and the output of the plant.

Comment 12. Given the region D of uncertainty in parameter space, it is not easy to choose $(n+1)$ parameters θ_i so that D lies in the convex hull of θ_i. This can be alleviated by use of a larger number of models (see Section 6).

3.4 Interactive/Evolutionary Adaptation

The third approach (chronologically) involving multiple identification models is distinctly different, both in philosophy and in details, from the three methods discussed thus far. In the late 1990s, the author attempted to formulate the identification problem in a time-varying setting, to be used in place of the adaptive identification model that had been in use for 3 decades. Since time variations can take a variety of forms, appropriate assumptions had to be made to render the problem both precise and tractable. While many modifications to that formulation were made later [23–25], the basic problem is stated in the following subsection.

3.4.1 Statement of the Problem

A plant to be controlled is defined by a time-varying parameter vector $\theta \in \mathbb{R}^n$. $\theta(t)$ can assume N distinct constant values $\theta_1, \theta_2, \ldots, \theta_N$ in parameter space. The parameter θ switches randomly between these values, governed by a Markov chain. The values of θ_i ($i \in \{1, 2, \ldots, N\} = \Omega$), the transition probabilities, and the instants at which switching occurs are unknown. After switching from one parameter value to another (ie, from one environment to another), the system retains its value for a minimum of $T_{\min}$ units of time. It is further assumed that all the models are visited infinitely often (or the Markov chain is ergodic).

On the basis of the observed output (and assuming that the system is discrete), the objective is to identify the parameters $\theta_i (i \in \Omega)$.

Further, assuming that a suitable adaptive controller is also designed online for each of the estimated models, the objective is to ensure a satisfactory response of the overall time-varying system as it attempts to track the output of a time-invariant reference model.

3.4.2 The Method

N adaptive identification models are used to identify the N unknown plant models. Both the plant and the models are described by difference equations, and the adaptive identification models have adjustable step sizes.

If $e_i(k)$ are the N identification errors, the parameters of the N models are adjusted in such a manner that smaller errors correspond to larger step sizes. In contrast to the methods of switching and tuning and second-level adaptation, all the models are adjusted at every instant.

If the plant and the models are described by difference equations, the output $y(k)$ of the plant and $\hat{y}_j(k)$ of the jth model are described by

$$y(k+1) = p^T(k)\phi(k), \quad p(k) \in S = \{p_1, p_2, \ldots, p_N\} \tag{24}$$

and

$$\hat{y}_j(k+1) = \hat{p}_j(k)\phi(k), \tag{25}$$

where $p_j(k), \phi(k) \in \mathbb{R}^n$, $\phi(k)$ is the regression vector and $\hat{p}_j(k)$ is the estimate of the jth model of the vector $p(k)$ at time k. All the models are adaptively adjusted with use of the law

$$\hat{p}_j(k+1) = \hat{p}_j(k) - \eta_j(k)\phi(k)e_j(k+1), \tag{26}$$

where $e_j(k) = \hat{y}_j(k) - y(k)$, and the step size $\eta_j(k)$ varies from model to model:

$$\eta_j(k) = \frac{\frac{1}{J_j(k)}}{\sum_{i=0}^{N} \frac{1}{J_i(k)}}, \tag{27}$$

where J is the performance index defined in Section 2.

Comment 13. As the plant changes, the directions in which all the parameters are adjusted also change, with the difference being that step sizes of models with smaller errors are larger. The theoretical question concerns the asymptotic behavior of all the models. More precisely, the objective is to show that all N models converge to the finite set S (to which the plant belongs), with exactly one model converging to one element of S.

Empirical Results

Over an extended period of several years, thousands of simulation studies were performed in which all N identification models converged to all N plant models. A typical simulation is shown in Fig. 10.

Simulation

A discrete-time plant P has a parameter vector $p(k)$ which assumes values in a finite set S with three elements. Each element is a constant vector in $\mathbb{R}^2$, and is indicated by a square in Fig. 10. Three adaptive vectors $\hat{p}_1(k)$,

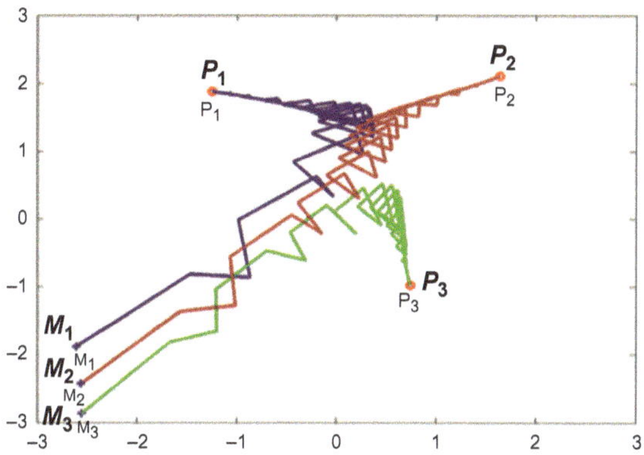

Fig. 10 Interactive adaptation

$\hat{p}_2(k)$, and $\hat{p}_3(k)$ are initialized at arbitrary values, denoted as M_1, M_2, and M_3 in parameter space. The plant and models have a common input $u(.)$. On the basis of the errors $e_i(k)$ $(= \hat{y}_i(k) - y(k))$, all the plant parameters are adjusted adaptively as given by Eq. (27). The location of $p(k)$ is governed by a Markov chain with transition probability $P_{ij} = \frac{1}{3}$. The models move in directions governed by the plant at that instant but ultimately converge to the set S, with each model converging to one unique element. It is not possible to predict which model converges to each of the elements of S.

Theoretical Results

Both static and dynamical systems have been investigated and numerous partial results have been obtained [23–25]. However, the basic problem is currently unresolved.

3.5 Summary

Among the four methods described in this section, switching using fixed models is the simplest. The computational overhead is minimal, and at every instant the model which represents the "best" approximation to the plant based on one of the criteria (see Section 2) is chosen. However, it requires a large number of models, which grows exponentially with the dimension of the unknown parameter vector. This is because the controller chosen to stabilize the final model must also stabilize the plant. Also, since the final model chosen is different from the plant, there is invariably a finite control error as $t \to \infty$. It is also worth noting that most of the models provide very little information for the final decision regarding the control input. Also, as mentioned earlier, there is always a residual error (for a general input) since the model chosen is different from the plant.

When all the models are adaptive, the computational overhead is large (since a $3n$-order differential equation is needed to represent an adaptively controlled plant of dimension n). However, the control input is continuous and relatively smooth. As mentioned in Section 3.1.1, since all the models converge to the plant, the method cannot be used in time-varying environments without re-initialization of the parameters.

Control using switching and tuning is significantly better than control using only switching, and adaptation ensures stability of the overall systems. It requires far fewer fixed models, and since it uses only two adaptive models, it is not computationally more intensive than switching control. Also, since one of the adaptive models is always reinitialized on the basis of performance, the method can be used without modifications in time-varying environments.

Second-level adaptation is an order of magnitude faster than the other two methods. However, locating $(n + 1)$ models so that the specified region of uncertainty lies in its convex hull is not a simple task. The problem is simplified to a great extent by locating the models at the corners of hypercubes, but this involves 2^n models. Since no switching is involved, the control input is invariably smooth, and with fixed models in the first level, is ideally suited when parameters vary with time.

The interactive/evolutionary method is ideally suited in those situations where the plant can change only in a finite number of ways that are unknown. Numerous modifications of the method have been simulated, and it has been shown to be ideally suited for exploring the regions in parameter space where the plant parameter might lie. Conceptually it is very elegant, and theorists would find it attractive since a theoretical proof of convergence has not been established despite years of effort.

4 EXTENSIONS

4.1 Linear Single-Input Single-Output Systems

In the discussions in the preceding section, the plant, the reference model, and the identification models in the switching, switching and tuning, and interactive/evolutionary methods (ie, 3.1, 3.2, 3.3) were all assumed to be described by state equations in a controllable canonical form with all state variables accessible. To make the methods more widely applicable and hence practically attractive, it is necessary to extend the concepts to more general systems. This has been done for switching, as well as for switching and tuning. More specifically, the two methods have been extended to the adaptive control of single-input single-output systems. Recently, a similar extension was proposed by Narendra et al. [26], who used second-level adaptation. In the case of the interactive/evolutionary method, the plant is assumed from the outset to be of a single-input single-output type, but since the representation is in discrete time, the counterparts of many of the problems that arise in the continuous case are substantially simpler.

4.2 Nonlinear Systems

Many of the concepts used in the identification of linear discrete-time systems by means of an autoregressive-moving average (ARMA) representation were extended in the 1990s to nonlinear systems with use of nonlinear autoregressive-moving average (NARMA) models. Yang et al. [27] recently argued that the multiple model approach provides a powerful tool for the identification and control of nonlinear systems. In particular, they consider plants described by the equation

$$y(k+d) = f_o(\phi(k)) + g_o(\phi(k))u(k), \tag{28}$$

where $\phi(k)$ is the regression vector without $u(k)$—that is, $\phi(k) = [y(k), y(k-1), \ldots, y(k-n), u(k-1), \ldots u(k-n)]^T$—so $u(k)$ appears linearly. The key idea of the multiple model approach is to partition $u(k)$ at time k into several intervals and to identify a local model that is linear in $u[k]$ within each interval. Since the equations are nonlinear, identification is carried out with neural networks. Both simulation studies and experimental results are presented in [27] to demonstrate the effectiveness of the control scheme.

Extensions to the adaptive control of nonlinear continuous time systems using neural networks and multiple models can also be found in [30] and [31].

4.3 Linear Stochastic Systems

In the early years of the last decade the results obtained for deterministic adaptive control included in Section 3 were extended to the stochastic case [28, 29]. The plant is described by the ARMA model with exogenous inputs model (ARMAX model):

$$A(q^{-1})y(k) = B(q^{-1})u(k) + C(q^{-1})w(t) \tag{29}$$

$$= p^T\psi(k), \tag{30}$$

where q^{-1} is a unit delay, A, B, and C are polynomials in q^{-1} with unknown parameters, p^T is an unknown plant parameter vector,

$$\psi(k)^T = [y(k), \ldots, y(k-n+1), u(k), \ldots, u(k-n+1), w(k), \ldots, w(k-n+1)] \tag{31}$$

is the regression vector, and $\{w(k)\}$ is a white noise sequence with $E[w(k)/(k-1)] = 0$ and $E[w^2(k)/(k-1)] = \sigma^2$. The objective is to estimate the parameters of the plant with use of multiple models. Since there is no estimation procedure which can be considered to be the best for stochastic systems, the multiple model approach was suggested to attempt all the different methods such as the extended least squares method, the recursive maximum likelihood method, and the output error with extended prediction model to decide which one would be the best.

5 SIMULATION STUDIES: TIME-VARYING SYSTEMS

The discussions thus far pertained mostly to time-invariant systems. The plant was assumed to be time invariant, and the performances of the different methods proposed were compared on the basis of speed and accuracy of the overall system. However, our interest in this chapter is in the performance of the different methods when the parameters of the plant vary with time. The difficulty in such a case is in characterizing the time variations. In this section, we consider three examples which are distinctly different from each other. These are (1) periodic continuous variations, (2) periodic piecewise constant variations, and (3) a random combination of periodic continuous variations and periodic piecewise constant variations as well as small-amplitude high-frequency variations and large-amplitude low-frequency variations. In all cases, we are interested in the parametric errors, as well as the identification and control errors.

The unknown plant and the reference model chosen for the simulation studies are described by the second-order differential equations

$$\text{(Plant)} \quad \ddot{y}(t) + \alpha(t)\dot{y}(t) + \beta(t)y(t) = u(t), \tag{32}$$

$$\text{(Reference model)} \quad \ddot{y}_{\mathrm{m}}(t) + \alpha_{\mathrm{m}}\dot{y}(t) + \beta_{\mathrm{m}}y_{\mathrm{m}}(t) = r(t), \tag{33}$$

where $\alpha(t)$ and $\beta(t)$ are the unknown time-varying parameters of the plant. The objective is to identify and control the plant with use of three of the methods described in Section 3 so that $y(t)$ follows the output $y_m(t)$ of the reference model asymptotically.

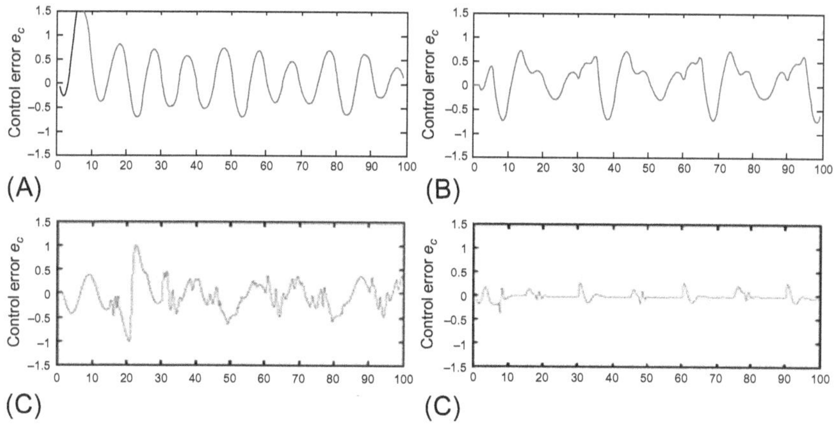

Fig. 11 Plant parameters switching periodically between two unstable values: (A) single model; (B) switching; (C) switching and tuning; (D) second-level adaptation.

In all three cases, the reference model is defined by $\alpha_m = 5$ and $\beta_m = 6$. The region of uncertainty of the plant parameters is large and is defined by the square

$$-15 \leq \alpha(t) \leq 15, \quad -15 \leq \beta(t) \leq 15$$

in parameter space.

Example 1. In this case, the parameter vector $[\alpha(t), \beta(t)]^T$ switches periodically, with period $T = 15$ units of time between two constants $[-10, -2]^T$ and $[-5, 10]^T$. With both constant values, the plant is seen to be unstable. The reference input $r(t)$ was chosen to be $6\sin(2\pi t)$. The performances of conventional adaptive control, switching, switching and tuning, and second-level adaptation were studied and are depicted in Fig. 11.

First-level adaptation using a single model (see Fig. 11A) is seen to result in very large errors. Switching and switching and tuning result in considerably better responses but the latter would not be considered satisfactory. Second-level adaptation results in an excellent response with small parameter and output errors. Only the control errors are shown in Fig. 11.

Example 2. In the second example, the parameter vector was assumed to vary periodically as $[\alpha(t), \beta(t)]^T = [12\sin(\omega t), 12(\cos\omega t)]^T$ so that the trajectory is a circle of radius 12 in parameter space. Only first-level and second-level adaptation were simulated, and the results are shown in Fig. 12.

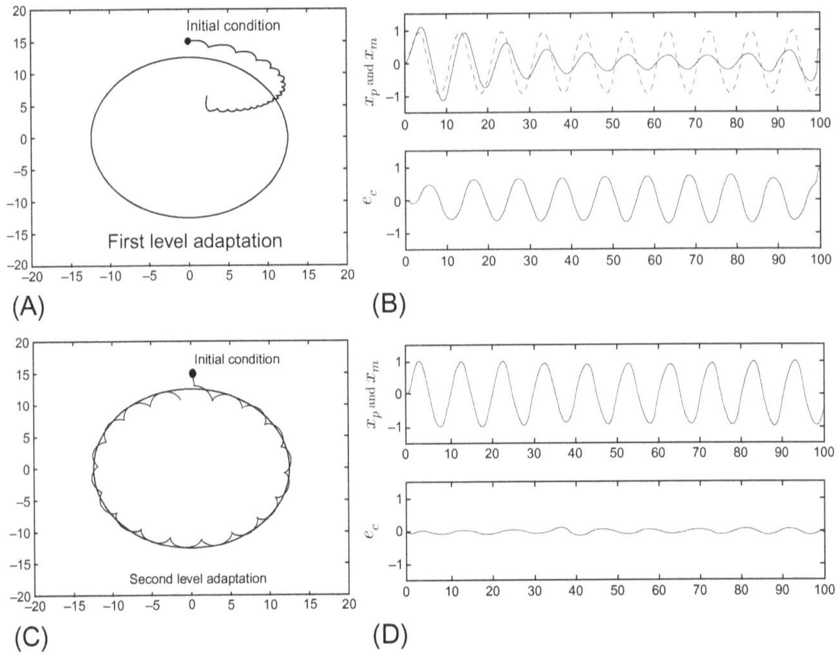

Fig. 12 Plant parameter sinusoidal variation: (A) single model—parameter space; (B) single model—response; (C) second-level adaptation—parameter space; (D) second-level adaptation—response.

While first-level adaptation fails entirely to track the parameter variations, second-level adaptation is seen to result in very satisfactory control with small parameter and control errors.

Example 3. As stated earlier, the time variations in this case are a random combination of different signals, as seen in Fig. 13. As expected, switching resulted in the chattering of parameter estimates, switching and tuning was considerably smoother, while with use of second-level adaptation, parameter tracking was excellent. The output errors are seen to be correspondingly large, medium, and small. These are seen in Figs. 14–16.

The poor response of adaptation based on switching may be attributed to the relatively "fast" variation of the parameters. In other simulations, with much faster variations, second-level adaptation may not fare much better (see Fig. 16). The research in progress at the present time, described in the following section, deals with such situations.

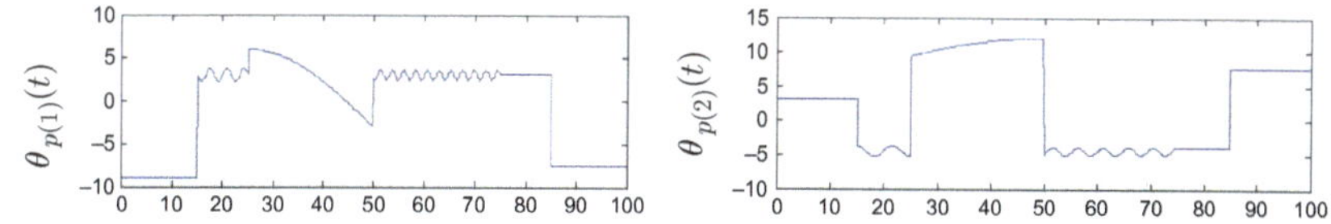

Fig. 13 Time variation of parameters.

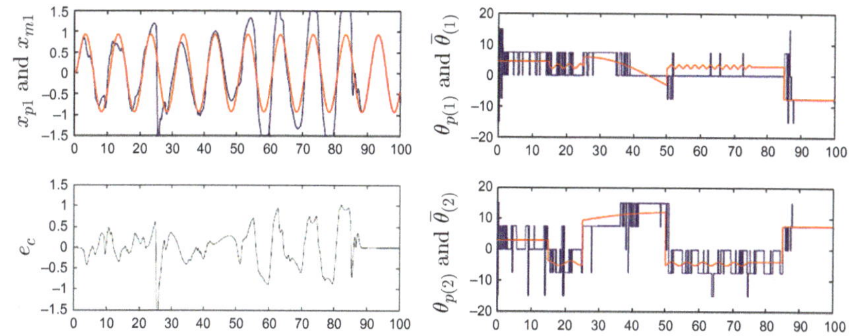

Fig. 14 Switching.

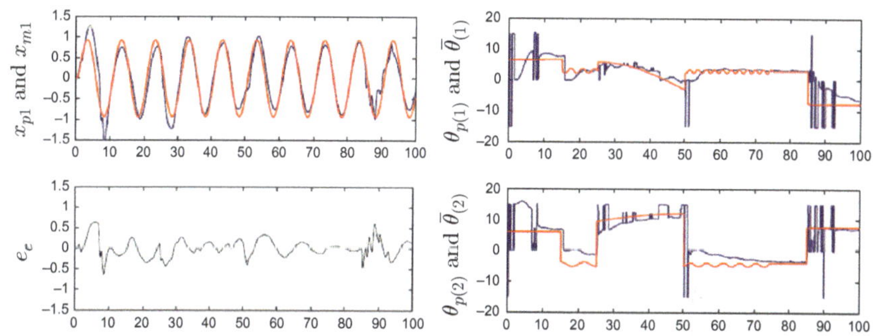

Fig. 15 Switching and tuning.

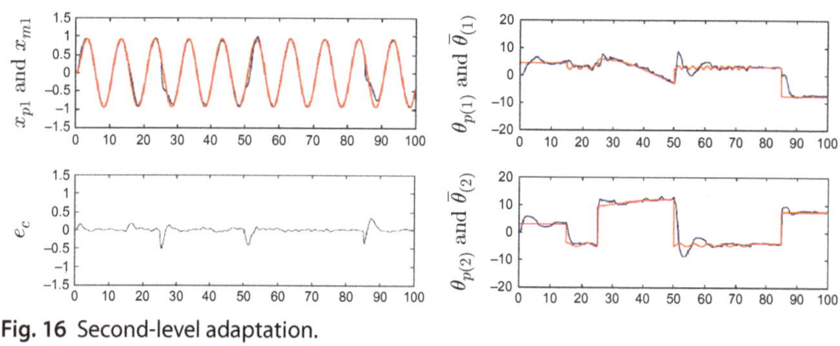

Fig. 16 Second-level adaptation.

6 WORK IN PROGRESS, PRINCIPAL ISSUES, AND FUTURE RESEARCH

The preceding sections summarized the work done during the past 2 decades on multiple model-based adaptive control by the author and his colleagues. The results obtained with each of the four constituent methods for adaptively controlling a time-varying plant were briefly described. Further, from Sections 3–5 it is clear that extensive simulation studies have been performed and considerable empirical knowledge has been gained. On the basis of this, the problems that are currently under investigation, the issues that they have given rise to, and possible directions for the future are briefly outlined in this section. All the work reported in this chapter was supported by the National Science Foundation under numerous grants. The problems proposed in this section were formulated only recently, are open-ended, and are continually being modified to be physically meaningful and mathematically tractable. As such, they should be of interest to students starting research in this area, as well as experienced engineers applying adaptive techniques to problems in industry.

The basic problem addressed in this chapter is the estimation of a parameter vector in a dynamical context. When the vector is a constant in parameter space, it can be estimated with use of powerful methods that currently exist. However, the primary interest of this chapter is in situations where the vector varies rapidly with time. When the problem is stated in such general terms, it shares many features with those that are currently being investigated in widely different fields, such as finance and security. In all cases, the central idea is to exploit the information from distributed sources.

Different assumptions regarding the parameter vector, the different domains where it can lie, the minimum time that it can spend in any specific region, its rate of variation, and the maximum change in amplitude all play important roles in the precise mathematical formulation of the problem. Equally important is the related information collected by each of the individual models scattered throughout the region of interest.

Every new formulation of the problem raises new sets of questions, and to classify these appropriately eight problems are embedded in the different subsections for future investigation. At first sight they may appear to be a random mix of unrelated problems. However, they help to highlight the complex nature of rapid adaptation, and are meant to serve as building blocks for achieving the overall objective.

6.1 The Need for a Hierarchy

For the solution of any problem, a set of assumptions have to be stated under which a solution is sought. Theoretical results related to stability, and convergence, are derived when these assumptions are valid. When the assumptions are relaxed, the scope of the problem changes and the theoretical and practical conclusions drawn have to be reevaluated.

All the methods treated earlier assumed that the plant parameter vector was constant, that the region of uncertainty was known, and that a large number of fixed or adaptive models were distributed throughout the region. For second-level adaptation, it was assumed that the unknown parameter vector lay in the convex hull of predetermined fixed models. By combination of the information from the different models, it was shown that fast and accurate convergence could be obtained.

When the space in which the unknown parameter $\theta_p(t)$ lies is of high dimension and when the variations in θ_p have large amplitudes and at the same time cannot be classified easily, many different models may be needed to generate the information required to make decisions. The location of the models, the information provided by them, and the manner in which they are combined all contribute to the success of the approach. At every instant not all models may be used. If resources are limited, the models may have to be redistributed. In all the above-mentioned cases a hierarchical approach is found to be an efficient and effective way for dealing with the problem.

Among many methods considered in earlier sections, second-level adaptation was found to be faster and more accurate than conventional adaptive control. Even the term "second level" was chosen to underscore the fact that decisions concerning the location of θ_p were based on information provided by models at a lower level. At the same time, it was assumed that it was known a priori that θ_p lay in the convex hull of the values of the parameters at a lower level. If $\theta_p(t)$ is rapidly varying, decisions have to be made at "higher levels of a hierarchy" to determine estimates of regions in which θ_p lies, the models that should be chosen to determine the control laws, the estimation criteria to be used, and when a new set of models are to be chosen.

6.2 Higher-Level Switching

Both switching and switching and tuning have proved to be efficient methods for rapidly controlling a time-invariant plant with unknown parameters. Both of them have been shown to be stable. It is not surprising that they have also been successful in adaptively controlling slowly varying plants, particularly when the parameters assume (almost) constant values over intervals of length greater than T, where T is empirically determined. When the plant parameter varies more rapidly (ie, T decreases), switching and switching and tuning have been used, not with errors from single models, but with errors from a cluster of models. This has enabled satisfactory control to be achieved in more rapidly varying environments. These empirical studies have led to many new theoretical questions which are included in Problem 1:

Problem 1. (a) How does the performance of switching algorithms degrade as the value of T decreases?

(b) If a conventional switching strategy is used, can sufficient conditions be derived on the value of T (ie, $T < T_1$) such that the resulting adaptive system is unstable?

(c) If modified switching using the average error of a cluster of models is attempted, can the improvement in stability be quantified?

Comment 14. While the stability of time-varying linear systems has been investigated extensively for more than 6 decades, the literature on the stability of time-varying adaptive systems is quite sparse.

6.3 Interactive Adaptation

Adjustment of multiple models at every instant in an interactive fashion was described in Section 3. The step sizes of all the models are governed by the models involved in the adaptive process. The author and his colleague Matthias Feiler have been working on the proof of convergence of the algorithm for more than 10 years. While every simulation has converged, the problem has resisted a theoretical solution. It has also raised numerous theoretical questions, some of which are included in Problems 2 and 3.

Problem 2. Under the conditions stated in Section 3, prove that N models converge to N values assumed by the plant, with exactly one model converging to one element of the set

Problem 3. As stated in Problem 2, every model is aware of the identification errors of all the models (ie, perfect communication exists). A natural modification would be to permit communication among neighbors in a graph. Such problems have been treated extensively in the literature on "consensus."

Problem 4. Second-level adaptation and interactive adaptation, while seemingly different, appear to share many common features. It is worth exploring how second-level adaptation could be modified to converge in a time-varying environment similar to that defined in the context of interactive adaptation.

6.4 Second-Level Adaptation

Extensive simulation studies in recent years have revealed that second-level adaptation has the fastest rate of convergence of all the methods that have been proposed. However, the method critically depends on the assumption that the true value of θ_p lies in the convex hull of some first-level model parameters. This has raised a number of questions in both the time-invariant and the time-varying cases which are included in Problems 5–7.

Problem 5. If θ_p is a constant vector and $\theta_p \in \Omega_1 \subset \Omega_2$, where Ω_1 is the convex hull of N_1 fixed models and Ω_2 is the convex hull of the N_2 fixed models, what advantages are gained (if any) by use of N_1 models instead of N_2 models? Are there any advantages in using N_2 models?

Problem 6. While second-level adaptation has been shown to be stable and robust when θ_p is a constant, the speed of convergence has not been demonstrated quantitatively. In particular, no quantitative comparisons are available regarding the convergence rates of first-level and second-level adaptation.

Problem 7. In a time-varying context, what considerations guide the choice of models when θ_p switches between constant values?

Thus far we have assumed that second-level adaptation is performed with no observation noise present. However, in almost all practical situations, observation noise is invariably present, at the output of the plant. However, the output of the models can be noisy or noise free, depending on the model structure used. This gives rise to the following problem.

Problem 8. When multiple adaptive identification models are used to identify the plant, the structure of each model is chosen to be of the series-parallel type to ensure stability. As described earlier, this has the form $\dot{\hat{x}}_p = A_m\hat{x} + [A(t) - A_m]x_p + bu$. However, for reasons described earlier, parallel models (ie, true models, with no input from the plant) described by equations of the form $\dot{\hat{x}} = \hat{A}_p\hat{x} + bu$ perform better than the former when a large amount of observation noise is present in $x(t)$. Methods for combining the two models are being investigated. The idea is to use parallel models when the error lies within precomputed limits, and to switch to series-parallel models outside these intervals, to ensure stability.

6.5 Combination of Methods

When it is known a priori that the plant parameter vector $\theta_p(t)$ can assume only a finite number of constant values, the interactive algorithm can be used to estimate it as t tends to infinity. At the same time, if the approximate regions where $\theta_p(t)$ may lie are known, different combinations of fixed models and second-level adaptation may be used to track $\theta_p(t)$. In those cases, where different approaches yield different information concerning $\theta_p(t)$ which is complementary, the designer can take advantage of it to improve the speed of convergence. Two simulation studies are included in this section to indicate the effect of combining different methods. In both cases, for simplicity, it is assumed that the plant parameter θ_p assumes four values, and that the plant is stable in each configuration. Further, our interest in these studies is only in identification (ie, estimation of the parameter vector $\theta_p(t) = [\theta_1(t), \theta_2(t)]^T$), rather than in control.

6.5.1 Simulation 1

A second-order plant in companion form ($\dot{x}_p = A_p x_p + bu$, with A_p in companion form), switches to four stable values—that is, $a_{p_1} = [-2.5, -2.5]$, $a_{p_2} = [-2.5, -3.5]$, $a_{p_3} = [-3.5, -3.5]$, and $a_{p_4} = [-3.5, -2.5]$. The switching is random, and is governed by a Markov chain with transition probability $P_{ij} = \frac{1}{4}$. The minimum time that $a_p(t)$ assumes at any specific value is 50 units. Our interest is in tracking the estimate of the parameter vector using the combination of methods discussed in this chapter. Because of space constraints, only the following three methods are shown in Fig. 17:

1. *Second-level adaptation.* A direct application of the method is shown in Fig. 17A. Since there is a discontinuous change in the value of $\alpha(t)$, the estimates are seen to have errors when the parameter value switches. However, for the given rate of variation, the response is satisfactory.
2. *Second-level adaptation and interactive adaptation combined.* Since the interactive approach attempts to determine the location of the constant values of the parameters asymptotically, the estimation is poor to begin with but is almost exact for large values of time (see Fig. 17B)
3. *Switching and second-level adaptation.* Assuming, a priori, that the parameter values could lie in specific regions, switching based on multiple models is used to select the region. Second-level adaptation is then used assuming the former to be true. This method is seen to be the best of the three considered, and is shown in Fig. 17C.

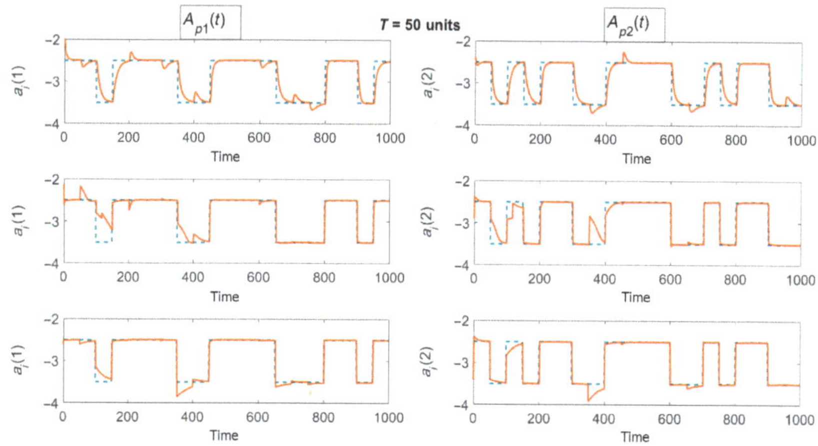

Fig. 17 $\theta_p(t)$ switching between four values (stable plant): (A) second-level adaptation; (B) second-level adaptation and interactive adaptation; (C) second-level adaptation and switching.

6.5.2 Simulation 2

The same experiments were repeated but the switching rate was increased substantially ($T = 20$ units). Combined switching and second-level adaptation was found to be far superior to regular second-level adaptation (see Fig. 18A and C).

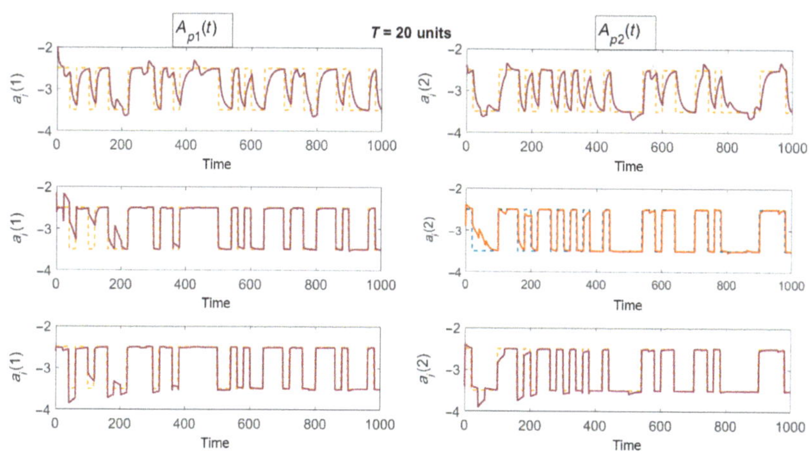

Fig. 18 Adaptation under faster switching: (A) second-level adaptation; (B) second-level adaptation and interactive adaptation; (C) second-level adaptation and switching.

7 CONCLUSION

In this chapter an attempt has been made to provide a broad overview of work done by the author and his colleagues on the adaptive control of time-varying systems using multiple models. The principal thesis of the chapter is that faster and more robust adaptation can be achieved by suitably combining the different existing methods, online, by using a hierarchical approach.

ACKNOWLEDGMENTS

The research reported here was supported by National Science Foundation grant 1503751.

It is a pleasure for the author to thank his former students Jeyendran Balakrishnan, Cheng Xiang, Osvaldo Driollet, Nicholas Oleng, Matthias Feiler, Zhuo Han, and Yu Wang, who worked with him for several years on different aspects of the problem. He is extremely grateful to his postdoctoral fellow Onur Cihan, who helped in many ways in the preparation of this chapter and without whose help this chapter would not have been completed. He is also thankful to Prashanth Harshangi, who helped in the preparation of all the simulations and provided valuable comments about parts of the chapter, and G. Mallesham, who also typed some parts of the chapter. The author is especially grateful to Snehasis Mukhopadhyay of Indiana University-Purdue University Indianapolis for his careful scrutiny of the chapter and for many valuable suggestions. All the work reported in this chapter was made possible by the continued support received from the National Science Foundation through grants 9521405, 0113239, 0400306, 0824118, and 1102178. Finally, the author thanks Sarangapani Jagannathan and Kyriakos G. Vamvoudakis for inviting him to contribute to this book.

REFERENCES

[1] D.T. Magill, Optimal adaptive estimation of sampled stochastic processes, IEEE Trans. Autom. Control 10 (1965) 434–439.
[2] D.G. Lainiotis, Partitioning: a unifying framework for adaptive systems. I. Estimation; II. Control, Proc. IEEE 64 (1976) 1126–1143, 1182–1197.
[3] M. Athans, D. Castanon, K. Dunn, C.S. Greene, W.H. Lee, N.R. Sandell, et al., The stochastic control of the F-8C aircraft using a multiple model adaptive control (MMAC) method: Part I. Equilibrium flight, IEEE Trans. Autom. Control 22 (1977) 768–780.
[4] D.W. Lane, P.S. Maybeck, Multiple model adaptive estimation applied to the lambda URV for failure detection and identification, in: Proceedings of the IEEE 33rd Conference on Decision and Control, Lake Buena Vista, FL, 1994, pp. 678–683.
[5] C. Yu, R.J. Roy, H. Kaufman, B.W. Bequette, Multiple-model adaptive predictive control of mean arterial pressure and cardiac output, IEEE Trans. Biomed. Eng. 39 (1992) 765–778.
[6] R.L. Moose, H.F.V. Landingham, D.H. McCabe, Modeling and estimation for tracking maneuvering targets, IEEE Trans. Aerosp. Electron. Syst. 15 (1979) 448–456.
[7] X.R. Li, Y. Bar-Shalom, Design of an interacting multiple model algorithm for air traffic control tracking, IEEE Trans. Control Syst. Technol. 1 (1993) 186–194.
[8] B. Martensson, Adaptive stabilization (Ph.D. dissertation), Lund Institute of Technology, Lund, Sweden, 1986.

[9] M. Fu, B.R. Barmish, Adaptive stabilization of linear systems via switching control, IEEE Trans. Autom. Control 31 (1986) 1097–1103.
[10] K. Poolla, S.J. Cusumano, A new approach to adaptive robust control: Parts I and II, Technical Report, Coordinated Science Laboratory, University of Illinois, Urbana, 1988.
[11] D.E. Miller, E.J. Davison, An adaptive controller which provides Lyapunov stability, IEEE Trans. Autom. Control 34 (1989) 599–609.
[12] D.E. Miller, Adaptive stabilization using a nonlinear time-varying controller, IEEE Trans. Autom. Control 39 (1994) 1347–1359.
[13] R.H. Middleton, G.C. Goodwin, D.J. Hill, D.Q. Mayne, Design issues in adaptive control, IEEE Trans. Autom. Control 33 (1988) 50–58.
[14] K.S. Narendra, J. Balakrishnan, Performance improvement in adaptive control systems using multiple models and switching, in: Proceedings of the Seventh Yale Workshop on Adaptive Learning System, Center for Systems Science, Yale University, New Haven, CT, 1992, pp. 27–33.
[15] A.S. Morse, Supervisory control of families of linear set point controllers, in: Proceedings of the IEEE 32nd Conference on Decision Control, San Antonio, TX, 1993.
[16] K.S. Narendra, J. Balakrishnan, Improving transient response of adaptive control systems using multiple models and switching, IEEE Trans. Autom. Control 39 (1994) 1861–1866.
[17] K.S. Narendra, J. Balakrishnan, Adaptive control using multiple models, IEEE Trans. Autom. Control 42 (1997) 171–187.
[18] A.S. Morse, Supervisory control of families of linear set-point controllers: Part I. Exact matching, IEEE Trans. Autom. Control 41 (1996) 1413–1431.
[19] Z. Han, K.S. Narendra, New concepts in adaptive control using multiple models, IEEE Trans. Autom. Control 57 (1) (2012) 78–89.
[20] K.S. Narendra, J. Balakrishnan, M.K. Ciliz, Adaptation and learning using multiple models, switching, and tuning, IEEE Control Syst. Mag. 15 (1995) 37–51.
[21] K.S. Narendra, Z. Han, A new approach to adaptive control using multiple models, Int. J. Adapt. Control Signal Process. 26 (2011) 778–799.
[22] K.S. Narendra, Z. Han, Adaptive control using collective information obtained from multiple models, in: Proceedings of the 18th IFAC World Congress, 2011, pp. 362–367.
[23] M.J. Feiler, K.S. Narendra, Adaptive control of rapidly time-varying systems, in: Proceedings of the 12th Yale Workshop on Adaptive and Learning Systems, Yale University, New Haven, CT, USA, 2003, pp. 33–42.
[24] M.J. Feiler, K.S. Narendra, Simultaneous Identification and Control of Linear Time Varying Plants Using Multiple Models, Technical Report No. 0401, Center for Systems Science, Yale University, New Haven, CT, USA, 2003.
[25] M.J. Feiler, K.S. Narendra, Simultaneous identification and control of time-varying systems, in: Proceedings of the 45th IEEE Conference on Decision and Control, 2006.
[26] K.S. Narendra, Y. Wang, W. Chen, Stability, Robustness, and Performance Issues in Second Level Adaptation, in: Proceedings of the 2014 American Control Conference, Portland, Oregon, 2014, pp. 2377–2382.
[27] Y. Yang, C. Xiang, T.H. Lee, Data-driven identification and control of nonlinear systems using multiple NARMA-L2 models, in: Proceedings of the 17th Yale Workshop on Adaptive and Learning Systems, Yale University, New Haven, CT, USA, 2015, pp. 78–83.
[28] K.S. Narendra, C. Xiang, Adaptive control of discrete-time systems using multiple models, IEEE Trans. Autom. Control 45 (9) (2000) 1669–1686.

[29] K.S. Narendra, O. Driollet, Stochastic adaptive control using multiple models in the presence of random disturbances, Int. J. Adapt. Control Signal Process. 15 (2001) 287–317.
[30] L. Chen, K.S. Narendra, Nonlinear adaptive control using neural networks and multiple models, Automatica 37 (8) (2001) 1245–1255.
[31] K.S. Narendra, K. George, Adaptive control of simple nonlinear systems using multiple models, in: Proceedings of the 2002 American Control Conference, Anchorage, Alaska, 2002, pp. 1779–1784.

CHAPTER THREE

Adaptive Stabilization of Uncertain Systems With Model-Based Control and Event-Triggered Feedback Updates

E. Garcia*, **P.J. Antsaklis**†
*Infoscitex Corporation, Wright-Patterson AFB, OH, United States
†University of Notre Dame, Notre Dame, IN, United States

1 INTRODUCTION

Networked control systems (NCSs) are control systems in which the system to be controlled and the different components such as actuators, controllers, and sensors are spatially distributed and communication between these components is achieved through the use of a digital communication network that is shared by other applications and control systems [1]. This type of implementation differ significantly from classical control systems, where all system components are attached directly to the control plant and exchange information using dedicated wiring. NCSs offer many advantages, such as cost-efficiency and improved functionality [2, 3], but new problems arise with use of this control implementation compared with the classical wired control systems.

One of the main problems in NCSs which is studied in this chapter is the design of control schemes accounting for the absence of feedback measurements for possibly long time intervals.

Many approaches have been introduced to reduce the amount of communication between sensor and controller nodes without compromising the stability of the control system. In particular, Walsh at al. [4] introduced a network control protocol called *try-once-discard* to allocate network resources to the different nodes in an NCS, all of which may access the network at any time assuming each access occurs before the maximum allowable transfer time. The minimum bit rate required to stabilize an NCS is a topic

Control of Complex Systems
http://dx.doi.org/10.1016/B978-0-12-805246-4.00003-3

of great interest. For example, Elia and Mitter [5] first characterize the least dense quantizer that stabilizes a linear discrete-time invariant system; then, in an extension, they search for the least dense scheme for sampling and quantization. In general, we need to quantize feedback signals more finely if the sampling time is large so as to maintain stability. The work in [6, 7] uses more efficiently the packet structure—that is, one obtains reduction in communication by sending packets of information using all data bits available (excluding overhead) in the structure of the packet.

Model uncertainties are especially important to consider when there are no feedback measurements for long time intervals. One of the attractive properties of classical closed-loop systems is that the appropriate design of closed-loop controllers reduces sensitivity to model uncertainties. Naturally, this property is lost as feedback measurements are no longer received at the controller node. An important approach that considers model uncertainties and that is able to reduce the sampling rate is that of model-based NCSs (MB-NCSs) [8, 9]. The MB-NCS approach incorporates an explicit model of the system to be controlled to provide estimates of the real state of the system during the time intervals when sensor measurements are unavailable. The length of the time intervals that an MB-NCS can operate in open loop depends on the accuracy of the model and of the estimated state compared with the real state. The results in this chapter provide a convenient way to obtain working models for MB-NCSs. Even when an accurate model is initially available, in many applications the parameters of a system may change over time because of the use and working conditions of the plant or of its components or because of the nature of the system; such is the case of switched systems, where the parameters switch values over time.

In this chapter we focus on applying parameter estimation algorithms in the MB-NCS context. Better knowledge of the plant dynamics will provide an improvement in the control action over the network—that is, we can operate over longer open-loop time intervals without the need for feedback. We use Kalman filters to estimate the unknown parameters of the system. A simple modification of the linear Kalman filter provides global convergence properties for deterministic systems. We estimate parameters in state-space noncanonical form and we do not require a persistently exciting input signal as most system identification algorithms do. The last two features represent important advantages compared with traditional closed-loop system identification algorithms. Since we do not estimate a canonical form of the state-space system, we do not need to know a priori the structure of the system and we only need to know its order or the dimension of the state. When the states of the system can all be measured, then this assumption is automatically fulfilled. With respect to MB-NCSs, we overcome the usual assumption that the controller is designed to stabilize the real system; this may be unrealistic since our knowledge of the plant dynamics is limited. As we will see, the identification process allows us to update not only the model but the controller itself so it can better respond to the dynamics of the real plant being controlled. In contrast to common adaptive techniques, we redefine the controller not at each time instant but only when the estimated parameter values are received in the controller node (ie, no closed-loop identification is performed).

To determine the time instants for the sensor node to transmit information to the controller node, we implement an event-triggered communication policy [10]. The main idea behind event-triggered control techniques is that the system transmits feedback measurements to the controller only when it is necessary—that is, only when a measure of the system state error is above a specified threshold. In particular, the model-based event-triggered (MB-ET) control framework uses the nominal model of the system to generate the control input between update time intervals and implements the event-triggered control paradigm to determine the update time instants. In this framework the state error that helps to determine the events is given by the difference between the plant state and the model state. The MB-ET control framework has also been used when it is not possible to measure the whole state but it is possible to measure an output of the system [11, 12]. This framework has also been used for stabilization of coupled subsystems [13] and for coordination of multiagent systems [14, 15].

In this chapter we introduce the adaptive MB-ET control framework where the system parameters are estimated online and the model of the system is upgraded. A system model that represents more accurately the real system dynamics provides a significant advantage in terms of reducing communication and improving the overall performance of the networked system.

This chapter is organized as follows. Section 2 presents stabilizing conditions for deterministic discrete-time systems using the MB-ET control framework. Parameter estimation using Kalman filters is addressed in Section 3. Stochastic systems are analyzed in Section 4. This chapter focuses on different applications where the adaptive MB-ET control scheme is implemented; these applications are investigated in Section 5. Conclusions are presented in Section 6.

2 STABILIZATION OF DETERMINISTIC DISCRETE-TIME SYSTEMS USING EVENT-TRIGGERED CONTROL

In this section we consider discrete-time uncertain systems described by

$$x(k+1) = Ax(k) + Bu(k). \tag{1}$$

The model of the system dynamics is represented by

$$\hat{x}(k+1) = \hat{A}\hat{x}(k) + \hat{B}u(k) \quad \text{for } k \in [k_i, k_{i+1}), \tag{2}$$

where $x \in \mathbb{R}^n$ represents the state of the system (or plant), $\hat{x} \in \mathbb{R}^n$ represents the state of the model, and $u \in \mathbb{R}^m$ is the control input. k_i represents the time instants at which the state of the model is updated with the current state of the real system for $i = 1, 2, \ldots$. In addition to the state of the model being updated at time instants k_i, the parameters of the model may or may not be updated at the same time instants depending on whether new estimated parameters are available. The matrices $\hat{A} \in \mathbb{R}^{n\times n}$ and $\hat{B} \in \mathbb{R}^{n\times m}$ are the parameters of the model and they represent estimated parameters of the real system matrices $A \in \mathbb{R}^{n\times n}$ and $B \in \mathbb{R}^{n\times m}$. The plant may be unstable (ie, one or more eigenvalues of A have magnitude greater than 1). We assume a state feedback control law given by $u(k) = K\hat{x}(k)$.

Model (2) represents a linear system, but if we introduce updates of its state it can be seen as an impulsive system when the same parameters are used and only the state is updated (state jump); this is a typical approach in [10, 16, 17]. In this chapter we follow the approach in [18] and we update the state of the model and we update the parameters of the model as well. So the model matrices switch when new parameter estimates are obtained. The real plant is a linear system and not a switched one as in model (2). In contrast to the model, the plant always operates with the same parameters and its state is never updated or reset.

In many applications it is convenient to drop the usual periodic update implementation in favor of one based on events. The event that the plant-model state error is equal to or greater than some predetermined threshold is typically used. A sensor node within the network broadcasts its local state only when it necessary (ie, when a measure of the local subsystem state error is above some predetermined threshold). In the MB-ET control framework the state error is defined as the difference between the state of the model and the state of the plant:

$$e(k) = \hat{x}(k) - x(k). \tag{3}$$

In this framework a copy of the model is needed in the sensor node to generate the model state and compute the state error (3).

In the adaptive MB-ET control framework we update both the parameters and the state of the model. In view of the fact that we need to be able to identify a system in a general state-space representation and not necessarily in canonical form as discussed in the literature [19–21], we will use the Kalman filter for identification of systems of the form (1). For the case we receive noisy measurements of the state, we will use the extended Kalman filter; see Section 4.

Theorem 1 provides stability bounds of the discrete-time MB-NCS when we update the model on the basis of error events. In the sensor node we measure the state of the plant and compare the norm of the state error (3) with a positive fixed threshold α. The plant state is transmitted to the controller and it is used to update the model state if the error is greater than the threshold—that is, when $\|e(k)\| > \alpha$.

Theorem 1. *Assume* $\|x(0)\| \leq \gamma$ for $0 < \gamma < \infty$. *Then system* (1) *with state feedback updates triggered when* $\|e(k)\| > \alpha$ *has a bounded state if the eigenvalues of* $A + BK$ *are within the unit circle of the complex plane.*

Proof. The system dynamics (1) can be written as

$$x(k+1) = (A + BK)x(k) + BKe(k) \tag{4}$$

after Eq. (3) and the control input $u = K\hat{x}$ have been used.

The response of the plant with initial time $k_0 = 0$ at any given time $k \geq 0$ is given by

$$x(k) = (A + BK)^k x(0) + \sum_{j=0}^{k-1} (A + BK)^{k-(j+1)} BKe(j), \tag{5}$$

where $e(k)$ is bounded by $\|e(k)\| \leq \alpha$ since when $\|e(k)\| > \alpha$ we send a measurement update and the error becomes $e(k_i) = 0$ at the update instant k_i. We can show that the state of the plant is bounded by evaluating its norm as follows:

$$\begin{aligned}\|x(k)\| &= \left\| (A + BK)^k x(0) + \sum_{j=0}^{k-1} (A + BK)^{k-(j+1)} BKe(j) \right\| \\ &\leq \left\| (A + BK)^k \right\| \|x(0)\| + \sum_{j=0}^{k-1} \left\| (A + BK)^{k-(j+1)} \right\| \|BK\| \left\| e(j) \right\| .\end{aligned}$$

Using the bound $\left\|(A+BK)^k\right\| \leq \beta\lambda^k, \lambda \in (0,1),\ \beta > 0$, we can write

$$\|x(k)\| \leq \gamma\beta\lambda^k + \alpha\beta\|BK\|\sum_{j=0}^{k-1}\lambda^{k-(j+1)}$$
$$\leq \gamma\beta\lambda^k + \frac{\alpha\beta\|BK\|(1-\lambda^k)}{1-\lambda}. \tag{6}$$

Note that $\lim_{k\to\infty}\|x(k)\| \leq \frac{\alpha\beta\|BK\|}{1-\lambda}$.

A choice of a stabilizing controller K, or in other words the fact that the eigenvalues of $A+BK$ lie strictly inside the unit circle, ensures that the first term on the right-hand side of Eq. (6) decreases exponentially with time and the second term is bounded for all time $k > 0$. □

Remark 1. Note that the design of the stabilizing gain K requires a robust-type controller since we have only the nominal parameters available. Note also that a specific location of the closed-loop eigenvalues is not needed in Theorem 1; they need only to have magnitude less than 1. The advantage of the framework in this section is that we can obtain accurate models that significantly reduce the initial uncertainty and stabilizing controllers can be easily obtained on the basis of these estimated parameters.

By applying a time-varying threshold, one can obtain asymptotic stability. From Eq. (6) we note that by varying the magnitude of the threshold value α in Theorem 1, we can obtain longer update intervals or a smaller output size as measured by the norm of the state as in Eq. (6). The norm of the state can be reduced by use of a smaller threshold α. A typical approach in event-triggered control is to reduce the threshold value as the system's state approaches its equilibrium point.

Consider again the plant and model described by Eqs. (1) and (2) and the control input $u = K\hat{x}$. Assume that the control input u renders the model (2) input-to-state stable (ISS) with respect to the measurement error e.

Definition 1. A continuous function $\gamma : [0,a) \to [0,\infty)$ is said to belong to class $\mathcal{K}$ if it is strictly increasing and $\gamma(0) = 0$. It is said to belong to class $\mathcal{K}_\infty$ if $a = \infty$ and $\gamma(r) \to \infty$ as $r \to \infty$.

For the definition of the ISS Lyapunov function for nonlinear discrete-time systems we use the following [22].

Definition 2. A continuous function $V : \mathbb{R}^n \to \mathbb{R}_0^+$ is called an ISS Lyapunov function for the dynamical system $x(k+1) = f(x(k), u(k))$, $x(k) \in \mathbb{R}^n, u(k) \in \mathbb{R}^m, k \in \mathbb{Z}^+$ if there exist class $\mathcal{K}_\infty$ functions $\alpha_1, \alpha_2, \alpha_3$, and γ satisfying

$$\alpha_1(\|x\|) \leq V(x) \leq \alpha_2(\|x\|), \tag{7}$$

$$V(f(x,u)) - V(x) \leq -\alpha_3(\|x\|) + \gamma(\|u\|). \tag{8}$$

The system $x(k+1) = f(x(k), u(k))$ is said to be ISS with respect to the input u if and only if there exists an ISS Lyapunov function for that system.

In this section we focus on linear systems and we choose a control law, $u = K\hat{x}$, that renders the closed-loop model (2) globally asymptotically stable. Any such K also renders the closed-loop model ISS with respect to the measurement errors. We proceed to choose a quadratic ISS Lyapunov function, $V = x^T Px$, where P is symmetric positive definite and is the solution of the closed-loop model discrete-time Lyapunov equation:

$$(\hat{A} + \hat{B}K)^T P(\hat{A} + \hat{B}K) - P = -Q, \tag{9}$$

where Q is a symmetric positive definite matrix.

Let us first analyze the case when $\hat{B} = B$ for simplicity and define the uncertainty$\tilde{A} = A - \hat{A}$; also assume that the following bounds on the uncertainty hold:

$$\begin{aligned} &\left\|\tilde{A}\right\| \leq \Delta_\mathrm{A}, \\ &\Delta_\mathrm{A} \|P\| \left(\Delta_\mathrm{A} + 2\left\|\hat{A} + BK\right\|\right) \leq \Delta < q, \end{aligned} \tag{10}$$

where $q = \underline{\sigma}(Q)$, the smallest singular value of Q in the Lyapunov equation (9). The parameters Δ and Δ_A represent given bounds on the norm of the uncertainty matrices. Theorem 2 provides stability conditions based on the error and the threshold value that is used to trigger a measurement update. The resulting (MB-ET) networked system is asymptotically stable. The error threshold is defined as a function of the norm of the state and the uncertainty bounds Δ and Δ_A. Similarly, the occurrence of an error event leads the sensor to send the current state measurement to the controller, which is used to update the state of the model. At the event time instants the error is equal to zero since the model state is made equal to the real state.

Theorem 2. *Assume that Eq. (10) holds. Then system* (1) *with* $u = K\hat{x}$ *and feedback based on error events generated when the relation*

$$\|e(k)\| > \alpha_\mathrm{r} \|x(k)\| \tag{11}$$

is first satisfied is globally asymptotically stable, where $0 < \sigma < 1$, $\alpha_\mathrm{r} = \min(\alpha_{\mathrm{r}1}, \alpha_{\mathrm{r}2})$, $\alpha_{\mathrm{r}1} = \sigma(q - \Delta)/2c_1$, $\alpha_{\mathrm{r}2} = (\sigma(q-\Delta)/2c_2)^{1/2}$, $c_1 = 2\|PBK\|\left(\Delta_A + \left\|\hat{A} + BK\right\|\right)$, *and* $c_2 = \left\|K^T B^T PBK\right\|$.

Proof. To prove this theorem we will set a bound on the first forward difference of $V = x^T Px$ along the trajectories of the system (4), which is equal to Eq. (1) when the input $u = K\hat{x}$ has already been substituted and expressed in terms of the state error. We can write

$$\begin{aligned}
&V(x(k+1)) - V(x(k)) \\
&= [x^T(A+BK)^T + e^T K^T B^T]P[(A+BK)x + BKe] - x^T Px \\
&= [x^T(\hat{A}+\tilde{A}+BK)^T + e^T K^T B^T]P[(\hat{A}+\tilde{A}+BK)x + BKe] - x^T Px \\
&= -x^T Qx + x^T[(\hat{A}+BK)^T P\tilde{A} + \tilde{A}^T P(\hat{A}+BK) + \tilde{A}^T P\tilde{A}]x \\
&\quad + 2x^T[(\hat{A}+\tilde{A}+BK)^T PBK]e + e^T K^T B^T PBKe.
\end{aligned}$$

We have just expressed $V(x(k+1)) - V(x(k))$ in terms of the model parameters and the uncertainty of the state matrix $\tilde{A}$. We now proceed to evaluate the contributions of each—the model, the uncertainty, and the error.

$$\begin{aligned}
&V(x(k+1)) - V(x(k)) \\
&\leq -q\|x\|^2 + \left\|(\hat{A}+BK)^T P\tilde{A} + \tilde{A}^T P(\hat{A}+BK) + \tilde{A}^T P\tilde{A}\right\| \|x\|^2 \\
&\quad + 2\left\|(\hat{A}+\tilde{A}+BK)^T PBK\right\| \|e\| \|x\| + \left\|K^T \hat{B}^T P\hat{B}K\right\| \|e\|^2 \\
&\leq (-q+\Delta)\|x\|^2 + c_1\|e\|\|x\| + c_2\|e\|^2,
\end{aligned}$$

where $e(k)$ is bounded by $\|e(k)\| \leq \alpha_r \|x(k)\|$. Then

$$V(x(k+1)) - V(x(k)) \leq (\sigma - 1)(q - \Delta)\|x\|^2, \tag{12}$$

which shows that V is guaranteed to decrease for any σ, $0 < \sigma < 1$. Here the state of the model is updated every time the error satisfies the condition imposed in Eq. (11). □

Remark 2. An important advantage of this approach is that we define the controller in terms of only the model parameters. The stabilizing threshold is designed with use of the model parameters $(\hat{A}, \hat{B})$ and some bounds on the uncertainty $\tilde{A}$, quantities that are known.

Remark 3. Also note that an accurate estimate of the plant parameters results in small values of Δ_A and Δ. It follows that a larger value for the threshold α_r can be obtained and longer update intervals can be achieved, which results in significant reduction in network communication.

3 PARAMETER ESTIMATION

In this section we will focus on parameter estimation of deterministic linear systems of the form (1) with no particular form of the matrices A and B. This identification problem can be solved with use of a linear Kalman filter; this implementation provides much better convergence properties than the extended Kalman filter, as was shown in [23].

In the particular case when the sensors provide noiseless measurements of the state, it is possible to modify the model that will be used for the Kalman filter equations so as to estimate the parameters A and B. To show this simple idea, let us focus on second-order autonomous systems (the idea can be easily extended to higher-order systems with deterministic inputs) with unknown time-invariant parameters a_{ij}:

$$\begin{bmatrix} x_1(k+1) \\ x_2(k+1) \end{bmatrix} = \begin{pmatrix} a_{11} & a_{12} \\ a_{21} & a_{22} \end{pmatrix} \begin{bmatrix} x_1(k) \\ x_2(k) \end{bmatrix}. \tag{13}$$

We do not know the values of the parameters and we receive only measurements of the states $x(0), \ldots, x(k)$. At any given step, because of the iterative nature of the Kalman filter, we need only $x(k)$ and $x(k-1)$. Now we rewrite Eq. (13) as

$$\begin{bmatrix} \bar{x}_1(k) \\ \bar{x}_2(k) \end{bmatrix} = \begin{bmatrix} x_1(k-1) & x_2(k-1) & 0 & 0 \\ 0 & 0 & x_1(k-1) & x_2(k-1) \end{bmatrix} \begin{bmatrix} \hat{a}_{11}(k) \\ \hat{a}_{12}(k) \\ \hat{a}_{21}(k) \\ \hat{a}_{22}(k) \end{bmatrix} = \hat{C}(k)\hat{a}(k), \tag{14a}$$

where $\hat{a}_{ij}$ represent the estimated values of the real parameters a_{ij}, and $\bar{x}_i$ are the estimates of the state based on the estimated parameters and on measurements of the real plant state x_i. Eq. (12) becomes the output equation of our filter. The state equation is described by

$$\begin{bmatrix} \hat{a}_{11}(k+1) \\ \hat{a}_{12}(k+1) \\ \hat{a}_{21}(k+1) \\ \hat{a}_{22}(k+1) \end{bmatrix} = \begin{pmatrix} 1 & 0 & 0 & 0 \\ 0 & 1 & 0 & 0 \\ 0 & 0 & 1 & 0 \\ 0 & 0 & 0 & 1 \end{pmatrix} \begin{bmatrix} \hat{a}_{11}(k) \\ \hat{a}_{12}(k) \\ \hat{a}_{21}(k) \\ \hat{a}_{22}(k) \end{bmatrix} = A_f\hat{a}(k). \tag{14b}$$

The state matrix of our filter model A_f is constant and the output matrix $\hat{C}(k)$ is time varying, but the overall model is linear. Therefore we can use a linear filter to obtain estimates of the parameters a_{ij} of the state matrix of the original system. We do not need any external input; we need only nonzero initial conditions on the state. We can also estimate the elements of matrices A and B if we receive measurements of the state and the deterministic input

$u(k)$. Any common inputs such as steps and sinusoidal inputs can be used for identification purposes; sinusoidal inputs do not need to have any particular frequency—that is, there is no requirement of a persistently exciting input, which makes this approach a suitable tool for adaptive stabilization. In this case we need to include estimates of the parameters of B in the state vector of the filter model and the input values in $\hat{C}(k)$. The output equation in this case is given by (for a second-order example with a single input):

$$\begin{bmatrix} \bar{x}_1(k) \\ \bar{x}_2(k) \end{bmatrix} = \begin{pmatrix} x_1(k-1) & x_2(k-1) & 0 & 0 & u(k-1) & 0 \\ 0 & 0 & x_1(k-1) & x_2(k-1) & 0 & u(k-1) \end{pmatrix} \begin{bmatrix} \hat{a}_{11}(k) \\ \hat{a}_{12}(k) \\ \hat{a}_{21}(k) \\ \hat{a}_{22}(k) \\ \hat{b}_1(k) \\ \hat{b}_2(k) \end{bmatrix}. \tag{15a}$$

The filter state matrix is expanded according to the number of additional states:

$$\begin{bmatrix} \hat{a}_{11}(k+1) \\ \hat{a}_{12}(k+1) \\ \hat{a}_{21}(k+1) \\ \hat{a}_{22}(k+1) \\ \hat{b}_1(k+1) \\ \hat{b}_2(k+1) \end{bmatrix} = \begin{pmatrix} 1 & 0 & 0 & 0 & 0 & 0 \\ 0 & 1 & 0 & 0 & 0 & 0 \\ 0 & 0 & 1 & 0 & 0 & 0 \\ 0 & 0 & 0 & 1 & 0 & 0 \\ 0 & 0 & 0 & 0 & 1 & 0 \\ 0 & 0 & 0 & 0 & 0 & 1 \end{pmatrix} \begin{bmatrix} \hat{a}_{11}(k) \\ \hat{a}_{12}(k) \\ \hat{a}_{21}(k) \\ \hat{a}_{22}(k) \\ \hat{b}_1(k) \\ \hat{b}_2(k) \end{bmatrix}. \tag{15b}$$

Both filter models, Eqs. (14) and (15), can be easily applied to higher-order systems by following the structures of Eqs. (14) and (15). A clear disadvantage is that the order of the filter is n^2, where n is the dimension of the state of the original system.

Song and Grizzle [24] have shown that the linear time-varying Kalman filter is a global asymptotic observer for the underlying deterministic system. Consider the deterministic system described by Eq. (14), and the associated noisy system:

$$\begin{aligned} \hat{a}(k+1) &= A_f \hat{a}(k) + N w_p(k), \\ \hat{y}(k) &= \hat{C}(k)\hat{a}(k) + M v_p(k), \end{aligned} \tag{16}$$

where the design parameters M and N are chosen to be positive definite matrices and the artificial noise processes w_p and v_p are white, zero-mean, uncorrelated processes and have known covariance matrices Q and R, respectively. Theorem 3 states the convergence of the estimation error.

Theorem 3 ([24]). *Consider the deterministic system* (14) *and the Kalman filter associated with Eq. (16). Suppose that the deterministic system is uniformly observable and* $A_f(k)$ *is invertible for all* k, *and that* $\|A\| := \sup\{\|A_f(k)\| : k \geq 0\}$ *and* $\|C\| := \sup\{\|C_{(k)}\| : k \geq 0\}$ *are bounded. Then the Kalman filter for the noisy system* (16) *is a global, uniform asymptotic observer for the deterministic system* (14).

Let us focus on the details pertaining to our specific model. From Eq. (14b) we can see that $A_f(k) = A_f = I$ is constant, bounded, and invertible for all k. The output matrix is built by use of the measurements of the deterministic system. For unstable systems it is required that the initial condition of the system is finite. The matrices M and N are chosen simply to be identity matrices of appropriate dimensions. The problem here is that the pair $(A_f, \hat{C}(k))$ is not observable. A simple solution is to increase the number of measurements used in the output equation (14a) as shown in [23], although this is not a necessary condition.

Remark 4. For systems of the form $x(k+1) = Ax(k)$, we do not need any external input to successfully identify all elements of the state matrix A; we need only nonzero initial conditions on the state.

Remark 5. In MB-NCSs we mainly focus on stabilization of systems of the form given by Eq. (1). For systems of this type we can estimate the elements of matrices A and B if measurements of the state and the deterministic input $u(k)$ are available. If we try to estimate the parameters online and at the same time try to stabilize the system, then the same input that is used for stabilization can be used for estimation purposes. This relaxation on the need for a persistently exciting input is a great advantage over traditional parameter estimation methods and it can be intuitively explained by our realizing that we are estimating the states, not the parameters, of a modified linear system. In this case we try to meet the conditions for state estimation such as those expressed in Theorem 3. In general, we try to find the best state estimate (the estimated parameters in our case) that minimize the second-order moment of the difference between the output of the system $x(k)$ and our predicted output $\bar{x}(k)$. It is also important to note that one particular case in which it is not possible to estimate the parameters is when the initial state of the system is zero. Since the main objective is stabilization of unstable linear systems, then this scenario does not represent a problem. Now, if a perturbation happens to move the state from its equilibrium point, then we are in a position to estimate the parameters and to compute a stabilizing control input.

Remark 6. We are not performing closed-loop identification since most of the time there is no communication between the sensor and the controller, and the model parameters are not updated until the sensor decides to broadcast a measurement; thus the overall system operates in open-loop mode.

4 ADAPTIVE STABILIZATION OF STOCHASTIC MODEL-BASED NETWORKED CONTROL SYSTEMS

In this section we study stochastic systems of the form

$$\begin{aligned} x(k+1) &= Ax(k) + Bu(k) + w(k), \\ y(k) &= x(k) + v(k), \end{aligned} \tag{17}$$

where the noise processes w and v are white, Gaussian, uncorrelated, zero-mean processes and have known covariance matrices Q and R, respectively. The model of the system is still given by Eq. (2) and, since we measure only $y(k)$, the error is now given by

$$e(k) = \hat{x}(k) - y(k). \tag{18}$$

Theorem 4. *Assume $x(0)$ is a random variable with a Gaussian distribution (μ_0, Σ_0). Then the state of the stochastic model-based networked system* (17) *with feedback based on error events generated when $\|e(k)\| > \alpha$, with $e(k)$ given by Eq. (18) and fixed threshold α, has finite mean and covariance for all k if the eigenvalues of $A + BK$ are within the unit circle of the complex plane.*

Proof. The state of Eq. (17) can be expressed by means of the linear system

$$x(k+1) = (A+BK)x(k) + BKe(k) + BKv(k) + w(k). \tag{19}$$

We can see that the state $x(k)$ is a Gaussian random variable for all k with mean and covariance given by

$$\mu_{(k)} = (A+BK)^k \mu_0 + \sum_{j=0}^{k-1} (A+BK)^{k-(j+1)} BKe(j), \tag{20}$$

$$\begin{aligned} \Sigma_{(k)} &= (A+BK)\Sigma_{(k-1)}(A+BK)^T + (BKR(BK)^T + Q) \\ &= (A+BK)^k \Sigma_0((A+BK)^k)^T \\ &\quad + \sum_{j=0}^{k-1} (A+BK)^j (BKR(BK)^T + Q)((A+BK)^j)^T. \end{aligned} \tag{21}$$

In Theorem 1 we showed that Eq. (20) is bounded if the eigenvalues of $A + BK$ lie strictly inside the unit circle. For the covariance $\Sigma_{(k)}$ to converge, we need the series (21) to be summable as $k \to \infty$, and we achieve this by making $(A + BK)^j$ converge to 0 (ie, if $A + BK$ is stable, then the covariance converges to a finite value). For a formal proof, we use Theorem 5. □

Theorem 5 ([25]). *Suppose $A + BK$ is stable. Then there is a positive semi-definite matrix Σ_∞ such that $\lim_{k\to\infty} \Sigma_{(k)} = \Sigma_\infty$. Moreover, Σ_∞ is the unique solution of the linear equation*

$$\Sigma_\infty = (A + BK)\Sigma_\infty(A + BK)^T + (BKR(BK)^T + Q). \tag{22}$$

By use of this theorem, the covariance $\Sigma_{(k)}$ of the state $x(k)$ in Eq. (17) is finite for all k if the condition in Theorem 4 is satisfied.

5 APPLICATIONS OF THE ADAPTIVE MODEL-BASED EVENT-TRIGGERED CONTROL FRAMEWORK

In Fig. 1 copies of the model and controller gain are contained in the sensor to generate the state $\hat{x}(k)$, which is used to compute the state error $e(k)$. The copies of the model and the controller are also used to generate the control input at the sensor node, which is needed to construct the filter equations. The parameters are estimated and the state error is evaluated at the sensor node at every sampling time of the system. At the update instants the sensor will transmit the measured state and the new value of the estimated parameters, or it can send a smaller packet containing only the state of the model if no significant variation in the parameter values has been detected.

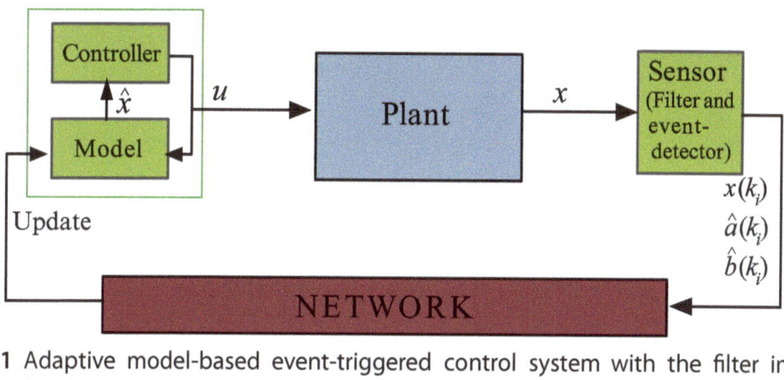

Fig. 1 Adaptive model-based event-triggered control system with the filter implemented in the sensor node.

5.1 Continuous-Time Systems

The adaptive MB-ET control framework described in this chapter can be used for stabilization of uncertain continuous-time systems. Consider an uncertain continuous-time system given by

$$\dot{x}(t) = A_c x(t) + B_c u(t),$$

where $u(t) = K\hat{x}(k)$ for $t \in [t_k, t_{k+1})$. The index t_k is used to determine the periodic sampling instants—that is, $h = t_{k+1} - t_k$ is the constant sampling interval. The state of the system is measured at every time instant t_k and the state error is computed and compared against the threshold. However, the state of the system is not transmitted at every sampling instant t_k. The event condition needs to be satisfied in order to trigger an event and send the current state measurement to the controller. We define the state error $e(k) = \hat{x}(k) - x(t_k)$. The events are triggered when $\|e(k)\| > \alpha$. Then we can write the following:

$$\dot{x}(t) = A_c x(t) + B_c K(x(t_k) + e(k)).$$

The response of the continuous-time system at any sampling instant t_k for $k = 1, 2, \ldots$ can be written as follows:

$$x(t_{k+1}) = Ax(t_k) + BKx(t_k) + BKe(k),$$

where $A = e^{A_c h}$ and $B = \int_0^h e^{A_c(h-s)} ds \cdot B_c$. Hence the state $x(t_k)$ can be bounded in a similar way as was shown in Theorem 1.

Example 1. Consider the model of a batch reactor [26]. In this example we consider the state feedback case and we also consider model uncertainties. The real matrices of the unstable continuous-time system are given by

$$A_c = \begin{bmatrix} 1.38 & -0.208 & 6.715 & -5.676 \\ -0.581 & -4.29 & 0 & 0.675 \\ 1.067 & 4.273 & -6.654 & 5.893 \\ 0.048 & 4.273 & 1.343 & -2.104 \end{bmatrix} \quad B_c = \begin{bmatrix} 0 & 0 \\ 5.679 & 0 \\ 1.136 & -3.146 \\ 1.136 & 0 \end{bmatrix}.$$

We use the notation A_c and B_c to emphasize the fact that these are the system parameters corresponding to the continuous-time system. All of the elements of the matrices A_c and B_c are unknown. The sampling period is $h = 0.005$ s. The Kalman filter implemented in the sensor location provides estimates of all parameters of the discretized system parameters. When the state error is greater than a predefined threshold, the sensor sends the latest

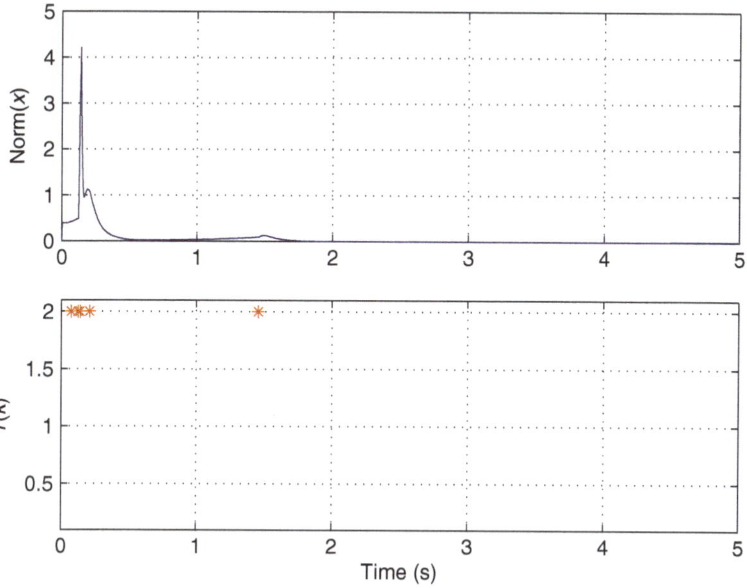

Fig. 2 Stabilization of the continuous-time batch reactor in Example 1 using adaptive model-based even-triggered control.

measured state to the controller. The fixed threshold used in this example was $\alpha = 0.09$. The most recent parameter estimates are sent only if there is a significant variation with respect to the previous updated parameters. For illustration purposes we construct the communication signal $r(k)$ as

$$r(k) = \begin{cases} 0 \text{ if no packet is sent,} \\ 1 \text{ if only the state is sent,} \\ 2 \text{ if both parameters and the state are sent.} \end{cases} \tag{23}$$

The control input obtained from the initial model did not stabilize the real plant. This can be seen at the beginning of the simulation in Fig. 2. After a few iterations the sensor is able to provide better estimates of the parameters. The controller is redesigned on the basis of the upgraded model and the state of the model is updated as well. The estimated discretized parameters that were obtained in this simulation are given by

$$\hat{A} = \begin{bmatrix} 1.007 & -0.001 & 0.0331 & -0.0279 \\ -0.0029 & 0.9788 & 0 & 0.0033 \\ 0.0052 & 0.021 & 0.9675 & 0.0287 \\ 0.0002 & 0.0211 & 0.0066 & 0.9896 \end{bmatrix} \quad \hat{B} = \begin{bmatrix} 0 & -0.0003 \\ 0.0281 & 0 \\ 0.006 & -0.0155 \\ 0.006 & -0.0001 \end{bmatrix}.$$

5.2 Actuator Fault Detection and Reconfiguration

The problem of fault detection and redesign of stabilizing controllers is studied in this section with use of the adaptive MB-ET control framework. We assume that multiple backup actuators are available. In the case that a stabilizing controller cannot be found (the fault makes the system not stabilizable), then the control scheme shuts down the faulty actuator configuration and turns on a healthy backup actuator in order to continue with the operation of the system and maintain closed-loop stability.

Consider an unknown linear discrete-time system represented by

$$x(k+1) = Ax(k) + B_s u(k), \tag{24}$$

where B_s represents the selected actuator configuration from the available choices B_i, where B_i represents an element of the backup actuator set B^{bp} with unknown backup actuator matrices:

$$B_i \in B^{\text{bp}}. \tag{25}$$

We implement a linear time-varying Kalman filter as described in Section 3 to estimate the current system parameters. In the case where we can obtain noise-free measurements of the state, we can estimate all parameters of the system, but here we focus on the elements of B_s. We also need to determine, after some change in the parameters of B_s has been detected, if the system remains stabilizable. A new controller needs to be calculated if the system can be stabilized under the partial fault detected; otherwise the faulty actuator has to be shut down for maintenance, a new actuator has to be activated, and its parameters have to be identified to compute the corresponding controller.

Example 2. Consider the unstable discrete-time system given by

$$A = \begin{bmatrix} 0.89 & 1.2 \\ 0.52 & 0.18 \end{bmatrix}$$

and the set of actuator configurations given by

$$B_1 = \begin{bmatrix} 0.7 \\ 0.5 \end{bmatrix}, \quad B_2 = \begin{bmatrix} -0.4 \\ 1 \end{bmatrix}, \quad B_3 = \begin{bmatrix} 2 \\ 0.4 \end{bmatrix}, \quad B_4 = \begin{bmatrix} -1 \\ -0.8 \end{bmatrix}.$$

A fault was introduced at time $k = 320$ and successfully detected moments later. The resulting configuration was not stabilizable, and configuration B_2 was activated. At time $k = 560$ a partial update was introduced and the new parameters were estimated. In this case the resulting configuration

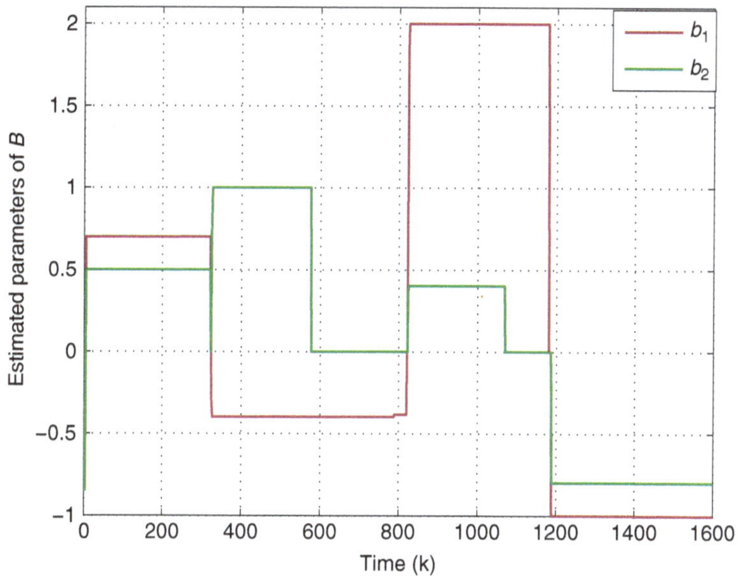

Fig. 3 Estimated parameters for the faulty actuator in Example 2.

could be stabilized and the new controller was designed. At time $k = 790$ the system experienced a fault and actuator configuration B_3 was activated. A partial fault was experienced at time $k = 1050$ and a new controller was computed. The last fault occurred at time $k = 1180$, and configuration B_4 was activated.

The estimated parameters corresponding to the elements of the matrix B_s (ie, of the activated actuator configuration) are shown in Fig. 3. The response of the system and the communication signal $r(k)$ as defined in Eq. (23) are shown in Fig. 4 for a threshold value of $\alpha = 0.1$. After a fault has occurred, an interval of time is needed in order to estimate the new parameters, to determine if the new pair (A, B_s) is stabilizable, and to activate a different actuator if necessary.

5.3 Identification and Rejection of Input Disturbances

A very interesting application of the modified Kalman filter that was discussed in Section 3 is related to the joint estimation of parameters and exogenous disturbances. The aim in this section is to stabilize systems with uncertain parameters that are affected by unknown piecewise constant external disturbances. Fig. 5 illustrates this scenario. The presence of a communication network does not allow the use of continuous feedback

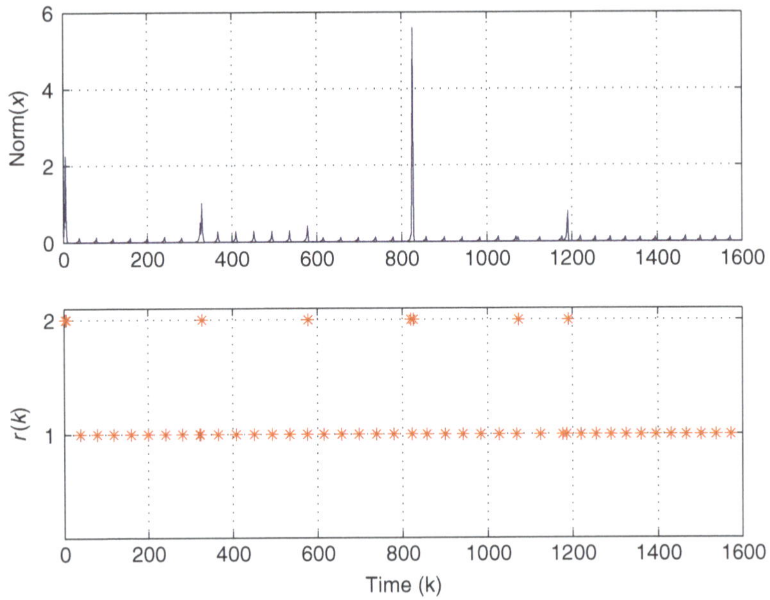

Fig. 4 Norm of the states of the system and communication signal in Example 2.

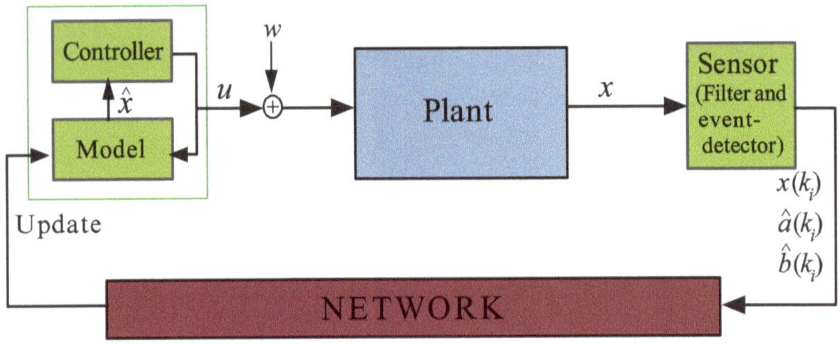

Fig. 5 Model-based networked system with external disturbance.

for control, and the networked system must operate in open-loop mode for possibly long time intervals. Because of these communication constraints we use the adaptive MB-ET control framework for stabilization of the uncertain system. In the following we present the details on how to estimate the system parameters and the unknown disturbance.

Consider the following type of uncertain discrete-time system:

$$x(k+1) = Ax(k) + B(u(k) + w(k)), \tag{26}$$

where $x \in \mathbb{R}^n$, $u \in \mathbb{R}^m$, and $w \in \mathbb{R}^m$ represent the state of the system, the control input, and the unknown input disturbance, respectively. We consider the particular case where B is known with certainty but the state matrix A and the input disturbance are unknown. Additionally, we restrict this work to the case when the input disturbances are given by piecewise constant signals and this represents all the information available about the external disturbance. The time instants at which the external disturbance changes its value are also unknown and they must be detected by our estimation algorithm.

Consider a second-order system with a single input for ease of exposition:

$$\begin{bmatrix} x_1(k+1) \\ x_2(k+1) \end{bmatrix} = \begin{pmatrix} a_{11} & a_{12} \\ a_{21} & a_{22} \end{pmatrix} \begin{bmatrix} x_1(k) \\ x_2(k) \end{bmatrix} + \begin{pmatrix} b_1 \\ b_2 \end{pmatrix} (u(k) + w(k)). \tag{27}$$

Similarly to Section 3, this approach can be easily generalized for any dimensions n and m, as long as the system is of the form (26). We rewrite Eq. (27) in the following form:

$$\begin{aligned} \begin{bmatrix} \bar{x}_1(k) \\ \bar{x}_2(k) \end{bmatrix} &= \begin{pmatrix} x_1(k-1) & x_2(k-1) & 0 & 0 & b_1 \\ 0 & 0 & x_1(k-1) & x_2(k-1) & b_2 \end{pmatrix} \begin{bmatrix} \hat{a}_{11}(k) \\ \hat{a}_{12}(k) \\ \hat{a}_{21}(k) \\ \hat{a}_{22}(k) \\ \hat{w}(k) \end{bmatrix} + \begin{pmatrix} b_1 \\ b_2 \end{pmatrix} u(k) \\ &= \hat{C}(k)\hat{a}(k) + \bar{u}(k). \end{aligned} \tag{28}$$

Eq. (28) represents the output equation for the filter. Since we consider linear time-invariant systems and piecewise constant input disturbances, the state equation for the filter is given by

$$\begin{bmatrix} \hat{a}_{11}(k+1) \\ \hat{a}_{12}(k+1) \\ \hat{a}_{21}(k+1) \\ \hat{a}_{22}(k+1) \\ \hat{w}(k+1) \end{bmatrix} = \begin{pmatrix} 1 & 0 & 0 & 0 & 0 \\ 0 & 1 & 0 & 0 & 0 \\ 0 & 0 & 1 & 0 & 0 \\ 0 & 0 & 0 & 1 & 0 \\ 0 & 0 & 0 & 0 & 1 \end{pmatrix} \begin{bmatrix} \hat{a}_{11}(k) \\ \hat{a}_{12}(k) \\ \hat{a}_{21}(k) \\ \hat{a}_{22}(k) \\ \hat{w}(k) \end{bmatrix} = \hat{A}\hat{a}(k), \tag{29}$$

where the scalars $\hat{a}_{ij}$ represent the estimates of the elements of the matrix A and $\hat{w}$ represents the estimate of the input disturbance.

The overall filter model is a linear time-varying system, similar to the filter model used in Section 3, and the same convergence properties apply to Eqs. (24) and (25).

The control input is now given by

$$u(k) = K\hat{x}(k) - \hat{w}_c(k), \tag{30}$$

where $\hat{x} \in \mathbb{R}^n$ represents the state of the model in the sensor and the controller nodes and $\hat{w}_c(k) = \hat{w}(k_i)$ for $k \in [k_i, k_{i+1})$. A copy of the model is implemented in the sensor to compute the state error and use an event-triggered control technique. The state of the model is also needed at the sensor node to compute the control signal, which in turn is needed by the estimation algorithm.

Example 3. Consider an unstable and uncertain system given by Eq. (26) with dimensions $n = 3$ and $m = 2$. The goal is to stabilize the system around its equilibrium point $x_e = [0\ 0\ 0]^T$ in the presence of unknown input disturbances. Matrix B is known and given by

$$B = \begin{pmatrix} 1 & 0 \\ 0 & -1 \\ -1 & 1 \end{pmatrix}.$$

The Kalman filter in the sensor node is in charge of estimating the parameters of matrix A and the piecewise constant external disturbance $w(k)$. The input disturbance changes values at time instants $k = 164$, $k = 314$, $k = 520$, $k = 690$, and $k = 804$. Fig. 6 shows the successful

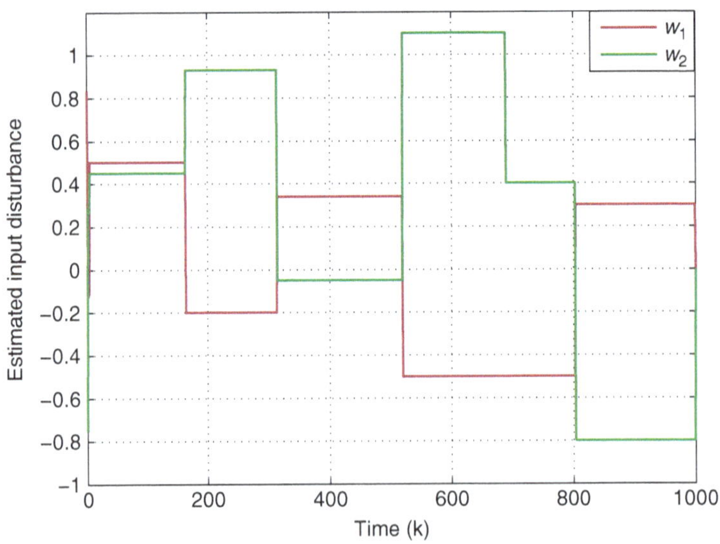

Fig. 6 Estimates of the external disturbance in Example 3.

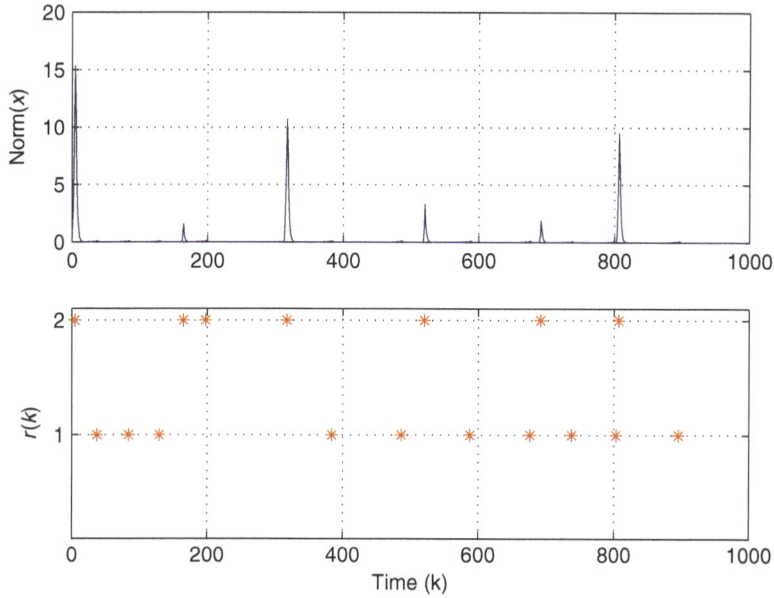

Fig. 7 States of the uncertain system (*top*) affected by an external disturbance, and communication signal for Example 3.

estimation of the input disturbance. Fig. 7 (top) shows the response of the system. When the disturbance changes value, there is an undesirable effect in the system, but after a few iterations we are able to estimate the new value of the disturbances in order to counteract this response and stabilize the system once again. The individual value of the disturbances need not change at the same instant for us to detect and estimate the new disturbance value. At time $k = 690$ only $w_2(k)$ changes its value, while $w_1(k)$ remains the same. The communication instants are shown in the second plot in Fig. 7.

This example shows that this approach offers good disturbance attenuation properties. In particular, we are able to reject the disturbance effect on the system a few iterations after the disturbance presents abrupt changes.

The adaptive MB-ET control framework, where a model-based controller and an event-triggered communication rule are combined with Kalman filter-based parameter and disturbance estimation, is able to adapt to the presence of unknown disturbances and to parameter uncertainties, while reducing the number of feedback measurements that need to be sent from the sensor to the controller.

5.4 Identification and Stabilization of Switched Systems With State Jumps

The application presented in this section considers the interesting problem of identification and stabilization of switched systems with state jumps and unknown switching sequences. Consider a discrete-time switched system:

$$x(k+1) = A_i x(k) + B_i u(k) \tag{31}$$

for each $i \in N$, where N denotes the set $\{1, 2, \dots N\}$ of N integers, where each $A_i \in \mathbb{R}^{n \times n}$ and $B_i \in \mathbb{R}^{n \times m}$ are unknown and each A_i may be unstable. In addition, the switching sequence is also unknown. This sequence of time instants is denoted by τ_{ij}, which represent the times k when the system switches from mode i to mode j for $i, j \in N$. Finally, when the system switches, the state is not restricted to keep the same values but it can be reset to a different finite value—that is, $x(\tau_{ij}) = f(\tau_{ij})$. The state jumps are dictated, in general, by a time-varying function, which can be nonlinear, stochastic, or any other function as long as $f(\tau_{ij}) < \infty$.

To identify the switching time instants we implement an event-triggered strategy using the state error. The sensor computes the norm of the state error (3) and compares it with a fixed threshold. In addition, the sensor node also checks for the difference between the current estimates of the system parameters with respect to the latest parameter updates, and it will send a parameter update if the current estimates are significantly different, which corresponds, in general, to a switch in the parameters of the real system. From time to time the sensor may send parameter updates even though the real system has not switched yet, depending on the convergence threshold we implement. These updates are basically refinements of the estimated parameters that were sent previously.

In addition to employing the adaptive MB-ET control framework, we also implement a learning approach. The estimated parameters are stored in the sensor node and used for future switching times. When a state jump is detected on the basis of the state error, the sensor uses past "learned" modes to predict the next states. The sensor node uses the next predicted error to select a past mode and send the corresponding parameters. The identification keeps executing as a backup option or to refine the estimated parameters. This type of implementation results, in general, in a better performance, as shown in the next example.

Example 4. Consider a switched system with state jumps that operates using the next set of unstable (unknown) modes:

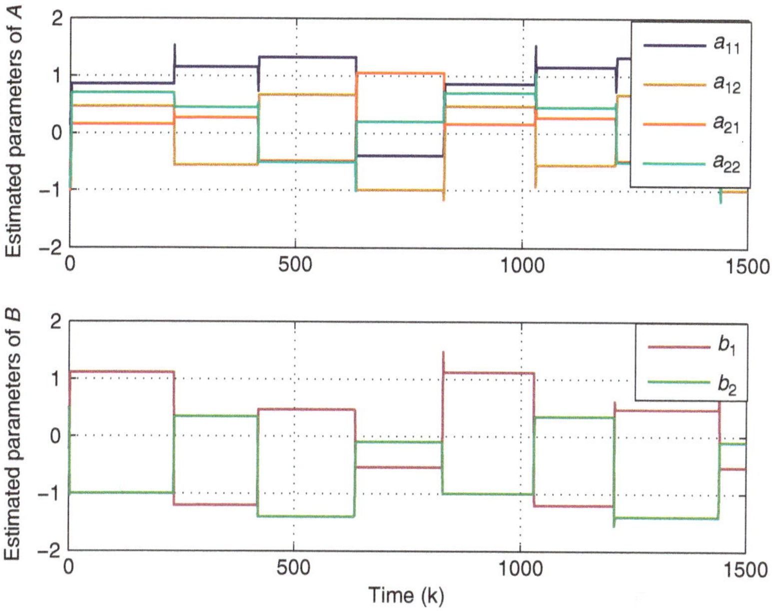

Fig. 8 Estimated parameters of the switched system in Example 4.

$$A_1 = \begin{bmatrix} 0.86 & 0.47 \\ 0.16 & 0.7 \end{bmatrix}, \quad B_1 = \begin{bmatrix} 1.12 \\ -1 \end{bmatrix},$$
$$A_2 = \begin{bmatrix} 1.15 & -0.56 \\ 0.27 & 0.45 \end{bmatrix}, \quad B_2 = \begin{bmatrix} -1.2 \\ 0.34 \end{bmatrix},$$
$$A_3 = \begin{bmatrix} 1.32 & 0.67 \\ -0.49 & -0.51 \end{bmatrix}, \quad B_3 = \begin{bmatrix} 0.47 \\ -1.4 \end{bmatrix},$$
$$A_4 = \begin{bmatrix} -0.4 & -1 \\ 1.05 & 0.2 \end{bmatrix}, \quad B_4 = \begin{bmatrix} -0.54 \\ -0.1 \end{bmatrix}.$$

The results of simulations are shown in Figs. 8 and 9. The switching time instants are random; however, we implement a fixed rule for the state jumps: $x(\tau_{ij}) = [-0.5 \;\; 0.3]^T$ for $i, j \in N$. We implement this fixed rule to illustrate the advantages of the learning approach.

Fig. 8 shows the estimated elements of A and B as they switch over time. Fig. 9 shows the response of the switched system and the communication signal $r(k)$ as defined in Eq. (23). This example shows that the switching parameters of the system can be successfully identified and the model can be updated to obtain better estimates of the state between feedback

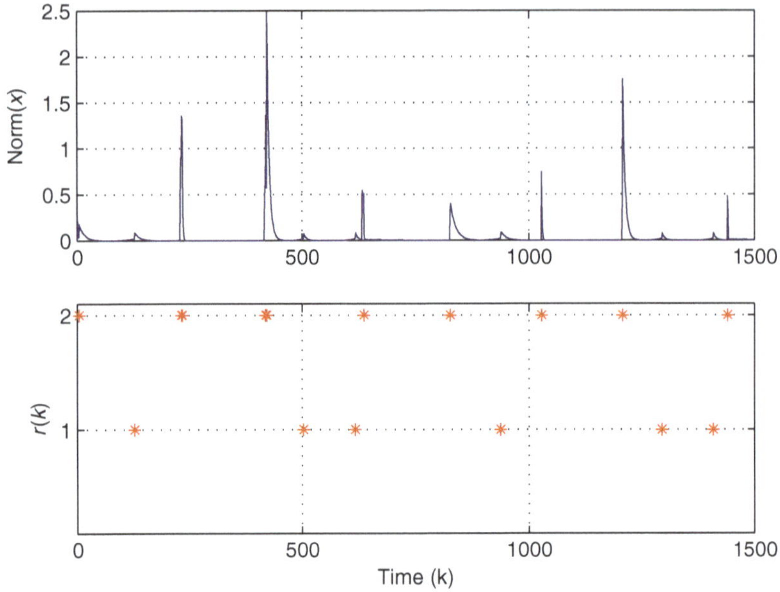

Fig. 9 Norm of the states of the system and communication signal in Example 4.

updates. Additionally, the switching sequence does not have to be known in advance since the sensor is able to detect the switching times and adjust the parameters of the new model. Similarly, a new control gain is designed to stabilize the current operating mode of the switched system. From Fig. 8,we can see that the system switches following the order 1, 2, 3, 4, 1, 2, 3, 4 during the simulation time shown. During the first cycle there is no previous knowledge of any system parameters (A_i, B_i); however, in the second cycle the sensor has stored estimated parameters for each system and is able improve the performance, as shown in Fig. 9, on the basis of learned system parameters.

6 CONCLUSIONS

We have described the adaptive MB-ET control framework for control of uncertain systems with limited feedback communication. This method employs an event-triggered communication policy to determine the time instants when the sensor transmits parameter and state updates to the controller. The Kalman filter was implemented in the sensor node to estimate the system parameters. We focused on important and practical control applications, such as stabilization of continuous-time systems using discrete-time models, fault detection, control of systems with external disturbance, and stabilization of switched systems. It was shown that the adaptive MB-ET control framework significantly reduces the number of transmissions from the sensor to the controller.

REFERENCES

[1] P.J. Antsaklis, J. Baillieul, Special issue on technology of networked control systems, Proc. IEEE 95 (1) (2007).

[2] F.L. Lian, J.R. Moyne, D.M. Tilbury, Performance evaluation of control networks: Ethernet, Controlnet, and Devicenet, IEEE Control Syst. Mag. (2001) 66–83.

[3] F.L. Lian, J.M. Moyne, D.M. Tilbury, Network design consideration for distributed control systems, IEEE Trans. Control Syst. Technol. 10 (2) (2002) 297–307.

[4] G.C. Walsh, H. Ye, L. Bushnell, Stability analysis of networked control systems, IEEE Trans. Control Syst. Technol. 10 (3) (2002) 438–446.

[5] N. Elia, S.K. Mitter, Stabilization of linear systems with limited information, IEEE Trans. Autom. Control 46 (9) (2001) 1384–1400.

[6] D. Georgiev, D.M. Tilbury, Packet-based control, in: Proceedings of American Control Conference, 2004, pp. 329–336.

[7] D.E. Quevedo, E.I. Silva, G.C. Goodwin, Packetized predictive control over erasure channels, in: Proceedings of American Control Conference, 2007, pp. 1003–1008.

[8] L.A. Montestruque, P.J. Antsaklis, On the model-based control of networked systems, Automatica 39 (10) (2003) 1837–1843.

[9] L.A. Montestruque, P.J. Antsaklis, Stability of model-based control of networked systems with time varying transmission times, IEEE Trans. Autom. Control 49 (9) (2004) 1562–1572.

[10] E. Garcia, P.J. Antsaklis, Model-based event-triggered control for systems with quantization and time-varying network delays, IEEE Trans. Autom. Control 58 (2) (2013) 422–434.

[11] E. Garcia, P.J. Antsaklis, Output feedback model-based control of uncertain discrete-time systems with network induced delays, in: Proceedings of the 51st IEEE Conference on Decision and Control, 2012, pp. 6647–6652.

[12] E. Garcia, P.J. Antsaklis, Event-triggered output feedback stabilization of networked systems with external disturbance, in: Proceedings of the 53rd IEEE Conference on Decision and Control, 2014, pp. 3566–3571.

[13] E. Garcia, P.J. Antsaklis, Decentralized model-based event-triggered control of networked systems, in: Proceedings of the American Control Conference, 2012, pp. 6485–6490.

[14] E. Garcia, Y. Cao, D.W. Casbeer, Cooperative control with general linear dynamics and limited communication: centralized and decentralized event-triggered control strategies, in: Proceedings of the 2014 American Control Conference, 2014, pp. 159–164.

[15] E. Garcia, Y. Cao, D.W. Casbeer, Decentralized event-triggered consensus with general linear dynamics, Automatica 50 (10) (2014) 2633–2640.

[16] E. Garcia, P.J. Antsaklis, Model-based event-triggered control with time-varying network delays, in: Proceedings of the 50th IEEE Conference on Decision and Control—European Control Conference, 2011, pp. 1650–1655.

[17] E. Garcia, P.J. Antsaklis, Optimal model-based control with limited communication, in: Proceedings of the 19th World Congress of the International Federation of Automatic Control, 2014, pp. 10908–10913.

[18] E. Garcia, P.J. Antsaklis, Adaptive stabilization of model-based networked control systems, in: Proceedings of the American Control Conference, 2011, pp. 1094–1099.
[19] K.J. Astrom, P. Eykhoff, System identification—a survey, Automatica 7 (2) (1971) 123–162.
[20] G. Kreisselmeier, Adaptive observers with exponential rate of convergence, IEEE Trans. Autom. Control AC-22 (1) (1977) 2–8.
[21] L.W. Nelson, E. Stear, The simultaneous on-line estimation of parameters and states in linear systems, IEEE Trans. Autom. Control AC-21 (1976).
[22] Z.P. Jiang, Y. Wang, Input-to-state stability for discrete-time nonlinear systems, Automatica 37 (6) (2001) 857–869.
[23] E. Garcia, P.J. Antsaklis, Parameter estimation and adaptive stabilization in time-triggered and event-triggered model-based control of uncertain systems, Int. J. Control 85 (9) (2012) 1327–1342.
[24] Y. Song, J.W. Grizzle, The extended Kalman filter as a local asymptotic observer for discrete-time nonlinear systems, J. Math. Syst. Estim. Control 5 (1) (1995).
[25] P.R. Kumar, P. Varaiya, Stochastic Systems: Estimation, Identification, and Adaptive Control, Prentice Hall, Englewood Cliffs, NJ, 1986.
[26] M.C.F. Donkers, W.P.M.H. Heemels, N. Wouw, L. Hetel, Stability analysis of networked control systems using a switched linear systems approach, IEEE Trans. Autom. Control 56 (2011) 2101–2115.

CHAPTER FOUR

A Neural Field Theory for Loss of Consciousness: Synaptic Drive Dynamics, System Stability, Attractors, Partial Synchronization, and Hopf Bifurcations Characterizing the Anesthetic Cascade

W.M. Haddad*, S.P. Hou†, J.M. Bailey‡, N. Meskin§

*Georgia Institute of Technology, Atlanta, GA, United States
†Singapore Institute of Manufacturing Technology, A*STAR, Singapore, Singapore
‡Northeast Georgia Medical Center, Gainesville, GA, United States
§Qatar University, Doha, Qatar

1 INTRODUCTION

The safe and effective administration of medications is based on both an understanding of the relationship between the drug dose and the resultant drug concentrations in various tissues (ie, *pharmacokinetics*), and the relationship between drug concentrations at the site of action and the subsequent physiological effect (ie, *pharmacodynamics*) [1]. A key concept of pharmacodynamic modeling is that of C_{50}, the drug concentration associated with 50% of the maximum pharmacological effect or achievement of a therapeutic effect in 50% of cases. In the case of anesthetic agents, the therapeutic effect is defined as a lack of response to a noxious stimulus, such as a skin incision, and hence is a binary variable. Usually anesthetic trials are conducted in populations of subjects, and in this case C_{50} refers to the drug concentration associated with a lack of response to noxious stimuli in 50% of the subjects.

There have been animal studies, and also a small number of human studies, in which the response of a single subject to various doses of an anesthetic agent is observed. In these studies it has been observed that the

http://dx.doi.org/10.1016/B978-0-12-805246-4.00004-5

concentration-response curve is very sharp [2]. In particular, as the dose or concentration of the anesthetic agent increases, there exists a point at which there is an abrupt transition from consciousness to unconsciousness. However, in these studies the concentration-response curve has not been found to be a strict step function. Rather, there is a narrow range of concentrations within which the therapeutic effect (response or lack of a response to a noxious stimulus) has been described as a probability of response. A common empirical model describing this effect is a variation on the Hill equation [3] in which the probability P of an anesthetic effect—that is, of no response to a noxious stimulus—is given by

$$P = \frac{C^{\gamma}}{(C_{50}^{\gamma} + C^{\gamma})}, \tag{1}$$

where C is the concentration of the anesthetic agent, C_{50} is the concentration of the anesthetic agent associated with a 50% probability of an anesthetic effect, and γ is a parameter that determines the steepness of the concentration-probability anesthetic curve. Typically for individual subjects γ ranges from 6 to 20 [4, 5].

The in vitro effects of anesthetic agents have been intensely investigated, and the actions of anesthetic agents on single neurons have been described [6, 7]. A key goal of anesthetic and neuroscience research is to understand how the effects on the single neuron translate into an abrupt transition from consciousness to unconsciousness as the concentration of the agent increases [8–10]. This requires consideration of the immense and complex neural network of the brain. Neurons, like all cells in the human body, maintain an electrochemical potential gradient between the inside of the cell and the surrounding milieu. However, neurons are characterized by excitability—that is, with a stimulus the neuron will discharge its electrochemical gradient and this discharge creates a change in the potential of neurons to which it is connected (the postsynaptic potential). And this postsynaptic potential, if excitatory, can serve as the stimulus for the discharge of the connected neuron or, if inhibitory, can hyperpolarize the connected neuron and inhibit firing [11, 12]. This leads to immense system complexity since the mammalian brain comprising approximately 10^{11} neurons each interconnecting with approximately 10^4–10^5 other neurons. To deal with this level of complexity, one can either make assumptions about neuronal connectivity or use mean field theories, or do both [10].

In this chapter we use dynamical system theory to develop a mechanistic mean field model for neural activity to study the anesthetic cascade.

Specifically, using the notions of synaptic drive—a measure of neuronal population activity that captures the influence of a given neuron population on the behavior of the neuronal network—and topological system stability, we show that this simplified model predicts interesting dynamic behavior. In particular, we show that brain activity decreases as anesthetic concentration increases. Even though this should be expected of any viable phenomenological model for addressing how anesthesia suppresses consciousness, the novelty of the model is the additional prediction of a conscious-unconscious transition as the implied anesthetic concentration increases, where excitatory neural activity is characterized by a stable limit cycle. This raises the interesting possibility that the probabilistic nature of the transition to the anesthetic state may reflect observations made randomly at various times during the cycle of excitatory neuron activity from high to low values. Then, as the anesthetic concentration increases further, the proposed dynamical system model undergoes a supercritical reverse Hopf bifurcation and transitions from a limit cycle behavior to an asymptotically stable equilibrium corresponding to an anesthetized state, where the excitatory neuron activity is very low.

Furthermore, we also extend our mean field theory to incorporate a more complex model for the postsynaptic potential. Specifically, by using an Erlang-type, time-multiplied exponential decay model for the postsynaptic potential rather than a simple exponential decay [11, 13], we can account for the delay in peak amplitude of the postsynaptic potential that occurs after a neuron discharges [14, Fig. 8.5]. In this case we see biphasic responses in the mean neural activity that seem to parallel the rise and decay of the postsynaptic potential.

Finally, we also address the more general question of synchronization of neural activity without mean field assumptions. There is extensive experimental verification that collections of neurons may function as oscillators and the synchronization of oscillators may play a key role in the transmission of information within the central nervous system [15–17]. This may be particularly relevant to understand the mechanism of action for general anesthesia [10]. More importantly, however, when patients lose consciousness other parts of the brain are still functional (heart rate control, ventilation, oxygenation, etc.), and hence the development of biological neural network models that exhibit *partial state synchronization* is critical. In particular, models that can account for synchronization of subsets of the brain with the nonsynchronized parts firing at normal levels are essential in capturing biophysical behavior.

We address partial state synchronization by focusing on a postulated subset of inhibitory neurons that are not themselves connected to other inhibitory neurons. In this case we provide sufficient conditions for global asymptotic and exponential partial synchronization for our excitatory and inhibitory cortical neuronal network. Specifically, we show that as the inhibitory time constants increase (one of the demonstrated in vitro effects of some anesthetic agents [6, 7]), the excitatory neurons that are coupled to inhibitory neurons all approach a zero synaptic drive state, and the inhibitory neurons that themselves are not coupled to inhibitory neurons achieve synchronization. Finally, several numerical experiments are presented to illustrate the different aspects of the proposed mechanistic anesthetic cascade model.

The notation used in this chapter is fairly standard. Specifically, for $x \in \mathbb{R}^n$, we write $x \geq\geq 0$ (or $x >> 0$) to indicate that every component x_i of x is nonnegative (or positive, respectively). In this case we say that x is *nonnegative* or *positive*, respectively. Likewise, $A \in \mathbb{R}^{n\times m}$ is *nonnegative* or *positive* if every entry of A is nonnegative or positive, respectively, which is written as $A \geq\geq 0$ or $A >> 0$, respectively. For $A \in \mathbb{R}^{n\times n}$, we write $A \geq 0$ (or $A > 0$) to indicate that A is nonnegative (or positive) definite. Furthermore, we write $\overline{\mathbb{R}}_+^n$ and $\mathbb{R}_+^n$ to denote the nonnegative and positive orthants of $\mathbb{R}^n$—that is, if $x \in \mathbb{R}^n$, then $x \in \overline{\mathbb{R}}_+^n$ and $x \in \mathbb{R}_+^n$ are equivalent to $x \geq\geq 0$ and $x >> 0$, respectively. In addition, we write $0_{n\times m}$ to denote the $n\times m$ zero matrix, I_n to denote the $n\times n$ identity matrix, and $\mathbf{e}_n \in \mathbb{R}^n$ to denote the 1's vector of order n—that is, $\mathbf{e}_n = [1, 1, \ldots, 1]^{\mathrm{T}}$; if the order of $\mathbf{e}_n$ is clear from the context, we simply write $\mathbf{e}$ for $\mathbf{e}_n$. Finally, we write $(\cdot)^{\mathrm{T}}$ to denote transpose, $\mathrm{tr}(\cdot)$ to denote trace, $\det(\cdot)$ to denote determinant, and $(\cdot)'$ to denote the Fréchet derivative.

2 A MODEL FOR EXCITATORY AND INHIBITORY NEURAL POPULATIONS

Biological neural network models predict a voltage in the receiving or postsynaptic neuron given by [10, 11]

$$v_i^{\mathrm{X}}(t) = \sum_{j=1}^{n_{\mathrm{E}}} A_{ij}^{\mathrm{XE}} \sum_k \alpha_j^{\mathrm{E}}(t - t_k) + \sum_{j'=1}^{n_{\mathrm{I}}} A_{ij'}^{\mathrm{XI}} \sum_{k'} \alpha_{j'}^{\mathrm{I}}(t - t_{k'}), \qquad (2)$$

where $v_i^{\mathrm{X}}(\cdot)$, $i = 1, \ldots, n_{\mathrm{X}}$, $\mathrm{X} \in \{\mathrm{E}, \mathrm{I}\}$, is the excitatory ($\mathrm{X} = \mathrm{E}$) and inhibitory ($\mathrm{X} = \mathrm{I}$) voltage in the ith receiving neuron, A_{ij}^{XY}, X, Y $\in$ {E,I}, are constants representing the coupling strengths (in volts) of the jth neuron on the ith neuron, k and $k' = 1, \ldots$ enumerate the action potential or firings of the excitatory and inhibitory transmitting (presynaptic) neurons at firing times t_k and $t_{k'}$, respectively, and $\alpha_j^{\mathrm{E}}(\cdot)$ and $\alpha_{j'}^{\mathrm{I}}(\cdot)$ are dimensionless functions describing the evolution of the excitatory and inhibitory postsynaptic potentials, respectively. With use of a (possibly discontinuous) function $f_i(\cdot)$ to represent the firing rate (in hertz) of the ith neuron that determines the instantaneous number of spikes that a neuron fires over an infinitesimal time interval $[t, t + \mathrm{d}t]$, and assuming that the firing rate is a function of the voltage $v_i^{\mathrm{E}}(\cdot)$ (or $v_i^{\mathrm{I}}(\cdot)$) across the membrane of the ith neuron given by $f_i(v_i^{\mathrm{E}})$ (or $f_i(v_i^{\mathrm{I}})$), it follows from Eq. (2) that

$$v_i^{\mathrm{E}}(t) = \sum_{j=1, j\neq i}^{n_{\mathrm{E}}} A_{ij}^{\mathrm{EE}} \int_{-\infty}^{t} \alpha_j^{\mathrm{E}}(t-\tau) f_j(v_j^{\mathrm{E}}(\tau))\mathrm{d}\tau + \sum_{j'=1}^{n_{\mathrm{I}}} A_{ij'}^{\mathrm{EI}} \int_{-\infty}^{t} \alpha_{j'}^{\mathrm{I}}(t-\tau) f_{j'}(v_{j'}^{\mathrm{I}}(\tau))\mathrm{d}\tau + v_{\mathrm{th}i}^{\mathrm{E}}(t), \quad i = 1, \ldots, n_{\mathrm{E}}, \tag{3}$$

$$v_i^{\mathrm{I}}(t) = \sum_{j=1}^{n_{\mathrm{E}}} A_{ij}^{\mathrm{IE}} \int_{-\infty}^{t} \alpha_j^{\mathrm{E}}(t-\tau) f_j(v_j^{\mathrm{E}}(\tau))\mathrm{d}\tau + \sum_{j'=1, j'\neq i}^{n_{\mathrm{I}}} A_{ij'}^{\mathrm{II}} \int_{-\infty}^{t} \alpha_{j'}^{\mathrm{I}}(t-\tau) f_{j'}(v_{j'}^{\mathrm{I}}(\tau))\mathrm{d}\tau + v_{\mathrm{th}i}^{\mathrm{I}}(t), \quad i = 1, \ldots, n_{\mathrm{I}}, \tag{4}$$

where $f_j(v_j^{\mathrm{X}}(t))\mathrm{d}t$, $\mathrm{X} \in \{\mathrm{E}, \mathrm{I}\}$, is the probability of a spike occurring in the time interval $[t, t+\mathrm{d}t]$, and the *neuronal connectivity matrix* A^{XY}, with units of volts, contains entries A_{ij}^{XY}, $\mathrm{X,Y} \in \{\mathrm{E,I}\}$, representing the coupling strength of the jth neuron on the ith neuron such that $A_{ij}^{\mathrm{XE}} > 0$ and $A_{ij}^{\mathrm{XI}} < 0$, $\mathrm{X} \in \{\mathrm{E}, \mathrm{I}\}$ if the jth neuron is connected (ie, contributes a postsynaptic potential) to the ith neuron and $A_{ij}^{\mathrm{XY}} = 0$ otherwise. Furthermore, $v_{\mathrm{th}i}^{\mathrm{E}}(\cdot)$ and $v_{\mathrm{th}i}^{\mathrm{I}}(\cdot)$ are continuous input voltages characterizing nerve impulses from sensory (pain) receptors, thermoreceptors (temperature-sensing receptors), or proprioceptive (motion-sensing) receptors. Alternatively, $v_{\mathrm{th}i}^{\mathrm{E}}(\cdot)$ and $v_{\mathrm{th}i}^{\mathrm{I}}(\cdot)$ can be thought of as inputs from the reticular activating system within the brainstem responsible for regulating arousal and sleep-wake transitions. Note that $A_{ii}^{\mathrm{EE}} \triangleq A_{ii}^{\mathrm{II}} \triangleq 0$ by definition.

If we define the *synaptic drive*—a dimensionless quantity—of each (excitatory or inhibitory) neuron as the convolution of the presynaptic firing rate with the postsynaptic potential given by [10, 11]

$$S_i^{(\mathrm{E,I})}(t) \triangleq \int_{-\infty}^{t} \alpha_i^{(\mathrm{E,I})}(t-\tau) f_i(v_i^{(\mathrm{E,I})}(\tau))\mathrm{d}\tau, \tag{5}$$

and assume an exponential decay of the synaptic voltages of the form [11]

$$\alpha_i^{(\mathrm{E,I})}(t) = B_i^{(\mathrm{E,I})} \mathrm{e}^{-\frac{t}{\lambda_i^{(\mathrm{E,I})}}}, \tag{6}$$

where the dimensionless gain $B_i^{(\mathrm{E,I})}$ is equal to B_i^{E} if the ith neuron is excitatory and B_i^{I} if the ith neuron is inhibitory, and similarly for $S_i^{(\mathrm{E,I})}$, $v_i^{\mathrm{E,I}}$, $\alpha_i^{(\mathrm{E,I})}$, and $\lambda_i^{(\mathrm{E,I})}$, it follows from Eqs. (5) and (6) that

$$\frac{\mathrm{d}S_i^{\mathrm{E}}(t)}{\mathrm{d}t} = -\frac{1}{\lambda_i^{\mathrm{E}}} S_i^{\mathrm{E}}(t) + B_i^{\mathrm{E}} f_i\left(\sum_{j=1, j\neq i}^{n_{\mathrm{E}}} A_{ij}^{\mathrm{EE}} S_j^{\mathrm{E}}(t) + \sum_{j'=1}^{n_{\mathrm{I}}} A_{ij'}^{\mathrm{EI}} S_{j'}^{\mathrm{I}}(t) + v_{\mathrm{th}i}^{\mathrm{E}}(t)\right),$$

$$i = 1, \ldots, n_{\mathrm{E}}, \tag{7}$$

$$\frac{\mathrm{d}S_i^{\mathrm{I}}(t)}{\mathrm{d}t} = -\frac{1}{\lambda_i^{\mathrm{I}}} S_i^{\mathrm{I}}(t) + B_i^{\mathrm{I}} f_i\left(\sum_{j=1}^{n_{\mathrm{E}}} A_{ij}^{\mathrm{IE}} S_j^{\mathrm{E}}(t) + \sum_{j'=1, j'\neq i}^{n_{\mathrm{I}}} A_{ij'}^{\mathrm{II}} S_{j'}^{\mathrm{I}}(t) + v_{\mathrm{th}i}^{\mathrm{I}}(t)\right),$$

$$i = 1, \ldots, n_{\mathrm{I}}. \tag{8}$$

A more general Erlang-type time-multiplied delay model for the postsynaptic potential is considered in Section 5.

The synaptic drive accounts for presynaptic and postsynaptic potentials in a neural network to give a measure of neural activity in the network. In particular, it follows from Eq. (5) that the synaptic drive quantifies the present activity (via the firing rate) along with all previous activity within a neural population appropriately scaled (via a temporal decay) when a particular firing event occurred. Hence the synaptic drive provides a measure of neuronal population activity that captures the influence of a given neuron population on the behavior of the network from the infinite past to the current time instant.

The above analysis reveals that a form for capturing the neuroelectric behavior of biological excitatory and inhibitory neuronal networks can be written as

$$\frac{\mathrm{d}S_i(t)}{\mathrm{d}t} = -\frac{1}{\tau_i}S_i(t) + B_i f_i\left(\sum_{j=1}^{n} A_{ij}S_j(t) + v_{\mathrm{th}i}(t)\right), \quad S_i(0) = S_{i0},$$

$$t \geq 0, \ i = 1, \ldots, n, \tag{9}$$

where $S_i(t) \in \mathcal{D} \subseteq \mathbb{R}$, $t \geq 0$, is the ith synaptic drive, $v_{\mathrm{th}i}(t) \in \mathbb{R}$, $t \geq 0$, denotes the input voltage to the ith neuron, A_{ij} is a constant representing the coupling strength of the jth neuron on the ith neuron, $\tau_i = \lambda_i$ is a time constant, B_i is a constant gain for the firing rate of the ith neuron, and $f_i(\cdot)$ is a nonlinear activation function describing the relationship between the synaptic drive and the firing rate of the ith neuron.

Since the neocortex contains on the order of 23 billion neurons, which have up to 20 trillion connections, the key challenge is whether it is possible to compute the neuronal connectivity matrix from patient or experimental data. Clearly this is a formidable task, and computational neuroscience has addressed this problem by reducing a large population of neurons to a distribution function describing their probabilistic evolution; that is, a function that captures the distribution of neuronal states at a given time [18, 19]. Specifically, a stochastic multiplicative uncertainty model can be used to capture modeling of a priori uncertainty in the coefficients of the neuronal connectivity matrix by means of state-dependent noise [10]. The philosophy of representing uncertain parameters by means of multiplicative white noise is motivated by the maximum entropy principle of Jaynes [20, 21] and statistical analysis [22].

Maximum entropy modeling is a form of stochastic modeling wherein stochastic integration is interpreted in the sense of Itô to provide a model for system parameter uncertainty. The use of stochastic theory to model system parameter uncertainty has been used within a modern information-theoretic interpretation of probability theory [20, 21, 23]. In particular, rather than regarding the probability of an event as an *objective* quantity such as the limiting frequency of outcomes of numerous repetitions, maximum entropy modeling adopts the view that the probability of an event is a *subjective* quantity which reflects the observer's certainty of a particular event occurring. This quantity corresponds to a measure of information. The validity of a stochastic model for a biological neural network does not rely on the existence of an ensemble model but rather lies in the interpretation that it expresses modeling certainty or uncertainty regarding the coefficients of the neuronal connectivity matrix. Hence a stochastic multiplicative uncertainty model uses state-dependent Gaussian white noise to represent parameter uncertainty by defining a measure of ignorance, in terms of an information-theoretic entropy, and then determining the probability distribution which *maximizes* this measure subject to agreement with a given model [10].

Even though the computation of the neural connectivity matrix and the graph topology of human brain networks are intractable, the phenomenological firing rate model given by Eqs. (7) and (8) involving the averaged behavior of the spiking rates of groups of neurons can be used to predict network system changes due to changes in a subset of the neuronal connectivity matrix A^{XY} containing the entries A_{ij}^{XY} so as to understand how the large neuron population changes qualitatively with the induction of anesthesia. The relevance of the model to realistic systems can be appraised by its prediction of salient aspects of anesthesia—in particular, predicting the abrupt transition from consciousness to unconsciousness resembling a phase transition as well as the biphasic response during induction of anesthesia leading to a paradoxical phase before the loss of consciousness [24].

In such population models the activity of a neuron—that is, the rate at which the neuron generates an action potential (ie, "fires")—is modeled as a function of the voltage (across the membrane). In this chapter we will assume continuous half-wave rectification activation functions as well as smooth sigmoidal functions. Specifically, for a typical neuron [12]

$$f_i(x) = [x]_+, \tag{10}$$

where $i \in \{1, \ldots, n\}$ and $[x]_+ = x$ if $x \geq 0$ and $[x]_+ = 0$ otherwise. The activation function (10) reflects the fact that as the voltage increases across the membrane of the ith neuron, the firing rate increases as well. Often the membrane potential firing rate curve exhibits a linear characteristic for a given range of the voltages. At higher voltages, however, a saturation phenomenon appears, indicating that the full effect of the firing rate has been reached. To capture this effect, $f_i(\cdot)$ can be modeled as a smooth (ie, infinitely differentiable) sigmoidal function

$$f_i(x) = \frac{f_{\max} e^{\gamma x}}{1 + e^{\gamma x}}, \tag{11}$$

where $i \in \{1, \ldots, n\}$, $\gamma \gg 0$, and $f_{\max} = \lim_{x \to \infty} f_i(x)$ denotes the maximum firing rate.

3 A TWO-CLASS MEAN EXCITATORY AND MEAN INHIBITORY SYNAPTIC DRIVE MODEL

To avoid the complexity of the larger-scale neural network model given by Eqs. (7) and (8), in this section we consider a mean field model. Specifically, the excitatory and inhibitory synaptic drive model given by Eqs. (7) and (8) can be reduced to a two-class mean excitatory and mean inhibitory model. In particular, with continuously differentiable $f_i(\cdot) = f(\cdot)$, $B_i^{\mathrm{E}} = B_i^{\mathrm{I}} = 1$, $\lambda_i^{\mathrm{E}} = \lambda^{\mathrm{E}}$, and $\lambda_i^{\mathrm{I}} = \lambda^{\mathrm{I}}$, Eqs. (7) and (8) collapse to (see [10] for details)

$$\frac{\mathrm{d}\overline{S}^{\mathrm{E}}(t)}{\mathrm{d}t} = \overline{f}_1(\overline{S}^{\mathrm{E}}(t), \overline{S}^{\mathrm{I}}(t)), \quad \overline{S}^{\mathrm{E}}(0) = \overline{S}_0^{\mathrm{E}}, \quad t \geq 0, \tag{12}$$

$$\frac{\mathrm{d}\overline{S}^{\mathrm{I}}(t)}{\mathrm{d}t} = \overline{f}_2(\overline{S}^{\mathrm{E}}(t), \overline{S}^{\mathrm{I}}(t)), \quad \overline{S}^{\mathrm{I}}(0) = \overline{S}_0^{\mathrm{I}}, \tag{13}$$

where

$$\overline{f}_1(\overline{S}^{\mathrm{E}}, \overline{S}^{\mathrm{I}}) = f(a\overline{S}^{\mathrm{E}} - b\overline{S}^{\mathrm{I}} + v_{\mathrm{th}}^{\mathrm{E}}) - \frac{1}{\lambda^{\mathrm{E}}}\overline{S}^{\mathrm{E}}, \tag{14}$$

$$\overline{f}_2(\overline{S}^{\mathrm{E}}, \overline{S}^{\mathrm{I}}) = f(c\overline{S}^{\mathrm{E}} - d\overline{S}^{\mathrm{I}} + v_{\mathrm{th}}^{\mathrm{I}}) - \frac{1}{\lambda^{\mathrm{I}}}\overline{S}^{\mathrm{I}}, \tag{15}$$

where $a \triangleq n_{\mathrm{E}}\overline{A}^{\mathrm{EE}}$, $b \triangleq -n_{\mathrm{I}}\overline{A}^{\mathrm{EI}}$, $c \triangleq n_{\mathrm{E}}\overline{A}^{\mathrm{IE}}$, and $d \triangleq -n_{\mathrm{I}}\overline{A}^{\mathrm{II}}$, where $\overline{A}^{\mathrm{EE}} = A_{ij}^{\mathrm{EE}} - \Delta_{ij}^{\mathrm{EE}}$, $\overline{A}^{\mathrm{EI}} = A_{ij}^{\mathrm{EI}} - \Delta_{ij}^{\mathrm{EI}}$, $\overline{A}^{\mathrm{IE}} = A_{ij}^{\mathrm{IE}} - \Delta_{ij}^{\mathrm{IE}}$, and $\overline{A}^{\mathrm{II}} = A_{ij}^{\mathrm{II}} - \Delta_{ij}^{\mathrm{II}}$, where $\overline{A}^{\mathrm{XY}} \triangleq \frac{1}{n_{\mathrm{X}} n_{\mathrm{Y}}} \sum_{i=1}^{n_{\mathrm{X}}} \sum_{j=1}^{n_{\mathrm{Y}}} A_{ij}^{\mathrm{XY}}$, X, Y $\in \{\mathrm{E}, \mathrm{I}\}$, denote the mean, $\Delta_{ij}^{\mathrm{XY}}$, X, Y $\in \{\mathrm{E}, \mathrm{I}\}$, are deviations from the mean, and $\overline{S}^{\mathrm{E}}(t) \triangleq \frac{1}{n_{\mathrm{E}}} \sum_{j=1}^{n_{\mathrm{E}}} S_j^{\mathrm{E}}(t)$ and $\overline{S}^{\mathrm{I}}(t) \triangleq \frac{1}{n_{\mathrm{I}}} \sum_{j=1}^{n_{\mathrm{I}}} S_j^{\mathrm{I}}(t)$ denote the mean excitatory synaptic drive and mean inhibitory synaptic drive in dimensionless units, respectively. The constants a, b, c, and d are nonnegative. Eqs. (12) and (13) represent the spatial average (mean) dynamics of the system given by Eqs. (7) and (8), and they are predicated on a mean field assumption that reduces the complex (approximately $10^{11} \times 10^{11}$) neuronal connectivity matrix to a 2×2 excitatory-inhibitory system. This is a drastic assumption but one which has been commonly used in theoretical neuroscience going back to the pioneering work of Wilson and Cowan [25].

To study the dynamic behavior of the mean field model given by Eqs. (12) and (13), we assume the postsynaptic activation function is given by Eq. (11). The next set of propositions and theorems present several results on the dynamic behavior of Eqs. (12) and (13). Here, we use the language of topological dynamics (ie, flows, equilibria, periodic orbits, limit sets) to study the dynamic behavior of Eqs. (12) and (13). Recall that a set $\mathcal{D} \subseteq \mathbb{R}^2$ is *positively invariant* with respect to Eqs. (12) and (13) if $\mathcal{D}$ contains the orbits of all its points. Moreover, a subset $\mathcal{D} \subseteq \mathbb{R}^2$ is *simply connected* if it can be continuously shrunk to a single point in $\mathcal{D}$. Finally, recall the standard *Lyapunov* and *asymptotic stability* definitions for an equilibrium point of Eqs. (12) and (13) given in [26], and recall that if all solutions of Eqs. (12) and (13) are bounded, then it follows from the Peano-Cauchy theorem [26, p. 76] that the maximal solution to Eqs. (12) and (13) exists on the semiinfinite interval $[0, \infty)$, and hence the mean field model given by Eqs. (12) and (13) is *forward complete.*

Proposition 1 ([10]). *Consider the two-class mean excitatory and mean inhibitory synaptic drive network given by Eqs. (12) and* (13). *If* $\overline{S}_0^{\mathrm{E}} \geq 0$ *and* $\overline{S}_0^{\mathrm{I}} \geq 0$, *then* $\overline{S}^{\mathrm{E}}(t) \geq 0$ *and* $\overline{S}^{\mathrm{I}}(t) \geq 0$ *for all* $t \geq 0$.

Proposition 2. *Consider the two-class mean excitatory and mean inhibitory synaptic drive network given by Eqs. (12) and* (13), *and let* $\mathcal{M} \triangleq \{(\overline{S}^{\mathrm{E}}, \overline{S}^{\mathrm{I}}) \in \mathbb{R}^2 : 0 \leq \overline{S}^{\mathrm{E}} \leq f_{\max}\lambda^{\mathrm{E}}$ *and* $0 \leq \overline{S}^{\mathrm{I}} \leq f_{\max}\lambda^{\mathrm{I}}\}$. *Then* $\mathcal{M}$ *is positively invariant with respect to Eqs. (12) and* (13).

Proof. Let $[1\ 0]^{\mathrm{T}}$ be a normal vector to the line $\overline{S}^{\mathrm{E}} = 0$ directed inward toward the region $\mathcal{M}$ and note that, since $f(x) \geq 0$, $x \in \mathbb{R}$,

$$[1\ 0]\begin{bmatrix}\overline{f}_1\\ \overline{f}_2\end{bmatrix} = \overline{f}_1 = f(a\overline{S}^{\mathrm{E}} - b\overline{S}^{\mathrm{I}} + v_{\mathrm{th}}^{\mathrm{E}}) - \frac{1}{\lambda^{\mathrm{E}}}\overline{S}^{\mathrm{E}} \geq 0, \quad \overline{S}^{\mathrm{E}} = 0.$$

Hence the vector field $\overline{f} \triangleq [\overline{f}_1\, \overline{f}_2]^{\mathrm{T}}$ along the line $\overline{S}^{\mathrm{E}} = 0$ is directed toward the region $\mathcal{M}$. Let $[-1\ 0]^{\mathrm{T}}$ be a normal vector to the line $\overline{S}^{\mathrm{E}} = f_{\max}\lambda^{\mathrm{E}}$ directed toward the region $\mathcal{M}$ and note that, since $f(x) \leq f_{\max}$, $x \in \mathbb{R}$,

$$[-1\ 0]\begin{bmatrix}\overline{f}_1\\ \overline{f}_2\end{bmatrix} = -\overline{f}_1 = -f(a\overline{S}^{\mathrm{E}} - b\overline{S}^{\mathrm{I}} + v_{\mathrm{th}}^{\mathrm{E}}) + \frac{1}{\lambda^{\mathrm{E}}}\overline{S}^{\mathrm{E}} \geq 0, \quad \overline{S}^{\mathrm{E}} = f_{\max}\lambda^{\mathrm{E}}.$$

Hence the vector field $\overline{f}$ along the line $\overline{S}^{\mathrm{E}} = f_{\max}\lambda^{\mathrm{E}}$ is directed inward toward the region $\mathcal{M}$.

Alternatively, let $[0\ 1]^{\mathrm{T}}$ be a normal vector along the line $\overline{S}^{\mathrm{I}} = 0$ directed toward the region $\mathcal{M}$ and note that, since $f(x) \geq 0$, $x \in \mathbb{R}$,

$$[0\ 1]\begin{bmatrix}\overline{f}_1\\ \overline{f}_2\end{bmatrix} = \overline{f}_2 = f(c\overline{S}^{\mathrm{E}} - d\overline{S}^{\mathrm{I}} + v_{\mathrm{th}}^{\mathrm{I}}) - \frac{1}{\lambda^{\mathrm{I}}}\overline{S}^{\mathrm{I}} \geq 0, \quad \overline{S}^{\mathrm{I}} = 0.$$

Hence the vector field $\overline{f}$ along the line $\overline{S}^{\mathrm{I}} = 0$ is directed inward toward the region $\mathcal{M}$. Finally, let $[0\ \ -1]^{\mathrm{T}}$ be a normal vector along the line

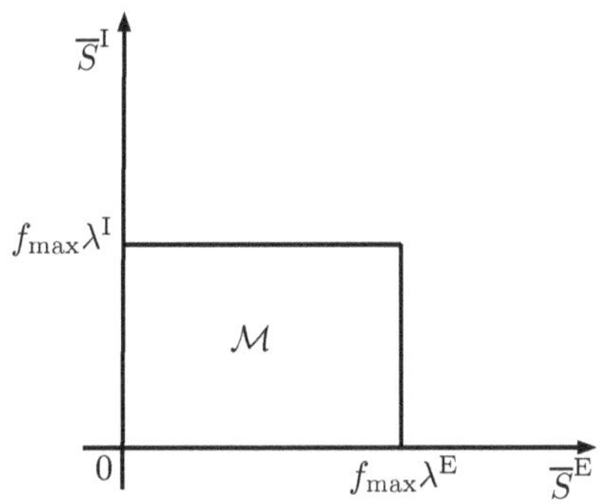

Fig. 1 Visualization of the region $\mathcal{M}$.

$\overline{S}^{\mathrm{I}} = f_{\max}\lambda^{\mathrm{I}}$ directed toward the region $\mathcal{M}$ and note that, since $f(x) \le f_{\max}$, $x \in \mathbb{R}$,

$$[0 \ -1]\begin{bmatrix}\overline{f}_1\\ \overline{f}_2\end{bmatrix} = -\overline{f}_2 = -f(c\overline{S}^{\mathrm{E}} - d\overline{S}^{\mathrm{I}} + v_{\mathrm{th}}^{\mathrm{I}}) + \frac{1}{\lambda^{\mathrm{I}}}\overline{S}^{\mathrm{I}} \ge 0, \quad \overline{S}^{\mathrm{I}} = f_{\max}\lambda^{\mathrm{I}}.$$

Hence the vector field $\overline{f}$ along the line $\overline{S}^{\mathrm{I}} = f_{\max}\lambda^{\mathrm{I}}$ is directed inward toward the region $\mathcal{M}$. Thus along the boundary $\partial\mathcal{M}$ of $\mathcal{M}$ the vector field $\overline{f}$ is directed inward toward the region $\mathcal{M}$, and hence $\mathcal{M}$ is positively invariant with respect to Eqs. (12) and (13). □

A visualization of the region $\mathcal{M}$ is shown in Fig. 1.

Proposition 3. *Consider the two-class mean excitatory and mean inhibitory synaptic drive network given by Eqs. (12) and (13). If $a \le \frac{4}{\gamma f_{\max}\lambda^{\mathrm{E}}}$, then the network given by Eqs. (12) and (13) has exactly one equilibrium point $(\overline{S}_{\mathrm{e}}^{\mathrm{E}}, \overline{S}_{\mathrm{e}}^{\mathrm{I}})$ and $(\overline{S}_{\mathrm{e}}^{\mathrm{E}}, \overline{S}_{\mathrm{e}}^{\mathrm{I}})$ is asymptotically stable.*

Proof. The equilibrium points $(\overline{S}_{\mathrm{e}}^{\mathrm{E}}, \overline{S}_{\mathrm{e}}^{\mathrm{I}})$ of Eqs. (12) and (13) are characterized by the solution to

$$f(a\overline{S}^{\mathrm{E}} - b\overline{S}^{\mathrm{I}} + v_{\mathrm{th}}^{\mathrm{E}}) = \frac{1}{\lambda^{\mathrm{E}}}\overline{S}^{\mathrm{E}}, \tag{16}$$

$$f(c\overline{S}^{\mathrm{E}} - d\overline{S}^{\mathrm{I}} + v_{\mathrm{th}}^{\mathrm{I}}) = \frac{1}{\lambda^{\mathrm{I}}}\overline{S}^{\mathrm{I}}. \tag{17}$$

First, note that it follows from Eq. (11) that $0 \le f(x) \le f_{\max}$, $x \in \mathbb{R}$. Thus it follows from Eqs. (12) and (13) that

$$\dot{\overline{S}}^{\mathrm{E}}(t) \le -\frac{1}{\lambda^{\mathrm{E}}}\overline{S}^{\mathrm{E}}(t) + f_{\max}, \quad \overline{S}^{\mathrm{E}}(0) = \overline{S}_0^{\mathrm{E}}, \quad t \ge 0, \tag{18}$$

$$\dot{\overline{S}}^{\mathrm{I}}(t) \leq -\frac{1}{\lambda^{\mathrm{I}}}\overline{S}^{\mathrm{I}}(t) + f_{\max}, \quad \overline{S}^{\mathrm{I}}(0) = \overline{S}^{\mathrm{I}}_0. \tag{19}$$

Now consider the dynamical system

$$\dot{x}(t) = -\frac{1}{\lambda^{\mathrm{E}}}x(t) + f_{\max}, \quad x(0) = \overline{S}^{\mathrm{E}}_0, \quad t \geq 0, \tag{20}$$

and note that

$$x(t) = (\overline{S}^{\mathrm{E}}_0 - f_{\max}\lambda^{\mathrm{E}})\mathrm{e}^{-\frac{1}{\lambda^{\mathrm{E}}}t} + f_{\max}\lambda^{\mathrm{E}}, \quad t \geq 0. \tag{21}$$

From the comparison principle [26, p. 126], it follows from Eqs. (18) and (21) that $\overline{S}^{\mathrm{E}}(t) \leq x(t)$, $t \geq 0$, and hence $\overline{S}^{\mathrm{E}}(t)$ is bounded for all $t \geq 0$. Similarly, it can be shown that $\overline{S}^{\mathrm{I}}(t)$ is bounded for all $t \geq 0$.

It follows from Eq. (11) that $f(x) \to 0$ if and only if $x \to -\infty$ and $f(x) \to f_{\max}$ if and only if $x \to +\infty$. Since a, b, c, d, $v^{\mathrm{E}}_{\mathrm{th}}$, $v^{\mathrm{I}}_{\mathrm{th}}$, $\overline{S}^{\mathrm{E}}(t)$, $t \geq 0$, and $\overline{S}^{\mathrm{I}}(t)$, $t \geq 0$, are bounded, it follows from Eq. (11) that $0 < f(a\overline{S}^{\mathrm{E}} - b\overline{S}^{\mathrm{I}} + v^{\mathrm{E}}_{\mathrm{th}}) < f_{\max}$ and $0 < f(c\overline{S}^{\mathrm{E}} - d\overline{S}^{\mathrm{I}} + v^{\mathrm{I}}_{\mathrm{th}}) < f_{\max}$. Thus it follows from Eqs. (16) and (17) that $0 < \overline{S}^{\mathrm{E}}_{\mathrm{e}} < f_{\max}\lambda^{\mathrm{E}}$ and $0 < \overline{S}^{\mathrm{I}}_{\mathrm{e}} < f_{\max}\lambda^{\mathrm{I}}$, and hence $(\overline{S}^{\mathrm{E}}_{\mathrm{e}}, \overline{S}^{\mathrm{I}}_{\mathrm{e}}) \in \mathcal{M}$.

Note that since $f(\cdot)$ is increasing, $f(\cdot)$ is invertible. Hence it follows from Eqs. (16) and (17) that

$$\overline{S}^{\mathrm{I}} = \frac{1}{b}\left(a\overline{S}^{\mathrm{E}} + v^{\mathrm{E}}_{\mathrm{th}} - f^{-1}\left(\frac{\overline{S}^{\mathrm{E}}}{\lambda^{\mathrm{E}}}\right)\right) \triangleq g(\overline{S}^{\mathrm{E}}), \tag{22}$$

$$\overline{S}^{\mathrm{E}} = \frac{1}{c}\left(d\overline{S}^{\mathrm{I}} - v^{\mathrm{I}}_{\mathrm{th}} + f^{-1}\left(\frac{\overline{S}^{\mathrm{I}}}{\lambda^{\mathrm{I}}}\right)\right) \triangleq h(\overline{S}^{\mathrm{I}}), \tag{23}$$

where $f^{-1}(x) = \frac{1}{\gamma}\log_{\mathrm{e}}\frac{x}{f_{\max}-x}$. It follows from Eqs. (22) and (23) that

$$\begin{aligned}\frac{\mathrm{d}g(\overline{S}^{\mathrm{E}})}{\mathrm{d}\overline{S}^{\mathrm{E}}} &= \frac{1}{b}\left(a - \frac{f_{\max}\lambda^{\mathrm{E}}}{\gamma\overline{S}^{\mathrm{E}}(f_{\max}\lambda^{\mathrm{E}} - \overline{S}^{\mathrm{E}})}\right)\\ &= \frac{a}{b\overline{S}^{\mathrm{E}}(f_{\max}\lambda^{\mathrm{E}} - \overline{S}^{\mathrm{E}})}\left(-\left(\overline{S}^{\mathrm{E}}\right)^2 + f_{\max}\lambda^{\mathrm{E}}\overline{S}^{\mathrm{E}} - \frac{f_{\max}\lambda^{\mathrm{E}}}{a\gamma}\right),\\ &= \frac{a}{b\overline{S}^{\mathrm{E}}(f_{\max}\lambda^{\mathrm{E}} - \overline{S}^{\mathrm{E}})}\left(-\left(\overline{S}^{\mathrm{E}} - \frac{f_{\max}\lambda^{\mathrm{E}}}{2}\right)^2 + \left(\frac{f_{\max}\lambda^{\mathrm{E}}}{2}\right)^2 - \frac{f_{\max}\lambda^{\mathrm{E}}}{a\gamma}\right),\end{aligned} \tag{24}$$

$$\frac{\mathrm{d}h(\overline{S}^{\mathrm{I}})}{\mathrm{d}\overline{S}^{\mathrm{I}}} = \frac{1}{c}\left(d + \frac{f_{\max}\lambda^{\mathrm{I}}}{\gamma\overline{S}^{\mathrm{I}}(f_{\max}\lambda^{\mathrm{I}} - \overline{S}^{\mathrm{I}})}\right) > 0. \tag{25}$$

Since h is increasing, h^{-1} exists and is an increasing function. Furthermore, it follows from Eq. (23) that $\overline{S}^{\mathrm{I}} = h^{-1}(\overline{S}^{\mathrm{E}})$. If $a \leq \frac{4}{\gamma f_{\max}\lambda^{\mathrm{E}}}$, it follows from Eq. (24) that $\frac{\mathrm{d}g(\overline{S}^{\mathrm{E}})}{\mathrm{d}\overline{S}^{\mathrm{E}}} \leq 0$, and hence g is a decreasing function. Thus the functions g and h^{-1} will intersect exactly once in $\overline{\mathbb{R}}_{+}^{2}$, and hence there exists only one equilibrium point of Eqs. (12) and (13) (see Fig. 2). In addition, note that $g(\cdot)$ approaches two vertical asymptotes as $\overline{S}^{\mathrm{E}} \to 0$ and $\overline{S}^{\mathrm{E}} \to f_{\max}\lambda^{\mathrm{E}}$, whereas $h^{-1}(\cdot)$ approaches two horizontal asymptotes as $\overline{S}^{\mathrm{I}} \to 0$ and $\overline{S}^{\mathrm{I}} \to f_{\max}\lambda^{\mathrm{I}}$ (see Fig. 2), both of whose characteristics are independent of the system parameter values.

To show the stability of the equilibrium point $(\overline{S}_{\mathrm{e}}^{\mathrm{E}}, \overline{S}_{\mathrm{e}}^{\mathrm{I}})$, note that $f'(x) = \gamma f_{\max}\frac{\mathrm{e}^{\gamma x}}{1+\mathrm{e}^{\gamma x}}\frac{1}{1+\mathrm{e}^{\gamma x}} = \gamma f(x)\left(1 - \frac{f(x)}{f_{\max}}\right)$. With use of Eqs. (16) and (17) it follows from Eqs. (12) and (13) that the Jacobian matrix of the system vector field evaluated at the equilibrium point $(\overline{S}_{\mathrm{e}}^{\mathrm{E}}, \overline{S}_{\mathrm{e}}^{\mathrm{I}})$ is given by

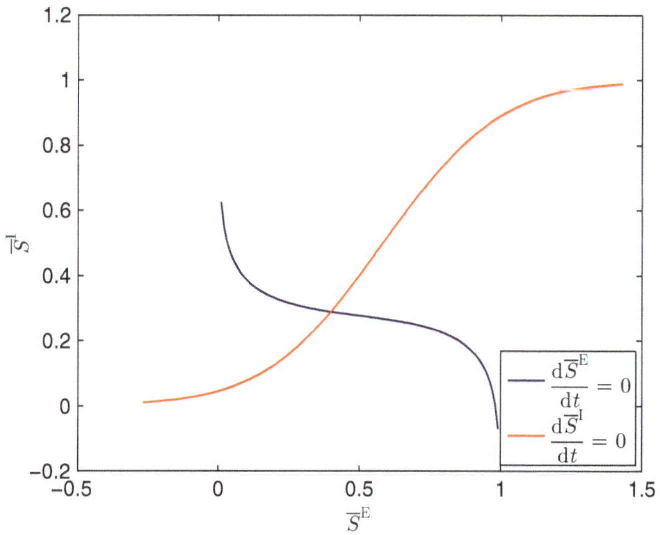

Fig. 2 Plots of $\frac{\mathrm{d}\overline{S}^{\mathrm{E}}}{\mathrm{d}t} = 0$ and $\frac{\mathrm{d}\overline{S}^{\mathrm{I}}}{\mathrm{d}t} = 0$ for $a = 3$, $b = 9$, $c = 6$, $d = 1$, $v_{\mathrm{th}}^{\mathrm{E}} = 1$, $v_{\mathrm{th}}^{\mathrm{I}} = -3$, $\lambda^{\mathrm{E}} = \lambda^{\mathrm{I}} = 1$, $f_{\max} = 1$, and $\gamma = 1$.

$$J(\overline{S}_{\mathrm{e}}^{\mathrm{E}}, \overline{S}_{\mathrm{e}}^{\mathrm{I}}) \triangleq \begin{bmatrix} J_{11}(\overline{S}_{\mathrm{e}}^{\mathrm{E}}, \overline{S}_{\mathrm{e}}^{\mathrm{I}}) & J_{12}(\overline{S}_{\mathrm{e}}^{\mathrm{E}}, \overline{S}_{\mathrm{e}}^{\mathrm{I}}) \\ J_{21}(\overline{S}_{\mathrm{e}}^{\mathrm{E}}, \overline{S}_{\mathrm{e}}^{\mathrm{I}}) & J_{22}(\overline{S}_{\mathrm{e}}^{\mathrm{E}}, \overline{S}_{\mathrm{e}}^{\mathrm{I}}) \end{bmatrix}$$
$$= \begin{bmatrix} -\frac{1}{\lambda^{\mathrm{E}}} + a\gamma \frac{\overline{S}_{\mathrm{e}}^{\mathrm{E}}}{\lambda^{\mathrm{E}}}\left(1 - \frac{\overline{S}_{\mathrm{e}}^{\mathrm{E}}}{f_{\max}\lambda^{\mathrm{E}}}\right) & -b\gamma \frac{\overline{S}_{\mathrm{e}}^{\mathrm{E}}}{\lambda^{\mathrm{E}}}\left(1 - \frac{\overline{S}_{\mathrm{e}}^{\mathrm{E}}}{f_{\max}\lambda^{\mathrm{E}}}\right) \\ c\gamma \frac{\overline{S}_{\mathrm{e}}^{\mathrm{I}}}{\lambda^{\mathrm{I}}}\left(1 - \frac{\overline{S}_{\mathrm{e}}^{\mathrm{I}}}{f_{\max}\lambda^{\mathrm{I}}}\right) & -\frac{1}{\lambda^{\mathrm{I}}} - d\gamma \frac{\overline{S}_{\mathrm{e}}^{\mathrm{I}}}{\lambda^{\mathrm{I}}}\left(1 - \frac{\overline{S}_{\mathrm{e}}^{\mathrm{I}}}{f_{\max}\lambda^{\mathrm{I}}}\right) \end{bmatrix}. \tag{26}$$

Since $0 \le \overline{S}_{\mathrm{e}}^{\mathrm{E}} < f_{\max}\lambda^{\mathrm{E}}$, $0 \le \overline{S}_{\mathrm{e}}^{\mathrm{I}} < f_{\max}\lambda^{\mathrm{I}}$, and a, b, c, and d are nonnegative constants, it follows that $J_{12}(\overline{S}_{\mathrm{e}}^{\mathrm{E}}, \overline{S}_{\mathrm{e}}^{\mathrm{I}}) < 0$, $J_{21}(\overline{S}_{\mathrm{e}}^{\mathrm{E}}, \overline{S}_{\mathrm{e}}^{\mathrm{I}}) > 0$, and $J_{22}(\overline{S}_{\mathrm{e}}^{\mathrm{E}}, \overline{S}_{\mathrm{e}}^{\mathrm{I}}) < 0$. It follows from Eq. (26) that

$$J_{11}(\overline{S}_{\mathrm{e}}^{\mathrm{E}}, \overline{S}_{\mathrm{e}}^{\mathrm{I}}) = \frac{a\gamma}{f_{\max}(\lambda^{\mathrm{E}})^2}\left[-\left(\overline{S}_{\mathrm{e}}^{\mathrm{E}}\right)^2 + f_{\max}\lambda^{\mathrm{E}}\overline{S}_{\mathrm{e}}^{\mathrm{E}} - \frac{f_{\max}\lambda^{\mathrm{E}}}{a\gamma}\right]$$
$$= \frac{a\gamma}{f_{\max}(\lambda^{\mathrm{E}})^2}\left[-\left(\overline{S}_{\mathrm{e}}^{\mathrm{E}} - \frac{f_{\max}\lambda^{\mathrm{E}}}{2}\right)^2 + \frac{(f_{\max}\lambda^{\mathrm{E}})^2}{4} - \frac{f_{\max}\lambda^{\mathrm{E}}}{a\gamma}\right]. \tag{27}$$

For $a \le \frac{4}{\gamma f_{\max}\lambda^{\mathrm{E}}}$ it follows from Eq. (27) that $J_{11}(\overline{S}_{\mathrm{e}}^{\mathrm{E}}, \overline{S}_{\mathrm{e}}^{\mathrm{I}}) \le 0$, and hence $\operatorname{tr} J(\overline{S}_{\mathrm{e}}^{\mathrm{E}}, \overline{S}_{\mathrm{e}}^{\mathrm{I}}) = J_{11}(\overline{S}_{\mathrm{e}}^{\mathrm{E}}, \overline{S}_{\mathrm{e}}^{\mathrm{I}}) + J_{22}(\overline{S}_{\mathrm{e}}^{\mathrm{E}}, \overline{S}_{\mathrm{e}}^{\mathrm{I}}) < 0$ and $\det J(\overline{S}_{\mathrm{e}}^{\mathrm{E}}, \overline{S}_{\mathrm{e}}^{\mathrm{I}}) > 0$, which, by Lyapunov's indirect theorem [26, p. 177], implies that $(\overline{S}_{\mathrm{e}}^{\mathrm{E}}, \overline{S}_{\mathrm{e}}^{\mathrm{I}})$ is an asymptotically stable equilibrium point. □

Theorem 1. *Consider the two-class mean excitatory and mean inhibitory synaptic drive network given by Eqs. (12) and* (13). *If* $a \le \frac{4}{\gamma f_{\max}\lambda^{\mathrm{E}}}$, *then the mean field model given by Eqs. (12) and* (13) *has no periodic orbits in* $\overline{\mathbb{R}}_+^2$, *and hence Eqs. (12) and* (13) *cannot exhibit a limit cycle behavior in* $\overline{\mathbb{R}}_+^2$.

Proof. Since $\mathcal{M}$ is positively invariant with respect to Eqs. (12) and (13), it follows that if the mean field model given by Eqs. (12) and (13) has a periodic orbit, then it must lie either entirely inside $\mathcal{M}$ or entirely outside $\mathcal{M}$. Since $f'(x) = \gamma f(x)\left(1 - \frac{f(x)}{f_{\max}}\right)$, it follows from Eqs. (14) and (15) that

$$\frac{\partial \bar{f}_1(\overline{S}^{\mathrm{E}}, \overline{S}^{\mathrm{I}})}{\partial \overline{S}^{\mathrm{E}}} = -\frac{1}{\lambda^{\mathrm{E}}} + a\gamma f(a\overline{S}^{\mathrm{E}} - b\overline{S}^{\mathrm{I}} + v_{\mathrm{th}}^{\mathrm{E}})\left(1 - \frac{f(a\overline{S}^{\mathrm{E}} - b\overline{S}^{\mathrm{I}} + v_{\mathrm{th}}^{\mathrm{E}})}{f_{\max}}\right), \tag{28}$$

$$\frac{\partial \bar{f}_2(\overline{S}^{\mathrm{E}}, \overline{S}^{\mathrm{I}})}{\partial \overline{S}^{\mathrm{I}}} = -\frac{1}{\lambda^{\mathrm{I}}} - d\gamma f(c\overline{S}^{\mathrm{E}} - d\overline{S}^{\mathrm{I}} + v_{\mathrm{th}}^{\mathrm{I}})\left(1 - \frac{f(c\overline{S}^{\mathrm{E}} - d\overline{S}^{\mathrm{I}} + v_{\mathrm{th}}^{\mathrm{I}})}{f_{\max}}\right). \tag{29}$$

Since $0 < f(c\overline{S}^{\mathrm{E}} - d\overline{S}^{\mathrm{I}} + v_{\mathrm{th}}^{\mathrm{I}}) < f_{\max}$ for $(\overline{S}^{\mathrm{E}}, \overline{S}^{\mathrm{I}}) \in \mathcal{M}$, it follows from Eq. (29) that $\frac{\partial \overline{f}_2(\overline{S}^{\mathrm{E}},\overline{S}^{\mathrm{I}})}{\partial \overline{S}^{\mathrm{I}}} < 0$ for $(\overline{S}^{\mathrm{E}}, \overline{S}^{\mathrm{I}}) \in \mathcal{M}$. Moreover, it follows from Eq. (28) that

$$\begin{aligned}\frac{\partial \overline{f}_1(\overline{S}^{\mathrm{E}}, \overline{S}^{\mathrm{I}})}{\partial \overline{S}^{\mathrm{E}}} &= \frac{a\gamma}{f_{\max}}\left[-f^2(a\overline{S}^{\mathrm{E}} - b\overline{S}^{\mathrm{I}} + v_{\mathrm{th}}^{\mathrm{E}}) + f_{\max} f(a\overline{S}^{\mathrm{E}} - b\overline{S}^{\mathrm{I}} + v_{\mathrm{th}}^{\mathrm{E}}) - \frac{f_{\max}}{a\gamma\lambda^{\mathrm{E}}}\right] \\ &= \frac{a\gamma}{f_{\max}}\left[-\left(f(a\overline{S}^{\mathrm{E}} - b\overline{S}^{\mathrm{I}} + v_{\mathrm{th}}^{\mathrm{E}}) - \frac{f_{\max}}{2}\right)^2 - \left(\frac{f_{\max}}{2}\right)^2 + \frac{f_{\max}}{a\gamma\lambda^{\mathrm{E}}}\right].\end{aligned} \tag{30}$$

It follows from Eq. (30) that $\frac{\partial \overline{f}_1(\overline{S}^{\mathrm{E}},\overline{S}^{\mathrm{I}})}{\partial \overline{S}^{\mathrm{E}}} \le 0$ for $a \le \frac{4}{\gamma f_{\max}\lambda^{\mathrm{E}}}$ and $(\overline{S}^{\mathrm{E}}, \overline{S}^{\mathrm{I}}) \in \mathcal{M}$. Thus

$$\nabla \overline{f}(\overline{S}^{\mathrm{E}}, \overline{S}^{\mathrm{I}}) \triangleq \frac{\partial \overline{f}_1(\overline{S}^{\mathrm{E}}, \overline{S}^{\mathrm{I}})}{\partial \overline{S}^{\mathrm{E}}} + \frac{\partial \overline{f}_2(\overline{S}^{\mathrm{E}}, \overline{S}^{\mathrm{I}})}{\partial \overline{S}^{\mathrm{I}}} < 0 \quad (\overline{S}^{\mathrm{E}}, \overline{S}^{\mathrm{I}}) \in \mathcal{M}. \tag{31}$$

Since $\nabla \overline{f}(\overline{S}^{\mathrm{E}}, \overline{S}^{\mathrm{I}})$ is not identically zero over any subdomain of $\mathcal{M}$ and does not change sign over $\mathcal{M}$, and $\mathcal{M}$ is a simply connected region in $\mathbb{R}^2$, it follows from Bendixson's theorem [26, p. 106] that $\mathcal{M}$ does not contain any nontrivial periodic solutions of Eqs. (12) and (13). It follows from Proposition 3 that for $a \le \frac{4}{\gamma f_{\max}\lambda^{\mathrm{E}}}$, the mean field model given by Eqs. (12) and (13) has exactly one equilibrium point and this equilibrium point is inside $\mathcal{M}$. Thus if there exists a periodic orbit outside $\mathcal{M}$, it must enclose $\mathcal{M}$. Since $\overline{S}^{\mathrm{E}}(t) \ge 0$ and $\overline{S}^{\mathrm{I}}(t) \ge 0$ for all $t \ge 0$, there does not exist any periodic orbit that encloses $\mathcal{M}$, and hence there does not exist any periodic orbit lying entirely outside $\mathcal{M}$. Thus the mean field model given by Eqs. (12) and (13) does not have periodic orbits in $\overline{\mathbb{R}}_+^2$, and hence Eqs. (12) and (13) cannot exhibit a limit cycle behavior in $\overline{\mathbb{R}}_+^2$. □

Proposition 4. *Consider the two-class mean excitatory and mean inhibitory synaptic drive network given by Eqs. (12) and (13). If $a > \frac{4}{\gamma f_{\max}\lambda^{\mathrm{E}}}$ and $\frac{4}{c\gamma f_{\max}\lambda^{\mathrm{I}}} > \frac{b\gamma f_{\max}\lambda^{\mathrm{E}}}{a\gamma f_{\max}\lambda^{\mathrm{E}}-4} - \frac{d}{c}$, then there exist input voltages $v_{\mathrm{th}}^{\mathrm{I}}$ and $v_{\mathrm{th}}^{\mathrm{E}}$ such that the mean field model given by Eqs. (12) and (13) has exactly three equilibrium points, two of which are asymptotically stable and one of which is unstable.*

Proof. If $a > \frac{4}{\gamma f_{\max}\lambda^{\mathrm{E}}}$, then $g(\overline{S}^{\mathrm{E}})$ has a maximum and a minimum value at $\overline{S}^{\mathrm{E}}_{\max}$ and $\overline{S}^{\mathrm{E}}_{\min}$, respectively. Furthermore, note that the values of $v_{\mathrm{th}}^{\mathrm{E}}$ and $v_{\mathrm{th}}^{\mathrm{I}}$ shift the curves $\frac{\mathrm{d}\overline{S}^{\mathrm{E}}}{\mathrm{d}t} = 0$ and $\frac{\mathrm{d}\overline{S}^{\mathrm{I}}}{\mathrm{d}t} = 0$ vertically and horizontally, respectively. If the maximum value of the gradient of $g(\cdot)$ is greater than

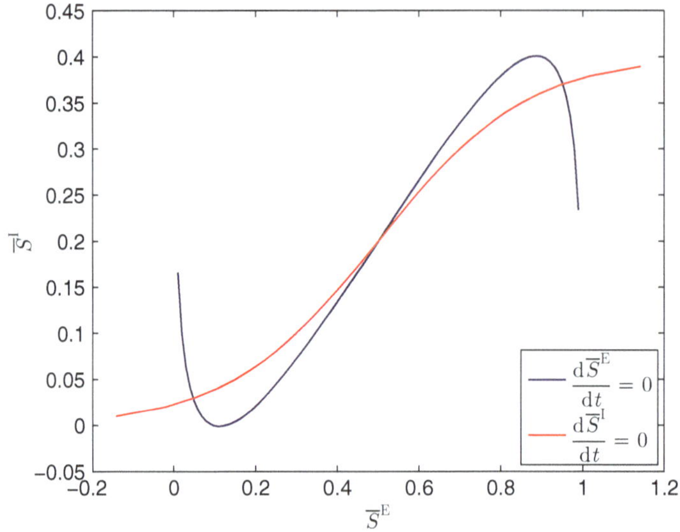

Fig. 3 Plots of $\frac{\mathrm{d}\overline{S}^{\rm E}}{\mathrm{d}t} = 0$ and $\frac{\mathrm{d}\overline{S}^{\rm I}}{\mathrm{d}t} = 0$ for $a = 10$, $b = 9$, $c = 6$, $d = 1$, $v_{\rm th}^{\rm E} = -3.2$, $v_{\rm th}^{\rm I} = -2.8$, $\lambda^{\rm E} = 1$, $\lambda^{\rm I} = 0.4$, $f_{\max} = 1$, and $\gamma = 1$.

the maximum value of the gradient of $h^{-1}(\cdot)$, then the curves $\frac{\mathrm{d}\overline{S}^{\rm E}}{\mathrm{d}t} = 0$ and $\frac{\mathrm{d}\overline{S}^{\rm I}}{\mathrm{d}t} = 0$ can be shifted to intersect at exactly three points; that is, there exist $v_{\rm th}^{\rm I}$ and $v_{\rm th}^{\rm E}$ such that Eqs. (12) and (13) have the three equilibrium points $(\overline{S}_{\rm e1}^{\rm E}, \overline{S}_{\rm e1}^{\rm I})$, $(\overline{S}_{\rm e2}^{\rm E}, \overline{S}_{\rm e2}^{\rm I})$, and $(\overline{S}_{\rm e3}^{\rm E}, \overline{S}_{\rm e3}^{\rm I})$ such that $\overline{S}_{\rm e1}^{\rm E} < \overline{S}_{\min}^{\rm E} < \overline{S}_{\rm e2}^{\rm E} < \overline{S}_{\max}^{\rm E} < \overline{S}_{\rm e3}^{\rm E}$, and the gradient of $g(\cdot)$ at $(\overline{S}_{\rm e2}^{\rm E}, \overline{S}_{\rm e2}^{\rm I})$ is greater than the gradient of $h^{-1}(\cdot)$ at $(\overline{S}_{\rm e2}^{\rm E}, \overline{S}_{\rm e2}^{\rm I})$ as shown in Fig. 3.

It follows from Eqs. (24) and (27) that $\left.\frac{\mathrm{d}g(\overline{S}^{\rm E})}{\mathrm{d}\overline{S}^{\rm E}}\right|_{(\overline{S}_{\rm e}^{\rm E}, \overline{S}_{\rm e}^{\rm I})}$ and $J_{11}(\overline{S}_{\rm e}^{\rm E}, \overline{S}_{\rm e}^{\rm I})$ have the same sign. Moreover, since $\overline{S}_{\rm e1}^{\rm E} < \overline{S}_{\min}^{\rm E}$, it follows from Eq. (24) that

$$\left.\frac{\mathrm{d}g(\overline{S}^{\rm E})}{\mathrm{d}\overline{S}^{\rm E}}\right|_{(\overline{S}_{\rm e1}^{\rm E}, \overline{S}_{\rm e1}^{\rm I})} < 0,$$

and thus $J_{11}(\overline{S}_{\rm e1}^{\rm E}, \overline{S}_{\rm e1}^{\rm I}) > 0$. Therefore, $\operatorname{tr} J(\overline{S}_{\rm e1}^{\rm E}, \overline{S}_{\rm e1}^{\rm I}) > 0$ and $\det J(\overline{S}_{\rm e1}^{\rm E}, \overline{S}_{\rm e1}^{\rm I}) > 0$, and hence the equilibrium point $(\overline{S}_{\rm e1}^{\rm E}, \overline{S}_{\rm e1}^{\rm I})$ is asymptotically stable. Similarly, for $\overline{S}_{\rm e3}^{\rm E} > \overline{S}_{\max}^{\rm E}$ it can be shown that $J_{11}(\overline{S}_{\rm e3}^{\rm E}, \overline{S}_{\rm e3}^{\rm I}) < 0$, and hence $\operatorname{tr} J(\overline{S}_{\rm e3}^{\rm E}, \overline{S}_{\rm e3}^{\rm I}) < 0$ and $\det J(\overline{S}_{\rm e3}^{\rm E}, \overline{S}_{\rm e3}^{\rm I}) > 0$. Thus the equilibrium point $(\overline{S}_{\rm e3}^{\rm E}, \overline{S}_{\rm e3}^{\rm I})$ is asymptotically stable.

If the gradient of $g(\cdot)$ at $(\overline{S}_{\rm e2}^{\rm E}, \overline{S}_{\rm e2}^{\rm I})$ is greater than the gradient of $h^{-1}(\cdot)$ at $(\overline{S}_{\rm e2}^{\rm E}, \overline{S}_{\rm e2}^{\rm I})$, then

$$\left.\frac{\mathrm{d}g(\overline{S}^{\mathrm{E}})}{\mathrm{d}\overline{S}^{\mathrm{E}}}\right|_{(\overline{S}^{\mathrm{E}}_{\mathrm{e}2},\overline{S}^{\mathrm{I}}_{\mathrm{e}2})} > \left(\left.\frac{\mathrm{d}h(\overline{S}^{\mathrm{I}})}{\mathrm{d}\overline{S}^{\mathrm{I}}}\right|_{(\overline{S}^{\mathrm{E}}_{\mathrm{e}2},\overline{S}^{\mathrm{I}}_{\mathrm{e}2})}\right)^{-1}.$$

It follows from Eqs. (24)–(26) that

$$\left.\frac{\mathrm{d}g(\overline{S}^{\mathrm{E}})}{\mathrm{d}\overline{S}^{\mathrm{E}}}\right|_{(\overline{S}^{\mathrm{E}}_{\mathrm{e}2},\overline{S}^{\mathrm{I}}_{\mathrm{e}2})} = -\frac{J_{11}(\overline{S}^{\mathrm{E}}_{\mathrm{e}2},\overline{S}^{\mathrm{I}}_{\mathrm{e}2})}{J_{12}(\overline{S}^{\mathrm{E}}_{\mathrm{e}2},\overline{S}^{\mathrm{I}}_{\mathrm{e}2})}$$

and

$$\left.\frac{\mathrm{d}h(\overline{S}^{\mathrm{I}})}{\mathrm{d}\overline{S}^{\mathrm{I}}}\right|_{(\overline{S}^{\mathrm{E}}_{\mathrm{e}2},\overline{S}^{\mathrm{I}}_{\mathrm{e}2})} = -\frac{J_{22}(\overline{S}^{\mathrm{E}}_{\mathrm{e}2},\overline{S}^{\mathrm{I}}_{\mathrm{e}2})}{J_{21}(\overline{S}^{\mathrm{E}}_{\mathrm{e}2},\overline{S}^{\mathrm{I}}_{\mathrm{e}2})},$$

and hence $-\frac{J_{11}(\overline{S}^{\mathrm{E}}_{\mathrm{e}2})}{J_{12}(\overline{S}^{\mathrm{E}}_{\mathrm{e}2})} > -\frac{J_{21}(\overline{S}^{\mathrm{I}}_{\mathrm{e}2})}{J_{22}(\overline{S}^{\mathrm{I}}_{\mathrm{e}2})}$. Therefore $\det J(\overline{S}^{\mathrm{E}}_{\mathrm{e}2},\overline{S}^{\mathrm{I}}_{\mathrm{e}2}) < 0$, and hence the equilibrium point $(\overline{S}^{\mathrm{E}}_{\mathrm{e}2},\overline{S}^{\mathrm{I}}_{\mathrm{e}2})$ is an unstable saddle equilibrium point.

Finally, note that the maximum values of the gradients of $g(\cdot)$ and $h^{-1}(\cdot)$ occur at the inflection points of $g(\cdot)$ and $h(\cdot)$, respectively, and from Eqs. (24) and (25) the inflection points of $g(\cdot)$ and $h(\cdot)$ correspond to $\overline{S}^{\mathrm{E}} = \frac{f_{\max}\lambda^{\mathrm{E}}}{2}$ and $\overline{S}^{\mathrm{I}} = \frac{f_{\max}\lambda^{\mathrm{I}}}{2}$, respectively. Thus the maximum values of the gradients of $g(\cdot)$ and $h^{-1}(\cdot)$ correspond to $\frac{1}{b}\left(a - \frac{4}{\gamma f_{\max}\lambda^{\mathrm{E}}}\right)$ and $\frac{1}{\frac{1}{c}\left(d+\frac{4}{\gamma f_{\max}\lambda^{\mathrm{I}}}\right)}$, respectively. The condition that the maximum value of the gradient of $g(\cdot)$ is greater than the maximum value of the gradient of $h^{-1}(\cdot)$ implies that $\frac{1}{b}\left(a - \frac{4}{\gamma f_{\max}\lambda^{\mathrm{E}}}\right) > \frac{1}{\frac{1}{c}\left(d+\frac{4}{\gamma f_{\max}\lambda^{\mathrm{I}}}\right)}$, and hence $\frac{4}{c\gamma f_{\max}\lambda^{\mathrm{I}}} > \frac{b\gamma f_{\max}\lambda^{\mathrm{E}}}{a\gamma\lambda^{\mathrm{E}}-4} - \frac{d}{c}$. □

Proposition 5. *Consider the two-class mean excitatory and mean inhibitory synaptic drive network given by Eqs. (12) and (13). If* $a > \frac{4}{\gamma f_{\max}\lambda^{\mathrm{E}}}$, $\frac{4}{c\gamma f_{\max}\lambda^{\mathrm{I}}} < \frac{b\gamma f_{\max}\lambda^{\mathrm{E}}}{a\gamma f_{\max}\lambda^{\mathrm{E}}-4} - \frac{d}{c}$, *and*

$$h^{-1}(\overline{S}^{\mathrm{E}}_{\max}) - h^{-1}(\overline{S}^{\mathrm{E}}_{\min}) < \frac{1}{b}\left[a\sqrt{(f_{\max}\lambda^{\mathrm{E}})^2 - \frac{4f_{\max}\lambda^{\mathrm{E}}}{a\gamma}}\right.$$
$$\left. -\frac{2}{\gamma}\log_e \frac{a\gamma\left(f_{\max}\lambda^{\mathrm{E}}+\sqrt{(f_{\max}\lambda^{\mathrm{E}})^2 - \frac{4f_{\max}\lambda^{\mathrm{E}}}{a\gamma}}\right)^2}{4f_{\max}\lambda^{\mathrm{E}}}\right],$$

then there exist input voltages $v_{\rm th}^{\rm I}$ and $v_{\rm th}^{\rm E}$ such that the mean field model given by Eqs. (12) and (13) has exactly five equilibrium points, two of which are asymptotically stable, two of which are unstable saddles, and one of which can be either asymptotically stable or unstable depending on the sign of the trace of the Jacobian matrix evaluated at that equilibrium point.

Proof. If $g(\overline{S}^{\rm E}_{\rm max}) - g(\overline{S}^{\rm E}_{\rm min}) > h^{-1}(\overline{S}^{\rm E}_{\rm max}) - h^{-1}(\overline{S}^{\rm E}_{\rm min})$ and the maximum value of the gradient of $h^{-1}(\cdot)$ is greater than the maximum value of the gradient of $g(\cdot)$, then the curves $\frac{{\rm d}\overline{S}^{\rm E}}{{\rm d}t} = 0$ and $\frac{{\rm d}\overline{S}^{\rm I}}{{\rm d}t} = 0$ can be shifted so that there exist exactly five equilibrium points $(\overline{S}^{\rm E}_{\rm e1}, \overline{S}^{\rm I}_{\rm e1})$, $(\overline{S}^{\rm E}_{\rm e2}, \overline{S}^{\rm I}_{\rm e2})$, $(\overline{S}^{\rm E}_{\rm e3}, \overline{S}^{\rm I}_{\rm e3})$, $(\overline{S}^{\rm E}_{\rm e4}, \overline{S}^{\rm I}_{\rm e4})$, and $(\overline{S}^{\rm E}_{\rm e5}, \overline{S}^{\rm I}_{\rm e5})$ such that $\overline{S}^{\rm E}_{\rm e1} < \overline{S}^{\rm E}_{\rm min} < \overline{S}^{\rm E}_{\rm e2} < \overline{S}^{\rm E}_{\rm e3} < \overline{S}^{\rm E}_{\rm e4} < \overline{S}^{\rm E}_{\rm max} < \overline{S}^{\rm E}_{\rm e5}$, and the gradient of $g(\cdot)$ at $(\overline{S}^{\rm E}_{\rm e2}, \overline{S}^{\rm I}_{\rm e2})$ and $(\overline{S}^{\rm E}_{\rm e4}, \overline{S}^{\rm I}_{\rm e4})$ is greater than the gradient of $h^{-1}(\cdot)$ at $(\overline{S}^{\rm E}_{\rm e2}, \overline{S}^{\rm I}_{\rm e2})$ and $(\overline{S}^{\rm E}_{\rm e4}, \overline{S}^{\rm I}_{\rm e4})$, and the gradient of $g(\cdot)$ at $(\overline{S}^{\rm E}_{\rm e3}, \overline{S}^{\rm I}_{\rm e3})$ is less than the gradient of $h^{-1}(\cdot)$ at $(\overline{S}^{\rm E}_{\rm e3}, \overline{S}^{\rm I}_{\rm e3})$ as shown in Fig. 4. It follows from Proposition 4 that $(\overline{S}^{\rm E}_{\rm e1}, \overline{S}^{\rm I}_{\rm e1})$ and $(\overline{S}^{\rm E}_{\rm e5}, \overline{S}^{\rm I}_{\rm e5})$ are asymptotically stable equilibrium points.

Since the gradient of $g(\cdot)$ at $(\overline{S}^{\rm E}_{\rm e2}, \overline{S}^{\rm I}_{\rm e2})$ and $(\overline{S}^{\rm E}_{\rm e4}, \overline{S}^{\rm I}_{\rm e4})$ is greater than the gradient of $h^{-1}(\cdot)$ at $(\overline{S}^{\rm E}_{\rm e2}, \overline{S}^{\rm I}_{\rm e2})$ and $(\overline{S}^{\rm E}_{\rm e4}, \overline{S}^{\rm I}_{\rm e4})$, it follows from Proposition 4 that $(\overline{S}^{\rm E}_{\rm e2}, \overline{S}^{\rm I}_{\rm e2})$ and $(\overline{S}^{\rm E}_{\rm e4}, \overline{S}^{\rm I}_{\rm e4})$ are unstable saddle equilibrium points.

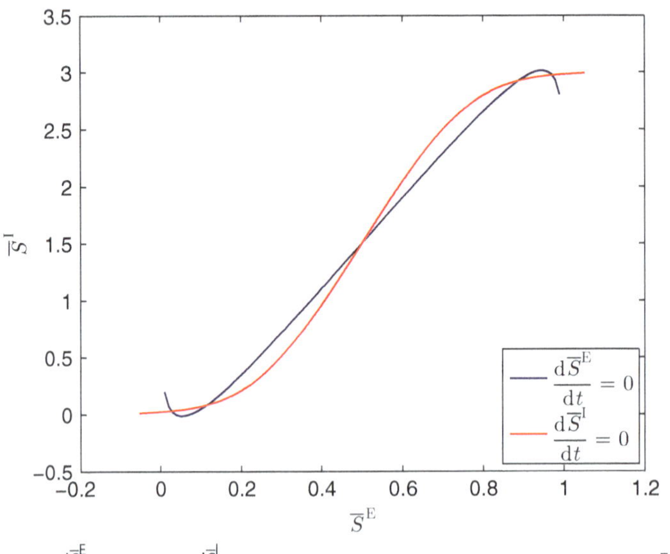

Fig. 4 Plots of $\frac{{\rm d}\overline{S}^{\rm E}}{{\rm d}t} = 0$ and $\frac{{\rm d}\overline{S}^{\rm I}}{{\rm d}t} = 0$ for $a = 20$, $b = 4$, $c = 13$, $d = 1$, $v^{\rm E}_{\rm th} = -4$, $v^{\rm I}_{\rm th} = -5$, $\lambda^{\rm E} = 1$, $\lambda^{\rm I} = 3$, $f_{\max} = 1$, and $\gamma = 1$.

Furthermore, since the gradient of $g(\cdot)$ at $(\overline{S}^{\rm E}_{\rm e3}, \overline{S}^{\rm I}_{\rm e3})$ is less than the gradient of $h^{-1}(\cdot)$ at $(\overline{S}^{\rm E}_{\rm e3}, \overline{S}^{\rm I}_{\rm e3})$, it follows from Eqs. (24)–(26) that $-\frac{J_{11}(\overline{S}^{\rm E}_{\rm e3})}{J_{12}(\overline{S}^{\rm E}_{\rm e3})} < -\frac{J_{21}(\overline{S}^{\rm I}_{\rm e3})}{J_{22}(\overline{S}^{\rm I}_{\rm e3})}$. Hence $\det J(\overline{S}^{\rm E}_{\rm e3}, \overline{S}^{\rm I}_{\rm e3}) > 0$. The stability of $(\overline{S}^{\rm E}_{\rm e3}, \overline{S}^{\rm I}_{\rm e3})$ depends on the sign of $\operatorname{tr} J(\overline{S}^{\rm E}_{\rm e3}, \overline{S}^{\rm I}_{\rm e3})$. Specifically, if $\operatorname{tr} J(\overline{S}^{\rm E}_{\rm e3}, \overline{S}^{\rm I}_{\rm e3}) < 0$, then $(\overline{S}^{\rm E}_{\rm e3}, \overline{S}^{\rm I}_{\rm e3})$ is an asymptotically stable equilibrium point, whereas if $\operatorname{tr} J(\overline{S}^{\rm E}_{\rm e3}, \overline{S}^{\rm I}_{\rm e3}) > 0$, then $(\overline{S}^{\rm E}_{\rm e3}, \overline{S}^{\rm I}_{\rm e3})$ is an unstable equilibrium point.

Finally, it follows from Eq. (24) that

$$\overline{S}^{\rm E}_{\max} = \frac{f_{\max}\lambda^{\rm E} + \sqrt{(f_{\max}\lambda^{\rm E})^2 - \frac{4f_{\max}\lambda^{\rm E}}{a\gamma}}}{2}, \tag{32}$$

$$\overline{S}^{\rm E}_{\min} = \frac{f_{\max}\lambda^{\rm E} - \sqrt{(f_{\max}\lambda^{\rm E})^2 - \frac{4f_{\max}\lambda^{\rm E}}{a\gamma}}}{2}, \tag{33}$$

and hence, using Eq. (22), we have

$$\begin{aligned} g(\overline{S}^{\rm E}_{\max}) - g(\overline{S}^{\rm E}_{\min}) &= \frac{1}{b}\left[a(\overline{S}^{\rm E}_{\max} - \overline{S}^{\rm E}_{\min}) - \frac{1}{\gamma}\log_e \frac{\overline{S}^{\rm E}_{\max}(f_{\max}\lambda^{\rm E} - \overline{S}^{\rm E}_{\min})}{\overline{S}^{\rm E}_{\min}(f_{\max}\lambda^{\rm E} - \overline{S}^{\rm E}_{\max})}\right] \\ &= \frac{1}{b}\left[a\sqrt{(f_{\max}\lambda^{\rm E})^2 - \frac{4f_{\max}\lambda^{\rm E}}{a\gamma}} - \frac{2}{\gamma}\log_e \frac{\overline{S}^{\rm E}_{\max}}{\overline{S}^{\rm E}_{\min}}\right] \\ &= \frac{1}{b}\left[a\sqrt{(f_{\max}\lambda^{\rm E})^2 - \frac{4f_{\max}\lambda^{\rm E}}{a\gamma}}\right. \\ &\qquad \left. - \frac{2}{\gamma}\log_e \frac{a\gamma\left(f_{\max}\lambda^{\rm E} + \sqrt{(f_{\max}\lambda^{\rm E})^2 - \frac{4f_{\max}\lambda^{\rm E}}{a\gamma}}\right)^2}{4f_{\max}\lambda^{\rm E}}\right], \end{aligned} \tag{34}$$

which proves the result. □

Proposition 6. *Consider the two-class mean excitatory and mean inhibitory synaptic drive network given by Eqs. (12) and (13). If* $a > \frac{4}{\gamma f_{\max}\lambda^{\rm E}}$, $\frac{4}{c\gamma f_{\max}\lambda^{\rm I}} < \frac{b\gamma f_{\max}\lambda^{\rm E}}{a\gamma f_{\max}\lambda^{\rm E} - 4} - \frac{d}{c}$, *and*

$$h^{-1}(\overline{S}^{\rm E}_{\max}) - h^{-1}(\overline{S}^{\rm E}_{\min}) > \frac{1}{b}\left[a\sqrt{(f_{\max}\lambda^{\rm E})^2 - \frac{4f_{\max}\lambda^{\rm E}}{a\gamma}}\right.$$
$$\left. - \frac{2}{\gamma}\log_e \frac{a\gamma\left(f_{\max}\lambda^{\rm E} + \sqrt{(f_{\max}\lambda^{\rm E})^2 - \frac{4f_{\max}\lambda^{\rm E}}{a\gamma}}\right)^2}{4f_{\max}\lambda^{\rm E}}\right],$$

then there exist input voltages $v^{\rm I}_{\rm th}$ *and* $v^{\rm E}_{\rm th}$ *such that the mean field model given by Eqs. (12) and* (13) *has exactly one equilibrium point* $(\overline{S}^{\rm E}_{\rm e}, \overline{S}^{\rm I}_{\rm e})$. *Moreover, if tr* $J(\overline{S}^{\rm E}_{\rm e}, \overline{S}^{\rm I}_{\rm e}) < 0$, *then* $(\overline{S}^{\rm E}_{\rm e}, \overline{S}^{\rm I}_{\rm e})$ *is an asymptotically stable equilibrium point, whereas if tr* $J(\overline{S}^{\rm E}_{\rm e}, \overline{S}^{\rm I}_{\rm e}) > 0$, *then* $(\overline{S}^{\rm E}_{\rm e}, \overline{S}^{\rm I}_{\rm e})$ *is an unstable equilibrium point.*

Proof. If $g(\overline{S}^{\rm E}_{\max}) - g(\overline{S}^{\rm E}_{\min}) < h^{-1}(\overline{S}^{\rm E}_{\max}) - h^{-1}(\overline{S}^{\rm E}_{\min})$ and the maximum value of the gradient of $h^{-1}(\cdot)$ is greater than the maximum value of the gradient of $g(\cdot)$, then $\frac{d\overline{S}^{\rm E}}{dt} = 0$ and $\frac{d\overline{S}^{\rm I}}{dt} = 0$ can be shifted so that there exists exactly one intersection $(\overline{S}^{\rm E}_{\rm e}, \overline{S}^{\rm I}_{\rm e})$ of $g(\cdot)$ and $h^{-1}(\cdot)$, and the gradient of $g(\cdot)$ at $(\overline{S}^{\rm E}_{\rm e}, \overline{S}^{\rm I}_{\rm e})$ is less than the gradient of $h^{-1}(\cdot)$ at $(\overline{S}^{\rm E}_{\rm e}, \overline{S}^{\rm I}_{\rm e})$ as shown in Fig. 5. Thus it follows from Proposition 5 that $\det J(\overline{S}^{\rm E}_{\rm e}, \overline{S}^{\rm I}_{\rm e}) > 0$. Hence if tr $J(\overline{S}^{\rm E}_{\rm e}, \overline{S}^{\rm I}_{\rm e}) < 0$, then $(\overline{S}^{\rm E}_{\rm e}, \overline{S}^{\rm I}_{\rm e})$ is an asymptotically stable equilibrium point, whereas if tr $J(\overline{S}^{\rm E}_{\rm e}, \overline{S}^{\rm I}_{\rm e}) > 0$, then $(\overline{S}^{\rm E}_{\rm e}, \overline{S}^{\rm I}_{\rm e})$ is an unstable equilibrium point. □

Theorem 2. *Consider the two-class mean excitatory and mean inhibitory synaptic drive network given by Eqs. (12) and* (13). *Assume that the conditions in Proposition 6 are satisfied and* tr $\mathrm{J}(\overline{\mathrm{S}}^{\rm E}_{\rm e}, \overline{\mathrm{S}}^{\rm I}_{\rm e}) > 0$. *Then there exist input voltages* $v^{\rm I}_{\rm th}$ *and* $v^{\rm E}_{\rm th}$ *such that the mean field model given by Eqs. (12) and* (13) *possesses a limit cycle. Moreover, the limit cycle is stable.*

Proof. It follows from Proposition 6 that the mean field model given by Eqs. (12) and (13) has exactly one equilibrium point $(\overline{S}^{\rm E}_{\rm e}, \overline{S}^{\rm I}_{\rm e})$ and that this equilibrium point $(\overline{S}^{\rm E}_{\rm e}, \overline{S}^{\rm I}_{\rm e})$ is unstable. Define $\mathcal{R} \triangleq \mathcal{M} \setminus (\overline{S}^{\rm E}_{\rm e}, \overline{S}^{\rm I}_{\rm e})$ and note that since $\mathcal{M}$ is positively invariant with respect to Eqs. (12) and (13), and $(\overline{S}^{\rm E}_{\rm e}, \overline{S}^{\rm I}_{\rm e})$ is an unstable equilibrium point, it follows that $\mathcal{R}$ is positively invariant with respect to Eqs. (12) and (13). Hence it follows from the Poincaré-Bendixson theorem [26, p. 109] that $\mathcal{R}$ contains a periodic orbit. Furthermore, since $\mathcal{R}$ is positively invariant and the equilibrium point $(\overline{S}^{\rm E}_{\rm e}, \overline{S}^{\rm I}_{\rm e})$ is unstable, the limit cycle is stable. □

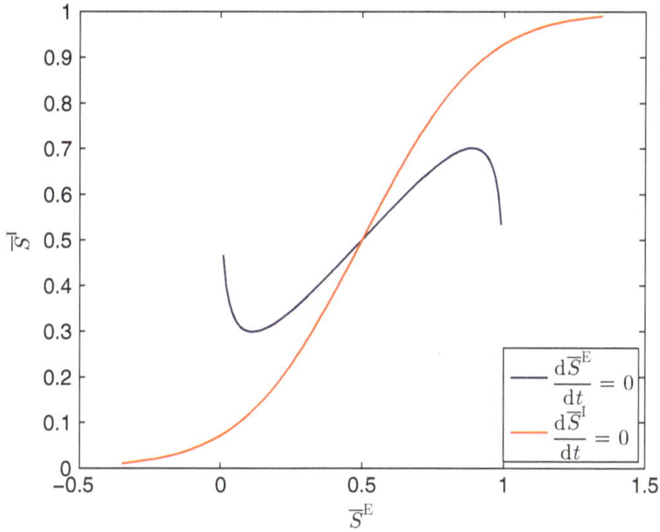

Fig. 5 Plots of $\frac{d\overline{S}^{\mathrm{E}}}{dt} = 0$ and $\frac{d\overline{S}^{\mathrm{I}}}{dt} = 0$ for $a = 10$, $b = 9$, $c = 6$, $d = 1$, $v_{\mathrm{th}}^{\mathrm{E}} = -0.5$, $v_{\mathrm{th}}^{\mathrm{I}} = -2.5$, $\lambda^{\mathrm{E}} = 1$, $\lambda^{\mathrm{I}} = 1$, $f_{\max} = 1$, and $\gamma = 1$.

4 LARGE NEURAL POPULATIONS, PARTIAL SYNCHRONIZATION, AND SUPPRESSION OF CONSCIOUSNESS

One of the most important questions in neuroscience is how do neurons, or collections of neurons, communicate. There is extensive experimental verification that collections of neurons may function as oscillators and the synchronization of oscillators may play a key role in the transmission of information within the central nervous system [15, 16, 27]. This may be particularly relevant to understanding the mechanism of action for general anesthesia.

It has been known for a long time that general anesthesia has profound effects on the spectrum of oscillations in the electroencephalograph [17, 28, 29]. More recently, John and Prichep [30] have suggested that thalamocortical circuits function as neural pacemakers and that alterations in the thalamic oscillations are associated with the induction of general anesthesia. Furthermore, it is well known that anesthetic drugs frequently induce epileptiform activity as part of the progression to the state of unconsciousness [31]. Multiple lines of evidence indicate that anesthetic agents impact neural oscillators. In addition, epileptiform activity implies synchronization of oscillators. This leads to the possibility that synchronization of these oscillators is involved in the transition to the anesthetic state.

Sufficient conditions for synchronization for an excitatory and inhibitory cortical neural network are given in [10]. It has been observed, however, that when patients lose consciousness some parts of the brain are still functional (eg, cardiac function), whereas other parts are suppressed. This can be captured by biological neural network models that exhibit *partial synchronization*, wherein part of the system's state is synchronized and the other parts fire at normal levels.

To address partial synchronization, we make a high-level assumption regarding neuronal connectivity by postulating the existence of a subset of inhibitory neurons that themselves do not receive inhibitory input. It is important to note that the existence of such a subset has not been demonstrated experimentally. However, it is not biologically implausible and, as we will demonstrate, this assumption leads to sufficient conditions for partial synchronization.

To address the problem of partial synchronization, let $S^{\mathrm{E}}(t) \triangleq [S_1^{\mathrm{E}}(t), \ldots, S_{n_{\mathrm{E}}}^{\mathrm{E}}(t)] \in \mathbb{R}^{n_{\mathrm{E}}}$, $S^{\mathrm{I}}(t) \triangleq [S_1^{\mathrm{I}}(t), \ldots, S_{n_{\mathrm{I}}}^{\mathrm{I}}(t)] \in \mathbb{R}^{n_{\mathrm{I}}}$, $\tau_i^{\mathrm{E}} = \lambda_i^{\mathrm{E}}$, $i = 1, \ldots, n_{\mathrm{E}}$, and $\tau_i^{\mathrm{I}} = \lambda_i^{\mathrm{I}}$, $i = 1, \ldots, n_{\mathrm{I}}$, and assume $B_i^{\mathrm{E}} = B_i^{\mathrm{I}} = 1$ so that the vector-matrix form of Eqs. (7) and (8) can be written as

$$\dot{S}^{\mathrm{E}}(t) = -L^{\mathrm{E}} S^{\mathrm{E}}(t) + f^{\mathrm{E}}(A^{\mathrm{EE}} S^{\mathrm{E}}(t) + A^{\mathrm{EI}} S^{\mathrm{I}}(t) + \tilde{v}_{\mathrm{th}}^{\mathrm{E}}(t)), \; S^{\mathrm{E}}(0) = S_0^{\mathrm{E}}, \quad t \geq 0, \tag{35}$$

$$\dot{S}^{\mathrm{I}}(t) = -L^{\mathrm{I}} S^{\mathrm{I}}(t) + f^{\mathrm{I}}(A^{\mathrm{IE}} S^{\mathrm{E}}(t) + A^{\mathrm{II}} S^{\mathrm{I}}(t) + \tilde{v}_{\mathrm{th}}^{\mathrm{I}}(t)), \quad S^{\mathrm{I}}(0) = S_0^{\mathrm{I}}, \tag{36}$$

where $L^{\mathrm{E}} \triangleq \mathrm{diag}\left[\frac{1}{\tau_1^{\mathrm{E}}}, \ldots, \frac{1}{\tau_{n_{\mathrm{E}}}^{\mathrm{E}}}\right] \in \mathbb{R}^{n_{\mathrm{E}} \times n_{\mathrm{E}}}$ and $L^{\mathrm{I}} \triangleq \mathrm{diag}\left[\frac{1}{\tau_1^{\mathrm{I}}}, \ldots, \frac{1}{\tau_{n_{\mathrm{I}}}^{\mathrm{I}}}\right] \in \mathbb{R}^{n_{\mathrm{I}} \times n_{\mathrm{I}}}$ are time constant matrices, $A^{\mathrm{EE}} \triangleq [A_{ij}^{\mathrm{EE}}] \in \mathbb{R}^{n_{\mathrm{E}} \times n_{\mathrm{E}}}$, $A^{\mathrm{EI}} \triangleq [A_{ij}^{\mathrm{EI}}] \in \mathbb{R}^{n_{\mathrm{E}} \times n_{\mathrm{I}}}$, $A^{\mathrm{IE}} \triangleq [A_{ij}^{\mathrm{IE}}] \in \mathbb{R}^{n_{\mathrm{I}} \times n_{\mathrm{E}}}$, and $A^{\mathrm{II}} \triangleq [A_{ij}^{\mathrm{II}}] \in \mathbb{R}^{n_{\mathrm{I}} \times n_{\mathrm{I}}}$ are matrices representing the strength of the synaptic interconnections, $\tilde{v}_{\mathrm{th}}^{\mathrm{E}} \triangleq [v_{\mathrm{th}1}^{\mathrm{E}}, \ldots, v_{\mathrm{th}n_{\mathrm{E}}}^{\mathrm{E}}]^{\mathrm{T}}$, $\tilde{v}_{\mathrm{th}}^{\mathrm{I}} \triangleq [v_{\mathrm{th}1}^{\mathrm{I}}, \ldots, v_{\mathrm{th}n_{\mathrm{I}}}^{\mathrm{I}}]^{\mathrm{T}}$, and

$$f^{\mathrm{E}}(A^{\mathrm{EE}} S^{\mathrm{E}} + A^{\mathrm{EI}} S^{\mathrm{I}} + \tilde{v}_{\mathrm{th}}^{\mathrm{E}}) \triangleq \left[f_1 \left(\sum_{j=1}^{n_{\mathrm{E}}} A_{1j}^{\mathrm{EE}} S_j^{\mathrm{E}} + \sum_{j'=1}^{n_{\mathrm{I}}} A_{1j'}^{\mathrm{EI}} S_{j'}^{\mathrm{I}} + v_{\mathrm{th}1}^{\mathrm{E}} \right), \ldots, \right.$$

$$f_{n_\mathrm{E}}\left(\sum_{j=1}^{n_\mathrm{E}} A^{\mathrm{EE}}_{n_\mathrm{E}j} S^\mathrm{E}_j + \sum_{j'=1}^{n_\mathrm{I}} A^{\mathrm{EI}}_{n_\mathrm{E}j'} S^\mathrm{I}_{j'} + v^\mathrm{E}_{\mathrm{th}n_\mathrm{E}}\right)\Bigg]^\mathrm{T}, \tag{37}$$

$$f^\mathrm{I}(A^{\mathrm{IE}}S^\mathrm{E} + A^{\mathrm{II}}S^\mathrm{I} + \tilde{v}^\mathrm{I}_{\mathrm{th}}) \triangleq \Bigg[f_1\left(\sum_{j=1}^{n_\mathrm{E}} A^{\mathrm{IE}}_{1j} S^\mathrm{E}_j + \sum_{j'=1}^{n_\mathrm{I}} A^{\mathrm{II}}_{1j'} S^\mathrm{I}_{j'} + v^\mathrm{I}_{\mathrm{th}1}\right), \ldots, \\ f_{n_\mathrm{I}}\left(\sum_{j=1}^{n_\mathrm{E}} A^{\mathrm{IE}}_{n_\mathrm{I}j} S^\mathrm{E}_j + \sum_{j'=1}^{n_\mathrm{I}} A^{\mathrm{II}}_{n_\mathrm{I}j'} S^\mathrm{I}_{j'} + v^\mathrm{I}_{\mathrm{th}n_\mathrm{I}}\right)\Bigg]^\mathrm{T} \tag{38}$$

denote the vector activation functions describing the relationship between the synaptic drives and the firing rates of neurons with $f_i(x)$, $x \in \mathbb{R}$, $i = 1, \ldots, n_\mathrm{E}$, and $f_k(x)$, $x \in \mathbb{R}$, $k = 1, \ldots, n_\mathrm{I}$, defined as in Eq. (10).

Define $S^\mathrm{I}(t) \triangleq [(\tilde{S}^\mathrm{I}_1(t))^\mathrm{T}, (\tilde{S}^\mathrm{I}_2(t))^\mathrm{T}]^\mathrm{T}$, where $\tilde{S}^\mathrm{I}_1(t) \triangleq [S^\mathrm{I}_1(t), \ldots, S^\mathrm{I}_q(t)]^\mathrm{T}$ denotes the vector of inhibitory synaptic drives that are coupled to both excitatory and inhibitory neurons and $\tilde{S}^\mathrm{I}_2(t) \triangleq [S^\mathrm{I}_{q+1}(t), \ldots, S^\mathrm{I}_{n_\mathrm{I}}(t)]^\mathrm{T}$ denotes the vector of inhibitory synaptic drives that are coupled to only excitatory neurons. In this case the neuronal connectivity matrix A can be partitioned as

$$A \triangleq \begin{bmatrix} A^{\mathrm{EE}} & A^{\mathrm{EI}} \\ A^{\mathrm{IE}} & A^{\mathrm{II}} \end{bmatrix} = \begin{bmatrix} A^{\mathrm{EE}} & A^{\mathrm{EI}}_1 & A^{\mathrm{EI}}_2 \\ A^{\mathrm{IE}}_1 & A^{\mathrm{II}}_1 & A^{\mathrm{II}}_2 \\ A^{\mathrm{IE}}_2 & 0_{(n_\mathrm{I}-q)\times q} & 0_{(n_\mathrm{I}-q)\times(n_\mathrm{I}-q)} \end{bmatrix}, \tag{39}$$

where

$$A^{\mathrm{EI}} = [A^{\mathrm{EI}}_1 \; A^{\mathrm{EI}}_2], \; A^{\mathrm{IE}} = \begin{bmatrix} A^{\mathrm{IE}}_1 \\ A^{\mathrm{IE}}_2 \end{bmatrix}, \quad A^{\mathrm{II}} = \begin{bmatrix} A^{\mathrm{II}}_1 & A^{\mathrm{II}}_2 \\ 0_{(n_\mathrm{I}-q)\times q} & 0_{(n_\mathrm{I}-q)\times(n_\mathrm{I}-q)} \end{bmatrix},$$

$A^{\mathrm{EI}}_1 \in \mathbb{R}^{n_\mathrm{E}\times q}$, $A^{\mathrm{EI}}_2 \in \mathbb{R}^{n_\mathrm{E}\times(n_\mathrm{I}-q)}$, $A^{\mathrm{IE}}_1 \in \mathbb{R}^{q\times n_\mathrm{E}}$, $A^{\mathrm{II}}_1 \in \mathbb{R}^{q\times q}$, $A^{\mathrm{II}}_2 \in \mathbb{R}^{q\times(n_\mathrm{I}-q)}$, and $A^{\mathrm{IE}}_2 \in \mathbb{R}^{(n_\mathrm{I}-q)\times n_\mathrm{E}}$.

We can write Eqs. (35) and (36) as

$$\dot{S}^\mathrm{E}(t) = -L^\mathrm{E} S^\mathrm{E}(t) + f^\mathrm{E}(A^{\mathrm{EE}}S^\mathrm{E}(t) + A^{\mathrm{EI}}_1\tilde{S}^\mathrm{I}_1(t) + A^{\mathrm{EI}}_2\tilde{S}^\mathrm{I}_2(t) + \tilde{v}^\mathrm{E}_{\mathrm{th}}(t)), \\ S^\mathrm{E}(0) = S^\mathrm{E}_0, \quad t \geq 0, \tag{40}$$

$$\dot{\tilde{S}}^\mathrm{I}_1(t) = -L^\mathrm{I}_1\tilde{S}^\mathrm{I}_1(t) + \tilde{f}^\mathrm{I}_1(A^{\mathrm{IE}}_1 S^\mathrm{E}(t) + A^{\mathrm{II}}_1\tilde{S}^\mathrm{I}_1(t) + A^{\mathrm{II}}_2\tilde{S}^\mathrm{I}_2(t) + \tilde{v}^\mathrm{I}_{\mathrm{th}1}(t)), \\ \tilde{S}^\mathrm{I}_1(0) = \tilde{S}^\mathrm{I}_{10}, \tag{41}$$

$$\dot{\tilde{S}}_2^{\rm I}(t) = -L_2^{\rm I}\tilde{S}_2^{\rm I}(t) + \tilde{f}_2^{\rm I}(A_2^{\rm IE}S^{\rm E}(t) + \tilde{v}_{\rm th2}^{\rm I}(t)), \quad \tilde{S}_2^{\rm I}(0) = \tilde{S}_{20}^{\rm I}, \tag{42}$$

where

$$L^{\rm I} \triangleq \begin{bmatrix} L_1^{\rm I} & 0_{q\times(n_{\rm I}-q)} \\ 0_{(n_{\rm I}-q)\times q} & L_2^{\rm I} \end{bmatrix}, \tag{43}$$

$L_1^{\rm I} \triangleq {\rm diag}\left[\frac{1}{\tau_1^{\rm I}},\ldots,\frac{1}{\tau_q^{\rm I}}\right] \in \mathbb{R}^{q\times q}$, $L_2^{\rm I} \triangleq {\rm diag}\left[\frac{1}{\tau_{q+1}^{\rm I}},\ldots,\frac{1}{\tau_{n_{\rm I}}^{\rm I}}\right] \in \mathbb{R}^{(n_{\rm I}-q)\times(n_{\rm I}-q)}$, $\tilde{v}_{\rm th1}^{\rm I} \triangleq [v_{\rm th1}^{\rm I},\ldots,v_{{\rm th}q}^{\rm I}]^{\rm T} \in \mathbb{R}^q$, $\tilde{v}_{\rm th2}^{\rm I} \triangleq [v_{{\rm th}(q+1)}^{\rm I},\ldots,v_{{\rm th}n_{\rm I}}^{\rm I}]^{\rm T} \in \mathbb{R}^{n_{\rm I}-q}$, and $f^{\rm I} = [(\tilde{f}_1^{\rm I})^{\rm T}\ (\tilde{f}_2^{\rm I})^{\rm T}]^{\rm T}$. Letting $S(t) \triangleq [(S^{\rm E}(t))^{\rm T}, (\tilde{S}_1^{\rm I}(t))^{\rm T}]^{\rm T} \triangleq [S_1(t),\ldots,S_{n_{\rm E}+q}(t)]^{\rm T}$, $\tilde{f} \triangleq [(f^{\rm E})^{\rm T}\ (\tilde{f}_1^{\rm I})^{\rm T}]^{\rm T}$, and $L \triangleq$ block-diag$[L^{\rm E}\ L_1^{\rm I}]$, we can write Eqs. (40), (41), and (42) as

$$\dot{S}(t) = -LS(t) + \tilde{f}(\tilde{A}S(t) + \tilde{B}\tilde{S}_2^{\rm I}(t) + \tilde{v}_{\rm th}(t)), \quad S(0) = S_0, \quad t \ge 0, \tag{44}$$

$$\dot{\tilde{S}}_2^{\rm I}(t) = -L_2^{\rm I}\tilde{S}_2^{\rm I}(t) + \tilde{f}_2^{\rm I}(A_2^{\rm IE}S^{\rm E}(t) + \tilde{v}_{\rm th2}^{\rm I}(t)), \quad \tilde{S}_2^{\rm I}(0) = \tilde{S}_{20}^{\rm I}, \tag{45}$$

where

$$\tilde{A} \triangleq \begin{bmatrix} A^{\rm EE} & A_1^{\rm EI} \\ A_1^{\rm IE} & A_1^{\rm II} \end{bmatrix} \in \mathbb{R}^{(n_{\rm E}+q)\times(n_{\rm E}+q)}, \quad \tilde{B} \triangleq \begin{bmatrix} A_2^{\rm EI} \\ A_2^{\rm II} \end{bmatrix} \in \mathbb{R}^{(n_{\rm E}+q)\times(n_{\rm I}-q)},$$

$$\tilde{v}_{\rm th} \triangleq \begin{bmatrix} \tilde{v}_{\rm th}^{\rm E} \\ \tilde{v}_{\rm th1}^{\rm I} \end{bmatrix} \in \mathbb{R}^{n_{\rm E}+q}.$$

The following proposition and definitions are needed for the statement of the main results of this section.

Proposition 7 ([10]). *Consider the excitatory-inhibitory network given by Eqs. (35) and* (36) *with the vector activation functions defined by Eqs. (37) and* (38), *where* $f_i(x)$, $x \in \mathbb{R}$, $i = 1,\ldots,n_E$, *and* $f_k(x)$, $x \in \mathbb{R}$, $k = 1,\ldots,n_I$, *defined as in Eqs. (10). If* $S_0^E \geq\geq 0$ *and* $S_0^I \geq\geq 0$, *then* $S^E(t) \geq\geq 0$ *and* $S^I(t) \geq\geq 0$ *for all* $t \ge 0$.

Definition 1. The biological neural network given by Eqs. (44) and (45) is said to be *globally exponentially stable with respect to* S *uniformly in* $\tilde{S}_{20}^I$ if there exist scalars α, $\beta > 0$ such that $\|S(t)\| \le \alpha\|S_0\|{\rm e}^{-\beta t}$, $t \ge 0$, for all $S_0 \in \overline{\mathbb{R}}_+^{n_{\rm E}+q}$ and $\tilde{S}_{20}^{\rm I} \in \overline{\mathbb{R}}_+^{n_{\rm I}-q}$.

Definition 2. The biological neural network given by Eqs. (44) and (45) is said to be *globally asymptotically partially synchronized* if

$$\lim_{t\to\infty} |S_i(t) - S_j(t)| = 0$$

for all $S_0 \in \overline{\mathbb{R}}_+^{n_{\rm E}+q}$, $\tilde{S}_{20}^{\rm I} \in \overline{\mathbb{R}}_+^{n_{\rm I}-q}$, and $i = 1,\ldots,n_{\rm E}+q$, $i \neq j$.

Definition 3. The biological neural network given by Eqs. (44) and (45) is said to be *globally exponentially partially synchronized* if there exist constants $p > 0$ and $\rho > 0$ such that

$$|S_i(t) - S_j(t)| \leq \rho e^{-pt}|S_{i0} - S_{j0}|, \quad t \geq 0,$$

for all $S_0 \in \overline{\mathbb{R}}_+^{n_E+q}$, $\tilde{S}_{20}^I \in \overline{\mathbb{R}}_+^{n_I-q}$, and $i = 1, \ldots, n_E + q$, $i \neq j$.

Remark 1. It follows from Definitions 1 and 3 that if the biological neural network given by Eqs. (44) and (45) is globally exponentially stable with respect to S uniformly in $\tilde{S}_{20}^I$, then the biological neural network given by Eqs. (44) and (45) is globally exponentially partially synchronized.

For the statement of the next theorem, $\mathcal{N}(X)$ denotes the null space of the matrix X.

Theorem 3. *Consider the excitatory-inhibitory network given by Eqs. (44) and (45) with the vector activation functions defined by Eqs. (37) and (38). If $\tilde{v}_{th}(t) \leq\leq -\tilde{B}w$, $t \geq 0$, where $w \triangleq [\min\{S_{q+1}^I(0), \tau_{q+1}^I\eta_{q+1}\}, \ldots, \min\{S_{n_I}^I(0), \tau_{n_I}^I\eta_{n_I}\}]^T \in \mathbb{R}^{n_I-q}$ and $\eta_i \triangleq \min_{t\geq 0} v_{th_i}^I(t)$, $i = q+1, \ldots, n_I$, and there exist positive-definite matrices P, $Q \in \mathbb{R}^{(n_E+q)\times(n_E+q)}$ and a diagonal positive-definite matrix $R \in \mathbb{R}^{(n_E+q)\times(n_E+q)}$ such that*

$$\begin{bmatrix} Q & -P \\ -P & R \end{bmatrix} \geq 0 \tag{46}$$

and $\Omega > 0$, where

$$\Omega \triangleq PL + LP - Q - \tilde{A}^T R \tilde{A}, \tag{47}$$

then Eqs. (44) and (45) is globally exponentially partially synchronized. Alternatively, if $\Omega \geq 0$ and $\mathcal{N}(\Omega) = \text{span}(\mathbf{e}_{n_E+q})$, then the biological neural network given by Eqs. (44) and (45) is globally asymptotically partially synchronized.

Proof. Consider the partial Lyapunov function candidate $V : \overline{\mathbb{R}}_+^{n_E+q} \to \mathbb{R}$ given by $V(S) = S^T PS$ and note that $\lambda_{min}(P)\|S\|_2^2 \leq V(S) \leq \lambda_{max}(P)\|S\|_2^2$, $S \in \overline{\mathbb{R}}_+^{n_E+q}$, where $\lambda_{max}(P)$ and $\lambda_{min}(P)$ denote the maximum and minimum eigenvalues of P, respectively. It follows that the derivative of $V(S)$ along the trajectories of Eq. (44) is given by

$$\dot{V}(S, \tilde{S}_2^I) = -S^T(PL + LP)S + 2S^T P\tilde{f}(\tilde{A}S + \tilde{B}\tilde{S}_2^I + \tilde{v}_{th}). \tag{48}$$

It follows from Eq. (46) that

$$[S^{\mathrm{T}}\ \tilde{f}^{\mathrm{T}}(\tilde{A}S+\tilde{B}\tilde{S}_2^{\mathrm{I}}+\tilde{v}_{\mathrm{th}})]\begin{bmatrix} Q & -P \\ -P & R \end{bmatrix}\begin{bmatrix} S \\ \tilde{f}(\tilde{A}S+\tilde{B}\tilde{S}_2^{\mathrm{I}}+\tilde{v}_{\mathrm{th}}) \end{bmatrix} \geq 0,$$

$$(S, \tilde{S}_2^{\mathrm{I}}) \in \overline{\mathbb{R}}_+^{n_{\mathrm{E}}+q} \times \overline{\mathbb{R}}_+^{n_{\mathrm{I}}-q}, \quad (49)$$

or, equivalently,

$$2S^{\mathrm{T}}P\tilde{f}(\tilde{A}S+\tilde{B}\tilde{S}_2^{\mathrm{I}}+\tilde{v}_{\mathrm{th}}) \leq S^{\mathrm{T}}QS+\tilde{f}^{\mathrm{T}}(\tilde{A}S+\tilde{B}\tilde{S}_2^{\mathrm{I}}+\tilde{v}_{\mathrm{th}})R\tilde{f}(\tilde{A}S+\tilde{B}\tilde{S}_2^{\mathrm{I}}+\tilde{v}_{\mathrm{th}}),$$
$$(S, \tilde{S}_2^{\mathrm{I}}) \in \overline{\mathbb{R}}_+^{n_{\mathrm{E}}+q} \times \overline{\mathbb{R}}_+^{n_{\mathrm{I}}-q}. \quad (50)$$

With use of Eq. (50) it follows from Eq. (48) that

$$\dot{V}(S, \tilde{S}_2^{\mathrm{I}}) \leq S^{\mathrm{T}}(-PL-LP+Q)S+\tilde{f}^{\mathrm{T}}(\tilde{A}S+\tilde{B}\tilde{S}_2^{\mathrm{I}}+\tilde{v}_{\mathrm{th}})R\tilde{f}(\tilde{A}S+\tilde{B}\tilde{S}_2^{\mathrm{I}}+\tilde{v}_{\mathrm{th}}),$$
$$(S, \tilde{S}_2^{\mathrm{I}}) \in \overline{\mathbb{R}}_+^{n_{\mathrm{E}}+q} \times \overline{\mathbb{R}}_+^{n_{\mathrm{I}}-q}. \quad (51)$$

Since $A^{\mathrm{IE}} \geq\geq 0$, $S^{\mathrm{E}} \geq\geq 0$, and $f_i(x)$, $x \in \mathbb{R}$, $i = q+1, \ldots, n_{\mathrm{I}}$, are nondecreasing functions, it follows that $\tilde{f}_2^{\mathrm{I}}(A_2^{\mathrm{IE}}S^{\mathrm{E}} + \tilde{v}_{\mathrm{th}2}^{\mathrm{I}}) \geq\geq \tilde{f}_2^{\mathrm{I}}(\tilde{v}_{\mathrm{th}2}^{\mathrm{I}})$, $S^{\mathrm{E}} \in \overline{\mathbb{R}}_+^{n_{\mathrm{E}}}$. And since $f_i(x) \geq x$, $x \in \mathbb{R}$, $i = q+1, \ldots, n_{\mathrm{I}}$, it follows that $\tilde{f}_2^{\mathrm{I}}(\tilde{v}_{\mathrm{th}2}^{\mathrm{I}}) \geq\geq \tilde{v}_{\mathrm{th}2}^{\mathrm{I}}$, and hence it follows from Eq. (45) that

$$\dot{\tilde{S}}_2^{\mathrm{I}}(t) \geq\geq -L_2^{\mathrm{I}}\tilde{S}_2^{\mathrm{I}}(t) + \tilde{v}_{\mathrm{th}2}^{\mathrm{I}}(t), \quad \tilde{S}_2^{\mathrm{I}}(0) = \tilde{S}_{20}^{\mathrm{I}}, \quad t \geq 0, \quad (52)$$

or, equivalently,

$$\dot{S}_i^{\mathrm{I}}(t) \geq -\frac{1}{\tau_i^{\mathrm{I}}}S_i^{\mathrm{I}}(t)+v_{\mathrm{th}i}^{\mathrm{I}}(t) \geq -\frac{1}{\tau_i^{\mathrm{I}}}S_i^{\mathrm{I}}(t)+\eta_i, \quad S_i^{\mathrm{I}}(0) = S_{i0}^{\mathrm{I}}, \quad t \geq 0, \quad i = q+1, \ldots, n_{\mathrm{I}}. \quad (53)$$

Now consider the dynamical system

$$\dot{y}_i(t) = -\frac{1}{\tau_i^{\mathrm{I}}}y_i(t) + \eta_i, \quad y(0) = S_{i0}^{\mathrm{I}}, \quad t \geq 0, \quad i = q+1, \ldots, n_{\mathrm{I}}, \quad (54)$$

and note that

$$y_i(t) = (S_{i0}^{\mathrm{I}} - \tau_i^{\mathrm{I}}\eta_i)e^{-\frac{1}{\tau_i^{\mathrm{I}}}t} + \tau_i^{\mathrm{I}}\eta_i, \quad i = q+1, \ldots, n_{\mathrm{I}}. \quad (55)$$

From the comparison principle [26, p. 126], it follows that $S_i^{\mathrm{I}}(t) \geq y_i(t) \geq \min\{S_{i0}^{\mathrm{I}}, \tau_i^{\mathrm{I}}\eta_i\}$, $t \geq 0$, $i = q+1, \ldots, n_{\mathrm{I}}$, or, equivalently,

$$\tilde{S}_2^{\mathrm{I}}(t) \geq\geq [\min\{S_{q+1}^{\mathrm{I}}(0), \tau_{q+1}^{\mathrm{I}}\eta_{q+1}\}, \ldots, \min\{S_{n_{\mathrm{I}}}^{\mathrm{I}}(0), \tau_{n_{\mathrm{I}}}^{\mathrm{I}}\eta_{n_{\mathrm{I}}}\}]^{\mathrm{T}}, \quad t \geq 0. \quad (56)$$

Since $\tilde{B} \leq\leq 0$, it follows that $\tilde{B}\tilde{S}_2^{\mathrm{I}}(t) \leq\leq \tilde{B}w$, $t \geq 0$, where $w = [\min\{S_{q+1}^{\mathrm{I}}(0), \tau_{q+1}^{\mathrm{I}}\eta_{q+1}\}, \ldots, \min\{S_{n_{\mathrm{I}}}^{\mathrm{I}}(0), \tau_{n_{\mathrm{I}}}^{\mathrm{I}}\eta_{n_{\mathrm{I}}}\}]^{\mathrm{T}}$, and hence $\tilde{v}_{\mathrm{th}}(t)+$

$\tilde{B}\tilde{S}_2^{\mathrm{I}}(t) \leq\leq \tilde{v}_{\mathrm{th}}(t) + \tilde{B}w \leq\leq 0$, $t \geq 0$. Thus $\tilde{f}(\tilde{A}S + \tilde{B}\tilde{S}_2^{\mathrm{I}} + \tilde{v}_{\mathrm{th}}) \leq\leq \tilde{f}(\tilde{A}S)$, $(S, \tilde{S}_2^{\mathrm{I}}) \in \overline{\mathbb{R}}_+^{n_{\mathrm{E}}+q} \times \overline{\mathbb{R}}_+^{n_{\mathrm{I}}-q}$, and since $R \in \mathbb{R}^{(n_{\mathrm{E}}+q)\times(n_{\mathrm{E}}+q)}$ is a diagonal positive-definite matrix, it follows that

$$\tilde{f}^{\mathrm{T}}(\tilde{A}S+\tilde{B}\tilde{S}_2^{\mathrm{I}}+\tilde{v}_{\mathrm{th}})R\tilde{f}(\tilde{A}S+\tilde{B}\tilde{S}_2^{\mathrm{I}}+\tilde{v}_{\mathrm{th}}) \leq \tilde{f}^{\mathrm{T}}(\tilde{A}S)R\tilde{f}(\tilde{A}S), \quad (S, \tilde{S}_2^{\mathrm{I}}) \in \overline{\mathbb{R}}_+^{n_{\mathrm{E}}+q} \times \overline{\mathbb{R}}_+^{n_{\mathrm{I}}-q}.$$

It follows from Eq. (51) that

$$\dot{V}(S, \tilde{S}_2^{\mathrm{I}}) \leq S^{\mathrm{T}}(-PL-LP+Q)S+\tilde{f}^{\mathrm{T}}(\tilde{A}S)R\tilde{f}(\tilde{A}S), \quad (S, \tilde{S}_2^{\mathrm{I}}) \in \overline{\mathbb{R}}_+^{n_{\mathrm{E}}+q} \times \overline{\mathbb{R}}_+^{n_{\mathrm{I}}-q}. \tag{57}$$

Since $f_i^2(x) \leq x^2$, $x \in \mathbb{R}$, for $f_i(\cdot)$, $i = 1, \ldots, n_{\mathrm{E}} + q$, given by Eq. (10), and $R \in \mathbb{R}^{(n_{\mathrm{E}}+q)\times(n_{\mathrm{E}}+q)}$ is a diagonal positive-definite matrix, it follows that

$$f^{\mathrm{T}}(\tilde{A}S)Rf(\tilde{A}S) \leq S^{\mathrm{T}}\tilde{A}^{\mathrm{T}}R\tilde{A}S, \quad S \in \overline{\mathbb{R}}_+^{n_{\mathrm{E}}+q}. \tag{58}$$

It follows from Eqs. (47) and (57) that

$$\dot{V}(S, \tilde{S}_2^{\mathrm{I}}) \leq -S^{\mathrm{T}}\Omega S \leq -\lambda_{\min}(\Omega)\|S\|_2^2, \quad (S, \tilde{S}_2^{\mathrm{I}}) \in \overline{\mathbb{R}}_+^{n_{\mathrm{E}}+q} \times \overline{\mathbb{R}}_+^{n_{\mathrm{I}}-q}. \tag{59}$$

Thus it follows from Theorem 4.1 in [26] that if $\Omega > 0$, then the biological neural network given by Eqs. (44) and (45) is globally exponentially stable with respect to S uniformly in $\tilde{S}_{20}^{\mathrm{I}}$, and hence the biological neural network given by Eqs. (44) and (45) is globally exponentially partially synchronized.

Alternatively, if $\Omega \geq 0$ and $\mathcal{N}(\Omega) = \mathrm{span}(\mathbf{e}_{n_{\mathrm{E}}+q})$ holds, then $\dot{V}(S(t), \tilde{S}_2^{\mathrm{I}}(t)) \leq 0$, $t \geq 0$, and hence $V(S(t)) \leq V(S_0)$ for all $t \geq 0$. Since P is positive definite and $\dot{V}(S(t), \tilde{S}_2^{\mathrm{I}}(t))$ is a nonincreasing function of time, it follows that $V(S(t))$ is bounded for all $t \geq 0$, and hence $S(t)$ is bounded for all $t \geq 0$, which further implies that $S^{\mathrm{E}}(t)$ is bounded for all $t \geq 0$. Thus $\tilde{f}_2^{\mathrm{I}}(A_2^{\mathrm{IE}}S^{\mathrm{E}}(t) + \tilde{v}_{\mathrm{th}2}^{\mathrm{I}}(t))$ is bounded for all $t \geq 0$, and hence there exists $\tilde{f}_{2\max}^{\mathrm{I}}$ such that $\tilde{f}_{2\max}^{\mathrm{I}} \geq\geq \tilde{f}_2^{\mathrm{I}}(A_2^{\mathrm{IE}}S^{\mathrm{E}}(t) + \tilde{v}_{\mathrm{th}2}^{\mathrm{I}}(t))$, $t \geq 0$. It follows from Eq. (45) that

$$\dot{\tilde{S}}_2^{\mathrm{I}}(t) \leq\leq -L_2^{\mathrm{I}}\tilde{S}_2^{\mathrm{I}}(t) + \tilde{f}_{2\max}^{\mathrm{I}}, \quad \tilde{S}_2^{\mathrm{I}}(0) = \tilde{S}_{20}^{\mathrm{I}}, \quad t \geq 0. \tag{60}$$

Next consider the dynamical system

$$\dot{x}(t) = -L_2^{\mathrm{I}}x(t) + \tilde{f}_{2\max}^{\mathrm{I}}, \quad x(0) = \tilde{S}_{20}^{\mathrm{I}}, \quad t \geq 0, \tag{61}$$

and note that since L_2^{I} is a diagonal positive-definite matrix, it follows that $x(t)$ is bounded for all $t \geq 0$. From the vector comparison principle [26], it follows from Eqs. (60) and (61) that $\tilde{S}_2^{\mathrm{I}}(t) \leq\leq x(t)$, $t \geq 0$, and hence $\tilde{S}_2^{\mathrm{I}}(t)$ is bounded for all $t \geq 0$. Since $S(t)$ and $\tilde{S}_2^{\mathrm{I}}(t)$ are bounded for all $t \geq 0$, it follows that $\ddot{V}(S(t), \tilde{S}_2^{\mathrm{I}}(t))$ is bounded for all $t \geq 0$. Thus $\dot{V}(S(t), \tilde{S}_2^{\mathrm{I}}(t))$ is uniformly continuous in t. Finally, it follows from Barbalat's lemma [26, p. 221] that $\dot{V}(S(t), \tilde{S}_2^{\mathrm{I}}(t)) \to 0$ as $t \to \infty$, which, since $\mathcal{N}(\Omega) = \text{span}(\mathbf{e}_{n_E+q})$, implies that the biological neural network given by Eqs. (44) and (45) is globally asymptotically partially synchronized. □

There is a body of experimental evidence that the mechanism of action of some anesthetic drugs is the prolongation of the inhibitory postsynaptic potential time constant [6, 7]. While Theorem 3 provides sufficient conditions for partial synchronization, it is not readily apparent how these conditions may be met as the inhibitory postsynaptic potential time constant increases. We address this question with the activation function [10]

$$f_i(x) = \begin{cases} 0, & x < 0, \\ x, & 0 \le x \le f_{\max}, \\ f_{\max}, & x > f_{\max}, \end{cases} \tag{62}$$

where $f_{\max}$ denotes the maximum firing rate.

Theorem 4. *Consider the excitatory-inhibitory network given by Eqs. (40), (41), and (42) with the vector activation functions defined by Eqs. (37) and (38), and $f_i(\cdot)$ given by Eq. (62). If $\tilde{v}^{\mathrm{I}}_{\mathrm{th}2}(t) >> 0$, $t \ge 0$, and the inhibitory time constants are such that $\tilde{S}^{\mathrm{I}}_{20} - (L^{\mathrm{I}}_2)^{-1}\min\{f_{\max}\mathbf{e}, \eta\} << 0$ and*

$$A^{\mathrm{EE}}\max\{S^{\mathrm{E}}_0, f_{\max}(L^{\mathrm{E}})^{-1}\mathbf{e}\} + A^{\mathrm{EI}}_2[(L^{\mathrm{I}}_2)^{-1}\min\{f_{\max}\mathbf{e}, \eta\} - \varepsilon\mathbf{e}] + \beta \le\le 0, \tag{63}$$

where $\eta \triangleq [\eta_{q+1}, \ldots, \eta_{n_{\mathrm{I}}}]$, $\eta_i = \min_{t\ge 0} v^{\mathrm{I}}_{\mathrm{th}i}(t)$, $i = q+1, \ldots, n_{\mathrm{I}}$, $\max\{x, y\} \triangleq [\max\{x_1, y_1\}, \ldots, \max\{x_n, y_n\}]^{\mathrm{T}}$, $x, y \in \mathbb{R}^n$, $\min\{x, y\} \triangleq [\min\{x_1, y_1\}, \ldots, \min\{x_n, y_n\}]^{\mathrm{T}}$, $x, y \in \mathbb{R}^n$, $\beta \triangleq [\beta_1, \ldots, \beta_{n_{\mathrm{E}}}]^{\mathrm{T}}$, $\beta_i \triangleq \max_{t\ge 0} v^{\mathrm{E}}_{\mathrm{th}i}(t)$, $i = 1, \ldots, n_{\mathrm{E}}$, and $\varepsilon > 0$ is a small positive constant, then $S^{\mathrm{E}}(t) \to 0$ as $t \to \infty$, and hence the excitatory-inhibitory network given by Eqs. (40)–(42) is globally asymptotically partially synchronized.

Proof. First, note that since $f_i(x) \le f_{\max}$, $x \in \mathbb{R}$, $i = 1, \ldots, n_{\mathrm{E}}$, it follows from Eq. (40) that

$$\dot{S}^{\mathrm{E}}_i(t) \le -\frac{1}{\tau^{\mathrm{E}}_i}S^{\mathrm{E}}_i(t) + f_{\max}, \quad S^{\mathrm{E}}_i(0) = S^{\mathrm{E}}_{i0}, \quad t \ge 0, \quad i = 1, \ldots, n_{\mathrm{E}}, \tag{64}$$

and hence

$$S^{\mathrm{E}}_i(t) \le (S^{\mathrm{E}}_{i0} - \tau^{\mathrm{E}}_i f_{\max})e^{-\frac{t}{\tau^{\mathrm{E}}_i}} + \tau^{\mathrm{E}}_i f_{\max}, \quad t \ge 0, \quad i = 1, \ldots, n_{\mathrm{E}}. \tag{65}$$

Thus $S^{\mathrm{E}}_i(t) \le \max\{S^{\mathrm{E}}_{i0}, \tau^{\mathrm{E}}_i f_{\max}\}$, $t \ge 0$, $i = 1, \ldots, n_{\mathrm{E}}$, or, equivalently,

$$S^{\mathrm{E}}(t) \le\le \max\{S^{\mathrm{E}}_0, f_{\max}(L^{\mathrm{E}})^{-1}\mathbf{e}\}, \quad t \ge 0. \tag{66}$$

Since $A^{\mathrm{IE}}_2 \ge\ge 0$, $S^{\mathrm{E}}(t) \ge\ge 0$, $t \ge 0$, and $f_i(x)$, $x \in \mathbb{R}$, $i = q+1, \ldots, n_{\mathrm{I}}$, are nondecreasing functions, it follows that $\tilde{f}^{\mathrm{I}}_2(A^{\mathrm{IE}}_2 S^{\mathrm{E}} + \tilde{v}^{\mathrm{I}}_{\mathrm{th}2}) \ge\ge \tilde{f}^{\mathrm{I}}_2(\tilde{v}^{\mathrm{I}}_{\mathrm{th}2})$,

$S^{\mathrm{E}} \in \overline{\mathbb{R}}_{+}^{n_{\mathrm{E}}}$. And since $f_i(x) \leq f_{\max}$, $x \in \mathbb{R}$, $i = q+1, \ldots, n_{\mathrm{I}}$, and $\tilde{v}_{\mathrm{th2}}^{\mathrm{I}}(t) >> 0$, $t \geq 0$, it follows that $\tilde{f}_2^{\mathrm{I}}(\tilde{v}_{\mathrm{th2}}^{\mathrm{I}}(t)) = \min\{f_{\max}\mathbf{e}, \tilde{v}_{\mathrm{th2}}^{\mathrm{I}}(t)\} \geq\geq \min\{f_{\max}\mathbf{e}, \eta\}$, $t \geq 0$. It follows from Eq. (42) that

$$\dot{\tilde{S}}_2^{\mathrm{I}}(t) \geq\geq -L_2^{\mathrm{I}}\tilde{S}_2^{\mathrm{I}}(t) + \min\{f_{\max}\mathbf{e}, \eta\}, \quad \tilde{S}_2^{\mathrm{I}}(0) = \tilde{S}_{20}^{\mathrm{I}}, \quad t \geq 0, \tag{67}$$

and hence

$$\tilde{S}_2^{\mathrm{I}}(t) \geq\geq (\tilde{S}_{20}^{\mathrm{I}} - (L_2^{\mathrm{I}})^{-1}\min\{f_{\max}\mathbf{e}, \eta\})\mathrm{e}^{-L_2^{\mathrm{I}}t} + (L_2^{\mathrm{I}})^{-1}\min\{f_{\max}\mathbf{e}, \eta\}, \quad t \geq 0. \tag{68}$$

Thus it follows from Eq. (68) that $\tilde{S}_2^{\mathrm{I}}(t) \geq\geq (L_2^{\mathrm{I}})^{-1}\min\{f_{\max}\mathbf{e}, \eta\}$ as $t \to \infty$, and since $\tilde{S}_{20}^{\mathrm{I}} - (L_2^{\mathrm{I}})^{-1}\min\{f_{\max}\mathbf{e}, \eta\} << 0$, there exists $t_1 > 0$ such that

$$\tilde{S}_2^{\mathrm{I}}(t) \geq\geq (L_2^{\mathrm{I}})^{-1}\min\{f_{\max}\mathbf{e}, \eta\} - \varepsilon\mathbf{e}, \quad t \geq t_1. \tag{69}$$

Since $A_1^{\mathrm{EI}} \leq\leq 0$ and $\tilde{S}_1^{\mathrm{I}}(t) \geq\geq 0$, $t \geq 0$, and $f_i(x)$, $x \in \mathbb{R}$, $i = 1, \ldots, n_{\mathrm{E}}$, are nondecreasing functions, it follows from Eq. (40) that

$$\dot{S}^{\mathrm{E}}(t) \leq\leq -L^{\mathrm{E}}S^{\mathrm{E}}(t) + f^{\mathrm{E}}(A^{\mathrm{EE}}S^{\mathrm{E}}(t) + A_2^{\mathrm{EI}}\tilde{S}_2^{\mathrm{I}}(t) + \tilde{v}_{\mathrm{th}}^{\mathrm{E}}(t)), \quad S^{\mathrm{E}}(0) = S_0^{\mathrm{E}}, \quad t \geq 0. \tag{70}$$

Since $A^{\mathrm{EE}} \geq\geq 0$ and $A_2^{\mathrm{EI}} \leq\leq 0$, it follows from Eqs. (66) and (69) that

$$\begin{aligned} A^{\mathrm{EE}}S^{\mathrm{E}}(t) + A_2^{\mathrm{EI}}\tilde{S}_2^{\mathrm{I}}(t) + \tilde{v}_{\mathrm{th}}^{\mathrm{E}}(t) &\leq\leq A^{\mathrm{EE}}\max\{S_0^{\mathrm{E}}, f_{\max}(L^{\mathrm{E}})^{-1}\mathbf{e}\} \\ &\quad + A_2^{\mathrm{EI}}[(L_2^{\mathrm{I}})^{-1}\min\{f_{\max}\mathbf{e}, \eta\} - \varepsilon\mathbf{e}] + \beta \\ &\leq\leq 0, \quad t \geq t_1. \end{aligned} \tag{71}$$

Finally, it follows from Eqs. (70) and (71) that

$$\dot{S}^{\mathrm{E}}(t) \leq\leq -L^{\mathrm{E}}S^{\mathrm{E}}(t), \quad S^{\mathrm{E}}(0) = S_0^{\mathrm{E}}, \quad t \geq t_1, \tag{72}$$

which implies that $S^{\mathrm{E}}(t) \to 0$ as $t \to \infty$, and hence the excitatory-inhibitory network given by Eqs. (40)–(42) is globally asymptotically partially synchronized. □

Condition (63) is satisfied if A_2^{EI} does not contain any row of zeros and L_2^{I} is small enough or, equivalently, the time constants of the inhibitory neurons that themselves do not receive any inhibitory inputs are large enough. This condition captures an important aspect of the effect of the anesthetic cascade. In particular, as the anesthetic drug concentration increases, resulting in an increase in the inhibitory time constants, the excitatory synaptic drives that receive inhibitory inputs from inhibitory neurons, which themselves do not receive inhibitory inputs, converge to zero, reflecting a state of unconsciousness. Condition (63) further implies that all the excitatory neurons receive inhibitory inputs from a subset $\mathcal{I}_1$

of inhibitory neurons that themselves do not receive inhibitory inputs. Alternatively, if only a subset $\mathcal{E}_1$ of excitatory neurons receive inhibitory inputs from the inhibitory neurons in $\mathcal{I}_1$, then it can be shown that only the synaptic drives of the excitatory neurons in $\mathcal{E}_1$ will converge to zero as the time constants of the inhibitory neurons in $\mathcal{I}_1$ increase.

5 GENERALIZED NEURAL POPULATION MODELS, NONMONOTONIC POSTSYNAPTIC POTENTIALS, AND DRUG BIPHASIC RESPONSE

The administration of increasing anesthetic doses can lead to a paradoxical state of excitement in the patient before decreases in the level of consciousness. This paradoxical boost in brain activity before hypnotic induction is known as *drug biphasic response* [24]. There is also a second biphasic surge (burst suppression) in the electroencephalograph power as the patient emerges from unconsciousness [29]. The model proposed in Section 2 does not capture such phenomena. Models that predict the aforementioned characteristics are of great clinical importance in providing the phenomenological trends of the anesthetic cascade. In this section we incorporate a more complex model for the postsynaptic potential to account for the delay in the peak amplitude of the postsynaptic potential that occurs after a neuron discharges [14, Fig. 8.5].

To capture an initial overshoot in the synaptic drive dynamics in an attempt to see biphasic responses in our neural field model, we assume the functional form of the postsynaptic potential is given by

$$\alpha_i^{(\mathrm{E,I})}(t) = B_i^{(\mathrm{E,I})} t \mathrm{e}^{-\frac{t}{\lambda_i^{(\mathrm{E,I})}}}. \tag{73}$$

In this case it follows from Eqs. (5) and (73) that

$$\begin{aligned}\frac{\mathrm{d}S_i^{(\mathrm{E,I})}(t)}{\mathrm{d}t} &= \int_{-\infty}^{t} B_i^{(\mathrm{E,I})} \mathrm{e}^{-\frac{t-\tau}{\lambda_i^{(\mathrm{E,I})}}} f_i(v_i^{(\mathrm{E,I})}(\tau))\mathrm{d}\tau - \int_{-\infty}^{t} \frac{t-\tau}{\lambda_i^{(\mathrm{E,I})}} B_i^{(\mathrm{E,I})} \mathrm{e}^{-\frac{t-\tau}{\lambda_i^{(\mathrm{E,I})}}} f_i(v_i^{(\mathrm{E,I})}(\tau))\mathrm{d}\tau \\ &= \int_{-\infty}^{t} B_i^{(\mathrm{E,I})} \mathrm{e}^{-\frac{t-\tau}{\lambda_i^{(\mathrm{E,I})}}} f_i(v_i^{(E,I)}(\tau))\mathrm{d}\tau - \frac{S_i^{(\mathrm{E,I})}(t)}{\lambda_i^{(\mathrm{E,I})}}.\end{aligned} \tag{74}$$

If we define

$$\overline{S_i}^{(\mathrm{E,I})}(t) \triangleq \int_{-\infty}^{t} B_i^{(\mathrm{E,I})} \mathrm{e}^{-\frac{t-\tau}{\lambda_i^{(\mathrm{E,I})}}} f_i(v_i^{(\mathrm{E,I})}(\tau))\mathrm{d}\tau, \tag{75}$$

it follows that

$$\frac{\mathrm{d}\overline{S}_i^{(\mathrm{E,I})}(t)}{\mathrm{d}t} = -\frac{1}{\lambda_i^{(\mathrm{E,I})}}\overline{S}_i^{(\mathrm{E,I})}(t) + B_i^{(\mathrm{E,I})} f_i(v_i^{(\mathrm{E,I})}(t)). \tag{76}$$

It follows from Eqs. (74) and (75) that

$$\frac{\mathrm{d}S_i^{(\mathrm{E,I})}(t)}{\mathrm{d}t} = \overline{S}_i^{(\mathrm{E,I})}(t) - \frac{S_i^{(\mathrm{E,I})}(t)}{\lambda_i^{(\mathrm{E,I})}}. \tag{77}$$

Differentiating Eq. (77) with respect to time, we obtain

$$\begin{aligned}
\frac{\mathrm{d}^2 S_i^{(\mathrm{E,I})}(t)}{\mathrm{d}t^2} &= \frac{\mathrm{d}\overline{S}_i^{(\mathrm{E,I})}(t)}{\mathrm{d}t} - \frac{1}{\lambda_i^{(\mathrm{E,I})}}\frac{\mathrm{d}S_i^{(\mathrm{E,I})}(t)}{\mathrm{d}t} \\
&= -\frac{1}{\lambda_i^{(E,I)}}\overline{S}_i^{(\mathrm{E,I})}(t) + B_i^{(\mathrm{E,I})} f_i(v_i^{(\mathrm{E,I})}(t)) - \frac{1}{\lambda_i^{(\mathrm{E,I})}}\frac{\mathrm{d}S_i^{(\mathrm{E,I})}(t)}{\mathrm{d}t} \\
&= -\frac{1}{\lambda_i^{(\mathrm{E,I})}}\left(\frac{\mathrm{d}S_i^{(\mathrm{E,I})}(t)}{\mathrm{d}t} + \frac{S_i^{(\mathrm{E,I})}(t)}{\lambda_i^{(\mathrm{E,I})}}\right) + B_i^{(\mathrm{E,I})} f_i(v_i^{(E,I)}(t)) - \frac{1}{\lambda_i^{(\mathrm{E,I})}}\frac{\mathrm{d}S_i^{(\mathrm{E,I})}(t)}{\mathrm{d}t} \\
&= -\frac{2}{\lambda_i^{(\mathrm{E,I})}}\frac{\mathrm{d}S_i^{(\mathrm{E,I})}(t)}{\mathrm{d}t} - \left(\frac{1}{\lambda_i^{(\mathrm{E,I})}}\right)^2 S_i^{(\mathrm{E,I})}(t) + B_i^{(\mathrm{E,I})} f_i(v_i^{(E,I)}(t)).
\end{aligned} \tag{78}$$

With use of the expressions for the excitatory and inhibitory voltage given by Eqs. (3) and (4), respectively, it follows that

$$\begin{aligned}
\frac{\mathrm{d}^2 S_i^{\mathrm{E}}(t)}{\mathrm{d}t^2} = &-\frac{2}{\lambda_i^{\mathrm{E}}}\frac{\mathrm{d}S_i^{\mathrm{E}}(t)}{\mathrm{d}t} - \left(\frac{1}{\lambda_i^{\mathrm{E}}}\right)^2 S_i^{\mathrm{E}}(t) + B_i^{\mathrm{E}} f_i\left(\sum_{j=1, j\neq i}^{n_{\mathrm{E}}} A_{ij}^{\mathrm{EE}} S_j^{\mathrm{E}}(t)\right. \\
&\left. + \sum_{j'=1}^{n_{\mathrm{I}}} A_{ij'}^{\mathrm{EI}} S_{j'}^{\mathrm{I}}(t) + v_{\mathrm{th}i}^{\mathrm{E}}(t)\right), \quad i = 1, \ldots, n_{\mathrm{E}},
\end{aligned} \tag{79}$$

$$\begin{aligned}
\frac{\mathrm{d}^2 S_i^{\mathrm{I}}(t)}{\mathrm{d}t^2} = &-\frac{2}{\lambda_i^{\mathrm{I}}}\frac{\mathrm{d}S_i^{\mathrm{I}}(t)}{\mathrm{d}t} - \left(\frac{1}{\lambda_i^{\mathrm{I}}}\right)^2 S_i^{\mathrm{I}}(t) + B_i^{\mathrm{I}} f_i\left(\sum_{j=1}^{n_{\mathrm{E}}} A_{ij}^{\mathrm{IE}} S_j^{\mathrm{E}}(t)\right. \\
&\left. + \sum_{j'=1, j'\neq i}^{n_{\mathrm{I}}} A_{ij'}^{\mathrm{II}} S_{j'}^{\mathrm{I}}(t) + v_{\mathrm{th}i}^{\mathrm{I}}(t)\right), \quad i = 1, \ldots, n_{\mathrm{I}}.
\end{aligned} \tag{80}$$

The above analysis reveals that an alternative form to Eq. (9) for capturing the neuroelectric behavior of biological excitatory and inhibitory neuronal networks can be given by the second-order differential equations

$$\frac{\mathrm{d}^2 S_i(t)}{\mathrm{d}t^2} = -\frac{2}{\tau_i}\frac{\mathrm{d}S_i(t)}{\mathrm{d}t} - \frac{1}{\tau_i^2}S_i(t) + B_i f_i\left(\sum_{j=1}^{n} A_{ij}S_j(t) + v_{\mathrm{th}i}(t)\right), \quad S_i(0) = S_{i0},$$
$$\dot{S}_i(0) = \dot{S}_{i0}, \quad t \geq 0, \quad i = 1, \ldots, n. \tag{81}$$

To address the problem of partial synchronization for the model given by Eqs. (79) and (80), the vector-matrix form of Eqs. (79) and (80) can be written as

$$\ddot{S}^{\mathrm{E}}(t) = -2L^{\mathrm{E}}\dot{S}^{\mathrm{E}}(t) - (L^{\mathrm{E}})^2 S^{\mathrm{E}}(t) + f^{\mathrm{E}}(A^{\mathrm{EE}}S^{\mathrm{E}}(t) + A^{\mathrm{EI}}S^{\mathrm{I}}(t) + \tilde{v}^{\mathrm{E}}_{\mathrm{th}}(t)),$$
$$S^{\mathrm{E}}(0) = S^{\mathrm{E}}_0, \quad \dot{S}^{\mathrm{E}}(0) = \dot{S}^{\mathrm{E}}_0, \quad t \geq 0, \tag{82}$$
$$\ddot{S}^{\mathrm{I}}(t) = -2L^{\mathrm{I}}\dot{S}^{\mathrm{I}}(t) - (L^{\mathrm{I}})^2 S^{\mathrm{I}}(t) + f^{\mathrm{I}}(A^{\mathrm{IE}}S^{\mathrm{E}}(t) + A^{\mathrm{II}}S^{\mathrm{I}}(t) + \tilde{v}^{\mathrm{I}}_{\mathrm{th}}(t)),$$
$$S^{\mathrm{I}}(0) = S^{\mathrm{I}}_0, \ \dot{S}^{\mathrm{I}}(0) = \dot{S}^{\mathrm{I}}_0, \tag{83}$$

where all vector-matrix notation is defined as in Section 4. Once again we partition the state vector $S^{\mathrm{I}}(t)$, $t \geq 0$, as in Section 4 so that the neuronal connectivity matrix is partitioned as in Eq. (39), and hence Eqs. (82) and (83) can be written as

$$\ddot{S}^{\mathrm{E}}(t) = -2L^{\mathrm{E}}\dot{S}^{\mathrm{E}}(t) - (L^{\mathrm{E}})^2 S^{\mathrm{E}}(t) + f^{\mathrm{E}}(A^{\mathrm{EE}}S^{\mathrm{E}}(t) + A_1^{\mathrm{EI}}\tilde{S}_1^{\mathrm{I}}(t) + A_2^{\mathrm{EI}}\tilde{S}_2^{\mathrm{I}}(t)$$
$$+ \tilde{v}^{\mathrm{E}}_{\mathrm{th}}(t)), \quad S^{\mathrm{E}}(0) = S^{\mathrm{E}}_0, \quad \dot{S}^{\mathrm{E}}(0) = \dot{S}^{\mathrm{E}}_0, \quad t \geq 0, \tag{84}$$
$$\ddot{\tilde{S}}_1^{\mathrm{I}}(t) = -2L_1^{\mathrm{I}}\dot{\tilde{S}}_1^{\mathrm{I}}(t) - (L_1^{\mathrm{I}})^2\tilde{S}_1^{\mathrm{I}}(t) + \tilde{f}_1^{\mathrm{I}}(A_1^{\mathrm{IE}}S^{\mathrm{E}}(t) + A_1^{\mathrm{II}}\tilde{S}_1^{\mathrm{I}}(t) + A_2^{\mathrm{II}}\tilde{S}_2^{\mathrm{I}}(t)$$
$$+ \tilde{v}^{\mathrm{I}}_{\mathrm{th}1}(t)), \quad \tilde{S}_1^{\mathrm{I}}(0) = \tilde{S}_{10}^{\mathrm{I}}, \quad \dot{\tilde{S}}_1^{\mathrm{I}}(0) = \dot{\tilde{S}}_{10}^{\mathrm{I}}, \tag{85}$$
$$\ddot{\tilde{S}}_2^{\mathrm{I}}(t) = -2L_2^{\mathrm{I}}\dot{\tilde{S}}_2^{\mathrm{I}}(t) - (L_2^{\mathrm{I}})^2\tilde{S}_2^{\mathrm{I}}(t) + \tilde{f}_2^{\mathrm{I}}(A_2^{\mathrm{IE}}S^{\mathrm{E}}(t) + \tilde{v}^{\mathrm{I}}_{\mathrm{th}2}(t)), \ \tilde{S}_2^{\mathrm{I}}(0) = \tilde{S}_{20}^{\mathrm{I}},$$
$$\dot{\tilde{S}}_2^{\mathrm{I}}(0) = \dot{\tilde{S}}_{20}^{\mathrm{I}}, \tag{86}$$

or, equivalently,

$$\ddot{S}(t) = -2L\dot{S}(t) - L^2 S(t) + \tilde{f}(\tilde{A}S(t) + \tilde{B}\tilde{S}_2^{\mathrm{I}}(t) + \tilde{v}_{\mathrm{th}}(t)), \quad S(0) = S_0,$$
$$\dot{S}(0) = \dot{S}_0, \quad t \geq 0, \tag{87}$$
$$\ddot{\tilde{S}}_2^{\mathrm{I}}(t) = -2L_2^{\mathrm{I}}\dot{\tilde{S}}_2^{\mathrm{I}}(t) - (L_2^{\mathrm{I}})^2\tilde{S}_2^{\mathrm{I}}(t) + \tilde{f}_2^{\mathrm{I}}(A_2^{\mathrm{IE}}S^{\mathrm{E}}(t) + \tilde{v}^{\mathrm{I}}_{\mathrm{th}2}(t)), \ \tilde{S}_2^{\mathrm{I}}(0) = \tilde{S}_{20}^{\mathrm{I}},$$
$$\dot{\tilde{S}}_2^{\mathrm{I}}(0) = \dot{\tilde{S}}_{20}^{\mathrm{I}}. \tag{88}$$

Finally, if we define $\hat{S}(t) \triangleq [S^{\mathrm{T}}(t) \ \dot{S}^{\mathrm{T}}(t)]^{\mathrm{T}}$, $\hat{S}_2^{\mathrm{I}}(t) \triangleq [(\tilde{S}_2^{\mathrm{I}}(t))^{\mathrm{T}} \ (\dot{\tilde{S}}_2^{\mathrm{I}}(t))^{\mathrm{T}}]^{\mathrm{T}}$, and $\hat{S}^{\mathrm{E}}(t) \triangleq [(\tilde{S}^{\mathrm{E}}(t))^{\mathrm{T}} \ (\dot{\tilde{S}}^{\mathrm{E}}(t))^{\mathrm{T}}]^{\mathrm{T}}$, Eqs. (87) and (88) can be written as

$$\dot{\hat{S}}(t) = -\Gamma\hat{S}(t) + \hat{f}(\hat{A}\hat{S}(t) + \hat{B}\hat{S}_2^{\mathrm{I}}(t) + \hat{v}_{\mathrm{th}}(t)), \quad \hat{S}(0) = S_0, \quad t \geq 0, \tag{89}$$

$$\dot{\hat{S}}_2^{\mathrm{I}}(t) = -\Gamma^{\mathrm{I}}\hat{S}_2^{\mathrm{I}}(t) + \hat{f}_2^{\mathrm{I}}(\hat{A}_2^{\mathrm{IE}}\hat{S}^{\mathrm{E}}(t) + \hat{v}_{\mathrm{th2}}^{\mathrm{I}}(t)), \quad \hat{S}_2^{\mathrm{I}}(0) = \hat{S}_{20}^{\mathrm{I}}, \tag{90}$$

where

$$\Gamma \triangleq \begin{bmatrix} 0_{(n_{\mathrm{E}}+q)\times(n_{\mathrm{E}}+q)} & -I_{(n_{\mathrm{E}}+q)\times(n_{\mathrm{E}}+q)} \\ (L^{\mathrm{E}})^2 & 2L^{\mathrm{E}} \end{bmatrix}, \quad \hat{A} \triangleq \begin{bmatrix} 0_{(n_{\mathrm{E}}+q)\times(n_{\mathrm{E}}+q)} & 0_{(n_{\mathrm{E}}+q)\times(n_{\mathrm{E}}+q)} \\ \tilde{A} & 0_{(n_{\mathrm{E}}+q)\times(n_{\mathrm{E}}+q)} \end{bmatrix},$$

$$\hat{B} \triangleq \begin{bmatrix} 0_{(n_{\mathrm{E}}+q)\times(n_{\mathrm{I}}-q)} & 0_{(n_{\mathrm{E}}+q)\times(n_{\mathrm{I}}-q)} \\ \tilde{B} & 0_{(n_{\mathrm{E}}+q)\times(n_{\mathrm{I}}-q)} \end{bmatrix}, \quad \hat{v}_{\mathrm{th}} \triangleq \begin{bmatrix} 0_{(n_{\mathrm{E}}+q)} \\ \tilde{v}_{\mathrm{th}} \end{bmatrix}, \quad \hat{f} \triangleq \begin{bmatrix} 0_{(n_{\mathrm{E}}+q)} \\ \tilde{f} \end{bmatrix},$$

$$\Gamma^{\mathrm{I}} \triangleq \begin{bmatrix} 0_{(n_{\mathrm{I}}-q)\times(n_{\mathrm{I}}-q)} & -I_{(n_{\mathrm{I}}-q)\times(n_{\mathrm{I}}-q)} \\ (L_2^{\mathrm{I}})^2 & 2L_2^{\mathrm{I}} \end{bmatrix}, \quad \hat{A}_2^{\mathrm{IE}} \triangleq \begin{bmatrix} 0_{(n_{\mathrm{I}}-q)\times n_{\mathrm{E}}} & 0_{(n_{\mathrm{I}}-q)\times n_{\mathrm{E}}} \\ A_2^{\mathrm{IE}} & 0_{(n_{\mathrm{I}}-q)\times n_{\mathrm{E}}} \end{bmatrix},$$

$$\hat{v}_{\mathrm{th2}}^{\mathrm{I}} \triangleq \begin{bmatrix} 0_{(n_{\mathrm{I}}-q)} \\ \tilde{v}_{\mathrm{th2}}^{\mathrm{I}} \end{bmatrix}, \quad \hat{f}_2^{\mathrm{I}} \triangleq \begin{bmatrix} 0_{(n_{\mathrm{I}}-q)} \\ \tilde{f}_2^{\mathrm{I}} \end{bmatrix}.$$

The following proposition and definitions are needed for the main results of this section.

Proposition 8. *Consider the excitatory-inhibitory network given by Eqs. (82) and (83) with the vector activation functions defined by Eqs. (37) and (38), and the synaptic drive defined by Eq. (5) with postsynaptic potential given by Eq. (73). Then,* $S^{\mathrm{E}}(t) \geq\geq 0$ *and* $S^{\mathrm{I}}(t) \geq\geq 0$ *for all* $t \geq 0$.

Proof. Since $B_i^{\mathrm{E}} = B_i^{\mathrm{I}} = 1$, it follows from Eqs. (5) and (73) that

$$S_i^{\mathrm{E}}(0) = S_{i0}^{\mathrm{E}} = \int_{-\infty}^{0} (-\tau)\mathrm{e}^{\frac{\tau}{\lambda_i^{\mathrm{E}}}} f_i(v_i^{\mathrm{E}}(\tau))\mathrm{d}\tau, \quad i = 1, \ldots, n_{\mathrm{E}}. \tag{91}$$

If we let $s \triangleq -\tau$, it follows from Eq. (91) that

$$S_{i0}^{\mathrm{E}} = \int_{\infty}^{0} s\mathrm{e}^{\frac{-s}{\lambda_i^{\mathrm{E}}}} f_i(v_i^{\mathrm{E}}(-s))(-\mathrm{d}s) = \int_{0}^{\infty} s\mathrm{e}^{\frac{-s}{\lambda_i^{\mathrm{E}}}} f_i(v_i^{\mathrm{E}}(-s))\mathrm{d}s, \quad i = 1, \ldots, n_{\mathrm{E}}. \tag{92}$$

Since $s\mathrm{e}^{\frac{-s}{\lambda_i^{\mathrm{E}}}} f_i(v_i^{\mathrm{E}}(-s)) \geq 0$ for $s \in [0, \infty)$ and $i = 1, \ldots, n_{\mathrm{E}}$, it follows from Eq. (92) that $S_{i0}^{\mathrm{E}} \geq 0$ for $i = 1, \ldots, n_{\mathrm{E}}$. It follows from Eq. (74) that

$$\dot{S}_{i0}^{\mathrm{E}} + \frac{1}{\lambda_i^{\mathrm{E}}} S_{i0}^{\mathrm{E}} = \int_{-\infty}^{0} \mathrm{e}^{\frac{\tau}{\lambda_i^{\mathrm{E}}}} f_i(v_i^{\mathrm{E}}(\tau))\mathrm{d}\tau \geq 0, \quad i = 1, \ldots, n_{\mathrm{E}}. \tag{93}$$

Since $f_i(x) \geq 0$, $x \in \mathbb{R}$, $i = 1, \ldots, n_{\mathrm{E}}$, it follows from Eq. (82) that

$$\ddot{S}_i^{\mathrm{E}}(t) + \frac{2}{\lambda_i^{\mathrm{E}}}\dot{S}_i^{\mathrm{E}}(t) + \frac{1}{(\lambda_i^{\mathrm{E}})^2} S_i^{\mathrm{E}}(t) \geq 0, \quad S_i^{\mathrm{E}}(0) = S_{i0}^{\mathrm{E}}, \quad \dot{S}_i^{\mathrm{E}}(0) = \dot{S}_{i0}^{\mathrm{E}},$$
$$t \geq 0, \quad i = 1, \ldots, n_{\mathrm{E}}. \tag{94}$$

If we let

$$u_i^{\mathrm{E}}(t) \triangleq \dot{S}_i^{\mathrm{E}}(t) + \frac{1}{\lambda_i^{\mathrm{E}}} S_i^{\mathrm{E}}(t), \quad t \geq 0, \quad i = 1, \ldots, n_{\mathrm{E}}, \tag{95}$$

it follows from Eqs. (94) and (95) that

$$\dot{u}_i^{\mathrm{E}}(t) + \frac{1}{\lambda_i^{\mathrm{E}}} u_i^{\mathrm{E}}(t) \geq 0, \quad u_i^{\mathrm{E}}(0) = \dot{S}_{i0}^{\mathrm{E}} + \frac{1}{\lambda_i^{\mathrm{E}}} S_{i0}^{\mathrm{E}}, \quad t \geq 0, \quad i = 1, \ldots, n_{\mathrm{E}}, \tag{96}$$

and since $u_i^{\mathrm{E}}(0) = \dot{S}_{i0}^{\mathrm{E}} + \frac{1}{\lambda_i^{\mathrm{E}}} S_{i0}^{\mathrm{E}} \geq 0$, $i = 1, \ldots, n_{\mathrm{E}}$, it follows from Eq. (96) that $u_i^{\mathrm{E}}(t) \geq 0$, $t \geq 0$, $i = 1, \ldots, n_{\mathrm{E}}$. Thus it follows from Eq. (95) that

$$\dot{S}_i^{\mathrm{E}}(t) + \frac{1}{\lambda_i^{\mathrm{E}}} S_i^{\mathrm{E}}(t) \geq 0, \quad S_i^{\mathrm{E}}(0) = S_{i0}^{\mathrm{E}}, \quad t \geq 0, \quad i = 1, \ldots, n_{\mathrm{E}}. \tag{97}$$

Finally, since $S_{i0}^{\mathrm{E}} \geq 0$, $i = 1, \ldots, n_{\mathrm{E}}$, it follows from Eq. (97) that $S_i^{\mathrm{E}}(t) \geq 0$, $t \geq 0$, $i = 1, \ldots, n_{\mathrm{E}}$. Similarly, it can be shown that $S_i^{\mathrm{I}}(t) \geq 0$, $t \geq 0$, $i = 1, \ldots, n_{\mathrm{I}}$, and hence $S^{\mathrm{E}}(t) \geq\geq 0$ and $S^{\mathrm{I}}(t) \geq\geq 0$ for all $t \geq 0$. □

Definition 4. The biological neural network given by Eqs. (89) and (90) is said to be *globally asymptotically partially synchronized* if

$$\lim_{t \to \infty} |S_i(t) - S_j(t)| = 0,$$
$$\lim_{t \to \infty} |\dot{S}_i(t) - \dot{S}_j(t)| = 0$$

for all $\hat{S}_0 \in \overline{\mathbb{R}}_+^{2(n_{\mathrm{E}}+q)}$, $\hat{S}_{20}^{\mathrm{I}} \in \overline{\mathbb{R}}_+^{2(n_{\mathrm{I}}-q)}$, and $i = 1, \ldots, n_{\mathrm{E}} + q$, $i \neq j$.

Definition 5. The biological neural network given by Eqs. (89) and (90) is said to be *globally exponentially partially synchronized* if there exist constants $p > 0$ and $\rho > 0$ such that

$$|S_i(t) - S_j(t)| \leq \rho e^{-pt} |S_{i0} - S_{j0}|, \quad t \geq 0,$$
$$|\dot{S}_i(t) - \dot{S}_j(t)| \leq \rho e^{-pt} |\dot{S}_{i0} - \dot{S}_{j0}|, \quad t \geq 0,$$

for all $\hat{S}_0 \in \overline{\mathbb{R}}_+^{2(n_{\mathrm{E}}+q)}$, $\hat{S}_{20}^{\mathrm{I}} \in \overline{\mathbb{R}}_+^{2(n_{\mathrm{I}}-q)}$, and $i = 1, \ldots, n_{\mathrm{E}} + q$, $i \neq j$.

Theorem 5. *Consider the excitatory-inhibitory network given by Eqs. (89) and (90) with the vector activation functions defined by Eqs. (37) and (38). If* $\tilde{v}_{\mathrm{th}}(t) \leq\leq -\tilde{B}\hat{w}$, $t \geq 0$, *where* $\hat{w} \triangleq [\min\{S_{q+1}^{\mathrm{I}}(0), \tau_{q+1}^{\mathrm{I}}\alpha_{q+1}\}, \ldots, \min\{S_{n_{\mathrm{I}}}^{\mathrm{I}}(0), \tau_{n_{\mathrm{I}}}^{\mathrm{I}}\alpha_{n_{\mathrm{I}}}\}]^T \in \mathbb{R}^{n_{\mathrm{I}}-q}$, $\alpha_i \triangleq \min\{\dot{S}_{i0}^{\mathrm{I}} + \frac{1}{\tau_i^{\mathrm{I}}} S_{i0}^{\mathrm{I}}, \tau_i^{\mathrm{I}}\eta_i\}$, $i = q+1, \ldots, n_{\mathrm{I}}$, $\eta_i = \min_{t \geq 0} v_{\mathrm{th}i}^{\mathrm{I}}(t)$, $i = q+1, \ldots, n_{\mathrm{I}}$, *and there exist positive-definite* *matrices* $\hat{P}$, $\hat{Q} \in \mathbb{R}^{2(n_E+q)\times 2(n_E+q)}$ *and a diagonal positive-definite matrix* $\hat{R} \in \mathbb{R}^{2(n_E+q)\times 2(n_E+q)}$ *such that*

$$\begin{bmatrix} \hat{Q} & -\hat{P} \\ -\hat{P} & \hat{R} \end{bmatrix} \geq 0 \qquad (98)$$

and $\hat{\Omega} > 0$, *where*

$$\hat{\Omega} \triangleq \hat{P}\hat{L} + \hat{L}\hat{P} - \hat{Q} - \hat{A}^{\mathrm{T}}\hat{R}\hat{A}, \qquad (99)$$

then Eqs. (89) and (90) *is globally exponentially partially synchronized. Alternatively, if* $\hat{\Omega} \geq 0$ *and* $\mathcal{N}(\hat{\Omega}) = \mathrm{span}([\mathbf{e}^{\mathrm{T}}_{(n_E+q)}\ 0^{\mathrm{T}}_{n_E+q}]^{\mathrm{T}}, [0^{\mathrm{T}}_{n_E+q}\ \mathbf{e}^{\mathrm{T}}_{(n_E+q)}]^{\mathrm{T}})$, *then the excitatory-inhibitory network given by Eqs. (89) and* (90) *is globally asymptotically partially synchronized.*

Proof. The proof follows as in the proof of Theorem 3 using the partial Lyapunov function candidate $\hat{V} : \overline{\mathbb{R}}_+^{2(n_E+q)} \to \mathbb{R}$ given by $\hat{V}(\hat{S}) = \hat{S}^{\mathrm{T}}\hat{P}\hat{S}$. Specifically, since $A^{\mathrm{IE}} \geq\geq 0$, $S^{\mathrm{E}} \geq\geq 0$, and $f_i(x)$, $x \in \mathbb{R}$, $i = q+1, \ldots, n_{\mathrm{I}}$, are nondecreasing functions, it follows that $\tilde{f}_2^{\mathrm{I}}(A_2^{\mathrm{IE}}S^{\mathrm{E}} + \tilde{v}^{\mathrm{I}}_{\mathrm{th}2}) \geq\geq \tilde{f}_2^{\mathrm{I}}(\tilde{v}^{\mathrm{I}}_{\mathrm{th}2})$, $S^{\mathrm{E}} \in \overline{\mathbb{R}}_+^{n_E}$. Since $f_i(x) \geq x$, $x \in \mathbb{R}$, $i = q+1, \ldots, n_{\mathrm{I}}$, it follows that $\tilde{f}_2^{\mathrm{I}}(\tilde{v}^{\mathrm{I}}_{\mathrm{th}2}) \geq\geq \tilde{v}^{\mathrm{I}}_{\mathrm{th}2}$, and hence it follows from Eq. (88) that

$$\ddot{\tilde{S}}_2^{\mathrm{I}}(t) \geq\geq -2L_2^{\mathrm{I}}\dot{\tilde{S}}_2^{\mathrm{I}}(t) - (L_2^{\mathrm{I}})^2\tilde{S}_2^{\mathrm{I}}(t) + \tilde{v}^{\mathrm{I}}_{\mathrm{th}2}(t),$$
$$\tilde{S}_2^{\mathrm{I}}(0) = \tilde{S}_{20}^{\mathrm{I}}, \quad \dot{\tilde{S}}_2^{\mathrm{I}}(0) = \dot{\tilde{S}}_{20}^{\mathrm{I}}, \quad t \geq 0, \qquad (100)$$

or, equivalently,

$$\begin{aligned} \ddot{S}_i^{\mathrm{I}}(t) &\geq -\frac{2}{\tau_i^{\mathrm{I}}}\dot{S}_i^{\mathrm{I}}(t) - \frac{1}{(\tau_i^{\mathrm{I}})^2}S_i^{\mathrm{I}}(t) + v^{\mathrm{I}}_{\mathrm{th}i}(t) \\ &\geq -\frac{2}{\tau_i^{\mathrm{I}}}\dot{S}_i^{\mathrm{I}}(t) - \frac{1}{(\tau_i^{\mathrm{I}})^2}S_i^{\mathrm{I}}(t) + \eta_i, \quad S_i^{\mathrm{I}}(0) = S_{i0}^{\mathrm{I}}, \quad \dot{S}_i^{\mathrm{I}}(0) = \dot{S}_{i0}^{\mathrm{I}}, \\ &\qquad t \geq 0, \quad i = q+1, \ldots, n_{\mathrm{I}}. \end{aligned} \qquad (101)$$

If we let

$$u_i^{\mathrm{I}}(t) \triangleq \dot{S}_i^{\mathrm{I}}(t) + \frac{1}{\tau_i^{\mathrm{I}}}S_i^{\mathrm{I}}(t), \quad t \geq 0, \quad i = 1, \ldots, n_{\mathrm{I}}, \qquad (102)$$

it follows from Eq. (101) that

$$\dot{u}_i^{\mathrm{I}}(t) + \frac{1}{\tau_i^{\mathrm{I}}}u_i^{\mathrm{I}}(t) \geq \eta_i, \quad u_i^{\mathrm{I}}(0) = \dot{S}_{i0}^{\mathrm{I}}(t) + \frac{1}{\tau_i^{\mathrm{I}}}S_{i0}^{\mathrm{I}}, \quad t \geq 0, \quad i = q+1, \ldots, n_{\mathrm{I}}, \qquad (103)$$

and hence

$$u_i^{\mathrm{I}}(t) \geq (u_i^{\mathrm{I}}(0) - \tau_i^{\mathrm{I}}\eta_i)\mathrm{e}^{-\frac{1}{\tau_i^{\mathrm{I}}}t} + \tau_i^{\mathrm{I}}\eta_i$$
$$\geq \min\left\{\dot{S}_{i0}^{\mathrm{I}} + \frac{1}{\tau_i^{\mathrm{I}}}S_{i0}^{\mathrm{I}}, \tau_i^{\mathrm{I}}\eta_i\right\} = \alpha_i, \quad t \geq 0, \quad i = q+1, \ldots, n_{\mathrm{I}}. \tag{104}$$

It follows from Eqs. (102) and (104) that

$$\dot{S}_i^{\mathrm{I}}(t) + \frac{1}{\tau_i^{\mathrm{I}}}S_i^{\mathrm{I}}(t) \geq \alpha_i, \quad S_i^{\mathrm{I}}(0) = S_{i0}^{\mathrm{I}}, \quad t \geq 0, \quad i = q+1, \ldots, n_{\mathrm{I}}, \tag{105}$$

and hence

$$S_i^{\mathrm{I}}(t) \geq (S_{i0}^{\mathrm{I}} - \tau_i^{\mathrm{I}}\alpha_i)\mathrm{e}^{-\frac{1}{\tau_i^{\mathrm{I}}}t} + \tau_i^{\mathrm{I}}\alpha_i \geq \min\{S_{i0}^{\mathrm{I}}, \tau_i^{\mathrm{I}}\alpha_i\},\ t \geq 0,\ i = q+1, \ldots, n_{\mathrm{I}}. \tag{106}$$

Thus

$$\tilde{S}_2^{\mathrm{I}}(t) \geq\geq [\min\{S_{q+1}^{\mathrm{I}}(0), \tau_{q+1}^{\mathrm{I}}\alpha_{q+1}\}, \ldots, \min\{S_{n_{\mathrm{I}}}^{\mathrm{I}}(0), \tau_{n_{\mathrm{I}}}^{\mathrm{I}}\alpha_{n_{\mathrm{I}}}\}]^{\mathrm{T}}, \quad t \geq 0, \tag{107}$$

where $\alpha_i = \min\{\dot{S}_{i0}^{\mathrm{I}}(t) + \frac{1}{\tau_i^{\mathrm{I}}}S_{i0}^{\mathrm{I}}, \tau_i^{\mathrm{I}}\eta_i\}$ and $\eta_i = \min_{t\geq 0}v_{\mathrm{th}i}^{\mathrm{I}}(t)$, $i = q+1, \ldots, n_{\mathrm{I}}$, $t \geq 0$. Since $\tilde{B} \leq\leq 0$, it follows that $\tilde{B}\tilde{S}_2^{\mathrm{I}}(t) \leq\leq \tilde{B}w$, $t \geq 0$, where $w = [\min\{S_{q+1}^{\mathrm{I}}(0), \tau_{q+1}^{\mathrm{I}}\alpha_{q+1}\}, \ldots, \min\{S_{n_{\mathrm{I}}}^{\mathrm{I}}(0), \tau_{n_{\mathrm{I}}}^{\mathrm{I}}\alpha_{n_{\mathrm{I}}}\}]^{\mathrm{T}}$, and hence $\tilde{v}_{\mathrm{th}}(t)+\tilde{B}\tilde{S}_2^{\mathrm{I}}(t) \leq\leq \tilde{v}_{\mathrm{th}}(t)+\tilde{B}w \leq\leq 0$, $t \geq 0$. Thus $\hat{B}\hat{S}_2^{\mathrm{I}}(t)+\hat{v}_{\mathrm{th}}(t) \leq\leq 0$, and hence $\hat{f}(\hat{A}\hat{S}(t) + \hat{B}\hat{S}_2^{\mathrm{I}}(t) + \hat{v}_{\mathrm{th}}(t)) \leq\leq \hat{f}(\hat{A}\hat{S})$, $(\hat{S}, \hat{S}_2^{\mathrm{I}}) \in \overline{\mathbb{R}}_+^{2(n_{\mathrm{E}}+q)} \times \overline{\mathbb{R}}_+^{2(n_{\mathrm{I}}-q)}$. The remainder of the proof now follows as in the proof of Theorem 3. □

Finally, we analyze the partial synchronization of the excitatory-inhibitory network given by Eqs. (84)–(86) with the activation function given by Eq. (62) in the face of increasing inhibitory time constants.

Theorem 6. *Consider the excitatory-inhibitory network given by Eqs. (84)–(86) with the vector activation functions defined by Eqs. (37) and (38), and $f_i(\cdot)$ given by Eq. (62). If $\tilde{v}_{\mathrm{th}2}^{\mathrm{I}}(t) >> 0$, $t \geq 0$, and the time constants of inhibitory neurons are such that $\dot{\tilde{S}}_{20}^{\mathrm{I}} + L_2^I\tilde{S}_{20}^{\mathrm{I}} - (L_2^{\mathrm{I}})^{-1}\min\{f_{\max}\mathbf{e}, \eta\} << 0$, $\tilde{S}_{20}^{\mathrm{I}} - (L_2^{\mathrm{I}})^{-1}((L_2^{\mathrm{I}})^{-1}\min\{f_{\max}\mathbf{e}, \eta\} - \varepsilon\mathbf{e}) << 0$, and*

$$A^{\mathrm{EE}}\max\{S_0^{\mathrm{E}}, (L^{\mathrm{E}})^{-1}\max\{\dot{S}_0^{\mathrm{E}} + L^{\mathrm{E}}S_0^{\mathrm{E}}, f_{\max}(L^{\mathrm{E}})^{-1}\mathbf{e}\}\}$$
$$+ A_2^{\mathrm{EI}}[(L_2^{\mathrm{I}})^{-1}[(L_2^{\mathrm{I}})^{-1}\min\{f_{\max}\mathbf{e}, \eta\} - \varepsilon\mathbf{e}] - \varepsilon\mathbf{e}] + \beta \leq\leq 0, \tag{108}$$

where $\eta = [\eta_{q+1}, \ldots, \eta_{n_{\mathrm{I}}}]$, $\eta_i = \min_{t\geq 0}v_{\mathrm{th}i}^{\mathrm{I}}(t)$, $i = q+1, \ldots, n_{\mathrm{I}}$, $\beta = [\beta_1, \ldots, \beta_{n_{\mathrm{E}}}]^{\mathrm{T}}$, $\beta_i = \max_{t\geq 0}v_{\mathrm{th}i}^{\mathrm{E}}(t)$, $i = 1, \ldots, n_{\mathrm{E}}$, and $\varepsilon > 0$ is a small *positive constant, then* $S^{\mathrm{E}}(t) \to 0$ *as* $t \to \infty$*, and hence the excitatory-inhibitory network given by Eqs. (84)–(86) is globally asymptotically partially synchronized.*

Proof. First, note that since $f_i(x) \le f_{\max}$, $x \in \mathbb{R}$, $i = 1, \ldots, n_{\mathrm{E}}$, it follows from Eq. (84) that

$$\ddot{S}_i^{\mathrm{E}}(t) + \frac{2}{\tau_i^{\mathrm{E}}}\dot{S}_i^{\mathrm{E}}(t) + \frac{1}{(\tau_i^{\mathrm{E}})^2}S_i^{\mathrm{E}}(t) \le f_{\max}, \quad S_i^{\mathrm{E}}(0) = S_{i0}^{\mathrm{E}},$$
$$\dot{S}_i^{\mathrm{E}}(0) = \dot{S}_{i0}^{\mathrm{E}}, \quad t \ge 0, \quad i = 1, \ldots, n_{\mathrm{E}}. \quad (109)$$

If we define $u_i^{\mathrm{E}}(t) \triangleq \dot{S}_i^{\mathrm{E}}(t) + \frac{1}{\tau_i^{\mathrm{E}}}S_i^{\mathrm{E}}(t)$, $t \ge 0$, $i = 1, \ldots, n_{\mathrm{E}}$, it can be shown that

$$S_i^{\mathrm{E}}(t) \le \max\left\{S_{i0}^{\mathrm{E}}, \tau_i^{\mathrm{E}} \max\left\{\dot{S}_{i0}^{\mathrm{E}} + \frac{1}{\tau_i^{\mathrm{E}}}S_{i0}^{\mathrm{E}}, \tau_i^{\mathrm{E}} f_{\max}\right\}\right\}, \quad t \ge 0, \quad i = 1, \ldots, n_{\mathrm{E}}, \quad (110)$$

or, equivalently,

$$S^{\mathrm{E}}(t) \le \max\{S_0^{\mathrm{E}}, (L^{\mathrm{E}})^{-1}\max\{\dot{S}_0^{\mathrm{E}} + L^{\mathrm{E}}S_0^{\mathrm{E}}, f_{\max}(L^{\mathrm{E}})^{-1}\mathbf{e}\}\}, \quad t \ge 0, \quad (111)$$

Since $A_2^{\mathrm{IE}} \geq\geq 0$, $S^{\mathrm{E}}(t) \geq\geq 0$, $t \ge 0$, and $f_i(x)$, $x \in \mathbb{R}$, $i = q + 1, \ldots, n_{\mathrm{I}}$, are nondecreasing functions, it follows that $\tilde{f}_2^{\mathrm{I}}(A_2^{\mathrm{IE}}S^{\mathrm{E}} + \tilde{v}_{\mathrm{th2}}^{\mathrm{I}}) \geq\geq \tilde{f}_2^{\mathrm{I}}(\tilde{v}_{\mathrm{th2}}^{\mathrm{I}})$, $S^{\mathrm{E}} \in \overline{\mathbb{R}}_+^{n_{\mathrm{E}}}$. And since $f_i(x) \le f_{\max}$, $x \in \mathbb{R}$, $i = q + 1, \ldots, n_{\mathrm{I}}$, and $\tilde{v}_{\mathrm{th2}}^{\mathrm{I}}(t) >> 0$, $t \ge 0$, it follows that $\tilde{f}_2^{\mathrm{I}}(\tilde{v}_{\mathrm{th2}}^{\mathrm{I}}(t)) = \min\{f_{\max}\mathbf{e}, \tilde{v}_{\mathrm{th2}}^{\mathrm{I}}(t)\} \geq\geq \min\{f_{\max}\mathbf{e}, \eta\}$, $t \ge 0$. It follows from Eq. (86) that

$$\ddot{\tilde{S}}_2^{\mathrm{I}}(t) \geq\geq -2L_2^{\mathrm{I}}\dot{\tilde{S}}_2^{\mathrm{I}}(t) - (L_2^{\mathrm{I}})^2\tilde{S}_2^{\mathrm{I}}(t) + \min\{f_{\max}\mathbf{e}, \eta\},$$
$$\tilde{S}_2^{\mathrm{I}}(0) = \tilde{S}_{20}^{\mathrm{I}}, \quad \dot{\tilde{S}}_2^{\mathrm{I}}(0) = \dot{\tilde{S}}_{20}^{\mathrm{I}}, \quad t \ge 0. \quad (112)$$

If we define

$$u^{\mathrm{I}}(t) \triangleq \dot{\tilde{S}}_2^{\mathrm{I}}(t) + L_2^{\mathrm{I}}\tilde{S}_2^{\mathrm{I}}(t), \quad t \ge 0, \quad (113)$$

it follows from Eq. (112) that

$$\dot{u}^{\mathrm{I}}(t) + L_2^{\mathrm{I}}u^{\mathrm{I}}(t) \geq\geq \min\{f_{\max}\mathbf{e}, \eta\}, \quad u^{\mathrm{I}}(0) = \dot{\tilde{S}}_{20}^{\mathrm{I}} + L_2^{\mathrm{I}}\tilde{S}_{20}^{\mathrm{I}}, \quad t \ge 0, \quad (114)$$

and hence

$$u^{\mathrm{I}}(t) \geq\geq (\dot{\tilde{S}}_{20}^{\mathrm{I}} + L_2^{\mathrm{I}}\tilde{S}_{20}^{\mathrm{I}} - (L_2^{\mathrm{I}})^{-1}\min\{f_{\max}\mathbf{e}, \eta\})e^{-L_2^{\mathrm{I}}t}$$
$$+ (L_2^{\mathrm{I}})^{-1}\min\{f_{\max}\mathbf{e}, \eta\}, \quad t \ge 0. \quad (115)$$

Thus it follows from Eq. (115) that $u^{\mathrm{I}}(t) \geq\geq (L_2^{\mathrm{I}})^{-1}\min\{f_{\max}\mathbf{e}, \eta\}$ as $t \to \infty$, and since $\dot{\tilde{S}}_{20}^{\mathrm{I}} + L_2^{\mathrm{I}}\tilde{S}_{20}^{\mathrm{I}} - (L_2^{\mathrm{I}})^{-1}\min\{f_{\max}\mathbf{e}, \eta\} << 0$, there exists $t_2 > 0$ such that

$$u^{\mathrm{I}}(t) \geq\geq (L_2^{\mathrm{I}})^{-1}\min\{f_{\max}\mathbf{e}, \eta\} - \varepsilon\mathbf{e}, \quad t \geq t_2. \tag{116}$$

It follows from Eqs. (113) and (116) that

$$\dot{\tilde{S}}_2^{\mathrm{I}}(t) + L_2^{\mathrm{I}}\tilde{S}_2^{\mathrm{I}}(t) \geq\geq (L_2^{\mathrm{I}})^{-1}\min\{f_{\max}\mathbf{e}, \eta\} - \varepsilon\mathbf{e}, \quad t \geq t_2, \tag{117}$$

and hence

$$\begin{aligned}\tilde{S}_2^{\mathrm{I}}(t) \geq\geq{} & (\tilde{S}_{20}^{\mathrm{I}} - (L_2^{\mathrm{I}})^{-1}((L_2^{\mathrm{I}})^{-1}\min\{f_{\max}\mathbf{e}, \eta\} - \varepsilon\mathbf{e}))\mathrm{e}^{-L_2^{\mathrm{I}}t} \\ & + (L_2^{\mathrm{I}})^{-1}((L_2^{\mathrm{I}})^{-1}\min\{f_{\max}\mathbf{e}, \eta\} - \varepsilon\mathbf{e}), \quad t \geq t_2. \end{aligned} \tag{118}$$

Thus it follows from Eq. (118) that $\tilde{S}_2^{\mathrm{I}}(t) \geq\geq (L_2^{\mathrm{I}})^{-1}((L_2^{\mathrm{I}})^{-1}\min\{f_{\max}\mathbf{e}, \eta\} - \varepsilon\mathbf{e})$ as $t \to \infty$, and since $\tilde{S}_{20}^{\mathrm{I}} - (L_2^{\mathrm{I}})^{-1}((L_2^{\mathrm{I}})^{-1}\min\{f_{\max}\mathbf{e}, \eta\} - \varepsilon\mathbf{e}) << 0$, there exist $t_3 > t_2 > 0$ such that

$$\tilde{S}_2^{\mathrm{I}}(t) \geq\geq (L_2^{\mathrm{I}})^{-1}[(L_2^{\mathrm{I}})^{-1}\min\{f_{\max}\mathbf{e}, \eta\} - \varepsilon\mathbf{e}] - \varepsilon\mathbf{e}, \quad t \geq t_3. \tag{119}$$

Since $A_1^{\mathrm{EI}} \leq\leq 0$ and $\tilde{S}_1^{\mathrm{I}}(t) \geq\geq 0$, $t \geq 0$, and $f_i(x)$, $x \in \mathbb{R}$, $i = 1, \ldots, n_{\mathrm{E}}$, are nondecreasing functions, it follows from Eq. (84) that

$$\begin{aligned}\ddot{S}^{\mathrm{E}}(t) &\leq\leq -2L^{\mathrm{E}}\dot{S}^{\mathrm{E}}(t) - (L^{\mathrm{E}})^2 S^{\mathrm{E}}(t) + f^{\mathrm{E}}(A^{\mathrm{EE}}S^{\mathrm{E}}(t) + A_2^{\mathrm{EI}}\tilde{S}_2^{\mathrm{I}} + \tilde{v}_{\mathrm{th}}^{\mathrm{E}}(t)), \\ S^{\mathrm{E}}(0) &= S_0^{\mathrm{E}}, \quad \dot{S}^{\mathrm{E}}(0) = \dot{S}_0^{\mathrm{E}}, \quad t \geq 0. \end{aligned} \tag{120}$$

Since $A^{\mathrm{EE}} \geq\geq 0$ and $A_2^{\mathrm{EI}} \leq\leq 0$, it follows from Eqs. (111) and (119) that

$$\begin{aligned}A^{\mathrm{EE}}S^{\mathrm{E}}(t) + A_2^{\mathrm{EI}}\tilde{S}_2^{\mathrm{I}}(t) + \tilde{v}_{\mathrm{th}}^{\mathrm{E}}(t) \leq\leq{} & A^{\mathrm{EE}}\max\{S_0^{\mathrm{E}}, (L^{\mathrm{E}})^{-1} \\ & \max\{\dot{S}_0^{\mathrm{E}} + L^{\mathrm{E}}S_0^{\mathrm{E}}, f_{\max}(L^{\mathrm{E}})^{-1}\mathbf{e}\}\} + A_2^{\mathrm{EI}}[(L_2^{\mathrm{I}})^{-1}[(L_2^{\mathrm{I}})^{-1} \\ & \min\{f_{\max}\mathbf{e}, \eta\} - \varepsilon\mathbf{e}] - \varepsilon\mathbf{e}] + \beta \leq\leq 0, \quad t \geq t_3. \end{aligned} \tag{121}$$

Finally, it follows from Eqs. (120) and (121) that

$$\begin{aligned}\ddot{S}^{\mathrm{E}}(t) \leq\leq -2L^{\mathrm{E}}\dot{S}^{\mathrm{E}}(t) - (L^{\mathrm{E}})^2 S^{\mathrm{E}}(t), \quad S^{\mathrm{E}}(0) = S_0^{\mathrm{E}}, \\ \dot{S}^{\mathrm{E}}(0) = \dot{S}_0^{\mathrm{E}}, \quad t \geq t_3, \end{aligned} \tag{122}$$

which implies that $S^{\mathrm{E}}(t) \to 0$ as $t \to \infty$, and hence the excitatory-inhibitory network given by Eqs. (84)–(86) is globally asymptotically partially synchronized. □

Condition (108) implies that all the excitatory neurons receive inhibitory inputs from the inhibitory neurons that themselves do not receive inhibitory inputs. As in the case of the simpler neural population model discussed in Section 4, if only a subset of excitatory neurons receive the inhibitory inputs from the inhibitory neurons that themselves do not receive inhibitory inputs, then only the synaptic drives of this subset of excitatory neurons will converge to zero as the inhibitory time constants increase.

6 A SECOND-ORDER MEAN EXCITATORY AND INHIBITORY SYNAPTIC DRIVE MODEL

In this section we reduce the model given in Eqs. (79) and (80) to a mean excitatory and mean inhibitory model. Consider Eqs. (79) and (80) with continuously differentiable $f_i(\cdot) = f(\cdot)$, $B_i^{\mathrm{E}} = B_i^{\mathrm{I}} = 1$, $\lambda_i^{\mathrm{E}} = \lambda^{\mathrm{E}}$, and $\lambda_i^{\mathrm{I}} = \lambda^{\mathrm{I}}$. In this case Eqs. (79) and (80) become

$$\frac{\mathrm{d}^2 S_i^{\mathrm{E}}(t)}{\mathrm{d}t^2} = -\frac{2}{\lambda^{\mathrm{E}}}\frac{\mathrm{d}S_i^{\mathrm{E}}(t)}{\mathrm{d}t} - \left(\frac{1}{\lambda^{\mathrm{E}}}\right)^2 S_i^{\mathrm{E}}(t) + f\left(\sum_{j=1}^{n_{\mathrm{E}}} A_{ij}^{\mathrm{EE}} S_j^{\mathrm{E}}(t) + \sum_{k=1}^{n_{\mathrm{I}}} A_{ik}^{\mathrm{EI}} S_k^{\mathrm{I}}(t) + v_{\mathrm{th}i}^{\mathrm{E}}(t)\right), \quad i = 1, \ldots, n_{\mathrm{E}}, \tag{123}$$

$$\frac{\mathrm{d}^2 S_i^{\mathrm{I}}(t)}{\mathrm{d}t^2} = -\frac{2}{\lambda^{\mathrm{I}}}\frac{\mathrm{d}S_i^{\mathrm{I}}(t)}{\mathrm{d}t} - \left(\frac{1}{\lambda^{\mathrm{I}}}\right)^2 S_i^{\mathrm{I}}(t) + f\left(\sum_{j=1}^{n_{\mathrm{E}}} A_{ij}^{\mathrm{IE}} S_j^{\mathrm{E}}(t) + \sum_{k=1}^{n_{\mathrm{I}}} A_{ik}^{\mathrm{II}} S_k^{\mathrm{I}}(t) + v_{\mathrm{th}i}^{\mathrm{I}}(t)\right), \quad i = 1, \ldots, n_{\mathrm{I}}, \tag{124}$$

where $A_{ii}^{\mathrm{EE}} = A_{ii}^{\mathrm{II}} = 0$. Let $A_{ij}^{\mathrm{EE}} = \overline{A}^{\mathrm{EE}} + \Delta_{ij}^{\mathrm{EE}}$, $A_{ij}^{\mathrm{EI}} = \overline{A}^{\mathrm{EI}} + \Delta_{ij}^{\mathrm{EI}}$, $A_{ij}^{\mathrm{IE}} = \overline{A}^{\mathrm{IE}} + \Delta_{ij}^{\mathrm{IE}}$, and $A_{ij}^{\mathrm{II}} = \overline{A}^{\mathrm{II}} + \Delta_{ij}^{\mathrm{II}}$, where $\overline{A}^{\mathrm{XY}} \triangleq \frac{1}{n_{\mathrm{X}} n_{\mathrm{Y}}} \sum_{i=1}^{n_{\mathrm{X}}} \sum_{j=1}^{n_{\mathrm{Y}}} A_{ij}^{\mathrm{XY}}$, X, Y $\in$ {E, I}, denote the mean and $\Delta_{ij}^{\mathrm{XY}}$, X, Y $\in$ {E, I}, are deviations from the mean. In this case it follows that

$$\sum_{i=1}^{n_{\mathrm{E}}}\sum_{j=1}^{n_{\mathrm{E}}} \Delta_{ij}^{\mathrm{EE}} = \sum_{i=1}^{n_{\mathrm{E}}}\sum_{j=1}^{n_{\mathrm{I}}} \Delta_{ij}^{\mathrm{EI}} = \sum_{i=1}^{n_{\mathrm{I}}}\sum_{j=1}^{n_{\mathrm{E}}} \Delta_{ij}^{\mathrm{IE}} = \sum_{i=1}^{n_{\mathrm{I}}}\sum_{j=1}^{n_{\mathrm{I}}} \Delta_{ij}^{\mathrm{II}} = 0. \tag{125}$$

Using the average and perturbed expression for A_{ij}^{XY}, X, Y $\in$ {E, I}, we can write Eqs. (123) and (124) as

$$\frac{\mathrm{d}^2 S_i^{\mathrm{E}}(t)}{\mathrm{d}t^2} = f\left(n_{\mathrm{E}}\overline{A}^{\mathrm{EE}}\overline{S}^{\mathrm{E}}(t) + \sum_{j=1}^{n_{\mathrm{E}}} \Delta_{ij}^{\mathrm{EE}} S_j^{\mathrm{E}}(t) + n_{\mathrm{I}}\overline{A}^{\mathrm{EI}}\overline{S}^{\mathrm{I}}(t) + \sum_{k=1}^{n_{\mathrm{I}}} \Delta_{ik}^{\mathrm{EI}} S_k^{\mathrm{I}}(t) + v_{\mathrm{th}i}^{\mathrm{E}}(t)\right)$$
$$- \frac{2}{\lambda^{\mathrm{E}}}\frac{\mathrm{d}S_i^{\mathrm{E}}(t)}{\mathrm{d}t} - \left(\frac{1}{\lambda^{\mathrm{E}}}\right)^2 S_i^{\mathrm{E}}(t), \quad i = 1, \ldots, n_{\mathrm{E}}, \tag{126}$$

$$\frac{\mathrm{d}^2 S_i^{\mathrm{I}}(t)}{\mathrm{d}t^2} = f\left(n_{\mathrm{E}}\overline{A}^{\mathrm{IE}}\overline{S}^{\mathrm{E}}(t) + \sum_{j=1}^{n_{\mathrm{E}}} \Delta_{ij}^{\mathrm{IE}} S_j^{\mathrm{E}}(t) + n_{\mathrm{I}}\overline{A}^{\mathrm{II}}\overline{S}^{\mathrm{I}}(t) + \sum_{k=1}^{n_{\mathrm{I}}} \Delta_{ik}^{\mathrm{II}} S_k^{\mathrm{I}}(t) + v_{\mathrm{th}i}^{\mathrm{I}}(t)\right)$$
$$- \frac{2}{\lambda^{\mathrm{I}}}\frac{\mathrm{d}S_i^{\mathrm{I}}(t)}{\mathrm{d}t} - \left(\frac{1}{\lambda^{\mathrm{I}}}\right)^2 S_i^{\mathrm{I}}(t), \quad i = 1, \ldots, n_{\mathrm{I}}, \tag{127}$$

where $\overline{S}^{\mathrm{E}}(t) \triangleq \frac{1}{n_{\mathrm{E}}}\sum_{j=1}^{n_{\mathrm{E}}} S_j^{\mathrm{E}}(t)$ and $\overline{S}^{\mathrm{I}}(t) \triangleq \frac{1}{n_{\mathrm{I}}}\sum_{j=1}^{n_{\mathrm{I}}} S_j^{\mathrm{I}}(t)$ denote the mean excitatory synaptic drive and mean inhibitory synaptic drive in dimensionless units, respectively. If we define $\delta_i^{\mathrm{E}}(t) \triangleq S_i^{\mathrm{E}}(t) - \overline{S}^{\mathrm{E}}(t)$ and $\delta_i^{\mathrm{I}}(t) \triangleq S_i^{\mathrm{I}}(t) - \overline{S}^{\mathrm{I}}(t)$, where $\delta_i^{\mathrm{E}}(t)$ and $\delta_i^{\mathrm{I}}(t)$ are deviations from the mean, Eqs. (126) and (127) become

$$\frac{\mathrm{d}^2 S_i^{\mathrm{E}}(t)}{\mathrm{d}t^2} = f\left(n_{\mathrm{E}}\overline{A}^{\mathrm{EE}}\overline{S}^{\mathrm{E}}(t) + \overline{S}^{\mathrm{E}}(t)\sum_{j=1}^{n_{\mathrm{E}}} \Delta_{ij}^{\mathrm{EE}} + \sum_{j=1}^{n_{\mathrm{E}}} \Delta_{ij}^{\mathrm{EE}}\delta_j^{\mathrm{E}}(t)\right.$$
$$\left. + n_{\mathrm{I}}\overline{A}^{\mathrm{EI}}\overline{S}^{\mathrm{I}}(t) + \overline{S}^{\mathrm{I}}(t)\sum_{k=1}^{n_{\mathrm{I}}} \Delta_{ik}^{\mathrm{EI}} + \sum_{k=1}^{n_{\mathrm{I}}} \Delta_{ik}^{\mathrm{EI}}\delta_k^{\mathrm{I}}(t) + v_{\mathrm{th}i}^{\mathrm{E}}(t)\right)$$
$$- \frac{2}{\lambda^{\mathrm{E}}}\frac{\mathrm{d}S_i^{\mathrm{E}}(t)}{\mathrm{d}t} - \left(\frac{1}{\lambda^{\mathrm{E}}}\right)^2 S_i^{\mathrm{E}}(t), \quad i = 1, \ldots, n_{\mathrm{E}}, \tag{128}$$

$$\frac{\mathrm{d}^2 S_i^{\mathrm{I}}(t)}{\mathrm{d}t^2} = f\left(n_{\mathrm{E}}\overline{A}^{\mathrm{IE}}\overline{S}^{\mathrm{E}}(t) + \overline{S}^{\mathrm{E}}(t)\sum_{j=1}^{n_{\mathrm{E}}} \Delta_{ij}^{\mathrm{IE}} + \sum_{j=1}^{n_{\mathrm{E}}} \Delta_{ij}^{\mathrm{IE}}\delta_j^{\mathrm{E}}(t) + n_{\mathrm{I}}\overline{A}^{\mathrm{II}}\overline{S}^{\mathrm{I}}(t)\right.$$
$$\left. + \overline{S}^{\mathrm{I}}(t)\sum_{k=1}^{n_{\mathrm{I}}} \Delta_{ik}^{\mathrm{II}} + \sum_{k=1}^{n_{\mathrm{I}}} \Delta_{ik}^{\mathrm{II}}\delta_k^{\mathrm{I}}(t) + v_{\mathrm{th}i}^{\mathrm{I}}(t)\right) - \frac{2}{\lambda^{\mathrm{I}}}\frac{\mathrm{d}S_i^{\mathrm{I}}(t)}{\mathrm{d}t}$$
$$- \left(\frac{1}{\lambda^{\mathrm{I}}}\right)^2 S_i^{\mathrm{I}}(t), \quad i = 1, \ldots, n_{\mathrm{I}}. \tag{129}$$

Assume that all terms with a factor $\Delta_{ij}^{\mathrm{XY}}$, X, Y $\in \{\mathrm{E}, \mathrm{I}\}$, $i = 1, \ldots, n_{\mathrm{X}}$ and $j = 1, \ldots, n_{\mathrm{Y}}$, in Eqs. (128) and (129) are small relative to the remaining terms in $f(\cdot)$. A first-order expansion of Eqs. (128) and (129) gives

$$
\begin{aligned}
\frac{\mathrm{d}^2 S_i^{\mathrm{E}}(t)}{\mathrm{d}t^2} =\, & f(n_{\mathrm{E}}\overline{A}^{\mathrm{EE}}\overline{S}^{\mathrm{E}}(t) + n_{\mathrm{I}}\overline{A}^{\mathrm{EI}}\overline{S}^{\mathrm{I}}(t) + v_{\mathrm{th}i}^{\mathrm{E}}(t)) + f'(n_{\mathrm{E}}\overline{A}^{\mathrm{EE}}\overline{S}^{\mathrm{E}}(t) \\
& + n_{\mathrm{I}}\overline{A}^{\mathrm{EI}}\overline{S}^{\mathrm{I}}(t) + v_{\mathrm{th}i}^{\mathrm{E}}(t)) \times \left[\overline{S}^{\mathrm{E}}(t)\sum_{j=1}^{n_{\mathrm{E}}} \Delta_{ij}^{\mathrm{EE}} + \sum_{j=1}^{n_{\mathrm{E}}} \Delta_{ij}^{\mathrm{EE}}\delta_j^{\mathrm{E}}(t) \right. \\
& \left. + \overline{S}^{\mathrm{I}}(t)\sum_{k=1}^{n_{\mathrm{I}}} \Delta_{ik}^{\mathrm{EI}} + \sum_{k=1}^{n_{\mathrm{I}}} \Delta_{ik}^{\mathrm{EI}}\delta_k^{\mathrm{I}}(t) \right] - \frac{2}{\lambda^{\mathrm{E}}}\frac{\mathrm{d}S_i^{\mathrm{E}}(t)}{\mathrm{d}t} \\
& - \left(\frac{1}{\lambda^{\mathrm{E}}}\right)^2 S_i^{\mathrm{E}}(t), \quad i = 1, \ldots, n_{\mathrm{E}},
\end{aligned} \tag{130}
$$

$$
\begin{aligned}
\frac{\mathrm{d}^2 S_i^{\mathrm{I}}(t)}{\mathrm{d}t^2} =\, & f(n_{\mathrm{E}}\overline{A}^{\mathrm{IE}}\overline{S}^{\mathrm{E}}(t) + n_{\mathrm{I}}\overline{A}^{\mathrm{II}}\overline{S}^{\mathrm{I}}(t) + v_{\mathrm{th}i}^{\mathrm{I}}(t)) + f'(n_{\mathrm{E}}\overline{A}^{\mathrm{IE}}\overline{S}^{\mathrm{E}}(t) \\
& + n_{\mathrm{I}}\overline{A}^{\mathrm{II}}\overline{S}^{\mathrm{I}}(t) + v_{\mathrm{th}i}^{\mathrm{I}}(t)) \times \left[\overline{S}^{\mathrm{E}}(t)\sum_{j=1}^{n_{\mathrm{E}}} \Delta_{ij}^{\mathrm{IE}} + \sum_{j=1}^{n_{\mathrm{E}}} \Delta_{ij}^{\mathrm{IE}}\delta_j^{\mathrm{E}}(t) \right. \\
& \left. + \overline{S}^{\mathrm{I}}(t)\sum_{k=1}^{n_{\mathrm{I}}} \Delta_{ik}^{\mathrm{II}} + \sum_{k=1}^{n_{\mathrm{I}}} \Delta_{ik}^{\mathrm{II}}\delta_k^{\mathrm{I}}(t) \right] - \frac{2}{\lambda^{\mathrm{I}}}\frac{\mathrm{d}S_i^{\mathrm{I}}(t)}{\mathrm{d}t} \\
& - \left(\frac{1}{\lambda^{\mathrm{I}}}\right)^2 S_i^{\mathrm{I}}(t), \quad i = 1, \ldots, n_{\mathrm{I}}.
\end{aligned} \tag{131}
$$

Assuming that the higher-order terms can be ignored, Eqs. (130) and (131) become

$$
\begin{aligned}
\frac{\mathrm{d}^2 S_i^{\mathrm{E}}(t)}{\mathrm{d}t^2} =\, & f(n_{\mathrm{E}}\overline{A}^{\mathrm{EE}}\overline{S}^{\mathrm{E}}(t) + n_{\mathrm{I}}\overline{A}^{\mathrm{EI}}\overline{S}^{\mathrm{I}}(t) + v_{\mathrm{th}i}^{\mathrm{E}}(t)) + f'(n_{\mathrm{E}}\overline{A}^{\mathrm{EE}}\overline{S}^{\mathrm{E}}(t) \\
& + n_{\mathrm{I}}\overline{A}^{\mathrm{EI}}\overline{S}^{\mathrm{I}}(t) + v_{\mathrm{th}i}^{\mathrm{E}}(t)) \times \left[\overline{S}^{\mathrm{E}}(t)\sum_{j=1}^{n_{\mathrm{E}}} \Delta_{ij}^{\mathrm{EE}} + \overline{S}^{\mathrm{I}}(t)\sum_{k=1}^{n_{\mathrm{I}}} \Delta_{ik}^{\mathrm{EI}} \right] \\
& - \frac{2}{\lambda^{\mathrm{E}}}\frac{\mathrm{d}S_i^{\mathrm{E}}(t)}{\mathrm{d}t} - \left(\frac{1}{\lambda^{\mathrm{E}}}\right)^2 S_i^{\mathrm{E}}(t), \quad i = 1, \ldots, n_{\mathrm{E}},
\end{aligned} \tag{132}
$$

$$
\begin{aligned}
\frac{\mathrm{d}^2 S_i^{\mathrm{I}}(t)}{\mathrm{d}t^2} =\, & f(n_{\mathrm{E}}\overline{A}^{\mathrm{IE}}\overline{S}^{\mathrm{E}}(t) + n_{\mathrm{I}}\overline{A}^{\mathrm{II}}\overline{S}^{\mathrm{I}}(t) + v_{\mathrm{th}i}^{\mathrm{I}}(t)) + f'(n_{\mathrm{E}}\overline{A}^{\mathrm{IE}}\overline{S}^{\mathrm{E}}(t) \\
& + n_{\mathrm{I}}\overline{A}^{\mathrm{II}}\overline{S}^{\mathrm{I}}(t) + v_{\mathrm{th}i}^{\mathrm{I}}(t)) \times \left[\overline{S}^{\mathrm{E}}(t)\sum_{j=1}^{n_{\mathrm{E}}} \Delta_{ij}^{\mathrm{IE}} + \overline{S}^{\mathrm{I}}(t)\sum_{k=1}^{n_{\mathrm{I}}} \Delta_{ik}^{\mathrm{II}} \right] \\
& - \frac{2}{\lambda^{\mathrm{I}}}\frac{\mathrm{d}S_i^{\mathrm{I}}(t)}{\mathrm{d}t} - \left(\frac{1}{\lambda^{\mathrm{I}}}\right)^2 S_i^{\mathrm{I}}(t), \quad i = 1, \ldots, n_{\mathrm{I}}.
\end{aligned} \tag{133}
$$

Finally, if we sum Eqs. (132) and (133) over $i = 1, \ldots, n_{\mathrm{E}}$ and $i = 1, \ldots, n_{\mathrm{I}}$, dividing by n_{E}, and n_{I}, respectively, using Eq. (125), and assume $\nu^{\mathrm{E}}_{\mathrm{th}i}(t) = \nu^{\mathrm{E}}_{\mathrm{th}}$ and $\nu^{\mathrm{I}}_{\mathrm{th}i}(t) = \nu^{\mathrm{I}}_{\mathrm{th}}$, $t \geq 0$, it follows that the average excitatory synaptic drive and the average inhibitory synaptic drive are given by

$$\frac{\mathrm{d}^2\overline{S}^{\mathrm{E}}(t)}{\mathrm{d}t^2} = f(n_{\mathrm{E}}\overline{A}^{\mathrm{EE}}\overline{S}^{\mathrm{E}}(t) + n_{\mathrm{I}}\overline{A}^{\mathrm{EI}}\overline{S}^{\mathrm{I}}(t) + \nu^{\mathrm{E}}_{\mathrm{th}}(t)) - \frac{2}{\lambda^{\mathrm{E}}}\frac{\mathrm{d}\overline{S}^{\mathrm{E}}(t)}{\mathrm{d}t} - \left(\frac{1}{\lambda^{\mathrm{E}}}\right)^2 \overline{S}^{\mathrm{E}}(t), \quad \overline{S}^{\mathrm{E}}(0) = \overline{S}^{\mathrm{E}}_0, \quad \dot{\overline{S}}^{\mathrm{E}}(0) = \dot{\overline{S}}^{\mathrm{E}}_0, \quad t \geq 0, \tag{134}$$

$$\frac{\mathrm{d}^2\overline{S}^{\mathrm{I}}(t)}{\mathrm{d}t^2} = f(n_{\mathrm{E}}\overline{A}^{\mathrm{IE}}\overline{S}^{\mathrm{E}}(t) + n_{\mathrm{I}}\overline{A}^{\mathrm{II}}\overline{S}^{\mathrm{I}}(t) + \nu^{\mathrm{I}}_{\mathrm{th}}(t)) - \frac{2}{\lambda^{\mathrm{I}}}\frac{\mathrm{d}\overline{S}^{\mathrm{I}}(t)}{\mathrm{d}t} - \left(\frac{1}{\lambda^{\mathrm{I}}}\right)^2 \overline{S}^{\mathrm{I}}(t), \quad \overline{S}^{\mathrm{I}}(0) = \overline{S}^{\mathrm{I}}_0, \quad \dot{\overline{S}}^{\mathrm{I}}(0) = \dot{\overline{S}}^{\mathrm{I}}_0. \tag{135}$$

An analysis similar to the analysis provided in Section 3 can be done for the coupled second-order differential equations (134) and (135). This can allow us to qualitatively study the abrupt transition from consciousness to unconsciousness as well as the biphasic response during induction of anesthesia. Initial quantitative numerical experiments with this model are presented in Section 7.

7 NUMERICAL SIMULATIONS OF THE ANESTHETIC CASCADE MODEL

In this section we use the different analysis results developed in this chapter to present several numerical experiments that illustrate the qualitative behavior of the proposed anesthetic cascade models.

7.1 Two-Class Mean Excitatory and Mean Inhibitory Synaptic Drive Model

For our first example we consider the two-class mean field model given by Eqs. (12) and (13) with a single asymptotically stable equilibrium point. Fig. 6 shows the synaptic drives converging to the equilibrium point, whereas Fig. 7 shows the phase portrait of Eqs. (12) and (13). Next we consider the case where the model possesses two asymptotically stable

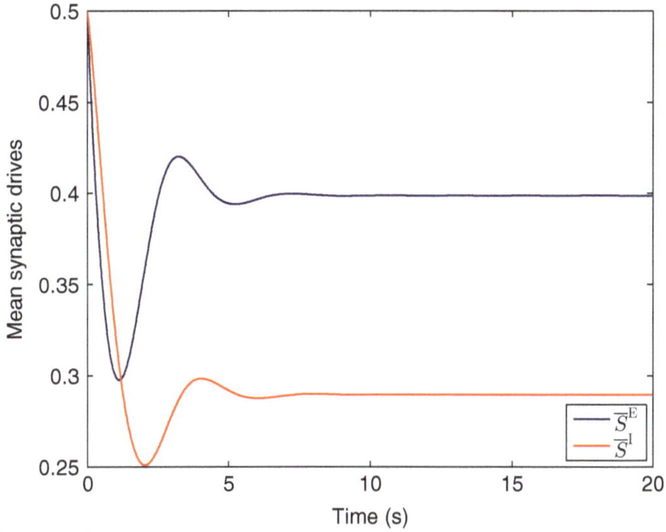

Fig. 6 $\overline{S}^{\mathrm{E}}(t)$ and $\overline{S}^{\mathrm{I}}(t)$ versus time for $a = 3$, $b = 9$, $c = 6$, $d = 1$, $v_{\mathrm{th}}^{\mathrm{E}} = 1$, $v_{\mathrm{th}}^{\mathrm{I}} = -3$, $\lambda^{\mathrm{E}} = \lambda^{\mathrm{I}} = 1$, $f_{\max} = 1$, and $\gamma = 1$ with $\overline{S}^{\mathrm{E}}(0) = \overline{S}^{\mathrm{I}}(0) = 0.5$.

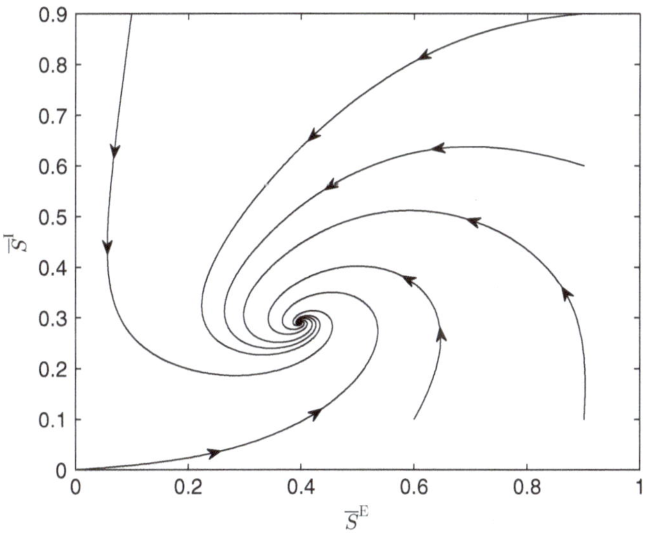

Fig. 7 Phase portrait of Eqs. (12) and (13) for $a = 3$, $b = 9$, $c = 6$, $d = 1$, $v_{\mathrm{th}}^{\mathrm{E}} = 1$, $v_{\mathrm{th}}^{\mathrm{I}} = -3$, $\lambda^{\mathrm{E}} = \lambda^{\mathrm{I}} = 1$, $f_{\max} = 1$, and $\gamma = 1$.

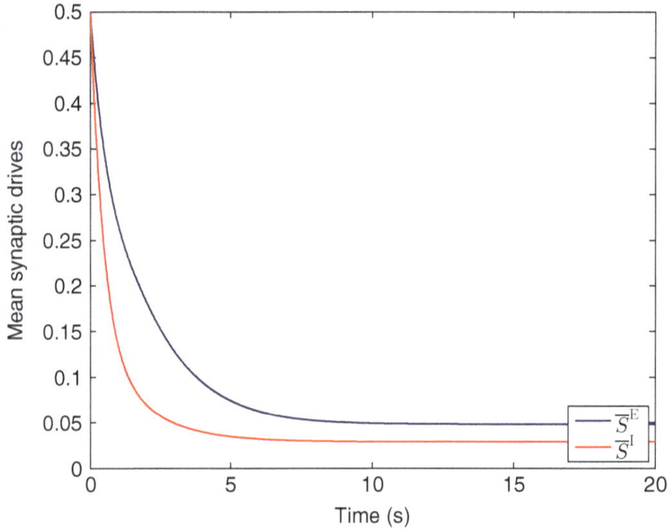

Fig. 8 $\overline{S}^{\mathrm{E}}(t)$ and $\overline{S}^{\mathrm{I}}(t)$ versus time for $a = 10, b = 9, c = 6, d = 1, v^{\mathrm{E}}_{\mathrm{th}} = -3.2, v^{\mathrm{I}}_{\mathrm{th}} = -2.8$, $\lambda^{\mathrm{E}} = 1, \lambda^{\mathrm{I}} = 0.4, f_{\max} = 1$, and $\gamma = 1$ with $\overline{S}^{\mathrm{E}}(0) = \overline{S}^{\mathrm{I}}(0) = 0.5$.

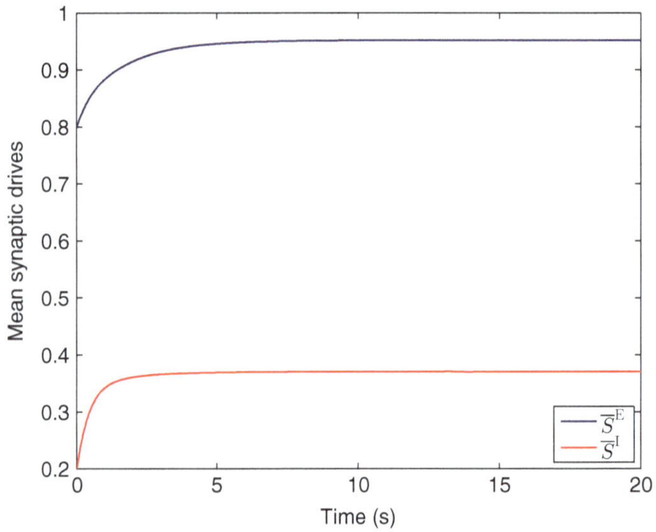

Fig. 9 $\overline{S}^{\mathrm{E}}(t)$ and $\overline{S}^{\mathrm{I}}(t)$ versus time for $a = 10, b = 9, c = 6, d = 1, v^{\mathrm{E}}_{\mathrm{th}} = -3.2, v^{\mathrm{I}}_{\mathrm{th}} = -2.8$, $\lambda^{\mathrm{E}} = 1, \lambda^{\mathrm{I}} = 0.4, f_{\max} = 1$, and $\gamma = 1$ with $\overline{S}^{\mathrm{E}}(0) = 0.8$ and $\overline{S}^{\mathrm{I}}(0) = 0.2$.

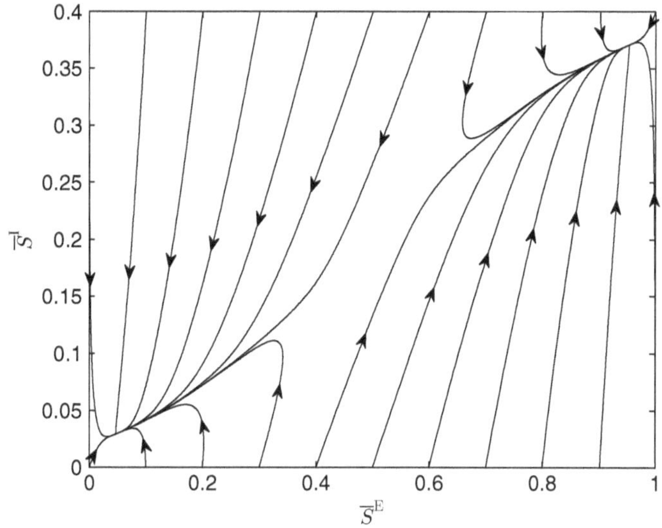

Fig. 10 Phase portrait of Eqs. (12) and (13) for $a = 10$, $b = 9$, $c = 6$, $d = 1$, $v_{\mathrm{th}}^{\mathrm{E}} = -3.2$, $v_{\mathrm{th}}^{\mathrm{I}} = -2.8$, $\lambda^{\mathrm{E}} = 1$, $\lambda^{\mathrm{I}} = 0.4$, $f_{\max} = 1$, and $\gamma = 1$.

equilibrium points and one unstable equilibrium point. Figs. 8 and 9 show the synaptic drives converging to two different points depending on the system's initial conditions, whereas Fig. 10 shows the corresponding phase portrait of Eqs. (12) and (13). Figs. 11 and 12 highlight the case where the model possesses five equilibrium points. In this case the synaptic drives converge to two different levels, indicating that there exist two asymptotically stable equilibrium points; the phase portrait for this case is shown in Fig. 13. Fig. 14 illustrates the case where there exists a single unstable equilibrium point. In this case the system possesses a limit cycle as can be seen in Fig. 15.

Next we illustrate the effect of the excitatory sensory input $v_{\mathrm{th}}^{\mathrm{E}}$ on the mean excitatory and mean inhibitory synaptic drives. We fix the parameters to be as in the simulation shown in Fig. 14 and vary $v_{\mathrm{th}}^{\mathrm{E}}$. It can be seen from Fig. 16 that for a value of $v_{\mathrm{th}}^{\mathrm{E}}$ below $v_{\mathrm{th1}}^{\mathrm{E}} = -1.6$, the two-class mean field model given by Eqs. (12) and (13) has exactly one asymptotically stable equilibrium point. As $v_{\mathrm{th}}^{\mathrm{E}}$ increases to $v_{\mathrm{th2}}^{\mathrm{E}} = -1.6$, the equilibrium point becomes unstable. At this point a supercritical Hopf bifurcation occurs, giving rise to a stable limit cycle. As $v_{\mathrm{th}}^{\mathrm{E}}$ increases beyond $v_{\mathrm{th2}}^{\mathrm{E}}$, the unstable equilibrium point reverts to an asymptotically stable equilibrium point and the stable limit cycle disappears. The value of $\overline{S}_{\mathrm{e}}^{\mathrm{E}}$ and that of $\overline{S}_{\mathrm{e}}^{\mathrm{I}}$ increases

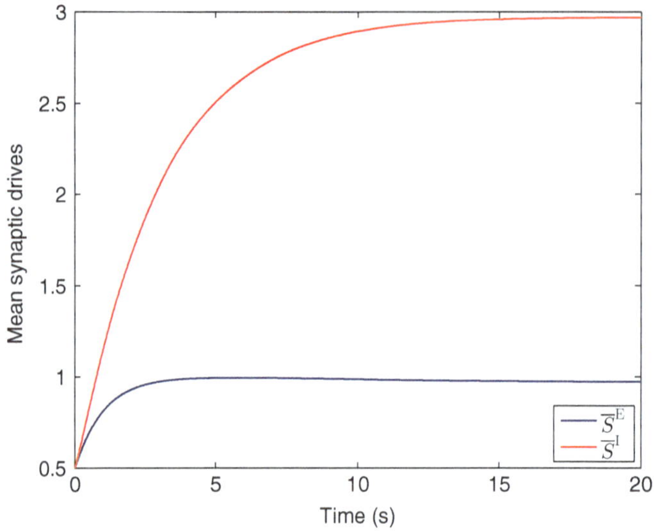

Fig. 11 $\overline{S}^{\mathrm{E}}(t)$ and $\overline{S}^{\mathrm{I}}(t)$ versus time for $a = 20, b = 4, c = 13, d = 1, v_{\mathrm{th}}^{\mathrm{E}} = -4, v_{\mathrm{th}}^{\mathrm{I}} = -5$, $\lambda^{\mathrm{E}} = 1, \lambda^{\mathrm{I}} = 3, f_{\mathrm{max}} = 1$, and $\gamma = 1$ with $\overline{S}^{\mathrm{E}}(0) = \overline{S}^{\mathrm{I}}(0) = 0.5$.

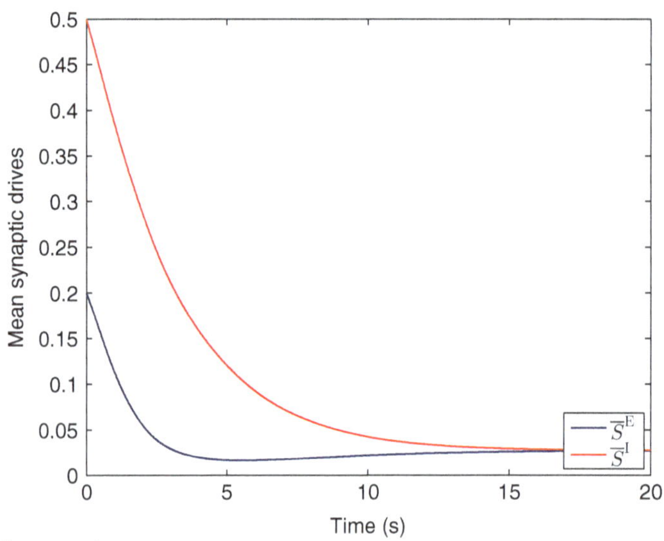

Fig. 12 $\overline{S}^{\mathrm{E}}(t)$ and $\overline{S}^{\mathrm{I}}(t)$ versus time for $a = 20, b = 4, c = 13, d = 1, v_{\mathrm{th}}^{\mathrm{E}} = -4, v_{\mathrm{th}}^{\mathrm{I}} = -5$, $\lambda^{\mathrm{E}} = 1, \lambda^{\mathrm{I}} = 3, f_{\mathrm{max}} = 1$, and $\gamma = 1$ with $\overline{S}^{\mathrm{E}}(0) = 0.2$ and $\overline{S}^{\mathrm{I}}(0) = 0.5$.

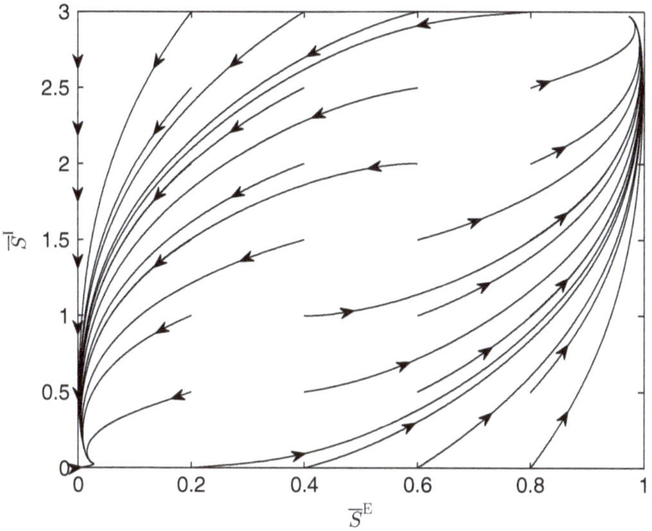

Fig. 13 Phase portrait of Eqs. (12) and (13) for $a = 20, b = 4, c = 13, d = 1, v_{\mathrm{th}}^{\mathrm{E}} = -4, v_{\mathrm{th}}^{\mathrm{I}} = -5, \lambda^{\mathrm{E}} = 1, \lambda^{\mathrm{I}} = 3, f_{\mathrm{max}} = 1$, and $\gamma = 1$.

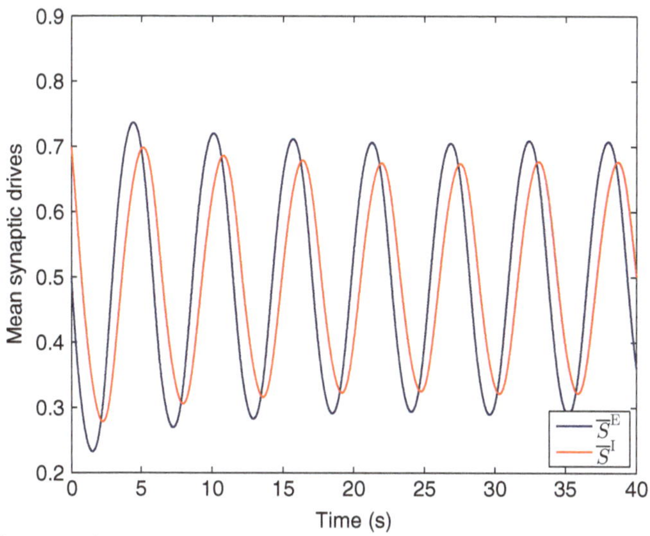

Fig. 14 $\overline{S}^{\mathrm{E}}(t)$ and $\overline{S}^{\mathrm{I}}(t)$ versus time for $a = 10, b = 9, c = 6, d = 1, v_{\mathrm{th}}^{\mathrm{E}} = -0.5, v_{\mathrm{th}}^{\mathrm{I}} = -2.5, \lambda^{\mathrm{E}} = 1, \lambda^{\mathrm{I}} = 1, f_{\mathrm{max}} = 1$, and $\gamma = 1$ with $\overline{S}^{\mathrm{E}}(0) = 0.5$ and $\overline{S}^{\mathrm{I}}(0) = 0.7$.

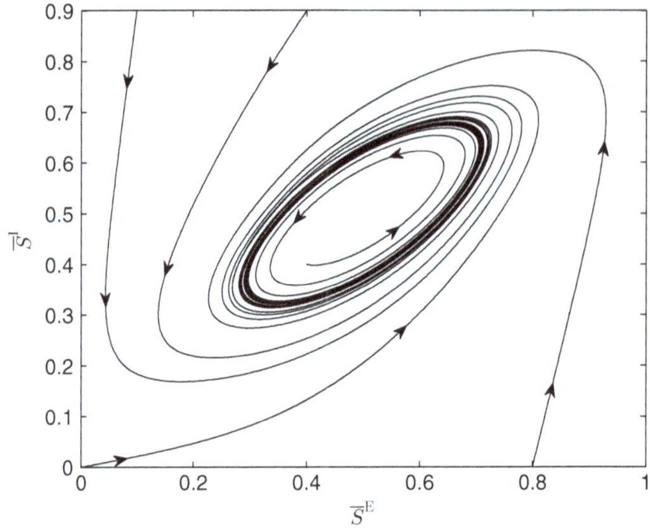

Fig. 15 Phase portrait of Eqs. (12) and (13) for $a = 10$, $b = 9$, $c = 6$, $d = 1$, $v_{\mathrm{th}}^{\mathrm{E}} = -0.5$, $v_{\mathrm{th}}^{\mathrm{I}} = -2.5$, $\lambda^{\mathrm{E}} = 1$, $\lambda^{\mathrm{I}} = 1$, $f_{\max} = 1$, and $\gamma = 1$.

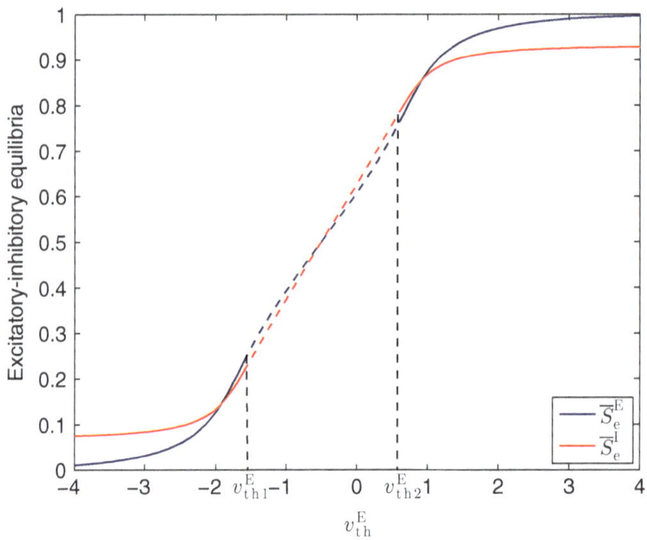

Fig. 16 Bifurcation diagram of Eqs. (12) and (13) for $a = 10$, $b = 9$, $c = 6$, $d = 1$, $v_{\mathrm{th}}^{\mathrm{I}} = -2.5$, $\lambda^{\mathrm{E}} = 1$, $\lambda^{\mathrm{I}} = 1$, $f_{\max} = 1$, and $\gamma = 1$ with $v_{\mathrm{th}}^{\mathrm{E}}$ as a bifurcation parameter. The solid line represents an asymptotically stable equilibrium point, whereas the dashed line represents an unstable equilibrium point. A supercritical Hopf bifurcation occurs at $v_{\mathrm{th}}^{\mathrm{E}} = -1.6$, where the single asymptotically stable equilibrium point becomes unstable and gives rise to a stable limit cycle. At $v_{\mathrm{th}}^{\mathrm{E}} = 0.6$ the unstable equilibrium point reverts to an asymptotically stable equilibrium point and the stable limit cycle disappears.

as $\nu_{\rm th}^{\rm E}$ increases; see Fig. 16. The effect of $\nu_{\rm th}^{\rm E}$ on the mean firing rates and mean synaptic drives of the excitatory and the inhibitory neurons is shown in Figs. 17 and 18, respectively. As $\nu_{\rm th}^{\rm E}$ increases, the mean firing rates and mean synaptic drives for the excitatory and the inhibitory neurons increase for the chosen parameter values.

The input voltage $\nu_{\rm th}^{\rm I}$ has a negative effect on the mean excitatory firing rate and synaptic drive. As can be seen from Fig. 19, when $\nu_{\rm th}^{\rm I}$ increases, the mean excitatory firing rate and synaptic drive decrease for the chosen parameter values.

Finally, we relate the effect of an anesthetic drug, corresponding to an increase in the inhibitory time constant $\lambda^{\rm I}$, to the mean excitatory and mean inhibitory synaptic drives. We fix the parameters to be as in the simulation shown in Fig. 14 and vary $\lambda^{\rm I}$, which corresponds to concentration variations in the anesthetic drug. As can be seen in Fig. 20, for a value of $\lambda^{\rm I}$ below $\lambda_1^{\rm I} = 0.85$, the two-class mean field model given by Eqs. (12) and (13) has exactly one asymptotically stable equilibrium point. As $\lambda^{\rm I}$ increases to $\lambda_1^{\rm I}$, the equilibrium point becomes unstable. At this point, a supercritical Hopf bifurcation occurs, giving rise to a stable limit cycle. As $\lambda^{\rm I}$ increases beyond $\lambda_2^{\rm I} = 2.3$, the unstable equilibrium point reverts to an asymptotically stable equilibrium point and the stable limit cycle disappears. The value of $\overline{S}_{\rm e}^{\rm E}$ decreases as $\lambda^{\rm I}$ increases and approaches zero as $\lambda^{\rm I}$ becomes large; see Fig. 20.

The mean excitatory synaptic drive $\overline{S}^{\rm E}(t)$, $t \geq 0$, also decreases as $\lambda^{\rm I}$ increases, and approaches zero with increasing $\lambda^{\rm I}$; see Fig. 21. The effect of an anesthetic drug can be explained by one observing that if $\lambda^{\rm I}$ is below $\lambda_1^{\rm I}$, corresponding to no drug effect, then the mean excitatory synaptic drive is at a high level, which corresponds to an awake state. The range of drug concentrations where we cannot predict with certainty whether the patient will respond or not respond to noxious stimuli (requiring a probabilistic model (1)) could certainly be explained by stochastic variation in $\nu_{\rm th}^{\rm E}$ and $\nu_{\rm th}^{\rm I}$; the postulated inputs from pain, sensorimotor, and proprioceptive sensors, or the reticular activating system.

However, an alternative explanation is that during the induction of anesthesia, the patient may or may not respond to a certain range of anesthetic drug concentration. This is captured in our model by the existence of limit cycle for a range of $\lambda^{\rm I}$ between $\lambda_1^{\rm I}$ and $\lambda_2^{\rm I}$. In this case the mean excitatory synaptic drive has a maximum and a minimum value as can be seen in Fig. 21. The probability of observing the response of

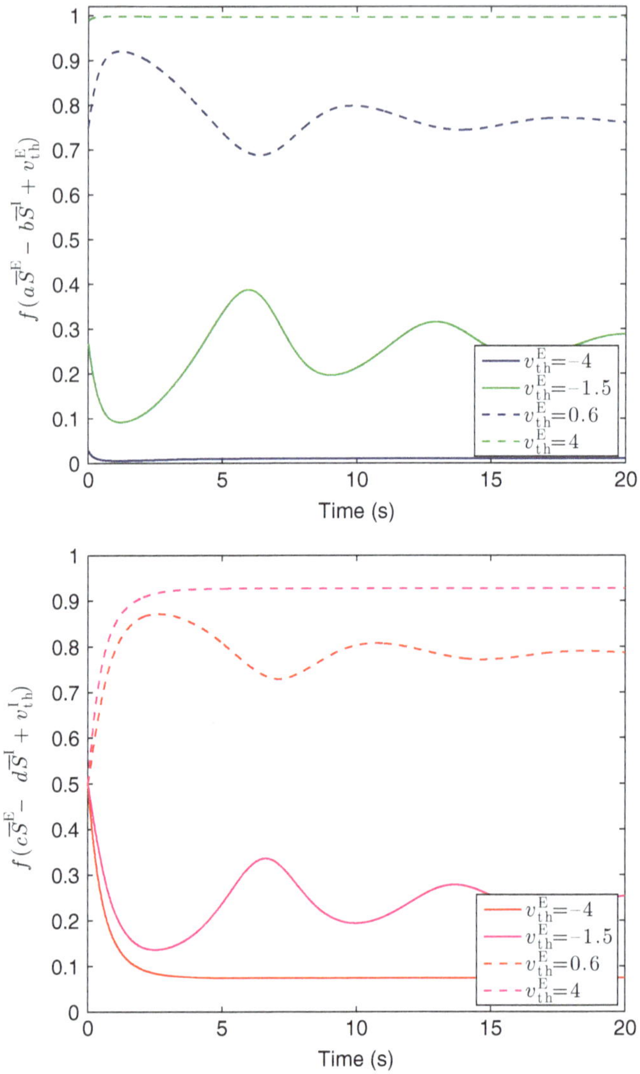

Fig. 17 Mean firing rates for $a = 10$, $b = 9$, $c = 6$, $d = 1$, $v_{\mathrm{th}}^{\mathrm{I}} = -2.5$, $\lambda^{\mathrm{E}} = 1$, $\lambda^{\mathrm{I}} = 1$, $f_{\max} = 1$, and $\gamma = 1$ with $\overline{S}^{\mathrm{E}}(0) = 0.5$ and $\overline{S}^{\mathrm{I}}(0) = 0.5$ and with different values of $v_{\mathrm{th}}^{\mathrm{E}}$.

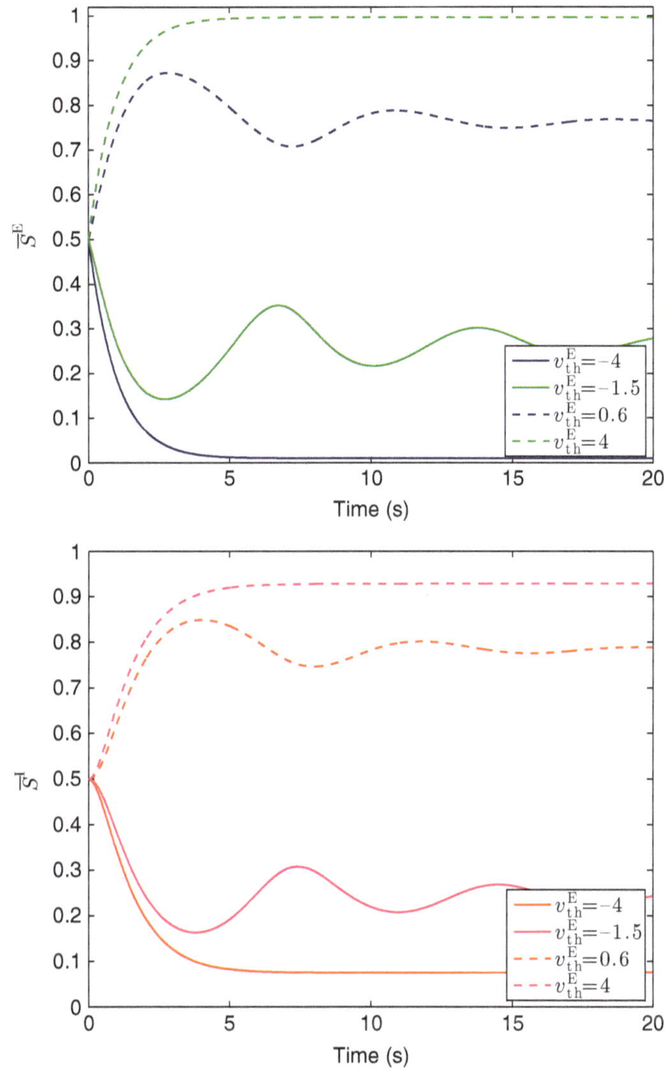

Fig. 18 Mean synaptic drives for $a = 10, b = 9, c = 6, d = 1, v_{th}^{I} = -2.5, \lambda^{E} = 1, \lambda^{I} = 1,$ $f_{max} = 1$, and $\gamma = 1$ with $\overline{S}^{E}(0) = 0.5$ and $\overline{S}^{I}(0) = 0.5$ and with different values of v_{th}^{E}.

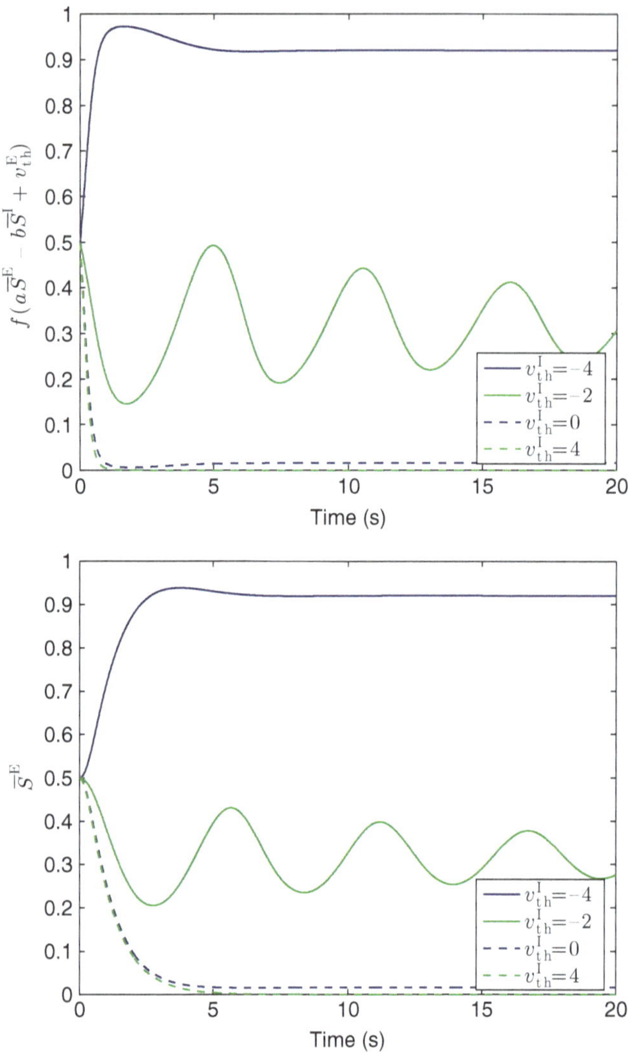

Fig. 19 Mean excitatory firing rates and mean excitatory synaptic drives for $a = 10$, $b = 9$, $c = 6$, $d = 1$, $v_{\mathrm{th}}^{\mathrm{E}} = -0.5$, $\lambda^{\mathrm{E}} = 1$, $\lambda^{\mathrm{I}} = 1$, $f_{\max} = 1$, and $\gamma = 1$ with $\overline{S}^{\mathrm{E}}(0) = 0.5$ and $\overline{S}^{\mathrm{I}}(0) = 0.5$ and with different values of $v_{\mathrm{th}}^{\mathrm{I}}$.

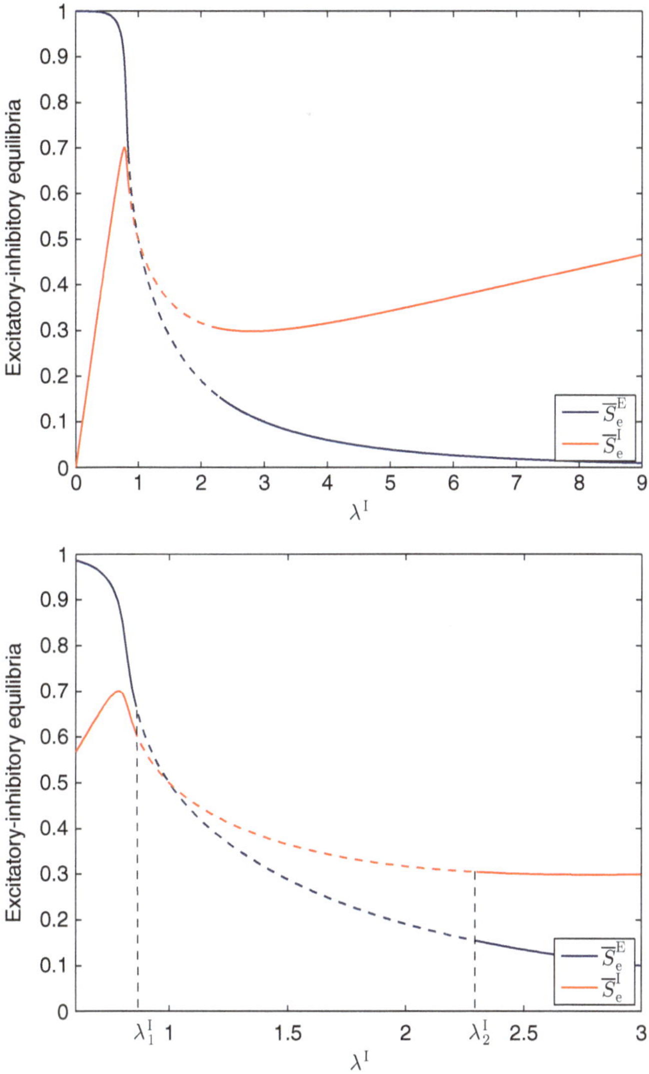

Fig. 20 Bifurcation diagram of Eqs. (12) and (13) for $a = 10$, $b = 9$, $c = 6$, $d = 1$, $v_{th}^E = -0.5$, $v_{th}^I = -2.5$, $\lambda^E = 1$, $f_{max} = 1$, and $\gamma = 1$ with λ^I as a bifurcation parameter. The solid line represents an asymptotically stable equilibrium point, whereas the dashed line represents an unstable equilibrium point. A supercritical Hopf bifurcation occurs at $\lambda_1^I = 0.85$, where the single asymptotically stable equilibrium point becomes unstable and gives rise to a stable limit cycle. At $\lambda_2^I = 2.3$ the unstable equilibrium point reverts to an asymptotically stable equilibrium point and the stable limit cycle disappears.

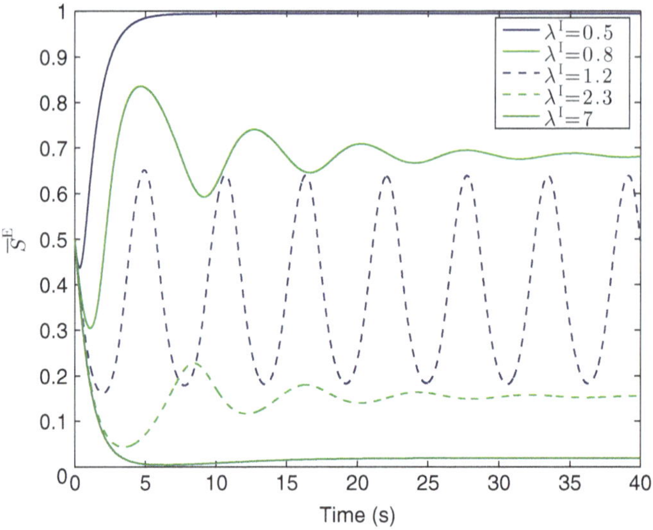

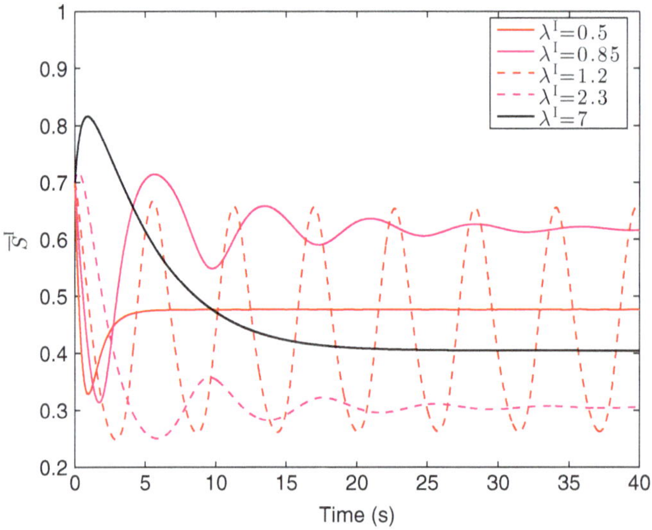

Fig. 21 Mean synaptic drives for $a = 10$, $b = 9$, $c = 6$, $d = 1$, $v_{th}^E = -0.5$, $v_{th}^I = -2.5$, $\lambda^E = 1$, $f_{max} = 1$, and $\gamma = 1$ with $\bar{S}^E(0) = 0.5$ and $\bar{S}^I(0) = 0.7$ and with different values of λ^I.

the patient depends on the time in the cycle at which the observation is made. The patient will respond if the observation occurs at the time when the mean excitatory synaptic drive is at its peak and will not respond if the mean excitatory synaptic drive is at its nadir. If the concentration of the anesthetic drug is further increased, corresponding to an increase in λ^{I}, then the mean excitatory synaptic drive approaches zero, as shown in Fig. 21. At this point the patient is deeply sedated and will not respond.

The mean excitatory synaptic drives for the parameters used in the different cases (ie, single stable equilibrium point, three equilibrium points, and five equilibrium points) are also shown for different values of λ^{I}. In all the cases, as λ^{I} increases, the mean excitatory synaptic drives decrease and approach zero with increasing λ^{I}; see Figs. 22–24.

7.2 Partial Synchronization of Synaptic Drive Dynamics

For our first example, we demonstrate partial synchronization for the model given by Eqs. (44) and (45) with three excitatory neurons $\mathrm{E}_1 - \mathrm{E}_3$ and three inhibitory neurons $\mathrm{I}_1 - \mathrm{I}_3$ as shown in Fig. 25. The neural connectivity matrix A is given by

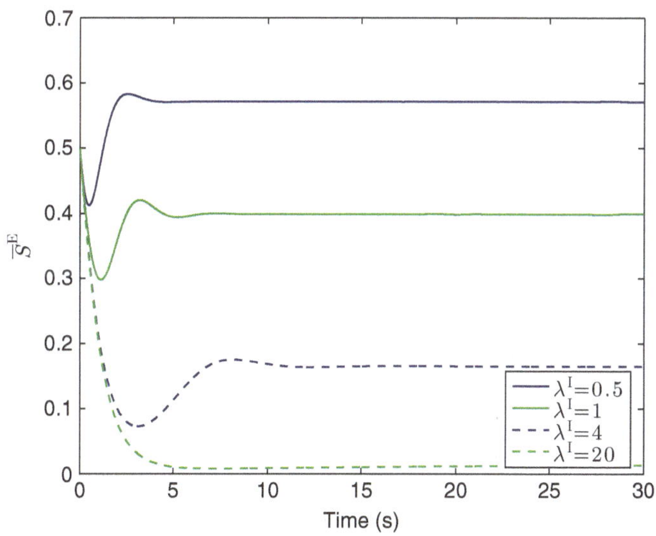

Fig. 22 Mean excitatory synaptic drives for $a = 3$, $b = 9$, $c = 6$, $d = 1$, $v_{\mathrm{th}}^{\mathrm{E}} = 1$, $v_{\mathrm{th}}^{\mathrm{I}} = -3$, $\lambda^{\mathrm{E}} = \lambda^{\mathrm{I}} = 1$, $f_{\mathrm{max}} = 1$, and $\gamma = 1$ with $\bar{S}^{\mathrm{E}}(0) = \bar{S}^{\mathrm{I}}(0) = 0.5$ and with different values of λ^{I}.

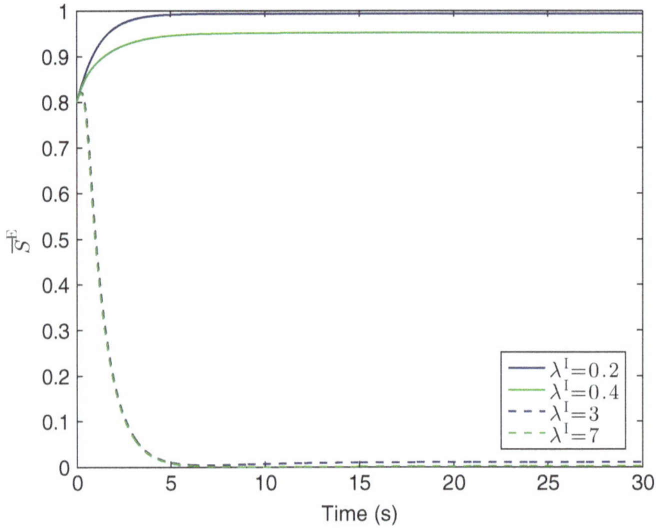

Fig. 23 Mean excitatory synaptic drives for $a = 10$, $b = 9$, $c = 6$, $d = 1$, $v_{\mathrm{th}}^{\mathrm{E}} = -3.2$, $v_{\mathrm{th}}^{\mathrm{I}} = -2.8$, $\lambda^{\mathrm{E}} = 1$, $\lambda^{\mathrm{I}} = 0.4$, $f_{\max} = 1$, and $\gamma = 1$ with $\bar{S}^{\mathrm{E}}(0) = 0.8$ and $\bar{S}^{\mathrm{I}}(0) = 0.2$ and with different values of λ^{I}.

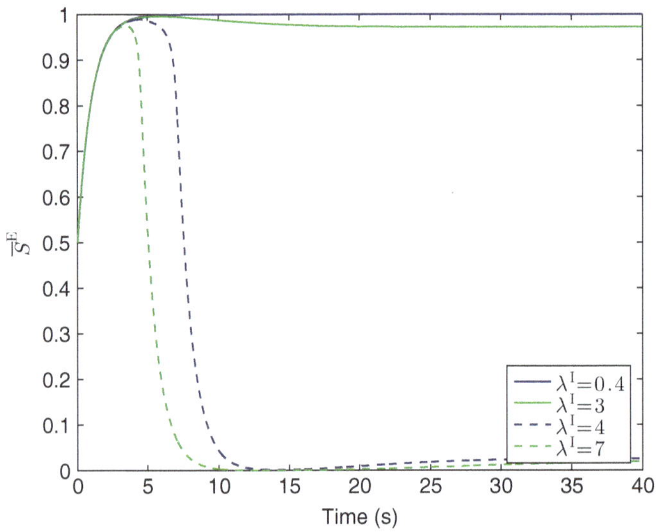

Fig. 24 Mean excitatory synaptic drives for $a = 20$, $b = 4$, $c = 13$, $d = 1$, $v_{\mathrm{th}}^{\mathrm{E}} = -4$, $v_{\mathrm{th}}^{\mathrm{I}} = -5$, $\lambda^{\mathrm{E}} = 1$, $\lambda^{\mathrm{I}} = 3$, $f_{\max} = 1$, and $\gamma = 1$ with $\bar{S}^{\mathrm{E}}(0) = \bar{S}^{\mathrm{I}}(0) = 0.5$ and with different values of λ^{I}.

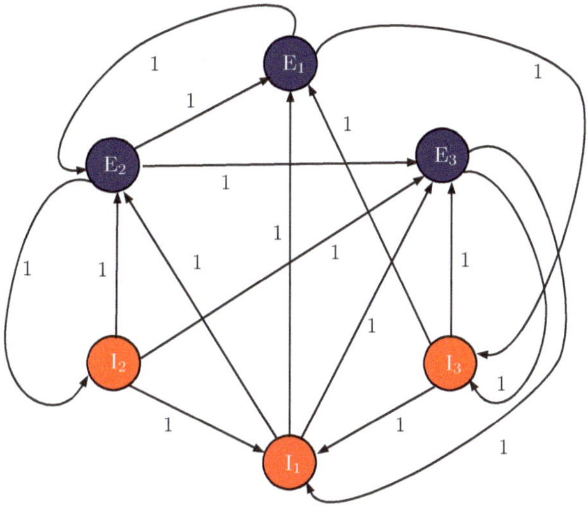

Fig. 25 A population of three excitatory neurons $E_1 - E_3$ and three inhibitory neurons $I_1 - I_3$. The synaptic weights shown on the arcs represent the coupling strength of the neural interconnection.

$$A = \begin{bmatrix} 0 & 1 & 0 & -1 & 0 & -1 \\ 1 & 0 & 0 & -1 & -1 & 0 \\ 0 & 1 & 0 & -1 & -1 & -1 \\ 0 & 0 & 1 & 0 & -1 & -1 \\ 0 & 1 & 0 & 0 & 0 & 0 \\ 1 & 0 & 1 & 0 & 0 & 0 \end{bmatrix},$$

which implies that the two inhibitory neurons I_2 and I_3 do not receive inhibitory inputs. Here, we assume that all the excitatory neurons have the same time constant $\lambda^{E} = 0.05$ s and all the inhibitory neurons have the same prolonged time constant $\lambda^{I} = 0.5$ s. Furthermore, we assume that $\tilde{\nu}_{\mathrm{th}}^{E} = [0.02,\ 0.02,\ 0.02]^{T}$, $\tilde{\nu}_{\mathrm{th}}^{I} = [0.02,\ 0.3,\ 0.5]^{T}$, and the activation functions $f_i(\cdot)$, $i = 1, 2, \ldots, 6$, are given by Eq. (10).

We use the MATLAB toolbox YALMIP for solving the linear matrix inequalities given by Eqs. (46) and (47). For the chosen values of A, λ^{E}, and λ^{I}, Eqs. (46) and (47) are satisfied with

$$P = \begin{bmatrix} 5.74 & 0 & 2.24 & -1.13 & -1.14 & -0.05 \\ 0 & 6.08 & 0 & -1.86 & -0.77 & -1.85 \\ 2.24 & 0 & 6.35 & -0.05 & -2.27 & -2.26 \\ -1.13 & -1.86 & -0.05 & 40.02 & 10.25 & 10.23 \\ -1.14 & -0.77 & -2.27 & 10.25 & 48.72 & 16.42 \\ -0.05 & -1.85 & -2.26 & 10.23 & 16.42 & 48.75 \end{bmatrix},$$

$$Q = \begin{bmatrix} 55.53 & 0.05 & -0.08 & -0.73 & -0.8 & -0.81 \\ 0.05 & 55.36 & 0.08 & -1.33 & -1.13 & -1.12 \\ -0.08 & 0.08 & 55.74 & -0.88 & -1.38 & -1.27 \\ -0.73 & -1.33 & -0.88 & 63.73 & 8.4 & 8.4 \\ -0.8 & -1.13 & -1.38 & 8.4 & 65.58 & 10.26 \\ -0.81 & -1.12 & -1.27 & 8.4 & 10.26 & 65.47 \end{bmatrix},$$

$$R = \mathrm{diag}[24.39 \;\; 24.55 \;\; 15.85 \;\; 49.42 \;\; 89.76 \;\; 91.31].$$

Hence the biological neural network given in Eqs. (44) and (45) is globally exponentially partially synchronized. As can be seen in Fig. 26, the synaptic drives of the excitatory neurons E_1–E_3 and the inhibitory neuron I_1 converge to zero, whereas the synaptic drives of the inhibitory neurons I_2 and I_3, which themselves do not receive inhibitory inputs, do not converge to zero.

Next we increase the number of neurons in the network to 12, with six excitatory neurons E_1–E_6 and six inhibitory neurons I_1–I_6. The connections between neurons are also increased as shown in the neural connectivity matrix A given by

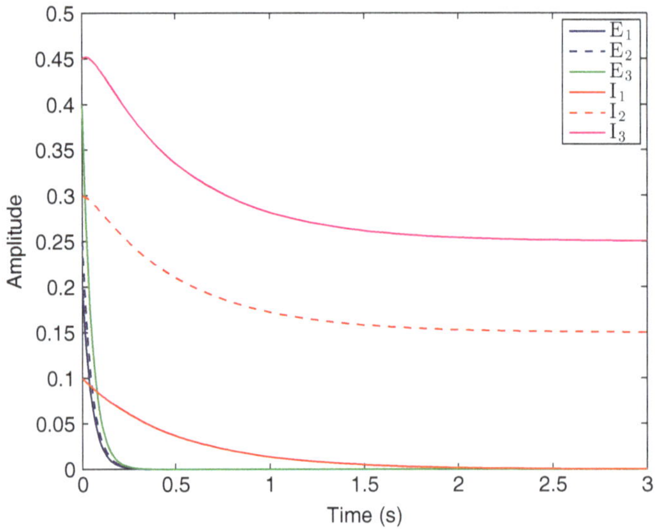

Fig. 26 Solutions to Eqs. (35) and (36) with initial conditions $S^E(0) = [0.2, 0.25, 0.4]^T$ and $S^I(0) = [0.1, 0.3, 0.45]^T$ for $\lambda^E = 0.05$ s and $\lambda^I = 0.5$ s. The synaptic drives of the excitatory neurons E_1–E_3 and the inhibitory neuron I_1 converge to zero, whereas the synaptic drives of the inhibitory neurons I_2 and I_3, which themselves do not receive inhibitory inputs, do not converge to zero.

$$A = \begin{bmatrix} 0 & 1 & 1 & 1 & 1 & 1 & -1 & 0 & -1 & -1 & -1 & 0 \\ 1 & 0 & 1 & 1 & 1 & 1 & 0 & -1 & -1 & -1 & 0 & -1 \\ 1 & 1 & 0 & 1 & 1 & 1 & -1 & 0 & -1 & 0 & -1 & -1 \\ 1 & 1 & 1 & 0 & 1 & 1 & 0 & -1 & 0 & -1 & 0 & -1 \\ 1 & 1 & 1 & 1 & 0 & 1 & -1 & 0 & -1 & 0 & -1 & -1 \\ 1 & 1 & 1 & 1 & 1 & 0 & 0 & -1 & 0 & -1 & 0 & -1 \\ 1 & 0 & 1 & 0 & 1 & 0 & 0 & -1 & -1 & 0 & -1 & 0 \\ 0 & 1 & 0 & 1 & 0 & 1 & -1 & 0 & -1 & 0 & -1 & 0 \\ 0 & 0 & 1 & 1 & 1 & 0 & -1 & -1 & 0 & -1 & -1 & 0 \\ 0 & 1 & 0 & 0 & 1 & 1 & 0 & 0 & 0 & 0 & 0 & 0 \\ 1 & 0 & 1 & 1 & 0 & 0 & 0 & 0 & 0 & 0 & 0 & 0 \\ 1 & 1 & 0 & 0 & 0 & 1 & 0 & 0 & 0 & 0 & 0 & 0 \end{bmatrix}, \quad (136)$$

which implies that the three inhibitory neurons I_4, I_5, and I_6 do not receive inhibitory inputs. We assume that all the excitatory neurons have the same time constant $\lambda^{\mathrm{E}} = 0.05$ s and all the inhibitory neurons have the same prolonged time constant $\lambda^{\mathrm{I}} = 0.35$ s. Furthermore, we assume that $\tilde{\nu}_{\mathrm{th}}^{\mathrm{E}} = [0.02,\ 0.02,\ 0.02,\ 0.02,\ 0.02,\ 0.02]^{\mathrm{T}}$, $\tilde{\nu}_{\mathrm{th}}^{\mathrm{I}} = [0.02,\ 0.02,\ 0.02,\ 0.1,\ 0.3,\ 0.5]^{\mathrm{T}}$, and the activation functions $f_i(\cdot)$, $i = 1, 2, \ldots, 12$, are given by Eq. (10).

The linear matrix inequalities given by Eqs. (46) and (47) are satisfied with

$$P = \begin{bmatrix} 0.24 & 0.08 & 0.11 & 0.11 & 0.15 & 0.08 & -0.07 & -0.24 & -0.06 \\ 0.08 & 0.2 & 0.07 & 0.11 & 0.08 & 0.11 & -0.16 & -0.08 & -0.12 \\ 0.11 & 0.07 & 0.19 & 0.1 & 0.11 & 0.07 & -0.06 & -0.17 & -0.09 \\ 0.11 & 0.11 & 0.1 & 0.23 & 0.11 & 0.11 & -0.16 & -0.13 & -0.15 \\ 0.15 & 0.08 & 0.11 & 0.11 & 0.24 & 0.08 & -0.07 & -0.24 & -0.06 \\ 0.08 & 0.11 & 0.07 & 0.11 & 0.08 & 0.2 & -0.16 & -0.07 & -0.15 \\ -0.07 & -0.16 & -0.06 & -0.16 & -0.07 & -0.16 & 1.13 & 0.03 & 0.49 \\ -0.24 & -0.08 & -0.17 & -0.13 & -0.24 & -0.07 & 0.03 & 1.42 & 0.13 \\ -0.06 & -0.12 & -0.09 & -0.15 & -0.06 & -0.15 & 0.49 & 0.13 & 1.07 \end{bmatrix},$$

$$Q = \begin{bmatrix} 1.45 & 0.08 & 0.09 & 0.1 & 0.11 & 0.07 & -0.13 & -0.31 & -0.14 \\ 0.08 & 1.41 & 0.06 & 0.09 & 0.08 & 0.07 & -0.18 & -0.16 & -0.18 \\ 0.09 & 0.06 & 1.41 & 0.08 & 0.09 & 0.06 & -0.12 & -0.23 & -0.13 \\ 0.1 & 0.09 & 0.08 & 1.44 & 0.1 & 0.08 & -0.2 & -0.22 & -0.2 \\ 0.11 & 0.08 & 0.09 & 0.1 & 1.45 & 0.07 & -0.13 & -0.31 & -0.14 \\ 0.07 & 0.07 & 0.06 & 0.08 & 0.07 & 1.41 & -0.19 & -0.15 & -0.19 \\ -0.13 & -0.18 & -0.12 & -0.2 & -0.13 & -0.19 & 1.9 & 0.15 & 0.47 \\ -0.31 & -0.16 & -0.23 & -0.22 & -0.31 & -0.15 & 0.15 & 2.22 & 0.21 \\ -0.14 & -0.18 & -0.13 & -0.2 & -0.14 & -0.19 & 0.47 & 0.21 & 1.86 \end{bmatrix},$$

$$R = \mathrm{diag}\,[0.63\ \ 0.63\ \ 93\ \ 0.84\ \ 0.63\ \ 0.80\ \ 1.51\ \ 1.36\ \ 1.39].$$

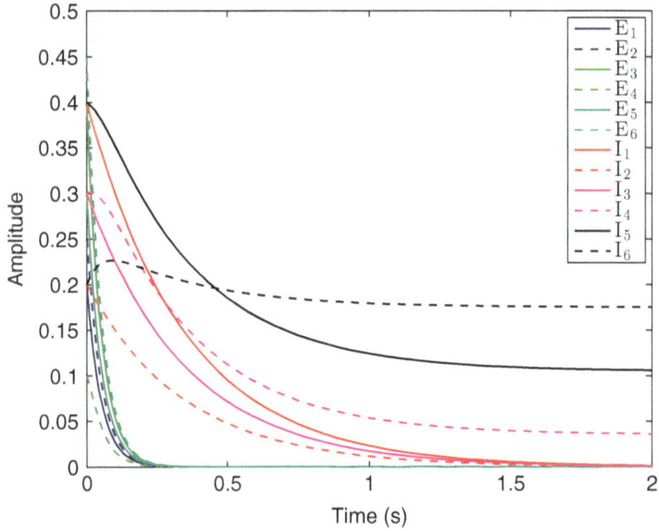

Fig. 27 Solutions to Eqs. (35) and (36) with initial conditions $S^{E}(0) = [0.2,\ 0.25,\ 0.4,\ 0.1,\ 0.3,\ 0.45]^{T}$ and $S^{I}(0) = [0.4,\ 0.2,\ 0.3,\ 0.3,\ 0.4,\ 0.2]^{T}$ for $\lambda^{E} = 0.05$ s and $\lambda^{I} = 0.35$ s. The synaptic drives of the excitatory neurons E_1–E_6 and three of the inhibitory neurons—I_1, I_2, and I_3—converge to zero, whereas the synaptic drives of the inhibitory neurons I_4, I_5, and I_6, which themselves do not receive inhibitory inputs, do not converge to zero.

Hence the biological neural network given in Eqs. (44) and (45) is globally exponentially partially synchronized. As can be seen in Fig. 27, the synaptic drives of the excitatory neurons E_1–E_6 and three of the inhibitory neurons—I_1, I_2, and I_3—converge to zero, whereas the synaptic drives of the inhibitory neurons I_4, I_5, and I_6, which themselves do not receive inhibitory inputs, do not converge to zero.

Next we apply Theorem 4 to the connectivity matrix given by Eq. (136). We assume that $\tilde{v}^{E}_{th} = [0.15,\ 0.15,\ 0.15,\ 0.15,\ 0.15,\ 0.15]^{T}$, $\tilde{v}^{I}_{th} = [0.15,\ 0.15,\ 0.15,\ 0.15,\ 0.15,\ 0.15]^{T}$, $L^{E} = \mathrm{diag}\,[2.1\ \ 1.9\ \ 1.85\ \ 2.15\ \ 2.05\ \ 2.5]$, $L^{I} = \mathrm{diag}\,[1.5\ \ 1.75\ \ 2.4\ \ 2.2\ \ 2.25\ \ 2]$, and the activation functions $f_i(\cdot)$, $i = 1, 2, \ldots, 12$, are given by Eq. (62) with $f_{\max} = 0.5$. As can be seen from Fig. 28, the synaptic drives of the excitatory neurons E_1—E_6 do not converge to zero. However, when the time constants of the inhibitory neurons are increased to $L^{I} = \mathrm{diag}\,[0.1\ \ 0.1\ \ 0.1\ \ 0.1\ \ 0.1\ \ 0.1]$, the synaptic drives of the excitatory neurons E_1–E_6 converge to zero; see Fig. 29,

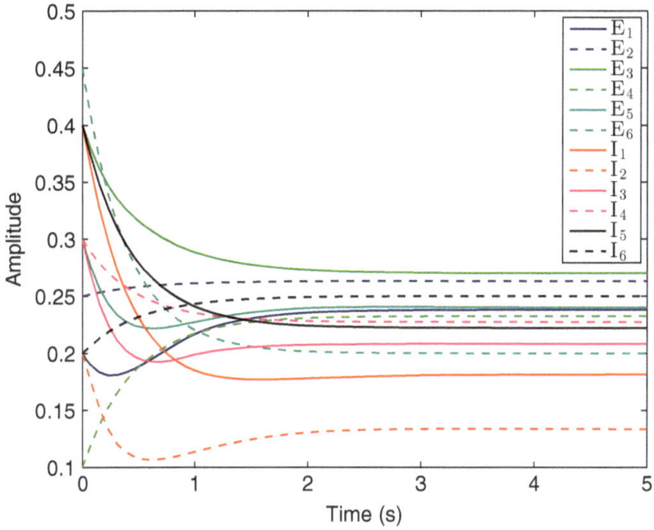

Fig. 28 Solutions to Eqs. (35) and (36) with initial conditions $S^{E}(0) = [0.2,\ 0.25,\ 0.4,\ 0.1,\ 0.3,\ 0.45]^{T}$ and $S^{I}(0) = [0.4,\ 0.2,\ 0.3,\ 0.3,\ 0.4,\ 0.2]^{T}$ for $L^{E} = \text{diag}[2.1\ 1.9\ 1.85\ 2.15\ 2.05\ 2.5]$, $L^{I} = \text{diag}\,[1.5\ 1.75\ 2.4\ 2.2\ 2.25\ 2]$, and $f_{\max} = 0.5$. The synaptic drives of the excitatory neurons E_1–E_6 do not converge to zero.

Next we present partial synchronization of the model presented in Eqs. (89) and (90) with four excitatory neurons E_1–E_4 and four inhibitory neurons I_1–I_4. The neural connectivity matrix A is given by

$$A = \begin{bmatrix} 0 & 1 & 1 & 1 & -1 & 0 & -1 & -1 \\ 1 & 0 & 1 & 1 & -1 & -1 & 0 & -1 \\ 1 & 1 & 0 & 1 & 0 & 0 & -1 & -1 \\ 1 & 1 & 1 & 0 & 0 & -1 & -1 & -1 \\ 1 & 1 & 1 & 1 & 0 & -1 & 0 & 0 \\ 1 & 1 & 1 & 1 & -1 & 0 & 0 & 0 \\ 1 & 0 & 1 & 0 & 0 & 0 & 0 & 0 \\ 0 & 1 & 0 & 1 & 0 & 0 & 0 & 0 \end{bmatrix}, \qquad (137)$$

which implies that the inhibitory neurons I_3 and I_4 do not receive inhibitory inputs. We assume that all the excitatory neurons have the time constant $\lambda^{E} = 0.05$ s and all the inhibitory neurons have the same prolonged time constant $\lambda^{I} = 0.3$ s. Furthermore, we assume that $\tilde{v}_{\text{th}}^{E} = [0.02,\ 0.02,\ 0.02,\ 0.02]^{T}$, $\tilde{v}_{\text{th}}^{I} = [0.02,\ 0.02,\ 0.4,\ 0.5]^{T}$, and the activation functions $f_i(\cdot)$, $i = 1, 2, \ldots, 8$, are given by Eq. (10).

The linear matrix inequalities given by Eqs. (98) and (99) are satisfied with

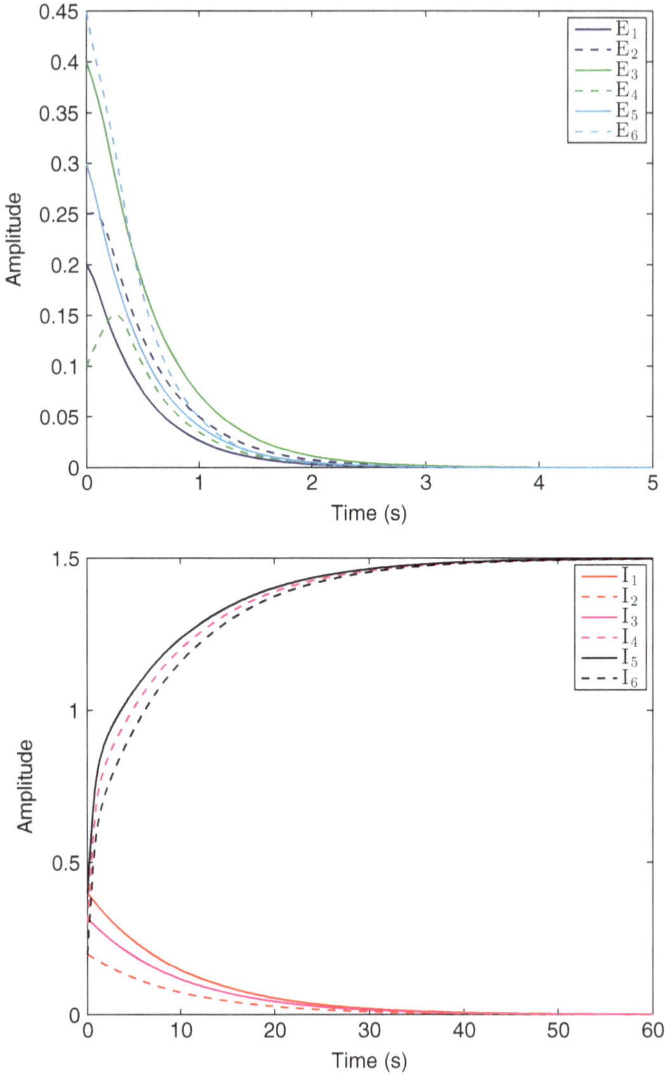

Fig. 29 Solutions to Eqs. (35) and (36) with initial conditions $S^{E}(0) = [0.2, 0.25, 0.4, 0.1, 0.3, 0.45]^{T}$ and $S^{I}(0) = [0.4, 0.2, 0.3, 0.3, 0.4, 0.2]^{T}$ for $L^{E} = \text{diag}\,[2.1\ 1.9\ 1.85\ 2.15\ 2.05\ 2.5]$, $L^{I} = \text{diag}\,[0.1\ 0.1\ 0.1\ 0.1\ 0.1\ 0.1]$, and $f_{max} = 0.5$. The synaptic drives of the excitatory neurons E_1–E_6 converge to zero.

$$\hat{P} = \begin{bmatrix} 0.2 & 0.04 & 0.04 & 0.04 & 0.03 & 0.03 & 0.11 & 0.06 & 0.06 & 0.06 & 0.03 & 0.03 \\ 0.04 & 0.2 & 0.04 & 0.04 & 0.03 & 0.03 & 0.07 & 0.11 & 0.07 & 0.07 & 0.03 & 0.03 \\ 0.04 & 0.04 & 0.2 & 0.04 & 0.03 & 0.03 & 0.06 & 0.06 & 0.11 & 0.06 & 0.03 & 0.03 \\ 0.04 & 0.04 & 0.04 & 0.2 & 0.03 & 0.03 & 0.06 & 0.06 & 0.06 & 0.11 & 0.03 & 0.03 \\ 0.03 & 0.03 & 0.03 & 0.03 & 4.8 & 0.73 & 0.04 & 0.04 & 0.04 & 0.04 & 2.97 & 1.62 \\ 0.03 & 0.03 & 0.03 & 0.03 & 0.73 & 4.8 & 0.04 & 0.04 & 0.04 & 0.04 & 1.62 & 2.97 \\ 0.11 & 0.07 & 0.06 & 0.06 & 0.04 & 0.04 & 0.19 & 0.05 & 0.05 & 0.05 & 0.04 & 0.04 \\ 0.06 & 0.11 & 0.06 & 0.06 & 0.04 & 0.04 & 0.05 & 0.19 & 0.05 & 0.05 & 0.04 & 0.04 \\ 0.06 & 0.07 & 0.11 & 0.06 & 0.04 & 0.04 & 0.05 & 0.05 & 0.19 & 0.05 & 0.04 & 0.04 \\ 0.06 & 0.07 & 0.06 & 0.11 & 0.04 & 0.04 & 0.05 & 0.05 & 0.05 & 0.19 & 0.04 & 0.04 \\ 0.03 & 0.03 & 0.03 & 0.03 & 2.97 & 1.62 & 0.04 & 0.04 & 0.04 & 0.04 & 3.29 & 1.35 \\ 0.03 & 0.03 & 0.03 & 0.03 & 1.62 & 2.97 & 0.04 & 0.04 & 0.04 & 0.04 & 1.35 & 3.29 \end{bmatrix},$$

$$\hat{Q} = \begin{bmatrix} 16.9 & 1.1 & 1 & 1 & 0.5 & 1.9 & -2.8 & -4.3 & -4.2 & -4.2 & -6.4 & -6.3 \\ 1.1 & 16.8 & 1 & 1.1 & 0.7 & 0.7 & -4.1 & -2.7 & -4.1 & -4.1 & -6.3 & -6.3 \\ 1 & 1 & 16.8 & 1 & 1.8 & 1.8 & -4.4 & -4.4 & -2.9 & -4.4 & -6.4 & -6.4 \\ 1 & 1.1 & 1 & 16.9 & 1.9 & 0.5 & -4.2 & -4.3 & -4.2 & -2.8 & -6.3 & -6.4 \\ 0.5 & 0.7 & 1.8 & 1.9 & 14.3 & 3.3 & -6.1 & -6.1 & -6.1 & -6.1 & 4.5 & 1.9 \\ 1.9 & 0.7 & 1.8 & 0.5 & 3.3 & 14.3 & -6.1 & -6.1 & -6.1 & -6.1 & 1.9 & 4.5 \\ -2.8 & -4.1 & -4.4 & -4.2 & -6.1 & -6.1 & 34.5 & 11.7 & 11.8 & 11.6 & -7.1 & -7.2 \\ -4.3 & -2.7 & -4.4 & -4.3 & -6.1 & -6.1 & 11.7 & 34.6 & 11.8 & 11.7 & -7.1 & -7.1 \\ -4.2 & -4.1 & -2.9 & -4.2 & -6.1 & -6.1 & 11.8 & 11.8 & 34.7 & 11.8 & -7.2 & -7.2 \\ -4.2 & -4.1 & -4.4 & -2.8 & -6.1 & -6.1 & 11.6 & 11.7 & 11.8 & 34.5 & -7.2 & -7.1 \\ -6.4 & -6.3 & -6.4 & -6.3 & 4.5 & 1.9 & -7.1 & -7.1 & -7.2 & -7.2 & 46.5 & 18.6 \\ -6.3 & -6.3 & -6.4 & -6.4 & 1.9 & 4.5 & -7.2 & -7.1 & -7.2 & -7.1 & 18.6 & 46.5 \end{bmatrix},$$

$$\hat{R} = \mathrm{diag}\,[19.8 \;\; 19.8 \;\; 19.8 \;\; 19.8 \;\; 22.73 \;\; 22.73 \;\; 3.55 \;\; 3.24 \;\; 3.34 \;\; 3.55 \;\; 3.87 \;\; 3.87].$$

Hence the biological neural network given in Eqs. (89) and (90) is globally exponentially partially synchronized. As can be seen in Fig. 30, the synaptic drives of the excitatory neurons E_1–E_4 and two of the inhibitory neurons—I_1 and I_2–converge to zero, whereas the synaptic drives of the inhibitory neurons I_3 and I_4 do not converge to zero.

Next we apply Theorem 6 to the connectivity matrix given by Eq. (137). We assume that $\tilde{v}_{\mathrm{th}}^{\mathrm{E}} = [0.12\ 0.12,\ 0.12,\ 0.12]^{\mathrm{T}}$, $\tilde{v}_{\mathrm{th}}^{\mathrm{I}} = [0.12,\ 0.12,\ 0.12,\ 0.12]^{\mathrm{T}}$, $L^{\mathrm{E}} = \mathrm{diag}\ [1 \quad 0.9 \quad 0.85 \quad 0.8]$, $L^{\mathrm{I}} = \mathrm{diag}\ [1.1 \quad 1.2 \quad 1.3 \quad 1.25]$, and the activation functions $f_i(\cdot)$, $i = 1, 2, \ldots, 8$, are given by Eq. (62) with $f_{\max} = 0.5$. As can be seen from Fig. 31, the synaptic drives of the excitatory neurons E_1–E_4 do not converge to zero. Once again, however, as we increase the time constants of the inhibitory neurons to $L^{\mathrm{I}} = \mathrm{diag}\ [0.2 \quad 0.2 \quad 0.2 \quad 0.2]$, the synaptic drives of the excitatory neurons E_1—E_4 converge to zero; see Fig. 32. It can also be seen from Figs. 30 and 32 that the synaptic drives of the excitatory neurons E_1—E_4 increase initially before converging to zero. This potentially captures the drug biphasic response, where there is a boost in brain activity before hypnotic induction.

Finally, we further explore the drug biphasic phenomena using the second-order mean excitatory and mean inhibitory synaptic drive model given by Eqs. (134) and (135) with the activation function $f(\cdot)$ given by Eq. (11). Once again we observe that there is an initial boost in the mean excitatory synaptic drive just before convergence to zero; see Fig. 33.

8 CONCLUSION

During the past decade there has been a remarkable explosion of our understanding of how anesthetic agents affect the properties of neurons. However, despite this advance, we still do not understand how molecular mechanisms translate into the induction of general anesthesia at the macroscopic level. In particular, there has been little focus on how the molecular properties of anesthetic agents lead to the observed macroscopic property that defines the anesthetic state—that is, lack of responsiveness to noxious stimuli.

In this chapter we used a synaptic drive model to derive a mechanistic neural field theory that can be employed for understanding the effects of anesthetic agents. We showed that even a simplified two-state mean field theory exhibits surprisingly varied behavior, including multiple equilibria (both stable and unstable) and limit cycles, as the system

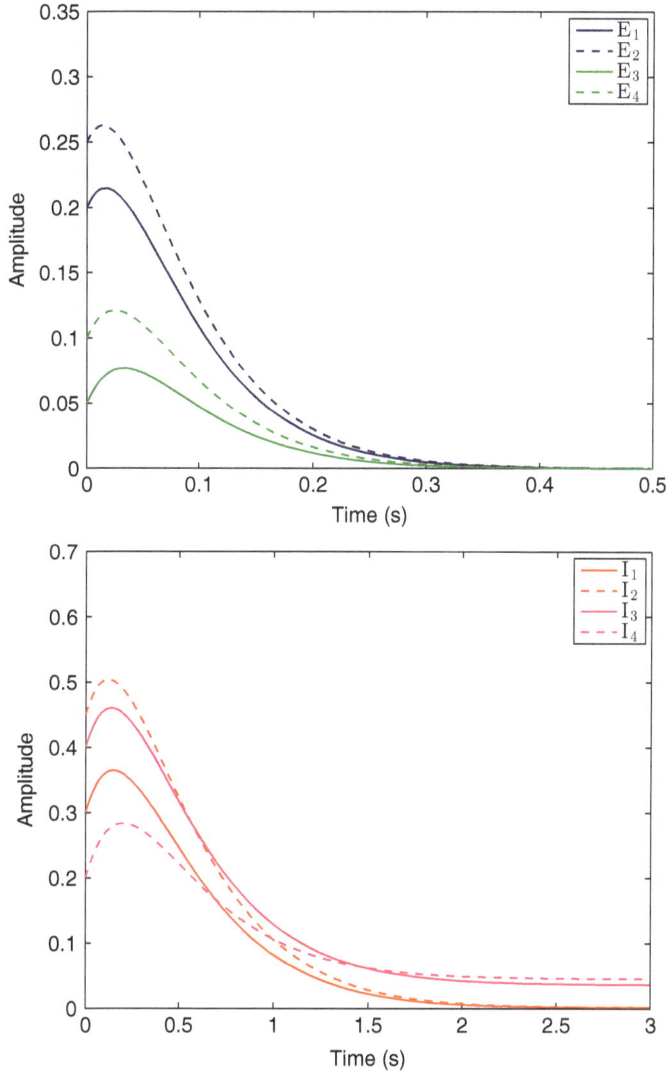

Fig. 30 Solutions to Eqs. (82) and (83) with initial conditions $S^{\mathrm{I}}(0) = [0.2,\ 0.25,\ 0.05,\ 0.1]^{\mathrm{T}}$, $S^{\mathrm{I}}(0) = [0.3,\ 0.45,\ 0.4,\ 0.2]^{\mathrm{T}}$, $\dot{S}^{\mathrm{E}}(0) = [2,\ 2,\ 2,\ 2]^{\mathrm{T}}$, and $\dot{S}^{\mathrm{I}}(0) = [1,\ 1,\ 1,\ 1]^{\mathrm{T}}$ for $\lambda^{\mathrm{E}} = 0.05$ s and $\lambda^{\mathrm{I}} = 0.3$ s. The synaptic drives of the excitatory neurons E_1–E_4 and two of the inhibitory neurons—I_1 and I_2—converge to zero, whereas the synaptic drives of the inhibitory neurons I_3 and I_4, which themselves do not receive inhibitory inputs, do not converge to zero.

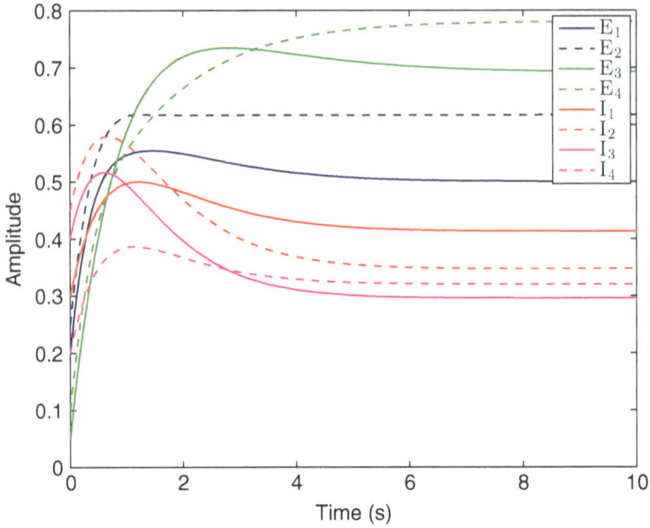

Fig. 31 Solutions to Eqs. (82) and (83) with initial conditions $S^{I}(0) = [0.2, 0.25, 0.05, 0.1]^{T}$, $S^{I}(0) = [0.3, 0.45, 0.4, 0.2]^{T}$, $\dot{S}^{E}(0) = [0.5, 0.5, 0.5, 0.5]^{T}$, and $\dot{S}^{I}(0) = [0.5, 0.5, 0.5, 0.5]^{T}$ for $L^{E} = \mathrm{diag}\,[1\ \ 0.9\ \ 0.85\ \ 0.8]$, $L^{I} = \mathrm{diag}\,[1.1\ \ 1.2\ \ 1.3\ \ 1.25]$, and $f_{\max} = 0.5$. The synaptic drives of the excitatory neurons E_1—E_4 do not converge to zero.

parameters are changed. We then extended our analysis without mean field assumptions by postulating the existence of a subset of inhibitory neurons that themselves do not receive inhibitory input. Using this assumption, we derived sufficient conditions for partial synchronization and showed that as the time constant of the inhibitory postsynaptic potential increases (an effect of anesthetic agents at the molecular level), partial synchronization occurs, with a subset of neurons converging to zero activity—a condition that can be identified with the anesthetic state. We further extended this analysis to consider a nonmonotonic postsynaptic potential as a possible model for the biphasic anesthetic response.

An assumption that is essential to a key finding of this chapter is the existence of a subset of inhibitory neurons that themselves do not receive inhibitory input. We are unaware of any experimental verification of such a subset. In the absence of such data, we view this assumption as defining a sufficient condition for partial synchronization and the anesthetic state. The proposed dynamical system framework can foster the development of new frameworks that can allow us to interpret experimental and clinical results, connect biophysical findings to psychophysical phenomena, explore

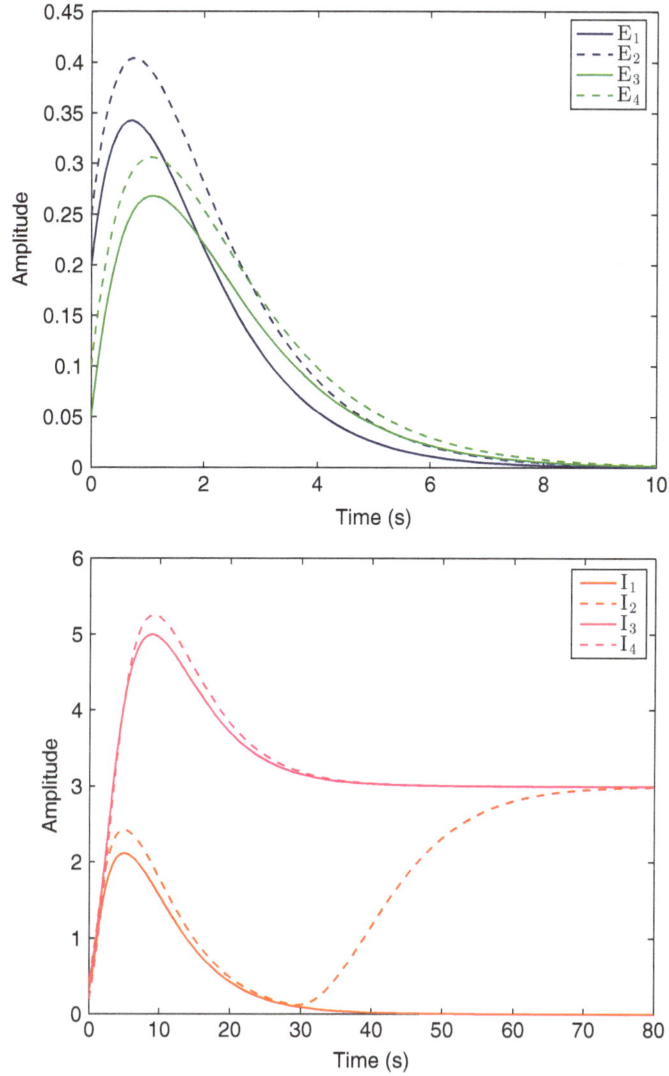

Fig. 32 Solutions to Eqs. (82) and (83) with initial conditions $S^{\mathrm{I}}(0) = [0.2,\ 0.25,\ 0.05,\ 0.1]^{\mathrm{T}}$, $S^{\mathrm{I}}(0) = [0.3,\ 0.45,\ 0.4,\ 0.2]^{\mathrm{T}}$, $\dot{S}^{\mathrm{E}}(0) = [0.5,\ 0.5,\ 0.5,\ 0.5]^{\mathrm{T}}$, and $\dot{S}^{\mathrm{I}}(0) = [0.5,\ 0.5,\ 0.5,\ 0.5]^{\mathrm{T}}$ for $L^{\mathrm{E}} = \mathrm{diag}\,[1\ \ 0.9\ \ 0.85\ \ 0.8]$, $L^{\mathrm{I}} = \mathrm{diag}\,[0.2\ \ 0.2\ \ 0.2\ \ 0.2]$, and $f_{\max} = 0.5$. The synaptic drives of the excitatory neurons E_1—E_4 converge to zero.

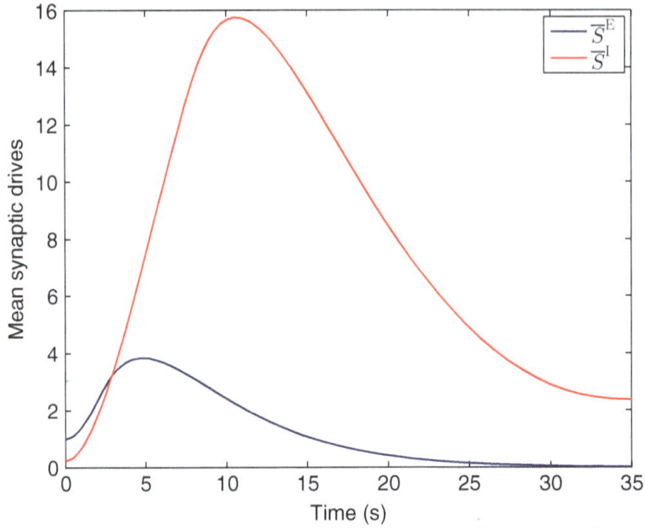

Fig. 33 Solutions to Eqs. (134) and (135) for $n_E\overline{A}^{EE} = 10, n_I\overline{A}^{II} = 12, n_E\overline{A}^{IE} = 4, n_I\overline{A}^{II} = 1$, $v^E_{th} = 0, v^I_{th} = 0, \lambda^E = 4$ s, $\lambda^I = 6$ s, $f_{max} = 1$, and $\gamma = 1$ with initial conditions $\overline{S}^E(0) = 1$, $\overline{S}^I(0) = 0.5$, and $\dot{\overline{S}}^E(0) = \dot{\overline{S}}^E(0) = 0$.

new hypotheses based on the cognitive neuroscience of consciousness and develop new assertions, and improve the reliability of general anesthesia. Dynamical systems theory is ideally suited for rigorously describing the behavior of large-scale networks of neurons and can potentially establish a scientific basis for new metrics of anesthetic depth by making the assessment of consciousness a mechanistically grounded tool.

ACKNOWLEDGMENTS

This publication was made possible by the Air Force Office of Scientific Research under grant FA9550-16-1-0100 and a National Priorities Research Program grant no. 4-187-2-060 from the Qatar National Research Fund (a member of the Qatar Foundation).

REFERENCES

[1] J.M. Bailey, M. Haddad, Drug dosing control in clinical pharmacology: paradigms, benefits, and challenges, IEEE Control Syst. Mag. 25 (2) (2005) 35–51.

[2] J. Vuyk, T. Lim, F. Engbers, A. Burm, A. Vletter, J. Bovill, Pharmacodynamics of alfentanil as a supplement to propofol or nitrous oxide for lower abdominal surgery in female patients, Anesthesiology 78 (6) (1993) 1036–1045.

[3] A.V. Hill, The possible effects of the aggregation of the molecules of haemoglobin on its dissociation curves, J. Physiol. 40 (1910) 4–7.

[4] E. Eger, D.M. Fisher, J.P. Dilger, J.M. Sonner, A. Evers, N.P. Franks, R.A. Harris, J.J. Kendig, W.R. Lieb, T. Yamakura, Relevant concentrations of inhaled anesthetics for in vitro studies of anesthetic mechanisms, Anesthesiology 94 (5) (2001) 915–921.
[5] J. Vuyk, T. Lim, F. Engbers, A. Burm, A.A. Vletter, J.G. Bovill, The pharmacodynamic interaction of propofol and alfentanil during lower abdominal surgery in women, Anesthesiology 83 (1) (1995) 8–22.
[6] A. Kitamura, W. Marszalec, J.Z. Yeh, T. Narahashi, Effects of halothane and propofol on excitatory and inhibitory synaptic transmission in rat cortical neurons, J. Pharmacol. Exp. Ther. 304 (1) (2002) 162–171.
[7] A. Hutt, A. Longtin, Effects of the anesthetic agent propofol on neural populations, Cogn. Neurodyn. 4 (1) (2009) 37–59.
[8] Q. Hui, W.M. Haddad, J.M. Bailey, Multistability, bifurcations, and biological neural networks: a synaptic drive firing model for cerebral cortex transition in the induction of general anesthesia, Nonlinear Anal. Hybrid Syst. 5 (3) (2011) 554–572.
[9] Q. Hui, W.M. Haddad, J.M. Bailey, T. Hayakawa, A stochastic mean field model for an excitatory and inhibitory synaptic drive cortical neuronal network, IEEE Trans. Neural Netw. Learn. Syst. 25 (4) (2014) 751–763.
[10] W.M. Haddad, Q. Hui, J.M. Bailey, Human brain networks: spiking neuron models, multistability, synchronization, thermodynamics, maximum entropy production, and anesthetic cascade mechanisms, Entropy 16 (7) (2014) 3939–4003.
[11] G.B. Ermentrout, D.H. Terman, Mathematical Foundations of Neuroscience, Springer-Verlag, New York, NY, 2010.
[12] P. Dayan, L.F. Abbott, Theoretical Neuroscience, MIT Press, Cambridge, MA, 2005.
[13] B. Gutkin, D. Pinto, B. Ermentrout, Mathematical neuroscience: from neurons to circuits to systems, J. Physiol. Paris 97 (2) (2003) 209–219.
[14] W. Gerstner, W.M. Kistler, Spiking Neuron Models: Single Neurons, Populations, Plasticity, Cambridge University Press, Cambridge, 2002.
[15] T. Bullock, M. McClune, J. Achimowicz, V. Iragui-Madoz, R. Duckrow, S. Spencer, Temporal fluctuations in coherence of brain waves, Proc. Natl. Acad. Sci. U. S. A. 92 (25) (1995) 11568–11572.
[16] G. Buzsaki, Rhythms of the Brain, Oxford University Press, Oxford, 2006.
[17] J.H. Tinker, F.W. Sharbrough, J.D. Michenfelder, Anterior shift of the dominant EEG rhythm during anesthesia in the Java monkey: correlation with anesthetic potency, Anesthesiology 46 (4) (1977) 252–259.
[18] G. Deco, V.K. Jirsa, P.A. Robinson, M. Breakspear, K. Friston, The dynamic brain: from spiking neurons to neural masses and cortical fields, PLoS Comput. Biol. 4 (8) (2008) e1000092.
[19] M.A. Buice, J.D. Cowan, Field-theoretic approach to fluctuation effects in neural networks, Phys. Rev. E 75 (5) (2007) 051919.
[20] E.T. Jaynes, Information theory and statistical mechanics, Phys. Rev. 106 (4) (1957) 620–630.
[21] E.T. Jaynes, Prior probabilities, IEEE Trans. Syst. Sci. Cybern. 4 (3) (1968) 227–241.
[22] R.H. Lyon, R. Lyon, Statistical Energy Analysis of Dynamical Systems: Theory and Applications, MIT Press, Cambridge, MA, 1975.
[23] R.D. Rosenkrantz, Papers on Probability, Statistics and Statistical Physics, Reidel, Boston, 1983.
[24] K. Kuizenga, J. Wierda, C. Kalkman, Biphasic EEG changes in relation to loss of consciousness during induction with thiopental, propofol, etomidate, midazolam or sevoflurane, Br. J. Anaesth. 86 (3) (2001) 354–360.
[25] H.R. Wilson, J.D. Cowan, Excitatory and inhibitory interactions in localized populations of model neurons, Biophys. J. 12 (1) (1972) 1–24.

[26] W.M. Haddad, V. Chellaboina, Nonlinear Dynamical Systems and Control: A Lyapunov-Based Approach, Princeton University Press, Princeton, NJ, 2008.
[27] C.M. Gray, Synchronous oscillations in neuronal systems: mechanisms and functions, J. Comput. Neurosci. 1 (1–2) (1994) 11–38.
[28] I.J. Rampil, A primer for EEG signal processing in anesthesia, Anesthesiology 89 (4) (1998) 980–1002.
[29] S. Ching, A. Cimenser, P.L. Purdon, E.N. Brown, N.J. Kopell, Thalamocortical model for a propofol-induced α-rhythm associated with loss of consciousness, Proc. Natl. Acad. Sci. U. S. A. 107 (52) (2010) 22665–22670.
[30] E.R. John, L.S. Prichep, The anesthetic cascade: a theory of how anesthesia suppresses consciousness, Anesthesiology 102 (2) (2005) 447–471.
[31] L.J. Voss, J.W. Sleigh, J.P. Barnard, H.E. Kirsch, The howling cortex: seizures and general anesthetic drugs, Anesth. Analg. 107 (5) (2008) 1689–1703.

SECTION 3

Adaptive Learning Algorithms

CHAPTER FIVE

Optimal Tracking Control of Uncertain Systems: On-Policy and Off-Policy Reinforcement Learning Approaches

B. Kiumarsi, H. Modares, F.L. Lewis
University of Texas at Arlington Research Institute, Fort Worth, TX, United States

1 INTRODUCTION

Although stability of the controlled system is the only concern of most nonlinear control design methods, it is a bare minimum requirement. To meet other design specifications, the optimal control theory tries to design a control input that not only stabilizes the system but also minimizes a predefined performance index. Optimal control problems can be divided into two major groups: optimal regulation and optimal tracking. In the optimal regulation problem, the goal is to design a control input to make the states or outputs of the system go to zero in an optimal manner. On the other hand, in the optimal tracking problem, the goal is to design a control input to make the states or outputs of the system track a reference trajectory.

It is well known that one can derive optimal control solutions for the regulation problem by solving the Hamilton-Jacobi-Bellman (HJB) equation. The HJB equation is nonlinear and extremely intractable to solve by analytical approaches. Thus several approximate methods have been presented in the literature to address the optimal control solutions. Traditional approaches for approximating the solution to the HJB equation are normally off-line approaches and require complete knowledge of the system dynamics.

Reinforcement learning (RL) [1–7] methods, inspired by learning mechanisms observed in mammals, have been widely used to learn the

Control of Complex Systems
http://dx.doi.org/10.1016/B978-0-12-805246-4.00005-7

solution to the HJB equation online in real time for unknown or partially unknown continuous-time systems [8–10] and discrete-time systems [11–14]. The optimal regulation control of nonlinear systems in the presence of input constraints has also been considered because of its importance in real-world applications [15–17].

In contrast to the solution to the optimal regulation problem, which consists of only a feedback term, the solution to the optimal tracking problem consists of two terms: a feedforward term which guarantees perfect tracking and a feedback term which stabilizes the error dynamics. Existing solutions to the nonlinear optimal tracking problem find the feedforward term using the dynamics inversion concept and find the feedback term by solving an HJB equation [18–23]. These methods require complete knowledge of the system dynamics. This is because, even if RL is used to find the feedback control term without any knowledge of the system dynamics, finding the feedforward term requires complete knowledge of the system dynamics. Moreover, the problem of optimal tracking control of nonlinear systems in the presence of the input constraints has not been considered yet. This is because, in standard methods, the feedback and feedforward control terms are obtained separately and only the feedback part of the control input appears in the performance function. Therefore it is not possible to encode input constraints into the performance function, as was done for the optimal regulation problem [15–17].

In this chapter the optimal tracking problem for nonlinear continuous-time and nonlinear discrete-time systems is solved with use of RL algorithms. The input constraints caused by actuator saturation are taken into account. First, the optimal tracking control problem is transformed into minimization of a nonquadratic performance function subject to an augmented system composed of the original system and the command generator system. Then, RL algorithms are developed to generate both feedback and feedforward parts of the control input simultaneously without complete knowledge of the augmented dynamics.

The rest of chapter is organized as follows. In Section 2 the optimal tracking control of input constraints for nonlinear continuous-time systems using on-policy RL and off-policy RL algorithms is presented. The solution of the optimal tracking control problem for the nonlinear discrete-time system using on-policy RL is considered in Section 3.

2 OPTIMAL TRACKING CONTROL OF CONSTRAINED-INPUT CONTINUOUS-TIME SYSTEMS

In this section the optimal tracking control for continuous-time systems is discussed. Both the standard formulation and its solution and the proposed formulation and its solution are presented.

The objective in the optimal tracking problem is to make the system states or outputs track a reference trajectory in an optimal manner by minimization of a performance function.

Consider the nonlinear continuous-time system

$$\dot{x} = f(x) + g(x)\,u, \tag{1}$$

where $x \in \mathbb{R}^n$ is the system state vector, $u \in \mathbb{R}^m$ is the control input, and $f(x) \in \mathbb{R}^{n\times n}$ and $g(x) \in \mathbb{R}^{n\times m}$ are the drift dynamics and input dynamics, respectively.

The reference trajectory $x_d(t)$ is assumed to be generated by a command generator function $h_d(x_d(t)) \in \mathbb{R}^n$ such that

$$\dot{x}_d(t) = h_d(x_d(t)). \tag{2}$$

2.1 Standard Formulation and Solution to the Optimal Tracking Control Problem

The standard solution to the optimal tracking control problem is composed of two parts: (1) a feedforward controller that guarantees perfect tracking of the reference trajectory and (2) an optimal feedback controller that finds the optimal stabilizer part of the control input. In the following, it is discussed how each of these parts are obtained with the standard method.

The steady-state feedforward part of the control input $u_d(t)$ is obtained by assuming that the reference trajectory satisfies

$$\dot{x}_d(t) = f(x_d(t)) + g(x_d(t))\,u_d(t). \tag{3}$$

If the drift and input dynamics of the system are known and $g^{-1}(x_d(t))$ exists, $u_d(t)$ becomes

$$u_d(t) = g^{-1}(x_d(t))\,(\dot{x}_d(t) - f(x_d(t)). \tag{4}$$

The feedback part of the control stabilizes the tracking error dynamics while minimizing the following performance function:

$$V(e_d(t)) = \int_t^{\infty} [e_d(\tau)^T Q\, e_d(\tau) + u_e^T R\, u_e]\; d\tau, \tag{5}$$

where $Q_e \in \mathbb{R}^{n\times n}$ and $R_e \in \mathbb{R}^{m\times m}$ are symmetric positive definite and

$$u_e(t) = u(t) - u_d(t) \tag{6}$$

is the feedback control input. We can obtain the optimal feedback control solution $u_e^*(t)$ which minimizes Eq. (5) by solving the HJB equation related to this performance function [1]. The standard optimal solution to the optimal tracking control problem is then constituted by the optimal feedback control $u_e^*(t)$ obtained.

Remark 1. The performance function (5) includes only the feedback part of the control input and ignores the feedforward part of the control input. This is because the feedforward part of the control input does not go to zero in general, which makes the value of the performance infinity for any given control policy. However, ignoring the feedforward part of the control input from the performance function leads to a suboptimal control solution and makes it impossible to encode the input constraints into the optimization problem by use of a nonquadratic performance function, as has been performed in the optimal regulation problem [15–17].

Remark 2. The optimal feedback controller $u_e^*(t)$ can be learned with use of RL techniques to obviate knowledge of the system dynamics [24]. However, exact knowledge of the system dynamics is required to find the feedforward controller given by Eq. (4), which cancels the usefulness of the RL techniques.

2.2 A New Optimal Tracker for Continuous-Time Systems

In this subsection a new formulation for solving the optimal tracking control problem of constrained continuous-time systems is presented. It is shown that defining an augmented system and a discounted performance function transforms the optimal tracking problem to an optimal regulation problem. It is also shown that a solution to the formulated problem can be found by solving an augmented tracking HJB equation and this solution stabilizes the error dynamics if the discount factor is chosen small enough. Both off-policy and on-policy RL algorithms are used to learn the solution to the HJB equation online in real time and without requiring complete knowledge of the system dynamics.

2.2.1 Tracking HJB Equation and On-Policy RL Technique

This subsection shows that the optimal solution to the tracking problem can be found by solving a tracking HJB equation. An on-policy RL algorithm is then developed to find the solution to the tracking HJB equation online and without knowledge of the drift dynamics of the system being required.

We define the tracking error as $e_d(t) = x(t) - x_d(t)$ and a general performance index as

$$V(e_d(t), x_d(t)) = \int_t^{\infty} e^{-\gamma(\tau - t)}[e_d(\tau)^T Q e_d(\tau) + U(u)]d\tau, \qquad (7)$$

where $Q \geq 0$, the discount factor $\gamma > 0$, and $U(u)$ is a positive definite penalty function for the control input. This function is defined such that minimization of the performance function (7) leads to a control input which never exceeds its permitted bounds. That is to ensure

$$|u_i(t)| \leq \lambda, \quad i = 1, \ldots, m, \qquad (8)$$

where λ is the permitted limit for the control input. To this end, $U(u)$ is defined as [15]

$$U(u) = 2\int_0^u (\lambda \beta^{-1}(v/\lambda))^T R\, dv. \qquad (9)$$

Remark 3. The use of the discount factor enables us to include the whole control input in the performance function (7), in contrast to the standard performance function (5). This allows us to define the non-quadratic performance function (7) such that the optimal control solution found by minimization of Eqs. (7) satisfies its permitted bounds given in Eqs. (8).

We have shown [25, 26] that if we define the augmented system state as

$$X(t) = [e_d(t)^T \; x_d(t)]^T \in \mathbb{R}^{2n}, \qquad (10)$$

the augmented system dynamics becomes

$$\dot{X}(t) = F(X(t)) + G(X(t))\, u(t), \qquad (11)$$

where $u(t) = u(X(t))$ and

$$F(X(t)) = \begin{bmatrix} f(e_d(t) + x_d(t)) - h_d(x_d(t)) \\ h_d(x_d(t)) \end{bmatrix},$$

$$G(X(t)) = \begin{bmatrix} g(e_d(t) + x_d(t)). \\ 0 \end{bmatrix}$$

Moreover, the performance function (7) in terms of the tracking error and the control input becomes [26]

$$V(X(t)) = \int_t^\infty e^{-\gamma(\tau - t)} [X(\tau)^T Q_T \, X(\tau) + U(u)] \, d\tau, \qquad (12)$$

where

$$Q_T = \begin{bmatrix} Q & \mathbf{0} \\ \mathbf{0} & \mathbf{0} \end{bmatrix}.$$

Therefore the tracking problem is transformed into a regulation problem with the augmented system (11) and the performance function (12). Using Leibniz's rule to differentiate V along the augmented system trajectories (11), one has

$$\nabla V^T(X) \, (F(X) + G(X) \, u(X)) + X^T Q_T X + 2 \int_0^u (\lambda \tanh^{-1}(v/\lambda))^T R$$

$$dv - \gamma V(X) = 0, \qquad (13)$$

where $\nabla V(X) = \partial V(X)/\partial X \in \mathbb{R}^{2n}$. We define the Hamiltonian function

$$\begin{aligned} H(X, u, \nabla V) = X^T Q_T X + 2 \int_0^u (\lambda \tanh^{-1}(v/\lambda))^T R \, dv \\ - \gamma V(X) + \nabla V^T(X) \, (F(X) + G(X) \, u(X)) = 0. \end{aligned} \qquad (14)$$

Now let $V^\star(X)$ be the optimal cost function defined as

$$V^\star(X(t)) = \min_{u \in \pi(\Omega)} \int_t^\infty e^{-\gamma(\tau - t)} [X^T Q_T X + U(u)] \, d\tau. \qquad (15)$$

Then, on the basis of Eq. (14), $V^\star(X)$ satisfies the tracking HJB equation:

$$\begin{aligned} H(X, u^*, \nabla V^\star) = X^T Q_T X + 2 \int_0^{u^*} (\lambda \tanh^{-1}(v/\lambda))^T R \, dv \\ - \gamma V^*(X) + \nabla V^{*T}(X) (F(X) + G(X) \, u^*(X)) = 0. \end{aligned} \qquad (16)$$

The optimal control input for the given problem is obtained by use of the stationarity condition [1] on the Hamiltonian (14). The result is

$$u^\star(X) = \arg\min_{u \in \pi(\Omega)} [H(X, u, \nabla V^\star)] = -\lambda \tanh((1/2\lambda) R^{-1}$$

$$G^T(X) \, \nabla V^\star(X)). \qquad (17)$$

This control is within its permitted bounds $\pm\lambda$. Substituting $u^{\star}(X)$ from Eq. (17) into Eq. (16) gives the following tracking HJB equation:

$$H(X, u^{*}, \nabla V^{\star}) \\ = X^{T} Q_{T} X - \gamma V^{*}(X) + \nabla V^{\star T}(X) F(X) + \lambda^{2} \bar{R} \ln (\underline{1} - \tanh^{2}(D^{\star})) = 0. \tag{18}$$

To solve the optimal tracking control problem, one solves the HJB equation (18) for the optimal value $V^{\star}$, then the optimal control is given as a feedback $u(V^{*})$ in terms of the HJB solution with use of Eq. (17).

Integral RL (IRL) [24] is used in the following to develop an on-policy RL algorithm for solving the tracking HJB equation online in real time using partial knowledge of the system dynamics. The idea of the IRL algorithm is to write the performance function (12) in the integral reinforcement form

$$V(X(t-T)) = \int_{t-T}^{t} e^{-\gamma(\tau-t+T)} \left[X(\tau)^{T} Q_{T} X(\tau) \; + U(u)\right] d\tau + e^{-\gamma T} \\ V(X(t)). \tag{19}$$

This is called the *IRL tracking Bellman equation*. With use of this Bellman equation, the following on-policy IRL-based algorithm can be used to solve the optimal tracking problem with only partial knowledge of the system dynamics.

Algorithm 5.1 On-Policy IRL Algorithm for Solving the Tracking HJB Equation

1. Given a control input $u_i(X)$, find $V_i(X)$ using the tracking IRL Bellman equation

$$V_i(X(t-T)) = \int_{t-T}^{t} e^{-\gamma(\tau-t+T)} \left[X(\tau)^{T} Q_{T} X(\tau) \; + \; U(u(\tau))\right] d\tau \\ + e^{-\gamma T} V_i(X(t)). \tag{20}$$

2. Update the control policy using

$$u_{i+1}(X) = -\lambda \, \tanh\left((1/2\lambda) R^{-1} G^{T}(X) \nabla V_i(X)\right). \tag{21}$$

The on-policy Algorithm 5.1 requires knowledge of the input dynamics $G(X)$ in the policy update law. To overcome this problem, an off-policy

RL algorithm is developed in the following which does not require any knowledge of the system dynamics.

2.2.2 Actor-Critic Structure for Solving the Nonlinear Tracking Problem

A critic neural network (NN) can be used to approximate the solution to the value function V_i in the tracking Bellman equation (20) and an actor NN can be used to update the control policy in Eq. (21). The weights of the NNs can be updated either by use of the least squares technique on collected data along the trajectory of the augmented system or by use of a gradient descent-based synchronous update law. A synchronous update law is designed in this subsection to update the NN weights continuously. To this end, assume the value function is a smooth function. Then, there exists a single-layer NN such that the value function solution can be approximated as

$$\hat{V}(X) = \hat{W}_1^T \phi(X), \tag{22}$$

where $\phi(X) \in \mathbb{R}^l$ is the basis function vector, $\hat{W}_1 \in \mathbb{R}^l$ is current estimation of the actual critic NN weights (ie, W_1), and l is the number of neurons. The critic NN (22) can be used to approximate the value function solution to the IRL tracking Bellman equation. Using the value approximation (22) in the tracking IRL (20), one has

$$\begin{aligned}\varepsilon_{\mathrm{B}}(t) \equiv \int_{t-T}^{t} & e^{-\gamma(\tau-t+T)} \Big[X(\tau)^T Q_T X(\tau) \\ & + 2\int_0^u (\lambda \tanh^{-1}(v/\lambda))^T R\, dv\Big] d\tau + \hat{W}_1^T \Delta\phi(X(t)),\end{aligned} \tag{23}$$

where

$$\Delta\phi(X(t)) = e^{-\gamma T}\phi(X(t)) - \phi(X(t-T)) \tag{24}$$

and ε_{B} is the tracking Bellman equation error due to the approximation error.

On the other hand, the actor NN is defined as

$$\hat{u}_1 = -\lambda \tanh((1/2\lambda)R^{-1}G^T \nabla\phi^T \hat{W}_2), \tag{25}$$

where $\hat{W}_2$ is the actor NN weights vector and it is considered as the current estimation of W_1. It was shown in [26] that if the actor and critic NN weights are updated as Eqs. (26) and (27), then the convergence to near optimal control and stability of the system during learning is guaranteed, provided that some mild assumptions are satisfied and the discount factor and RL interval are chosen small enough.

$$\dot{\hat{W}}_1 = -\frac{-\alpha_1 \Delta\phi}{(1+\Delta\phi^T\Delta\phi)^2}\,\hat{e}_{\mathrm{B}}, \tag{26}$$

$$\dot{\hat{W}}_2 = -\alpha_2(\nabla\phi G e_u + \nabla\phi G \tanh^2(\hat{D}) e_u + Y\hat{W}_2), \tag{27}$$

where $Y > 0$ is a design parameter, $\hat{D} = \frac{1}{2\lambda} R^{-1} G^T \nabla\phi^T \hat{W}_2$,

$$\hat{e}_{\mathrm{B}}\,(t) \triangleq \int_{t-T}^{t} e^{-\gamma(\tau-t+T)}\left[X(\tau)^T Q_T\, X(\tau) + \hat{U}\right] d\tau + \hat{W}_1^T \Delta\phi(X(t)), \tag{28}$$

and

$$e_u = \hat{u}_1 - u_1 = \lambda\left(\tanh\left(\frac{1}{2\lambda} R^{-1} G^T \nabla\phi^T \hat{W}_2\right) - \tanh\left(\frac{1}{2\lambda} R^{-1} G^T \nabla\phi^T \hat{W}_1\right)\right). \tag{29}$$

Knowledge of the input dynamics is required to update the actor NN. In the following, a least squares-based off-policy RL algorithm is developed to solve the tracking HJB equation online and without any knowledge of the system dynamics being required.

2.2.3 Off-Policy RL for Solving the Optimal Tracking Problem

In this subsection a model-free off-policy RL algorithm is developed to solve the optimal tracking problem. In the off-policy RL algorithm, the policy which is applied to the system to generate the data required for learning can be different from the policy which is updated and improved. The off-policy IRL algorithm presented in [27–30] to solve the optimal regulation problem can be extended to the discounted optimal control to avoid the requirement of knowledge of $G(X)$. To this end, inspired by [27, 28], we first write the system dynamics (11) as

$$\dot{X} = F(X) + G(X)\, u_i + G(X)(u - u_i). \tag{30}$$

Differentiating $V_i(X)$ along with the system dynamics (30) and using Eqs. (14) and (21), one obtains

$$\begin{aligned}\dot{V}_i &= V_{X_i}{}^T(F + G u_i) + \Delta V_i^T G\,(u - u_i)\\ &= \gamma\, V_i - X^T Q_T\, X - U(u_i) - 2\lambda v_{i+1}{}^T R\,(\tanh(v) - \tanh(v_i)),\end{aligned} \tag{31}$$

where

$$\nu_{i+1} = 0.5\,\lambda^{-1} R^{-1} G^T \nabla V_i, \tag{32}$$

and thus on the basis of Eq. (21),

$$u_i = -\lambda \tanh(\nu_i). \tag{33}$$

Multiplying both sides of Eq. (31) by $e^{-\gamma(\tau-t)}$and integrating from both sides, one obtains the following off-policy IRL Bellman equation:

$$\begin{aligned} & e^{-\gamma T} V_i(X(t+T)) - V_i(X(t)) \\ & = -\int_t^{t+T} e^{-\gamma(\tau-t)} (X^T Q_T X + U(u_i))\, d\tau \\ & \quad - 2\lambda \int_t^{t+T} e^{-\gamma(\tau-t)} {\nu_{i+1}}^T R\, (\tanh(\nu) - \tanh(\nu_i))\, d\tau. \end{aligned} \tag{34}$$

This off-policy tracking Bellman equation can be solved for V_i and ν_{i+1} simultaneously without any knowledge of the system dynamics being required. Note that ν_{i+1} gives u_{i+1} because of Eq. (33). Therefore, both the value function and the control inputs can be found simultaneously by our solving the Bellman equation (34). The following algorithm uses this Bellman equation to solve the optimal tracking problem without requiring any knowledge of the system dynamics.

Algorithm 5.2 Off-Policy RL Algorithm for Solving the Tracking HJB Equation

1. Solve the following Bellman equation for V_i and ν_{i+1} (and consequently u_{i+1}) simultaneously:

$$\begin{aligned} & e^{-\gamma T} V_i(X(t+T)) - V_i(X(t)) \\ & = -\int_t^{t+T} e^{-\gamma(\tau-t)} (X^T Q_T X + U(u_i))\, d\tau \\ & \quad - 2\lambda \int_t^{t+T} e^{-\gamma(\tau-t)} {\nu_{i+1}}^T R\, (\tanh(\nu) - \tanh(\nu_i))\, d\tau. \end{aligned} \tag{35}$$

Stop if a stopping criterion is met, otherwise set $i = i + 1$ and go to step 1.

Remark 4. Like the on-policy-based RL algorithm developed in the previous subsection, two NNs can be used to approximate the solution to the value function V_i and v_{i+1} in the tracking Bellman equation (35) simultaneously. The weights of the NNs can be updated by use of the least squares technique on collected data along the trajectory of the augmented system.

Remark 5. The existence of the solution to the tracking HJB equation (18) is discussed in [25] for linear systems. The same results can be derived for affine nonlinear systems. Optimality of the solution to the tracking HJB equation is also shown in [26]. Moreover, convergence of on-policy IRL Algorithm 5.1 to the solution of the tracking HJB equation is shown in [26]. The same proof can be derived for the convergence of off-policy RL Algorithm 5.2.

3 OPTIMAL TRACKING CONTROL OF CONSTRAINED-INPUT DISCRETE-TIME SYSTEMS

In this section the optimal tracking control for discrete-time systems is discussed. Both the standard formulation and its solution and the proposed formulation and its solution are presented.

Consider the deterministic nonlinear affine discrete-time system given by

$$x(k+1) = f(x(k)) + g(x(k))\, u(k), \tag{36}$$

where $k > 0$, $x(k) \in \mathbb{R}^n$ represents the state vector of the system, $u(k) \in \mathbb{R}^m$ represents the control vector, and $f(x(k)) \in \mathbb{R}^{n\times n}$ and $g(x(k)) \in \mathbb{R}^{n\times m}$ are the drift dynamics and the input dynamics, respectively.

The goal in the infinite-horizon optimal tracking problem is to design an optimal controller for the system (36) which ensures that the state $x(k)$ tracks the reference trajectory $r(k)$ in an optimal manner.

The reference trajectory dynamics for the nonlinear tracking problem is generated by the command generator model

$$r(k+1) = \psi(r(k)), \tag{37}$$

where $r(k) \in \mathbb{R}^n$.

3.1 Standard Formulation and Solution to the Optimal Tracking Control Problem

The standard solution to the tracking problem of nonlinear discrete-time systems is composed of two parts: (1) a feedforward controller to ensure perfect tracking and (2) a feedback controller to make the system stable [1]. In the following it is discussed how each of these parts are obtained with the standard method.

The steady-state feedforward part of the control input $u_d(k)$ is obtained by assuming that the reference trajectory satisfies

$$r(k+1) = f(r(k)) + g(r(k))\, u_d(k). \tag{38}$$

If the system dynamics is known, $u_d(k)$ is obtained by

$$u_d(k) = g(r(k))^{+}\,(r(k+1) - f(r(k))), \tag{39}$$

where $g(r(k))^{+} = (g(r(k))^{T} g(r(k)))^{-1} g(r(k))^{T}$ is the generalized inverse of $g(r(k))$, with $g(r(k))^{+} g(r(k)) = I$.

The feedback control input is designed to stabilize the tracking error dynamics in an optimal manner by minimizing the following performance function:

$$J(e(k), u_e(k)) = \sum_{i=k}^{\infty} \left[e^{T}(i)\, Q_e e(i) + u_e^{T}(i) R_e u_e(i) \right] \tag{40}$$

where $Q_e \in \mathbb{R}^{n\times n}$ and $R_e \in \mathbb{R}^{m\times m}$ are symmetric positive definite.

Then the standard optimal control input is given by

$$u^{\star}(k) = u_d(k) + u_e^{\star}(k), \tag{41}$$

where u_d is given by Eq. (39) and $u_e^{\star}$ is obtained by our solving the HJB equation related to the performance function (40) [1].

Remark 6. The RL algorithms for finding the feedback part of the control input (41) could be implemented to obviate knowledge of the system dynamics [4]. In contrast, obtaining the feedforward part of the control input needs complete knowledge of the system dynamics and the reference trajectory dynamics.

Remark 7. Since the feedforward part of the control input (39) is found separately and is not involved in the optimization of the performance (40), it is not possible to encode the constraints of the control input u into the optimization problem. In the following, a new formulation of the problem is given that gives both feedback and feedforward parts of the control input simultaneously by minimizing a predefined performance function.

3.2 Optimal Tracker for Discrete-Time Systems

In this subsection an alternative approach for formulating and solving the nonlinear discrete-time tracking problem is presented that gives both parts of the control input simultaneously and also takes into account the input constraints. First, an augmented system composed of the original system and the reference trajectory is constructed [31]. On the basis of this augmented system a new performance function is presented that consists of both feedback and feedforward parts of the control input. Then, the solution to the formulated problem is obtained by use of the actor-critic structure.

3.2.1 New Formulation for the Tracking Problem of Nonlinear Discrete-Time Systems

The tracking error is defined as $e(k) = x(k) - r(k)$. The tracking error dynamics in terms of the system dynamics (36) and the reference trajectory dynamics (37) is given by

$$\begin{aligned} e(k+1) &= x(k+1) - r(k+1) \\ &= f(e(k)+r(k)) - \psi(r(k)) + g(e(k)+r(k))u(k). \end{aligned} \tag{42}$$

On the basis of Eqs. (37) and (42), the augmented system is constructed in terms of the tracking error $e(k)$ and the reference trajectory $r(k)$ as

$$\begin{aligned} \begin{bmatrix} e(k+1) \\ r(k+1) \end{bmatrix} &= \begin{bmatrix} f(e(k)+r(k)) - \psi(r(k)) \\ \psi(r(k)) \end{bmatrix} + \begin{bmatrix} g(e(k)+r(k)) \\ 0 \end{bmatrix} u(k) \\ &\equiv F(X(k)) + G(X(k))u(k), \end{aligned} \tag{43}$$

where the augmented state is

$$X(k) = \begin{bmatrix} e(k) \\ r(k) \end{bmatrix} \in \mathbb{R}^{2n} \tag{44}$$

and

$$u(k) \in \mathbb{R}^{m} \big| \quad |u_j(k)| \le \bar{u}_j, \quad j = 1, \ldots, m, \tag{45}$$

where $\bar{u}_j$ is the permitted limit for the control input.

On the basis of the augmented system (43), define the value function for the tracking problem as

$$V(X(k)) = \sum_{i=k}^{\infty} \gamma^{i-k} U_i = \sum_{i=k}^{\infty} \gamma^{i-k} \left[X(i)^T Q_1 X(i) + W(u(i)) \right], \tag{46}$$

where

$$Q_1 = \begin{bmatrix} Q & 0 \\ 0 & 0 \end{bmatrix}. \tag{47}$$

Here $0 < \gamma \leq 1$ is the discount factor, Q is positive definite, and $W(u(i)) \in \mathbb{R}$ is a positive definite penalty function for the control input.

To deal with the constrained control input [15–17, 32–34], a non-quadratic function is defined as

$$W(u(i)) = 2\int_0^{u(i)} \varphi^{-T}(\bar{U}^{-1}s)\, \bar{U}\, R\, ds, \tag{48}$$

where $\bar{U} \in \mathbb{R}^{m\times m}$ is the constant diagonal matrix given by $\bar{U} = \mathrm{diag}\{\bar{u}_1, \bar{u}_2, \ldots, \bar{u}_m\}$ and R is positive definite and assumed to be diagonal. $\varphi(.)$ is a bounded one-to-one function satisfying $|\varphi(.)| < 1$.

Eqs. (46) and (48) yield

$$\begin{aligned} V(X(k)) = {} & X(k)^T Q_1\, X(k) + 2\int_0^{u(k)} \varphi^{-T}(\bar{U}^{-1}s)\, \bar{U}\, R\, ds \\ & + \sum_{i=k+1}^{\infty} \gamma^{i-k}\left[X(i)^T Q_1\, X(i) + 2\int_0^{u(i)} \varphi^{-T}(\bar{U}^{-1}s)\, \bar{U}\, R\, ds\right], \end{aligned} \tag{49}$$

which yields the nonlinear tracking Bellman equation:

$$V(X(k)) = X(k)^T Q_1\, X(k) + 2\int_0^{u(k)} \varphi^{-T}(\bar{U}^{-1}s)\, \bar{U}\, R\, ds + \gamma\, V(X(k+1)). \tag{50}$$

On the basis of Eq. (50), define the Hamiltonian function

$$\begin{aligned} H(X(k), u(k), V) = {} & X(k)^T Q_1\, X(k) \\ & + 2\int_0^{u(k)} \varphi^{-T}(\bar{U}^{-1}s)\bar{U}R\, ds \\ & + \gamma\, V(X(k+1)) - V(X(k)). \end{aligned} \tag{51}$$

$V^{\star}(X(k))$ is the optimal cost function and is defined as

$$\begin{aligned} V^{\star}(X(k)) = \min_{u(k)} \Big[& X(k)^T Q_1\, X(k) \\ & + 2\int_0^{u(k)} \varphi^{-T}(\bar{U}^{-1}s)\, \bar{U}\, R\, ds + \gamma\, V^{\star}(X(k+1))\Big]. \end{aligned} \tag{52}$$

Then, on the basis of Eq. (51), $V^{\star}(X(k))$ satisfies the tracking HJB equation:

$$\begin{aligned} &H(X(k), u^{\star}(k), V^{\star}) \\ &= X(k)^T Q_1 X(k) + 2\int_0^{u^{\star}(k)} \varphi^{-T}(\bar{U}^{-1}s)\,\bar{U}\,R\,ds + \gamma V^{\star}(X(k+1)) \\ &\quad - V^{\star}(X(k)). \end{aligned} \tag{53}$$

A necessary condition for optimality is the stationarity condition [1]:

$$\begin{aligned} &\frac{\partial H(X(k), u(k), V^{\star})}{\partial u(k)} \\ &= \frac{\partial \left(X(k)^T Q_1 X(k) + 2\int_0^{u(k)} \varphi^{-T}(\bar{U}^{-1}s)\,\bar{U}\,R\,ds \right)}{\partial u(k)} \\ &\quad + \gamma \left(\frac{\partial X(k+1)}{\partial u(k)} \right)^T \frac{\partial V^{\star}(X(k+1))}{\partial X(k+1)} = 0. \end{aligned} \tag{54}$$

Then the optimal control input is

$$u^{\star}(k) = \bar{U}\varphi\left(-\frac{\gamma}{2}(\bar{U}R)^{-T} G(X)^T \frac{\partial V^{\star}(X(k+1))}{\partial X(k+1)} \right). \tag{55}$$

If we substitute Eq. (55) in the Bellman equation, the discrete-time tracking HJB equation (52) becomes

$$V^{\star}(X(k)) = X(k)^T Q_1 X(k) + W(u^{\star}(k)) + \gamma V^{\star}(X(k+1)), \tag{56}$$

where $V^{\star}(X(k))$ is the value function corresponding to the optimal control input.

To find the optimal control solution, first the discrete-time tracking HJB equation (56) is solved and then the optimal control solution is given by Eq. (55) with use of the solution of the discrete-time tracking HJB equation. However, in general, the discrete-time tracking HJB equation cannot be solved analytically. In the subsequent sections, it is shown how to solve the discrete-time tracking HJB equation online in real time without complete knowledge of the augmented system dynamics.

Remark 8. It is clear from Eq. (55) that the optimal control input never exceeds its permitted bounds. This is a result of reformulation of the nonlinear tracking problem to minimization of the nonquadratic performance function (46) subject to the augmented system (43). In this formulation, both feedback and feedforward parts of the control input are obtained simultaneously by minimization of the performance function (46) and the control input stays in its permitted bounds because of the use of the nonquadratic function (48) for penalizing the control input.

3.2.2 Actor-Critic Structure for Solving the Nonlinear Tracking Problem

In this section an RL algorithm on the actor-critic structure is used to solve the discrete-time tracking HJB equation. The proposed algorithm is an online synchronized algorithm and does not require knowledge of the drift system dynamics or the reference trajectory dynamics. This method has been developed for continuous-time systems [35] and for discrete-time systems with a one-layer NN [14]. The value function and the control input are approximated with two separate two-layer perceptron NNs [36–39] and are updated simultaneously.

A two-layer NN with one hidden layer is used as the actor NN to approximate the control input for the nonlinear tracking problem. The output of the actor NN is given as

$$\begin{aligned} &\hat{u}(X(k), W_a^{(1)}(k), W_a^{(2)}(k)) = [\hat{u}_1, \ldots, \hat{u}_m], \\ &\hat{u}_i = \sigma_i(W_{a_i}^{(2)}(k)\,\phi_a(W_a^{(1)}(k)X(k))) \\ &\quad = \sigma_i\left(\sum_{l=1}^{N_a} \hat{w}_{a_{il}}^{(2)}(k)\,\phi_{a_l}\left(\sum_{j=1}^{2n} \hat{w}_{a_{lj}}^{(1)}(k)X_j(k)\right)\right), \end{aligned} \tag{57}$$

where $\phi(k) = [\phi_1(k), \ldots, \phi_{N_a}(k)]^T \in \mathbb{R}^{N_a \times 1}$ is the vector of hidden-layer activation functions with $\phi(.) = \tanh(.)$, and $\sigma_i(.) = \bar{u}_i \tanh(.)$, $i = 1, \ldots, m$ is the ith output layer activation function.

The weight matrix between the neurons in the input and the hidden layers is defined as

$$W_a^{(1)} = \begin{bmatrix} w_{a_{11}}^{(1)} & \cdots & w_{a_{1(2n)}}^{(1)} \\ \vdots & \vdots & \vdots \\ w_{a_{N_a1}}^{(1)} & \cdots & w_{a_{N_a(2n)}}^{(1)} \end{bmatrix} \in \mathbb{R}^{N_a \times 2n},$$

where N_a is number of the hidden layer neurons and each element $w_{a_{ij}}^{(1)}$ denotes the weight from the jth input to the ith hidden neurons. Also, we define the matrix of the weights between the neurons in the hidden and the output layers as

$$W_a^{(2)} = \begin{bmatrix} w_{a_{11}}^{(2)} & \cdots & w_{a_{1N_a}}^{(2)} \\ \vdots & \vdots & \vdots \\ w_{a_{m1}}^{(2)} & \cdots & w_{a_{mN_a}}^{(2)} \end{bmatrix} \in \mathbb{R}^{m \times N_a},$$

where $\hat{w}_{a_{ij}}^{(2)}(k)$ is the weight between the jth hidden layer and the ith output layer.

The output of the actor NN satisfies the input bounds defined in Eq. (45).

A two-layer NN with one hidden layer is used as a critic NN to approximate the value function. The output of the critic NN is given as

$$\hat{V}(X(k)) = W_c^{(2)}(k)\,\phi_c(W_c^{(1)}(k)X(k)) = \sum_{i=1}^{N_c} \hat{w}_{c_i}^{(2)}(k)\,\phi_{c_i}\left(\sum_{j=1}^{2n} \hat{w}_{c_{ij}}^{(1)}(k)X_j(k)\right), \tag{58}$$

where $\phi(k) = [\phi_1(k), \ldots, \phi_{N_c}(k)]^T \in \mathbb{R}^{N_c \times 1}$ is the vector of hidden layer activation functions, with $\phi(.) = \tanh(.)$ and the matrix of the weights between the neurons in the input and the hidden layers is defined as

$$W_c^{(1)} = \begin{bmatrix} w_{c_{11}}^{(1)} & \cdots & w_{c_{1(2n)}}^{(1)} \\ \vdots & \vdots & \vdots \\ w_{c_{N_c1}}^{(1)} & \cdots & w_{c_{N_c(2n)}}^{(1)} \end{bmatrix} \in \mathbb{R}^{N_c \times 2n},$$

where N_c is number of the hidden layer neurons and each element $w_{c_{ij}}^{(1)}$ denotes the weight from the jth input to ith hidden neurons. Also, we define the vector of the weights between the neurons in the hidden and the output layer as

$$W_c^{(2)} = [w_{c_1}^{(2)}, w_{c_2}^{(2)}, \ldots, w_{c_{N_c}}^{(2)}] \in \mathbb{R}^{1 \times N_c},$$

where $\hat{w}_{c_i}^{(2)}(k)$ is the weight between the ith hidden layer and output layer.

3.2.3 Learning Rules for Actor and Critic NNs

The critic NN is updated to minimize the tracking Bellman equation error and the actor NN is updated to minimize the value function. In this section, approximate gradient descent rules for updating the critic and actor NNs are developed. To obviate the need for $V(X(k+1))$, which requires the system model to predict $X(k+1)$, the previous values of the system state

$X(k)$ and the state value $V(X(k))$ are stored and used for updating the actor and critic NN weights.

The prediction error of the tracking Bellman equation is defined as

$$e_c(k) = \gamma \hat{V}(X(k)) + U_k - \hat{V}(X(k-1)), \tag{59}$$

where

$$U_k = X(k)^T Q_1 X(k) + 2 \int_0^{u(k)} \varphi^{-T}(\bar{U}^{-1}s)\, \bar{U} R\, ds. \tag{60}$$

It is desired to select the weights of the critic NN to minimize the square Bellman error:

$$E_c(k) = \frac{1}{2} e_c^2(k). \tag{61}$$

The tuning laws for the critic weights are selected as the normalized approximation to the gradient descent algorithm—that is,

$$\hat{w}_{c_{ij}}^{(1)}(k+1) = \hat{w}_{c_{ij}}^{(1)}(k) - l_c \frac{\partial E_c(k)}{\partial \hat{w}_{c_{ij}}^{(1)}(k)}, \tag{62}$$

$$\hat{w}_{c_i}^{(2)}(k+1) = \hat{w}_{c_i}^{(2)}(k) - l_c \frac{\partial E_c(k)}{\partial \hat{w}_{c_i}^{(2)}(k)}, \tag{63}$$

where $l_c > 0$ is the learning rate.

Using the chain rule, one has

$$\begin{aligned} \frac{\partial E_c(k)}{\partial \hat{w}_{c_{ij}}^{(1)}(k)} &= \frac{\partial E_c(k)}{\partial e_c(k)} \frac{\partial e_c(k)}{\partial \hat{V}(k)} \frac{\partial \hat{V}(k)}{\partial \phi_{c_i}(k)} \frac{\partial \phi_{c_i}(k)}{\partial \hat{w}_{c_{ij}}^{(1)}(k)} \\ &= \gamma\, e_c(k) \hat{w}_{c_i}^{(2)}(k)(1 - \phi_{c_i}^2(k)) X_j(k), \end{aligned} \tag{64}$$

$$\frac{\partial E_c(k)}{\partial \hat{w}_{c_i}^{(2)}(k)} = \frac{\partial E_c(k)}{\partial \hat{V}(k)} \frac{\partial \hat{V}(k)}{\partial \hat{w}_{c_i}^{(2)}(k)} = \gamma\, e_c(k)\, \phi_{c_i}(k). \tag{65}$$

The actor (57) is updated to minimize the approximated value function. One does this by minimizing the error between a target control input which is obtained by minimization of the value function and the actual control input which is applied to the system. Let the current estimation of the value function be $\hat{V}$. Then, on the basis of the policy update law (55), the target control input at time k is

$$\tilde{u}(X) = \bar{U}\varphi \left(-\frac{\gamma}{2} (\bar{U}R)^{-1} G(X(k))^T \frac{\partial \hat{V}(X(k+1))}{\partial X(k+1)} \right). \tag{66}$$

However, to obtain the value at time $k+1$, the states are required to be predicted by use of a model network. But we do not use a model network to predict the future value. Rather, we store the previous value of the system state and the state value and try to minimize the error between the target control input and the actual control input given current actor and critic estimate weights while the previous stored state $X(k-1)$ is used as the input to the actor and critic. That is, we try to minimize

$$e_a(k) = \tilde{u}(X(k-1)) - \hat{u}(k, k-1), \tag{67}$$

where $\hat{u}(k,k-1) = \hat{u}(X(k-1), W_a^{(1)}(k), W_a^{(2)}(k))$ is the output of the actor NN at time $(k-1)$ if the current NN weights are used. It is desired to select the weights of the actor NN to minimize the square actor NN error:

$$E_a(k) = \frac{1}{2} e_a^2(k). \tag{68}$$

The tuning laws for the actor weights are selected as the normalized gradient descent algorithm—that is,

$$\hat{w}_{alj}^{(1)}(k+1) = \hat{w}_{ailj}^{(1)}(k) - l_a \frac{\partial E_a(k)}{\partial \hat{w}_{alj}^{(1)}(k)}, \tag{69}$$

$$\hat{w}_{aij}^{(2)}(k+1) = \hat{w}_{aij}^{(2)}(k) - l_a \frac{\partial E_a(k)}{\partial \hat{w}_{aij}^{(2)}(k)}, \tag{70}$$

where $l_a > 0$ is learning rate.

The chain rule for the weights between the input layer and the hidden layer yields

$$\begin{aligned}\frac{\partial E_a(k)}{\partial \hat{w}_{aij}^{(1)}(k)} &= \frac{\partial E_a(k)}{\partial e_a(k)} \frac{\partial e_a(k)}{\partial \hat{u}(k,k-1)} \frac{\partial \hat{u}(k,k-1)}{\partial \hat{w}_{aij}^{(1)}(k)} \\ &= e_a(k) \times \left(-\sum_{l=1}^{m} \left[\bar{u}_l \left(1 - \frac{1}{\bar{u}_l^2} \sigma_l^2(X(k-1), W_a^{(1)}(k), W_a^{(2)}(k)) \right) \right.\right. \\ &\quad \left.\left. \times \sum_{i=1}^{N_a} \hat{w}_{ali}^{(2)}(k)\,(1 - \phi_{a_l}(k)^2) \right] \right) \times \sum_{j=1}^{2n} X_j(k-1),\end{aligned} \tag{71}$$

and for the weights between the hidden layers and the output layers yields

$$\frac{\partial E_a(k)}{\partial \hat{w}_{a_{ij}}^{(2)}(k)} = \frac{\partial E_a(k)}{\partial e_a(k)} \frac{\partial e_a(k)}{\partial \hat{u}(k,k-1)} \frac{\partial \hat{u}(k,k-1)}{\partial \hat{w}_{a_{ij}}^{(2)}(k)}$$
$$= -e_a(k) \times \bar{u}_i \left(1 - \frac{1}{\bar{u}_i^2}\sigma_i^2(X(k-1), W_a^{(1)}(k), W_a^{(2)}(k))\right) \times \phi_j(k). \quad (72)$$

Remark 9. The stability and optimality of the solution found by our solving the discrete-time HJB equation are proved in [40]. Moreover, the stability proof and convergence of the actor-critic network during learning are provided in [14, 41].

4 CONCLUSION

Optimal tracking control of both continuous-time and discrete-time systems was considered. In both cases an augmented system composed of the original system dynamics and the command generator dynamics was formed. A nonquadratic performance function was defined to take into account the input constraints caused by actuator saturation. It was shown that the solving of the optimal tracking problems is equivalent to minimization of the nonquadratic performance function subject to the augmented system. Then, an augmented HJB equation was derived which gives both feedback and feedforward control input simultaneously. RL algorithms were developed to find the solution to the HJB equation online and without complete knowledge of the system dynamics being required.

ACKNOWLEDGMENTS

This work was supported in part by National Science Foundation grant ECCS-1405173, Office of Naval Research grant N00014-13-1-0562, Army Research Office grant W911NF-11-D-0001, China Natural National Science Foundation grant 61120106011, and China Education Ministry Project 111 (no. B08015).

REFERENCES

[1] F.L. Lewis, D. Vrabie, V. Syrmos (Eds.), Optimal Control, Wiley, New York, 2012.
[2] D.P. Bertsekas, J.N. Tsitsiklis, Neuro-dynamic Programming, Athena Scientific, Belmont, MA, 1996.
[3] W.B. Powell, Approximate Dynamic Programming: Solving the Curses of Dimensionality, Wiley, New York, 2011.
[4] R.S. Sutton, A.G. Barto, Reinforcement Learning an Introduction, MIT Press, Cambridge, MA, 1998.
[5] F.L. Lewis, D. Vrabie, K.G. Vamvoudakis, Reinforcement learning and feedback control, IEEE Control Syst. Mag. 32 (6) (2012) 76–105.

[6] F.L. Lewis, D. Liu, Reinforcement Learning and Approximate Dynamic Programming for Feedback Control, Wiley, Hoboken, NJ, 2013.
[7] D. Vrabie, K.G. Vamvoudakis, F.L. Lewis, Optimal Adaptive Control and Differential Games by Reinforcement Learning Principles, The Institution of Engineering and Technology, London, United Kingdom, 2013.
[8] S. Bhasin, R. Kamalapurkar, M. Johnson, K.G. Vamvoudakis, F.L. Lewis, W. Dixon, A novel actor-critic-identifier architecture for approximate optimal control of uncertain nonlinear systems, Automatica 49 (1) (2013) 82–92.
[9] D. Liu, X. Yang, H. Li, Adaptive optimal control for a class of continuous-time affine nonlinear systems with unknown internal dynamics, Neural Comput. Appl. 23 (2012) 1843–1850.
[10] D. Vrabie, F.L. Lewis, Neural network approach to continuous-time direct adaptive optimal control for partially unknown nonlinear systems, Neural Netw. 22 (2009) 237–246.
[11] D. Prokhorov, D. Wunsch, Adaptive critic designs, IEEE Trans. Neural Netw. 8 (5) (1997) 997–1007.
[12] A. Al-Tamimi, F.L. Lewis, M. Abu-Khalaf, Discrete-time nonlinear HJB solution using approximate dynamic programming: convergence proof, IEEE Trans. Syst. Man Cybern. B 38 (4) (2008) 943–949.
[13] Y. Luo, H. Zhang, Approximate optimal control for a class of nonlinear discrete-time systems with saturating actuators, Prog. Nat. Sci. 18 (8) (2008) 1023–1029.
[14] T. Dierks, S. Jagannathan, Online optimal control of nonlinear discrete-time systems using approximate dynamic programming, J. Control Theory Appl. 9 (3) (2011) 361–369.
[15] M. Abou-Khalaf, F.L. Lewis, Nearly optimal control laws for nonlinear systems with saturating actuators using a neural network HJB approach, Automatica 41 (2005) 779–791.
[16] H. Modares, M.B. Naghibi-Sistani, F.L. Lewis, A policy iteration approach to online optimal control of continuous-time constrained-input systems, ISA Trans. 52 (2013) 611–621.
[17] H. Modares, M.B. Naghibi-Sistani, F.L. Lewis, Integral reinforcement learning and experience replay for adaptive optimal control of partially unknown constrained-input continuous-time systems, Automatica 50 (2014) 193–202.
[18] H. Zhang, Q. Wei, Y. Luo, A novel infinite-time optimal tracking control scheme for a class of discrete-time nonlinear systems via the greedy HDP iteration algorithm, IEEE Trans. Syst. Man Cybern. B: Cybern. 38 (4) (2008) 937–942.
[19] T. Dierks, S. Jagannathan, Optimal tracking control of affine nonlinear discrete-time systems with unknown internal dynamics, in: Proceedings of the 48th IEEE CDC, 2009, pp. 6750–6755.
[20] Y. Huang, D. Liu, Neural-network-based optimal tracking control scheme for a class of unknown discrete-time nonlinear systems using iterative ADP algorithm, Neurocomputing 125 (2014) 46–56.
[21] H. Zhang, L. Cui, X. Zhang, X. Luo, Data-driven robust approximate optimal tracking control for unknown general nonlinear systems using adaptive dynamic programming method, IEEE Trans. Neural Netw. 22 (12) (2011) 2226–2236.
[22] Q. Wei, D. Liu, Adaptive dynamic programming for optimal tracking control of unknown nonlinear systems with application to coal gasification, IEEE Trans. Autom. Sci. Eng. 11 (4) (2014) 1020–1036.
[23] A. Mannava, S.N. Balakrishnan, L. Tang, R.G. Landers, Optimal tracking control of motion systems, IEEE Trans. Control Syst. Technol. 20 (6) (2012) 1548–1556.
[24] D. Vrabie, O. Pastravanu, M. Abou-Khalaf, F.L. Lewis, Adaptive optimal control for continuous-time linear systems based on policy iteration, Automatica 45 (2009) 477–484.

[25] H. Modares, F.L. Lewis, Linear quadratic tracking control of partially-unknown continuous-time systems using reinforcement learning, IEEE Trans. Autom. Control 59 (11) (2014) 3051–3056.
[26] H. Modares, F.L. Lewis, Optimal tracking control of nonlinear partially-unknown constrained-input systems using integral reinforcement learning, Automatica 50 (2014) 1780–1792.
[27] Y. Jiang, Z.P. Jiang, Computational adaptive optimal control for continuous-time linear systems with completely unknown dynamics, Automatica 48 (2012) 2699–2704.
[28] Y. Jiang, Z.P. Jiang, Robust adaptive dynamics programming feedback stabilization of nonlinear systems, IEEE Trans. Neural Netw. Learn. Syst. 25 (5) (2014) 882–893.
[29] B. Luo, H. Wu, T. Huang, Off-policy H_∞ reinforcement learning for control design, IEEE Trans. Cybern. 45 (1) (2015) 65–76.
[30] H. Modares, F.L. Lewis, Z.P. Jiang, H_∞ tracking control of completely unknown continuous-time systems via off-policy reinforcement learning, IEEE Trans. Neural Netw. Learn. Syst. 26 (10) (2015) 2550–2562.
[31] B. Kiumarsi, F.L. Lewis, H. Modares, A. Karimpour, M.B. Naghibi-Sistani, Reinforcement Q-learning for optimal tracking control of linear discrete-time systems with unknown dynamics, Automatica 50 (2014) 1167–1175.
[32] H. Modares, F.L. Lewis, M.B. Naghibi-Sistani, Adaptive optimal control of unknown constrained-input systems using policy iteration and neural networks, IEEE Trans. Neural Netw. Learn. Syst. 24 (10) (2013) 1513–1525.
[33] S.E. Lyshevski, Optimal control of nonlinear continuous-time systems: design of bounded controllers via generalized nonquadratic functionals, in: Proceedings of the IEEE ACC, 1998, pp. 205–209.
[34] K. Doya, Reinforcement learning in continuous time and space, Neural Comput. 12 (1) (2000) 219–245.
[35] K.G. Vamvodakis, F.L. Lewis, Online actor-critic algorithm to solve the continuous-time infinite horizon optimal control problem, Automatica 46 (5) (2010) 878–888.
[36] P.J. Werbos, A Menu of Designs for Reinforcement Learning Over Time, MIT Press, Cambridge, MA, 1991.
[37] P.J. Werbos, Approximate Dynamic Programming for Real-Time Control and Neural Modeling, Reinhold, New York, 1992.
[38] P.J. Werbos, Neural networks for control and system identification, in: Proceedings of the 28th IEEE CDC, 1989, pp. 260–265.
[39] P.J. Werbos, Stable adaptive control using new critic designs, 1998, [Online]. Available: http://arxiv.org/abs/adap-org/9810001.
[40] B. Kiumarsi, F.L. Lewis, Actor-critic-based optimal tracking for partially unknown nonlinear discrete-time systems, IEEE Trans. Neural Netw. Learn. Syst. 26 (1) (2014) 140–151.
[41] X. Luo, J. Si, Stability of direct heuristic dynamic programming for nonlinear tracking control using PID neural network, in: Proceedings of the IJCNN, 2013, pp. 1–7.

CHAPTER SIX

Addressing Adaptation and Learning in the Context of Model Predictive Control With Moving-Horizon Estimation

D.A. Copp, J.P. Hespanha
University of California, Santa Barbara, CA, United States

1 INTRODUCTION

Having an accurate model of a system to be controlled is often vital for effective control of that system. This is certainly true for a model predictive control (MPC) approach in which a finite-horizon online optimization problem is solved to determine an optimal control input given the system's dynamics and a desired control objective [1]. However, in most practical applications there are unknown parameters in the model of a system or, at least, uncertain parameters that are known only to be within some set of values. These uncertainties may include uncertain model parameters, input disturbances, and measurement noise. Because of this, much work on MPC approaches has involved investigation of robustness to model parameter uncertainty, input disturbances, and measurement noise. This work is known as robust MPC [2, 3], which also includes worst-case or min-max MPC [4].

An attractive, and perhaps less conservative, approach to controlling systems with parameter uncertainties is to update the model of the system with new estimates of the parameters as they become available, which is the underlying idea behind indirect adaptive control [5, 6]. Very little work has been done on adaptive MPC, but there are a few proposed approaches. Mayne and Michalska [7] proposed an adaptive MPC scheme that uses a standard estimator and certainty equivalence to update the model with the current estimates of the parameters. Adetola et al. [8]

Control of Complex Systems
http://dx.doi.org/10.1016/B978-0-12-805246-4.00006-9

investigated nonlinear systems that are affine with respect to unknown parameters and performed adaptive control by combining a parameter adjustment mechanism with robust MPC algorithms such as min-max MPC. A cost function is minimized with respect to feedback control policies and maximized with respect to the unknown parameters so that the MPC approach is robust to the worst-case values of the unknown parameters. For both of these approaches, it is assumed that the full state is available for feedback. This is often the case for MPC approaches so as to alleviate issues that arise from uncertainties, noise, and disturbances.

Unfortunately, in most practical applications the full state is not known or necessarily available for feedback. Because of this, output-feedback MPC should be considered, and an independent algorithm for estimating the state is needed. A convenient estimation algorithm for use with MPC is moving-horizon estimation (MHE). MHE can be used for estimation of the state of constrained nonlinear systems and similarly involves the solution of a finite-horizon online optimization problem where a criterion based on a finite number of past output measurements is minimized to find the best estimate of the state [9, 10]. It is straightforward to also incorporate parameter estimation into the formulation of MHE, so the state and parameters can both be estimated with the same estimator [11].

Copp and Hespanha [12] presented a framework for solving the output-feedback MPC problem with MHE that solves the control and estimation problems as a single min-max optimization problem. This framework already incorporates input disturbances and measurement noise. In this chapter we further incorporate uncertain model parameters into this framework and obtain parameter estimates by including the uncertain parameters as optimization variables. In this way we solve simultaneously the control problem and the state and parameter estimation problems, resulting in effective control of uncertain systems. Our approach can be likened to an indirect model reference adaptive control approach as described in the adaptive control literature [5, 6] in that, at each time step, new estimates of the uncertain parameters are computed and used to update the model and a new sequence of future control inputs that minimize an objective criterion is also simultaneously computed.

Because MPC and MHE involve the solution of an online optimization problem, this approach lends itself to adaptation to both constant and time-varying parameters because, at each time step, a new estimate is computed, and the model can be updated accordingly. We show in examples that when the system is sufficiently excited, this MPC with MHE approach is able to learn the true values of the uncertain parameters, but even if the system is not sufficiently excited to learn the true values of the parameters, this approach finds estimates that are consistent with the dynamics and often still enables effective control and disturbance rejection.

The main assumption for this work is that a saddle-point solution exists for the min-max optimization problem at each sampling time. This assumption presumes an appropriate form of observability for the closed-loop system and is a common requirement in game-theoretical approaches to control design [13]. For controllability, we additionally require that there exists a terminal cost that is an input-to-state stability (ISS) control Lyapunov function with respect to a disturbance input, which is a common assumption in MPC [14].

The rest of the chapter is organized as follows. In Section 2 we formulate the adaptive MPC with MHE problem that we would like to solve. Stability results that can be used to prove state boundedness and reference tracking are given in Section 3. In Section 4 we discuss several linear and nonlinear systems with parameter uncertainty and show that using our MPC with MHE scheme, we are able to not only stabilize the system but also estimate the correct values of the uncertain parameters. Finally, we provide some conclusions and directions for future work in Section 5.

2 PROBLEM FORMULATION

In the formulation of standard MPC and MHE problems, a time-varying nonlinear discrete-time process of the form

$$x_{t+1} = f_t(x_t, u_t, d_t), \quad y_t = g_t(x_t) + n_t, \quad \forall t \in \mathbb{Z}_{\geq 0} \tag{1}$$

is considered, with *state* x_t taking values in a set $\mathcal{X} \subset \mathbb{R}^{n_x}$. The inputs to this system are the *control input* u_t, which must be restricted to the set $\mathcal{U} \subset \mathbb{R}^{n_u}$, the *unmeasured disturbance* d_t, which is assumed to belong to the set $\mathcal{D} \subset \mathbb{R}^{n_d}$, and the *measurement noise* $n_t \in \mathbb{R}^{n_n}$. The signal $y_t \in \mathbb{R}^{n_y}$ denotes the *measured output* that is available for feedback.

We investigate MPC and MHE of processes with uncertain model parameters. These uncertain parameters are denoted by the vector θ, whose elements are known to belong to the set $\Theta \subset \mathbb{R}^{n_\theta}$. In this formulation the process dynamics depend explicitly on the uncertain parameter θ, so we redefine the process dynamics in Eq. (1) to include the uncertain parameters as

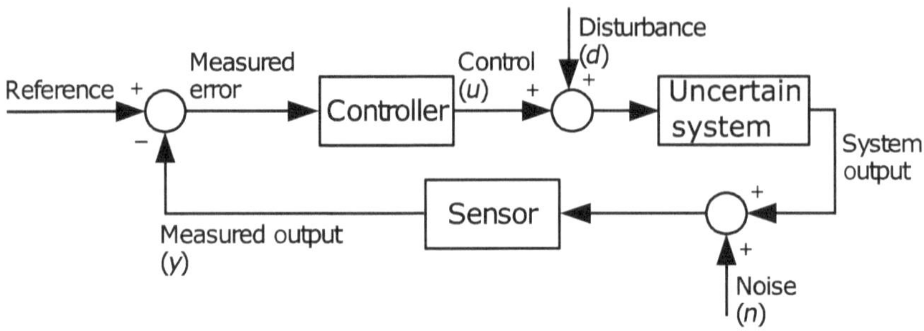

Fig. 1 The process given in Eq. (2).

$$x_{t+1} = f_t(x_t, \theta, u_t, d_t), \quad y_t = g_t(x_t, \theta) + n_t, \quad \forall t \in \mathbb{Z}_{\geq 0}. \tag{2}$$

We assume that θ is a constant parameter—that is, $\theta = \theta_t$ for all $t \in \mathbb{Z}_{\geq 0}$—but as will be shown later, we are still able to adapt to and learn changing parameter values.

A block diagram depicting the process given in Eq. (2) is shown in Fig. 1.

2.1 Moving-Horizon Estimation

In MHE the current state of the system x_t at time t is estimated by the solving of a finite-horizon online optimization problem with use of a finite number of past measurements [9]. If we consider a finite horizon of L time steps, then the objective of the MHE problem is to find an estimate of the current state x_t so as to minimize a criterion of the form

$$\sum_{s=t-L}^{t} \eta_s \left(y_s - g_s(x_s)\right) + \sum_{s=t-L}^{t-1} \rho_s(d_s), \tag{3}$$

given the system dynamics (1). The functions $\eta_s(\cdot)$ and $\rho_s(\cdot)$ are assumed to take nonnegative values. This is similar to the MHE criterion considered in [9, 10].

If the system dynamics also include uncertain model parameters, as in Eq. (2), the MHE problem can be formulated so as to estimate both the current state x_t and the uncertain parameter θ. Then the MHE problem can be written as

$$\min_{\hat{x}_{t-L} \in \mathcal{X}, \hat{d}_{t-L:t-1} \in \mathcal{D}, \hat{\theta} \in \Theta} \sum_{s=t-L}^{t} \eta_s \left(y_s - g_s(\hat{x}_s, \hat{\theta})\right) + \sum_{s=t-L}^{t-1} \rho_s(\hat{d}_s), \tag{4}$$

where the initial state x_{t-L} is constrained to belong to the set $\mathcal{X}$, each element of the input disturbance sequence $d_{t-L:t-1}$ is assumed to belong to the set $\mathcal{D}$, and the uncertain parameter θ is known to belong to the set Θ. Throughout this chapter, given two times t_1 and t_2 with $t_1 < t_2$, we use the notation $x_{t_1:t_2}$ to denote the time series $x_{t_1}, x_{t_1+1}, \ldots, x_{t_2-1}, x_{t_2}$. An estimate of the current state is then determined from the dynamics (2) given the known past control inputs applied $u_{t-L:t-1}$ and estimates of the initial state $\hat{x}_{t-L}$, the input disturbance sequence $\hat{d}_{t-L:t-1}$, and the uncertain parameter $\hat{\theta}$. The optimization (4) is solved again at each time t in a receding-horizon fashion.

2.2 Model Predictive Control

In MPC a sequence of future control inputs that achieve a desired control objective is computed by the solving of a finite-horizon online optimization problem with use of an estimate of the current state $\hat{x}_t$ and the system dynamics [1]. If we consider a finite-horizon of T time steps, then the objective of the MPC problem is to find a sequence of future control inputs $u_{t:t+T-1}$ that minimizes a criterion of the form

$$\sum_{s=t}^{t+T-1} (c_s(x_s, u_s, d_s) - \rho_s(d_s)) + q_{t+T}(x_{t+T}), \tag{5}$$

given the system dynamics (1). The functions $c_s(\cdot)$, $\rho_s(\cdot)$, and $q_{t+T}(\cdot)$ are all assumed to take nonnegative values. The negative sign in front of $\rho_t(\cdot)$ penalizes the maximizer for using large values of d_t. The function $q_{t+T}(x_{t+T})$ is a terminal cost that penalizes the "final" state x_{t+T} and is needed to prove stability (see Assumption 3). Criterion (5) is similar to the closed-loop min-max MPC criterion considered in [15, 16].

If the system dynamics also include uncertain model parameters, as in Eq. (2), the MPC criterion (5) can be reformulated so as to incorporate worst-case values of the uncertain parameters θ, and the MPC problem can be written as

$$\min_{\hat{u}_{t:t+T-1}\in\mathcal{U}} \max_{\hat{d}_{t:t+T-1}\in\mathcal{D},\hat{\theta}\in\Theta} \sum_{s=t}^{t+T-1} \left(c_s(x_s, \hat{u}_s, \hat{d}_s) - \rho_s(\hat{d}_s)\right) + q_{t+T}(x_{t+T}), \tag{6}$$

where each element of the future control input sequence $u_{t:t+T-1}$ is constrained to belong to the set $\mathcal{U}$, each element of the future disturbance sequence $d_{t:t+T-1}$ is assumed to belong to the set $\mathcal{D}$, and the uncertain parameter θ is known to belong to the set Θ. To overcome the conservativeness of open-loop control, at each time step t the first element $\hat{u}_t^*$ of the future control input sequence $\hat{u}_{t:t+T-1}^*$ that is the solution to Eq. (6) is applied to the system, and the optimization (6) is solved again at each time step in a receding-horizon fashion. This is similar to the adaptive MPC problem with exogenous inputs considered in [17].

2.3 Adaptive MPC Combined with MHE

Next we show how both the MPC problem (6) and the MHE problem (4) can be formulated and solved simultaneously as a single min-max optimization problem.

Taking criterion (5) and subtracting criterion (3) gives a criterion of the form

$$J_t := \sum_{s=t}^{t+T-1} c_s(x_s, u_s, d_s) + q_{t+T}(x_{t+T}) - \sum_{s=t-L}^{t} \eta_s(n_s) - \sum_{s=t-L}^{t+T-1} \rho_s(d_s), \quad (7)$$

which contains $T \in \mathbb{Z}_{\geq 1}$ terms of the running cost $c_s(x_s, u_s, d_s)$, which recede as the current time t advances, $L + 1 \in \mathbb{Z}_{\geq 1}$ terms of the measurement cost $\eta_s(n_s)$, and $L + T \in \mathbb{Z}_{\geq 1}$ terms of the cost on the input disturbance $\rho_s(d_s)$. Again the function $q_{t+T}(x_{t+T})$ acts as a terminal cost so as to penalize the "final" state at time $t + T$. The functions $c_t(\cdot)$, $q_{t+T}(\cdot)$, $\eta_t(\cdot)$, and $\rho_t(\cdot)$ in Eq. (7) are all assumed to take nonnegative values. We use finite horizons into the past and into the future to decrease the computational complexity of the optimization problem, and we use online optimization to generate closed-loop solutions.

The *control objective* is to select the control signal $u_t \in \mathcal{U}$, $\forall t \in \mathbb{Z}_{\geq 0}$, so as to minimize criterion (7) under worst-case assumptions on the *unknown* system's initial condition $x_{t-L} \in \mathcal{X}$, unmeasured disturbances $d_t \in \mathcal{D}$, measurement noise $n_t \in \mathbb{R}^{n_n}$, and uncertain parameter $\theta \in \Theta$, for all $t \in \mathbb{Z}_{\geq 0}$, subject to the constraints imposed by the system dynamics (2) and the measurements $y_{t-L:t}$ collected up to the current time t.

Because the objective is to optimize criterion (7) at the current time t so as to compute control inputs u_s for times $s \geq t$, there is no reason to penalize other irrelevant terms. For instance, the first summation in Eq. (7) starts at time t because there is no reason to penalize the running cost $c_s(x_s, u_s, d_s)$ for past time instants $s < t$. There is also no reason to consider the values of future measurement noise at times $s > t$ as they will not affect choices made at time t. Thus the second summation in Eq. (7) ends at time t. However, all values of the unmeasured disturbance d_s for $t - L \leq s \leq t + T - 1$ need to be considered because past values affect the (unknown) current state x_t, and future values affect the future values of the running cost.

Boundedness of Eq. (7) by a constant γ guarantees that

$$\sum_{s=t}^{t+T-1} c_s(x_s, u_s, d_s) + q_{t+T}(x_{t+T}) \leq \gamma + \sum_{s=t-L}^{t} \eta_s(n_s) + \sum_{s=t-L}^{t+T-1} \rho_s(d_s). \quad (8)$$

This shows that we can bound the running and final costs involving the future states $x_{t:t+T}$ in terms of bounds on the noise and disturbance.

Remark 1 (Quadratic case). While the results presented here are general, it may be easier to gain intuition on the results if we consider a quadratic criterion for Eq. (7) such as $c_t(x_t, u_t, d_t) := \|x_t\|^2 + \|u_t\|^2$, $\eta_t(n_t) := \|n_t\|^2$, $\rho_t(d_t) := \|d_t\|^2$. For this choice of the criterion, boundedness of Eq. (7) guarantees that the state x_t and input u_t are ℓ_2 provided that the disturbance d_t and noise n_t are also ℓ_2 (cf. Eq. 8).

With the objective of optimizing criterion (7) at a given time $t \in \mathbb{Z}_{\geq 0}$ for the future control inputs $u_{t:t+T-1}$ and worst-case estimates of x_{t-L}, $d_{t-L:t+T-1}$, and θ, the combined adaptive MPC with MHE problem amounts to the solving of the following min-max optimization:

$$J_t^* = \min_{\hat{u}_{t:t+T-1} \in \mathcal{U}} \max_{\hat{x}_{t-L} \in \mathcal{X}, \hat{d}_{t-L:t+T-1} \in \mathcal{D}, \hat{\theta} \in \Theta}$$

$$\sum_{s=t}^{t+T-1} c_s(\hat{x}_s, \hat{u}_s, \hat{d}_s) + q_{t+T}(\hat{x}_{t+T}) - \sum_{s=t-L}^{t} \eta_s\left(y_s - g_s(\hat{x}_s, \hat{\theta})\right) - \sum_{s=t-L}^{t+T-1} \rho_s(\hat{d}_s), \quad (9)$$

with the understanding that

$$\hat{x}_{s+1} = \begin{cases} f_s(\hat{x}_s, \hat{\theta}, u_s, \hat{d}_s) & \text{for } t - L \leq s < t, \\ f_s(\hat{x}_s, \hat{\theta}, \hat{u}_s, \hat{d}_s) & \text{for } t \leq s < t + T. \end{cases}$$

We can view the optimization variables $\hat{x}_{t-L}$, $\hat{d}_{t-L:t+T-1}$, and $\hat{\theta}$ as (worst-case) estimates of the initial state, disturbances, and uncertain parameter, respectively, based on the past inputs $u_{t-L:t-1}$ and outputs $y_{t-L:t}$ available at time t.

Just as in the MPC problem, after solving this optimization problem at each time t, we use as the control input the first element of the sequence

$$\hat{u}^*_{t:t+T-1} = \{\hat{u}^*_t, \hat{u}^*_{t+1}, \hat{u}^*_{t+2}, \ldots, \hat{u}^*_{t+T-1}\} \in \mathcal{U}$$

that minimizes Eq. (9), leading to the control law

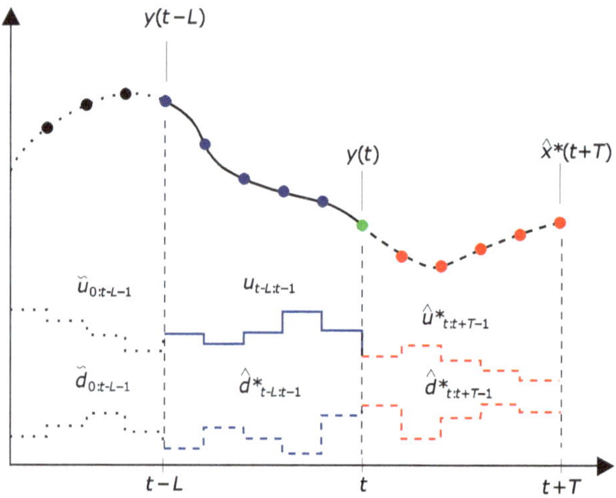

Fig. 2 Example solution to the combined finite-horizon control and estimation problem. The elements from $t-L$ to t correspond to the moving-horizon estimation problem, and the elements from t to $t+T$ correspond to the model predictive control problem. The variables denoted as $\hat{\cdot}$ are optimization variables, the variables denoted as $\tilde{\cdot}$ are not relevant for the optimization, and the other variables are known.

$$u_t = \hat{u}_t^*, \qquad \forall t \geq 0. \tag{10}$$

A depiction of an example solution to the combined finite-horizon control and estimation problem is shown in Fig. 2.

3 STABILITY RESULTS

Next we discuss under what appropriate assumptions the control law (10) leads to boundedness of the state of the closed-loop system resulting from the finite-horizon optimization introduced in Section 2.3.

To implement the control law (10), the outer minimization in Eq. (9) must lead to a finite value for the optimum. For the stability results in this section, we require the existence of a finite-valued saddle-point solution to the min-max optimization in Eq. (9), which is a common requirement in game-theoretical approaches to control design [13].

Assumption 1 (Saddle point). The min-max optimization (9) always has a finite-valued saddle-point solution for which the min and max commute. Specifically, for all $t \in \mathbb{Z}_{\geq 0}$, $u_{t-L:t-1} \in \mathcal{U}$, $y_{t-L:t}$, there exist $J_t^* \in \mathbb{R}$, $\hat{x}^*_{t-L} \in \mathcal{X}$, $\hat{u}^*_{t:t+T-1} \in \mathcal{U}$, $\hat{d}^*_{t-L:t+T-1} \in \mathcal{D}$, and $\hat{\theta}^* \in \Theta$ such that

$$\begin{aligned} J_t^* &= \min_{\hat{u}_{t:t+T-1}\in\mathcal{U}} \max_{\hat{x}_{t-L}\in\mathcal{X},\hat{d}_{t-L:t+T-1}\in\mathcal{D},\hat{\theta}\in\Theta} J_t \\ &= \max_{\hat{x}_{t-L}\in\mathcal{X},\hat{d}_{t-L:t+T-1}\in\mathcal{D},\hat{\theta}\in\Theta} \min_{\hat{u}_{t:t+T-1}\in\mathcal{U}} J_t \\ &< \infty. \end{aligned}$$

Assumption 1 presumes an appropriate form of *observability/detectability* adapted to the criterion $\sum_{s=t}^{t+T-1} c_s(x_s, u_s, d_s)$. In particular, it implies that the size of the *current* state can be bounded with use of past outputs and past/future input disturbances, regardless of the value of $\theta \in \Theta$.

To ensure controllability and to establish state boundedness under the control (10) defined by the finite-horizon optimization (9), we require additional assumptions regarding the dynamics and the terminal cost $q_t(\cdot)$.

Assumption 2 (Reversible dynamics). For every $t \in \mathbb{Z}_{\geq 0}$, $x_{t+1} \in \mathcal{X}$, $\theta \in \Theta$, and $u_t \in \mathcal{U}$, there exists a state $\tilde{x}_t \in \mathcal{X}$ and a disturbance $\tilde{d}_t \in \mathcal{D}$ such that

$$x_{t+1} = f_t(\tilde{x}_t, \theta, u_t, \tilde{d}_t). \tag{11}$$

Assumption 3 (ISS control Lyapunov function). The terminal cost $q_t(\cdot)$ is an *ISS control Lyapunov function*, in the sense that, for every $t \in \mathbb{Z}_{\geq 0}$, $x \in \mathcal{X}$, $d \in \mathcal{D}$, and $\theta \in \Theta$, there exists a control $u \in \mathcal{U}$ such that

$$q_{t+1}\left(f_t(x, \theta, u, d)\right) - q_t(x) \leq -c_t(x, u, d) + \rho_t(d). \tag{12}$$

The mild Assumption 2 essentially implies that the sets of disturbances $\mathcal{D}$ and past states $\mathcal{X}$ are sufficiently rich to allow for a jump to any future state in $\mathcal{X}$. In fact, for linear dynamics, Assumption 2 is satisfied if the state-space A matrix has no eigenvalues at the origin (eg, if it results from the time discretization of a continuous-time system).

Assumption 3 plays the role of the common assumption in MPC that the terminal cost must be a control Lyapunov function for the closed loop [14]. Without the disturbance d_t, Eq. (12) would imply that $q_t(\cdot)$ could be viewed as a control Lyapunov function that decreases along system trajectories for an appropriate control input u_t [18]. With the disturbance d_t, $q_t(\cdot)$ should be viewed as an ISS control Lyapunov function that satisfies an ISS condition for the disturbance input d_t and an appropriate control input u_t [19]. In the case of linear dynamics and a quadratic cost function, a terminal cost $q_t(\cdot)$ can typically be found by the solving of a system of linear matrix inequalities.

3.1 State Boundedness

The following theorem provides a bound that can be used to prove boundedness of the state when the control signal is computed by the solving of the finite-horizon optimization (9).

Theorem 1 (Finite-horizon cost-to-go bound). *Suppose that Assumptions 1, 2, and 3 hold. Then there exists a finite constant* J_0^* *and vectors* $\tilde{d}_s \in \mathcal{D}$, $\tilde{n}_s \in \mathbb{R}^{n_n}$, $\forall s \in \{0, 1, \dots, t-L-1\}$ *for which*

$$\eta_s(\tilde{n}_s), \rho_s(\tilde{d}_s) < \infty, \quad \forall s \in \{0, 1, t-L-1\},$$

and the trajectories of the process (2) *with control* (10) *defined by the finite-horizon optimization* (9) *satisfy*

$$c_t(x_t, u_t, d_t) \leq J_0^* + \sum_{s=0}^{t-L-1} \eta_s(\tilde{n}_s) + \sum_{s=0}^{t-L-1} \rho_s(\tilde{d}_s) + \sum_{s=t-L}^{t} \eta_s(n_s) + \sum_{s=t-L}^{t} \rho_s(d_s), \quad \forall t \in \mathbb{Z}_{\geq 0}. \tag{13}$$

Proof. This result is an extension of Theorem 1 presented in [12]. If the state is augmented such that $\bar{x}_t = [x_t\ \theta]^\top$, and the process is defined as $\bar{x}_{t+1} = [\bar{f}_t(\bar{x}_t, u_t, d_t)\ \theta]^\top$, then the same proof used for Theorem 1 in [12] can be applied here with use of $\bar{x}_t$ in place of x_t. □

We refer the reader to [12] for a discussion on how this result can be used to ensure state boundedness and reference tracking.

4 SIMULATION STUDY

In this section we consider several examples of systems with parametric uncertainty and present closed-loop simulations using the control approach described in Section 2. For all of the following examples, we use a cost function of the form

$$J_t = \sum_{s=t}^{t+T-1} \|h_s(x_s)\|_2^2 + \lambda_u \sum_{s=t}^{t+T-1} \|u_s\|_2^2 - \lambda_n \sum_{s=t-L}^{t} \|n_s\|_2^2 - \lambda_d \sum_{s=t-L}^{t+T-1} \|d_s\|_2^2, \tag{14}$$

where $h_s(x_s)$ is a function of the state x_s that is especially relevant for the example under consideration, and λ_u, λ_n, and λ_d are positive weighting constants.

Given the optimization criterion (14), the following examples involve optimization of this criterion with respect to the future control inputs $u_{t:t+T-1}$ under worst-case assumptions on x_{t-L}, $d_{t-L:t+T-1}$, and θ by the solving of the following min-max optimization problem:

$$\min_{\hat{u}_{t:t+T-1}\in\mathcal{U}} \quad \max_{\hat{x}_{t-L}\in\mathcal{X},\hat{d}_{t-L:t+T-1},\hat{\theta}\in\Theta} J_t. \tag{15}$$

The first time this optimization is solved, guesses for the initial values of the uncertain parameter θ, initial state x_{t-L}, past control inputs $u_{t-L:t-1}$, and input disturbances $d_{t-L:t-1}$ need to be made. Then values for the past states $x_{t-L+1:t}$ that are consistent with the dynamics are picked. These states can be used to determine the output measurements $y_{t-L:t}$, and then the optimization (15) can be solved for the first time. At subsequent times, all of the variables from the solution of Eq. (15) at the previous time step (after moving the variables away from the constraints) can be used as a "warm start" for solving Eq. (15) at the current time step.

To solve the optimization (15) we have developed a primal-dual-like interior-point method that finds the saddle-point solution. Details of this method can be found in [20]. Under appropriate convexity assumptions, this method is guaranteed to terminate at a global solution. However, the simulation results show that it also converges in problems that are severely nonconvex, such as most of the examples below.

Example 1 (Linear system—uncertain gain and poles). Consider a discrete-time linear system described by the transfer function

$$G(z) = \frac{b}{(z-p_1)(z-p_2)}, \tag{16}$$

where $p = [p_1\ p_2]^\top$ denotes the uncertain pole locations p_1 and p_2, which are assumed to belong to the set $\mathcal{P} := \{p \in \mathbb{R}^2 : 0 \leq p_i \leq 2, i = 1, 2\}$, so they may be stable or unstable. The parameter b is an uncertain gain assumed to belong in the interval $\mathcal{B} := \{b \in \mathbb{R} : 1 \leq b \leq 5\}$.

The transfer function (16) can be rewritten in state-space controllable canonical form as

$$\begin{aligned} x_{t+1} &= \begin{bmatrix} p_1 + p_2 & -p_1 p_2 \\ 1 & 0 \end{bmatrix} x_t + \begin{bmatrix} 1 \\ 0 \end{bmatrix} (u_t + d_t), \\ y_t &= \begin{bmatrix} 0 & b \end{bmatrix} x_t + n_t, \quad \forall t \in \mathbb{Z}_{\geq 0}, \end{aligned} \tag{17}$$

where y_t is the measured output at time t with noise n_t, and d_t is an additive input disturbance. For all $t \in \mathbb{Z}_{\geq 0}$, the control input u_t is constrained to belong in the set $\mathcal{U} := \{u_t \in \mathbb{R} : \|u_t\|_\infty \leq 8\}$, and the input disturbance d_t is assumed to belong to the set $\mathcal{D} := \{d_t \in \mathbb{R} : \|d_t\|_\infty \leq 0.1\}$.

By defining $a_1 = p_1 + p_2$ and $a_2 = p_1 p_2$, we can reparametrize the state-space model (17) as

$$\begin{aligned} x_{t+1} &= \begin{bmatrix} a_1 & -a_2 \\ 1 & 0 \end{bmatrix} x_t + \begin{bmatrix} 1 \\ 0 \end{bmatrix} (u_t + d_t), \\ y_t &= \begin{bmatrix} 0 & b \end{bmatrix} x_t + n_t, \quad \forall t \in \mathbb{Z}_{\geq 0}. \end{aligned} \tag{18}$$

Letting $a = [a_1\ a_2]^\top$, we assume the uncertain parameter a belongs to $\mathcal{A} := \{a \in \mathbb{R}^2 : 0 \leq a_i \leq 4, i = 1, 2\}$. This set $\mathcal{A}$ is conservative, and a tighter nonconvex set could be used. Now the model (18) is linear in the uncertain parameters. This is a standard problem that can be solved with classical adaptive control techniques. We will show that our MPC with MHE approach can solve this problem, and in following examples we will see that our approach does not require reparameterization such that the system is linear in the uncertain parameters.

The uncertain parameters (a and b) can be estimated by our including them as optimization variables in the following problem:

$$\min_{\hat{u}_{t:t+T-1} \in \mathcal{U}} \quad \max_{\hat{x}_{t-L} \in \mathcal{X}, \hat{d}_{t-L:t+T-1} \in \mathcal{D}, \hat{a} \in \mathcal{A}, \hat{b} \in \mathcal{B}}$$

$$\sum_{s=t}^{t+T} \|y_s - r_s\|_2^2 + \lambda_u \sum_{s=t}^{t+T-1} \|\hat{u}_s\|_2^2 - \lambda_n \sum_{s=t-L}^{t} \|n_s\|_2^2 - \lambda_d \sum_{s=t-L}^{t+T-1} \|\hat{d}_s\|_2^2, \tag{19}$$

where r_t is a desired reference signal for the output of the system to follow. Figs. 3 and 4 show simulations of the resulting closed-loop system for a square-wave reference defined as $r_t = 10\text{sgn}(\sin(0.4t))$ and the backward and forward horizon lengths chosen as $L = 10$ and $T = 10$, respectively. The weights in the cost function are chosen to be $\lambda_u = 0.1$, $\lambda_n = 1000$, and $\lambda_d = 1000$. In this simulation the actual input disturbance d_t and measurement noise n_t are unmeasured Gaussian independently and identically distributed (i.i.d.) random variables with zero mean and standard deviations of 0.001 and 0.005, respectively.

Fig. 3 shows the combined MPC and MHE method successfully regulating the output of the system to the given square-wave reference trajectory. The system is initialized with incorrect guesses for the initial values of the uncertain parameters, and zero control input (ie, $u_t = 0$) is applied for the first $L = 10$ time steps. After that point, starting at time $t = 11$, the optimization problem (19) is solved at each time step, and the computed control input $\hat{u}_t^*$ is applied in a receding-horizon fashion.

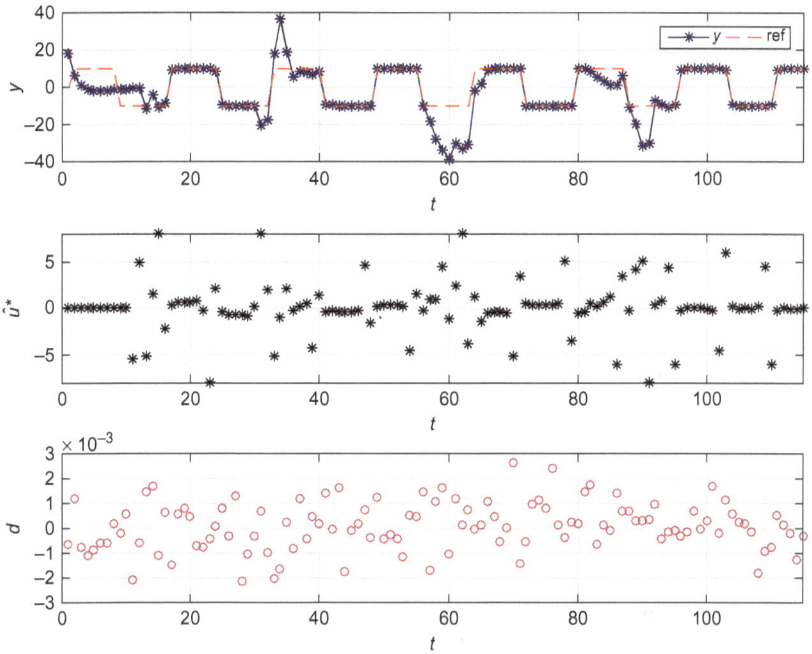

Fig. 3 Linear system: output and inputs. The top plot shows the measured output (*asterisks*) and the reference signal (*dashes*). The second plot shows the input $\hat{u}_t^*$ applied to the system, and the third plot shows the unmeasured disturbance input d_t that the system is subjected to.

At several times throughout the simulation the true model of the system (17) is altered by the changing of the value of the gain or the poles (which can be seen in Fig. 4). The estimates of the gain and pole locations shown in Fig. 4 are obtained from the estimates of the uncertain parameters a and b from the solution of the optimization (19). Fig. 4 shows that the estimates of the gain and pole locations converge to their true values. Even after the true values of the gain and pole locations have been changed during the simulation, the combined MPC and MHE scheme is able to adapt to the changing system, effectively regulating the system to the reference trajectory and correctly learning the new parameters of the model.

Example 2 (Inverted pendulum—uncertain mass and friction). Consider an inverted pendulum actuated by a torque at the base as shown in Fig. 5 and described by the model

$$ml^2\ddot{\phi} = mgl\sin(\phi) - b\dot{\phi} + \tau,$$

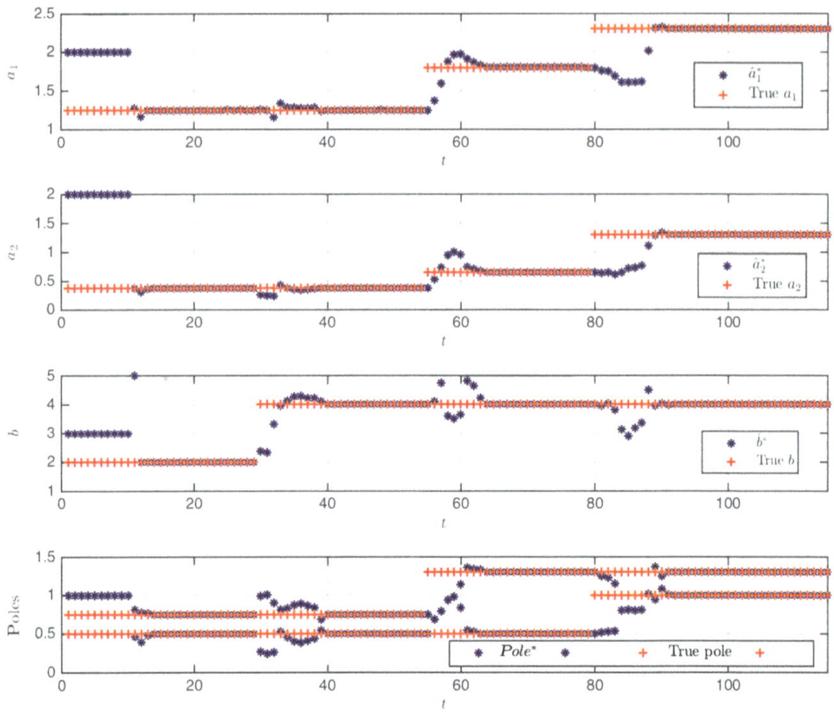

Fig. 4 Linear system: parameters. The top two plots show the true values of a_1 and a_2 (*crosses*) and their estimated values (*asterisks*). The third plot shows the true value of the gain b (*crosses*) and its estimated value (*asterisks*). The bottom plot shows the true values of the poles p_1 and p_2 (*crosses*) and their estimated values (*asterisks*) computed from $\hat{a}_1^*$ and $\hat{a}_2^*$.

where m is the mass at the end of the pendulum, l is the length of the link, ϕ is the angle from vertical, g is the gravitational constant, b is the coefficient of friction, and τ is the torque applied at the base.

We can rewrite this model in state-space form as

$$\begin{aligned} \dot{x}_1 &= x_2, \\ \dot{x}_2 &= \frac{g}{l}\sin(x_1) - \frac{b}{ml^2}x_2 + \frac{1}{ml^2}u, \end{aligned} \tag{20}$$

where $x_1 = \phi$, $x_2 = \dot{\phi}$, and $u = \tau$.

By letting $l = 1$, $g = 9.81$, and $a = 1/m$, adding an input disturbance d, and discretizing the model using Euler's method with time step Δt, we find the system (20) becomes

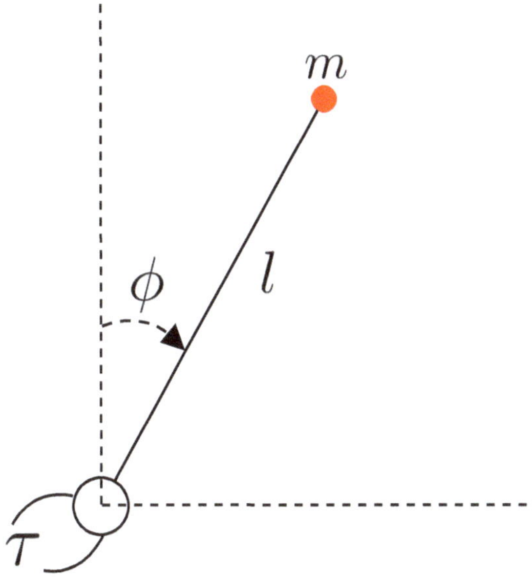

Fig. 5 The pendulum considered in Examples 2 and 3.

$$\begin{aligned} x_{1,t+1} &= x_{1,t} + \Delta t\, x_{2,t}, \\ x_{2,t+1} &= x_{2,t} + \Delta t \left(9.81 \sin(x_{1,t}) - abx_{2,t} + a(u_t + d_t)\right), \\ y_t &= x_{1,t} + n_t, \quad \forall t \in \mathbb{Z}_{\geq 0}, \end{aligned} \tag{21}$$

where y_t is the measurement available at time t with noise n_t. According to this model, a noisy measurement of the angle x_1 is available at each time t.

The inverse of the mass and the coefficient of friction (a and b, respectively) are uncertain but are assumed to belong to the sets $\mathcal{A} := \{a \in \mathbb{R} : 1/2 \leq a \leq 1\}$ and $\mathcal{B} := \{b \in \mathbb{R} : 0.2 \leq b \leq 0.7\}$, respectively. The control input u_t is constrained to the set $\mathcal{U} := \{u_t \in \mathbb{R} : \|u_t\|_\infty \leq 5\}$, and the disturbance input d_t is assumed to belong to $\mathcal{D} := \{d_t \in \mathbb{R} : \|d_t\|_\infty \leq 0.3\}$ for all $t \in \mathbb{Z}_{\geq 0}$.

The control objective is to regulate the output (the noisy measurement of the angle ϕ) to a desired reference. The uncertain mass and coefficient of friction can be determined with use of estimates of the parameters a and b in Eq. (21). We can estimate these parameters by including them as optimization variables in the following problem:

$$\min_{\hat{u}_{t:t+T-1} \in \mathcal{U}} \quad \max_{\hat{x}_{t-L} \in \mathcal{X}, \hat{d}_{t-L:t+T-1} \in \mathcal{D}, \hat{a} \in \mathcal{A}, \hat{b} \in \mathcal{B}}$$

$$\sum_{s=t}^{t+T} \|y_s - r_s\|_2^2 + \lambda_u \sum_{s=t}^{t+T-1} \|\hat{u}_s\|_2^2 - \lambda_n \sum_{s=t-L}^{t} \|n_s\|_2^2 - \lambda_d \sum_{s=t-L}^{t+T-1} \|\hat{d}_s\|_2^2. \tag{22}$$

A noteworthy challenge in this nonlinear problem is that the unknown parameters a and b appear multiplied by the unmeasurable state x_2 in the system dynamics (21).

Fig. 6 shows a simulation of the closed-loop system with the discrete time step chosen as $\Delta t = 0.2$ and a square-wave reference given as $r_t = 5(\pi/180)\text{sgn}(\sin(0.5t))$. The backward and forward horizon lengths are chosen to be $L = 6$, and $T = 7$, respectively. The weights in the cost function are chosen to be $\lambda_u = 0$ (ie, the control input is not penalized), $\lambda_n = 1000$, and $\lambda_d = 10$. In this simulation the actual input disturbance d_t and measurement noise n_t are unmeasured Gaussian i.i.d. random variables

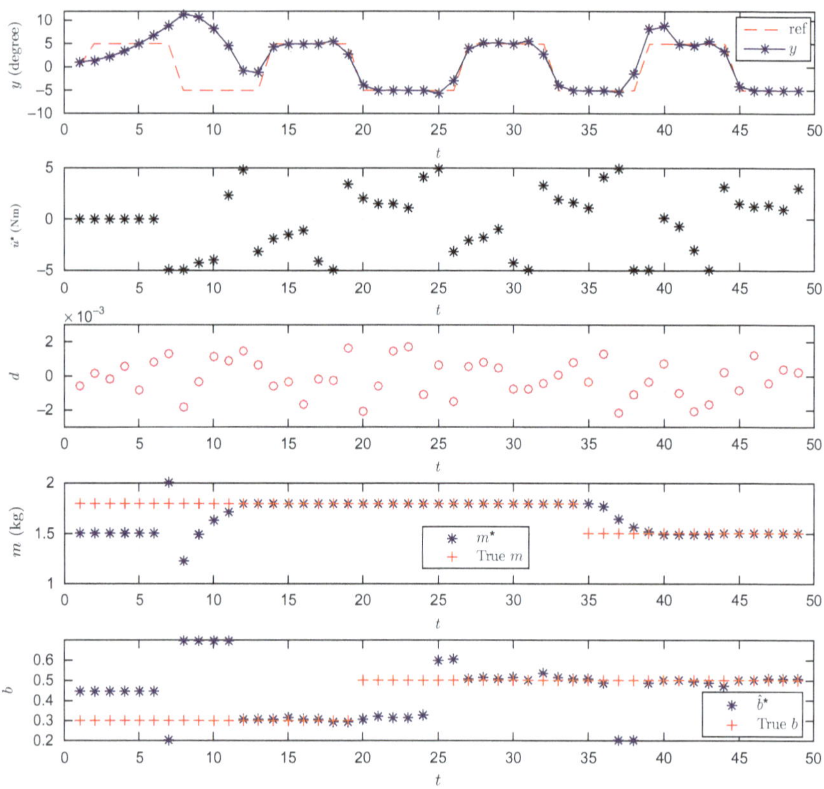

Fig. 6 Inverted pendulum: uncertain mass and friction. The top plot shows the measured output (*asterisks*) tracking the square reference signal (*dashes*). The second plot shows the control input $\hat{u}_t^*$ that is applied. The third plot shows the unmeasured input disturbance d_t that the system is subjected to. The bottom two plots show the true values of the mass and coefficient of friction (*crosses*) and the estimated values of the mass m^* (computed from $\hat{a}^*$) and coefficient of friction $\hat{b}^*$ (*asterisks*).

with zero mean and standard deviations of 0.001 and 0.0001, respectively. The system is initialized with incorrect guesses for the initial values of the uncertain parameters, and zero control input (ie, $u_t = 0$) is applied for the first $L = 6$ time steps. After that point, starting at time $t = 7$, the optimization problem (22) is solved at each time step, and the computed control input $\hat{u}_t^*$ is applied in a receding-horizon fashion.

As in the previous example the true values of the uncertain parameters (the mass and coefficient of friction) are changed several times throughout the simulation. Even as the true values of the mass and coefficient of friction change, the control input $\hat{u}_t^*$, computed by the solving of the optimization (22), is able to successfully regulate the output of the system to the reference trajectory, and the estimates of the uncertain mass m and coefficient of friction b converge to their true values.

Example 3 (Inverted pendulum—stabilization and disturbance rejection without learning true parameter values). This example shows that this adaptive MPC with MHE approach can stabilize uncertain systems even when the systems are not persistently excited and the true values of the uncertain parameters are not learned. Furthermore, the results of this example show that this estimation and control approach is not only robust to the model uncertainty but is also able to reject large input disturbances.

Again we consider an inverted pendulum as depicted in Fig. 5 and described by a discretized model of the form (21). This time, rather than following a reference trajectory, the control objective is to stabilize the system at the unstable equilibrium $x_1 = 0$. This means that the same optimization given in Eq. (22) is solved but with $r_t = 0$. Fig. 7 shows a simulation of the resulting closed-loop system.

For this example the parameters in the optimization (22) are chosen to be the same as in Example 2 except with respect to the input disturbance d_t. In this example the unmeasured disturbance d_t is larger and assumed to belong to the set $\mathcal{D} := \{d_t \in \mathbb{R} : \|d_t\|_\infty \leq 0.5\}$. The weight on the input disturbance in Eq. (22) is chosen as $\lambda_d = 1$. The actual input disturbance that the system is subjected to is a Gaussian i.i.d. random variable with zero mean and a standard deviation of 0.15. Again the system is initialized with guesses for the initial values of the uncertain parameters, and zero control input (ie, $u_t = 0$) is applied for the first $L = 6$ time steps. Starting at time $t = 7$, the optimization problem (22) is solved at each time step, and the computed control input $\hat{u}_t^*$ is applied in a receding-horizon fashion.

Fig. 7 shows that the system is not sufficiently excited to correctly learn the true values of the mass m and coefficient of friction b. However,

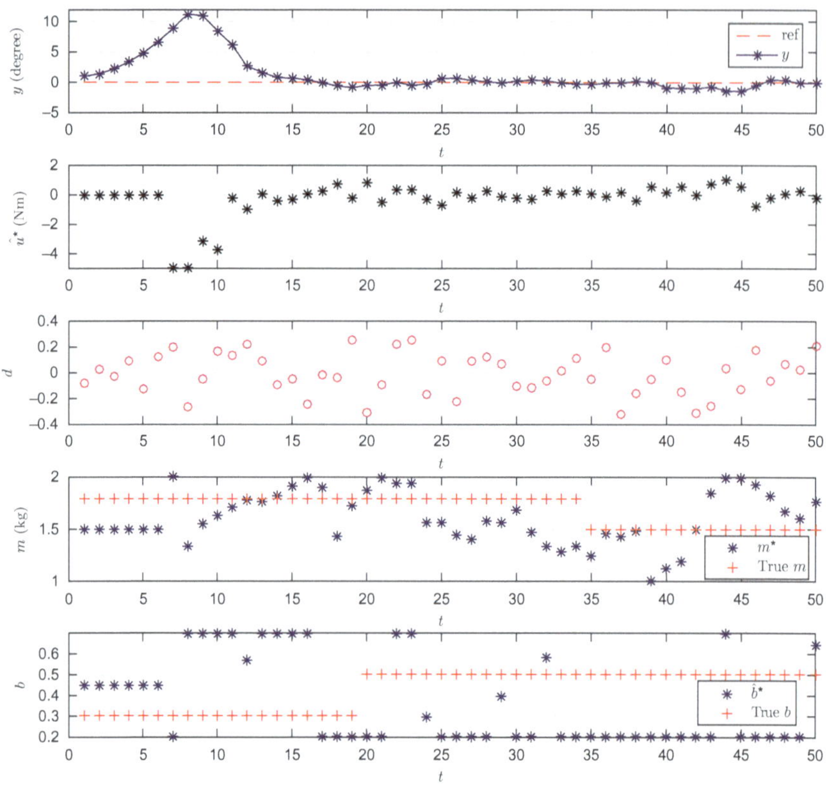

Fig. 7 Inverted pendulum: stabilization and disturbance rejection. The top plot shows the output (*asterisks*) converging to the unstable equilibrium $\phi = 0$ (*dashes*). The second plot shows the control input $\hat{u}_t^*$ that is applied. The third plot shows the large unknown input disturbance d_t that the system is subjected to. The fourth and fifth plots show the true values of the mass and coefficient of friction (*crosses*) and the estimated values of the mass m^* (computed from $\hat{a}^*$) and coefficient of friction $\hat{b}^*$ (*asterisks*).

the control input $\hat{u}_t^*$ computed by the solving of the optimization (22) is nonetheless able to stabilize the system (even as the true values of the uncertain parameters change) and reject the large unmeasured disturbance input.

Example 4 (Nonlinear pursuit-evasion—uncertain wind). In this example, we consider a two-player pursuit-evasion game where the pursuer is modeled as a unicycle vehicle and the evader is modeled as a double integrator. The pursuer is an aerial vehicle that is subject to wind disturbances and the evader is a ground vehicle that is not susceptible to the wind. A nonlinear discrete-time model of the overall system is given as follows:

$$x_{t+1} = x_t + \begin{bmatrix} v\cos\phi_t + w_1 \\ v\sin\phi_t + w_2 \\ u_t \\ d_t \end{bmatrix}, \qquad y_t = x_t + n_t, \quad \forall t \in \mathbb{Z}_{\geq 0}. \tag{23}$$

The state of the system is denoted by $x_t = [p_t\ \phi_t\ z_t]^\top$, where $p_t = [p_1\ p_2]_t^\top$ denotes the planar position of the pursuer, ϕ_t denotes the orientation of the pursuer, and $z_t = [z_1\ z_2]_t^\top$ denotes the planar position of the evader. The planar wind speed is denoted by $w = [w_1\ w_2]^\top$, where w_1 is the component of the wind speed in the x-direction, w_2 is the component of the wind speed in the y-direction, and w is known to belong to the set $\mathcal{W} := \{w \in \mathbb{R}^2 : \|w\|_\infty \leq 0.05\}$. The control input u_t is constrained to belong in the set $\mathcal{U} := \{u_t \in \mathbb{R} : \|u_t\|_\infty \leq 0.35\}$. The evader's velocity is given by $d_t = \left[d_1\ d_2\right]_t^\top$ and is constrained to the set $\mathcal{D} := \{d_t \in \mathbb{R}^2 : \|d_t\|_\infty \leq 0.05\}$, and $n_t \in \mathbb{R}^{n_n}$ is measurement noise.

The pursuer's objective is to make the distance between its position p_t and the position of the evader z_t as small as possible, so the pursuer wants to minimize the value of $\|p_t - z_t\|$. The evader's objective is to do the opposite—namely, maximize the value of $\|p_t - z_t\|$. The pursuer and evader try to achieve these objectives by choosing appropriate values for u_t and d_t, respectively. The wind speed is unknown, but both the pursuer and evader would benefit from learning the wind speed. Therefore the optimal solution will involve each player adapting his or her action (choice of u_t and d_t) to the current estimate of the wind speed. These considerations motivate our solving the following problem:

$$\min_{\hat{u}_{t:t+T-1} \in \mathcal{U}} \quad \max_{\hat{x}_{t-L} \in \mathcal{X}, \hat{d}_{t-L:t+T-1} \in \mathcal{D}, \hat{w} \in \mathcal{W}}$$

$$\sum_{s=t}^{t+T} \|p_s - z_s\|_2^2 + \lambda_u \sum_{s=t}^{t+T-1} \|\hat{u}_s\|_2^2 - \lambda_n \sum_{s=t-L}^{t} \|n_s\|_2^2 - \lambda_d \sum_{s=t-L}^{t+T-1} \|\hat{d}_s\|_2^2, \tag{24}$$

where the pursuer's future actions $u_{t:t+T-1}$, the unknown evader's actions $d_{t-L:t+T-1}$, the unknown initial state x_{t-L}, and the unknown wind speed w are included as optimization variables. A simulation of the resulting closed-loop system is shown in Figs. 8–10.

The parameters chosen for the model (23) and the optimization (24) are given as follows. The pursuer moves with constant velocity $v = 0.1$. The backward and forward horizon lengths are chosen to be $L = 8$ and $T = 12$, respectively. The weights in the cost function in Eq. (24) are chosen to be $\lambda_u = 10$, $\lambda_n = 10{,}000$, and $\lambda_d = 100$. The actual measurement noise

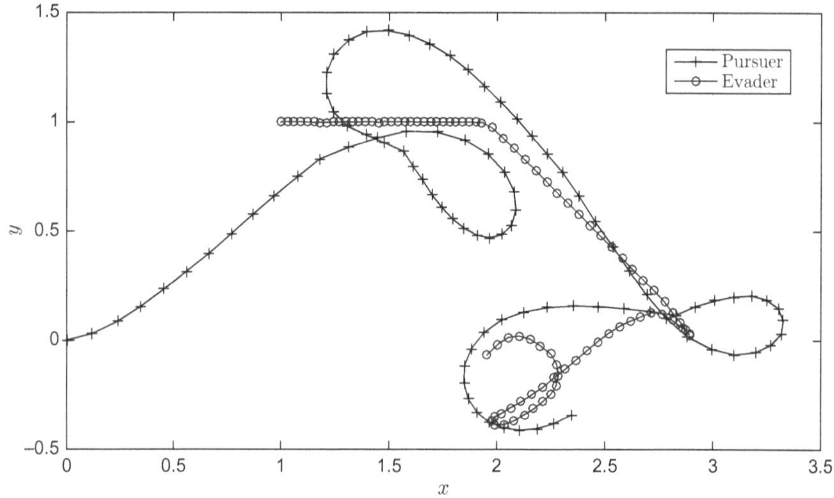

Fig. 8 Pursuit-evasion: trajectories. The pursuer's trajectory (*crosses*) begins at (0,0) and the evader's trajectory (*circles*) begins at (1,1).

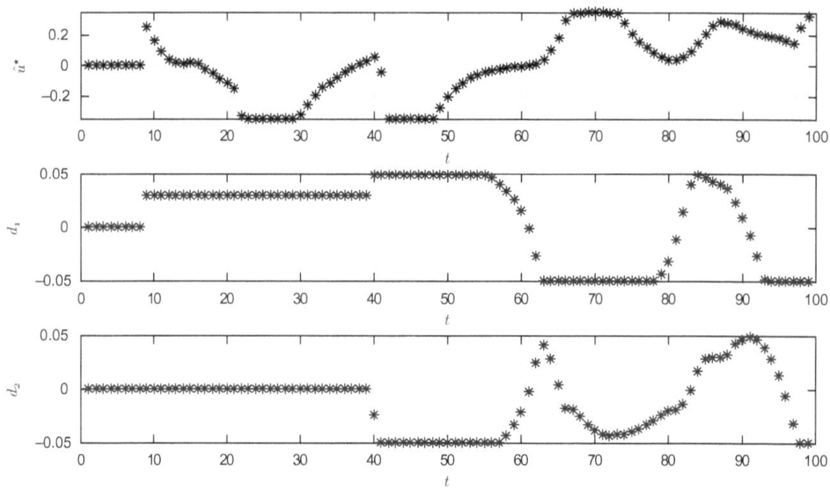

Fig. 9 Pursuit-evasion: inputs. The top plot shows the pursuer's input $\hat{u}_t^*$ that is applied. The lower plots show the evader's input d_t that is applied. The evader applies constant velocity until time $t = 40$, at which time the optimal $\hat{d}_t^*$ is applied for the remainder of the simulation.

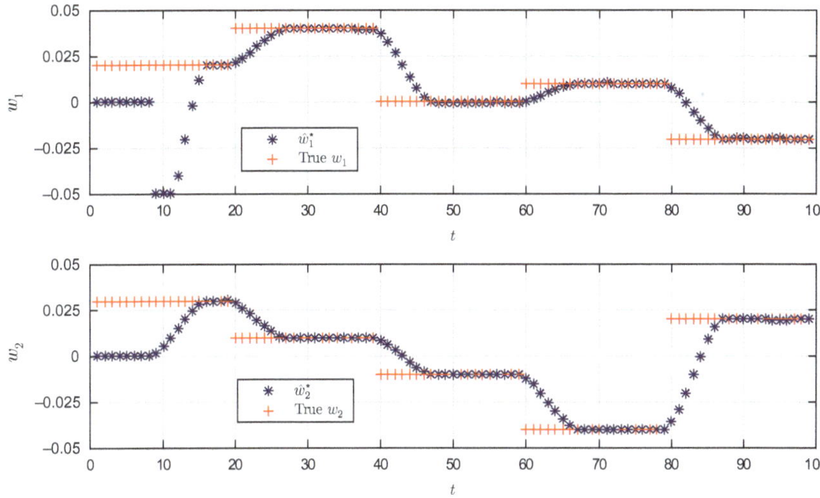

Fig. 10 Pursuit-evasion: wind. The plots show the true values of the wind speed (*crosses*) and the estimated values $\hat{w}_1^*$ and $\hat{w}_2^*$ of the wind speed (asterisks) in the *x*-direction (*top plot*) and *y*-direction (*bottom plot*).

n_t is an unmeasured Gaussian i.i.d. random variable with zero mean and a standard deviation of 0.001.

The trajectories that each player follow are shown in Fig. 8. The evader moves with constant velocity in the positive x-direction until time $t = 40$, when the optimal $\hat{d}_t^*$ begins to be applied. The pursuer applies $\hat{u}_t^*$ throughout the entire simulation. Rapidly the pursuer catches up to the evader and is forced to make a loop because of its nonholonomic dynamics. The evader, on the other hand, is able to make sharp maneuvers because of its double-integrator dynamics. The inputs that each player applies are shown in Fig. 9. Fig. 10 shows that the estimates of the uncertain wind speed converge to their true values even as they change throughout the simulation.

5 CONCLUSIONS

In this chapter we addressed adaptation and learning in the context of output-feedback MPC with MHE. Often the MPC and MHE problems are formulated with a known model of the dynamics. However, in this chapter we investigated the solving of the MPC and MHE problems using a model with uncertain parameters. We did this by simultaneously solving the MPC and MHE problems as a single min-max optimization problem and including the uncertain model parameters as optimization variables to be estimated. Under appropriate assumptions ensuring controllability and observability, Theorem 1 provides bounds on the state of the system.

In a simulation study we showed that the combined control and estimation approach effectively controls linear and nonlinear systems with model parameter uncertainty, adapts to changing model parameters, and also learns the uncertain model parameters when the system is sufficiently excited. However, even when the system is not sufficiently excited to learn the true values of the uncertain model parameters, the computed control law is still able to effectively reject disturbances and stabilize the system. Using a primal-dual-like interior point method, one can find solutions to this MPC with MHE approach even for severely nonconvex examples.

Future work may involve investigation of under what specific conditions the estimates of the uncertain parameters are guaranteed to converge to their true values.

REFERENCES

[1] J.B. Rawlings, D.Q. Mayne, Model Predictive Control: Theory and Design, Nob Hill Publishing, Madison, WI, 2009.
[2] A. Bemporad, M. Morari, Robust model predictive control: a survey, in: Robustness in Identification and Control, Springer, Berlin, 1999, pp. 207–226.
[3] L. Magni, G. De Nicolao, R. Scattolini, F. Allgöwer, Robust model predictive control for nonlinear discrete-time systems, Int. J. Robust Nonlinear Control 13 (3–4) (2003) 229–246.
[4] J.H. Lee, Z. Yu, Worst-case formulations of model predictive control for systems with bounded parameters, Automatica 33 (5) (1997) 763–781.
[5] P.A. Ioannou, J. Sun, Robust Adaptive Control, Dover Publications, Inc., Mineola, NY, 2012.
[6] K.J. Åström, B. Wittenmark, Adaptive Control, Dover Publications, Inc., Mineola, NY, 2013.
[7] D.Q. Mayne, H. Michalska, Adaptive receding horizon control for constrained nonlinear systems, in: Proceedings of the 32nd IEEE Conference on Decision and Control, 1993, pp. 1286–1291.
[8] V. Adetola, D. DeHaan, M. Guay, Adaptive model predictive control for constrained nonlinear systems, Syst. Control Lett. 58 (5) (2009) 320–326.
[9] C.V. Rao, J.B. Rawlings, D.Q. Mayne, Constrained state estimation for nonlinear discrete-time systems: stability and moving horizon approximations, IEEE Trans. Autom. Control 48 (2) (2003) 246–258.
[10] A. Alessandri, M. Baglietto, G. Battistelli, Moving-horizon state estimation for nonlinear discrete-time systems: new stability results and approximation schemes, Automatica 44 (7) (2008) 1753–1765.
[11] D.G. Robertson, J.H. Lee, J.B. Rawlings, A moving horizon-based approach for least-squares estimation, AIChE J. 42 (8) (1996) 2209–2224.
[12] D.A. Copp, J.P. Hespanha, Nonlinear output-feedback model predictive control with moving horizon estimation, in: IEEE Conference on Decision and Control, 2014, pp. 3511–3517.

[13] T. Başar, G.J. Olsder, Dynamic Noncooperative Game Theory, Academic Press, London, 1995.
[14] D.Q. Mayne, J.B. Rawlings, C.V. Rao, P.O. Scokaert, Constrained model predictive control: stability and optimality, Automatica 36 (6) (2000) 789–814.
[15] L. Magni, R. Scattolini, Robustness and robust design of MPC for nonlinear discrete-time systems, in: Assessment and Future Directions of Nonlinear Model Predictive Control, Springer, Berlin, 2007 pp. 239–254.
[16] D.M. Raimondo, D. Limon, M. Lazar, L. Magni, E. Camacho, Min-max model predictive control of nonlinear systems: a unifying overview on stability, Eur. J. Control 15 (1) (2009) 5–21.
[17] D. DeHaan, M. Guay, Adaptive robust MPC: a minimally-conservative approach, in: American Control Conference, IEEE, New York, 2007, pp. 3937–3942.
[18] E.D. Sontag, Control-Lyapunov functions, in: Open Problems in Mathematical Systems and Control Theory, Springer, Berlin, 1999, pp. 211–216.
[19] D. Liberzon, E.D. Sontag, Y. Wang, Universal construction of feedback laws achieving ISS and integral-ISS disturbance attenuation, Syst. Control Lett. 46 (2) (2002) 111–127.
[20] D.A. Copp, J.P. Hespanha, Nonlinear Output-Feedback Model Predictive Control With Moving Horizon Estimation, Tech. Rep., University of California, Santa Barbara, 2014, http://www.ece.ucsb.edu/hespanha/techrep.html.

CHAPTER SEVEN

Stochastic Adaptive Dynamic Programming for Robust Optimal Control Design

T. Bian, Z.-P. Jiang
New York University, Brooklyn, NY, United States

1 INTRODUCTION

Dynamic programming (DP) [1] aims at solving the optimal control problem for dynamic systems using Bellman's principle of optimality. However, the direct implementation of DP in real-world applications is usually prohibited by the "curse of dimensionality" [2] and the "curse of modeling" [3]. One way of solving these two shortcomings is to use approximate DP, which was proposed by Werbos [4, 5] to bring reinforcement learning and DP together to approximate the (generally intractable) solution to Bellman's equation via online learning. Inspired by Werbos's original work, numerous approximate DP methods, such as Q-learning [6, 7], temporal-difference learning [8, 9], and actor-critic algorithms [10–12], have been widely adopted to solve the optimal control problem for Markov decision processes (MDPs) in the past two decades. Similar ideas have been proposed by Bertsekas and Tsitsiklis under the name of *neuro-DP* [3]. The interested reader can consult [3, 13, 14] for a nice tutorial. Despite their popularity in real-world applications, the above-mentioned methods usually overlooked the stability issue. Moreover, the underlying state and action spaces are assumed to be either finite or countable, which is restrictive for many real-world applications.

Different from the early approximate DP methods, adaptive dynamic programming (ADP) was introduced to find a stabilizing optimal controller for continuous-state space control systems. Over the past decade, ADP-based control design has been extensively studied by several research groups for both continuous-time systems [15–26] and discrete-time [27–32]

Control of Complex Systems
http://dx.doi.org/10.1016/B978-0-12-805246-4.00007-0

systems. See also [33–35] for three review papers, and [36] for a recent book. Unfortunately, most existing ADP methods are devised for deterministic systems. Although preliminary results [37, 38] have been obtained for a class of continuous-time stochastic systems, additive noise is excluded from the model in question. Since the quadratic cost in the classic linear-quadratic regulator problem may become ill-posed because of the presence of the additive noise, we need to consider new types of cost functionals for stochastic optimal control problems in the presence of additive noise. In the first part of this chapter (Sections 2–5) we investigate the optimal control problem for linear stochastic systems subject to both multiplicative noise and additive noise. In Section 2, two cost functionals—the discounted cost and the biased cost—are presented. The optimal control problems corresponding to these two types of cost functionals are solved in Sections 3 and 4, respectively, with rigorous convergence and stability analysis provided. In Section 5, ADP techniques are developed to design stochastic adaptive optimal controllers through online successive approximations.

In the second part of this chapter (Sections 6 and 7) we further develop a stochastic robust ADP (RADP) method to solve the stochastic robust optimal control problem for continuous-time linear stochastic systems subject to nonlinear dynamic uncertainties. Different from the traditional ADP, RADP can address the presence of dynamic uncertainties caused by the modeling error in linear and nonlinear dynamic systems (see, eg, [35, 39–42], and references therein). By use of the RADP algorithm, precise information on the system dynamics and order is no longer required, and robust optimal control laws are designed directly on the basis of the real-time data. In Section 6 we conduct robust optimality and stability analysis for a class of continuous-time partially linear stochastic systems, by employing the stochastic H^{∞} theory [43, 44] and the small-gain theory [45–47]. It can be shown that if the input-dependent noise is small enough, then the robust stability of the closed-loop system under the optimal control policy is guaranteed by tuning Q and R matrices properly. On the basis of the robust optimality results obtained, a stochastic RADP algorithm with convergence and stability analysis is given in Section 7.

To illustrate the results obtained, we present three practical examples from the engineering and computational neuroscience fields in Section 8. In the first example the two-joint human arm movement in a divergent force field [48] is considered, where an undiscounted cost is used. This example shows that our stochastic ADP method appears to be a suitable candidate for a computational learning mechanism in the central nervous system to coordinate movements. Then we apply the stochastic ADP algorithms to solve a vehicle suspension control problem, where stochastic additive noise is involved. Both discount-optimal control and bias-optimal control are derived. Finally, to illustrate the stochastic RADP algorithm, a partially linear single-joint human arm movement model is presented in the third example. These three examples demonstrate that the stochastic ADP method presented in this chapter serves as a powerful tool to solve data-driven, non-model-based robust optimal control problems for stochastic systems.

Throughout this chapter we use $\mathbb{R}$ to denote the set of real numbers. I_n denotes the identity matrix of dimension n. $|\cdot|$ denotes the Euclidean norm for vectors, or the induced matrix norm for matrices. For a matrix $A \in \mathbb{R}^{n\times m}$, $\text{vec}(A) = [a_1^T \; a_2^T \; \cdots \; a_m^T]^T$, where $a_i \in \mathbb{R}^n$ is the ith column of A. $\otimes$ indicates the Kronecker product. We denote by $(\Omega, \mathscr{F}, P)$ the underlying probability space [49, Chapter 5.2], where Ω is a sample space, P is a probability measure, and $\mathscr{F}$ is a σ-field of Borel sets equipped with a nature filtration $\{\mathscr{F}_t\}$, $\mathscr{F}_s \subseteq \mathscr{F}_t \subseteq \mathscr{F}$ for $0 \leq s \leq t$. $\mathcal{E}$ and $\mathcal{L}$ denote the expectation operator and the differential generator, respectively.

2 MODEL AND PROBLEM FORMULATION

2.1 System Description

Suppose $z := [\eta_1, \eta_2, \ldots, \eta_{q_1}, w_1, w_2, \ldots, w_{q_2}, \zeta_1, \zeta_2, \ldots, \zeta_{q_3}]^T$ is an $\{\mathscr{F}_t\}$-adapted $(q_1+q_2+q_3)$-dimensional standard Brownian motion. Denote $\mathscr{F}_t^z$ as the σ-field generated by $z(s)$, $0 \leq s \leq t$, which contains all the P null sets. Consider the system described by the following stochastic differential equation:

$$dx = (Ax + Bu)dt + Bdv + \sum_{i=1}^{q_3} E_i d\zeta_i, \tag{1}$$

where $dv = \sum_{i=1}^{q_1} F_i x d\eta_i + \sum_{i=1}^{q_2} G_i u dw_i$ represents the state- and control-dependent noise received by the plant from the input channel; $x \in \mathbb{R}^n$ and $u \in \mathbb{R}^m$ are $\{\mathcal{F}_t^z\}$-adapted random processes representing the state and the input of the system, respectively; $x(0) = \xi$ is a constant; $A \in \mathbb{R}^{n\times n}$, $B \in \mathbb{R}^{n\times m}$, $F_i \in \mathbb{R}^{m\times n}$, and $G_i \in \mathbb{R}^{m\times m}$ are unknown constant matrices; and $E_i \in \mathbb{R}^m$ are unknown constant vectors.

Following the definitions in [50], we give the following definition:

Definition 1. The uncontrolled nominal system of Eq. (1) (with $u = 0$ and $E_i = 0$ for all $i = 1, \ldots, q_3$) is said to be *globally asymptotically stable (GAS) at the origin in the mean square sense* if all initial states $x(0)$ yield $\lim_{t\to\infty} \mathcal{E}|x(t)|^2 = 0$. The controlled nominal system of Eq. (1) is said to be *mean square stabilizable* if there exists a controller u such that the system is GAS at the origin in the mean square sense.

Throughout this chapter we assume the nominal system of Eq. (1) is mean square stabilizable.

2.2 Problem Formulation

In this subsection we present three stochastic optimal control problems.

Given $\xi \in \mathbb{R}^n$, denote the *average cost* as

$$\rho(u) = \limsup_{\tau\to\infty} \frac{1}{\tau} \int_0^\tau \mathcal{E}_\xi^u \left[x^{\mathrm{T}} Qx + u^{\mathrm{T}} Ru\right] dt, \tag{2}$$

where $Q = Q^{\mathrm{T}} > 0$, $R = R^{\mathrm{T}} > 0$, and $\mathcal{E}_\xi^u$ denotes the expectation for the processes corresponding to system (1) with initial state $x(0) = \xi$ and control input u.

It has been shown [51–54] that if $\sum_{i=1}^{q_3} E_i E_i^{\mathrm{T}} > 0$, then ρ is minimized under the unique *average-optimal* controller $u^\star = -K^\star x$, where $K^\star = (\Sigma(P^\star) + R)^{-1} B^{\mathrm{T}} P^\star$, where $P^\star \in \mathbb{R}^{n\times n}$ is the unique symmetric positive definite solution to the following algebraic Riccati equation:

$$0 = A^{\mathrm{T}} P + PA - PB(\Sigma(P) + R)^{-1} B^{\mathrm{T}} P + \Pi(P) + Q, \tag{3}$$

with $\Pi(P) = \sum_{i=1}^{q_1} F_i^{\mathrm{T}} B^{\mathrm{T}} PBF_i$ and $\Sigma(P) = \sum_{i=1}^{q_2} G_i^{\mathrm{T}} B^{\mathrm{T}} PBG_i$. Moreover, $\rho(u^\star) = E(P^\star) := \sum_{i=1}^{q_3} E_i^{\mathrm{T}} P^\star E_i$.

Note that the results in [51–55] rely heavily on the nondegeneracy assumption $\sum_{i=1}^{q_3} E_i E_i^{\mathrm{T}} > 0$. If this condition is violated, then ρ may not have a unique minimum. One such example is when $E_i = 0$ for all $i = 1, \ldots, q_3$, all the stabilizing controllers for system (1) are average-optimal, and $\inf_u \rho(u) = 0$. Thus, in the degeneracy case, the averaged optimal control problem may become ill-posed. Moreover, Eq. (2) cannot characterize the transient performance of the control system. It is possible that two different controllers may lead to the same average cost, as long as they give the same steady-state distribution. To tackle these problems we introduce another two cost functionals that are well established even in the degeneracy case and can reflect the transient performance of the closed-loop system.

Given $\xi \in \mathbb{R}^n$, denote the *discounted cost* as

$$\mathcal{J}_\lambda(\xi; u) = \mathcal{E}_\xi^u \left[\int_0^\infty e^{-\lambda t}(x^T Qx + u^T Ru)dt \right], \tag{4}$$

where $\lambda > 0$ is a constant scalar.

It has been shown [56, Theorem 4] that under certain conditions $\mathcal{J}_\lambda(\xi; u)$ is minimized under the *discount-optimal* controller $u_\lambda^\star = -K_\lambda^\star x$, where $K_\lambda^\star = (\Sigma(P_\lambda^\star) + R)^{-1}B^T P_\lambda^\star$, where $P_\lambda^\star \in \mathbb{R}^{n\times n}$ is the unique symmetric positive definite solution to the following algebraic Riccati equation:

$$\lambda P = A^T P + PA - PB(\Sigma(P) + R)^{-1}B^T P + \Pi(P) + Q. \tag{5}$$

Moreover, $\mathcal{J}_\lambda(\xi; u_\lambda^\star) = \xi^T P_\lambda^\star \xi + \lambda^{-1}E(P_\lambda^\star)$.

Note that if $E_i = 0$ for all i, we can pick $\lambda = 0$, and then Eq. (4) becomes the undiscounted cost in the standard stochastic linear-quadratic regulator problem [57].

One drawback with the main results of [56] is that the stability issue is overlooked. Indeed, $\mathcal{E}_\xi^{u_\lambda^\star} x(t)$ may not converge as t goes to infinity. Furthermore, the discount-optimal controller may not provide good steady-state performance, since the long-term cost is "discounted" by the discounting factor.

The stability issue is conquered in Section 3 by adding one more condition on λ and Q. The stability of the closed-loop system is guaranteed in the sense that $x(t)$ converges to a stationary process [58, Chapter 2.1]. To improve the steady-state performance, we next introduce the biased cost to system (1). Given $\xi \in \mathbb{R}^n$, denote the *biased cost* as

$$\mathcal{J}_b(\xi; u) = \mathcal{E}_\xi^u \left[\int_0^\infty (x^T Qx + u^T Ru - \rho(u))dt \right]. \tag{6}$$

Borrowing the definition in [59] for MDPs, we call an average-optimal linear controller $u = -Kx$ *bias optimal* if it minimizes Eq. (6).

$\mathcal{J}_b(\xi; u)$ characterizes the relative difference in the total reward gained from starting in state ξ as opposed to some other states [59]. Indeed, for ξ_1, $\xi_2 \in \mathbb{R}^n$,

$$\begin{aligned}
&\mathcal{J}_b(\xi_1; u) - \mathcal{J}_b(\xi_2; u) \\
&= \mathcal{E}_{\xi_1}^u \left[\int_0^\infty (x^T Qx + u^T Ru)dt \right] - \mathcal{E}_{\xi_2}^u \left[\int_0^\infty (x^T Qx + u^T Ru)dt \right].
\end{aligned}$$

As we will show in Sections 4 and 8, bias-optimal controllers provide improved steady-state performance, since the future cost is not discounted. Because of this advantage, the biased problem has been extensively studied for MDPs [59–64]. An extension to nondegenerate diffusions can be found in [65, Chapter 3] and [66]. In Section 4 we provide a detailed analysis on the bias-optimal control problem for linear continuous-time stochastic systems. The nondegeneracy assumption on the stochastic process is dropped. Several important results on the biased problem for MDPs [61, 63] are extended to the continuous-time setting. An ADP algorithm is also developed in Section 5 to find the bias-optimal controller through online learning.

3 DISCOUNT-OPTIMAL CONTROL

In this section the discount-optimal control design for system (1) is developed. First, consider a simplified model of system (1) with $E_i = 0$, $\forall i = 1, \ldots, q_3$:

$$dx = (Ax + Bu)dt + Bdv. \tag{7}$$

Next a stability result is given for system (7). Since Theorem 1 is a direct result of Corollary 11.4.14 in [67], the proof is omitted.

Theorem 1. *A necessary (or sufficient) condition for system (7) under the controller $u = -Kx$ to be GAS at the origin in the mean square sense is that for every (or for some) real symmetric positive definite matrix Q, the matrix equation*

$$0 = (A - BK)^{\mathrm{T}} P + P(A - BK) + K^{\mathrm{T}} \Sigma(P) K + \Pi(P) + Q$$

has a unique solution $P \in \mathbb{R}^{n \times n}$ that is symmetric positive definite.

Now we define the admissible feedback gain matrix for system (1).

Definition 2. A feedback gain matrix $K \in \mathbb{R}^{m \times n}$ is *admissible* to system (1) with respect to the cost (4) if

- system (7) is GAS at the origin in the mean square sense under $u = -Kx$;
- $\lim_{t \to \infty} \mathcal{E}_{\xi}^{u}[x(t)x^{\mathrm{T}}(t)]$ is finite;
- $\mathcal{J}_{\lambda}(\xi; u) < \infty$ for all $\xi \in \mathbb{R}^n$.

The following result is a direct extension of the result in [53]:

Lemma 1. *If system (7) is GAS at the origin in the mean square sense under $u = -Kx$, then*

$$\lim_{t \to \infty} \mathcal{E}_{\xi}^{u}[x(t)x^{\mathrm{T}}(t)] = \bar{X},$$

where $\bar{X} \in \mathbb{R}^{n\times n}$ *is the unique symmetric positive semidefinite solution to the following matrix equation:*

$$0 = (A - BK)\bar{X} + \bar{X}(A - BK)^{\mathrm{T}} + \sum_{i=1}^{q_1} BF_i\bar{X}F_i^{\mathrm{T}}B^{\mathrm{T}} + \sum_{i=1}^{q_2} BG_iK\bar{X}K^{\mathrm{T}}G_i^{\mathrm{T}}B^{\mathrm{T}} + \sum_{i=1}^{q_3} E_iE_i^{\mathrm{T}}.$$

Now we state the main result of this section:

Theorem 2. *Choose* $K_0 \in \mathbb{R}^{m\times n}$ *that is admissible to system* (1). *Let* $P_k \in \mathbb{R}^n$ *be the unique symmetric positive definite solution to the following linear matrix equation:*

$$\lambda P_k = A_k^{\mathrm{T}}P_k + P_kA_k + K_k^{\mathrm{T}}(\Sigma(P_k) + R)K_k + \Pi(P_k) + Q, \tag{8}$$

where $A_k = A - BK_k$, *and*

$$K_{k+1} = (\Sigma(P_k) + R)^{-1}B^{\mathrm{T}}P_k, \quad k = 0, 1, \ldots. \tag{9}$$

If $\lambda P_0 < Q$, *then:*

- $P_\lambda^\star \leq \cdots \leq P_{k+1} \leq P_k \cdots \leq P_0$;
- $\lim_{k\to\infty} P_k = P_\lambda^\star$, $\lim_{k\to\infty} K_k = K_\lambda^\star$;
- $K_\lambda^\star$ *and* K_k, $k = 0, 1, \ldots$, *are admissible to system* (1).

Remark 1. From [56, Theorem 4] and Theorem 1, it is easily seen that if a pair (P_k, K_k), where $P_k = P_k^{\mathrm{T}} \in \mathbb{R}^{n\times n}$ and $K_k \in \mathbb{R}^{n\times m}$, satisfies Eq. (8) with $0 < \lambda P_k < Q$, then K_k is admissible to system (1), and

$$\mathcal{E}_\xi^u\left[\int_0^\infty \mathrm{e}^{-\lambda t}x^{\mathrm{T}}(Q + K_k^{\mathrm{T}}RK_k)x\,dt\right] = \xi^{\mathrm{T}}P_k\xi + \frac{1}{\lambda}E(P_k), \tag{10}$$

where $u = -K_kx$. On the other hand, if Eq. (10) holds for a pair (P_k, K_k), then Eq. (8) holds.

Proof of Theorem 2. The proofs of the first two conclusions in Theorem 2 are similar to the proofs in [53, Theorem 1] and thus are omitted. Suppose K_k is admissible to system (1). From Eqs. (8) and (9), it is directly checked that

$$\begin{aligned} &A_{k+1}^{\mathrm{T}}P_k + P_kA_{k+1} + \Pi(P_k) + K_{k+1}^{\mathrm{T}}\Sigma(P_k)K_{k+1} \\ &= \lambda P_k - (K_{k+1} - K_k)^{\mathrm{T}}(R + \Sigma(P_k))(K_{k+1} - K_k) - Q - K_{k+1}^{\mathrm{T}}RK_{k+1}. \end{aligned} \tag{11}$$

Since $\lambda P_k \leq \lambda P_0 < Q$, the right-hand side of Eq. (11) is negative definite. Hence one concludes from Theorem 1 that system (7) is GAS at the origin in the mean square sense under $u = -K_{k+1}x$, and thus by Lemma 1 that $\lim_{t\to\infty} \mathcal{E}_\xi^u[x(t)x^T(t)]$ is finite.

Moreover, by Eq. (11),

$$\mathcal{E}_\xi^u\left[\int_0^\infty e^{-\lambda t}x^T(Q+K_{k+1}^T RK_{k+1})xdt\right] \le \xi^T P_k\xi + \frac{1}{\lambda}E(P_k), \quad \forall\xi \in \mathbb{R}^n,$$

where $u = -K_{k+1}x$. Therefore K_{k+1} is admissible to system (1). Since K_0 is admissible, by induction, K_k, $k = 0, 1, \ldots,$ are admissible to system (1). Finally, by definition, $K_\lambda^\star$ is admissible. This completes the proof.

Theorem 2 shows that the policy iteration scheme given by Eqs. (8) and (9) can be used to successively approximate the optimal controller and the minimum cost value for system (1) with respect to Eq. (4).

Remark 2. If we let k go to infinity, Eqs. (5) and (10) imply that

$$\mathcal{E}_\xi^{u_\lambda^\star}\left[\int_0^\infty e^{-\lambda t}x^T(Q+(K_\lambda^\star)^T RK_\lambda^\star)xdt\right] = \xi^T P_\lambda^\star\xi + \frac{1}{\lambda}E(P_\lambda^\star).$$

Remark 3. Different from [56, Theorem 4], by addition of one more condition $\lambda P_0 < Q$, the stability of system (1) is guaranteed in the sense that

$$\lim_{t\to\infty} |\mathcal{E}_\xi^{u_\lambda^\star}x(t)| = 0, \quad \lim_{t\to\infty} |\mathcal{E}_\xi^{u_\lambda^\star}[x(t)x^T(t)]| < \infty, \quad \forall\xi \in \mathbb{R}^n.$$

This implies that system (1) maintains a stationary distribution [58, Chapter 2.1].

4 BIAS-OPTIMAL CONTROL

One problem with the discounted cost is that the cost generated by the future trajectories of the system is discounted in time. Moreover, it is not easy to select Q, R, and λ in reality, such that $\lambda P_0 < Q$. To avoid these difficulties we investigate the bias-optimal control problem in this section.

Lemma 2. *If a pair (P, K), where $P = P^T > 0$, satisfies*

$$0 = (A-BK)^T P + P(A-BK) + K^T\Sigma(P)K + \Pi(P) + Q + K^T RK, \tag{12}$$

then

$$\mathcal{E}_\xi^u\left[\int_0^\infty (x^T(Q+K^T RK)x - E(P))dt\right] = \xi^T P\xi - E(\tilde{P}), \tag{13}$$

where $\tilde{P} = \tilde{P}^{\mathrm{T}} > 0$ *is the unique solution to*

$$0 = (A - BK)^{\mathrm{T}}\tilde{P} + \tilde{P}(A - BK) + K^{\mathrm{T}}\Sigma(\tilde{P})K + \Pi(\tilde{P}) + P. \tag{14}$$

Conversely, if Eqs. (13) and (14) *hold for a pair* (P, K), *then Eq. (12) holds.*

Proof. By the tower property [49, Problem 1.2.17] and Ito's lemma [68, p. 16], following a procedure similar to that in the proof of Theorem 8.1 in [58], one easily has the necessary condition.

Next we prove the sufficient part. If Eq. (12) holds for some pair (P, K), then

$$\mathcal{L}(x^{\mathrm{T}}Px) = E(P) - x^{\mathrm{T}}(Q + K^{\mathrm{T}}RK)x, \quad \forall x \in \mathbb{R}^n,$$

where $\mathcal{L}$ is taken along the solutions to system (1) with $u = -Kx$. Using Dynkin's formula [68, p. 10], we have

$$\xi^{\mathrm{T}}P\xi - \lim_{t\to\infty} \mathcal{E}_\xi^u\left[x^{\mathrm{T}}(t)Px(t)\right] = \mathcal{E}_\xi^u\left[\int_0^\infty (x^{\mathrm{T}}(Q + K^{\mathrm{T}}RK)x - E(P))dt\right]. \tag{15}$$

Note from Lemma 1 that $\lim_{t\to\infty} \mathcal{E}_\xi^u\left[x(t)x^{\mathrm{T}}(t)\right] = \bar{X}$. Following [53, Eq. (18)], we have

$$\lim_{t\to\infty} \mathcal{E}_\xi^u\left[x^{\mathrm{T}}(t)Px(t)\right] = tr\{\lim_{t\to\infty} \mathcal{E}_\xi^u\left[Px(t)x^{\mathrm{T}}(t)\right]\} = tr\{P\bar{X}\} = E(\tilde{P}).$$

From the above equation and Eq. (15), we have Eq. (13). This completes the proof. □

Remark 4. Note from Eq. (13) that

$$\mathcal{J}_b(\xi_1; u) - \mathcal{J}_b(\xi_2; u) = \xi_1^{\mathrm{T}}P\xi_1 - \xi_2^{\mathrm{T}}P\xi_2, \quad \forall \xi_1, \xi_2 \in \mathbb{R}^n.$$

Thus the biased cost represents the relative different cost value caused by different initial states. This also matches the observation in Section 2.2.

To derive the policy iteration algorithm for the biased problem, we first study the relationship between the biased cost and the discounted cost. We fix a feedback gain matrix K such that the following two matrix equations have unique symmetric positive definite solutions P_λ and P, respectively:

$$\lambda P_\lambda = (A - BK)^{\mathrm{T}}P_\lambda + P_\lambda(A - BK) + K^{\mathrm{T}}\Sigma(P_\lambda)K + \Pi(P_\lambda) + K^{\mathrm{T}}RK + Q, \tag{16}$$

$$0 = (A - BK)^{\mathrm{T}}P + P(A - BK) + K^{\mathrm{T}}\Sigma(P)K + \Pi(P) + K^{\mathrm{T}}RK + Q. \tag{17}$$

Next we give two lemmas. The proof of Lemma 3 is omitted, as it can be easily obtained by our following the procedure in [53].

Lemma 3. *For system* (1) *with* $u = -Kx$, *we have* $\rho(u) = E(P)$.

Lemma 4. *Consider system* (1) *with* $u = -Kx$. *We have*

$$\lim_{\lambda \to 0} \left| \mathcal{J}_\lambda(\xi; u) - \frac{1}{\lambda}\rho(u) - \mathcal{J}_b(\xi; u) \right| = 0. \tag{18}$$

Proof. First we show that $\lim_{\lambda \to 0}(P - P_\lambda)/\lambda = \tilde{P}$, where $\tilde{P}$ is defined as in Eq. (14). By the definitions of P and P_λ,

$$\begin{aligned} 0 = {} & (A - BK)^{\mathrm{T}}(P - P_\lambda) + (P - P_\lambda)(A - BK) + K^{\mathrm{T}}\Sigma(P - P_\lambda)K \\ & + \Pi(P - P_\lambda) + \lambda P_\lambda. \end{aligned}$$

Then we easily see from the above equation that $\lim_{\lambda \to 0} P_\lambda = P$ and $\lim_{\lambda \to 0} \lambda^{-1}(P - P_\lambda) = \tilde{P}$.

Therefore from Lemma 3, Eqs. (10) and (13), we have

$$\begin{aligned} & \lim_{\lambda \to 0} \left| \mathcal{J}_\lambda(\xi; u) - \frac{1}{\lambda}\rho(u) - \mathcal{J}_b(\xi; u) \right| \\ & = \lim_{\lambda \to 0} \left| \xi^{\mathrm{T}} P_\lambda \xi + \frac{1}{\lambda}E(P_\lambda) - \frac{1}{\lambda}E(P) - \xi^{\mathrm{T}} P \xi + E(\tilde{P}) \right| = 0. \end{aligned}$$

This completes the proof. □

It is of interest to notice that Lemma 4 is similar to Corollary 8.2.4 in [61] developed for MDPs over a finite state space. Our result can be considered as an extension of Corollary 8.2.4 in [61] to the continuous-time case.

Now we give the optimality result.

Theorem 3. $u^\star$ *is bias optimal.*

Proof. By definition, $u^\star$ is average optimal. For another controller $u = -Kx$ that is optimal with respect to the averaged cost (2), by Eqs. (10) and (18), and the fact that $\lim_{\lambda \to 0} P_\lambda = P$,

$$\mathcal{J}_b(\xi; u) - \mathcal{J}_b(\xi; u^\star) = \lim_{\lambda \to 0}(\mathcal{J}_\lambda(\xi; u) - \mathcal{J}_\lambda(\xi; u^\star)) \geq 0.$$

Thus $u^\star$ minimizes both $\mathcal{J}_b(\xi; u)$ and $\rho(u)$—that is, $u^\star$ is bias optimal. □

Remark 5. From Theorem 3,

$$\mathcal{J}_b(\xi; u^\star) = \xi^{\mathrm{T}} P^\star \xi - E(\tilde{P}^\star), \quad \forall \xi \in \mathbb{R}^n,$$

where $\tilde{P}^\star$ is the unique symmetric positive definite solution to

$$0 = (A - BK^\star)^{\mathrm{T}}\tilde{P}^\star + \tilde{P}^\star(A - BK^\star) + (K^\star)^{\mathrm{T}}\Sigma(\tilde{P}^\star)K^\star + \Pi(\tilde{P}^\star) + P^\star,$$

and $K^\star$ and $P^\star$ are defined as in Section 2.2.

Remark 6. Following the definition in [61, Chapter 5.4.3], we say a controller u' is *n-discount optimal*, where $n = -1, 0, 1, 2, \ldots$, if

$$\liminf_{\lambda \to 0} \lambda^{-n}(\mathcal{J}_\lambda(\xi; u) - \mathcal{J}_\lambda(\xi; u')) \geq 0, \quad \forall \xi \in \mathbb{R}^n, \ u \neq u'.$$

From Eq. (18),

$$\lim_{\lambda \to 0} (\mathcal{J}_\lambda(\xi; u') - \mathcal{J}_\lambda(\xi; u)) = \lim_{\lambda \to 0} \left(\frac{1}{\lambda} \rho(u') + \mathcal{J}_b(\xi; u') - \frac{1}{\lambda} \rho(u) - \mathcal{J}_b(\xi; u) \right).$$

Then we easily see that all the average-optimal controllers are equivalent to (-1)-discount optimal controllers, and the bias-optimal controller $u^\star$ is equivalent to 0-discount optimal controller. This also matches the results in [59].

Remark 7. If $E_i = 0$ for $i = 1, 2, \ldots, q_3$, we know from Eq. (13) and Remark 5 that

$$\mathcal{E}_\xi^{u^\star} \left[\int_0^\infty x^{\mathrm{T}} (Q + K^{\star \mathrm{T}} R K^\star) x dt \right] = \xi^{\mathrm{T}} P^\star \xi.$$

Now we can give the policy iteration algorithm to approximate the bias-optimal control policy.

Theorem 4. *Choose $K_0 \in \mathbb{R}^{m \times n}$ such that system (7) under $u = -K_0 x$ is GAS at the origin in the mean square sense. Let P_k and $\tilde{P}_k$ be the unique symmetric positive definite solutions to the following two linear matrix equations:*

$$0 = A_k^{\mathrm{T}} P_k + P_k A_k + K_k^{\mathrm{T}} (\Sigma(P_k) + R) K_k + \Pi(P_k) + Q,$$
$$0 = A_k^{\mathrm{T}} \tilde{P}_k + \tilde{P}_k A_k + K_k^{\mathrm{T}} \Sigma(\tilde{P}_k) K_k + \Pi(\tilde{P}_k) + P_k,$$

where K_k is defined by Eq. (9). Then we have:

- $P^\star \leq \cdots \leq P_{k+1} \leq P_k \cdots \leq P_0$, *and system (7) under $u = -K_k x$ is GAS at the origin in the mean square sense.*
- $\lim_{k\to\infty} P_k = P^\star$, $\lim_{k\to\infty} K_k = K^\star$, *and $\mathcal{J}_b(\xi; u_k) = \xi^{\mathrm{T}} P_k \xi - E(\tilde{P}_k)$ is the biased cost corresponding to system (1) with $u_k = -K_k x$.*
- *$u = -K^\star x$ is bias optimal.*

Proof. The first result is easily obtained by following the proof of Theorem 1 in [53]. The existence and the uniqueness of $\{\tilde{P}_k\}_{k=0}^\infty$ and $\tilde{P}^\star$ are guaranteed by Theorem 1. The value of $\mathcal{J}_b(\xi; u_k)$ is given by Eq. (13). Finally, the optimality of $u^\star$ is guaranteed by Theorem 3 and Remark 5. □

5 STOCHASTIC ADP DESIGN

In this section stochastic ADP algorithms are developed to approximate the discount-optimal and the bias-optimal control policies investigated in Sections 3 and 4.

5.1 Stochastic ADP Algorithms

First we derive a stochastic ADP design with respect to the discounted problem.

Consider system (1) under an $\{\mathcal{F}_t^z\}$-adapted locally essentially bounded input $u = -K_0 x + \varepsilon$, where K_0 is an initial control gain matrix satisfying the conditions in Theorem 2 and ε is a bounded exploration noise, such as sinusoidal signals or random noise. A detailed explanation for adding ε can be found in Remark 9. From Ito's lemma [68, p. 16], we have

$$\begin{aligned} d(x^{\mathrm{T}} P_k x) &= 2x^{\mathrm{T}} P_k A x dt + x^{\mathrm{T}} \Pi(P_k) x dt + E(P_k) dt + u^{\mathrm{T}} \Sigma(P_k) u dt \\ &\quad + 2x^{\mathrm{T}} P_k d\zeta + 2x^{\mathrm{T}} P_k B d\hat{u}, \end{aligned} \tag{19}$$

where P_k is defined in Theorem 2, $d\zeta = \sum_{i=1}^{q_3} E_i d\zeta_i$, and $d\hat{u} = u dt + \sum_{i=1}^{q_1} F_i x d\eta_i + \sum_{i=1}^{q_2} G_i u dw_i$. Substituting Eq. (8) into Eq. (19) leads to

$$\begin{aligned} d(x^{\mathrm{T}} P_k x) &= 2x^{\mathrm{T}} P_k B K_k x dt + E(P_k) dt + u^{\mathrm{T}} \Sigma(P_k) u dt \\ &\quad - x^{\mathrm{T}} K_k^{\mathrm{T}} (\Sigma(P_k) + R) K_k x dt \\ &\quad + 2x^{\mathrm{T}} P_k d\zeta + 2x^{\mathrm{T}} P_k B d\hat{u} - x^{\mathrm{T}} Q x dt + \lambda x^{\mathrm{T}} P_k x dt. \end{aligned}$$

Integrating both sides of the above equation along the solutions of Eq. (1), we have

$$\begin{aligned} &\left(x^{\mathrm{T}} \otimes x^{\mathrm{T}} \,|_t^{t+\delta t} - \lambda \int_t^{t+\delta t} x^{\mathrm{T}} \otimes x^{\mathrm{T}} ds - 2 \int_t^{t+\delta t} x^{\mathrm{T}} \otimes d\zeta^{\mathrm{T}} \right) \mathrm{vec}(P_k) \\ &= 2 \left(\int_t^{t+\delta t} x^{\mathrm{T}} \otimes d\hat{u}^{\mathrm{T}} \right) \mathrm{vec}(B^{\mathrm{T}} P_k) \\ &\quad + 2 \left(\int_t^{t+\delta t} x^{\mathrm{T}} \otimes x^{\mathrm{T}} ds \right) (I_n \otimes K_k^{\mathrm{T}}) \mathrm{vec}(B^{\mathrm{T}} P_k) \\ &\quad + \left(\int_t^{t+\delta t} u^{\mathrm{T}} \otimes u^{\mathrm{T}} ds \right) \mathrm{vec}(\Sigma(P_k)) \\ &\quad - \left(\int_t^{t+\delta t} x^{\mathrm{T}} \otimes x^{\mathrm{T}} ds \right) (K_k^{\mathrm{T}} \otimes K_k^{\mathrm{T}}) \mathrm{vec}(\Sigma(P_k)) \\ &\quad - \left(\int_t^{t+\delta t} x^{\mathrm{T}} \otimes x^{\mathrm{T}} ds \right) \mathrm{vec}(Q + K_k^{\mathrm{T}} R K_k) + \delta t E(P_k), \end{aligned} \tag{20}$$

where $\delta t > 0$ is a known constant. Here the noisy signals $d\hat{u}$ and $d\zeta$ are assumed to be known and are exploited for online learning. Of course, they are not used for feedback control design.

Next, we define five matrices:

$$\delta_{xx} = \left[x \otimes x \,|_{t_0}^{t_1}, x \otimes x \,|_{t_1}^{t_2}, \ldots, x \otimes x \,|_{t_{l-1}}^{t_l}\right]^{\mathrm{T}},$$

$$I_{xx} = \left[\int_{t_0}^{t_1} x \otimes x ds, \int_{t_1}^{t_2} x \otimes x ds, \ldots, \int_{t_{l-1}}^{t_l} x \otimes x ds\right]^{\mathrm{T}},$$

$$I_{uu} = \left[\int_{t_0}^{t_1} u \otimes u ds, \int_{t_1}^{t_2} u \otimes u ds, \ldots, \int_{t_{l-1}}^{t_l} u \otimes u ds\right]^{\mathrm{T}},$$

$$I_{xd\hat{u}} = \left[\int_{t_0}^{t_1} x \otimes d\hat{u}, \int_{t_1}^{t_2} x \otimes d\hat{u}, \ldots, \int_{t_{l-1}}^{t_l} x \otimes d\hat{u}\right]^{\mathrm{T}},$$

$$I_{xd\zeta} = \left[\int_{t_0}^{t_1} x \otimes d\zeta, \int_{t_1}^{t_2} x \otimes d\zeta, \ldots, \int_{t_{l-1}}^{t_l} x \otimes d\zeta\right]^{\mathrm{T}},$$

where l is a positive integer and $0 \leq t_0 < t_1 < \cdots < t_l$ are known constants standing for learning time instants.

Using the above notation, we can rewrite Eq. (20) in a compact form:

$$\Theta_k^{\lambda} \begin{bmatrix} \mathrm{vec}(P_k) \\ \mathrm{vec}(B^{\mathrm{T}} P_k) \\ \mathrm{vec}(\Sigma(P_k)) \\ E(P_k) \end{bmatrix} = \Xi_k \in \mathbb{R}^l, \tag{21}$$

where

$$\begin{aligned} \Theta_k^{\lambda} &= [\delta_{xx} - \lambda I_{xx} - 2I_{xd\zeta}, -2I_{xx}(I_n \otimes K_k^{\mathrm{T}}) \\ &\quad - 2I_{xd\hat{u}}, I_{xx}(K_k^{\mathrm{T}} \otimes K_k^{\mathrm{T}}) - I_{uu}, -\Delta t], \\ \Xi_k &= -I_{xx}\mathrm{vec}(Q + K_k^{\mathrm{T}} R K_k), \quad \Delta t = [t_1 - t_0, t_2 - t_1, \ldots, t_l - t_{l-1}]^{\mathrm{T}}. \end{aligned}$$

If Eq. (21) has unique solutions of symmetric matrices P_k and $\Sigma(P_k)$, matrix $B^{\mathrm{T}} P_k$, and scalar $E(P_k)$, then K_{k+1} can be obtained directly from Eq. (9).

On the basis of the above analysis, we are in a position to formulate the ADP algorithm with respect to the discounted cost.

Algorithm 7.1 Stochastic ADP for the Discounted Problem

1. Choose $K_0 \in \mathbb{R}^{n \times m}$ satisfying the conditions in Theorem 2. Set $k = 0$.
2. Apply $u = -K_0 x + \varepsilon$ to system (1) to generate the online data.

3. Compute Θ_k^λ and Ξ_k. Then solve P_k, $E(P_k)$, and K_{k+1} from Eqs. (21) and (9), respectively. If $|P_k - P_{k+1}| < \bar{\varepsilon}$, where $\bar{\varepsilon} > 0$ is a sufficiently small threshold, go to the next step. Otherwise set $k+1 \to k$ and repeat step 3.
4. Denote $\hat{u}_\lambda^\star = -K_k x$ as an approximation to $u_\lambda^\star$. Then $\mathcal{J}_\lambda(\xi; \hat{u}_\lambda^\star) = \xi^{\mathrm{T}} P_k \xi + \lambda^{-1} E(P_k)$.

Remark 8. We can also update the control input u at each learning iteration to generate the online data. In this case, $u = -K_k x + \varepsilon$ at the kth iteration. Moreover, in some systems (eg, the human sensorimotor system) the learning process does not terminate, and thus there is no need to check $|P_k - P_{k+1}| < \bar{\varepsilon}$.

Next we give the stochastic ADP design with respect to the biased cost. Similarly to the discounted case, one easily has from Theorem 4 that

$$\Theta_k^0 \begin{bmatrix} \mathrm{vec}(P_k) \\ \mathrm{vec}(B^{\mathrm{T}} P_k) \\ \mathrm{vec}(\Sigma(P_k)) \\ E(P_k) \end{bmatrix} = \Xi_k, \quad \Theta_k^0 \begin{bmatrix} \mathrm{vec}(\tilde{P}_k) \\ \mathrm{vec}(B^{\mathrm{T}} \tilde{P}_k) \\ \mathrm{vec}(\Sigma(\tilde{P}_k)) \\ E(\tilde{P}_k) \end{bmatrix} = \tilde{\Xi}_k, \tag{22}$$

where $\tilde{\Xi}_k = -I_{xx}\mathrm{vec}(P_k)$. If Eq. (22) has unique solutions of symmetric matrices P_k, $\tilde{P}_k$ and $\Sigma(P_k)$, matrix $B^{\mathrm{T}} P_k$, and scalar $E(P_k)$, then K_{k+1} can be given directly from Eq. (9).

Algorithm 7.2 Stochastic ADP for the Biased Problem

1. Choose $K_0 \in \mathbb{R}^{n \times m}$ satisfying the conditions in Theorem 4. Set $k = 0$.
2. Apply $u = -K_0 x + \varepsilon$ to system (1) to generate the online data.
3. Compute Θ_k^0, Ξ_k, and $\tilde{\Xi}_k$. Then solve P_k, $\tilde{P}_k$, and K_{k+1} from Eqs. (22) and (9), respectively. If $|P_k - P_{k+1}| < \bar{\varepsilon}$, go to the next step. Otherwise set $k+1 \to k$ and repeat step 3.
4. Denote $\hat{u}^\star = -K_k x$ as an approximation to $u^\star$. Then, $\mathcal{J}(\xi; \hat{u}^\star) = \xi^{\mathrm{T}} P_k \xi - E(\tilde{P}_k)$ and $\rho(\hat{u}^\star) = E(P_k)$.

5.2 Convergence Analysis

5.2.1 Convergence Proof for Algorithm 7.1

First we show that the two sequences $\{P_k\}_{k=0}^\infty$ and $\{K_k\}_{k=0}^\infty$ derived from Algorithm 7.1 converge to $P_\lambda^\star$ and $K_\lambda^\star$, respectively. To begin with, a rank condition is given.

Lemma 5. *Suppose there exists a positive scalar $\bar{l} > 0$ such that for $l \geq \bar{l}$*

$$\text{rank}([I_{xx}, I_{xd\hat{u}}, I_{uu}, \Delta t]) = \frac{n(n+1)}{2} + mn + \frac{m(m+1)}{2} + 1.$$

Then Eq. (21) has unique solutions of P_k, $B^T P_k$, $\Sigma(P_k)$, and $E(P_k)$.

Proof. To show Eq. (21) has unique solutions, it is sufficient to show that the null space of Θ_k^λ is $\{0\}$. From the definition, for any matrices $S = S^T \in \mathbb{R}^{n\times n}$, $L \in \mathbb{R}^{m\times n}$, $H = H^T \in \mathbb{R}^{m\times m}$, and scalar $U > 0$, we have

$$\begin{aligned}\Theta_k^\lambda \begin{bmatrix} \text{vec}(S) \\ \text{vec}(L) \\ \text{vec}(H) \\ \text{vec}(U) \end{bmatrix} &= I_{xx}\text{vec}(X_k) + 2I_{xd\hat{u}}\text{vec}(SB - L^T) + I_{uu}\text{vec}(Y) \\ &\quad + \Delta t(E(S) - U) \\ &= I_{\bar{x}}\text{vec}(\bar{X}_k) + 2I_{xd\hat{u}}\text{vec}(SB - L^T) + I_{\bar{u}}\text{vec}(\bar{Y}) \\ &\quad + \Delta t(E(S) - U),\end{aligned}$$

where

$$\begin{aligned} X_k &= A^T S + SA + K_k^T H K_k + \Pi(S) \\ &\quad - L^T K_k - K_k^T L - \lambda S, \quad Y = \Sigma(S) - H; \\ I_{\bar{x}} &= \left[\int_{t_0}^{t_1} \bar{x} ds, \int_{t_1}^{t_2} \bar{x} ds, \ldots, \int_{t_{l-1}}^{t_l} \bar{x} ds\right]^T; \\ I_{\bar{u}} &= \left[\int_{t_0}^{t_1} \bar{u} ds, \int_{t_1}^{t_2} \bar{u} ds, \ldots, \int_{t_{l-1}}^{t_l} \bar{u} ds\right]^T; \\ \bar{x} &= [x_1^2, x_1x_2, \ldots, x_1x_n, x_2^2, x_2x_3, \ldots, x_{n-1}x_n, x_n^2]^T; \\ \bar{u} &= [u_1^2, u_1u_2, \ldots, u_1u_m, u_2^2, u_2u_3, \ldots, u_{m-1}u_m, u_m^2]^T; \\ \bar{X}_k &= [X_{11}, 2X_{12}, \ldots, 2X_{1n}, X_{22}, 2X_{23}, \ldots, 2X_{(n-1)n}, X_{nn}]^T; \\ \bar{Y} &= [Y_{11}, 2Y_{12}, \ldots, 2Y_{1m}, Y_{22}, 2Y_{23}, \ldots, 2Y_{(m-1)m}, Y_{mm}]^T; \end{aligned}$$

x_j and u_j denote the jth element of x and u, respectively; and X_{ij} and Y_{ij} denote the element of X_k and Y on the ith row and the jth column, respectively.

Since the rank condition implies that $[I_{\bar{x}}, I_{xd\hat{u}}, I_{\bar{u}}, \Delta t]$ has full column rank, the solutions to

$$\Theta_k^\lambda \begin{bmatrix} \text{vec}(S) \\ \text{vec}(L) \\ \text{vec}(H) \\ \text{vec}(U) \end{bmatrix} = 0$$

must be $X_k = 0$, $SB - L^{\mathrm{T}} = 0$, $Y = 0$, and $E(S) - U = 0$, which implies

$$SA_k + A_k^{\mathrm{T}} S + K_k^{\mathrm{T}} \Sigma(S) K_k + \Pi(S) = \lambda S, \quad E(S) = U.$$

Since K_k is admissible to system (1), we know Eq. (8) has a unique symmetric positive definite solution, and thus $S = 0$. Then we have $S = 0$, $L = 0$, $H = 0$, and $U = 0$. This implies that only the zero vector is in the null space of Θ_k^{λ}. This completes the proof. □

Remark 9. To ensure the rank condition in Lemma 5 is satisfied, an exploration noise ε can be added to the system input during the learning process. The role of the rank condition is similar to that of the persistent excitation condition in adaptive control [69].

Next the convergence result is given for Algorithm 7.1.

Theorem 5. *If the conditions in Theorem 2 and Lemma 5 hold, then* $\lim_{k\to\infty} P_k = P_\lambda^\star$ *and* $\lim_{k\to\infty} K_k = K_\lambda^\star$, *where* $\{P_k\}_{k=0}^{\infty}$ *and* $\{K_k\}_{k=0}^{\infty}$ *are obtained from Algorithm 7.1.*

Proof. By Lemma 5, Eq. (21) has unique solutions. Therefore P_k and K_{k+1} are the unique solutions to Eqs. (8) and (9), respectively. Otherwise there is another pair (P_k', K_{k+1}') satisfying Eq. (20) and thus Eq. (21). We know from Theorem 2 that $\{P_k\}_{k=0}^{\infty}$ and $\{K_k\}_{k=0}^{\infty}$ converge to $P_\lambda^\star$ and $K_\lambda^\star$, respectively. This completes the proof. □

5.2.2 Convergence Proof for Algorithm 7.2

Now we analyze the convergence property for Algorithm 7.2. The proofs are similar to the proofs for the discounted cost case and are thus omitted.

Lemma 6. *If the conditions in Lemma 5 hold, then Eq. (22) has unique solutions.*

Theorem 6. *If the conditions in Theorem 4 and Lemma 6 hold, then* $\lim_{k\to\infty} P_k = P^\star$, $\lim_{k\to\infty} \tilde{P}_k = \tilde{P}^\star$, *and* $\lim_{k\to\infty} K_k = K^\star$, *where* $\{P_k\}_{k=0}^{\infty}$, $\{\tilde{P}_k\}_{k=0}^{\infty}$, *and* $\{K_k\}_{k=0}^{\infty}$ *are obtained from Algorithm 7.2.*

6 STOCHASTIC ROBUST OPTIMAL CONTROL

In the following two sections we consider the stochastic robust optimal control problem for a class of partially linear stochastic systems. Some of the results in these two sections were presented in [70]. For simplicity, additive noise is not considered.

6.1 System Description

Suppose $w := (w_1, w_2, \eta) \in \mathbb{R}^{q_1} \times \mathbb{R}^{q_2} \times \mathbb{R}^{q_3}$ is an $\{\mathscr{F}_t\}$-adapted standard Brownian motion. Denote $\mathscr{F}_t^w$ as the σ-field generated by $w(s)$, $0 \leq s \leq t$, and containing all the P null sets. Consider the class of dynamically perturbed stochastic systems described by the following stochastic differential equation:

$$dx = (Ax + B(u + \Delta(x, z)))dt + Bdv, \tag{23}$$

$$dz = f(x, z)dt + g(x, z)d\eta, \tag{24}$$

where $dv = \sum_{i=1}^{q_1} F_i x dw_{1i} + \sum_{i=1}^{q_2} G_i(u + \Delta(x, z))dw_{2i}$ represents the state- and input-dependent noise received by the plant from the input channel, (x, z) and u are $\{\mathscr{F}_t^w\}$-adapted random processes representing the state and the input of the system, respectively, $\Delta : \mathbb{R}^n \times \mathbb{R}^{m_z} \to \mathbb{R}^m$ is a continuous function satisfying $|\Delta(x, z)|^2 \leq C_1|x|^2 + C_2|z|^2$ for some known constants $C_1,\ C_2 > 0$, $(x(0), z(0)) = \xi \in \mathbb{R}^n \times \mathbb{R}^{m_z}$ is a constant initial value, $A \in \mathbb{R}^{n\times n}$, $B \in \mathbb{R}^{n\times m}$, $F_i \in \mathbb{R}^{m\times n}$, and $G_i \in \mathbb{R}^{m\times m}$ are unknown constant matrices, the Borel measurable functions $f : \mathbb{R}^n \times \mathbb{R}^{m_z} \to \mathbb{R}^m$ and $g : \mathbb{R}^n \times \mathbb{R}^{m_z} \to \mathbb{R}^{m_z \times q_3}$ are locally essentially bounded and locally Lipschitz continuous, $f(0, 0) = 0$, and $g(0, 0) = 0$. It is also assumed that the nominal system of Eq. (23) with $\Delta = 0$ is mean square stabilizable.

In this subsection we consider the following undiscounted cost functional:

$$\mathcal{J}(\xi; u) = \mathcal{E}_\xi^u \left[\int_0^\infty (x^{\mathrm{T}} Qx + u^{\mathrm{T}} Ru)dt \right], \tag{25}$$

where $\mathcal{E}_\xi^u$ denotes the expectation for the process corresponding to the system given by Eqs. (23) and (24) with initial state $(x(0), z(0)) = \xi$ and controller u, and Q and R follow the definitions in Section 2.2.

Definition 3. A feedback gain matrix $K \in \mathbb{R}^{m\times n}$ is called *admissible* to system (23) if

- system (23) with $u = -Kx$ and $\Delta \equiv 0$ is GAS at the origin in the mean square sense;
- $\mathcal{J}(\xi; u) < \infty$ for all $\xi \in \mathbb{R}^n$.

$\mathcal{J}(\xi; u)$ is a special case of the biased cost and discounted cost defined in Section 2.2. Thus $\mathcal{J}(\xi; u)$ is minimized under $u = -K^\star x$, where $K^\star = (\Sigma(P^\star) + R)^{-1} B^{\mathrm{T}} P^\star$, where $P^\star$ is the unique symmetric positive definite solution to Eq. (3).

6.2 Robust Optimal Control Design

In this subsection the robust optimal control problem for the system given by Eqs. (23) and (24) is considered. The objective is to find a robust optimal controller that minimizes $\mathcal{J}(\xi; u)$ when the dynamic uncertainty characterized by the z-subsystem (24) is absent while ensuring the robust stability of the system given by Eqs. (23) and (24).

The following assumption is imposed on the dynamic uncertainty (24).

Assumption 1. There exist constant scalars $\gamma > 0$, $\alpha_2 \geq \alpha_1 > 0$ and a twice continuously differentiable positive definite function $W : \mathbb{R}^{m_z} \to \mathbb{R}$ such that

$$\alpha_1|z|^2 \leq W(z) \leq \alpha_2|z|^2, \quad \mathcal{L}W(z) \leq \gamma^2|x|^2 - |z|^2, \quad \forall z \in \mathbb{R}^{m_z}, \tag{26}$$

where $\mathcal{L}$ is the differential generator of system (24).

Remark 10. Since

$$\begin{aligned}\lim_{\tau\to\infty} \mathcal{E}\left[W(z(\tau)) \mid \mathscr{F}_0^w\right] - W(z(0)) \\ \leq \lim_{\tau\to\infty} \mathcal{E}\left[\int_0^\tau (\gamma^2|x|^2 - |z|^2)dt \mid \mathscr{F}_0^w\right],\end{aligned}$$

Assumption 1 suggests that the dynamic uncertainty (24) has a finite linear L^2 gain γ [43, Definition 2.2] when x is considered as the input. Moreover, since $|\Delta|^2 \leq C_1|x|^2 + C_2|z|^2$, system (24) also admits a finite linear L^2 gain $(C_2\gamma^2 + C_1)^{1/2}$, with x as the input and Δ as the output. It is possible to relax this assumption by means of Sontag's input-to-state stability [71, 72].

Next a robust optimal control result is given for the system represented by Eqs. (23) and (24).

Theorem 7. *Suppose Assumption 1 holds. The closed-loop system given by Eqs. (23) and (24) with the robust optimal controller $u = -K^\star x$ is GAS at the origin in the mean square sense if the following matrix inequalities are satisfied:*

$$\left(\gamma^2 + \frac{C_1}{C_2}\right) I_n < Q, \quad \begin{bmatrix} R & R + 2\Sigma(P^\star) \\ R + 2\Sigma(P^\star) & \frac{1}{C_2} I_m - \Sigma(P^\star) \end{bmatrix} > 0.$$

Proof. Choose the Lyapunov function candidate as $V(x) = x^{\mathrm{T}} P^\star x$. Then it follows from Eq. (3) that

$$\begin{aligned}\mathcal{L}V(x) &= 2x^{\mathrm{T}}P^\star(A - BK^\star)x + x^{\mathrm{T}}\Pi(P^\star)x + x^{\mathrm{T}}(K^\star)^{\mathrm{T}}\Sigma(P^\star)K^\star x \\ &\quad + 2\Delta^{\mathrm{T}}\Sigma(P^\star)K^\star x \\ &\quad + 2x^{\mathrm{T}}P^\star B\Delta + \Delta^{\mathrm{T}}\Sigma(P^\star)\Delta \\ &= -x^{\mathrm{T}}(Q + (K^\star)^{\mathrm{T}}RK^\star)x + 2\Delta^{\mathrm{T}}\Sigma(P^\star)K^\star x + 2x^{\mathrm{T}}P^\star B\Delta \\ &\quad + \Delta^{\mathrm{T}}\Sigma(P^\star)\Delta, \quad \forall x \in \mathbb{R}^n.\end{aligned} \tag{27}$$

Denote $\bar{V}(x,z) = V(x) + W(z)$. From Eqs. (26) and (27), and the fact that $|\Delta(x,z)|^2 \le C_1|x|^2 + C_2|z|^2$, we have for all $(x,z) \in \mathbb{R}^n \times \mathbb{R}^{m_z}$ that

$$\begin{aligned}
\mathcal{L}\bar{V}(x,z) &\le -x^{\mathrm{T}}(Q+(K^\star)^{\mathrm{T}}RK^\star)x + 2\Delta^{\mathrm{T}}\Sigma(P^\star)K^\star x + 2x^{\mathrm{T}}P^\star B\Delta \\
&\quad + \Delta^{\mathrm{T}}\Sigma(P^\star)\Delta + \gamma^2|x|^2 - |z|^2 \\
&= -x^{\mathrm{T}}(Q-\gamma^2 I_n)x - (K^\star x)^{\mathrm{T}}RK^\star x + 2\Delta^{\mathrm{T}}\Sigma(P^\star)K^\star x \\
&\quad + 2(K^\star x)^{\mathrm{T}}(\Sigma(P^\star)+R)\Delta + \Delta^{\mathrm{T}}\Sigma(P^\star)\Delta - |z|^2 \\
&= -x^{\mathrm{T}}(Q-\gamma^2 I_n)x - |z|^2 \\
&\quad - \begin{bmatrix} K^\star x \\ -\Delta \end{bmatrix}^{\mathrm{T}} \begin{bmatrix} R & R+2\Sigma(P^\star) \\ R+2\Sigma(P^\star) & -\Sigma(P^\star) \end{bmatrix} \begin{bmatrix} K^\star x \\ -\Delta \end{bmatrix} \\
&\le -x^{\mathrm{T}}(Q-(\gamma^2+\frac{C_1}{C_2})I_n)x \\
&\quad - \begin{bmatrix} K^\star x \\ -\Delta \end{bmatrix}^{\mathrm{T}} \begin{bmatrix} R & R+2\Sigma(P^\star) \\ R+2\Sigma(P^\star) & \frac{1}{C_2}I_m - \Sigma(P^\star) \end{bmatrix} \begin{bmatrix} K^\star x \\ -\Delta \end{bmatrix}.
\end{aligned}$$

[−14pt]

By the conditions in Theorem 7, we know there exists a $\delta > 0$ such that

$$\mathcal{L}\bar{V}(x,z) \le -\delta|x|^2.$$

Using Eq. (26) and the above inequality, we easily obtain

$$\mathcal{L}\bar{V}_\gamma(x,z) \le -\frac{\delta}{2}|x|^2 - \frac{\delta}{2\gamma^2}|z|^2, \quad \bar{V}_\gamma(x,z) := \left(1+\frac{\delta}{2\gamma^2}\right)W(z) + V(x).$$

Thus from stochastic stability theory [58, Theorem 5.13], one concludes that the closed-loop system given by Eqs. (23) and (24) is GAS at the origin in the mean square sense under the controller $u = -K^\star x$. □

The following corollary provides easily checkable, although slightly more restrictive, conditions for the robust stability of the closed-loop system given by Eqs. (23) and (24).

Corollary 1. *The system given by Eqs. (23) and (24) is GAS at the origin in the mean square sense under the optimal controller $u = -K^\star x$ if there exists $\alpha \in (0,1)$ such that*

$$\begin{aligned}
&Q > \left(\gamma^2 + \frac{C_1}{C_2}\right)I_n, \quad 0 < R < \frac{\alpha}{C_2}I_m, \\
&\Sigma(P^\star) < \min\left\{\frac{1-\alpha}{2\alpha}R,\ \frac{\alpha/C_2 I_m - R}{\alpha+2}\right\}.
\end{aligned}$$

Proof. For any $\zeta,\xi\in\mathbb{R}^m$, $(\zeta,\xi)\neq(0,0)$, by completing the squares, it holds that

$$\begin{aligned}
&\begin{bmatrix}\zeta\\ \xi\end{bmatrix}^{\mathrm{T}}\begin{bmatrix}R & R+2\Sigma(P^\star)\\ R+2\Sigma(P^\star) & \frac{1}{C_2}I_m-\Sigma(P^\star)\end{bmatrix}\begin{bmatrix}\zeta\\ \xi\end{bmatrix}\\
&=\zeta^{\mathrm{T}}R\zeta+2\zeta^{\mathrm{T}}(R+2\Sigma(P^\star))\xi+\xi^{\mathrm{T}}\left(\frac{1}{C_2}I_m-\Sigma(P^\star)\right)\xi\\
&\geq\zeta^{\mathrm{T}}R\zeta-\alpha\zeta^{\mathrm{T}}(R+2\Sigma(P^\star))\zeta-\alpha^{-1}\xi^{\mathrm{T}}(R+2\Sigma(P^\star))\xi\\
&\quad+\xi^{\mathrm{T}}\left(\frac{1}{C_2}I_m-\Sigma(P^\star)\right)\xi\\
&=\zeta^{\mathrm{T}}(R-\alpha(R+2\Sigma(P^\star)))\zeta\\
&\quad+\xi^{\mathrm{T}}\left(\frac{1}{C_2}I_m-\Sigma(P^\star)-\alpha^{-1}(R+2\Sigma(P^\star))\right)\xi>0,
\end{aligned}$$

where the last inequality follows from the conditions in Corollary 1. Thus

$$\begin{bmatrix}R & R+2\Sigma(P^\star)\\ R+2\Sigma(P^\star) & \frac{1}{C_2}I_m-\Sigma(P^\star)\end{bmatrix}>0.$$

Moreover, since $Q>(\gamma^2+C_1/C_2)I_n$, we can conclude from Theorem 7 that the system given by Eqs. (23) and (24) under $u=-K^\star x$ is GAS at the origin in the mean square sense. □

Remark 11. The conditions in Theorem 7 and Corollary 1 imply that as long as $|\sum_{i=1}^{q_2}G_i^{\mathrm{T}}G_i|$ is small enough, one can choose Q and R properly such that the system given by Eqs. (23) and (24) under $u=-K^\star x$ is GAS at the origin in the mean square sense.

To further illustrate the idea behind Theorem 7, we next study the linear input-to-output L^2 gain of system (23) using stochastic H^∞ theory [43, 44]. Before proceeding, we first give a lemma.

Lemma 7 (Theorem 7.7.6 in [73]). *For any real symmetric matrix M,*

$$M=\begin{bmatrix}M_{11} & M_{12}\\ M_{12}^{\mathrm{T}} & M_{22}\end{bmatrix}>0$$

is equivalent to $M_{22}>0$ *and* $M_{11}-M_{12}M_{22}^{-1}M_{12}^{\mathrm{T}}>0$, *where* M_{11} *and* M_{22} *are square matrices.*

Theorem 8. *If the conditions in Theorem 7 hold, then system* (23) *under* $u=-K^\star x$ *admits a finite linear* L^2 *gain* $(C_2\gamma^2+C_1)^{-1/2}$ *with* Δ *as the input and* x *as the output.*

Proof. From the conditions in Theorem 7, using Lemma 7, we easily have

$$0 > -Q + \left(\gamma^2 + \frac{C_1}{C_2}\right) I_n,$$

$$0 < (K^\star)^{\mathrm{T}} R K^\star - (\Sigma(P^\star)K^\star + B^{\mathrm{T}} P^\star)^{\mathrm{T}} \left(\frac{1}{C_2} I_m - \Sigma(P^\star)\right)^{-1} (\Sigma(P^\star)K^\star + B^{\mathrm{T}} P^\star).$$

Substituting the above inequalities into Eq. (3), one can derive

$$\begin{aligned}
0 &> (A - BK^\star)^{\mathrm{T}} P^\star + P^\star (A - BK^\star) + \Pi(P^\star) \\
&\quad + K^{\star\mathrm{T}} \Sigma(P^\star) K^\star + \left(\gamma^2 + \frac{C_1}{C_2}\right) I_n \\
&\quad + (\Sigma(P^\star)K^\star + B^{\mathrm{T}} P^\star)^{\mathrm{T}} \left(\frac{1}{C_2} I_m - \Sigma(P^\star)\right)^{-1} (\Sigma(P^\star)K^\star + B^{\mathrm{T}} P^\star) \\
&= (A - BK^\star)^{\mathrm{T}} M + M(A - BK^\star) + \Pi(M) + K^{\star\mathrm{T}} \Sigma(M) K^\star + I_n \\
&\quad + (\Sigma(M)K^\star + B^{\mathrm{T}} M)^{\mathrm{T}} ((C_2\gamma^2 + C_1)^{-1} I_m \\
&\quad - \Sigma(M))^{-1} (\Sigma(M)K^\star + B^{\mathrm{T}} M),
\end{aligned}$$

where $M = C_2/(C_2\gamma^2 + C_1)P^\star$. From the above inequality, and with use of Corollary 2.14 in [43], it follows that system (23) under $u = -K^\star x$ admits a finite linear L^2 gain $(C_2\gamma^2 + C_1)^{-1/2}$ with Δ as the input and x as the output. Indeed, if we choose $\hat{V}(x) = x^{\mathrm{T}} M x$ as the Lyapunov function candidate, then

$$\mathcal{L}\hat{V}(x) \leq (C_2\gamma^2 + C_1)^{-1} |\Delta|^2 - |x|^2, \quad \forall x \in \mathbb{R}^n,$$

where $\mathcal{L}$ is the differential generator of Eq. (23) under $u = -K^\star x$. This completes the proof. □

Remark 12. Theorem 8 implies that by tuning Q and R, we can assign the linear L^2 gain for stochastic system (23) using $u = -K^\star x$.

Remark 13. From Theorem 8, Assumption 1, and Remark 10, it follows from Theorem 1 in [46] that the system given by Eqs. (23) and (24) is GAS at the origin in the mean square sense. This also matches the result we obtained in Theorem 7.

7 STOCHASTIC RADP DESIGN

In this section a stochastic RADP method is developed for the system given by Eqs. (23) and (24).

7.1 Stochastic RADP Design Algorithm

Using a method similar to that used in Section 5.1, we easily have

$$
\begin{aligned}
& x^{\mathrm{T}} \otimes x^{\mathrm{T}}\big|_t^{t+\delta t}\operatorname{vec}(P_k) \\
& = 2\left(\int_t^{t+\delta t} x^{\mathrm{T}} \otimes d\hat{u}^{\mathrm{T}}\right)\operatorname{vec}(B^{\mathrm{T}}P_k) \\
& + 2\left(\int_t^{t+\delta t} x^{\mathrm{T}} \otimes x^{\mathrm{T}} ds\right)(I_n \otimes K_k^{\mathrm{T}})\operatorname{vec}(B^{\mathrm{T}}P_k) \\
& + \left(\int_t^{t+\delta t} \bar{u}^{\mathrm{T}} \otimes \bar{u}^{\mathrm{T}} ds\right)\operatorname{vec}(\Sigma(P_k)) \\
& - \left(\int_t^{t+\delta t} x^{\mathrm{T}} \otimes x^{\mathrm{T}} ds\right)(K_k^{\mathrm{T}} \otimes K_k^{\mathrm{T}})\operatorname{vec}(\Sigma(P_k)) \\
& - \left(\int_t^{t+\delta t} x^{\mathrm{T}} \otimes x^{\mathrm{T}} ds\right)\operatorname{vec}(Q + K_k^{\mathrm{T}}RK_k),
\end{aligned}
$$

where x is the solution to system (23) under $u = -K_0x + \varepsilon$, $\bar{u} = u + \Delta$, and $d\hat{u} = \bar{u}dt + \sum_{i=1}^{q1} F_i x dw_{1i} + \sum_{i=1}^{q2} G_i \bar{u} dw_{2i}$. Then, similarly to Eq. (21),

$$
\Theta_k \begin{bmatrix} \operatorname{vec}(P_k) \\ \operatorname{vec}(B^{\mathrm{T}}P_k) \\ \operatorname{vec}(\Sigma(P_k)) \end{bmatrix} = \Xi_k, \qquad (28)
$$

where

$$
\Theta_k = [\delta_{xx}, -2I_{xx}(I_n \otimes K_k^{\mathrm{T}}) - 2I_{xd\hat{u}}, I_{xx}(K_k^{\mathrm{T}} \otimes K_k^{\mathrm{T}}) - I_{uu}],
$$

and the definitions of Ξ_k, I_{xx}, I_{uu}, and $I_{xd\hat{u}}$ are the same as the definitions in Section 5.1, except that u is replaced by $\bar{u}$ here.

If Eq. (28) has unique solutions of symmetric matrices P_k and $\Sigma(P_k)$ and matrix $B^{\mathrm{T}}P_k$, then K_{k+1} can be solved directly from Eq. (9).

On the basis of the above analysis, a stochastic RADP algorithm is given in Algorithm 7.3:

Algorithm 7.3 Stochastic RADP Algorithm

1. Choose $K_0 \in \mathbb{R}^{n\times m}$ that is admissible to system (23). Set $k = 0$.
2. Apply $u = -K_0x + \varepsilon$ to the system given by Eqs. (23) and (24), and collect the online data.
3. Solve P_k and K_{k+1} from Eqs. (28) and (9), respectively. If $|P_k - P_{k+1}| < \bar{\varepsilon}$, go to the next step. Otherwise set $k \leftarrow k + 1$ and repeat step 3.
4. Denote $\hat{u}^\star = -K_kx$ as an approximation to the optimal controller.

7.2 Convergence and Stability Analysis

In this subsection we show that the two sequences $\{P_k\}_{k=0}^{\infty}$ and $\{K_k\}_{k=0}^{\infty}$ obtained from Algorithm 7.3 converge to $P^\star$ and $K^\star$, respectively. The proofs of Lemma 8 and Theorem 9 are similar to the proofs in Section 5.1 and are thus omitted.

Lemma 8. *Suppose there exists a positive scalar $\bar{l} > 0$ such that for all $l \geq \bar{l}$*

$$rank([I_{xx}, I_{xd\hat{u}}, I_{uu}]) = \frac{n(n+1)}{2} + mn + \frac{m(m+1)}{2}.$$

Then, Eq. (28) has unique solutions.

Theorem 9. *If the conditions in Theorem 7 and Lemma 8 hold, then* $\lim_{k\to\infty} P_k = P^\star$ *and* $\lim_{k\to\infty} K_k = K^\star$, *where $\{P_k\}_{k=0}^{\infty}$ and $\{K_k\}_{k=0}^{\infty}$ are obtained from Algorithm 7.3.*

In the following theorem the robust stability analysis is conducted for the system given by Eqs. (23) and (24) under the controller $\hat{u}^\star$ obtained from Algorithm 7.3.

Theorem 10. *There exists a threshold $\bar{\varepsilon} > 0$ such that the system given by Eqs. (23) and (24) is GAS at the origin in the mean square sense under the controller $\hat{u}^\star$ obtained from Algorithm 7.3.*

Proof. From the conditions on Q and R in Theorem 7, it is easy to see that there exists a $\delta > 0$ such that

$$Q - \left(\gamma^2 + \frac{C_1}{C_2}\right) I_n > \delta I_n, \quad \begin{bmatrix} R & R + 2\Sigma(P^\star) \\ R + 2\Sigma(P^\star) & \frac{1}{C_2} I_m - \Sigma(P^\star) \end{bmatrix} > \delta I_{2m}.$$

Since $\lim_{k\to\infty} P_k = P^\star$ and $\lim_{k\to\infty} K_k = K^\star$, there exists a small $\bar{\varepsilon} > 0$ such that

$$\begin{aligned} \delta \geq |2\delta^{-1}|\Sigma(P^\star)|(K^\star - K_k)^{\mathrm{T}}\Sigma(P^\star)(K^\star - K_k) \\ + 2K_k^{\mathrm{T}}\Sigma(P^\star)K_k - 2(K^\star)^{\mathrm{T}}\Sigma(P^\star)K^\star \\ + 2(K^\star - K_k)^{\mathrm{T}}RK^\star + 2(K^\star)^{\mathrm{T}}R(K^\star - K_k)|, \end{aligned}$$

as long as $|P_k - P_{k+1}| \leq \bar{\varepsilon}$.

Then it is easy to derive that

$$\begin{aligned} \mathcal{L}\bar{V}(x,z) \leq & -x^{\mathrm{T}}\left(Q - \left(\gamma^2 + \frac{C_1}{C_2}\right) I_n\right) x + \frac{\delta}{2}|x|^2 + \delta|\Delta|^2 \\ & - \begin{bmatrix} K^\star x \\ -\Delta \end{bmatrix}^{\mathrm{T}} \begin{bmatrix} R & R + 2\Sigma(P^\star) \\ R + 2\Sigma(P^\star) & \frac{1}{C_2} I_m - \Sigma(P^\star) \end{bmatrix} \begin{bmatrix} K^\star x \\ -\Delta \end{bmatrix} \\ \leq & -\frac{\delta}{2}|x|^2, \quad \forall (x,z) \in \mathbb{R}^n \times \mathbb{R}^m, \end{aligned}$$

where $\mathcal{L}$ is taken along the solutions to the system given by Eqs. (23) and (24), with $u = \hat{u}^\star$. Thus using a technique similar to that used in the proof of Theorem 7, we know the system given by Eqs. (23) and (24) is GAS at the origin in the mean square sense. This completes the proof. □

8 ILLUSTRATIVE EXAMPLES

8.1 Two-Joint Arm Movements in a Divergent Force Field

In this example the stochastic ADP method proposed in Section 5 is used to study the learning mechanism of human arm movements in a divergent force field. Since the additive noise is not considered, the undiscounted cost (25) is used.

The model is given as follows [74]:

$$\begin{aligned} dp &= v dt, \\ m dv &= (a - bv + f) dt, \\ \tau\, da &= (u - a) dt + d\zeta, \end{aligned}$$

where $p = [p_x\ p_y]^T$, $v = [v_x\ v_y]^T$, and $a = [a_x\ a_y]^T$ denote the distance between the hand position and the origin, the hand velocity, and the actuator state, respectively; $u = [u_x\ u_y]^T$ is the control input; $m = 1.3$kg is the hand mass; $b = 10\,\mathrm{N\,s/m}$ is viscosity constant; $\tau = 0.05\,\mathrm{s}$ is the time constant; and $d\zeta$ is the signal-dependent noise [75]:

$$d\zeta = \begin{bmatrix} c_1 & 0 \\ c_2 & 0 \end{bmatrix} \begin{bmatrix} u_x \\ u_y \end{bmatrix} dw_1 + \begin{bmatrix} 0 & -c_2 \\ 0 & c_1 \end{bmatrix} \begin{bmatrix} u_x \\ u_y \end{bmatrix} dw_2,$$

where w_i are independent standard Brownian motions, and $c_1 = 0.075$ and $c_2 = 0.025$ are noise magnitudes. The force generated by the divergent force field is $f = 150 p_x$.

The weighting matrices in the cost are chosen as in [38]:

$$Q = \begin{bmatrix} Q_0 & 0 & 0 \\ 0 & 10^{-2} Q_0 & 0 \\ 0 & 0 & 5 \times 10^{-5} Q_0 \end{bmatrix}, \quad Q_0 = \mathrm{diag}\{8, 10\} \times 10^4, \quad R = I_2.$$

The movement trajectories, the velocity curves, and the endpoint force curves are given in Figs. 1 and 2. Fig. 1A shows the optimal trajectories in the null field. When the subject was first exposed to the divergent force field, the variations were amplified by the divergence force, and thus the system is no longer stable. This can be seen from Fig. 1B. To regain stable

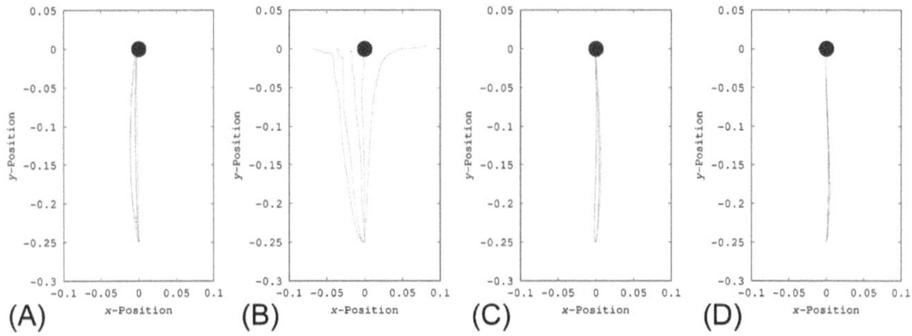

Fig. 1 Movement trajectories in the divergent force field (*DF*). (A) Five independent movement trajectories in the null filed (*NF*) with the initial control policy. (B) Five independent movement trajectories in the DF with the initial control policy. (C) Five independent movement trajectories in the DF after adaptive dynamic programming learning. (D) Five independent movement trajectories when the DF was removed.

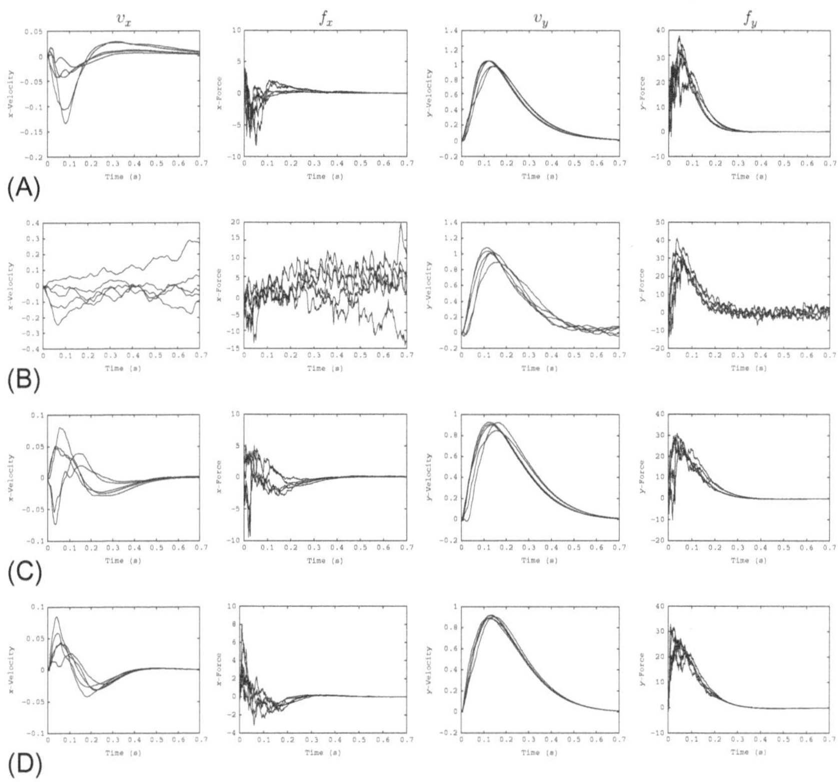

Fig. 2 Velocity and endpoint force curves. (A) Five trials in NF. (B) First five trials in DF. (C) Five after learning trials in DF. (D) Five after effect trials in NF. *DF*, divergent field; *NF*, null field.

behavior, the central nervous system will increase the stiffness along the direction of the divergence force [76]. Here we increased the first entry in the first row of the feedback gain matrix by 300 and set the resultant matrix to be K_0, which is stabilizing. Then the proposed stochastic ADP algorithm is applied with this K_0 as the initial stabilizing feedback gain matrix. After 30 learning trials, a new feedback gain matrix is obtained. The simulation for the system under the new control policy is given in Fig. 1C. To test the aftereffects, the divergent force field is then unexpectedly removed. The aftereffects of the motor learning are shown in Fig. 1D. The results obtained are consistent with the experimental results in [48, 77].

This example shows that our stochastic ADP method appears to be a suitable candidate for computational learning mechanism in the central nervous system to coordinate movements.

8.2 Vehicle Suspension System With Two Degrees of Freedom

In this example a vehicle suspension system with two degrees of freedom is considered to illustrate the stochastic ADP results for the discounted and biased problems. The model of the vehicle suspension system with two degrees of freedom is given in Fig. 3 [78].

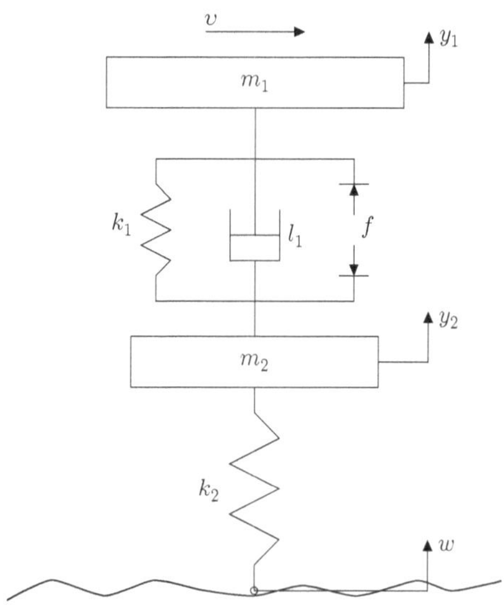

Fig. 3 Vehicle suspension system with two degrees of freedom.

The system dynamics can be represented by the following stochastic differential equations:

$$
\begin{aligned}
0 &= m_1\ddot{y}_1 + m_1 g + l_1(\dot{y}_1 - \dot{y}_2) + k_1(y_1 - y_2) - f,\\
0 &= m_2\ddot{y}_2 + m_2 g + l_1(\dot{y}_2 - \dot{y}_1) + k_1(y_2 - y_1) + k_2(y_2 - w) + f,\\
dw &= -avwdt + d\zeta,
\end{aligned}
$$

where m_1 and m_2 represent the masses of the body and the wheels of the vehicle, respectively, y_1 and y_2 represent the absolute vertical displacements of the body and the wheels, respectively, k_1 and l_1 are the linear spring and damping elements, respectively, f denotes the suspension force, k_2 is the tire stiffness, w denotes road irregularities, v is the vehicle velocity, a is a coefficient depending on the type of road surface, and ζ is the stochastic noise satisfying $\mathcal{E}[d\zeta^2(t)] = 2\sigma^2 avdt$, where σ^2 denotes the variance of the road irregularities.

It is easy to see that the equilibriums of the system are $y_{2,0} = (m_1 + m_2)/k_2 g$, $y_{1,0} = y_{2,0} - m_1/k_1 g$, and $w_0 = 0$. Suppose the equilibriums are given. Then the state-space representation of the error system is given as

$$dx = (Ax + Bu)dt + Ed\zeta,$$

where $u = f$,

$$
x = \begin{bmatrix} \Delta y_1 \\ \Delta \dot{y}_1 \\ \Delta y_2 \\ \Delta \dot{y}_2 \\ w \end{bmatrix}, \quad
A = \begin{bmatrix}
0 & 1 & 0 & 0 & 0\\
-\frac{k_1}{m_1} & -\frac{l_1}{m_1} & \frac{k_1}{m_1} & \frac{l_1}{m_1} & 0\\
0 & 0 & 0 & 1 & 0\\
\frac{k_1}{m_2} & \frac{l_1}{m_2} & -\frac{k_1+k_2}{m_2} & -\frac{l_1}{m_2} & \frac{k_2}{m_2}\\
0 & 0 & 0 & 1 & -\frac{a}{v}
\end{bmatrix},
$$

$$B = [0, m_1^{-1}, 0, -m_2^{-1}, 0]^{\mathrm{T}}, \quad E = [0, 0, 0, 0, \sqrt{2\sigma^2 av}]^{\mathrm{T}},$$

$\Delta y_1 = y_1 - y_{10}$, and $\Delta y_2 = y_2 - y_{20}$.

Following the definition in [78], we can define the one-step cost as

$$
\begin{aligned}
[x^{\mathrm{T}}\ u^{\mathrm{T}}]\begin{bmatrix} Q & N \\ N^{\mathrm{T}} & R \end{bmatrix}\begin{bmatrix} x \\ u \end{bmatrix} &= \left(-\frac{k_1}{m_1}x_1 - \frac{l_1}{m_1}x_2 + \frac{k_1}{m_1}x_3 + \frac{l_1}{m_1}x_4 + \frac{1}{m_1}u\right)^2\\
&\quad + \rho_1(x_1 - x_3)^2\\
&\quad + \rho_2(x_3 - x_5)^2 + \rho_3 u^2 + s_1 x_2^2 + s_2 x_4^2 + s_3 x_5^2.
\end{aligned}
$$

Note that differently from the cost in [78], we add s_1, s_2, and s_3 to characterize the influence of $\Delta\dot{y}_1$, $\Delta\dot{y}_2$, and w.

The parameters of the system are given in Table 1.

Table 1 Parameters of the vehicle suspension system

Parameter	Value	Parameter	Value
m_1	500 kg	m_2	80 kg
k_1	18 kN/m	l_1	3 kN s/m
k_2	200 kN/m	σ^2	300×10^{-6} m^2
v	20 m/s	a	0.45/m
ρ_1	10^3	ρ_2	10^4
ρ_3	10^{-2}	s_1	10
s_2	10	s_3	10

With a simple mathematical manipulation, the one-step cost is rewritten as $x^{\mathrm{T}}\bar{Q}x + v^{\mathrm{T}}Rv$, where $\bar{Q} = Q - NRN^{\mathrm{T}}$. The control input is given as $u = -R^{-1}N^{\mathrm{T}}x + v$. Then the original problem is transformed to the optimal control problem for the following system:

$$dx = (A - BR^{-1}N^{\mathrm{T}})xdt + Bvdt + Ed\zeta,$$

with respect to the one-step cost $x^{\mathrm{T}}\bar{Q}x + v^{\mathrm{T}}Rv$.

Algorithms 7.1 and 7.2 are used to approximate the optimal feedback gain matrices for the discounted and biased problems, respectively. To illustrate the effectiveness of the proposed ADP methods, choose $K_0 = [30, 80, -200, -40, 20]$, which is far from $K_\lambda^\star$ and $K^\star$. Suppose there is an obstacle on the ground represented by an impulse input in w at $t = 0$ such that the initial state is $x(0) = [0.01, 0.01, 0.02, 0.04, 0.3]^{\mathrm{T}}$. The initial input is chosen as $u = -K_0x + 130\sin(200t) + 30\sin(20t)$. The online information from $t = 0$ s to $t = 4$ s is used in the stochastic ADP algorithms.

Both the discounted cost and the biased cost are considered. Choose $\lambda = 0.3$. The control policies obtained ($\hat{u}_\lambda^\star$ and $\hat{u}^\star$) are applied separately to the system at $t = 5$ s. The system trajectories are given in Fig. 4. To further illustrate the effectiveness of the methods obtained, we solve the optimal feedback gain matrix $K_\lambda^\star$ and $K^\star$ directly from Eqs. (5) and (3). It is easy to see that the results obtained from Algorithms 7.1 and 7.2 are very close to the optimal values:

$$\begin{aligned}
K_\lambda^\star &= [184.785, 495.4953, -851.997, -287.0263, 33.32046],\\
\hat{K}_\lambda^\star &= [188.4489, 479.1928, -841.3594, -289.1891, 34.55221],\\
K^\star &= [298.684, 1176.283, -987.723, -219.847, 55.257],\\
\hat{K}^\star &= [299.8287, 1138.114, -974.0073, -224.8217, 55.84284].
\end{aligned}$$

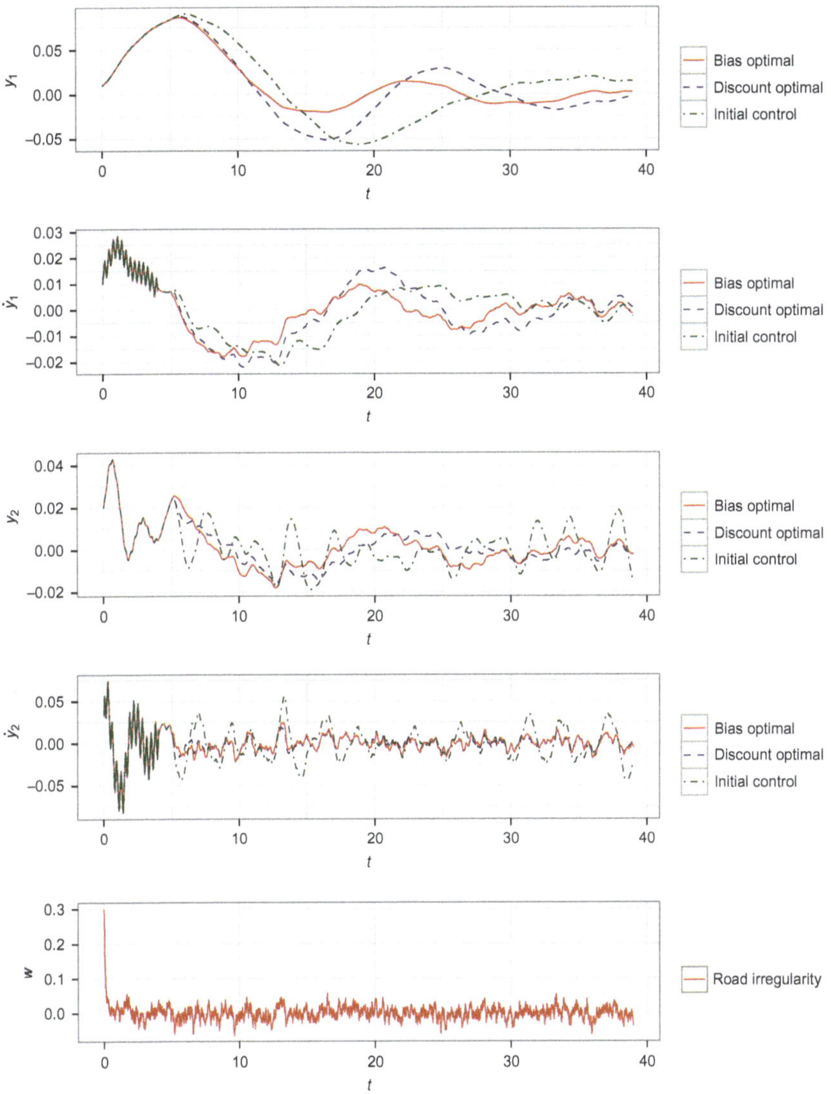

Fig. 4 Vehicle suspension optimal control.

It is easy to see that in both cases the control policies obtained from the ADP algorithms give better performance. Moreover, one can easily see from Fig. 4 that the system maintains a better steady-state performance under the control policy obtained from Algorithm 7.2. This is not surprising, since the bias-optimal control is designed to minimize both the average cost and the biased cost.

8.3 Single-Joint Arm Movement

In this section a numerical simulation of the single-joint human arm movement is given to illustrate the stochastic RADP method obtained in Section 7.

The dynamic model of the single-joint human arm movement is [79]

$$I\ddot{\theta} = -mgl\cos\theta + n + T_{\mathrm{m}},$$

where $m = 1.65\,\mathrm{kg}$ is the mass of the segment, $I = 0.0779\,\mathrm{kg\,m^2}$ is the inertia, $g = 9.81\,\mathrm{m/s^2}$ is the gravitational constant, $l = 0.179\,\mathrm{m}$ is the distance of the center of mass from the joint, θ is the joint angular position, T_{m} is the input to the muscle from the motoneurons, and n denotes the inputs from the neural integrator satisfying

$$\dot{n} = -\frac{n}{\tau_{\mathrm{N}}} + T_{\mathrm{m}},$$

where $\tau_{\mathrm{N}} = 2$ is a time constant.

Define

$$x_1 = \theta - \theta_0, \quad x_2 = \dot{\theta}, \quad z = n - \frac{\tau_{\mathrm{N}} mgl\cos\theta_0}{\tau_{\mathrm{N}} + 1} - Ix_2,$$
$$u = T_{\mathrm{m}} - \frac{mgl\cos\theta_0}{\tau_{\mathrm{N}} + 1},$$

where $\theta_0 = \pi/4$ is the desired endpoint angular position. Then we can write the following linearized model:

$$dz = -\left(\frac{1+\tau_{\mathrm{N}}}{\tau_{\mathrm{N}}}(z + Ix_2) + 2mgl\sin\left(\frac{x_1}{2} + \theta_0\right)\sin\left(\frac{x_1}{2}\right)\right)dt,$$
$$dx_1 = x_2 dt,$$
$$dx_2 = \left(\frac{mgl}{I}\sin(\theta_0)x_1 + x_2\right)dt + \frac{1}{I}((u+z)dt + cudw),$$

where $cudw$ represents the signal-dependent noise; $c = 0.05$. Note that the z-subsystem is regarded as the nonlinear dynamic uncertainty.

The initial state is $[x_1(0), x_2(0), z(0)]^{\mathrm{T}} = [-\pi/4, 0, 1]^{\mathrm{T}}$, and the initial control gain matrix is $K_0 = [5, 0.5]$. First we simulate the system under the initial control policy. After 10 trials of learning using Algorithm 7.3, a new controller

$$u = -9.746359x_1 - 1.716147x_2$$

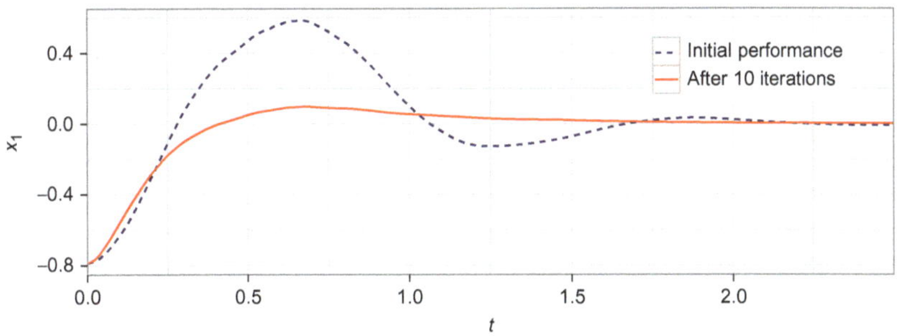

Fig. 5 Trajectory of the angle error.

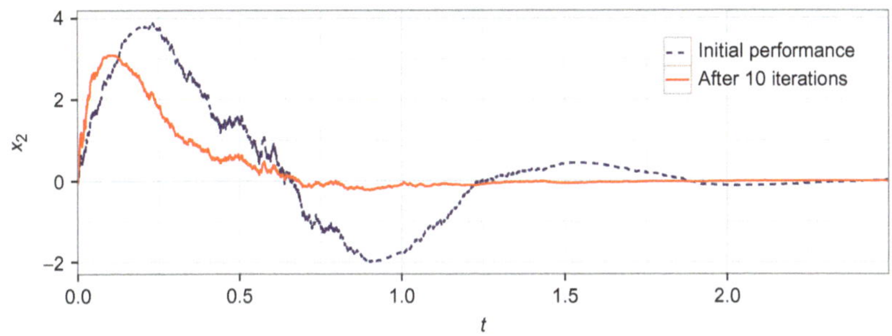

Fig. 6 Trajectory of the angle velocity.

is obtained. The trajectories of x_1 and x_2 are given in Figs. 5 and 6, respectively. The results obtained match the experimental observations in [80].

9 SUMMARY

We have presented preliminary results of stochastic ADP-based control design for continuous-time linear stochastic systems with unknown dynamics. Both stochastic optimal control and stochastic robust optimal problems were investigated. The results obtained were illustrated by three practical examples arising from the sensorimotor control and vehicle suspension control problems.

The proposed methods can be extended in several directions. For example, the stochastic noise in systems (1) and (23) is not necessarily independent. Moreover, for the stochastic robust optimal control problem, one can consider high-order systems by a combined application of recursive backstepping and small-gain techniques, or by relaxation of the linear growth condition on dynamic uncertainties by nonlinear growth conditions.

ACKNOWLEDGMENTS

This work was partially supported by the National Science Foundation under grants ECCS-1501044 and ECCS-1230040.

REFERENCES

[1] R.E. Bellman, Dynamic Programming, Princeton University Press, Princeton, NJ, 1957.
[2] R.E. Bellman, Adaptive Control Processes: A Guided Tour, Princeton University Press, Princeton, NJ, 1961.
[3] D.P. Bertsekas, J.N. Tsitsiklis, Neuro-Dynamic Programming, Athena Scientific, Belmont, MA, 1996.
[4] P.J. Werbos, The elements of intelligence, Cybernetica (Namur) 3 (1968).
[5] P.J. Werbos, Intelligence in the brain: a theory of how it works and how to build it, Neural Netw. 22 (3) (2009) 200–212.
[6] C.J.C.H. Watkins, P. Dayan, Q-learning, Mach. Learn. 8 (3–4) (1992) 279–292.
[7] J.N. Tsitsiklis, Asynchronous stochastic approximation and Q-learning, Mach. Learn. 16 (3) (1994) 185–202.
[8] R.S. Sutton, Learning to predict by the methods of temporal differences, Mach. Learn. 3 (1) (1988) 9–44.
[9] J.N. Tsitsiklis, B. Van Roy, An analysis of temporal-difference learning with function approximation, IEEE Trans. Autom. Control 42 (5) (1997) 674–690.
[10] A.G. Barto, R.S. Sutton, C.W. Anderson, Neuronlike adaptive elements that can solve difficult learning control problems, IEEE Trans. Syst. Man Cybern. SMC-13 (5) (1983) 834–846.
[11] D.V. Prokhorov, D.C. Wunsch, Adaptive critic designs, IEEE Trans. Neural Netw. 8 (5) (1997) 997–1007.
[12] V.R. Konda, Actor-critic algorithms (Ph.D. thesis), Department of Electrical Engineering and Computer Science, Massachusetts Institute of Technology, 2002.
[13] R.S. Sutton, A.G. Barto, Reinforcement Learning: An Introduction, MIT Press, Cambridge, MA, 1998.
[14] W.B. Powell, Approximate Dynamic Programming: Solving the Curses of Dimensionality, second ed., John Wiley & Sons, Inc., Hoboken, NJ, 2007.
[15] J.J. Murray, C.J. Cox, G.G. Lendaris, R. Saeks, Adaptive dynamic programming, IEEE Trans. Syst. Man Cybern. C: Appl. Rev. 32 (2) (2002) 140–153.
[16] D. Vrabie, O. Pastravanu, M. Abu-Khalaf, F.L. Lewis, Adaptive optimal control for continuous-time linear systems based on policy iteration, Automatica 45 (2) (2009) 477–484.
[17] D. Vrabie, F.L. Lewis, Neural network approach to continuous-time direct adaptive optimal control for partially unknown nonlinear systems, Neural Netw. 22 (3) (2009) 237–246.
[18] Y. Jiang, Z.P. Jiang, Computational adaptive optimal control for continuous-time linear systems with completely unknown dynamics, Automatica 48 (10) (2012) 2699–2704.

[19] T. Bian, Y. Jiang, Z.P. Jiang, Adaptive dynamic programming and optimal control of nonlinear nonaffine systems, Automatica 50 (10) (2014) 2624–2632.
[20] W. Gao, Y. Jiang, Z.P. Jiang, T. Chai, Adaptive and optimal output feedback control of linear systems: an adaptive dynamic programming approach, in: Proceeding of the 11th World Congress on Intelligent Control and Automation, 2014, pp. 2085–2090.
[21] W. Gao, Z.P. Jiang, Adaptive dynamic programming and adaptive optimal output regulation of linear systems, IEEE Trans. Autom. Control (2016) accepted.
[22] H. Modares, F.L. Lewis, Linear quadratic tracking control of partially-unknown continuous-time systems using reinforcement learning, IEEE Trans. Autom. Control 59 (11) (2014) 3051–3056.
[23] H. Li, D. Liu, D. Wang, Integral reinforcement learning for linear continuous-time zero-sum games with completely unknown dynamics, IEEE Trans. Autom. Sci. Eng. 11 (3) (2014) 706–714.
[24] R. Song, F.L. Lewis, Q. Wei, H. Zhang, Z.P. Jiang, D. Levine, Multiple actor-critic structures for continuous-time optimal control using input-output data, IEEE Trans. Neural Netw. Learn. Syst. 26 (4) (2015) 851–865.
[25] H. Zargarzadeh, T. Dierks, S. Jagannathan, Optimal control of nonlinear continuous-time systems in strict-feedback form, IEEE Trans. Neural Netw. Learn. Syst. 26 (10) (2015) 2535–2549.
[26] T. Bian, Z.P. Jiang, Value iteration and adaptive optimal control for linear continuous-time systems, in: IEEE 7th International Conference on Cybernetics and Intelligent Systems (CIS) and IEEE Conference on Robotics, Automation and Mechatronics (RAM), 2015, pp. 53–58.
[27] A. Al-Tamimi, F.L. Lewis, M. Abu-Khalaf, Discrete-time nonlinear HJB solution using approximate dynamic programming: convergence proof, IEEE Trans. Syst. Man Cybern. B: Cybern. 38 (4) (2008) 943–949.
[28] H. Zhang, Y. Luo, D. Liu, Neural-network-based near-optimal control for a class of discrete-time affine nonlinear systems with control constraints, IEEE Trans. Neural Netw. 20 (9) (2009) 1490–1503.
[29] F.L. Lewis, K.G. Vamvoudakis, Reinforcement learning for partially observable dynamic processes: adaptive dynamic programming using measured output data, IEEE Trans. Syst. Man Cybern. B: Cybern. 41 (1) (2011) 14–25.
[30] D. Wang, D. Liu, Q. Wei, D. Zhao, N. Jin, Optimal control of unknown nonaffine nonlinear discrete-time systems based on adaptive dynamic programming, Automatica 48 (8) (2012) 1825–1832.
[31] A. Heydari, Revisiting approximate dynamic programming and its convergence, IEEE Trans. Cybern. 44 (12) (2014) 2733–2743.
[32] Z. Ni, H. He, X. Zhong, D.V. Prokhorov, Model-free dual heuristic dynamic programming, IEEE Trans. Neural Netw. Learn. Syst. 26 (8) (2015) 1834–1839.
[33] F.L. Lewis, D. Vrabie, Reinforcement learning and adaptive dynamic programming for feedback control, IEEE Circuits Syst. Mag. 9 (3) (2009) 32–50.
[34] F.L. Lewis, D. Vrabie, K.G. Vamvoudakis, Reinforcement learning and feedback control: using natural decision methods to design optimal adaptive controllers, IEEE Control Syst. 32 (6) (2012) 76–105.
[35] Z.P. Jiang, Y. Jiang, Robust adaptive dynamic programming for linear and nonlinear systems: an overview, Eur. J. Control 19 (5) (2013) 417–425.
[36] F.L. Lewis, D. Liu, Reinforcement Learning and Approximate Dynamic Programming for Feedback Control, John Wiley & Sons, Inc., Piscataway, NJ, 2013.
[37] Y. Jiang, Z.P. Jiang, Approximate dynamic programming for optimal stationary control with control-dependent noise, IEEE Trans. Neural Netw. 22 (12) (2011) 2392–2398.
[38] Y. Jiang, Z.P. Jiang, Adaptive dynamic programming as a theory of sensorimotor control, Biol. Cybern. 108 (4) (2014) 459–473.

[39] Y. Jiang, Z.P. Jiang, Robust adaptive dynamic programming for large-scale systems with an application to multimachine power systems, IEEE Trans. Circuits Syst. II: Express Briefs 59 (10) (2012) 693–697.
[40] Y. Jiang, Z.P. Jiang, Robust adaptive dynamic programming with an application to power systems, IEEE Trans. Neural Netw. Learn. Syst. 24 (7) (2013) 1150–1156.
[41] Y. Jiang, Z.P. Jiang, Robust adaptive dynamic programming and feedback stabilization of nonlinear systems, IEEE Trans. Neural Netw. Learn. Syst. 25 (5) (2014) 882–893.
[42] T. Bian, Y. Jiang, Z.P. Jiang, Decentralized adaptive optimal control of large-scale systems with application to power systems, IEEE Trans. Ind. Electron. 62 (4) (2015) 2439–2447.
[43] D. Hinrichsen, A. Pritchard, Stochastic H^{∞}, SIAM J. Control Optim. 36 (5) (1998) 1504–1538.
[44] N. Berman, U. Shaked, H_{∞}-like control for nonlinear stochastic systems, Syst. Control Lett. 55 (3) (2006) 247–257.
[45] Z.P. Jiang, A.R. Teel, L. Praly, Small-gain theorem for ISS systems and applications, Math. Control Signals Syst. 7 (2) (1994) 95–120.
[46] V. Drăgan, A. Halanay, A. Stoica, A small gain theorem for linear stochastic systems, Syst. Control Lett. 30 (5) (1997) 243–251.
[47] Z.-J. Wu, X.-J. Xie, S.-Y. Zhang, Adaptive backstepping controller design using stochastic small-gain theorem, Automatica 43 (4) (2007) 608–620.
[48] E. Burdet, R. Osu, D.W. Franklin, T.E. Milner, M. Kawato, The central nervous system stabilizes unstable dynamics by learning optimal impedance, Nature 414 (6862) (2001) 446–449.
[49] I. Karatzas, S.E. Shreve, Brownian Motion and Stochastic Calculus, second ed., Springer, New York, 1991.
[50] J.L. Willems, J.C. Willems, Feedback stabilizability for stochastic systems with state and control dependent noise, Automatica 12 (3) (1976) 277–283.
[51] W. Wonham, A Liapunov method for the estimation of statistical averages, J. Differ. Equ. 2 (4) (1966) 365–377.
[52] W. Wonham, Optimal stationary control of a linear system with state-dependent noise, SIAM J. Control 5 (3) (1967) 486–500.
[53] D.L. Kleinman, Optimal stationary control of linear systems with control-dependent noise, IEEE Trans. Autom. Control 14 (6) (1969) 673–677.
[54] U. Haussmann, Optimal stationary control with state control dependent noise, SIAM J. Control 9 (2) (1971) 184–198.
[55] P.R. Kumar, P. Varaiya, Stochastic Systems: Estimation, Identification, and Adaptive Control, Prentice Hall, Englewood Cliffs, NJ, 1986.
[56] H. Kushner, Optimal discounted stochastic control for diffusion processes, SIAM J. Control 5 (4) (1967) 520–531.
[57] P. McLane, Optimal stochastic control of linear systems with state- and control-dependent disturbances, IEEE Trans. Autom. Control 16 (6) (1971) 793–798.
[58] R. Khas'minskii, Stochastic Stability of Differential Equations, Springer, Berlin, Heidelberg, 2012.
[59] S. Mahadevan, Average reward reinforcement learning: foundations, algorithms, and empirical results, Mach. Learn. 22 (1–3) (1996) 159–195.
[60] A. Schwartz, A reinforcement learning method for maximizing undiscounted rewards, in: Proceedings of the 10th International Conference on Machine Learning, Amherst, MA, 1993, pp. 298–305.
[61] M.L. Puterman, Markov Decision Processes: Discrete Stochastic Dynamic Programming, John Wiley & Sons, Inc., New York, NY, 1994.
[62] J.N. Tsitsiklis, B. Van Roy, Average cost temporal-difference learning, Automatica 35 (11) (1999) 1799–1808.

[63] J.N. Tsitsiklis, B. Van Roy, On average versus discounted reward temporal-difference learning, Mach. Learn. 49 (2–3) (2002) 179–191.
[64] H. Yu, D.P. Bertsekas, Convergence results for some temporal difference methods based on least squares, IEEE Trans. Autom. Control 54 (7) (2009) 1515–1531.
[65] A. Arapostathis, V.S. Borkar, M.K. Ghosh, Ergodic Control of Diffusion Processes, Cambridge University Press, New York, NY, 2012.
[66] A. Arapostathis, V.S. Borkar, K. Kumar, Convergence of the relative value iteration for the ergodic control problem of nondegenerate diffusions under near-monotone costs, SIAM J. Control Optim. 52 (1) (2014) 1–31.
[67] L. Arnold, Stochastic Differential Equations: Theory and Applications, John Wiley & Sons, Inc., New York, 1974.
[68] H.J. Kushner, Stochastic Stability and Control, Academic Press, London, UK, 1967.
[69] G. Tao, Multivariable adaptive control: a survey, Automatica 50 (11) (2014) 2737–2764.
[70] T. Bian, Z.P. Jiang, Robust adaptive dynamic programming for continuous-time linear stochastic systems, in: 2014 IEEE International Symposium on Intelligent Control (ISIC), 2014, pp. 536–541.
[71] E.D. Sontag, On the input-to-state stability property, Eur. J. Control 1 (1) (1995) 24–36.
[72] E.D. Sontag, Mathematical Control Theory: Deterministic Finite Dimensional Systems, second ed., Springer, New York, 1998.
[73] R.A. Horn, C.R. Johnson, Matrix Analysis, second ed., Cambridge University Press, New York, NY, 2013.
[74] D. Liu, E. Todorov, Evidence for the flexible sensorimotor strategies predicted by optimal feedback control, J. Neurosci. 27 (35) (2007) 9354–9368.
[75] C.M. Harris, D.M. Wolpert, Signal-dependent noise determines motor planning, Nature 394 (6695) (1998) 780–784.
[76] N. Hogan, The mechanics of multi-joint posture and movement control, Biol. Cybern. 52 (5) (1985) 315–331.
[77] D.W. Franklin, E. Burdet, R. Osu, M. Kawato, T. Milner, Functional significance of stiffness in adaptation of multijoint arm movements to stable and unstable dynamics, Exp. Brain Res. 151 (2) (2003) 145–157.
[78] A. Hać, Suspension optimization of a 2-DOF vehicle model using a stochastic optimal control technique, J. Sound Vib. 100 (3) (1985) 343–357.
[79] R. Shadmehr, S.P. Wise, The Computational Neurobiology of Reaching and Pointing: A Foundation for Motor Learning, MIT Press, Cambridge, MA, 2005.
[80] R. Shadmehr, S. Mussa-Ivaldi, Biological Learning and Control: How the Brain Builds Representations, Predicts Events, and Makes Decisions, MIT Press, Cambridge, MA, 2012.

CHAPTER EIGHT

Model-Based Reinforcement Learning for Approximate Optimal Regulation

R. Kamalapurkar, P. Walters, W.E. Dixon
University of Florida, Gainesville, FL, United States

1 INTRODUCTION

Reinforcement learning (RL) enables a cognitive agent to learn desirable behavior from interactions with its environment. In control theory the desirable behavior is typically quantified using a cost function, and the control problem is formulated as the desire to find the optimal policy that minimizes a cumulative cost. In recent years various RL-based techniques have been developed to approximately solve optimal control problems for continuous-time and discrete-time deterministic systems [1–34]. The approximate solution is facilitated via value function approximation, where the optimal policy is computed on the basis of an estimate of the value function.

Methods that seek online solutions to optimal control problems are comparable to adaptive control [3, 8, 12, 14, 35, 36]. In adaptive control the estimates for the uncertain parameters in the plant model are updated using the current tracking error as a performance metric, whereas in online RL-based techniques, estimates for the uncertain parameters in the value function are updated using the Bellman error (BE) as a performance metric. Typically, to establish regulation or tracking, adaptive control methods do not require the adaptive estimates to convergence to the true values. However, convergence of the RL-based controller to a neighborhood of the optimal controller requires convergence of the parameter estimates to a neighborhood of their ideal values.

Parameter convergence has been a focus of research in adaptive control for several decades. It is common knowledge that least squares-based and

Control of Complex Systems
http://dx.doi.org/10.1016/B978-0-12-805246-4.00008-2

gradient descent-based update laws generally require persistence of excitation (PE) in the system state for convergence of the parameter estimates. Modification schemes such as projection algorithms, σ-modification, and e-modification are used to guarantee boundedness of parameter estimates and overall system stability; however, these modifications do not guarantee parameter convergence unless the PE condition, which is generally impossible to verify online, is satisfied [37–40].

In RL-based approximate online optimal control, the Hamilton-Jacobi-Bellman (HJB) equation along with an estimate of the state derivative [7, 12] or an integral form of the HJB equation [41] is used to approximately evaluate the BE at each visited state along the system trajectory. The BE provides an indirect measure of the quality of the current estimate of the value function at each visited state along the system trajectory. Hence the unknown value function parameters are updated on the basis of evaluation of the BE along the system trajectory. Such weight update strategies create two challenges for analysis of convergence. The system states need to satisfy PE, and the system trajectory needs to visit enough points in the state space to generate a good approximation to the value function over the entire operating domain: that is, exploration versus exploitation. These challenges are typically addressed in related literature [5, 8, 12, 19, 25–27, 42–44] by the addition of an exploration signal to the control input to ensure sufficient exploration in the desired region of the state space. However, no analytical methods exist to compute the appropriate exploration signal when the system dynamics are nonlinear.

The aforementioned challenges arise from the restriction that the BE can be evaluated only along the system trajectories. In particular, the integral BE is only meaningful as a measure of the quality of the value function if it is evaluated along the system trajectories, and state derivative estimators can only generate estimates of the state derivative along the system trajectories using numerical smoothing. Modares et al. [26] recently demonstrated that experience replay can be used to improve data efficiency in online approximate optimal control by reuse of recorded data. However, since the data need to be recorded along the system trajectory, the system trajectory under the designed approximate optimal controller needs to provide enough excitation for learning. In general, such excitation is not available; hence the simulation results in [26] were generated with an added probing signal.

In this chapter a different approach is used to improve data efficiency by observing that if the system dynamics are known, the state derivative, and hence the BE, can be evaluated at any desired point in the state space. Unknown parameters in the value function can therefore be adjusted on the basis of least-squares minimization of the BE evaluated at any number of desired points in the state space. For example, in an infinite-horizon regulation problem, the BE can be computed at sampled points uniformly distributed in a neighborhood around the origin of the state space. The results in this chapter indicate that convergence of the unknown parameters in the value function is guaranteed provided the selected points satisfy a rank condition. Since the BE can be evaluated at any desired point in the state space, sufficient exploration can be achieved by appropriate selection of the points in a desired neighborhood.

If each new evaluation of the BE along the system trajectory is interpreted as gaining experience via exploration, the use of a model to evaluate the BE at an unexplored point in the state space can be interpreted as a simulation of experience. Learning based on simulation of experience has been investigated for stochastic model-based RL [45–50]; however, the optimal control problem was solved offline in the sense that repeated learning trials need to be performed before the algorithm learns the controller, and system stability during the learning phase is not analyzed.

In this chapter a novel implementation of simulation of experience is presented for deterministic nonlinear systems using BE extrapolation. A detailed stability analysis is presented to establish online approximate learning of the optimal controller while maintaining system stability during the learning phase. The stability analysis shows that provided an exact model of the system is available, simulation of experience based on the model implemented via BE extrapolation can be used to approximately solve an infinite horizon optimal regulation problem online.

Exact model knowledge is assumed in this chapter for ease of exposition. Using techniques similar to those in [15, 20, 22, 51], the developed method can be easily extended to establish set-point regulation and convergence to optimality when a system identifier is employed instead of an exact model of the system. Using techniques similar to those in [22, 30], the developed method can also be extended to optimally track a class of desired trajectories.

Simulation results are provided that demonstrate the efficacy of the system identification-based extension of the developed method for uncertain inherently unstable control affine nonlinear systems without the addition of a probing signal. The performance of the developed controller is demonstrated through approximate solution of an optimal control problem that has a known analytical solution. On the basis of the known solution, an exact polynomial basis is used for value function approximation. Another simulation demonstrates the applicability of the developed technique in the case where the analytical solution, and hence the basis for value function approximation, is unknown. In this case, since the optimal solution is unknown, the optimal trajectories obtained with the developed technique are compared with optimal trajectories obtained through numerical optimal control techniques.

The performance of the developed controller is demonstrated via experiments conducted on an autonomous underwater vehicle (AUV) at a spring head. The developed approximate optimal controller is used to regulate three degrees of freedom of the AUV (ie, surge, sway, and yaw) to a set point. All the computations required to implement the controller are performed onboard using an embedded processor. The experimental results demonstrate the capability of the developed method to concurrently identify, optimize, and control a real-world nonlinear autonomous system.

2 PROBLEM FORMULATION

Consider a control affine nonlinear dynamical system of the form

$$\dot{x}(t) = f(x(t)) + g(x(t))\,u(t), \tag{1}$$

$t \in \mathbb{R}_{\geq t_0}$,[1] where t_0 denotes the initial time, $x : \mathbb{R}_{\geq t_0} \to \mathbb{R}^n$ denotes the system state $f : \mathbb{R}^n \to \mathbb{R}^n$ and $g : \mathbb{R}^n \to \mathbb{R}^{n\times m}$ denote the drift dynamics and the control effectiveness, respectively, and $u : \mathbb{R}_{\geq t_0} \to \mathbb{R}^m$ denotes the control input. The functions f and g are assumed to be locally Lipschitz continuous. Furthermore, $f(\mathbf{0}_{n\times 1}) = \mathbf{0}_{n\times 1}$ and $\nabla f : \mathbb{R}^n \to \mathbb{R}^{n\times n}$ are continuous.[2,3] In the following the notation $\phi^u(t; t_0, x^o)$ denotes the trajectory of the system in Eq. (1) under the control signal u with the initial condition[4] $x^o \in \mathbb{R}^n$ and initial time $t_0 \in \mathbb{R}_{\geq 0}$.

The control objective is to solve the infinite-horizon optimal regulation problem online—that is, to design a control signal u online to minimize the cost functional

[1] $\mathbb{R}_{\geq a}$ denotes the interval $[a, \infty)$, and $\mathbb{R}_{>a}$ denotes the interval (a, ∞).
[2] $\nabla f(x, y, \ldots)$ denotes the partial derivative of f with respect to the first argument.
[3] $\mathbf{0}_{n\times m}$ and I_n denote an $n \times m$ matrix of zeros and an $n \times n$ identity matrix, respectively.
[4] $(\cdot)^o$ is used to denote an arbitrary variable.

$$J(x, u) \triangleq \int_{t_0}^{\infty} r(x(\tau), u(\tau))\, d\tau \tag{2}$$

under the dynamic constraint in Eq. (1) while regulating the system state to the origin. In Eq. (2), $r : \mathbb{R}^n \times \mathbb{R}^m \to \mathbb{R}_{\geq 0}$ denotes the instantaneous cost defined as

$$r(x^{\mathrm{o}}, u^{\mathrm{o}}) \triangleq Q(x^{\mathrm{o}}) + u^{\mathrm{o}T} R u^{\mathrm{o}}, \tag{3}$$

for all $x^{\mathrm{o}} \in \mathbb{R}^n$ and $u^{\mathrm{o}} \in \mathbb{R}^m$, where $Q : \mathbb{R}^n \to \mathbb{R}_{\geq 0}$ is a positive definite function and $R \in \mathbb{R}^{m \times m}$ is a constant positive definite symmetric matrix.

Assuming an optimal controller exists, a closed-form solution to the optimal control problem is formulated in terms of the optimal value function $V^{\star} : \mathbb{R}^n \to \mathbb{R}_{\geq 0}$ defined as

$$V^{\star}(x^{\mathrm{o}}) \triangleq \min_{u(\tau) \in U | \tau \in \mathbb{R}_{\geq t}} \int_{t}^{\infty} r(\phi^u(\tau; t, x^{\mathrm{o}}), u(\tau))\, d\tau. \tag{4}$$

Assuming that the optimal value function is continuously differentiable, it can be obtained by solving the corresponding HJB equation [52]

$$\nabla V^{\star}(x^{\mathrm{o}}) \left(f(x^{\mathrm{o}}) + g(x^{\mathrm{o}}) u^{\star}(x^{\mathrm{o}}) \right) + Q(x^{\mathrm{o}}) + u^{\star T}(x^{\mathrm{o}}) R u^{\star}(x^{\mathrm{o}}) = 0, \tag{5}$$

for all $x^{\mathrm{o}} \in \mathbb{R}^n$, with the boundary condition $V^{\star}(0) = 0$. The optimal control law can be determined with use of the optimal value function as $u^{\star}(x^{\mathrm{o}}) = -\frac{1}{2} R^{-1} g^T(x^{\mathrm{o}}) \left(\nabla V^{\star}(x^{\mathrm{o}}) \right)^T$ [52].

An analytical solution of the HJB equation is generally infeasible; hence an approximate solution is sought. An approximate solution of the HJB equation is facilitated by replacing $V^{\star}$ and $u^{\star}$ in Eq. (5) by their respective subsequently defined parametric estimates $\hat{V}\left(x^{\mathrm{o}}, \hat{W}_{\mathrm{c}}^{\mathrm{o}}\right)$ and $\hat{u}\left(x^{\mathrm{o}}, \hat{W}_{\mathrm{a}}^{\mathrm{o}}\right)$ to compute the BE $\delta : \mathbb{R}^{n+2L} \to \mathbb{R}$ as

$$\begin{aligned} \delta\left(x^{\mathrm{o}}, \hat{W}_{\mathrm{c}}^{\mathrm{o}}, \hat{W}_{\mathrm{a}}^{\mathrm{o}}\right) = {} & \nabla \hat{V}\left(x^{\mathrm{o}}, \hat{W}_{\mathrm{c}}^{\mathrm{o}}\right) \left(f(x^{\mathrm{o}}) + g(x^{\mathrm{o}}) \hat{u}\left(x^{\mathrm{o}}, \hat{W}_{\mathrm{a}}^{\mathrm{o}}\right) \right) + x^{\mathrm{o}T} Q x^{\mathrm{o}} \\ & + \hat{u}^T\left(x^{\mathrm{o}}, \hat{W}_{\mathrm{a}}^{\mathrm{o}}\right) R \hat{u}\left(x^{\mathrm{o}}, \hat{W}_{\mathrm{a}}^{\mathrm{o}}\right), \end{aligned} \tag{6}$$

where $\hat{W}_{\mathrm{c}}^{\mathrm{o}} \in \mathbb{R}^L$ and $\hat{W}_{\mathrm{a}}^{\mathrm{o}} \in \mathbb{R}^L$ denote the estimates of the unknown parameters in the approximation of the value function and the policy, respectively. The control objective is achieved by simultaneous adjustment of the estimates $\hat{W}_{\mathrm{c}}^{\mathrm{o}}$ and $\hat{W}_{\mathrm{a}}^{\mathrm{o}}$ to minimize the BE.

Since the BE depends on the drift dynamics, f is assumed to be known. The focus of this chapter is a novel implementation of simulation of experience for online approximate optimal control of deterministic nonlinear systems. If a system model is available, then the approximate optimal control technique can be implemented with the model. However, if an exact model of the system is unavailable, then parametric system identification can be employed to generate an estimate of the system model. A possible approach is to use parameters that are estimated offline in a separate experiment. A more useful approach is to use the offline estimate as an initial guess and employ an adaptive system identification technique capable of refining the initial guess on the basis of input-output data. The proposed technique can be easily extended to incorporate an online adaptive system identifier [15, 20, 22, 51].

3 APPROXIMATE OPTIMAL CONTROL

3.1 Value Function Approximation

Approximations to the optimal value function $V^{\star}$ and the optimal policy $u^{\star}$ are designed on the basis of neural network (NN)-based representations. Given any compact set $\chi \subset \mathbb{R}^n$ and positive constants $\bar{\epsilon}, \bar{\epsilon}' \in \mathbb{R}$, the universal approximation property of NNs can be exploited to represent the optimal value function $V^{\star}$ as $V^{\star}(x^{\mathrm{o}}) = W^T \sigma(x^{\mathrm{o}}) + \epsilon(x^{\mathrm{o}})$, for all $x^{\mathrm{o}} \in \chi$, where $W \in \mathbb{R}^L$ is the ideal weight matrix, which is bounded above by a known positive constant $\bar{W}$ in the sense that $\|W\| \leq \bar{W}$, $\sigma : \mathbb{R}^n \to \mathbb{R}^L$ is a continuously differentiable nonlinear activation function such that $\sigma(0) = 0$ and $\sigma'(0) = 0$, $L \in \mathbb{N}$ is the number of neurons, and $\epsilon : \mathbb{R}^n \to \mathbb{R}$ is the function reconstruction error such that $\sup_{x^{\mathrm{o}} \in \chi} |\epsilon(x^{\mathrm{o}})| \leq \bar{\epsilon}$ and $\sup_{x^{\mathrm{o}} \in \chi} |\nabla\epsilon(x^{\mathrm{o}})| \leq \bar{\epsilon}'$.

On the basis of the NN representation of the value function, an NN-based representation of the optimal controller is derived as $u^{\star}(x^{\mathrm{o}}) = -\frac{1}{2} R^{-1} g^T(x^{\mathrm{o}}) \left(\nabla\sigma^T(x^{\mathrm{o}}) W + \nabla\epsilon^T(x^{\mathrm{o}})\right)$. The NN-based approximations $\hat{V} : \mathbb{R}^n \times \mathbb{R}^L \to \mathbb{R}$ and $\hat{u} : \mathbb{R}^n \times \mathbb{R}^L \to \mathbb{R}^m$ are defined as

$$\hat{V}\left(x^{\mathrm{o}}, \hat{W}_{\mathrm{c}}^{\mathrm{o}}\right) \triangleq \hat{W}_{\mathrm{c}}^{\mathrm{o}T} \sigma(x^{\mathrm{o}}),$$

$$\hat{u}\left(x^{\mathrm{o}}, \hat{W}_{\mathrm{a}}^{\mathrm{o}}\right) \triangleq -\frac{1}{2} R^{-1} g^T(x^{\mathrm{o}}) \nabla\sigma^T(x^{\mathrm{o}}) \hat{W}_{\mathrm{a}}^{\mathrm{o}}, \tag{7}$$

where $\hat{W}_{\mathrm{c}}^{\mathrm{o}} \in \mathbb{R}^L$ and $\hat{W}_{\mathrm{a}}^{\mathrm{o}} \in \mathbb{R}^L$ denote the estimates of the ideal weights W. The use of two sets of weights to estimate the same set of ideal weights is motivated by the stability analysis and the fact that it enables a formulation of the BE that is linear in the value function weight estimates $\hat{W}_c^o$, enabling a least squares-based adaptive update law.

3.2 Simulation of Experience via BE Extrapolation

In traditional RL-based algorithms the value function estimate and the policy estimate are updated on the basis of observed data. The use of observed data to learn the value function naturally leads to a sufficient exploration condition, which demands sufficient richness in the observed data. In stochastic systems, this is achieved by a randomized stationary policy [7, 53, 54], whereas in deterministic systems a probing noise is added to the derived control law [8, 12, 55–57]. The technique developed in this chapter implements simulation of experience in a model-based RL scheme by using the system model to extrapolate the approximate BE to unexplored areas of the state space.

In the following, $\delta_t : \mathbb{R}_{\geq t_0} \to \mathbb{R}$ denotes the BE in Eq. (6) evaluated along the trajectories of Eqs. (1), (8), and (10) as $\delta_t(t) \triangleq \delta\big(x(t), \hat{W}_c(t), \hat{W}_a(t)\big)$ and $\delta_{ti} : \mathbb{R}_{\geq t_0} \to \mathbb{R}$ denotes the BE extrapolated along the trajectories of Eqs. (8) and (10), and a predefined set of trajectories $\left\{x_i : \mathbb{R}_{\geq t_0} \to \mathbb{R}^n \mid i = 1, \ldots, N\right\}$ as $\delta_{ti} \triangleq \delta\left(x_i(t), \hat{W}_c(t), \hat{W}_a(t)\right)$. A least-squares update law for the value function weights is designed on the basis of the subsequent stability analysis as

$$\dot{\hat{W}}_c = -k_{c1}\Gamma(t)\frac{\omega(t)}{\rho(t)}\delta_t(t) - \frac{k_{c2}}{N}\Gamma(t)\sum_{i=1}^{N}\frac{\omega_i(t)}{\rho_i(t)}\delta_{ti}(t), \tag{8}$$

$$\dot{\Gamma}(t) = \beta\Gamma(t) - k_{c1}\Gamma(t)\frac{\omega(t)\omega^T(t)}{\rho^2(t)}\Gamma(t) - \frac{k_{c2}}{N}\Gamma(t)\sum_{i=1}^{N}\frac{\omega_i(t)\omega_i^T(t)}{\rho_i^2(t)}\Gamma(t), \tag{9}$$

$\Gamma(t_0) = \Gamma_0$, where $\Gamma : \mathbb{R}_{\geq t_0} \to \mathbb{R}^{L\times L}$ is a time-varying least-squares gain matrix, $\omega(t) \triangleq \sigma'(x(t))\left(Y(x(t))\hat{\theta}(t) + g(x(t))\hat{u}\left(x(t), \hat{W}_a(t)\right)\right)$, $\omega_i(t) \triangleq \sigma'(x_i(t))\left(Y(x_i(t))\hat{\theta}(t) + g(x_i(t))\hat{u}\left(x_i(t), \hat{W}_a(t)\right)\right)$, $\rho(t) \triangleq 1 + \gamma_1\omega^T(t)\omega(t)$, and $\rho_i(t) \triangleq 1 + \gamma_1\omega_i^T(t)\omega_i(t)$, where $\gamma_1 \in \mathbb{R}$ is a constant positive normalization gain, $\mathbf{1}_{\{\cdot\}}$ denotes the indicator function, $\overline{\Gamma} > 0 \in \mathbb{R}$ is a saturation constant, $\beta > 0 \in \mathbb{R}$ is a constant forgetting factor, and $k_{c1}, k_{c2} > 0 \in \mathbb{R}$ are constant adaptation gains.

The policy weights are updated to follow the value function weights as[5]

$$\dot{\hat{W}}_a(t) = -k_{a1}\left(\hat{W}_a(t) - \hat{W}_c(t)\right) - k_{a2}\hat{W}_a(t) + \frac{k_{c1}G_{\sigma}^T(t)\hat{W}_a(t)\omega^T(t)}{4\rho(t)}\hat{W}_c(t)$$
$$+\sum_{i=1}^{N}\frac{k_{c2}G_{\sigma i}^T(t)\hat{W}_a(t)\omega_i^T(t)}{4N\rho_i(t)}\hat{W}_c(t), \quad (10)$$

where k_{a1}, $k_{a2} \in \mathbb{R}$ are positive constant adaptation gains, $G_{\sigma}(t) \triangleq \sigma'(x(t))g(x(t))R^{-1}g^T(x(t))\sigma'^T(x(t))$, and $G_{\sigma i}(t) \triangleq \nabla\sigma_i(t)g_i(t)R^{-1}g_i^T(t)\nabla\sigma_i^T(t) \in \mathbb{R}^{L\times L}$, where $g_i(t) = g(x_i(t))$ and $\nabla\sigma_i(t) = \nabla\sigma(x_i(t))$. Using the weight estimates $\hat{W}_a$, the controller for the system in Eq. (1) is designed as

$$u(t) = \hat{u}\left(x(t), \hat{W}_a(t)\right). \quad (11)$$

The following rank condition facilitates the subsequent stability analysis.

Assumption 1. *There exists a finite set of trajectories* $\{x_i : \mathbb{R}_{\geq t_0} \to \mathbb{R}^n \mid i = 1, \ldots, N\}$ *and a constant* $T \in \mathbb{R}_{>0}$ *such that*

$$\underline{c}_1 I_L \leq \int_t^{t+T}\left(\frac{\omega(\tau)\omega^T(\tau)}{\rho^2(\tau)}\right)d\tau, \quad \forall t \in \mathbb{R}_{\geq t_0}, \quad (12)$$

$$\underline{c}_2 I_L \leq \inf_{t\in\mathbb{R}_{\geq t_0}}\left(\frac{1}{N}\sum_{i=1}^{N}\frac{\omega_i(t)\omega_i^T(t)}{\rho_i^2(t)}\right), \quad (13)$$

$$\underline{c}_3 I_L \leq \frac{1}{N}\int_t^{t+T}\left(\sum_{i=1}^{N}\frac{\omega_i(\tau)\omega_i^T(\tau)}{\rho_i^2(\tau)}\right)d\tau, \quad \forall t \in \mathbb{R}_{\geq t_0}, \quad (14)$$

where at least one of the nonnegative constants $\underline{c}_1, \underline{c}_2$, *and* $\underline{c}_3$ *is strictly positive.*

The rank conditions in Eqs. (12)–(14) depend on the estimates $\hat{W}_a$; hence, in general, they are impossible to guarantee a priori. However, unlike the traditional adaptive dynamic programming literature that assumes ω is persistently exciting, Assumption 1 only requires either the regressor ω or the regressor ω_i to be persistently exciting. The regressor ω is completely determined by the system state x and the weights $\hat{W}_a$. Hence excitation in ω vanishes as the system states and the weights converge. Hence, in general,

[5] Using the fact that the ideal weights are bounded, a projection-based [58] update law $\dot{\hat{W}}_a = \text{proj}\left\{-k_{a1}\left(\hat{W}_a - \hat{W}_c\right)\right\}$ can be used to update the policy weights. Since the policy weights are bounded a priori by the projection algorithm, a less complex stability analysis can be used to establish the result in Theorem 1.

it is unlikely that $\underline{c}_1 > 0$. However, the regressor ω_i depends on x_i, which can be designed independently of the system state x. Hence $\underline{c}_3$ can be made strictly positive if the signal x_i contains enough frequencies, and $\underline{c}_2$ can be made strictly positive by selection of a sufficient number of extrapolation functions (ie, $N \gg L$).

The update law in Eq. (8) is fundamentally different from the concurrent-learning or concurrent learning adaptive update in [59, 60], in the sense that the trajectories $\{x_i\}$ are selected a priori on the basis of prior information about the desired behavior of the system. Given the system dynamics, or an estimate of the system dynamics, the approximate BE can be extrapolated to any desired point in the state space, whereas in adaptive control the prediction error is used as a metric which can be evaluated only at observed data points along the state trajectory.

4 STABILITY ANALYSIS

4.1 Boundedness of the Least-Squares Gain Under Persistent Excitation

Intuitively, the selection of time-varying trajectories for BE extrapolation results in virtual excitation. That is, instead of our using input-output data from a persistently excited system, the dynamic model is used to simulate persistent excitation to facilitate parameter convergence. In the following, upper and lower bounds on the eigenvalues of the least-squares learning gain matrix Γ are established. Bounds on the eigenvalues of Γ are traditionally established under PE. The following lemma extends the traditional PE-based result to incorporate the generalized excitation conditions in Assumption 1.

Lemma 1. *Provided Assumption 1 holds and $\lambda_{\min}\left\{\Gamma_0^{-1}\right\} > 0$, the update law in Eq. (9) ensures that the least-squares gain matrix satisfies*

$$\underline{\Gamma} I_L \leq \Gamma(t) \leq \overline{\Gamma} I_L, \tag{15}$$

$\forall t \in \mathbb{R}_{\geq 0}$, *where* $\overline{\Gamma} = \frac{1}{\min\left\{k_{c1}\underline{c}_1 + k_{c2}\max\left\{\underline{c}_2 T, \underline{c}_3\right\}, \lambda_{\min}\left\{\Gamma_0^{-1}\right\}\right\} e^{-\beta T}}$ *and* $\underline{\Gamma} = \frac{1}{\lambda_{\max}\left\{\Gamma_0^{-1}\right\} + \frac{(k_{c1}+k_{c2})}{\beta\gamma_1}}$. *Furthermore,* $\overline{\Gamma} > 0$.

Proof. The proof closely follows the proof of Corollary 4.3.2 in [40]. The update law in Eq. (9) implies that $\frac{d}{dt}\Gamma^{-1}(t) = -\beta\Gamma^{-1}(t) + k_{c1}\frac{\omega(t)\omega^T(t)}{\rho^2(t)} + \frac{k_{c2}}{N}\sum_{i=1}^{N}\frac{\omega_i(t)\omega_i^T(t)}{\rho_i^2(t)}$. Hence

$$\Gamma^{-1}(t) = \mathrm{e}^{-\beta t}\Gamma_0^{-1} + k_{c1}\int_0^t \mathrm{e}^{-\beta(t-\tau)}\frac{\omega(\tau)\,\omega^T(\tau)}{\rho^2(\tau)}d\tau$$
$$+ \frac{k_{c2}}{N}\int_0^t \mathrm{e}^{-\beta(t-\tau)}\sum_{i=1}^{N}\frac{\omega_i(\tau)\,\omega_i^T(\tau)}{\rho_i^2(\tau)}d\tau.$$

To facilitate the proof, let $t < T$. Then

$$\Gamma^{-1}(t) \geq \mathrm{e}^{-\beta t}\Gamma_0^{-1} \geq \mathrm{e}^{-\beta T}\Gamma_0^{-1} \geq \lambda_{\min}\left\{\Gamma_0^{-1}\right\}\mathrm{e}^{-\beta T}I_L.$$

If $t \geq T$, then since the integrands are positive, Γ^{-1} can be bounded as

$$\Gamma^{-1}(t) \geq k_{c1}\int_{t-T}^t \mathrm{e}^{-\beta(t-\tau)}\frac{\omega(\tau)\,\omega^T(\tau)}{\rho^2(\tau)}d\tau$$
$$+ \frac{k_{c2}}{N}\int_{t-T}^t \mathrm{e}^{-\beta(t-\tau)}\sum_{i=1}^{N}\frac{\omega_i(\tau)\,\omega_i^T(\tau)}{\rho_i^2(\tau)}d\tau.$$

Hence

$$\Gamma^{-1}(t) \geq k_{c1}\mathrm{e}^{-\beta T}\int_{t-T}^t \frac{\omega(\tau)\,\omega^T(\tau)}{\rho^2(\tau)}d\tau + \frac{k_{c2}}{N}\mathrm{e}^{-\beta T}\int_{t-T}^t \sum_{i=1}^{N}\frac{\omega_i(\tau)\,\omega_i^T(\tau)}{\rho_i^2(\tau)}d\tau.$$

Using Assumption 1,

$$\frac{1}{N}\int_{t-T}^t \sum_{i=1}^{N}\frac{\omega_i(\tau)\,\omega_i^T(\tau)}{\rho_i^2(\tau)}d\tau \geq \max\left\{\underline{c}_2 T, \underline{c}_3\right\}I_L, \quad \int_{t-T}^t \frac{\omega(\tau)\,\omega^T(\tau)}{\rho^2(\tau)}d\tau \geq \underline{c}_1 I_L.$$

Hence a lower bound for Γ^{-1} is obtained as

$$\Gamma^{-1}(t) \geq \min\left\{k_{c1}\underline{c}_1 + k_{c2}\max\left\{\underline{c}_2 T, \underline{c}_3\right\}, \lambda_{\min}\left\{\Gamma_0^{-1}\right\}\right\}\mathrm{e}^{-\beta T}I_L. \quad (16)$$

Provided Assumption 1 holds, the lower bound in Eq. (16) is strictly positive. Furthermore, using the facts that $\frac{\omega(t)\omega^T(t)}{\rho^2(t)} \leq \frac{1}{\gamma_1}$ and $\frac{\omega_i(t)\omega_i^T(t)}{\rho_i^2(t)} \leq \frac{1}{\gamma_1}$ for all $t \in \mathbb{R}_{\geq t_0}$,

$$\Gamma^{-1}(t) \leq \int_0^t \mathrm{e}^{-\beta(t-\tau)}\left(k_{c1}\frac{1}{\gamma_1} + \frac{k_{c2}}{N}\sum_{i=1}^{N}\frac{1}{\gamma_1}\right)I_L d\tau + \mathrm{e}^{-\beta t}\Gamma_0^{-1}$$
$$\leq \left(\lambda_{\max}\left\{\Gamma_0^{-1}\right\} + \frac{(k_{c1}+k_{c2})}{\beta\gamma_1}\right)I_L.$$

Since the inverse of the lower and upper bounds on Γ^{-1} are the upper and lower bounds on Γ, respectively, the proof is complete. □

4.2 Regulation and Convergence to Optimality

For notational brevity the dependence of all the functions on the system states and time is suppressed hereafter unless it is required for clarity of exposition. To facilitate the subsequent stability analysis, the approximate BE is expressed in terms of the weight estimation errors $\tilde{W}_{\mathrm{c}} \triangleq W - \hat{W}_{\mathrm{c}}$ and $\tilde{W}_{\mathrm{a}} \triangleq W - \hat{W}_{\mathrm{a}}$. Subtracting Eq. (5) from Eq. (6), an unmeasurable form of the instantaneous BE can be expressed as

$$\delta_t = -\omega^T \tilde{W}_{\mathrm{c}} + \frac{1}{4} \tilde{W}_{\mathrm{a}}^T G_\sigma \tilde{W}_{\mathrm{a}} + \Delta, \tag{17}$$

where $G \triangleq gR^{-1}g^T \in \mathbb{R}^{n\times n}$, $\Delta \triangleq \frac{1}{2} W^T \nabla\sigma G \nabla\epsilon^T + \frac{1}{4} G_\epsilon - \nabla\epsilon f \in \mathbb{R}$, $G_\epsilon \triangleq \nabla\epsilon G \nabla\epsilon^T \in \mathbb{R}$, and G_σ was introduced in Eq. (10). Similarly, the approximate BE evaluated along the selected trajectories $\{x_i \mid i = 1, \ldots, N\}$ can be expressed as

$$\delta_{ti} = -\omega_i^T \tilde{W}_{\mathrm{c}} + \frac{1}{4} \tilde{W}_{\mathrm{a}}^T G_{\sigma i} \tilde{W}_{\mathrm{a}} + \Delta_i, \tag{18}$$

where $\nabla\epsilon_i = \nabla\epsilon(x_i)$, $f_i = f(x_i)$, $G_i \triangleq g_i R^{-1} g_i^T \in \mathbb{R}^{n\times n}$, $\Delta_i \triangleq \frac{1}{2} W^T \nabla\sigma_i G_i \nabla\epsilon_i^T + \frac{1}{4} G_{\epsilon i} - \nabla\epsilon_i f_i \in \mathbb{R}$ is a constant, $G_{\epsilon i} \triangleq \nabla\epsilon_i G_i \nabla\epsilon_i^T \in \mathbb{R}$, and $G_{\sigma i}$ was introduced in Eq. (10).

Let $B_\zeta \subset \mathbb{R}^{n+2L}$ denote a closed ball with radius ζ centered at the origin. Let $\chi \triangleq B_\zeta \cap \mathbb{R}^n$. Let the notation $\overline{\|(\cdot)\|}$ be defined as $\overline{\|h\|} \triangleq \sup_{x^{\mathrm{o}} \in \chi} \|h(x^{\mathrm{o}})\|$ for some continuous function $h : \mathbb{R}^n \to \mathbb{R}^k$. To facilitate the analysis, let $\left\{\varpi_j \in \mathbb{R}_{>0} \mid j = 1, \ldots, 7\right\}$ be constants such that $\varpi_1 + \varpi_2 + \varpi_3 = 1$ and $\varpi_4 + \varpi_5 + \varpi_6 + \varpi_7 = 1$. Let $\underline{c} \in \mathbb{R}_{>0}$ be a constant defined as

$$\underline{c} \triangleq \frac{\beta}{2\overline{\Gamma} k_{c2}} + \frac{\underline{c}_2}{2} \tag{19}$$

and let $\iota \in \mathbb{R}$ be a positive constant defined as

$$\iota \triangleq \frac{(k_{c1} + k_{c2})^2 \overline{\|\Delta\|}^2}{4k_{c2}\underline{c}\varpi_3} + \frac{1}{4(k_{a1} + k_{a2})\varpi_6} \left(\frac{\overline{W}\overline{\|G_\sigma\|}}{2} + \frac{(k_{c1} + k_{c2}) \overline{W}^2 \overline{\|G_\sigma\|}}{4} + \frac{\overline{\|\nabla\epsilon G^T \nabla\sigma^T\|}}{2} + k_{a2}\overline{W} \right)^2 + \frac{1}{4}\overline{\|G_\epsilon\|}. \tag{20}$$

Let $V_L : \mathbb{R}^{n+2L} \times \mathbb{R}_{\geq 0} \to \mathbb{R}_{\geq 0}$ be a continuously differentiable candidate Lyapunov function defined as

$$V_L(Z^{\mathrm{o}}, t) \triangleq V^\star(x^{\mathrm{o}}) + \frac{1}{2} \tilde{W}_{\mathrm{c}}^{\mathrm{o}T} \Gamma^{-1} \tilde{W}_{\mathrm{c}}^{\mathrm{o}} + \frac{1}{2} \tilde{W}_{\mathrm{a}}^{\mathrm{o}T} \tilde{W}_{\mathrm{a}}^{\mathrm{o}}, \tag{21}$$

where $V^\star$ is the optimal value function and $Z^{\mathrm{o}} \triangleq \left[x^{oT},\ \tilde{W}_{\mathrm{c}}^{oT},\ \tilde{W}_{\mathrm{a}}^{oT}\right]^T$. Using the fact that $V^\star$ is positive definite, Eq. (15), Lemma 1, and Lemma 4.3 in [61] yield

$$\underline{\nu_l}\left(\|Z^{\mathrm{o}}\|\right) \leq V_L\left(Z^{\mathrm{o}}, t\right) \leq \overline{\nu_l}\left(\|Z^{\mathrm{o}}\|\right), \tag{22}$$

for all $t \in \mathbb{R}_{\geq t_0}$ and for all $Z^{\mathrm{o}} \in \mathbb{R}^{n+2L}$, where $\underline{\nu_l}, \overline{\nu_l} : \mathbb{R}_{\geq 0} \rightarrow \mathbb{R}_{\geq 0}$ are class $\mathcal{K}$ functions. Let $\nu_l : \mathbb{R}_{\geq 0} \rightarrow \mathbb{R}_{\geq 0}$ be a class $\mathcal{K}$ function such that $\nu_l\left(\|Z^{\mathrm{o}}\|\right) \geq \frac{Q(x^{\mathrm{o}})}{2} + \frac{k_{c2}\underline{c}\varpi_1}{2}\left\|\tilde{W}_{\mathrm{c}}^{\mathrm{o}}\right\|^2 + \frac{(k_{a1}+k_{a2})\varpi_4}{2}\left\|\tilde{W}_{\mathrm{a}}^{\mathrm{o}}\right\|^2$.

Let $Z : \mathbb{R}_{\geq t_0} \rightarrow \mathbb{R}^{n+2L}$ denote the concatenated trajectories of Eqs. (1), (8), and (10), defined as $Z(t) \triangleq \left[x^T(t),\ \tilde{W}_{\mathrm{c}}^T(t),\ \tilde{W}_{\mathrm{a}}^T(t)\right]^T$. The sufficient conditions for ultimate boundedness of Z are derived on the basis of the subsequent stability analysis as

$$k_{c2}\underline{c}\left(k_{a1}+k_{a2}\right)\varpi_5\varpi_2 \geq \left(k_{a1} + \frac{1}{4}\left(k_{c1}+k_{c2}\right)\overline{W\|G_\sigma\|}\right), \tag{23}$$

$$\left(k_{a1}+k_{a2}\right)\varpi_7 \geq \frac{1}{4}\left(k_{c1}+k_{c2}\right)\overline{W\|G_\sigma\|}, \tag{24}$$

$$\nu_l^{-1}(\iota) < \overline{\nu_l}^{-1}\left(\underline{\nu_l}(\zeta)\right). \tag{25}$$

The bound on the function f and the NN function approximation errors depend on the underlying compact set; hence ι is a function of ζ. Even though, in general, ι increases with increasing ζ, the sufficient condition in Eq. (25) can be satisfied provided the points for BE extrapolation are selected such that the constant $\underline{c}$, introduced in Eq. (19) is large enough and that the basis for value function approximation is selected such that $\overline{\|\epsilon\|}$ and $\overline{\|\nabla\epsilon\|}$ are small enough. The main result of this chapter can now be stated as follows.

Theorem 1. *Provided Assumption 1 holds and the sufficient gain conditions in Eqs. (23)–(25) are satisfied, the controller in Eq. (11) along with the adaptive update laws in Eqs. (8)–(10) ensure that the state x, the value function weight estimation error $\tilde{W}_{\mathrm{c}}$, and the policy weight estimation error $\tilde{W}_{\mathrm{a}}$ are uniformly ultimately bounded.*

Proof. The time derivative of Eq. (21) along the trajectories of Eq. (1) and Eqs. (8)–(10) is given by

$$\dot{V}_L = \dot{V}^\star - \tilde{W}_{\mathrm{c}}^T\Gamma^{-1}\dot{\hat{W}}_{\mathrm{c}} - \frac{1}{2}\tilde{W}_{\mathrm{c}}^T\dot{\Gamma}^{-1}\tilde{W}_{\mathrm{c}} - \tilde{W}_{\mathrm{a}}^T\dot{\hat{W}}_{\mathrm{a}},$$

$$= \nabla V^\star\left(f + gu\right) - \tilde{W}_{\mathrm{c}}^T\left(-k_{c1}\frac{\omega}{\rho}\delta_t - \frac{k_{c2}}{N}\sum_{i=1}^{N}\frac{\omega_i}{\rho_i}\delta_{ti}\right)$$

$$
\begin{aligned}
&-\frac{1}{2}\tilde{W}_c^T\Gamma^{-1}\left(\beta\Gamma - k_{c1}\left(\Gamma\frac{\omega\omega^T}{\rho^2}\Gamma\right) - \frac{k_{c2}}{N}\Gamma\sum_{i=1}^{N}\frac{\omega_i\omega_i^T}{\rho_i^2}\Gamma\right)\Gamma^{-1}\tilde{W}_c \\
&-\tilde{W}_a^T\left(-k_{a1}\left(\hat{W}_a - \hat{W}_c\right) - k_{a2}\hat{W}_a\right) \\
&-\tilde{W}_a^T\left(\frac{k_{c1}G_\sigma^T\hat{W}_a\omega^T}{4\rho} + \sum_{i=1}^{N}\frac{k_{c2}G_{\sigma i}^T\hat{W}_a\omega_i^T}{4N\rho_i}\right)\hat{W}_c.
\end{aligned}
$$

By substitution of the approximate BEs from Eqs. (17) and (18) and using the inequality $\frac{\omega\omega^T}{\rho^2} \leq \frac{\omega\omega^T}{\rho}$, the Lyapunov derivative can be bounded as

$$
\begin{aligned}
\dot{V}_L \leq\ & -Q(x) - k_{c2}\tilde{W}_c^T\left(\frac{\beta\Gamma^{-1}}{2k_{c2}} + \sum_{i=1}^{N}\frac{\omega_i\omega_i^T}{2N\rho_i}\right)\tilde{W}_c - (k_{a1}+k_{a2})\tilde{W}_a^T\tilde{W}_a \\
& + \left(\frac{W^TG_\sigma + \nabla\epsilon G^T\nabla\sigma^T}{2} + k_{a2}W^T - \frac{k_{c1}W^T\omega W^TG_\sigma}{4\rho}\right. \\
& \left. - \sum_{i=1}^{N}\frac{k_{c2}W^T\omega_iW^TG_{\sigma i}}{4N\rho_i}\right)\tilde{W}_a \\
& + \tilde{W}_c^T\left(k_{a1} + \frac{k_{c1}\omega W^TG_\sigma}{4\rho} + \sum_{i=1}^{N}\frac{k_{c2}\omega_iW^TG_{\sigma i}}{4N\rho_i}\right)\tilde{W}_a \\
& + W^T\frac{k_{c1}\omega}{4\rho}\tilde{W}_a^TG_\sigma\tilde{W}_a + W^T\sum_{i=1}^{N}\frac{k_{c2}\omega_i}{4N\rho_i}\tilde{W}_a^TG_{\sigma i}\tilde{W}_a \\
& + \tilde{W}_c^T\left(k_{c1}\frac{\omega}{\rho}\Delta + \frac{1}{N}\sum_{i=1}^{N}\frac{k_{c2}\omega_i}{\rho_i}\Delta_i\right) + \frac{1}{4}G_\epsilon.
\end{aligned}
$$

Provided the gains are selected on the basis of the sufficient conditions in Eqs. (23)–(25), the Lyapunov derivative can be upper bounded as

$$
\dot{V}_L \leq -v_l(\|Z\|), \quad \forall \|Z\| > v_l^{-1}(\iota), \tag{26}
$$

for all $t \geq 0$ and $\forall Z \in B_\zeta$. Using Eqs. (22), (25), and (26), Theorem 4.18 in [61] can now be invoked to conclude that Z is uniformly ultimately bounded in the sense that $\limsup_{t\to\infty}\|Z(t)\| \leq \underline{v_l}^{-1}\left(\overline{v_l}\left(v_l^{-1}(\iota)\right)\right)$. Furthermore, the concatenated state trajectories are bounded such that $\|Z(t)\| \in B_\zeta$ for all $t \in \mathbb{R}_{\geq t_0}$. Since the estimates $\hat{W}_a$ approximate the ideal weights W, the policy $\hat{u}$ approximates the optimal policy $u^\star$. □

5 SIMULATION

This section presents two simulations to demonstrate the performance and the applicability of the developed technique. First, the performance of the developed controller is demonstrated through approximate solution of an optimal control problem that has a known analytical solution. On the basis of the known solution, an exact polynomial basis is selected for value function approximation. The second simulation demonstrates the applicability of the developed technique in the case where the analytical solution, and hence the basis for value function approximation, is unknown. In this case, since the optimal solution is unknown, the optimal trajectories obtained with the developed technique are compared with optimal trajectories obtained through offline numerical optimal control techniques.

5.1 Problem with a Known Basis

The performance of the developed controller is demonstrated by simulation of a nonlinear, control affine system with a two-dimensional state $x = [x_1, x_2]^T$. The system dynamics are described by Eq. (1), where [8]

$$f = \begin{bmatrix} x_1 & x_2 & 0 & 0 \\ 0 & 0 & x_1 & x_2\left(1 - (\cos(2x_1) + 2)^2\right) \end{bmatrix} \begin{bmatrix} a \\ b \\ c \\ d \end{bmatrix},$$

$$g = \begin{bmatrix} 0 \\ \cos(2x_1) + 2 \end{bmatrix}. \tag{27}$$

In Eq. (27), $a, b, c, d \in \mathbb{R}$ are positive unknown parameters.[6] The parameters are selected as[7] $a = -1$, $b = 1$, $c = -0.5$, and $d = -0.5$. The control objective is to minimize the cost in Eq. (4), where $Q = I_2$ and $R = 1$. The optimal value function and optimal control for the system in Eq. (27) are given by $V^\star(x) = \frac{1}{2}x_1^2 + x_2^2$ and $u^\star(x) = -(\cos(2x_1) + 2)x_2$ [8].

[6] To relax the requirement of exact model knowledge, the simulations and the experiment employ a concurrent learning-based system identifier [15, 20, 22, 51]. Using techniques similar to those in [15, 20, 22, 51], the analysis in Section 4 can be easily extended to establish set-point regulation and convergence to optimality when a system identifier is employed instead of an exact model of the system.

[7] The origin is an unstable equilibrium point of the unforced system $\dot{x} = f(x)$.

To facilitate the adaptive dynamic programming-based controller, the basis function $\sigma : \mathbb{R}^2 \to \mathbb{R}^3$ for value function approximation is selected as $\sigma = \begin{bmatrix} x_1^2, & x_1 x_2, & x_2^2 \end{bmatrix}$. On the basis of the analytical solution, the ideal weights are $W = [0.5,\ 0,\ 1]^T$. The data points for the concurrent learning-based update law in Eq. (8) are selected to be on a 5×5 grid around the origin. The initial condition for the system state is selected as $x(0) = [-1,\ -1]^T$.

Fig. 1 demonstrates that the system state is regulated to the origin, the unknown parameters in the drift dynamics are identified, and the value function and the policy weights converge to their true values. Furthermore, unlike previous results, a probing signal to ensure PE is not required. Fig. 2 demonstrates the satisfaction of Assumption 1.

5.2 Problem with an Unknown Basis

To demonstrate the applicability of the developed controller, a nonlinear, control affine system with a four-dimensional state $x = [x_1,\ x_2,\ x_3,\ x_4]^T$ is simulated. The system dynamics are described by Eq. (1), where

$$f(x) = \begin{bmatrix} x_3 \\ x_4 \\ -M^{-1} V_m \begin{bmatrix} x_3 \\ x_4 \end{bmatrix} \end{bmatrix} + \begin{bmatrix} 0, & 0, & 0, & 0 \\ 0, & 0, & 0, & 0 \\ \left[M^{-1}, \right. & & \left. M^{-1} \right] D & \end{bmatrix} \begin{bmatrix} f_{d1} \\ f_{d2} \\ f_{s1} \\ f_{s2} \end{bmatrix},$$

$$g(x) = \left[[0,\ 0]^T,\ [0,\ 0]^T,\ \left(M^{-1}\right)^T \right]^T. \tag{28}$$

In Eq. (28), $x \triangleq [x_1,\ x_2,\ x_3,\ x_4]^T$, $D \triangleq \operatorname{diag}[x_3,\ x_4,\ \tanh x_3,\ \tanh x_4]$, and the matrices $M,\ V_m,\ F_d,\ F_s \in \mathbb{R}^{2\times 2}$ are defined as $M \triangleq \begin{bmatrix} p_1 + 2p_3 c_2, & p_2 + p_3 c_2 \\ p_2 + p_3 c_2, & p_2 \end{bmatrix}$, $V_m \triangleq \begin{bmatrix} -p_3 s_2 x_4, & -p_3 s_2 (x_3 + x_4) \\ p_3 s_2 x_3, & 0 \end{bmatrix}$, $F_d \triangleq \begin{bmatrix} f_{d1}, & 0 \\ 0, & f_{d2} \end{bmatrix}$, and $F_s \triangleq \begin{bmatrix} f_{s1} \tanh x_3, & 0 \\ 0, & f_{s2} \tanh x_3 \end{bmatrix}$, where $c_2 = \cos x_2$, $s_2 = \sin x_2$, $p_1 = 3.473$, $p_2 = 0.196$, and $p_3 = 0.242$, and $f_{d1}, f_{d2}, f_{s1}, f_{s2} \in \mathbb{R}$ are positive unknown parameters.[8] The parameters are selected as $f_{d1} = 5.3$, $f_{d2} = 1.1$, $f_{s1} = 8.45$, and $f_{s2} = 2.35$. The control objective is to minimize the cost in Eq. (4), where $Q = \operatorname{diag}([10,\ 10,\ 1,\ 1])$ and $R = \operatorname{diag}([1,\ 1])$.

[8] See footnote 6.

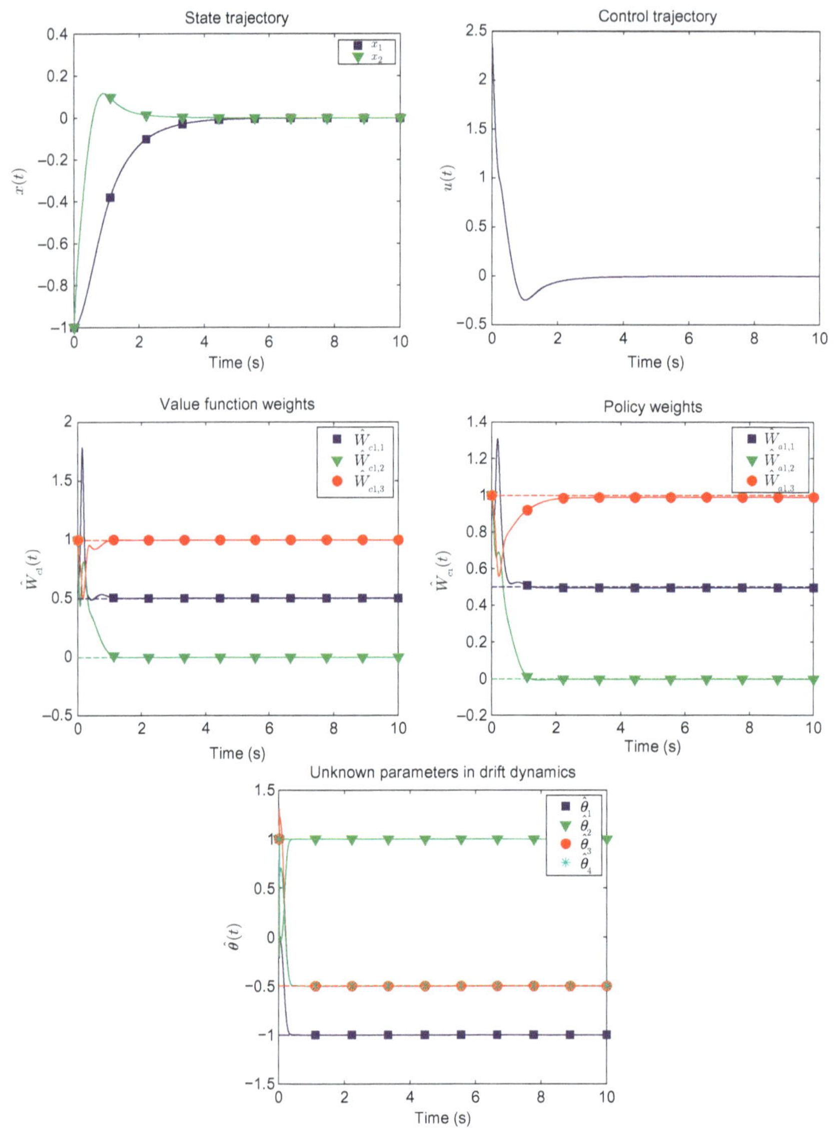

Fig. 1 System trajectories generated with the proposed method, and compared with the analytical solution.

To facilitate the adaptive dynamic programming-based controller, the basis function $\sigma : \mathbb{R}^4 \rightarrow \mathbb{R}^{10}$ for value function approximation is selected as $\sigma(x) = \left[x_1x_3, \ \ x_2x_4, \ \ x_3x_2, \ \ x_4x_1, \ \ x_1x_2, \ \ x_4x_3, \ \ x_1^2, \ \ x_2^2, \ \ x_3^2, \ \ x_4^2\right]$. The points for the concurrent learning-based update law in Eq. (8) are selected to be on a $3 \times 3 \times 3 \times 3$ grid around the origin, and the

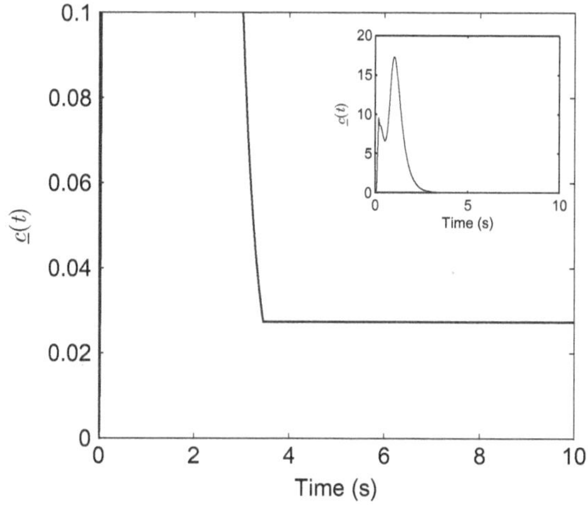

Fig. 2 Satisfaction of Assumption 1 for the simulation with known basis.

policy weights are updated using a projection-based update law. The initial condition for the system state is selected as $x(0) = [1,\ 1,\ 0,\ 0]^T$.

Fig. 3 demonstrates that the system state is regulated to the origin, the unknown parameters in the drift dynamics are identified, and the value function and the policy weights converge. Fig. 4 demonstrates the satisfaction of Assumption 1. The value function and the policy weights converge to the following values:

$$\hat{W}_{\mathrm{c}}^{\star} = \hat{W}_{\mathrm{a}}^{\star} = [24.7,\ 1.19,\ 2.25,\ 2.67,\ 1.18,\ 0.93,\ 44.34,\ 11.31,\ 3.81\ ,0.10]^T. \tag{29}$$

Since the true values of the value function weights are unknown, the weights in Eq. (29) cannot be compared with their true values. However, a measure of the proximity of the weights in Eq. (29) to the ideal weights W can be obtained by comparison of the system trajectories resulting from application of the feedback control policy $\hat{u}^{\star}(x) = -\frac{1}{2}R^{-1}g^T(x)\nabla\sigma^T(x)\hat{W}_{\mathrm{a}}^{\star}$ to the system against numerically computed optimal system trajectories. In Fig. 5 the numerical optimal solution is obtained with an infinite-horizon Gauss pseudospectral method [62] with 45 collocation points. Fig. 5 indicates that the weights in Eq. (29) generate state and control trajectories that closely match the numerically computed optimal trajectories.

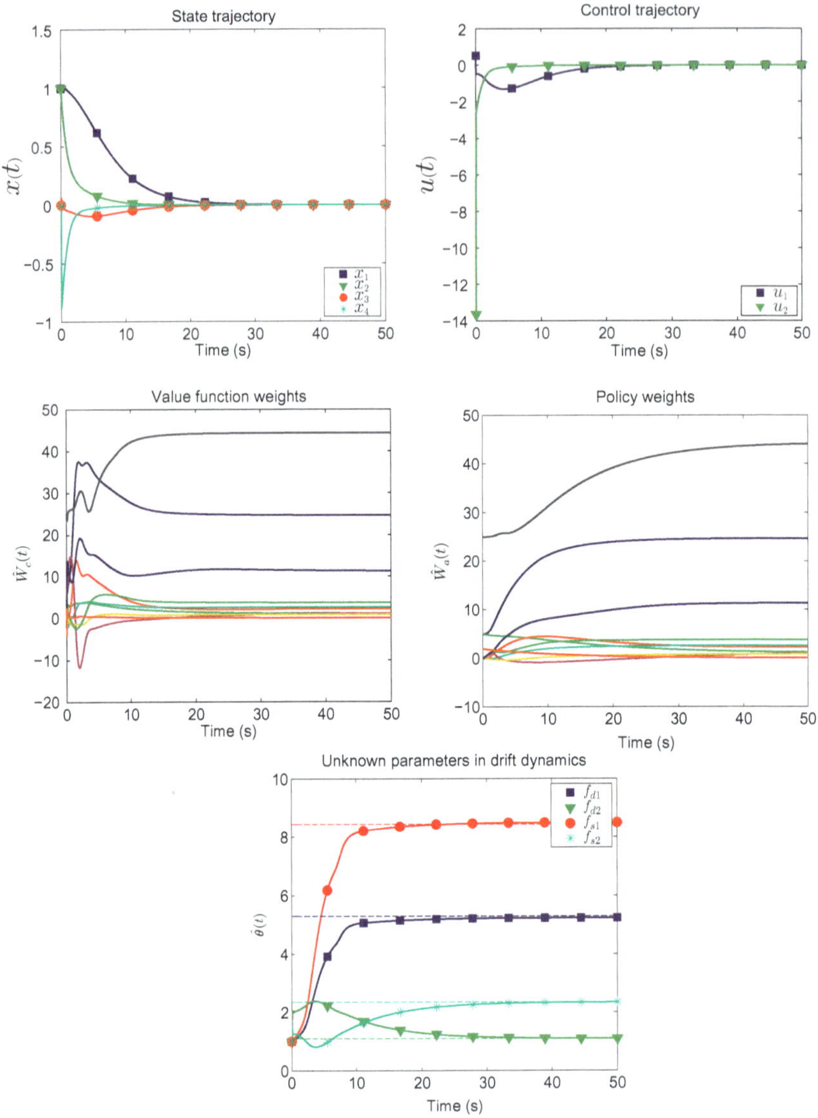

Fig. 3 System trajectories generated with the proposed method, where the drift parameter estimates are compared with the actual drift parameters.

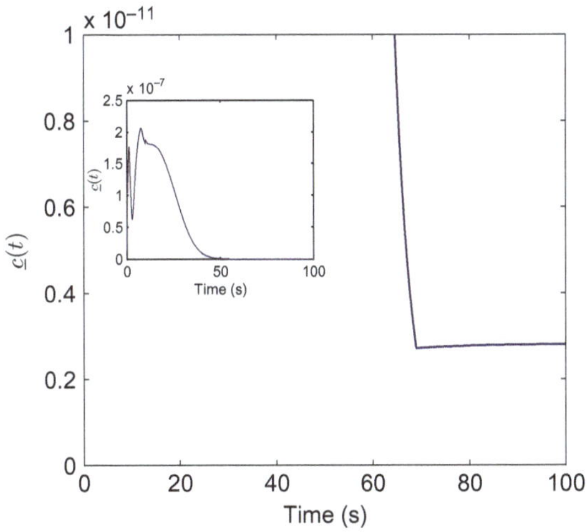

Fig. 4 Satisfaction of Assumption 1 for the simulation with unknown basis.

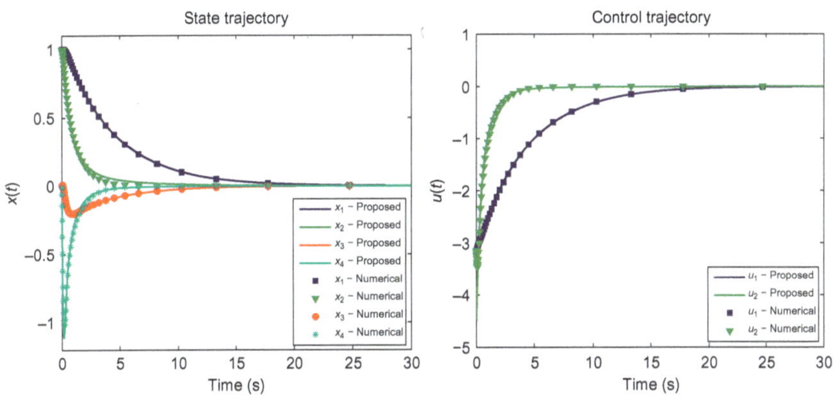

Fig. 5 State and control trajectories generated with feedback policy $\hat{u}^*(x)$ compared with a numerical optimal solution.

6 EXPERIMENTAL VALIDATION

The performance of the developed controller is demonstrated by experiments conducted at Ginnie Springs in High Springs, Florida. Ginnie Springs is a second-magnitude spring discharging 142×10^6 L of freshwater daily with a spring pool measuring 27.4 m in diameter and 3.7 m deep [63]. Ginnie Springs is selected to validate the proposed controller because of its relatively high flow rate and clear waters for vehicle observation. For clarity of exposition[9] and to remain within the vehicle's depth limitations,[10] the proposed method is implemented on three degrees of freedom of an AUV (ie, surge, sway, and yaw).

6.1 Experimental Platform

Experiments are conducted on an AUV, SubjuGator 7, developed at the University of Florida. The AUV, shown in Fig. 6, is a small two-man portable AUV with a mass of 40.8 kg. The vehicle is overactuated with eight bidirectional thrusters.

Designed to be modular, the vehicle has multiple specialized pressure vessels that house computational capabilities, sensors, batteries, and mission-specific payloads. The central pressure vessel houses the vehicle's motor controllers, network infrastructure, and core computing capability. The core computing capability services the vehicle's environmental sensors (eg, visible light cameras, scanning sonar), the vehicle's high-level mission planning, and low-level command and control software. A standard

Fig. 6 SubjuGator 7 autonomous underwater vehicle operating at Ginnie Springs, Florida.

[9] The number of basis functions and weights required to support a six degrees of freedom model greatly increases from the set required for the three degrees of freedom model. The increased number of parameters and complexity reduces the clarity of this proof-of-principle experiment.

[10] The vehicle's Doppler velocity log has a minimum height over the bottom of approximately 3 m that is required to measure water velocity. A minimum depth of approximately 0.5 m is required to remove the vehicle from surface effects. With the depth of the spring nominally 3.7 m, a narrow window of about 20 cm is left to operate the vehicle in heave.

small-form factor computer makes up the computing capability and consists of a 2.13-GHz server-grade quad-core processor. Located near the front of the vehicle, the navigation vessel houses the vehicle's basic navigation sensors. The suite of navigation sensors include an inertial measurement unit, a Doppler velocity log, a depth sensor, and a digital compass. The navigation vessel also includes an embedded 720-MHz processor for preprocessing and packaging navigation data. Along the sides of the central pressure vessel, two vessels house batteries of 44 Ah for propulsion and the electronics.

The vehicle's software runs within the Robot Operating System framework in the central pressure vessel. For the experiment, three main software nodes were used: navigation, control, and thruster mapping nodes. The navigation node receives packaged navigation data from the navigation pressure vessel, where an unscented Kalman filter estimates the vehicle's full state at 50 Hz. The desired force and moment produced by the controller are mapped to the eight thrusters by means of a least-squares minimization algorithm. The controller node contains the proposed controller and system identifier.

6.2 Controller Implementation

Implementation of the developed controller is divided into three parts: system identification, value function iteration, and control iteration. Implementation of the system identifier requires a recorded data set. The data set is collected in a swimming pool. The vehicle is commanded to track an exciting trajectory with a robust integral of the sign of the error (RISE) controller [64] while the state-action pairs were recorded. The recorded data are trimmed to a subset of 40 sampled points that are selected to maximize the minimum singular value of the history stack as in Section 6.2 in [65]. The system identifier is updated at 50 Hz.

Eqs. (6) and (8) form the value function iteration. Evaluating the extrapolated BE (6) with each control iteration is computationally expensive. Because of the limited computational resources available onboard the AUV, the update of the value function weights is calculated at 5 Hz.

For the experiments the proposed controller was restricted to three degrees of freedom (ie, surge, sway, and yaw). The RISE controller is used to regulate the remaining degrees of freedom (ie, heave, roll, and pitch) so as to maintain the implied assumption that roll and pitch remain at zero and the depth remains constant. Implementation of the proposed controller requires Eqs. (10) and (11). The RISE controller in conjunction with the proposed controller runs at a rate of 50 Hz.

The vehicle uses water profiling data from the Doppler velocity log to measure the relative water velocity near the vehicle in addition to bottom tracking data for the state estimator. Between the state estimator, water profiling data, and recorded data, the equations used to implement the proposed controller contain only known or measurable quantities.

6.3 Results

The vehicle is commanded to hold a station near the vent of Ginnie Spring. An initial condition of $x(t_0) = \begin{bmatrix} 4\,\text{m} & 4\,\text{m} & \frac{\pi}{4}\,\text{rad} & 0\,\text{m/s} & 0\,\text{m/s} & 0\,\text{rad/s} \end{bmatrix}^T$ is given to demonstrate the controller's ability to regulate the state. The optimal control weighting matrices are selected to be $Q = \text{diag}([20, 50, 20, 10, 10, 10])$ and $R = I_3$. The NN weights are initialized to match the ideal values for the linearized optimal control problem, which one obtains by solving the algebraic Riccati equation with the dynamics linearized about the station. The BE is extrapolated to 2025 points in a grid about the station.

Fig. 7 illustrates the ability of the generated policy to regulate the state. Fig. 8 illustrates the total control effort applied to the body of the vehicle. Fig. 9 illustrates the convergence of the parameters of the system identifier and Fig. 10 illustrates convergence of the NN weights representing the value function.

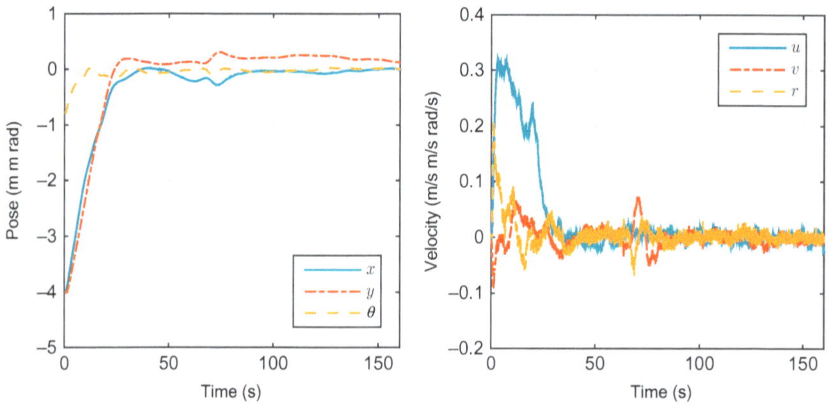

Fig. 7 Inertial position error $\eta \triangleq [x, y, \theta]^T$ (*left*) and body-fixed velocity error $\nu \triangleq [u, v, r]^T$ (*right*) of the autonomous underwater vehicle.

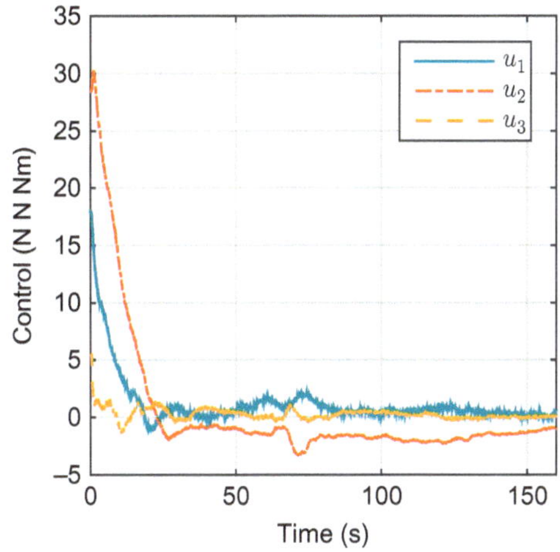

Fig. 8 Body-fixed optimal control effort commanded about the center of mass of the vehicle.

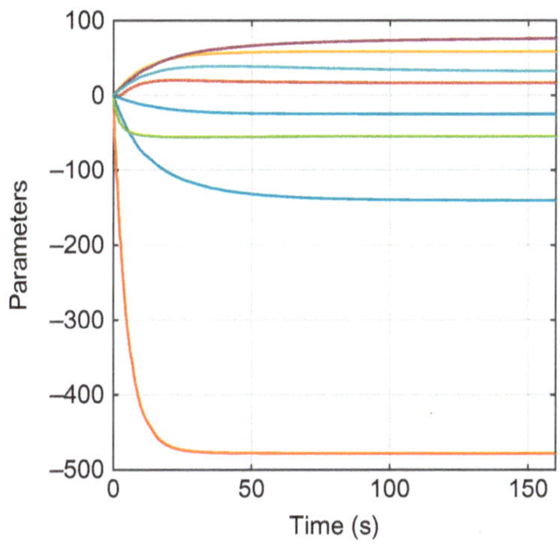

Fig. 9 Identified system parameters determined for the vehicle online. The parameter definitions may be found in Example 6.2 and Equation 6.100 in [66].

7 CONCLUSION

An online approximate optimal controller is developed, where the value function is approximated without PE in the system states via novel use of a model to evaluate the BE over unexplored areas of the state space. The PE condition along the system trajectories is replaced by an excitation condition that needs to be satisfied along virtual trajectories selected a priori. Uniformly ultimately bounded regulation of the system

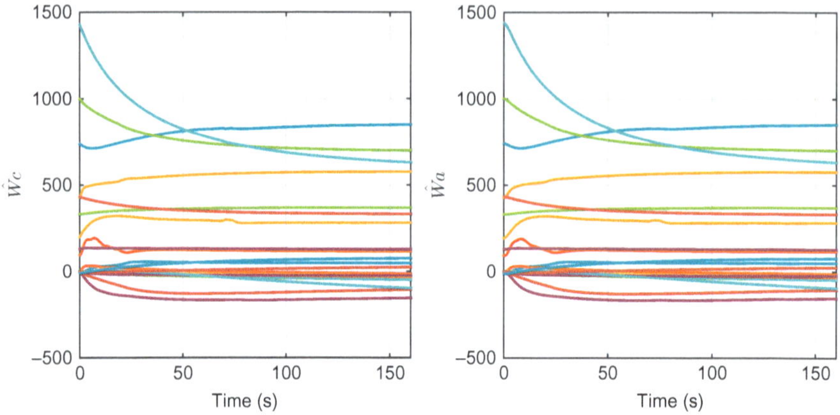

Fig. 10 Online convergence of critic $\hat{W}_c$ (*left*) and actor $\hat{W}_a$ (*right*) neural network weight estimates.

states to a neighborhood of the origin and convergence of the policy to a neighborhood of the optimal policy are established using a Lyapunov-based analysis. Simulations demonstrate that the developed technique generates an approximation to the optimal controller online, while maintaining system stability, without the use of a probing signal. Experiments demonstrate the ability to concurrently identify the uncertainties in the dynamics and generate an approximate optimal policy using the identified model. The vehicle follows the generated policy to achieve its station-keeping objective in the presence of external disturbances using industry standard navigation and environmental sensors.

REFERENCES

[1] K. Doya, Reinforcement learning in continuous time and space, Neural Comput. 12 (1) (2000) 219–245.

[2] R. Padhi, S. Balakrishnan, T. Randolph, Adaptive-critic based optimal neuro control synthesis for distributed parameter systems, Automatica 37 (8) (2001) 1223–1234.

[3] R. Padhi, N. Unnikrishnan, X. Wang, S. Balakrishnan, A single network adaptive critic (SNAC) architecture for optimal control synthesis for a class of nonlinear systems, Neural Netw. 19 (10) (2006) 1648–1660.

[4] A. Al-Tamimi, F.L. Lewis, M. Abu-Khalaf, Discrete-time nonlinear HJB solution using approximate dynamic programming: convergence proof, IEEE Trans. Syst. Man Cybern. B Cybern. 38 (2008) 943–949.

[5] F.L. Lewis, D. Vrabie, Reinforcement learning and adaptive dynamic programming for feedback control, IEEE Circuits Syst. Mag. 9 (3) (2009) 32–50.

[6] T. Dierks, B. Thumati, S. Jagannathan, Optimal control of unknown affine nonlinear discrete-time systems using offline-trained neural networks with proof of convergence, Neural Netw. 22 (5–6) (2009) 851–860.

[7] P. Mehta, S. Meyn, Q-learning and Pontryagin's minimum principle, in: Proc. IEEE Conf. Decis. Control, December 2009, 2009, pp. 3598–3605.

[8] K. Vamvoudakis, F. Lewis, Online actor-critic algorithm to solve the continuous-time infinite horizon optimal control problem, Automatica 46 (5) (2010) 878–888.

[9] H. Zhang, L. Cui, X. Zhang, Y. Luo, Data-driven robust approximate optimal tracking control for unknown general nonlinear systems using adaptive dynamic programming method, IEEE Trans. Neural Netw. 22 (12) (2011) 2226–2236.

[10] F.L. Lewis, D. Vrabie, V.L. Syrmos, Optimal Control, Wiley, New York, 2012.

[11] D. Wang, D. Liu, Q. Wei, Finite-horizon neuro-optimal tracking control for a class of discrete-time nonlinear systems using adaptive dynamic programming approach, Neurocomputing 78 (1) (2012) 14–22.

[12] S. Bhasin, R. Kamalapurkar, M. Johnson, K. Vamvoudakis, F.L. Lewis, W. Dixon, A novel actor-critic-identifier architecture for approximate optimal control of uncertain nonlinear systems, Automatica 49 (1) (2013) 89–92.

[13] H. Zhang, L. Cui, Y. Luo, Near-optimal control for nonzero-sum differential games of continuous-time nonlinear systems using single-network ADP, IEEE Trans. Cybern. 43 (1) (2013) 206–216.

[14] H. Zhang, D. Liu, Y. Luo, D. Wang, Adaptive dynamic programming for control algorithms and stability, in: Communications and Control Engineering, Springer-Verlag, London, 2013.

[15] R. Kamalapurkar, P. Walters, W.E. Dixon, Concurrent learning-based approximate optimal regulation, in: Proc. IEEE Conf. Decis. Control, Florence, Italy, December 2013, 2013, pp. 6256–6261.

[16] Q. Wei, D. Liu, Optimal tracking control scheme for discrete-time nonlinear systems with approximation errors, in: C. Guo, Z.-G. Hou, Z. Zeng (Eds.), Advances in Neural Networks—ISNN 2013, Lecture Notes in Computer Science, vol. 7952, Springer, Berlin, Heidelberg, 2013, pp. 1–10.

[17] A. Heydari, S. Balakrishnan, Finite-horizon control-constrained nonlinear optimal control using single network adaptive critics, IEEE Trans. Neural Netw. Learn. Syst. 24 (1) (2013) 145–157.

[18] A. Heydari, S.N. Balakrishnan, Fixed-final-time optimal control of nonlinear systems with terminal constraints, Neural Netw. 48 (2013) 61–71.

[19] H. Modares, F. Lewis, M.-B. Naghibi-Sistani, Adaptive optimal control of unknown constrained-input systems using policy iteration and neural networks, IEEE Trans. Neural Netw. Learn. Syst. 24 (10) (2013) 1513–1525.

[20] R. Kamalapurkar, J. Klotz, W. Dixon, Concurrent learning-based online approximate feedback Nash equilibrium solution of N-player nonzero-sum differential games, IEEE/CAA J. Autom. Sin. 1 (2014) 239–247.

[21] R. Kamalapurkar, J. Klotz, W.E. Dixon, Model-based reinforcement learning for on-line feedback-Nash equilibrium solution of n-player nonzero-sum differential games, in: Proceedings of the American Control Conference, 2014, pp. 3000–3005.

[22] R. Kamalapurkar, L. Andrews, P. Walters, W.E. Dixon, Model-based reinforcement learning for infinite-horizon approximate optimal tracking, in: Proc. IEEE Conf. Decis. Control, 2014, pp. 5083–5088.

[23] X. Yang, D. Liu, Q. Wei, Online approximate optimal control for affine non-linear systems with unknown internal dynamics using adaptive dynamic programming, IET Control Theory Appl. 8 (16) (2014) 1676–1688.

[24] X. Yang, D. Liu, D. Wang, Reinforcement learning for adaptive optimal control of unknown continuous-time nonlinear systems with input constraints, Int. J. Control 87 (3) (2014) 553–566.

[25] H. Modares, F.L. Lewis, Optimal tracking control of nonlinear partially-unknown constrained-input systems using integral reinforcement learning, Automatica 50 (7) (2014) 1780–1792.
[26] H. Modares, F.L. Lewis, M.-B. Naghibi-Sistani, Integral reinforcement learning and experience replay for adaptive optimal control of partially-unknown constrained-input continuous-time systems, Automatica 50 (1) (2014) 193–202.
[27] B. Kiumarsi, F.L. Lewis, H. Modares, A. Karimpour, M.-B. Naghibi-Sistani, Reinforcement Q-learning for optimal tracking control of linear discrete-time systems with unknown dynamics, Automatica 50 (2014) 1167–1175.
[28] A. Heydari, S.N. Balakrishnan, Adaptive critic-based solution to an orbital rendezvous problem, J. Guid. Control Dyn. 37 (2014) 344–350.
[29] T. Bian, Y. Jiang, Z.-P. Jiang, Adaptive dynamic programming and optimal control of nonlinear nonaffine systems, Automatica 50 (10) (2014) 2624–2632.
[30] R. Kamalapurkar, H. Dinh, S. Bhasin, W.E. Dixon, Approximate optimal trajectory tracking for continuous-time nonlinear systems, Automatica 51 (2015) 40–48.
[31] R. Kamalapurkar, J. Rosenfeld, W.E. Dixon, State following (StaF) kernel functions for function approximation Part II: Adaptive dynamic programming, in: Proceedings of the American Control Conference, 2015, pp. 521–526.
[32] T. Bian, Y. Jiang, Z.-P. Jiang, Decentralized adaptive optimal control of large-scale systems with application to power systems, IEEE Trans. Ind. Electron. 62 (2015) 2439–2447.
[33] H. Modares, F.L. Lewis, Z.-P. Jiang, H∞ tracking control of completely unknown continuous-time systems via off-policy reinforcement learning, IEEE Trans. Neural Netw. Learn. Syst. 26 (10) (2015) 2550–2562.
[34] R. Song, F. Lewis, Q. Wei, H.-G. Zhang, Z.-P. Jiang, D. Levine, Multiple actor-critic structures for continuous-time optimal control using input-output data, IEEE Trans. Neural Netw. Learn. Syst. 26 (2015) 851–865.
[35] P. He, S. Jagannathan, Reinforcement learning neural-network-based controller for nonlinear discrete-time systems with input constraints, IEEE Trans. Syst. Man Cybern. B Cybern. 37 (2) (2007) 425–436.
[36] H. Zhang, Q. Wei, Y. Luo, A novel infinite-time optimal tracking control scheme for a class of discrete-time nonlinear systems via the greedy HDP iteration algorithm, IEEE Trans. Syst. Man Cybern. B: Cybern. 38 (4) (2008) 937–942.
[37] K.S. Narendra, A.M. Annaswamy, A new adaptive law for robust adaptive control without persistent excitation, IEEE Trans. Autom. Control 32 (1987) 134–145.
[38] K. Narendra, A. Annaswamy, Stable Adaptive Systems, Prentice Hall, Englewood Cliffs, NJ, 1989.
[39] S. Sastry, M. Bodson, Adaptive Control: Stability, Convergence, and Robustness, Prentice Hall, Upper Saddle River, NJ, 1989.
[40] P. Ioannou, J. Sun, Robust Adaptive Control, Prentice Hall, Upper Saddle River, NJ, 1996.
[41] D. Vrabie, Online adaptive optimal control for continuous-time systems (Ph.D. thesis), University of Texas at Arlington, 2010.
[42] A. Al-Tamimi, F.L. Lewis, M. Abu-Khalaf, Model-free Q-learning designs for linear discrete-time zero-sum games with application to H_∞ control, Automatica 43 (2007) 473–481.
[43] K. Vamvoudakis, F. Lewis, Multi-player non-zero-sum games: online adaptive learning solution of coupled Hamilton-Jacobi equations, Automatica 47 (2011) 1556–1569.
[44] K.G. Vamvoudakis, F.L. Lewis, G.R. Hudas, Multi-agent differential graphical games: online adaptive learning solution for synchronization with optimality, Automatica 48 (8) (2012) 1598–1611.

[45] S.P. Singh, Reinforcement learning with a hierarchy of abstract models, in: AAAI Natl. Conf. Artif. Intell., vol. 92, 1992, pp. 202–207.
[46] C.G. Atkeson, S. Schaal, Robot learning from demonstration, in: Int. Conf. Mach. Learn., vol. 97, 1997, pp. 12–20.
[47] P. Abbeel, M. Quigley, A.Y. Ng, Using inaccurate models in reinforcement learning, in: Int. Conf. Mach. Learn., ACM, New York, NY, USA, 2006, pp. 1–8.
[48] M.P. Deisenroth, Efficient Reinforcement Learning Using Gaussian Processes, KIT Scientific Publishing, Karlsruhe, 2010.
[49] D. Mitrovic, S. Klanke, S. Vijayakumar, Adaptive optimal feedback control with learned internal dynamics models, in: O. Sigaud, J. Peters (Eds.), From Motor Learning to Interaction Learning in Robots, Studies in Computational Intelligence, vol. 264, Springer, Berlin, Heidelberg, 2010, pp. 65–84.
[50] M.P. Deisenroth, C.E. Rasmussen, Pilco: a model-based and data-efficient approach to policy search, in: Int. Conf. Mach. Learn, 2011, pp. 465–472.
[51] R. Kamalapurkar, Model-based reinforcement learning for online approximate optimal control (Ph.D. thesis), University of Florida, 2014.
[52] D. Kirk, Optimal Control Theory: An Introduction, Dover, Mineola, NY, 2004.
[53] R.S. Sutton, A.G. Barto, Reinforcement Learning: An Introduction, MIT Press, Cambridge, MA, USA, 1998.
[54] V. Konda, J. Tsitsiklis, On actor-critic algorithms, SIAM J. Control Optim. 42 (4) (2004) 1143–1166.
[55] T. Dierks, S. Jagannathan, Optimal tracking control of affine nonlinear discrete-time systems with unknown internal dynamics, in: Proc. IEEE Conf. Decis. Control, 2009, pp. 6750–6755.
[56] K. Vamvoudakis, F. Lewis, Online synchronous policy iteration method for optimal control, in: W. Yu (Ed.), Recent Advances in Intelligent Control Systems, Springer, Berlin, 2009, pp. 357–374.
[57] T. Dierks, S. Jagannathan, Optimal control of affine nonlinear continuous-time systems, in: Proceedings of the American Control Conference, 2010, pp. 1568–1573.
[58] W.E. Dixon, A. Behal, D.M. Dawson, S. Nagarkatti, Nonlinear Control of Engineering Systems: A Lyapunov-Based Approach, Birkhauser, Boston, 2003.
[59] G.V. Chowdhary, E.N. Johnson, Theory and flight-test validation of a concurrent-learning adaptive controller, J. Guid. Control Dyn. 34 (2011) 592–607.
[60] G. Chowdhary, T. Yucelen, M. Mühlegg, E.N. Johnson, Concurrent learning adaptive control of linear systems with exponentially convergent bounds, Int. J. Adapt. Control Signal Process. 27 (4) (2013) 280–301.
[61] H.K. Khalil, Nonlinear Systems, third ed., Prentice Hall, Upper Saddle River, NJ, USA, 2002.
[62] D. Garg, W.W. Hager, A.V. Rao, Pseudospectral methods for solving infinite-horizon optimal control problems, Automatica 47 (4) (2011) 829–837.
[63] W. Schmidt, Springs of Florida 2004, bulletin 66, Florida Geological Survey.
[64] N. Fischer, D. Hughes, P. Walters, E. Schwartz, W.E. Dixon, Nonlinear RISE-based control of an autonomous underwater vehicle, IEEE Trans. Robot. 30 (2014) 845–852.
[65] G. Chowdhary, Concurrent learning adaptive control for convergence without persistency of excitation (Ph.D. thesis), Georgia Institute of Technology, 2010.
[66] T.I. Fossen, Handbook of Marine Craft Hydrodynamics and Motion Control, Wiley, New York, 2011.

CHAPTER NINE

Continuous-Time Distributed Adaptive Dynamic Programming for Heterogeneous Multiagent Optimal Synchronization Control

Q. Wei*, R. Song†, D. Liu†, B. Luo*

*The State Key Laboratory of Management and Control for Complex Systems, Institute of Automation, Chinese Academy of Sciences, Beijing, China

†School of Automation and Electrical Engineering, University of Science and Technology Beijing, Beijing, China

1 INTRODUCTION

In many real-world control systems, there are multicontrollers in the systems and the performance index functions for these controllers are often operated in a group with coupled performance index functions as a game, such as the multiagent graphical game. Synchronization behavior, found in the flocking of birds, schooling of fish, and other natural systems, is a primary behavior in multiagent systems. Distributed synchronization control, which drives all the agents to a desired synchronization behavior, is an important research topic for the control of multiagent systems [1–3]. However, in the previous literature on distributed synchronization control problems for multiagent systems, the stability of multiagent systems was focused on, while the optimality property, which is also a primary property for multiagent systems, was scarcely investigated. The main difficulty lies in solving the coupled Hamilton-Jacobi (HJ) equations. Hence, intelligent and adaptive optimal control has always been the key focus for solving these multicontroller systems [4–7].

Adaptive dynamic programming (ADP) [8, 9], characterized by strong abilities of self-learning and adaptivity, is receiving significantly increased attention and is becoming an important brain-like intelligent optimal control method [10–14]. Several synonyms for "adaptive dynamic programming" (ADP) [15–18] have been used, including "adaptive critic design" [19],

Control of Complex Systems
http://dx.doi.org/10.1016/B978-0-12-805246-4.00009-4

"approximate dynamic programming" [20], "neurodynamic programming" [21], "neural dynamic programming" [22], and "reinforcement learning" [23]. Iterative methods, including value and policy iterations [24], are widely used in ADP to obtain the optimal performance index function indirectly [25–31]. Value iteration for optimal control was first given in [21]. The convergence of the value iteration ADP method was proven in [32]. Recently, an optimal adaptive neural control method with finite approximation errors was developed [33], where a new design method of convergence criterion was established which made a suitable approximation error be adaptively designed to guarantee the convergence of the algorithm. Value iteration algorithms were proposed in [34, 35] to solve two-person zero-sum games for discrete-time linear systems. With the system disturbance as a control input, the optimal control problem with disturbances was successfully solved in [18] by a new data-based multicontroller zero-sum adaptive neural control method, without knowledge of the system model, which was applied to a coal-based process. Despite the reconstruction errors of neural networks and the system disturbances, the proposed data-based ADP methods [18, 36] guaranteed the properties in both convergence and stability efficiently. A novel Q-learning-based value iteration algorithm was developed in [37] to solve the optimal battery management and control problem in smart residential environments. For the first time, two iteration indices—internal and external iteration indices—were employed in a Q-learning algorithm. It was proven that the iterative Q function reaches the optimal solution of the optimal battery management problem. A novel distributed value iteration algorithm was developed in [38] to solve the multibattery optimal coordination control problems for home energy management systems. In each iteration of the distributed iterative ADP algorithm, only a single-input optimization problem is implemented, which effectively avoids the solution of high-dimensional optimization for multibattery systems. For value iteration algorithms the stability of the system under the iterative control law cannot be guaranteed. Policy iteration algorithms are a class of the primary and most important iterative ADP algorithms [39–41]. Policy iteration for optimal control of continuous-time systems was given in [42, 43]. A policy iteration algorithm was successfully applied in [44] to solve continuous-time complex-valued systems. Discrete-time policy iteration with convergence and stability proofs was developed in [45]. A novel generalized policy iteration algorithm was described and analyzed in [46], which made it possible for all discrete-time iterative reinforcement learning optimal control algorithms to be constructed and analyzed by a unified generalized policy iteration algorithm. Policy iteration was successfully applied in [47] to achieve optimal control laws for multiperson zero-sum differential games. A policy iteration algorithm for non-zero-sum games was proposed in [48, 49]. A near-optimal control for non-zero-sum differential games was solved in [50] with use of a single-network policy iteration algorithm. In previous policy iteration ADP algorithms for multiplayer games, it was always desired that the system's states for each agent converge to the equilibrium of the systems. In many real-world games, it is required that the states of each agent track a desired dynamics (i.e., to achieve synchronization control). Synchronization behavior of the multiagent optimal control based on ADP was pioneered by Vamvoudakis et al. [51]. They developed a policy iteration algorithm for multiagent differential graphical games, where the Nash equilibrium of the game was effectively achieved. However, to guarantee the convergence of the performance index function of the agent for each node, it was required that all the dynamics of autonomous systems for each node of the multiagent graphical games are the same, which is equal to the desired dynamics. If the autonomous systems of each agent are not the same (i.e., for heterogeneous multiagent differential graphical games), the convergence analysis in [51] is invalid. This is a disadvantage of the policy iteration algorithm proposed in [51]. To the best of the authors' knowledge, when the dynamics of the graphical game are different, the optimal synchronization control for the synchronization of heterogeneous multiagent differential graphical games is not discussed. This is the motivation for our research.

In this chapter, inspired by the pioneering work of Vamvoudakis et al. [51], an optimal distributed synchronization control for heterogeneous multiagent differential graphical games is investigated. We emphasize that in the heterogeneous differential graphical games developed, the autonomous system of the agent for each node and the desired dynamics can be different from each other. By means of system transformations, the optimal synchronization control problem is transformed into an optimal multiagent regulation problem. The main contribution of this chapter is the development of an effective cooperative policy iteration algorithm to obtain the optimal distributed synchronization control law for heterogeneous multiagent differential graphical games. Convergence properties of the cooperative policy iteration algorithm are developed, and these guarantee that the iterative value function of the agent for each node converges to the Nash equilibrium of the games. Two simulation examples are presented to show the effectiveness of the algorithm developed.

2 GRAPHS AND SYNCHRONIZATION OF MULTIAGENT SYSTEMS

In this section a background review of communication graphs is given and the problem of synchronization of heterogeneous multiagent systems is formulated.

2.1 Graph Theory

Let $\mathcal{G} = (\mathcal{V}, \mathcal{E}, \mathcal{A})$ be a weighted graph of N nodes with the nonempty finite set of nodes $\mathcal{V} = \{v_1, \ldots, v_N\}$, where the set of edges $\mathcal{E}$ belongs to the product space of $\mathcal{V}$ (i.e., $\mathcal{E} \subseteq \mathcal{V} \times \mathcal{V}$), an edge of $\mathcal{G}$ is denoted by $\rho_{ij} = (v_j, v_i)$, which is a direct path from node j to node i, and $\mathcal{A} = [\rho_{ij}]$ is a weighted adjacency matrix with nonnegative adjacency elements—that is, $\rho_{ij} \geq 0$, $\rho_{ij} \in \mathcal{E} \Leftrightarrow \rho_{ij} > 0$; otherwise $\rho_{ij} = 0$. The node index i belongs to a finite index set $\mathcal{I} = \{1, 2, \ldots, N\}$.

Definition 1 (Laplacian matrix). The graph Laplacian matrix $L = [l_{ij}]$ is defined as $\mathcal{D} - \mathcal{A}$, with $\mathcal{D} = \text{diag}\{d_i\} \in \mathbb{R}^{N \times N}$ being the in-degree matrix of the graph, where $d_i = \Sigma_{j=1}^{N} \rho_{ij}$ is the in-degree of node v_i in the graph.

We assume that the graph is simple—for example, no repeated edges and no self-loops. The set of neighbors of node v_i is denoted by $N_i = \{v_j \in \mathcal{V}; (v_j, v_i) \in \mathcal{E}\}$. A graph is referred to as a *spanning tree* if there is a node v_i (called the root) such that there is a directed path from the root to any other nodes in the graph. A graph is said to be strongly connected if there is a directed path from node i to node j for all distinct nodes $v_i, v_j \in \mathcal{V}$. A digraph has a spanning tree if it is strongly connected, but not vice versa.

2.2 Synchronization and Tracking Error Dynamic Systems

Consider a heterogeneous multiagent system with N agents in the form of communication network $\mathcal{G}$. For an arbitrary node $i \in \mathcal{I}$, the heterogeneous agent system is expressed as

$$\dot{x}_i = A_i x_i + B_i u_i, \tag{1}$$

where $x_i(t) \in \mathbb{R}^n$ is the state of node i and $u_i(t) \in \mathbb{R}^{m_i}$ is the input coordination control. Let A_i and B_i, $\forall\, i \in \mathcal{I}$, be constant matrices. We assume that the control gain matrix B_i satisfies $\text{rank}\{B_i\} \geq n$ for the convenience of our analysis. For any $i \in \mathcal{I}$, assume that the state x_i of the agent for each node is controllable. In this chapter the synchronization control problem is to design local control protocols for all the nodes in $\mathcal{G}$ to synchronize to a control node or leader. The local neighborhood tracking error for all $i \in \mathcal{I}$ is defined as

$$\delta_i = \sum_{j \in N_i} \rho_{ij}(x_i - x_j) + \sigma_i(x_i - x_0), \tag{2}$$

where $\sigma_i \geq 0$ is the pinning gain. We let $\sigma_i = 0$ if and only if there is not a direct path from the control node to the ith node in $\mathcal{G}$. Otherwise we have $\sigma_i > 0$. Note that $\sigma_i > 0$ for at least one i.

Let $x_0(t) \in \mathbb{R}^n$ be the target state of the control node, which satisfies the dynamics

$$\dot{x}_0 = A_0 x_0. \tag{3}$$

According to Eq. (1), the global network dynamics can be expressed as

$$\dot{x} = Ax + Bu, \tag{4}$$

where $x = [x_1^T, x_2^T, \ldots, x_N^T]^T \in \mathbb{R}^{Nn}$, $u = [u_1^T, u_2^T, \ldots, u_N^T]^T \in \mathbb{R}^{Nm}$, $A = \mathrm{diag}\{A_i\}$, and $B = \mathrm{diag}\{B_i\}$, $i = 1, 2, \ldots, N$. According to Eqs. (2) and (4) the global error vector for the network $\mathcal{G}$ is

$$\begin{aligned}\delta &= (L \otimes I_n)x + (G \otimes I_n)(x - \bar{x}_0)\\ &= ((L + G) \otimes I_n)(x - \bar{x}_0)\\ &= \mathcal{L}(x - \bar{x}_0),\end{aligned} \tag{5}$$

where $\mathcal{L} = (L + G) \otimes I_n$, I_n is an $n \times n$ identity matrix, L is the Laplacian matrix, $\otimes$ denotes the Kronecker product operator, and $G = [\sigma_{ij}] \in \mathbb{R}^{N \times N}$ is a diagonal matrix with diagonal entries σ_i, $\bar{x}_0 = \underline{I} \otimes x_0$, where $\underline{I} = \underline{1} \otimes I_n$ and $\underline{1}$ is an N-dimensional vector of 1's.

In [51], $\forall i \in \mathcal{I}$, the system matrices, satisfy that $A_i = A_0$. This makes the control $u_i \equiv 0$ as x_i reaches the target state x_0. In this chapter, according to Eqs. (1) and (3), $\forall i \in \mathcal{I}$, the control u_i may not be zero, even when $x_i = x_0$. Thus we should first solve the desired control at the target state. Let u_{di} be the desired control that satisfies the following equation:

$$\dot{x}_0 = A_i x_0 + B_i u_{di}. \tag{6}$$

According to Eqs. (3) and (6) we have

$$u_{di} = B_i^{+}(A_0 - A_i)x_0, \tag{7}$$

where B_i^+ is the Moore-Penrose pseudo inverse of B_i. Let $x_{ei} = x_i - x_0$ and $u_{ei} = u_i - u_{di}$. Then, according to Eq. (7), system (1) can be rewritten as

$$\begin{aligned}\dot{x}_{ei} &= A_i x_{ei} + B_i u_{ei} + (A_i - A_0)x_0 + B_i u_{di} \\ &= A_i x_{ei} + B_i u_{ei}.\end{aligned} \tag{8}$$

According to Eq. (2), the tracking error vector for the network $\mathcal{G}$ can be expressed as

$$\begin{aligned}\dot{\delta}_i &= \sum_{j\in N_i} \rho_{ij}(\dot{x}_i - \dot{x}_j) + \sigma_i(\dot{x}_i - \dot{x}_0) \\ &= \sum_{j\in N_i} \rho_{ij}(\dot{x}_{ei} - \dot{x}_{ej}) + \sigma_i \dot{x}_{ei} \\ &= \sum_{j\in N_i} \rho_{ij}(A_i x_{ei} + B_i u_{ei} - A_j x_{ej} + B_j u_{ej}) + \sigma_i(A_i x_{ei} + B_i u_{ei}) \\ &= \sum_{j\in N_i} \rho_{ij} A_i (x_{ei} - x_{ej}) + \sum_{j\in N_i} \rho_{ij}(A_i - A_j)x_{ej} + \sum_{j\in N_i} \rho_{ij}(B_i u_{ei} - B_j u_{ej}) \\ &\quad + \sigma_i A_i x_{ei} + \sigma_i B_i u_{ei} \\ &= A_i \delta_i + \sum_{j\in N_i} \rho_{ij}(A_i - A_j)x_{ej} + (d_i + \sigma_i) B_i u_{ei} - \sum_{j\in N_i} \rho_{ij} B_j u_{ej}.\end{aligned} \tag{9}$$

From Eq. (9), we can see that the tracking error vector is a function with respect to δ_i and x_{ej}. This means that the distributed control u_i should be designed by using the information of δ_i and x_{ej}. A system transformation is necessary.

Let $j_1^i, \ldots, j_{N_i}^i$ be the nodes in N_i. Define the new state and control vectors as $x_{-ei} = \left[x_{ej_1^i}^T, \ldots, x_{ej_{N_i}^i}^T\right]^T$ and $u_{-ei} = \left[u_{ej_1^i}^T, \ldots, u_{ej_{N_i}^i}^T\right]^T$, respectively. Letting $z_i = \left[\delta_i^T, x_{-ei}^T\right]^T$ and $\bar{u}_{ei} = \left[u_{ei}^T, u_{-ei}^T\right]^T$, then we can obtain the following expression:

$$\dot{z}_i = \bar{A}_i z_i + \bar{B}_i \bar{u}_{ei}, \tag{10}$$

where $\bar{A}_i$ and $\bar{B}_i$ are expressed as

$$\bar{A}_i = \begin{bmatrix} A_i & e_{ij_1^i}(A_i - A_{j_1^i}) & \cdots & e_{ij_{N_i}^i}(A_i - A_{j_{N_i}^i}) \\ 0 & A_{j_1^i} & \cdots & 0 \\ \vdots & \vdots & \ddots & \vdots \\ 0 & 0 & \cdots & A_{j_{N_i}^i} \end{bmatrix} \tag{11}$$

and

$$\bar{B}_i = \begin{bmatrix} (d_i + \sigma_i)B_i & -e_{ij_1^i} B_{j_1^i} & \cdots & -e_{ij_{N_i}^i} B_{j_{N_i}^i} \\ 0 & B_{j_1^i} & \cdots & 0 \\ \vdots & \vdots & \ddots & \vdots \\ 0 & 0 & \cdots & B_{j_{N_i}^i} \end{bmatrix}. \tag{12}$$

We can see that system (10) is a multi-input system. The controls are u_{ei} and all the u_{ej} from its neighbors, and the controls for agent i are coupled with the agents of its neighbors. This makes the controller design very difficult. In the next section, an optimal distributed control by an iterative ADP algorithm is developed to make all agents reach the target state.

3 OPTIMAL DISTRIBUTED SYNCHRONIZATION CONTROL FOR HETEROGENEOUS MULTIAGENT DIFFERENTIAL GRAPHICAL GAMES

In this section our goal is to design an optimal distributed synchronization control to reach the desired dynamics and simultaneously to minimize the local performance index function for system (10).

3.1 Cooperative Performance Index Function

For $i \in \mathcal{I}$, let Q_{ii}, R_{ii}, and R_{ij}, $j \in N_i$, be constant and symmetric matrices which satisfy $Q_{ii} \geq 0$, $R_{ii} > 0$, and $R_{ij} \geq 0$. For $i \in \mathcal{I}$, define the utility function as

$$\begin{aligned} U_i(z_i, u_{ei}, u_{-ei}) &= z_i^T \bar{Q}_{ii} z_i + u_{ei}^T R_{ii} u_{ei} + \sum_{j \in N_i} u_{ej}^T R_{ij} u_{ej} \\ &= \delta_i^T Q_{ii} \delta_i + u_{ei}^T R_{ii} u_{ei} + \sum_{j \in N_i} u_{ej}^T R_{ij} u_{ej}, \end{aligned} \tag{13}$$

where

$$\begin{aligned} z_i^T \bar{Q}_{ii} z_i &= \begin{bmatrix} \delta_i \\ x_{ej_1^i} \\ \vdots \\ x_{ej_{N_i}^i} \end{bmatrix}^T \begin{bmatrix} Q_{ii} & 0 & \cdots & 0 \\ 0 & 0 & \cdots & 0 \\ \vdots & \vdots & \ddots & \vdots \\ 0 & 0 & \cdots & 0 \end{bmatrix} \begin{bmatrix} \delta_i \\ x_{ej_1^i} \\ \vdots \\ x_{ej_{N_i}^i} \end{bmatrix} \\ &= \delta_i^T Q_{ii} \delta_i. \end{aligned} \tag{14}$$

Then we can define the local performance index functions as

$$J_i(z_i(0), u_{ei}, u_{-ei}) = \int_0^\infty \left(\delta_i^T Q_{ii}\delta_i + u_{ei}^T R_{ii}u_{ei} + \sum_{j\in N_i} u_{ej}^T R_{ij}u_{ej} \right) \mathrm{d}t$$
$$= \int_0^\infty U_i\left(z_i(t), u_{ei}(t), u_{-ei}(t)\right) \mathrm{d}t. \tag{15}$$

From Eq. (15), we can see that the performance index functions include only the information about the inputs of node i and its neighbors. The goal of this chapter is to design the local optimal distributed control to minimize the local performance index functions (15) subject to Eq. (10) and make all nodes (agents) reach consensus at the target state x_0.

Definition 2 (Admissible coordination control law). [51] The control law u_{ei}, $i \in \mathcal{I}$ is said to be an admissible coordination control law if u_{ei} is continuous, $u_{ei}(0) = 0$, u_{ei} stabilizes agent (10) locally, and the performance index function (15) is finite.

If u_{ei} and u_{-ei} are state-dependent admissible coordination control laws, then we can define the local performance index function $V_i(z_i)$ as

$$V_i(z_i) = \int_t^\infty U_i\left(z_i, u_{ei}, u_{-ei}\right) \mathrm{d}t. \tag{16}$$

For admissible coordination control laws u_{ei} and u_{-ei}, the Hamiltonian $H_i\left(z_i, \frac{\partial V_i}{\partial z_i}, u_{ei}, u_{-ei}\right)$ satisfies the following cooperative HJ equation:

$$H_i\left(z_i, \frac{\partial V_i}{\partial z_i}, u_{ei}, u_{-ei}\right) \equiv \frac{\partial V_i^T}{\partial z_i}\left(\bar{A}_i z_i + \bar{B}_i \bar{u}_{ei}\right) + \delta_i^T Q_{ii}\delta_i + u_{ei}^T R_{ii}u_{ei}$$
$$+ \sum_{j\in N_i} u_{ej}^T R_{ij}u_{ej} = 0 \tag{17}$$

with boundary condition $V_i(0) = 0$. If $V_i^\star(z_i)$ is the local performance index function, then $V_i^\star(z_i)$ satisfies the coupled HJ equations

$$\min_{u_{ei}} H_i\left(z_i, \frac{\partial V_i^\star}{\partial z_i}, u_{ei}, u_{-ei}\right) = 0 \tag{18}$$

and the local optimal control law $u_{ei}^\star$ satisfies

$$u_{ei}^{\star} = \arg\min_{u_{ei}} H_i\left(z_i, \frac{\partial V_i^{\star}}{\partial z_i}, u_{ei}, u_{-ei}\right). \tag{19}$$

3.2 Nash Equilibrium

In this subsection the optimality of multiagent systems will be developed. The corresponding property will also be presented. According to [51], we introduce the Nash equilibrium definition of multiagent differential games.

Definition 3 (Global Nash equilibrium). A sequence of control laws $\left\{u_{e1}^{\star}, u_{e2}^{\star}, \ldots, u_{eN}^{\star}\right\}$ is said to be a Nash equilibrium solution for an N multiagent game for all $i \in \mathcal{I}$ if $\forall\, i \in \mathcal{I}$,

$$J_i^{\star} \stackrel{\Delta}{=} J_i(u_{e1}^{\star}, \ldots, u_{ei}^{\star}, \ldots, u_{eN}^{\star}) \leq J_i(u_{e1}^{\star}, \ldots, u_{ei}, \ldots, u_{eN}^{\star}), \tag{20}$$

for $u_{ei} \neq u_{ei}^{\star}$.

The N-tuple of the local performance index function $\{J_1^{\star}, J_2^{\star}, \ldots, J_N^{\star}\}$ is known as the Nash equilibrium of the multiagent differential graphical game in $\mathcal{G}$.

Theorem 1. *Let $J_i^{\star}$ be the Nash equilibrium solution that satisfies Eq. (20). If $\left\{u_{e1}^{\star}, u_{e2}^{\star}, \ldots, u_{eN}^{\star}\right\}$ are optimal distributed synchronization control laws, then for all $i \in \mathcal{I}$ we have*

$$u_{ei}^{\star} = \arg\min_{u_{ei}} H_i\left(z_i, \frac{\partial J_i^{\star}}{\partial z_i}, u_{ei}, u_{-ei}^{\star}\right). \tag{21}$$

Proof. The conclusion can be proven by contradiction. Assume that the conclusion is false. Let $\left\{u_{e1}^{\star}, u_{e2}^{\star}, \ldots, u_{eN}^{\star}\right\}$ be optimal control laws. Then there must exist a $u_{el}^{\star}$ such that

$$u_{el}^{\star} \neq \arg\min_{u_{el}} H_l\left(z_l, \frac{\partial J_l^{\star}}{\partial z_l}, u_{el}, u_{-el}^{\star}\right). \tag{22}$$

As $\left\{u_{e1}^{\star}, u_{e2}^{\star}, \ldots, u_{eN}^{\star}\right\}$ is the optimal control law and $J_l^{\star}$ is the Nash equilibrium solution, for $l = 1, 2, \ldots, N$, we can get

$$H_l\left(z_l, \frac{\partial J_l^{\star}}{\partial z_l}, u_{el}^{\star}, u_{-el}^{\star}\right) = \frac{\partial J_l^{\star}}{\partial z_l}\left(\bar{A}_l z_l + \bar{B}_l \bar{u}_{el}^{\star}\right) + \delta_l^T Q_{ll} \delta_l + u_{el}^{\star T} R_{ll} u_{el}^{\star}$$
$$+ \sum_{j \in N_l} u_{ej}^{\star T} R_{lj} u_{ej}^{\star} = 0. \tag{23}$$

If Eq. (23) holds, then we can let u_{el}^{o} satisfy

$$u_{el}^{o} = \arg\min_{u_{el}} H_l\left(z_l, \frac{\partial J_l^{\star}}{\partial z_l}, u_{el}, u_{-el}^{\star}\right). \tag{24}$$

According to Eq. (24), we have

$$H_l\left(z_l, \frac{\partial J_l^{\star}}{\partial z_l}, u_{el}^{o}, u_{-el}^{\star}\right) = \min_{u_{el}} H_l\left(z_l, \frac{\partial J_l^{\star}}{\partial z_l}, u_{el}, u_{-el}^{\star}\right). \tag{25}$$

Hence, we can obtain

$$\begin{aligned} H_l\left(z_l, \frac{\partial J_l^{\star}}{\partial z_l}, u_{el}^{o}, u_{-el}^{\star}\right) &\leq H_l\left(z_l, \frac{\partial J_l^{\star}}{\partial z_l}, u_{el}^{\star}, u_{-el}^{\star}\right) \\ &= 0. \end{aligned} \tag{26}$$

If we can find a performance index function $J_l^{o}(z_l(0), u_{el}^{o}, u_{-el}^{\star})$ that satisfies

$$\begin{aligned} H_l\left(z_l, \frac{\partial J_l^{o}}{\partial z_l}, u_{el}^{o}, u_{-el}^{\star}\right) &= \frac{\partial J_l^{o}}{\partial z_l}\left(\bar{A}_l z_l + \bar{B}_l \bar{u}_{el}^{o}\right) + \delta_l^T Q_{ll} \delta_l + u_{el}^{\star T} R_{ll} u_{el}^{\star} \\ &\quad + \sum_{j \in N_l} u_{ej}^{\star T} R_{lj} u_{ej}^{\star} = 0, \end{aligned} \tag{27}$$

then we can obtain

$$\dot{J}_l^{o} = -\left(\delta_l^T Q_{ll} \delta_l + u_{el}^{oT} R_{ll} u_{el}^{o} + \sum_{j \in N_l} u_{ej}^{\star T} R_{lj} u_{ej}^{\star}\right). \tag{28}$$

According to Eqs. (27) and (28), we have

$$\begin{aligned} \dot{J}_l^{\star} &= H_l\left(z_l, \frac{\partial J_l^{\star}}{\partial z_l}, u_{el}^{o}, u_{-el}^{\star}\right) - \left(\delta_l^T Q_{ll} \delta_l + u_{el}^{oT} R_{ll} u_{el}^{o} + \sum_{j \in N_l} u_{ej}^{\star T} R_{lj} u_{ej}^{\star}\right) \\ &\leq \dot{J}_l^{o}. \end{aligned} \tag{29}$$

Hence, $J_l^{\star} \geq J_l^{o}$ for all $l = 1, 2, \ldots N$. This is a contradiction, and so the conclusion holds. □

From Theorem 1, we can see that if $\forall i \in \mathcal{I}$, all the control from its neighbors is optimal (i.e., $u_{ej} = u_{ej}^{\star}, j \in N_i$), and then the optimal distributed synchronization control $u_{ei}^{\star}$ can be obtained by Eq. (21). The following results can be derived according to the properties of $u_{ei}^{\star}$, $i \in \mathcal{I}$.

Lemma 1 ([51]). *Let $V^{\star}(z_i) > 0$, $i \in \mathcal{I}$ be a solution to the HJ equation* (18) *and let the optimal distributed control policies $u_{ei}^{\star}$, $i \in \mathcal{I}$ be given by Eq. (19) in terms of $V^{\star}(z_i)$. Then*

(i) *The local neighborhood consensus error* (10) *converges to zero.*

(ii) *The local performance index functions $J_i^{\star}(z_i, u_{ei}^{\star}, u_{-ei}^{\star})$ are equal to $V_i^{\star}(z_i)$, $i \in \mathcal{I}$, and $u_{ei}^{\star}$ and $u_{ej}^{\star}$, $j \in N_i$, are in Nash equilibrium.*

4 HETEROGENEOUS MULTIAGENT DIFFERENTIAL GRAPHICAL GAMES BY AN ITERATIVE ADP ALGORITHM

From the previous section, we know that if the optimal performance index function is obtained, then the optimal distributed synchronization control law can be obtained by our solving the HJ equation (18). However, Eq. (18) is difficult or impossible to solve directly by dynamic programming. In this section, a cooperative policy iteration algorithm of ADP is developed which solves the HJ equation (18) iteratively. Convergence properties of the developed algorithm will also be presented in this section.

4.1 Derivation of the Cooperative Policy Iteration Algorithm for Heterogeneous Multiagent Differential Graphical Games

In this subsection, we will derive the cooperative policy iteration algorithm for heterogeneous multiagent differential graphical games (heterogeneous multiagent policy iteration algorithm in brief). In the policy iteration algorithm to be developed, the value function and control law are updated by iterations, with the iteration index k increasing from 0 to infinity. For all $i \in \mathcal{I}$, let u_i^0 be an arbitrary admissible control law. Let V_i^0 be the iterative value function constructed by u_i^0, which satisfies

$$H_i\left(z_i, \frac{\partial V_i^0}{\partial z_i}, u_{ei}^0, u_{-ei}^0\right) = 0. \tag{30}$$

According to Eq. (30), we can update the iterative control law u_{ei}^1 by

$$u_{ei}^1 = \arg\min_{u_{ei}} H_i\left(z_i, \frac{\partial V_i^0}{\partial z_i}, u_{ei}, u_{-ei}^0\right). \tag{31}$$

For $k = 1, 2, \ldots$, solve for V_i^k that satisfies the following Hamiltonian:

$$H_i\left(z_i, \frac{\partial V_i^k}{\partial z_i}, u_{ei}^k, u_{-ei}^k\right) = \frac{\partial V_i^k}{\partial z_i}\left(\bar{A}_i z_i + \bar{B}_i \bar{u}_{ei}^k\right) + \delta_i^T Q_{ii} \delta_i + u_{ei}^{kT} R_{ii} u_{ei}^k$$

$$+ \sum_{j \in N_i} u_{ej}^{kT} R_{ij} u_{ej}^k = 0, \tag{32}$$

and update the iterative control law by

$$u_{ei}^{k+1} = \arg\min_{u_{ei}} H_i\left(z_i, \frac{\partial V_i^k}{\partial z_i}, u_{ei}, u_{-ei}^k\right). \tag{33}$$

Remark 1. The heterogeneous multiagent policy iteration algorithm has been shown in Eqs. (30)–(33). In Eqs. (30) and (32), given a current control law u_{ei}^k, $k = 0, 1, \ldots$, an iterative value function can be determined by our solving the Hamiltonian (32). This procedure is known as "policy evaluation." In this iteration procedure the iterative value function V_i^k is updated. Based on the iterative value function we can always use it to find another iterative control law u_{ei}^{k+1} that is better, or at least no worse. This iteration procedure is known as "policy improvement." In this iteration procedure, the iterative control law is updated.

From the heterogeneous multiagent policy iteration algorithm given by Eqs. (30)–(33), we can see that V_i^k is used to approximate $J_i^\star$ and u_i^k is used to approximate $u_i^\star$. Therefore, when $i \to \infty$, it is expected that the algorithm makes V_i^k and u_i^k converge to the optimal values. In the next subsection, we will show such properties of the above policy iteration algorithm.

4.2 Properties of the Cooperative Policy Iteration Algorithm for Heterogeneous Multiagent Differential Graphical Games

In this subsection, we will first present the convergence property of the heterogeneous multiagent policy iteration algorithm. Initialized by an admissible coordination control law, the monotonicity of the iterative performance index functions is discussed. The properties of the iterative value function were analyzed in [51] for the multiagent system, which is expressed by $\dot{x}_i = Ax_i + B_i u_i$. In this chapter, inspired by Vamvoudakis et al. [51], the properties of the iterative value function are analyzed for the heterogeneous multiagent system $\dot{x}_i = A_i x_i + B_i u_i$.

Lemma 2 (Solution for the best iterative control law). *Given fixed neighbor policies* $u_{-ei} \triangleq \{u_j : j \in N_i\}$, *the best iterative control law can be expressed by*

$$u_{ei}^{k+1} = -\frac{1}{2} R_{ii}^{-1} (d_i + \sigma_i) B_i^T \frac{\partial V_i^k}{\partial \delta_i}. \tag{34}$$

Proof. According to Eqs. (32) and (33), taking the derivative of u_{ei}, we have

$$\begin{aligned} u_{ei}^{k+1} &= \arg\min_{u_{ei}} \left\{ \left(\frac{\partial V_i^k}{\partial z_i} \right)^T \left(\bar{A}_i z_i + \bar{B}_i \bar{u}_{ei}^k \right) + \delta_i^T Q_{ii} \delta_i \right. \\ &\quad \left. + u_{ei}^{kT} R_{ii} u_{ei}^k + \sum_{j \in N_i} u_{ej}^{kT} R_{ij} u_{ej}^k \right\}, \\ &= -\frac{1}{2} R_{ii}^{-1} \left(\frac{\partial \bar{u}_{ei}^k}{\partial u_{ei}} \right)^T \begin{bmatrix} (d_i + \sigma_i) B_i^T & 0 & \cdots & 0 \\ -e_{ij_1^i} B_{j_1^i}^T & B_{j_1^i}^T & & 0 \\ \vdots & & \ddots & \\ -e_{ij_{N_i}^i} B_{j_{N_i}^i}^T & 0 & & B_{j_{N_i}^i}^T \end{bmatrix} \\ &\quad \times \left[\left(\frac{\partial V_i^k}{\partial \delta_i} \right)^T, \left(\frac{\partial V_i^k}{\partial x_{ej_1^i}} \right)^T, \ldots, \left(\frac{\partial V_i^k}{\partial x_{j_{N_i}^i}} \right)^T \right]^T \\ &= -\frac{1}{2} (d_i + \sigma_i) R_{ii}^{-1} B_i^T \frac{\partial V_i^k}{\partial \delta_i}. \end{aligned} \tag{35}$$

The proof is completed. □

Remark 2. In [51], with the system matrices $A_i = A_0$, $\forall i \in \mathcal{I}$, an iterative ADP algorithm for multiagent differential graphical games was proposed. As $A_i = A_0$, in [51] the desired control $u_{di} \equiv 0$. Compared with [51], in this chapter, as $A_i \neq A_0$, we have $u_{di} \neq 0$. Thus, when x_i reaches x_0, we have the control u_i is not zero. Second, in [51], it is shown that the iterative control law and the iterative value function are both functions of δ_i. In this chapter, the iterative value function is a function of z_i, that is, $V_i^k(z_i)$. Hence, the heterogeneous multiagent differential graphical game in this chapter is obviously different from the one in [51]. On the other hand, we can see that the multiagent differential graphical games in [51] are special cases of the heterogeneous multiagent differential graphical games in this chapter. Thus, the heterogeneous multiagent policy iteration algorithm in this chapter possess more potential for applications.

Next, inspired by Vamvoudakis et al. [51], the convergence properties of the heterogeneous multiagent policy iteration algorithm will be developed by the following theorems.

Theorem 2 (Convergence of the heterogeneous multiagent policy iteration algorithm when only one agent updates its policy and all players u_{-ei} in its neighborhood do not change). *Given fixed neighbor policies u^o_{-ei}, let the iterative value function V_i^k and the iterative control law u_{ei}^{k+1} be updated by the heterogeneous multiagent policy iteration algorithm given by Eqs. (30)–(33). Then, the iterative value function V_i^k is convergent as $k \to \infty$,—that is, $\lim_{k\to\infty} V_i^k = V_i^\infty$, where V_i^∞ satisfies the following HJ equation:*

$$\min_{u_{ei}} H_i\left(z_i, \frac{\partial V_i^k}{\partial z_i}, u_{ei}, u^o_{-ei}\right) = 0. \tag{36}$$

Proof. Let $\hat{u}_{ei}^k = \left[u_{ei}^{kT}, u_{-ei}^{oT}\right]^T$. Taking the derivative of V_i^k along the trajectory of z_i, we obtain

$$\begin{aligned} \dot{V}_i^k &= \frac{\partial V_i^k}{\partial z_i}(\bar{A}_i z_i + \bar{B}_i \hat{u}_{ei}^{k+1}) \\ &= H_i\left(z_i, \frac{\partial V_i^k}{\partial z_i}, u_{ei}^{k+1}, u^o_{-ei}\right) \\ &\quad - \left(\delta_i^T Q_{ii}\delta_i + u_{ei}^{(k+1)T} R_{ii} u_{ei}^{k+1} + \sum_{j\in N_i} u_{ej}^{oT} R_{ij} u_{ej}^o\right). \end{aligned} \tag{37}$$

According to Eqs. (32) and (33), we can get

$$\begin{aligned} H_i\left(z_i, \frac{\partial V_i^k}{\partial z_i}, u_{ei}^{k+1}, u_{-ei}^o\right) &= \min_{u_{ei}} H_i\left(z_i, \frac{\partial V_i^k}{\partial z_i}, u_{ei}, u_{-ei}^o\right) \\ &\le H_i\left(z_i, \frac{\partial V_i^k}{\partial z_i}, u_{ei}^k, u_{-ei}^o\right) \\ &= 0. \end{aligned} \tag{38}$$

Substituting the iteration index $k+1$ into the HJ equation (32), we have

$$\begin{aligned} H_i\left(z_i, \frac{\partial V_i^{k+1}}{\partial z_i}, u_{ei}^{k+1}, u_{-ei}^o\right) &= \frac{\partial V_i^{k+1}}{\partial z_i}(\bar{A}_i z_i + \bar{B}_i \hat{u}_{ei}^{k+1}) \\ &+ \left(\delta_i^T Q_{ii}\delta_i + u_{ei}^{(k+1)T} R_{ii} u_{ei}^{k+1} + \sum_{j\in N_i} u_{ej}^{oT} R_{ij} u_{ej}^o\right) = 0, \end{aligned} \tag{39}$$

which means

$$\dot{V}_i^{k+1} = -\left(\delta_i^T Q_{ii}\delta_i + u_{ei}^{(k+1)T} R_{ii} u_{ei}^{k+1} + \sum_{j\in N_i} u_{ej}^{oT} R_{ij} u_{ej}^o\right). \tag{40}$$

According to Eqs. (37), (38) and (40), we can obtain $\dot{V}_i^{k+1} \ge \dot{V}_i^k$. By integrating both sides of $\dot{V}_i^{k+1} \ge \dot{V}_i^k$, we can obtain $V_i^{k+1} \le V_i^k$. As V_i^k is lower bounded, V_i^k is convergent as $k \to \infty$,—that is, $\lim_{k\to\infty} V_i^k = V_i^\infty$. According to Eqs. (32) and (33) and letting $k \to \infty$, we can obtain Eq. (36). The proof is completed. □

From Theorem 2, we can see that for $i = 1, 2, \ldots, N_i$, if the neighboring control laws u_{-ei}^o are fixed, then the iterative value functions and iterative control laws are convergent. In the next theorem, the convergence property for the iterative value function and iterative control laws when all nodes update their control laws will be developed.

Theorem 3 (Convergence of the heterogeneous multiagent policy iteration algorithm when all agents update their policies). *Assume all nodes update their policies at each iteration of the policy iteration algorithm given by Eqs. (30)–(33). Define $\varsigma(R_{ij}R_{jj}^{-1})$ as the maximum singular value of $R_{ij}R_{jj}^{-1}$. For a small $\varsigma(R_{ij}R_{jj}^{-1})$, the iterative value function V_i^k converges to the optimal $J^\star$—that is, $\lim_{k\to\infty} V_i^k = J_i^\star$.*

Proof. According to Eqs. (33) and (35), letting $\breve{u}_{ei}^{k+1} = \left[u_{ei}^{(k+1)T}, u_{-ei}^{kT}\right]^T$, we have

$$\dot{V}_i^k = \frac{\partial V_i^k}{\partial z_i}(\bar{A}_i z_i + \bar{B}_i \breve{u}_{ei}^{k+1}). \tag{41}$$

According to Hamiltonian (32), we have

$$\frac{\partial V_i^k}{\partial z_i}\bar{A}_i z_i = -\frac{\partial V_i^k}{\partial z_i}\bar{B}_i \bar{u}_{ei}^k - \delta_i^T Q_{ii}\delta_i - u_{ei}^{kT} R_{ii} u_{ei}^k - \sum_{j\in N_i} u_{ej}^{kT} R_{ij} u_{ej}^k. \tag{42}$$

According to Hamiltonian (32), for $k+1$, we can obtain

$$\begin{aligned}
\dot{V}_i^{k+1} - \dot{V}_i^k &= \frac{\partial V_i^k}{\partial z_i}\bar{B}_i(\bar{u}_{ei}^k - \breve{u}_{ei}^{k+1}) - u_{ei}^{(k+1)T} R_{ii} u_{ei}^{k+1} \\
&\quad - \sum_{j\in N_i} u_{ej}^{(k+1)T} R_{ij} u_{ej}^{k+1} + u_{ei}^{kT} R_{ii} u_{ei}^k + \sum_{j\in N_i} u_{ej}^{kT} R_{ij} u_{ej}^k \\
&= (d_i + \sigma_i)\frac{\partial V_i^k}{\partial \delta_i} B_i (u_{ei}^k - u_{ei}^{k+1}) - u_{ei}^{(k+1)T} R_{ii} u_{ei}^{k+1} \\
&\quad - \sum_{j\in N_i} u_{ej}^{(k+1)T} R_{ij} u_{ej}^{k+1} + u_{ei}^{kT} R_{ii} u_{ei}^k + \sum_{j\in N_i} u_{ej}^{kT} R_{ij} u_{ej}^k \\
&= 2u_{ei}^{(k+1)T} R_{ii}(u_{ei}^{k+1} - u_{ei}^k) - u_{ei}^{(k+1)T} R_{ii} u_{ei}^{k+1} + u_{ei}^{kT} R_{ii} u_{ei}^k \\
&\quad + \sum_{j\in N_i} (u_{ej}^{k+1} - u_{ej}^k)^T R_{ij}(u_{ej}^{k+1} - u_{ej}^k) - 2\sum_{j\in N_i} (u_{ej}^{k+1} - u_{ej}^k) R_{ij} u_{ej}^{k+1} \\
&= (u_{ei}^{k+1} - u_{ei}^k)^T R_{ii}(u_{ei}^{k+1} - u_{ei}^k) + \sum_{j\in N_i} (u_{ej}^{k+1} - u_{ej}^k)^T \\
&\quad R_{ij}(u_{ej}^{k+1} - u_{ej}^k) - 2\sum_{j\in N_i} (u_{ej}^{k+1} - u_{ej}^k) R_{ij} u_{ej}^{k+1}.
\end{aligned} \tag{43}$$

A sufficient condition for $\dot{V}_i^{k+1} - \dot{V}_i^k \geq 0$ is

$$\begin{aligned}
\Delta u_{ei}^{kT} R_{ii} \Delta u_{ei}^k + \sum_{j\in N_i} \Delta u_{ej}^{kT} R_{ij} \Delta u_{ej}^k &\geq 2\sum_{j\in N_i} \Delta u_{ej}^{kT} R_{ij} u_{ej}^{k+1} \\
&= -\sum_{j\in N_i} (d_j + \sigma_j) \Delta u_{ej}^{kT} R_{ij} R_{jj}^{-1} B_j \frac{\partial V_j^k}{\partial \delta_j},
\end{aligned} \tag{44}$$

where $\Delta u_{ei} = u_{ei}^{k+1} - u_{ei}^{k}, i \in \{i, N_i\}$. Let $\varsigma(R_{ij})$ be the maximum singular value of R_{ij}. We can see that for all i, if $\varsigma(R_{ij}R_{jj}^{-1})$, ρ_{ij}, $j \in N_i$ and σ_i are small, then inequality (44) holds, which means $\dot{V}_i^{k+1} \geq \dot{V}_i^{k}$. By integrating $\dot{V}_i^{k+1} \geq \dot{V}_i^{k}$, we can obtain $V_i^{k+1} \leq V_i^{k}$. Hence, the iterative value function V_i^k is monotonically nonincreasing and lower bounded. As such, V_i^k is convergent as $k \to \infty$,—that is, $\lim_{k\to\infty} V_i^k = V_i^{\infty}$.

It is obvious that $V_i^{\infty} \geq J^{\star}$. On the other hand, let $\{u_{e1}^{\ell}, u_{e2}^{\ell}, \ldots, u_{eN}^{\ell}\}$ be a set of arbitrary admissible coordination control laws. For all i, there must exist a value function V_i^{ℓ} that satisfies

$$H_i\left(z_i, \frac{\partial V_i^{\ell}}{\partial z_i}, u_{ei}^{\ell}, u_{-ei}^{\ell}\right) = 0. \tag{45}$$

Let $u_{ei}^{\ell+1} = \arg\min_{u_{ei}} H_i\left(z_i, \frac{\partial V_i^{\ell}}{\partial z_i}, u_{ei}, u_{-ei}^{\ell}\right)$ and then we have $V_i^{\infty} \leq V_i^{\ell+1} \leq V_i^{\ell}$. As the set $\{u_{e1}^{\ell}, u_{e2}^{\ell}, \ldots, u_{eN}^{\ell}\}$ is arbitrary, we can let

$$\{u_{e1}^{\ell}, u_{e2}^{\ell}, \ldots, u_{eN}^{\ell}\} = \left\{u_{e1}^{\star}, u_{e2}^{\star}, \ldots, u_{eN}^{\star}\right\} \tag{46}$$

and then we obtain $V_i^{\infty} \leq J^{\star}$. Therefore $\lim_{k\to\infty} V_i^k = J_i^{\star}$. The proof is completed. □

Remark 3. In [51] it is shown for multiagent system $\dot{x}_i = Ax_i + B_i u_i$ that if the edge weights ρ_{ij} and $\varsigma(R_{ij}R_{jj}^{-1})$ are small, then the iterative value function converges to the optimum. From Theorem 3, we see that for heterogeneous multiagent system (1) if $\varsigma(R_{ij}R_{jj}^{-1})$ is small, then iterative value function V_i^k will also converge to the optimum as $k \to \infty$, while the constraint for ρ_{ij} in [51] is avoided. This is an advantage of the heterogeneous multiagent policy iteration algorithm developed in this chapter.

4.3 Heterogeneous Multiagent Policy Iteration Algorithm

On the basis of the previous preparation, we summarize the heterogeneous multiagent policy iteration algorithm in Algorithm 9.1.

Algorithm 9.1 Heterogeneous Multiagent Policy Iteration Algorithm

Initialization:

Choose an admissible coordination control law u_i^0, $\forall\, i \in \mathcal{I}$.
Choose a computation precision ε.
Choose a constant $0 < \zeta < 1$.
Choose positive definite matrices Q_{ii}, R_{ii}, and R_{ij}. Obtain the desired control u_{di}, $\forall\, i \in \mathcal{I}$ by $u_{di} = B_i^{+}(A_0 - A_i)x_0$.
Transform the agent into Eq. (10)—that is, $\dot{z}_i = \bar{A}_i z_i + \bar{B}_i \bar{u}_{ei}$.

Iteration:

1: Construct the utility function $U_i(z_i, u_{ei}, u_{-ei})$ by Eq. (13).
2: Let the iteration index $k = 0$. Construct an iterative value function V_i^0 to satisfy Eq. (30).
3: For $i \in \mathcal{I}$, let $k = k + 1$. Do **policy improvement**

$$u_{ei}^k = \arg\min_{u_{ei}} H_i\left(z_i, \frac{\partial V_i^{k-1}}{\partial z_i}, u_{ei}, u_{-ei}^{k-1}\right).$$

4: Do **policy evaluation**

$$H_i\left(z_i, \frac{\partial V_i^k}{\partial z_i}, u_{ei}^k, u_{-ei}^k\right) = 0.$$

5: If $V_i^k \leq V_i^{k-1}$, go to the next step. Else let $R_{ij} = \zeta R_{ij}$, $j \in N_i$ and go to step 2.
6: If $V_i^{k-1} - V_i^k \leq \varepsilon$, then the optimal value function and optimal control law are obtained. Go to step 7. Else go to step 3.
7: **return** u_{ei}^k and V_i^k. Let $u_i^{\star} = u_{ei}^k + u_{di}$.

5 SIMULATION STUDY

To evaluate the performance of our heterogeneous multiagent policy iteration algorithm, two examples are given for numerical experiments to solve the optimal distributed synchronization control problems.

Example 1. Our first example is chosen as a multiagent system in [51] with modifications. The structure of the five-node digraph is shown in Fig. 1 with a leader node connected to node 3. The edge weights are chosen as $\rho_{13} = 0.1$, $\rho_{14} = 0.1$, $\rho_{21} = 0.5$, $\rho_{31} = 0.4$, $\rho_{32} = 0.3$,

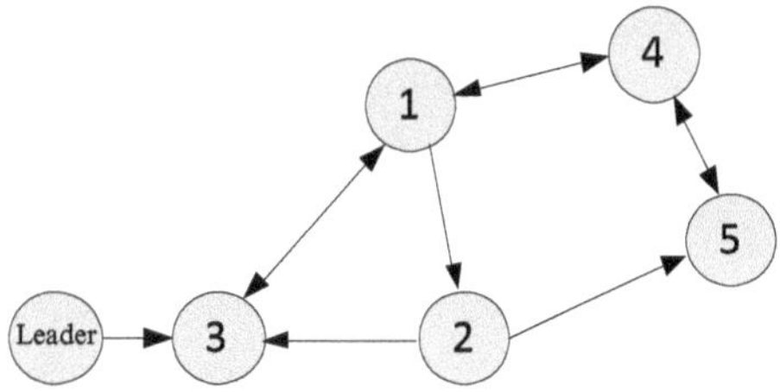

Fig. 1 The structure of the five-node digraph.

$\rho_{41} = 0.2$, $\rho_{45} = 0.1$, $\rho_{52} = 0.8$, and $\rho_{54} = 0.7$, respectively. The pinning gains are taken as $\sigma_1 = 0.5$, $\sigma_2 = 0.3$, $\sigma_3 = 0.2$, $\sigma_4 = 0.1$, and $\sigma_5 = 0.1$, respectively.

For the structure in Fig. 1, the dynamics of the nodes is considered as

Node 1:

$$\begin{bmatrix} \dot{x}_{11} \\ \dot{x}_{12} \end{bmatrix} = \begin{bmatrix} 0 & 1 \\ -1 & 0 \end{bmatrix} \begin{bmatrix} x_{11} \\ x_{12} \end{bmatrix} + \begin{bmatrix} 2 & 0 \\ 0 & 1 \end{bmatrix} \begin{bmatrix} u_{11} \\ u_{12} \end{bmatrix}, \tag{47}$$

Node 2:

$$\begin{bmatrix} \dot{x}_{21} \\ \dot{x}_{22} \end{bmatrix} = \begin{bmatrix} 0 & 1 \\ -1 & 0 \end{bmatrix} \begin{bmatrix} x_{21} \\ x_{22} \end{bmatrix} + \begin{bmatrix} 2 & 0 \\ 0 & 3 \end{bmatrix} \begin{bmatrix} u_{21} \\ u_{22} \end{bmatrix}, \tag{48}$$

Node 3:

$$\begin{bmatrix} \dot{x}_{31} \\ \dot{x}_{32} \end{bmatrix} = \begin{bmatrix} 0 & 1 \\ -1 & 0 \end{bmatrix} \begin{bmatrix} x_{31} \\ x_{32} \end{bmatrix} + \begin{bmatrix} 2 & 0 \\ 0 & 2 \end{bmatrix} \begin{bmatrix} u_{31} \\ u_{32} \end{bmatrix}, \tag{49}$$

Node 4:

$$\begin{bmatrix} \dot{x}_{41} \\ \dot{x}_{42} \end{bmatrix} = \begin{bmatrix} 0 & 1 \\ -1 & 0 \end{bmatrix} \begin{bmatrix} x_{41} \\ x_{42} \end{bmatrix} + \begin{bmatrix} 1 & 0 \\ 0 & 1 \end{bmatrix} \begin{bmatrix} u_{41} \\ u_{42} \end{bmatrix}, \tag{50}$$

Node 5:

$$\begin{bmatrix} \dot{x}_{51} \\ \dot{x}_{52} \end{bmatrix} = \begin{bmatrix} 0 & 1 \\ -1 & 0 \end{bmatrix} \begin{bmatrix} x_{51} \\ x_{52} \end{bmatrix} + \begin{bmatrix} 3 & 0 \\ 0 & 2 \end{bmatrix} \begin{bmatrix} u_{51} \\ u_{52} \end{bmatrix}. \tag{51}$$

Let the desired dynamics be expressed as

$$\begin{bmatrix} \dot{x}_{01} \\ \dot{x}_{02} \end{bmatrix} = \begin{bmatrix} 0 & 1 \\ -1 & 0 \end{bmatrix} \begin{bmatrix} x_{01} \\ x_{02} \end{bmatrix}. \tag{52}$$

Define the utility function as in Eq. (13), where the weight matrices are expressed as

$$\begin{aligned} &Q_{11} = Q_{22} = Q_{33} = Q_{44} = Q_{55} = I, \\ &R_{11} = 4I, \quad R_{12} = I, \quad R_{13} = I, \quad R_{14} = I, \\ &R_{21} = I, \quad R_{22} = 4I, \\ &R_{31} = I, \quad R_{32} = I, \quad R_{33} = 5I, \\ &R_{41} = I, \quad R_{44} = 9I, \quad R_{45} = I, \\ &R_{52} = I, \quad R_{54} = I, \quad R_{55} = 9I, \end{aligned} \tag{53}$$

where I denotes the identity matrix with suitable dimension. Let the initial states of each agent in the game be

$$\begin{aligned} &x_1(0) = \begin{bmatrix} 1 \\ -1 \end{bmatrix}, \quad x_2(0) = \begin{bmatrix} -1 \\ 1 \end{bmatrix}, \quad x_3(0) = \begin{bmatrix} 1 \\ 1 \end{bmatrix}, \\ &x_4(0) = \begin{bmatrix} 2 \\ -2 \end{bmatrix}, \quad x_5(0) = \begin{bmatrix} -2 \\ 2 \end{bmatrix}, \quad x_0(0) = \begin{bmatrix} 0.5 \\ -0.5 \end{bmatrix}. \end{aligned} \tag{54}$$

Define $x_{ei} = x_i - x_0$, $i = 1, 2, 3, 4, 5$. Implement Algorithm 9.1 for $k = 20$ iterations. The tracking errors of every agent for nodes 1–5 are shown in Figs. 2 and 3A. The control errors u_{ei}, $i = 1, 2, \ldots, 5$ are shown in Figs. 3B–D and 4.

From the multiagent system given by Eqs. (47)–(51), we can see that the autonomous systems in each node are the same, that is, $A_i = A_j$, $i \neq j$, $i, j = 1, 2, \ldots, 5$. Furthermore, all the autonomous systems in each node equal the desired dynamics, that is, $A_i = A_0$, $\forall i = 1, 2, \ldots, 5$. This makes the desired control $u_{di} \equiv 0$. In this situation the multiagent system is reduced to the one in [51]. From the simulation results in Figs. 2–4, we can see that after 20 iterations the iterative states and iterative control laws are convergent to the optimum. For implementation of the achieved optimal control laws to the multiagent system for 300 s, the optimal state trajectories are shown in Fig. 5 and the corresponding optimal control trajectories for each node are shown in Fig. 6.

From Figs. 5 and 6 we can see that with use of the heterogeneous multiagent policy iteration algorithm developed, optimal control can be obtained to make the system states in each node track the desired state trajectories. Hence, we can say that the graphical game in [51] is a

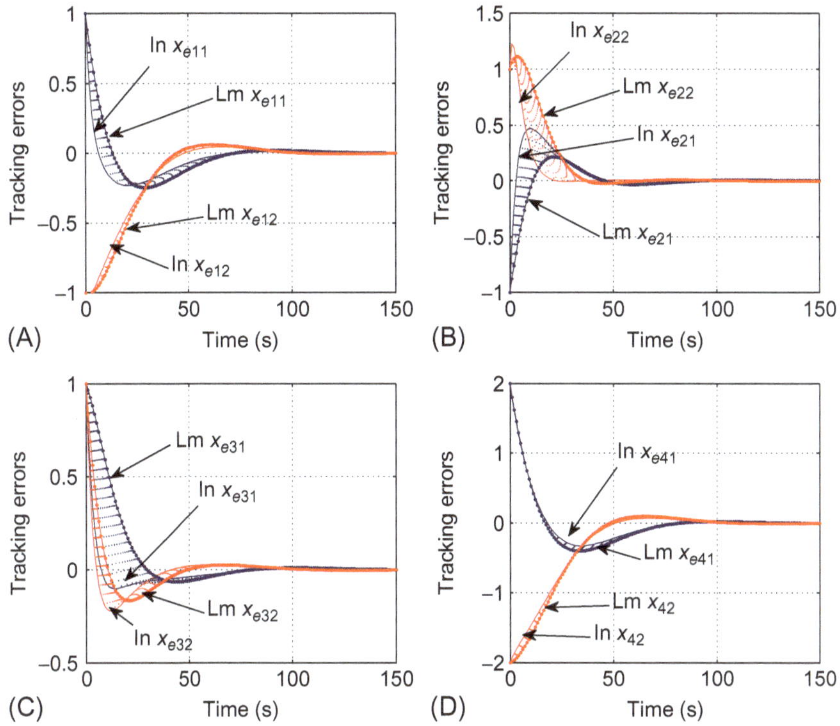

Fig. 2 Tracking errors x_{ei}, $i = 1, 2, 3, 4$: (A) tracking error x_{e1}; (B) tracking error x_{e2}; (C) tracking error x_{e3}; (D) tracking error x_{e4}.

special case of the heterogeneous multiagent differential graphical game in this chapter and the effectiveness of the heterogeneous multiagent policy iteration algorithm can be verified. On the other hand, if $A_i \neq A_j$, for $i \neq j$, $\forall i, j = 1, 2, \ldots, 5$, then the method for the graphical game in [51] is invalid. In the next example we will show that for the situation that $A_i \neq A_j$, for $i \neq j$, $\forall i, j = 1, 2, \ldots, 5$, the heterogeneous multiagent policy iteration algorithm developed is also effective to obtain the optimal distributed synchronization control law.

Example 2. A heterogeneous multiagent differential graphical game is considered. Let the structure of the five-node digraph be as shown in Fig. 1, which is the same as for Example 1. For the structure in Fig. 1, each node dynamics is expressed by

Node 1:

$$\begin{bmatrix} \dot{x}_{11} \\ \dot{x}_{12} \end{bmatrix} = \begin{bmatrix} -1 & 1 \\ -3 & -1 \end{bmatrix} \begin{bmatrix} x_{11} \\ x_{12} \end{bmatrix} + \begin{bmatrix} 2 & 0 \\ 0 & 1 \end{bmatrix} \begin{bmatrix} u_{11} \\ u_{12} \end{bmatrix}, \tag{55}$$

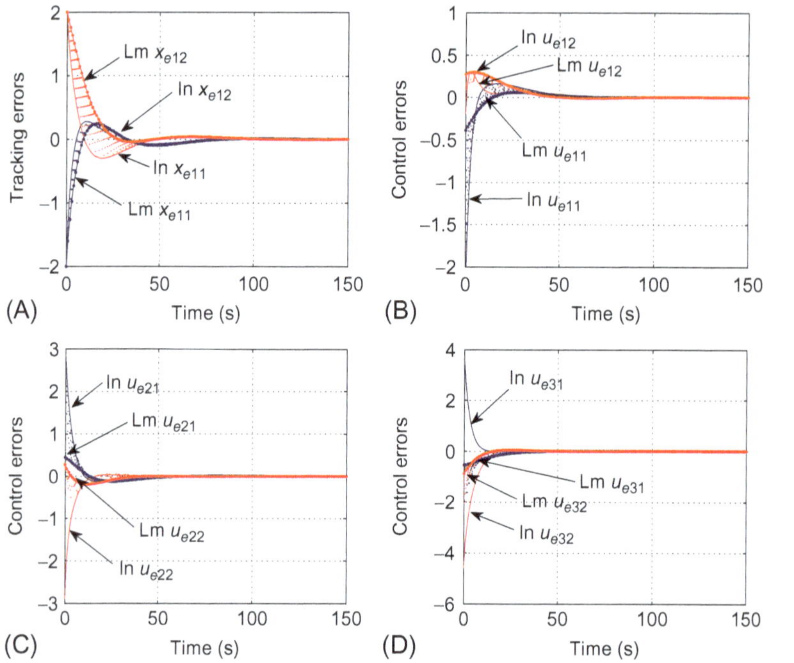

Fig. 3 Tracking error x_{e5} and the control errors u_{ei}, $i = 1, 2, 3$: (A) tracking errors x_{e5}; (B) control error u_{e1}; (C) control error u_{e2}; (D) control error u_{e3}.

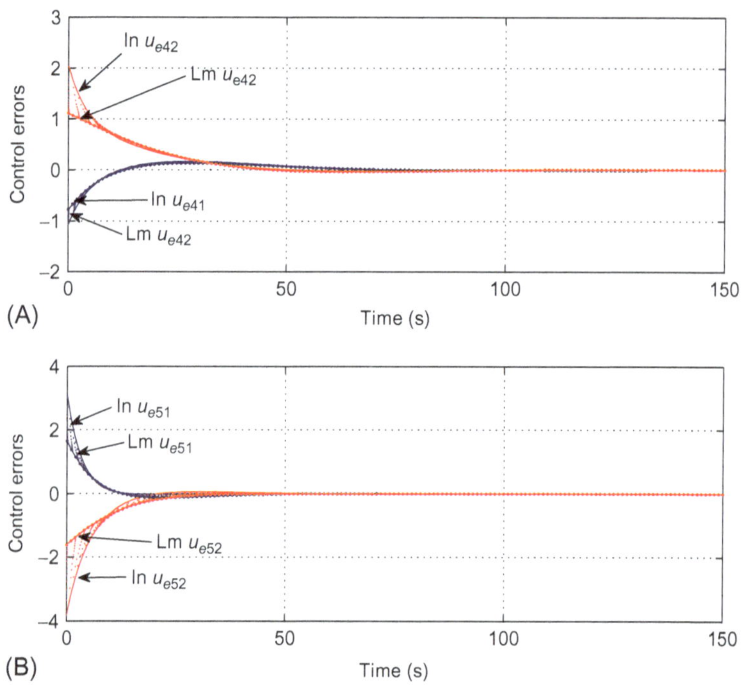

Fig. 4 Control errors u_{ei}, $i = 4, 5$: (A) control error u_{e4}; (B) control error u_{e5}.

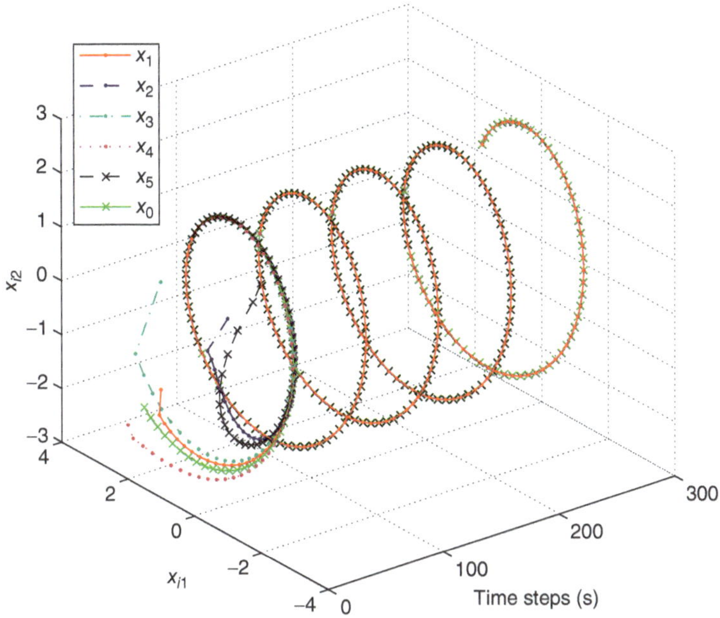

Fig. 5 The evolution of the agent states.

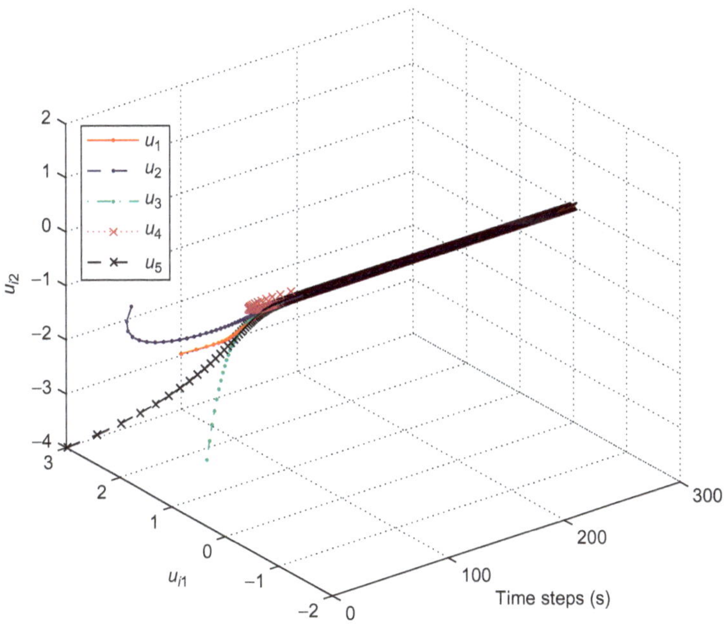

Fig. 6 The evolution of the agent controls.

Node 2:

$$\begin{bmatrix} \dot{x}_{21} \\ \dot{x}_{22} \end{bmatrix} = \begin{bmatrix} -1 & 1 \\ -4 & -1 \end{bmatrix} \begin{bmatrix} x_{21} \\ x_{22} \end{bmatrix} + \begin{bmatrix} 2 & 0 \\ 0 & 3 \end{bmatrix} \begin{bmatrix} u_{21} \\ u_{22} \end{bmatrix}, \tag{56}$$

Node 3:

$$\begin{bmatrix} \dot{x}_{31} \\ \dot{x}_{32} \end{bmatrix} = \begin{bmatrix} -2 & 1 \\ -1 & -1 \end{bmatrix} \begin{bmatrix} x_{31} \\ x_{32} \end{bmatrix} + \begin{bmatrix} 2 & 0 \\ 0 & 2 \end{bmatrix} \begin{bmatrix} u_{31} \\ u_{32} \end{bmatrix}, \tag{57}$$

Node 4:

$$\begin{bmatrix} \dot{x}_{41} \\ \dot{x}_{42} \end{bmatrix} = \begin{bmatrix} -1 & 1 \\ -2 & -1 \end{bmatrix} \begin{bmatrix} x_{41} \\ x_{42} \end{bmatrix} + \begin{bmatrix} 1 & 0 \\ 0 & 1 \end{bmatrix} \begin{bmatrix} u_{41} \\ u_{42} \end{bmatrix}, \tag{58}$$

Node 5:

$$\begin{bmatrix} \dot{x}_{51} \\ \dot{x}_{52} \end{bmatrix} = \begin{bmatrix} -2 & 1 \\ -3 & -1 \end{bmatrix} \begin{bmatrix} x_{51} \\ x_{52} \end{bmatrix} + \begin{bmatrix} 3 & 0 \\ 0 & 2 \end{bmatrix} \begin{bmatrix} u_{51} \\ u_{52} \end{bmatrix}. \tag{59}$$

Let the desired dynamics be expressed as Eq. (52). Define the utility function as in Eq. (53) and let the initial states of each agent in the games be the same as the ones in Eq. (54). The heterogeneous multiagent policy iteration algorithm (Algorithm 9.1) is implemented for $k = 20$ iterations. The tracking errors of every agent for nodes 1–5 are shown in Figs. 7 and 8A. The control errors u_{ei}, $i = 1, 2, \ldots, 5$ are shown in Figs. 8B–D and 9.

From Eqs. (55)–(59), we can see that the autonomous systems in each node are the not same, that is, $A_i \neq A_j$, $i \neq j$, $i, j = 1, 2, \ldots, 5$. Moreover, the autonomous systems in each node does not equal the desired dynamics, that is, $A_i \neq A_0$, $\forall i = 1, 2, \ldots, 5$. This makes the policy iteration algorithm in [51] invalid. From Figs. 7–9, we can see that with use of the heterogeneous multiagent policy iteration algorithm developed in this chapter, the optimal control law can be obtained. We emphasize that this is an obvious advantage of the heterogeneous multiagent policy iteration algorithm developed in this chapter compared with the algorithm in [51].

From Figs. 7–9, we can see that after 20 iterations the iterative states and iterative control laws are convergent to the optimum. For implementation of the achieved optimal control laws with each agent for 300 s, the state trajectories are shown in Fig. 10 and the corresponding optimal control trajectories for each node are shown in Fig. 11. From Figs. 10 and 11, we can see that the states for each node are convergent to the desired trajectory x_0. In [51], as $A_i = A_0$, $\forall i = 1, 2, \ldots, 5$, the optimal control

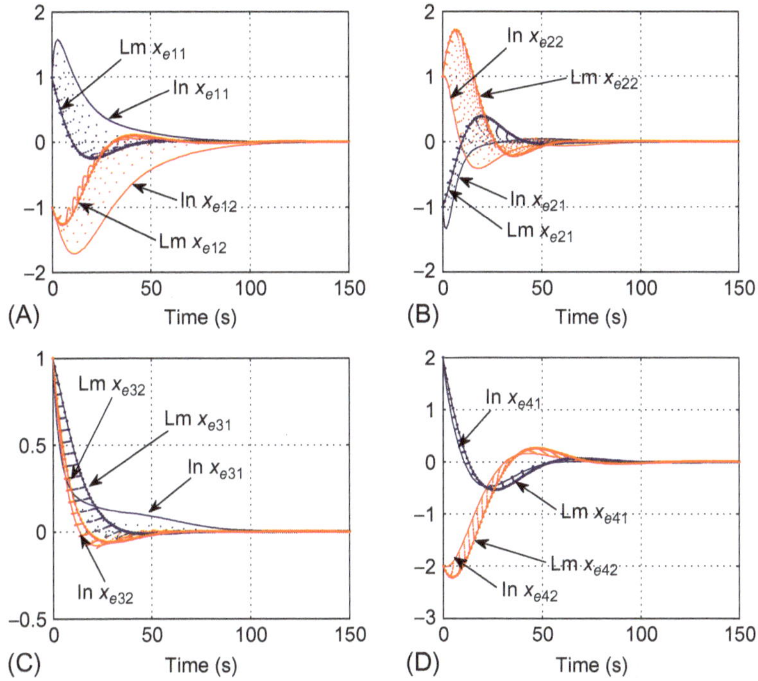

Fig. 7 Tracking errors x_{ei}, $i = 1, 2, 3, 4$: (A) tracking error x_{e1}; (B) tracking error x_{e2}; (C) tracking error x_{e3}; (D) tracking error x_{e4}.

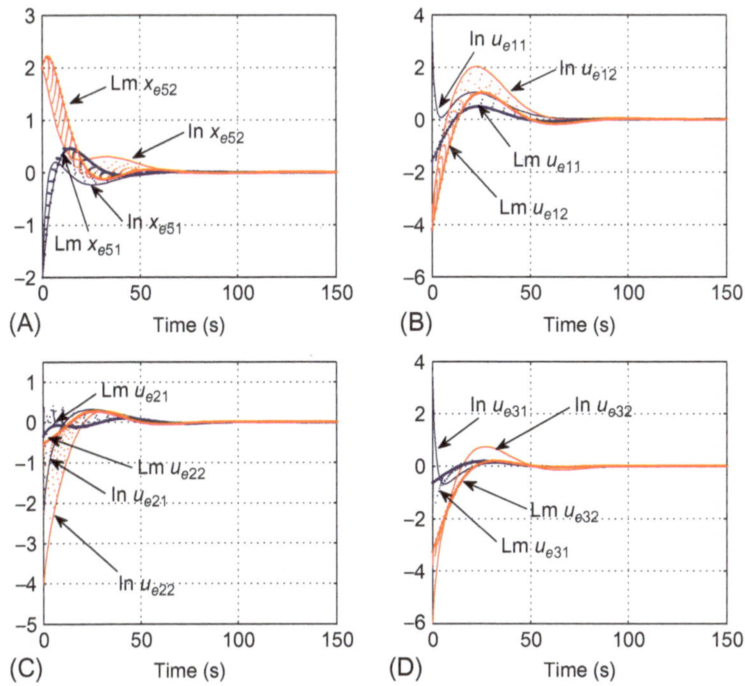

Fig. 8 Tracking error x_{e5} and the control errors u_{ei}, $i = 1, 2, 3$: (A) tracking errors x_{e5}; (B) control error u_{e1}; (C) control error u_{e2}; (D) control error u_{e3}.

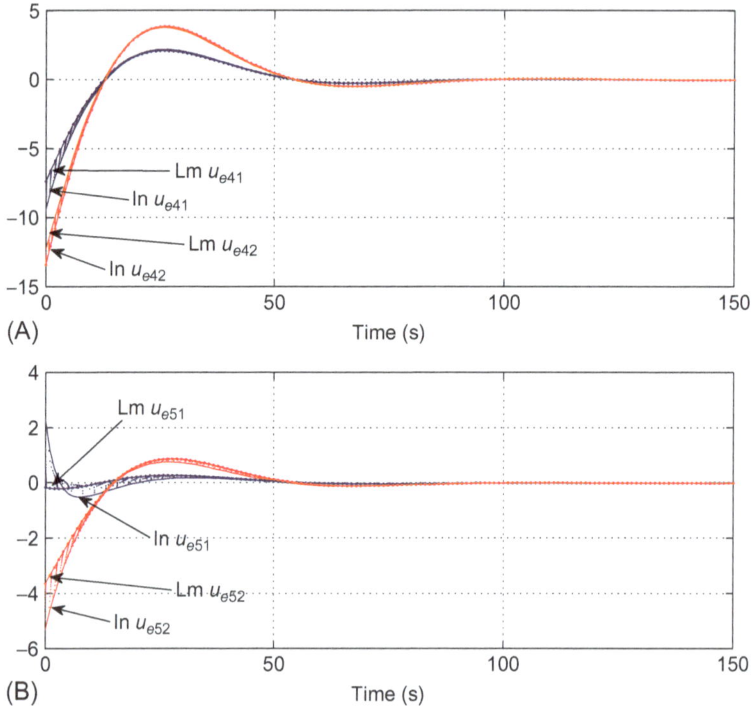

Fig. 9 Control errors u_{ei}, $i = 4, 5$: (A) control error u_{e4}; (B) control error u_{e5}.

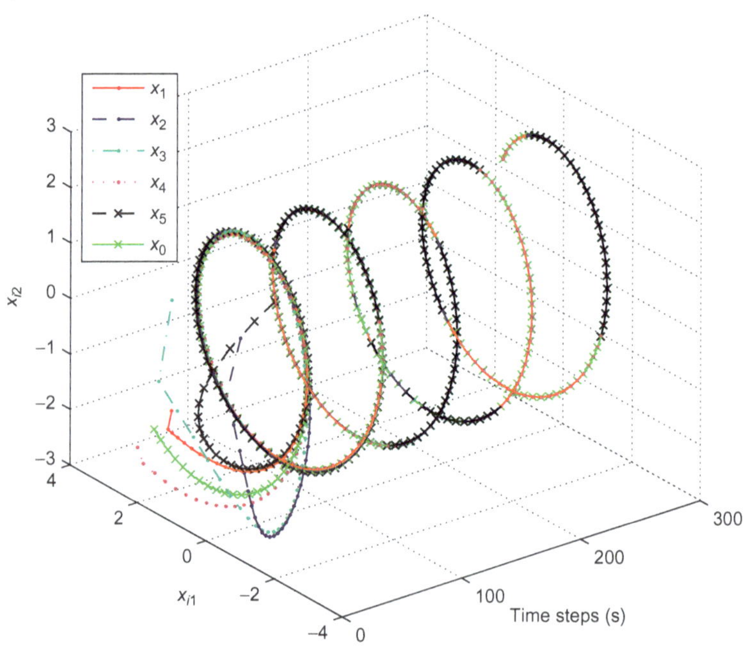

Fig. 10 The evolution of the agent states.

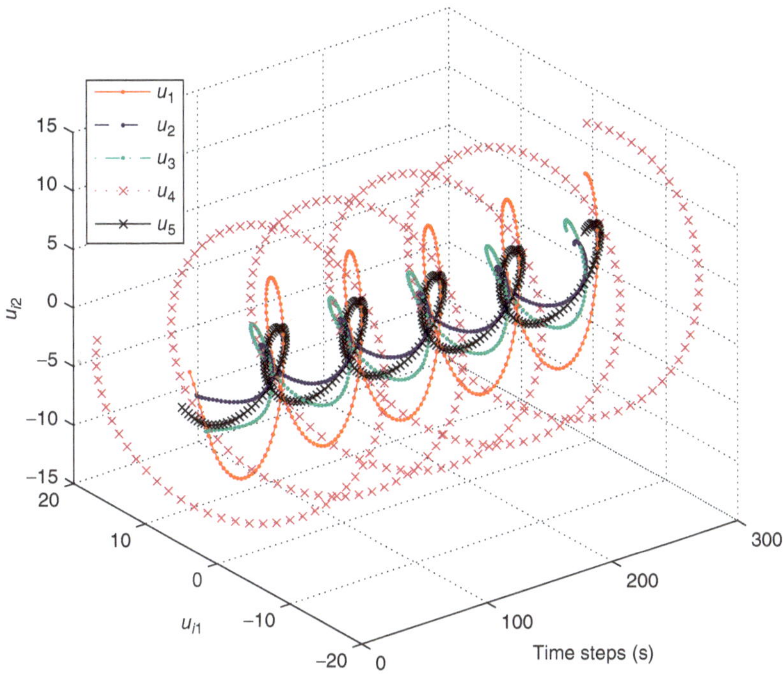

Fig. 11 The evolution of the agent controls.

converges to zero, as the implementation time increases to infinity. For the heterogeneous multiagent system in this chapter, from Fig. 11 we can see that the control for each node is not zero and the achieved optimal control can make the system states in each node track the desired state trajectories. Hence, the effectiveness of the heterogeneous multiagent policy iteration algorithm developed can be verified.

6 CONCLUSION

An effective heterogeneous policy iteration-based ADP algorithm was developed to solve the optimal distributed coordination control problems for heterogeneous multiagent differential graphical games. The heterogeneous differential graphical games developed permit the agent dynamics of each node to be different from that of other nodes. A heterogeneous multiagent policy iteration algorithm was developed to achieve the optimal distributed coordination control law for the agent for each node to guarantee that the dynamics of each node can track the desired one. Convergence analysis was performed to show that the iterative value functions of heterogeneous multiagent differential graphical games can converge to the Nash equilibrium. Finally, simulation results show the effectiveness of the optimal control scheme developed.

REFERENCES

[1] H.L. Trentelman, K. Takaba, N. Monshizadeh, Robust synchronization of uncertain linear multi-agent systems, IEEE Trans. Autom. Control 58 (2013) 1511–1523.

[2] W. Wang, Carrier frequency synchronization in distributed wireless sensor networks, IEEE Syst. J. 9 (3) (2015) 703–713.

[3] Y. Tang, H. Gao, W. Zou, J. Kurths, Distributed synchronization in networks of agent systems with nonlinearities and random switchings, IEEE Trans. Cybern. 43 (2013) 358–370.

[4] X. Su, L. Wu, P. Shi, Y. Song, A novel approach to output feedback control of fuzzy stochastic systems, Automatica 50 (2014) 3268–3275.

[5] M. Abu-Khalaf, F.L. Lewis, J. Huang, Policy iterations on the Hamilton-Jacobi-Isaacs equation for state feedback control with input saturation, IEEE Trans. Autom. Control 51 (2006) 1989–1995.

[6] M. Abu-Khalaf, F.L. Lewis, J. Huang, Neurodynamic programming and zero-sum games for constrained control systems, IEEE Trans. Neural Netw. 19 (2008) 1243–1252.

[7] T. Basar, P. Bernhard, H_∞ Optimal Control and Related Minimax Design Problems, Birkhäuser Press, Boston, MA, 1995.

[8] P.J. Werbos, Gen. Syst. Yearb. 22 (1977) 25–38.

[9] P.J. Werbos, A menu of designs for reinforcement learning over time, in: W.T. Miller, R.S. Sutton, P.J. Werbos (Eds.), Neural Networks for Control, MIT Press, Cambridge, MA, 1990, pp. 67–95.

[10] H. Zhang, Q. Wei, D. Liu, An iterative adaptive dynamic programming method for solving a class of nonlinear zero-sum differential games, Automatica 47 (2011) 207–214.

[11] D. Liu, Y. Zhang, H. Zhang, A self-learning call admission control scheme for CDMA cellular networks, IEEE Trans. Neural Netw. 16 (2005) 1219–1228.

[12] A. Heydari, S.N. Balakrishnan, Finite-horizon control-constrained nonlinear optimal control using single network adaptive critics, IEEE Trans. Neural Netw. Learn. Syst. 24 (2013) 145–157.

[13] Z. Ni, H. He, J. Wen, Adaptive learning in tracking control based on the dual critic network design, IEEE Trans. Neural Netw. Learn. Syst. 24 (2013) 913–928.

[14] H. Zhang, Q. Wei, Y. Luo, A novel infinite-time optimal tracking control scheme for a class of discrete-time nonlinear systems via the greedy HDP iteration algorithm, IEEE Trans. Syst. Man Cybern. B: Cybern. 38 (2008) 937–942.

[15] D. Liu, D. Wang, X. Yang, An iterative adaptive dynamic programming algorithm for optimal control of unknown discrete-time nonlinear systems with constrained inputs, Inform. Sci. 220 (2013) 331–342.

[16] Q. Wei, D. Liu, An iterative ε-optimal control scheme for a class of discrete-time nonlinear systems with unfixed initial state, Neural Netw. 32 (2012) 236–244.

[17] Q. Wei, D. Liu, A novel iterative θ-adaptive dynamic programming for discrete-time nonlinear systems, IEEE Trans. Autom. Sci. Eng. 11 (2014) 1176–1190.

[18] Q. Wei, D. Liu, Adaptive dynamic programming for optimal tracking control of unknown nonlinear systems with application to coal gasification, IEEE Trans. Autom. Sci. Eng. 11 (2014) 1020–1036.

[19] D.V. Prokhorov, D.C. Wunsch, Adaptive critic designs, IEEE Trans. Neural Netw. 8 (1997) 997–1007.
[20] D. Molina, G.K. Venayagamoorthy, J. Liang, R.G. Harley, Intelligent local area signals based damping of power system oscillations using virtual generators and approximate dynamic programming, IEEE Trans. Smart Grid 4 (2013) 498–508.
[21] D.P. Bertsekas, J.N. Tsitsiklis, Neuro-Dynamic Programming, Athena Scientific, Belmont, MA, 1996.
[22] R. Enns, J. Si, Helicopter trimming and tracking control using direct neural dynamic programming, IEEE Trans. Neural Netw. 14 (2003) 929–939.
[23] B. Kiumarsi, F.L. Lewis, H. Modares, A. Karimpour, M.B. Naghibi-Sistani, Reinforcement-learning for optimal tracking control of linear discrete-time systems with unknown dynamics, Automatica 50 (2014) 1167–1175.
[24] F.L. Lewis, D. Vrabie, K.G. Vamvoudakis, Reinforcement learning and feedback control: using natural decision methods to design optimal adaptive controllers, IEEE Control Syst. 32 (2012) 76–105.
[25] M. Aurangzeb, F.L. Lewis, Internal structure of coalitions in competitive and altruistic graphical coalitional games, Automatica 50 (2014) 335–348.
[26] D. Liu, Q. Wei, Finite-approximation-error-based optimal control approach for discrete-time nonlinear systems, IEEE Trans. Cybern. 43 (2013) 779–789.
[27] Q. Wei, D. Liu, Numerical adaptive learning control scheme for discrete-time nonlinear systems, IET Control Theory Appl. 7 (2013) 1472–1486.
[28] Q. Wei, D. Liu, Neural-network-based adaptive optimal tracking control scheme for discrete-time nonlinear systems with approximation errors, Neurocomputing 149 (2015) 106–115.
[29] Q. Wei, D. Liu, Stable iterative adaptive dynamic programming algorithm with approximation errors for discrete-time nonlinear systems, Neural Comput. Appl. 24 (2014) 1355–1367.
[30] Q. Wei, D. Wang, D. Zhang, Dual iterative adaptive dynamic programming for a class of discrete-time nonlinear systems with time-delays, Neural Comput. Appl. 23 (2013) 1851–1863.
[31] R. Song, F.L. Lewis, Q. Wei, H. Zhang, Z.P. Jiang, D. Levine, Multiple actor-critic structures for continuous-time optimal control using input-output data, IEEE Trans. Neural Netw. Learn. Syst. 26 (2015) 851–865.
[32] A. Al-Tamimi, F.L. Lewis, M. Abu-Khalaf, Discrete-time nonlinear HJB solution using approximate dynamic programming: convergence proof, IEEE Trans. Syst. Man Cybern. B: Cybern. 38 (2008) 943–949.
[33] Q. Wei, F. Wang, D. Liu, X. Yang, Finite-approximation-error based discrete-time iterative adaptive dynamic programming, IEEE Trans. Cybern. 44 (2014) 2820–2833.
[34] A. Al-Tamimi, M. Abu-Khalaf, F.L. Lewis, Adaptive critic designs for discrete-time zero-sum games with application to H-infinity control, IEEE Trans. Syst. Man Cybern. B: Cybern. 37 (2007) 240–247.
[35] A. Al-Tamimi, F.L. Lewis, M. Abu-Khalaf, Model-free Q-learning designs for linear discrete-time zero-sum games with application to H-infinity control, Automatica 43 (2007) 473–481.
[36] Q. Wei, D. Liu, Data-driven neuro-optimal temperature control of water gas shift reaction using stable iterative adaptive dynamic programming, IEEE Trans. Ind. Electron. 61 (2014) 6399–6408.
[37] Q. Wei, D. Liu, G. Shi, A novel dual iterative Q-learning method for optimal battery management in smart residential environments, IEEE Trans. Ind. Electron. 62 (2015) 2509–2518.
[38] Q. Wei, D. Liu, G. Shi, Y. Liu, Optimal multi-battery coordination control for home energy management systems via distributed iterative adaptive dynamic programming, IEEE Trans. Ind. Electron. 42 (2015) 4203–4214.

[39] H. Modares, F.L. Lewis, Optimal tracking control of nonlinear partially-unknown constrained-input systems using integral reinforcement learning, Automatica 50 (2014) 1780–1792.
[40] H. Modares, F.L. Lewis, M.B. Naghibi-Sistani, Integral reinforcement learning and experience replay for adaptive optimal control of partially-unknown constrained-input continuous-time systems, Automatica 50 (2014) 193–202.
[41] Y. Jiang, Z. Jiang, Robust adaptive dynamic programming and feedback stabilization of nonlinear systems, IEEE Trans. Neural Netw. Learn. Syst. 25 (2014) 882–893.
[42] M. Abu-Khalaf, F.L. Lewis, Nearly optimal control laws for nonlinear systems with saturating actuators using a neural network HJB approach, Automatica 41 (2005) 779–791.
[43] J.J. Murray, C.J. Cox, G.G. Lendaris, R. Saeks, Adaptive dynamic programming, IEEE Trans. Syst. Man Cybern. C: Appl. Rev. 32 (2002) 140–153.
[44] R. Song, W. Xiao, H. Zhang, C. Sun, Adaptive dynamic programming for a class of complex-valued nonlinear systems, IEEE Trans. Neural Netw. Learn. Syst. 25 (2014) 1733–1739.
[45] D. Liu, Q. Wei, Policy iteration adaptive dynamic programming algorithm for discrete-time nonlinear systems, IEEE Trans. Neural Netw. Learn. Syst. 25 (2014) 621–634.
[46] Q. Wei, D. Liu, X. Yang, Infinite horizon self-learning optimal control of nonaffine discrete-time nonlinear systems, IEEE Trans. Neural Netw. Learn. Syst. 26 (2015) 866–879.
[47] D. Liu, Q. Wei, Multi-person zero-sum differential games for a class of uncertain nonlinear systems, Int. J. Adapt. Control Signal Process. 28 (2014) 205–231.
[48] D. Vrabie, K.G. Vamvoudakis, F.L. Lewis, Optimal Adaptive Control and Differential Games by Reinforcement Learning Principles, The Institution of Engineering and Technology, London, United Kingdom, 2013.
[49] K.G. Vamvoudakis, F.L. Lewis, Multi-player non-zero-sum games: online adaptive learning solution of coupled Hamilton-Jacobi equations, Automatica 47 (2011) 1556–1569.
[50] H. Zhang, L. Cui, Y. Luo, Near-optimal control for nonzero-sum differential games of continuous-time nonlinear systems using single-network ADP, IEEE Trans. Cybern. 43 (2013) 206–216.
[51] K.G. Vamvoudakis, F.L. Lewis, G.R. Hudas, Multi-agent differential graphical games: online adaptive learning solution for synchronization with optimality, Automatica 48 (2012) 1598–1611.

CHAPTER TEN

Model-Free Learning of Nash Games With Applications to Network Security

K.G. Vamvoudakis
Virginia Tech, VA, United States

1 INTRODUCTION

Cyber-physical systems usually comprise a network of physically distributed embedded sensors and actuators equipped with computing and communicating capabilities. Although each individual device is fairly inept at monitoring or regulating the physical substratum, the coordinated action of the individual network nodes has the potential for unprecedented capabilities. Several military and civilian applications, including sensor networks and mobile robots, can be described as networked teams that are concerned with reaching an agreement on a common value by exchanging only information available in the neighborhood over communication channels. Consensus protocols became popular after Tsitsiklis's Ph.D. thesis [1] and since then have been fundamental for multiagent coordination [2–5]—for example, time synchronization, rendezvous, formation control, and distributed estimation. But all wireless communications suffer from the problems of interference, collision, or noise at the receiver side or from attack. To understand this problem, one needs to think that the devices of contemporary local wireless networks and microwave ovens operate on the same frequency. Moreover, wireless networks also suffer interference from other sources, such as electrical high-voltage cables and generators.

Because of the highly uncertain and dynamic nature of military conflicts, enabling autonomous agents to gracefully adapt to mission and environmental changes is a very challenging task. These capabilities are necessary in the asymmetric battles waged against insurgencies, where enemy combatants quickly adapt to Army strategies and tactics. The ability

Control of Complex Systems
http://dx.doi.org/10.1016/B978-0-12-805246-4.00010-0

to synchronize activities provides an important strategic capability in a wide variety of military missions. The US Army, Air Force, and Navy have recently shown interest in the cyber security aspect of unmanned aerial vehicles and unmanned ground vehicles, which rely heavily on their onboard autopilots and controllers to function. Most of the currently available autopilot systems were built without cyber security considerations, and are thus vulnerable to cyber attacks. This chapter provides knowledge of the problem of commanding multiple tactical assets in military and adversarial environments, attending the Army's expectations that networked teams will perform in a reliable manner especially when being attacked by advanced and persistent threats.

Game theory is a very useful mathematical tool used for years in networks, initially to describe resource allocation and routing in competitive environments. The limited resources impose a conflict of interest, and every player decides in a distributed way for the strategy that will optimize his or her payoff. The users of a network are considered rational, which means that they will always try to optimize their own utility [6]. This can be interpreted in our problem as if the jammer fails, the transmitter will transmit successfully and the payoff of the jammer will be zero; on the other hand, if the jammer succeeds, the payoff of the transmitter will be zero. So it is straightforward to see that the game should be described as zero sum. This game looks similar to a pursuit-evasion game, where the game ends when the pursuer captures the evader and which has been studied extensively in the literature [7–9]. For multiple evaders or pursuers the game becomes extremely difficult because the computation is expensive and thus general theory is beginning to be developed and most of current theory assumes knowledge of each other's objective function [10]. So there is a need for efficient algorithms to solve such optimization problems.

A pursuit-evasion game between a plane and a missile by use of a minimax Q-learning algorithm was investigated in [11]. A survey of existing cyber threats in multiagent systems and models of realistic and rational adversary models and presented in [12, 13]. Consensus in the presence of persistent adversaries has been focused on detecting, identifying, and isolating the failing nodes [14]. These algorithms are computationally expensive, and mostly they use global information and specific graph connectivity. The adversaries can easily drive the system to instability and make the system operate with an undesired behavior. *Thus it is better to fight adversaries than guard against them as was done in* [31]. Most of the consensus algorithms proposed in the literature either are do not optimize some performance criteria [15] or optimize global ones and solve the problem off-line with complicated Riccati matrix equations [16, 17]. Bauso et al. [18] evaluated the cost of every agent by considering constant states for the other agents. Yucelen and Egerstedt [19] proposed a controller that suppresses the effect of constant and time-varying disturbances by using information on the agent's and neighbors' states. Zhu and Martinez [20] proposed a distributed resilient formation control algorithm that consists of a formation control block that is adapted online to minimize a local formation error function and an online learning block to collect information in real time and update the estimates of adversaries. The discussion in [21] states the attack vulnerability of the US electrical grid and how important it is to focus on the development of intelligent and robust resilient algorithms.

This chapter is motivated by emerging networking applications characterized by actuation attacks, time-varying connectivity, jamming attacks, and lack of centralized control and access to information. Although the agents in the team can be endowed with behaviors designed a priori, they often need to learn new behaviors, such as how the agent, or the whole team, will gradually improve its performance. Also, in a time-changing environment, hardwired behavior is inappropriate. The agents interact with their neighborhood environment through sensing data. The algorithms proposed have to be robust against changes in actuators and topology and have to guarantee optimal performance at all times. For this reason the cost functions are defined for the case of actuation attacks as a zero-sum game problem for each agent and a non-zero-sum game for all the agents in the team, and for the jamming attacks as a max min time problem, where there is a team of agents who want to maximize the time they are connected and a jammer who wants to minimize this time. For the latter case, to avoid using bang-bang control, which despite its simplicity is rarely used since it is very sensitive to modeling errors in the dynamics and in implementation issues, we will base our cost index on [22], which is based on time energy and precision terms and uses the solution to the Riccati equation to compute the optimal feedback gain and the optimal time. Elalami and Znaidi [23] extended the results of [22] to discrete-time systems. This zero-sum linear-quadratic optimal time game requires the solution of the Hamilton-Jacobi-Isaacs equation, which is infeasible and to the best of my knowledge there is no descent work on this in the literature with guaranteed convergence and stability. Hence approximation theory combined with learning ideas is an ideal approach to solve these two problems. The solutions assume that the value functions are differentiable. There are several stories of success in applying approximate techniques in networked control systems, transportation systems, and power systems. One of the most fundamental frameworks of learning is animal behavior-inspired actor-critic algorithms [24], which use an actor to select the control policies to improve the performance and a critic to evaluate the actor's decisions. Several online actor-critic learning techniques along with Lyapunov convergence proofs were proposed in [25].

Notation

For the ease of readers, we provide here a partial notation list which will be explained in more detail later throughout the chapter. $\mathbb{R}$ denotes the set of real numbers, $\mathbb{R}^+$ denotes the set $\{x \in \mathbb{R} : x > 0\}$, $\mathbb{R}^{n_1 \times n_2}$ is the set of $n_1 \times n_2$ real matrices, $\|.\|$ denotes the Euclidean norm, $\|.\|_{\mathcal{P}}$ denotes the weighted Euclidean norm, with $\mathcal{P}$ a matrix of appropriate dimensions, $|S|$ is the cardinality of S, and $(.)^T$ denotes the transpose. Also, $\mathbf{0}_{m\times m}$ denotes an $m \times m$ matrix of zeros, $\mathbf{I}_{m\times m}$ denotes an identity $m \times m$ matrix, col(.) describes a column vector, diag(.) describes a diagonal matrix, the symbol $\otimes$ denotes the Kronecker product, and the column vector with m 1's $\mathbf{1}_m = [1, 1, \ldots 1]^T$. Moreover, we write $\underline{\lambda}(M)$ for the minimum eigenvalue of matrix M. Also, a continuous function $\alpha : [0, \alpha) \to [0, \infty)$ is said to belong to class $\mathcal{K}$ functions if it is strictly increasing and $\alpha(0) = 0$, and a scalar continuous function $\beta(r, s)$ defined for $r \in [0, a)$ and $s \in [0, \infty)$ is said to belong to class $\mathcal{KL}$ if for each fixed s the mapping $\beta(r, s)$ belongs to class $\mathcal{K}$ with respect to r and for fixed r the mapping $\beta(r, s)$ is decreasing with respect to s and $\beta(r, s) \to 0$ as $s \to \infty$. Furthermore, $\dot{x}(t) = \frac{dx}{dt}$ with $x \in \mathbb{R}^n$.

2 ACTUATION CORRUPTION ATTACKS

2.1 Background on Networks and Graphs

In this section multiagent system interactions are modeled by a fixed strongly connected *graph* $\mathcal{G} = (\mathcal{V}_{\mathcal{G}}, \mathcal{E}_{\mathcal{G}})$ defined by a finite set $\mathcal{V}_{\mathcal{G}} = \{n_1, \ldots, n_N\}$ of N nodes that represent agents and a set of edges $\mathcal{E}_{\mathcal{G}} \subseteq \mathcal{V}_{\mathcal{G}} \times \mathcal{V}_{\mathcal{G}}$ that represent interagent information exchange links. The set of *neighbors* of a node n_i is defined as the set of nodes with edges coming in to n_i and is denoted by $\mathcal{N}_i = \left\{n_j : (n_j, n_i) \in \mathcal{E}_{\mathcal{G}}\right\}$. We assume the graph is simple (ie, it does not have repeated edges) and has no self-loops—that is, $(n_i, n_i) \notin \mathcal{E}_\mathcal{G}$ for all $n_i \in \mathcal{V}_\mathcal{G}$. The *connectivity matrix* $\mathcal{A}_\mathcal{G} = [a_{ij}]_{N\times N}$ of the graph $\mathcal{G}$ is an $N \times N$ matrix of the form

$$a_{ij} \begin{cases} > 0 & \text{if } (n_j, n_i) \in \mathcal{E}_\mathcal{G}, \\ = 0 & \text{otherwise,} \end{cases}$$

where a_{ij} is the *weight* associated with the edge $(n_j, n_i) \in \mathcal{E}_\mathcal{G}$. The *degree* matrix of the graph $\mathcal{G}$ is an $N \times N$ diagonal matrix whose ith entry of the main diagonal is the weighted degree $d_i = \sum_{j\in\mathcal{N}_i} a_{ij}$ of node i (ie, ith row sum of $\mathcal{A}_\mathcal{G}$). The number of neighbors of agent i is denoted by $|\mathcal{N}_i|$, which is equal to d_i.

We consider a system consisting of N two-input dynamics, each modeled as

$$\dot{x}_i = Fx_i + B_i u_i + D_i v_i, \quad \forall i \in \mathcal{N} :- \{1, \ldots, N\}, \tag{1}$$

where $x_i \in \mathbb{R}^n$ is the state of each agent, $u_i, v_i \in \mathbb{R}^m$ are the two inputs to agent i, which are controlled by players with opposite goals, $F \in \mathbb{R}^{n\times n}$ is the system matrix, and $B_i \in \mathbb{R}^{n\times m}$ and $D_i \in \mathbb{R}^{n\times m}$ are the input and the "attacker corruption" matrix, respectively. We thus have a total of $2N$ players, N of them—which we call *controllers*—select values for $u_i(t)$, $t \geq 0$, $i \in \mathcal{N}$ within appropriate sets $\mathcal{U}_i$ and the other N—which we call *attackers*—select values for the $v_i(t)$, $t \geq 0$, $i \in \mathcal{N}$ within appropriate sets $\mathcal{V}_i$.

For each $i \in \mathcal{N}$ the criteria of the players (ie, agents) associated with the inputs u_i and v_i are symmetric and depend on their state $[x_i']$, their inputs $[u_i' \; v_i']$, and also on the states and inputs of a subset $\mathcal{N}_i \subset \mathcal{N}$ of the other agents. Specifically, u_i and v_i want to minimize and maximize, respectively, the following criteria:

$$J_i(x_i, x_{\mathcal{N}_i}(0); u_i, u_{\mathcal{N}_i}, v_i, v_{\mathcal{N}_i}) = \frac{1}{2}\int_0^\infty r_i(e_i, u_i, u_{\mathcal{N}_i}, v_i, v_{\mathcal{N}_i})dt$$
$$\equiv \frac{1}{2}\int_0^\infty \left(\|e_i\|^2_{H_i} + \|u_i\|^2_{R_{ii}} - \gamma_{ii}^2\|v_i\|^2 + \sum_{j\in\mathcal{N}_i}\|u_j\|^2_{R_{ij}} - \sum_{j\in\mathcal{N}_i}\gamma_{ij}^2\|v_j\|^2 \right) dt, \tag{2}$$

where

$$e_i := \sum_{j\in\mathcal{N}_i}(x_i - x_j) + g_i(x_i - x_0), \quad \forall i \in \mathcal{N}, \tag{3}$$

where $g_i \in \mathbb{R}^+$ is the pinning gain that shows if an agent is pinned to the leader node (ie, $g_i \neq 0$) and it is nonzero for at least one node. We will now proceed to the design of the user-defined *distributed performances*.

The dynamics of Eq. (3) are given by

$$\begin{aligned}\dot{e}_i &= Fe_i + (d_i + g_i)(B_i u_i + D_i v_i) \\ &\quad - \sum_{j \in \mathcal{N}_i} (B_j u_j + D_j v_j), \quad \forall i \in \mathcal{N},\end{aligned} \tag{4}$$

with $e_i \in \mathbb{R}^n$.

The cost functionals associated with each agent $i \in \mathcal{N}$, which depend on the tracking error e_i, the control u_i, the controls in the neighborhood of agent i given as $u_{\mathcal{N}_i} := \{u_j : j \in \mathcal{N}_i\}$, the adversarial input v_i, and the adversarial inputs in the neighborhood of agent i given as $v_{\mathcal{N}_i} := \{v_j : j \in \mathcal{N}_i\}$, have the following form:

$$\begin{aligned}\mathcal{J}_i(e_i(0); u_i, u_{\mathcal{N}_i}, v_i, v_{\mathcal{N}_i}) = \frac{1}{2}\int_0^\infty \Bigg(&\|e_i\|^2_{H_i} + \|u_i\|^2_{R_{ii}} - \gamma_{ii}^2 \|v_i\|^2 \\ &+ \sum_{j \in \mathcal{N}_i} \|u_j\|^2_{R_{ij}} - \sum_{j \in \mathcal{N}_i} \gamma_{ij}^2 \|v_j\|^2 \Bigg) dt, \quad \forall i \in \mathcal{N},\end{aligned} \tag{5}$$

with user-defined matrices $H_i \succeq 0$, $R_{ii} \succ 0$, and $R_{ij} \succeq 0$, for all $i, j \in \mathcal{N}$ of appropriate dimensions, and $\gamma_{ii}, \gamma_{ij} \in \mathbb{R}^+$ for all $i \in \mathcal{N}$.

Hence, given a strongly connected graph $\mathcal{G}$, we are interested in finding a graphical Nash equilibrium [7, 26] that is translated to a saddle point $u_i^\star, v_i^\star$ for every agent $i \in \mathcal{N}$ in the sense that

$$\begin{aligned}&\mathcal{J}_i(e_i(0); u_i^\star, u_{\mathcal{N}_i}^\star, v_i, v_{\mathcal{N}_i}^\star) \le \mathcal{J}_i(e_i(0); u_i^\star, u_{\mathcal{N}_i}^\star, v_i^\star, v_{\mathcal{N}_i}^\star) \\ &\le \mathcal{J}_i(e_i(0); u_i, u_{\mathcal{N}_i}^\star, v_i^\star, v_{\mathcal{N}_i}^\star), \ \forall u_i, v_i \ i \in \mathcal{N}.\end{aligned} \tag{6}$$

This can be expressed by the following coupled distributed optimization problems:

$$\mathcal{J}_i(e_i(0); u_i^\star, u_{\mathcal{N}_i}^\star, v_i^\star, v_{\mathcal{N}_i}^\star) = \min_{u_i} \max_{v_i} \mathcal{J}_i(e_i(0); u_i, u_{\mathcal{N}_i}^\star, v_i, v_{\mathcal{N}_i}^\star), \quad \forall i \in \mathcal{N},$$

given the dynamics in Eq. (4).

Thus the ultimate goal is to find the distributed optimal value functions $V_i^\star$ for all $i \in \mathcal{N}$ defined by

$$V_i^\star(e_i(t)) := \min_{u_i} \max_{v_i} \int_t^\infty \frac{1}{2} \left(\|e_i\|_{H_i}^2 + \|u_i\|_{R_{ii}}^2 - \gamma_{ii}^2 \|v_i\|^2 + \sum_{j \in \mathcal{N}_i} \|u_j\|_{R_{ij}}^2 - \sum_{j \in \mathcal{N}_i} \gamma_{ij}^2 \|v_j\|^2 \right) dt, \quad \forall t, \forall i \in \mathcal{N}, \tag{7}$$

but without any information on the system matrices F, B_i, and D_i for all $i \in \mathcal{N}$ and pinning gains g_i for all $i \in \mathcal{N}$. First we will define the Hamiltonian associated with each agent's neighborhood tracking error (4) and each $V_i^\star$ given in Eq. (7) as follows:

$$\mathcal{H}_i\left(e_i, u_i, u_{\mathcal{N}_i}, v_i, v_{\mathcal{N}_i}, \frac{\partial V_i^\star}{\partial e_i}\right) = \frac{\partial V_i^\star}{\partial e_i}^T \left(Fe_i + (d_i + g_i)(B_i u_i + D_i v_i) - \sum_{j \in \mathcal{N}_i} (B_j u_j + D_j v_j) \right) + \frac{1}{2} \left(\|e_i\|_{H_i}^2 + \|u_i\|_{R_{ii}}^2 - \gamma_{ii}^2 \|v_i\|^2 + \sum_{j \in \mathcal{N}_i} \|u_j\|_{R_{ij}}^2 - \sum_{j \in \mathcal{N}_i} \gamma_{ij}^2 \|v_j\|^2 \right), \quad \forall e_i, u_i, v_i \; \forall i \in \mathcal{N}. \tag{8}$$

After employing the stationarity conditions, in the Hamiltonian (8), we can find the saddle-point solution—that is, $\frac{\partial \mathcal{H}_i(\cdot)}{\partial u_i} = 0$, and $\frac{\partial \mathcal{H}_i(\cdot)}{\partial v_i} = 0$. Hence the optimal control for each $i \in \mathcal{N}$ can be found to be

$$u_i^\star(e_i) = \arg\min_{u_i} \mathcal{H}_i\left(e_i, u_i, u_{\mathcal{N}_i}, v_i, v_{\mathcal{N}_i}, \frac{\partial V_i^\star}{\partial e_i}\right) = -(d_i + g_i) R_{ii}^{-1} B_i^T \frac{\partial V_i^\star}{\partial e_i}, \quad \forall e_i, \tag{9}$$

and the worst-case adversarial input can be found to be

$$v_i^\star(e_i) = \arg\max_{v_i} \mathcal{H}_i\left(e_i, u_i, u_{\mathcal{N}_i}, v_i, v_{\mathcal{N}_i}, \frac{\partial V_i^\star}{\partial e_i}\right) = \frac{(d_i + g_i)}{\gamma_{ii}^2} D_i^T \frac{\partial V_i^\star}{\partial e_i}, \quad \forall e_i. \tag{10}$$

The saddle-point solution given by Eqs. (9) and (10) should satisfy the appropriate coupled Hamilton-Jacobi equations:

$$\mathcal{H}_i\left(e_i, u_i^\star, u_{\mathcal{N}_i}^\star, v_i^\star, v_{\mathcal{N}_i}^\star, \frac{\partial V_i^\star}{\partial e_i}\right) = 0, \quad \forall i \in \mathcal{N}. \tag{11}$$

The value functions can be represented as quadratic in the neighborhood tracking error—that is, $V_i^\star(e_i) : \mathbb{R}^n \to \mathbb{R}$:

$$V_i^\star(e_i) = \frac{1}{2}e_i^T P_i e_i, \quad \forall e_i, \ \forall i \in \mathcal{N}, \tag{12}$$

where $P_i \in \mathbb{R}^{n\times n}$ for all $i \in \mathcal{N}$ are the unique symmetric positive definite matrices that solve the following complicated distributed coupled equations:

$$\begin{aligned}
& e_i^T P_i \left(Fe_i - (d_i + g_i)^2 \left(B_i R_{ii}^{-1} B_i^T - \frac{1}{\gamma_{ii}^2} D_i D_i^T\right) P_i e_i\right. \\
& \left. + \sum_{j\in\mathcal{N}} (d_j + g_j) \left(B_j R_{jj}^{-1} B_j^T - \frac{1}{\gamma_{ii}^2} D_j D_j^T\right) P_j e_j\right) \\
& + \left(Ae_i - (d_i + g_i)^2 \left(B_i R_{ii}^{-1} B_i^T - \frac{1}{\gamma_{ii}^2} D_i D_i^T\right) P_i e_i\right. \\
& \left. + \sum_{j\in\mathcal{N}} (d_j + g_j) \left(B_j R_{jj}^{-1} B_j^T - \frac{1}{\gamma_{ii}^2} D_j D_j^T\right) P_j e_j\right)^T P_i e_i \\
& + \sum_{j\in\mathcal{N}_i} (d_j + g_j)^2 e_j^T P_j \left(B_j R_{jj}^{-T} R_{ij} R_{jj}^{-1} B_j^T - \frac{\gamma_{ij}^2}{\gamma_{jj}^4} D_j D_j^T\right) P_j e_j \\
& + (d_i + g_i)^2 e_i^T P_i \left(B_i R_{ii} B_i^T - \frac{1}{\gamma_{ii}^2} D_i D_i^T\right) P_i e_i \\
& + e_i^T H_i e_i = 0, \quad \forall i \in \mathcal{N}.
\end{aligned} \tag{13}$$

By use of Eq. (12), the optimal control (9) for every agent $i \in \mathcal{N}$ can be written as

$$u_i^\star(e_i) = -(d_i + g_i) R_{ii}^{-1} B_i^T P_i e_i, \quad \forall e_i, \tag{14}$$

and the worst-case adversarial input (10) for every agent $i \in \mathcal{N}$ can be written as

$$v_i^\star(e_i) = \frac{(d_i + g_i)}{\gamma_{ii}^2} D_i^T P_i e_i, \quad \forall e_i. \tag{15}$$

Eqs. (13)–(15) are highly coupled, difficult to solve because of the cross terms $e_i^T e_j$, and require complete knowledge of the system matrix F, the input matrices B_i and D_i, $i \in \mathcal{N}$, and leader connection information g_i. We will show a new *cooperative* Q-*learning*-based approach to solve the graphical Nash game problem without any information on the system dynamics while attenuating persistent adversarial inputs, by adjusting parameters in an adaptive way.

2.2 *Q*-Learning-Based Approach

The value functions (12) need to be parameterized as functions of the neighborhood tracking error e_i, the controls u_i and $u_{\mathcal{N}_i}$, and the adversarial inputs v_i and $v_{\mathcal{N}_i}$ to represent the distributed Q-function (ie, cooperative learning) for each agent in the game. The optimal value given by Eq. (12) after addition of the Hamiltonian from Eq. (8) can be written as the following distributed Q-function or *action-dependent value* $Q_i(e_i, u_i, u_{\mathcal{N}_i}, v_i, v_{\mathcal{N}_i}) : \mathbb{R}^{n+m_i+l_i+\sum_{j\in\mathcal{N}_i}(m_j+l_j)} \to \mathbb{R}$:

$$\begin{aligned}
Q_i(e_i, u_i, u_{\mathcal{N}_i}, v_i, v_{\mathcal{N}_i}) &:= V_i^\star(e_i) + \mathcal{H}_i\left(e_i, u_i, u_{\mathcal{N}_i}, v_i, v_{\mathcal{N}_i}, \frac{\partial V_i^\star}{\partial e_i}\right) \\
&= V_i^\star(e_i) + \frac{\partial V_i^{\star\,T}}{\partial e_i}\left(Fe_i + (d_i + g_i)(B_i u_i + D_i v_i) - \sum_{j\in\mathcal{N}_i}(B_j u_j + D_j v_j)\right) \\
&+ \frac{1}{2}\left(e_i^T H_i e_i + (u_i^T R_{ii} u_i - \gamma_{ii}^2 v_i^T v_i) + \sum_{j\in\mathcal{N}_i}(u_j^T R_{ij} u_j - \gamma_{ij}^2 v_j^T v_j)\right), \\
&\forall e_i, u_i, u_{\mathcal{N}_i}, v_i, v_{\mathcal{N}_i}, \ \forall i \in \mathcal{N},
\end{aligned} \tag{16}$$

where $\mathcal{H}_i\left(e_i, u_i, u_{\mathcal{N}_i}, v_i, v_{\mathcal{N}_i}, \frac{\partial V_i^\star}{\partial e_i}\right)$ is given by Eq. (8) and the optimal cost is $V_i^\star(e_i) = \frac{1}{2} e_i^T P_i e_i$ for all $i \in \mathcal{N}$. Each agent's distributed Q-function (16) can be written in a compact distributed quadratic from with respect to the neighborhood error e_i, controls u_i and u_{N_i} and adversarial inputs v_i and v_{N_i} as in:

$$
Q_i(e_i, u_i, u_{\mathcal{N}_i}, v_i, v_{\mathcal{N}_i}) = \frac{1}{2} z_i^T
\begin{bmatrix}
P_i + H_i + P_i F + F^T P_i & (d_i+g_i)P_iB_i & -\mathrm{row}(P_iB_j) & (d_i+g_i)P_iD_i & -\mathrm{row}\left(P_iD_j\right)_{j\in\mathcal{N}_i} \\
(d_i+g_i)B_i^T P_i & R_{ii} & \mathbf{0} & \mathbf{0} & \mathbf{0} \\
-\mathrm{col}\left(B_j^T P_i\right)_{j\in\mathcal{N}_i} & \mathbf{0} & \mathrm{diag}\left(R_{ij}\right)_{j\in\mathcal{N}_i} & \mathbf{0} & \mathbf{0} \\
(d_i+g_i)D_i^T P_i & \mathbf{0} & \mathbf{0} & \gamma_{ii}^2 & \mathbf{0} \\
-\mathrm{col}\left(D_j^T P_i\right)_{j\in\mathcal{N}_i} & \mathbf{0} & \mathbf{0} & \mathbf{0} & \mathrm{diag}\left(\gamma_{ij}^2\right)_{j\in\mathcal{N}_i}
\end{bmatrix} z_i
$$

$$
:= \frac{1}{2} z_i^T
\begin{bmatrix}
Q^i_{e_ie_i} & Q^i_{e_iu_i} & Q^i_{e_iu_{\mathcal{N}_i}} & Q^i_{e_iv_i} & Q^i_{e_iv_{\mathcal{N}_i}} \\
Q^i_{u_ie_i} & Q^i_{u_iu_i} & Q^i_{u_iu_{\mathcal{N}_i}} & Q^i_{u_iv_i} & Q^i_{u_iv_{\mathcal{N}_i}} \\
Q^i_{u_{\mathcal{N}_i}e_i} & Q^i_{u_{\mathcal{N}_i}u_i} & Q^i_{u_{\mathcal{N}_i}u_{\mathcal{N}_i}} & Q^i_{u_{\mathcal{N}_i}v_i} & Q^i_{u_{\mathcal{N}_i}v_{\mathcal{N}_i}} \\
Q^i_{v_ie_i} & Q^i_{v_iu_i} & Q^i_{v_iu_{\mathcal{N}_i}} & Q^i_{v_iv_i} & Q^i_{v_iv_{\mathcal{N}_i}} \\
Q^i_{v_{\mathcal{N}_i}e_i} & Q^i_{v_{\mathcal{N}_i}u_i} & Q^i_{v_{\mathcal{N}_i}u_{\mathcal{N}_i}} & Q^i_{v_{\mathcal{N}_i}v_i} & Q^i_{v_{\mathcal{N}_i}v_{\mathcal{N}_i}}
\end{bmatrix} z_i := \frac{1}{2} z_i^T \bar{Q}_i z_i, \ \forall z_i, \ \forall i \in \mathcal{N}, \tag{17}
$$

where $\mathbf{0}$ are zero matrices of appropriate dimensions, $\mathrm{diag}\left(R_{ij}\right)_{j\in\mathcal{N}_i}$ and $\mathrm{diag}\left(\gamma_{ij}\right)_{j\in\mathcal{N}_i}$ are stacked diagonal matrices, $\mathrm{col}\left(B_j^T P_i\right)_{j\in\mathcal{N}_i}$ and $\mathrm{col}\left(D_j^T P_i\right)_{j\in\mathcal{N}_i}$ are stacked column matrices, and $\mathrm{row}\left(P_iB_j\right)_{j\in\mathcal{N}_i}$ and $\mathrm{row}\left(P_iD_j\right)_{j\in\mathcal{N}_i}$ are stacked row matrices.

In Eq. (17) the equivalences of $Q^i_{(\cdot)}$ are straightforward—for example, $Q^i_{e_ie_i} = P_i + H_i + P_iF + F^TP_i$, $Q^i_{e_iu_i} = (d_i+g_i)P_iB_i$, $Q^i_{e_iv_i} = (d_i+g_i)P_iD_i$, $Q^i_{u_iu_i} = R_{ii}$, and $Q^i_{v_iv_i} = \gamma_{ii}^2$, etc., and positive definite matrices $\bar{Q}^i \in \mathbb{R}^{(n+m_i+l_i+\sum_{j\in\mathcal{N}_i}(m_j+l_j))\times(n+m_i+l_i+\sum_{j\in\mathcal{N}_i}(m_j+l_j))}$ for all $i \in \mathcal{N}$. A model-free formulation of Eq. (14) can be found by the solving of $\frac{\partial Q_i(e_i,u_i,u_{\mathcal{N}_i},v_i,v_{\mathcal{N}_i})}{\partial u_i} = 0$ to write

$$
\begin{aligned}
u_i^\star(e_i) &= \arg\min_{u_i} Q_i(e_i, u_i, u^\star_{\mathcal{N}_i}, v_i, v^\star_{\mathcal{N}_i}) \\
&= -(Q^i_{u_iu_i})^{-1} Q^i_{u_ie_i} e_i, \ \forall i \in \mathcal{N},
\end{aligned} \tag{18}
$$

and a model-free formulation of Eq. (15) can be found by the solving of $\frac{\partial Q_i(e_i,u_i,u_{\mathcal{N}_i},v_i,v_{\mathcal{N}_i})}{\partial v_i} = 0$ to write

$$
\begin{aligned}
v_i^\star(e_i) &= \arg\max_{v_i} Q_i(e_i, u_i, u^\star_{\mathcal{N}_i}, v_i, v^\star_{\mathcal{N}_i}) \\
&= (Q^i_{v_iv_i})^{-1} Q^i_{v_ie_i} e_i, \ \forall i \in \mathcal{N}.
\end{aligned} \tag{19}
$$

We will use machine intelligence and especially an actor-critic framework to find Eqs. (18) and (19).

The critic neural network will approximate the distributed Q-function (17), the control actor neural network will approximate the optimal controller (18), and the adversarial input actor neural network will approximate the worst-case adversarial input (19) of each agent $i \in \mathcal{N}$. Specifically, $Q_i^\star(e_i, u_i^\star, u_{\mathcal{N}_i}^\star, v_i^\star, v_{\mathcal{N}_i}^\star)$ can be written as

$$Q_i^\star(e_i, u_i^\star, u_{\mathcal{N}_i}^\star, v_i^\star, v_{\mathcal{N}_i}^\star) = \frac{1}{2} z_i^T \bar{Q}^i z_i, \quad \forall i \in \mathcal{N} \tag{20}$$

or

$$Q_i^\star(e_i, u_i^\star, u_{\mathcal{N}_i}^\star, v_i^\star, v_{\mathcal{N}_i}^\star) = \frac{1}{2} \mathrm{vech}\left(\bar{Q}^i\right)^T (z_i \otimes z_i), \quad \forall i \in \mathcal{N},$$

where $z_i := \begin{bmatrix} e_i^T & u_i^T & u_{\mathcal{N}_i}^T & v_i^T & v_{\mathcal{N}_i}^T \end{bmatrix}^T$.

We denote $\mathrm{vech}\left(\bar{Q}^i\right) \in \mathbb{R}^{\frac{1}{2}(n+m_i+l_i+\sum_{j\in\mathcal{N}_i}(m_j+l_j))(n+m_i+l_i+\sum_{j\in\mathcal{N}_i}(m_j+l_j)+1)}$ as a half-vectorization of the matrix $\bar{Q}^i$ that returns a column vector by stacking the elements of the diagonal and upper triangular part of the symmetric matrix into a vector where the off-diagonal elements are taken as $2\bar{Q}^i_{\kappa_1\kappa_2}$. The Kronecker product quadratic polynomial basis vector is denoted by $\otimes$.

By denoting $W_{ic} := \frac{1}{2}\mathrm{vech}\left(\bar{Q}^i\right)$, we can write Eq. (20) in a compact form as

$$Q_i^\star(e_i, u_i^\star, u_{\mathcal{N}_i}^\star, v_i^\star, v_{\mathcal{N}_i}^\star) = W_{ic}^T (z_i \otimes z_i), \quad \forall i \in \mathcal{N},$$

where W_{ic} are the ideal weights. Next we estimate $Q_i^\star$, $u_i^\star$, and $v_i^\star$ with the following actual values for the critic neural network with $\hat{W}_{ic} := \frac{1}{2}\mathrm{vech}\left(\hat{\bar{Q}}^i\right)$,

$$\hat{Q}_i(e_i, u_i, u_{\mathcal{N}_i}, v_i, v_{\mathcal{N}_i}) = \hat{W}_{ic}^T (z_i \otimes z_i), \quad \forall i \in \mathcal{N}, \tag{21}$$

where $\hat{W}_{ic}$ are the estimated critic weights.

The control actor neural network is given as

$$\hat{u}_i(e_i) = \hat{W}_{ia}^T e_i, \quad \forall i \in \mathcal{N}, \tag{22}$$

where $\hat{W}_{ia} \in \mathbb{R}^{n \times m_i}$ are the estimated actor weights; note also that the neighborhood tracking error, e_i, serves as an activation function for the action neural network. Note that the optimal value for each control actor neural network is given by Eq. (18).

Finally for the adversarial actor neural network we have

$$\hat{v}_i(e_i) = \hat{W}_{id}^T e_i, \quad \forall i \in \mathcal{N}, \tag{23}$$

where $\hat{W}_{id} \in \mathbb{R}^{n \times l_i}$ are the estimated actor weights; note also that the neighborhood tracking error, e_i, serves as an activation function for the action neural network. Note that the optimal value for each adversarial actor neural network is given by Eq. (19).

By using integral reinforcement learning (see also the previous section), we can write

$$
\begin{aligned}
&Q_i^\star(e_i(t), u_i^\star(t), u_{\mathcal{N}_i}^\star(t), v_i^\star(t), v_{\mathcal{N}_i}^\star(t)) \\
&= Q^\star(e_i(t-T), u_i^\star(t-T), u_{\mathcal{N}_i}^\star(t-T), v_i^\star(t-T), v_{\mathcal{N}_i}^\star(t-T)) \\
&\quad - \frac{1}{2}\int_{t-T}^{t} \Bigg(\|e_i\|_{H_i}^2 + \|u_i\|_{R_{ii}}^2 - \gamma_{ii}^2 \|v_i\|^2 \\
&\quad + \sum_{j \in \mathcal{N}_i} \|u_j\|_{R_{ij}}^2 - \sum_{j \in \mathcal{N}_i} \gamma_{ij}^2 \|v_j\|^2 \Bigg) d\tau, \quad \forall i \in \mathcal{N},
\end{aligned} \tag{24}
$$

where $T \in \mathbb{R}^+$ is a small fixed time interval that defines how fast one measures the neighborhood tracking error e_i, the control u_i, the adversarial input v_i, and the adversarial inputs and controls in the neighborhood $v_{\mathcal{N}_i}$ and $u_{\mathcal{N}_i}$.

Now we will find tuning updates for $\hat{W}_{ic}$, $\hat{W}_{ia}$, and $\hat{W}_{id}$. By following adaptive control techniques as in [27], we can find the gradient descent estimate of $\hat{W}_{ic}$ for the critic weights of each agent, as in Eq. (25):

$$
\dot{\hat{W}}_{ic} = -\alpha_{ic} \frac{(z_i(t) \otimes z_i(t) - z_i(t-T) \otimes z_i(t-T))}{\big(1 + (z_i(t) \otimes z_i(t) - z_i(t-T) \otimes z_i(t-T))^T (z_i(t) \otimes z_i(t) - z_i(t-T) \otimes z_i(t-T))\big)^2} E_i^T, \quad \forall i \in \mathcal{N}, \tag{25}
$$

where

$$
\begin{aligned}
E_i := &\ \hat{Q}_i(e_i(t), \hat{u}_i(t), \hat{u}_{\mathcal{N}_i}(t), \hat{v}_i(t), \hat{v}_{\mathcal{N}_i}(t)) \\
&- \hat{Q}_i(e_i(t-T), \hat{u}_i(t-T), \hat{u}_{\mathcal{N}_i}(t-T), \hat{v}_i(t-T), \hat{v}_{\mathcal{N}_i}(t-T)) \\
&+ \frac{1}{2}\int_{t-T}^{t} \left(e_i^T H_i e_i + \left(\hat{u}_i^T R_{ii} \hat{u}_i - \gamma_{ii}^2 \hat{v}_i^T \hat{v}_i\right) + \sum_{j \in \mathcal{N}_i} \left(\hat{u}_j^T R_{ij} \hat{u}_j - \gamma_{ij}^2 \hat{v}_j^T \hat{v}_j\right) \right) d\tau
\end{aligned}
$$

$$
\begin{aligned}
&= \hat{W}_{ic}^{T}\left(z_i(t) \otimes z_i(t)\right) \\
&\quad + \frac{1}{2}\int_{t-T}^{t}\left(e_i^T H_i e_i + \left(\hat{u}_i^T R_{ii}\hat{u}_i - \gamma_{ii}^2 \hat{v}_i^T \hat{v}_i\right) + \sum_{j\in\mathcal{N}_i}\left(\hat{u}_j^T R_{ij}\hat{u}_j - \gamma_{ij}^2 \hat{v}_j^T \hat{u}_j\right)\right) d\tau \\
&\quad - \hat{W}_{ic}^{T}\left(z_i(t-T) \otimes z_i(t-T)\right)
\end{aligned}
$$

and $\alpha_{ic} \in \mathbb{R}^+$ is a constant gain that determines the speed of critic neural network convergence. Similarly, the gradient descent estimate of $\hat{W}_{ia}$ for the control actor weights can be constructed as

$$
\dot{\hat{W}}_{ia} = -\alpha_{ia} e_i \left(\hat{W}_{ia}^T e_i + (\hat{Q}^i_{u_i u_i})^{-1} \hat{Q}^i_{u_i e_i} e_i\right)^T, \quad \forall i \in \mathcal{N}, \tag{26}
$$

where $\alpha_{ia} \in \mathbb{R}^+$ is a constant gain that determines the speed of actor neural network convergence. Finally, the gradient descent estimate of $\hat{W}_{id}$ for the adversarial actor weights can be constructed as

$$
\dot{\hat{W}}_{id} = -\alpha_{id} e_i \left(\hat{W}_{id}^T e_i - (\hat{Q}^i_{v_i v_i})^{-1} \hat{Q}^i_{v_i e_i} e_i\right)^T, \quad \forall i \in \mathcal{N}, \tag{27}
$$

where $\alpha_{id} \in \mathbb{R}^+$ is a constant gain that determines the speed of actor neural network convergence.

As one can see, to enable full autonomy and resiliency without any offline computations or exhaustive modeling, one can simply plug Eqs. (21), (22), (23), (25), (26), and (27) in every agent.

The main theorem is presented next and provides a Lyapunov-based closed-loop stability of each distributed subsystem.

Theorem 1. *Assume that the agents are strongly connected in the networked system $\mathcal{G}$. Consider the neighborhood tracking error dynamics given by Eq. (4), the critic neural network for each agent $i \in \mathcal{N}$ given by Eq. (21), the optimal control for each agent $i \in \mathcal{N}$ given by Eq. (22), and the worst-case adversarial input for each agent $i \in \mathcal{N}$ given by Eq. (23). The tuning law for the weights of the critic is given by Eq. (25), for the weights of the control actor is given by Eq. (26), and for the weights of the adversarial actor is given by Eq. (27). Then the equilibrium point of the system $\psi_i := \begin{bmatrix} e_i^T & \tilde{W}_{ic}^T & \tilde{W}_{ia}^T & \tilde{W}_{\mathcal{N}_i a}^T & \tilde{W}_{id}^T & \tilde{W}_{\mathcal{N}_i d}^T \end{bmatrix}^T$, $i \in \mathcal{N}$ (where $\tilde{W}_{\mathcal{N}_i a}^T$ and $\tilde{W}_{\mathcal{N}_i d}^T$ are the control actor and adversarial error estimation weights of the neighbors) for all initial conditions $\psi_i(0)$ is asymptotically stable given $\alpha_{ic} \gg \alpha_{ia}$ for all $i \in \mathcal{N}$.*

Proof. The proof will appear elsewhere because of limited space. □

2.3 Simulation

To show the effectiveness of the proposed distributed Q-learning model-free approach, we will use a connected graph $\mathcal{G}$ with 10 agents as shown in Fig. 1. The system dynamics are given in the form of Eq. (1) with a marginally stable $F = \begin{bmatrix} 0 & 2 \\ -2 & 0 \end{bmatrix}$ and $B_1 = \begin{bmatrix} -1 \\ 2 \end{bmatrix}$, $B_2 = \begin{bmatrix} 3 \\ -2 \end{bmatrix}$, $B_4 = \begin{bmatrix} 1 \\ -4 \end{bmatrix}$, $B_5 = \begin{bmatrix} 1 \\ 1 \end{bmatrix}$, $B_6 = \begin{bmatrix} -3 \\ 2 \end{bmatrix}$, $B_7 = \begin{bmatrix} 1 \\ 3 \end{bmatrix}$, $B_8 = \begin{bmatrix} 5 \\ 1 \end{bmatrix}$, $B_9 = \begin{bmatrix} 0.5 \\ 1 \end{bmatrix}$, and $B_{10} = \begin{bmatrix} 1 \\ 0.5 \end{bmatrix}$ with random initial conditions and a leader given as $\dot{x}_0 = Ax_0$. The user-defined matrices in the distributed performances are given as $H_i = 10I_2$ for all $i \in \{1, 2, \ldots, 10\}$, $R_{ii} = 5$ for all $i \in \{1, 2, \ldots, 10\}$, and $R_{ij} = 2$ for all $i \neq j \in \{1, 2, \ldots, 10\}$, and the tuning gains of the algorithm described by Eqs. (21), (22), (23), (25), (26), and (27) are picked as $\alpha_{ic} = 50$ and $\alpha_{ia} = 0.3$ for all $i \in \{1, 2, \ldots, 10\}$ and $T = 0.02\,\text{s}$. For $0 \leq t \leq 0.5\,\text{s}$ of the simulation, we add exponentially decreasing exploration noise in the

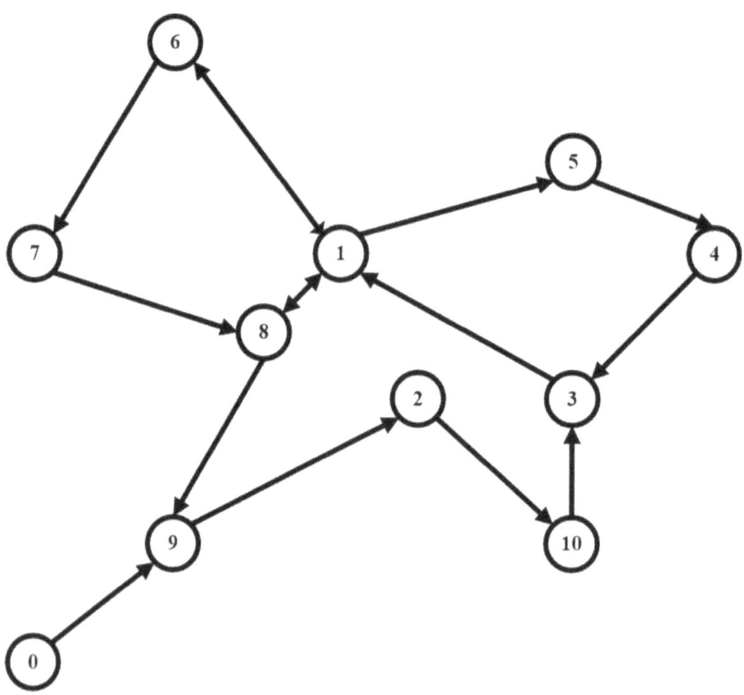

Fig. 1 A networked system $\mathcal{G}$ with 10 agents and a leader node 0 pinned to agent 9.

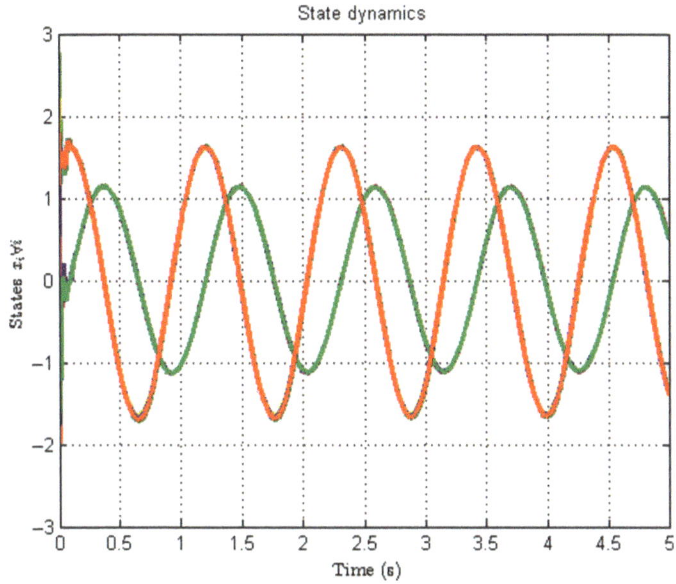

Fig. 2 Evolution of the agents' states and the leader state. It is obvious that all the agents' dynamics are synchronized to the first and second states of the leader.

control inputs to ensure persistence of excitation and state exploration. The evolution of the agents' and leader's states are shown in Fig. 2, where it is clear that synchronization occurs.

3 COMMUNICATION NETWORK CORRUPTION (JAMMING) ATTACKS

We consider a multiagent system consisting of N agents with integrated communication capabilities networked together. We also consider a jammer to be the $(N + 1)$th agent with an objective to break the communication links of the *networked team*. Suppose that the dynamics of every agent are modeled as

$$\dot{x}_i = Fx_i + B_i u_i, \quad x_{i0} = x_i(0), \quad \forall i = \{1, 2, \ldots N + 1\}, \tag{28}$$

where $x_i \in \mathbb{R}^n$ for each $t \in [t_0, T]$ ($T \in \mathbb{R}^+$ unknown final time when the motion ceases) are the states and $t \mapsto u_i \in \mathbb{R}^m$ is the control input of the ith agent, and $F \in \mathbb{R}^{n\times n}$ and $B_i \in \mathbb{R}^{n\times m}$ are the plant and input matrices, respectively The control input u_i for all $i \in \mathcal{N}$ belongs to some control set $U_i \subset \mathbb{R}^m$ for all $i \in \mathcal{N}$ and is chosen by each agent. The control input u_{N+1}

belongs to some control set $U_{N+1} \subset \mathbb{R}^m$ and is chosen by the jammer. We thus have two players with opposite goals—the networked team and the jammer. We formulate a control optimization problem for reaching a target set C_{stop}. The target C_{stop} is a subset of the state space where the agents in the networked team try to maintain connectivity as long as possible outside this set and the jammer aims to reach C_{stop} as soon as possible. The goal is to define an optimization scheme to select u_i for all $i \in \mathcal{N}$ to keep the agents connected so they can flock and select u_{N+1} such that the adversary can jam the communication.

3.1 Background on Dynamic Graphs

A dynamic graph model $\mathcal{G}(t) = \{\mathcal{V}, \mathcal{E}_\mathcal{G}(t), \mathcal{A}_\mathcal{G}(t)\}$ is used to model the communication between the N agents and consists of a set of vertices $\mathcal{V} = \{v_1, v_2, \ldots, v_N\}$, a time-varying set of edges $\mathcal{E}_\mathcal{G}(t) \subseteq \mathcal{V}\times\mathcal{V}$ constructed on the basis of k-proximity graphs, where edges are established between nodes v_i, v_j if and only if they are within a disagreement k of each other (ie, $\|q_i(t) - q_j(t)\| < k$) as described in [28], and a time-varying $N \times N$ adjacency matrix $\mathcal{A}_\mathcal{G}(t) = [a_{ij}]^{N\times N}$ with elements $a_{ij} \in [0, 1]$ for all $i, j \in \mathcal{N}$.

Assumption 1. $\mathcal{E}_\mathcal{G}(t)$ satisfies the following conditions for all $i, j \in \mathcal{N}$:

- The graph is assumed to be *connected* at $t = 0$ and the initial links are given by

$$\mathcal{E}_\mathcal{G}(0) = \left\{(i,j) \middle| \left\|x_i(0) - x_j(0)\right\| < k\right\}.$$

- If $0 < \|x_i(t) - x_j(t)\| < k$, then $(i, j) \in \mathcal{E}_\mathcal{G}(t)$.
- If $\|x_i(t) - x_j(t)\| \geq k$, then $(i, j) \notin \mathcal{E}_\mathcal{G}(t)$, and the time-varying adjacency matrix will be constructed by a procedure described next.

Assumption 2. The graph is simple—for example, no repeated edges and $(v_i, v_i) \notin \mathcal{E}_\mathcal{G}(t)$ for all $i \in \mathcal{N}$ no self-loops ($a_{ii}(q) = 0$, where $x \equiv \text{col}(x_1, x_2, \ldots, x_N)$ is the position vector of all the agents in the team). Also we denote as a directed path of length κ of the graph $\mathcal{G}(t)$ the sequence of $\kappa + 1$ nodes $v_1, v_2, \ldots, v_\kappa \in \mathcal{V}$ such that $v_q, v_{q+1} \in \mathcal{E}_\mathcal{G}(t)$ for all $1 \leq q \leq \kappa$. A directed graph has a directed spanning tree if there exists at least one node having a direct path to all other nodes.

The communication channel between two nodes $i, j \in \mathcal{N}$ is jammed if

$$\eta \geq \frac{\min\left\{\left\|x_{N+1}(t) - x_i(t)\right\|, \left\|x_j(t) - x_{N+1}(t)\right\|\right\}}{\left\|x_j(t) - x_i(t)\right\|}, \; j \neq i,$$

where $\eta \in \mathbb{R}^+$ depends on the strength of the jamming signal (for an explicit function, see [29]). *Hence the nodes can communicate if they lie in each other's perception range.*

Now one can write the condition that the graph stays connected for all $i \in \mathcal{N}$:

$$\mathcal{N}_i(t) = \Big\{j \in \mathcal{N} : \left\|x_j(t) - x_i(t)\right\| \leq \min\left\{k, \eta \left\|x_{N+1}(t) - x_i(t)\right\|, \eta \left\|x_j(t) - x_{N+1}(t)\right\|\right\},\ \mathcal{N} = \{1, 2, \ldots, N\}, j \neq i\Big\},$$

where all the agents have a uniform cluster k, and $\|.\|$ is the Euclidean distance. We define the time-varying *degree* matrix as $\mathcal{D}(t) = \mathrm{diag}(d_i(x))$, where $d_i(x) = \sum_{j \in \mathcal{N}_i} a_{ij}(x)$ is the weighted degree of node i (ie, ith row sum of $\mathcal{A}_{\mathcal{G}}(t)$). We define the graph Laplacian matrix as $\mathcal{L}_{\mathcal{G}}(t) = \mathcal{D}(t) - \mathcal{A}_{\mathcal{G}}(t)$, which has all row sums equal to zero. Let the ordered eigenvalues for all t of $\mathcal{L}_{\mathcal{G}}(t)$ be given by $\lambda_1(\mathcal{L}_{\mathcal{G}}(t)) \leq \lambda_2(\mathcal{L}_{\mathcal{G}}(t)) \leq \cdots \leq \lambda_N(\mathcal{L}_{\mathcal{G}}(t))$. Then $\mathcal{L}_{\mathcal{G}}(t)$ has the following properties:

- $\lambda_1(\mathcal{L}_{\mathcal{G}}(t)) = 0$ with corresponding eigenvector **1** and the algebraic multiplicity of this eigenvalue is equal to the number of connected agents in $\mathcal{G}(t)$.
- The Fiedler eigenvalue $\lambda_2(\mathcal{L}_{\mathcal{G}}(t)) > 0$ if and only if the graph $\mathcal{G}(t)$ is connected for all t and satisfies $\lambda_2(\mathcal{L}_{\mathcal{G}}(t)) = \min_{x \neq 0, \mathbf{1}^T x = 0} \frac{x^T \mathcal{L}_{\mathcal{G}}(t) x}{x^T x}$. In other words, the agents want to maintain (this will be interpreted with an edge tension function that will be defined later) $\lambda_2(\mathcal{L}_{\mathcal{G}}(t)) > 0$ and the jammer wants to make $\lambda_2(\mathcal{L}_{\mathcal{G}}(t)) = 0$.

Because of the discontinuities, the elements $a_{ij}(x)$ of the adjacency matrix $\mathcal{A}_{\mathcal{G}}(t)$ can be constructed as position $x \equiv \mathrm{col}(x_1, x_2, \ldots, x_N)$ dependent for all $i, j \in \mathcal{N}$ with the following assignment:

$$a_{ij}(x) = \begin{cases} \Psi\left(\left\|x_j - x_i\right\|_\sigma / k\right), & i \neq j, \\ 0, & i = j, \end{cases}$$

where $\left\|x_j - x_i\right\|_\sigma$ is the σ-norm [3] defined as $\left\|x_j - x_i\right\|_\sigma = \frac{1}{\zeta}\left[\sqrt{1 + \zeta \left\|x_j - x_i\right\|} - 1\right]$ with $\zeta > 0$, which is differentiable everywhere (compared with the Euclidean norm, which is not differentiable at the point where $x_j = x_i$), and $\Psi(z) : \mathbb{R} \to [0, 1]$ is a *smooth and compactly supported* projection function (eg, logistic function) with the following properties:

$$\Psi(z) = \begin{cases} 1, & z \leq \min\left\{k, \eta \left\|x_{N+1}(t) - x_i(t)\right\|, \eta \left\|x_j(t) - x_{N+1}(t)\right\|\right\}, \\ 0, & z > \min\left\{k, \eta \left\|x_{N+1}(t) - x_i(t)\right\|, \eta \left\|x_j(t) - x_{N+1}(t)\right\|\right\}, \\ \in (0, 1), & \text{otherwise} \end{cases}$$

and

$$\frac{d\Psi(z)}{dz} = \begin{cases} 0, & z \leq \min\left\{k, \eta \left\|x_{N+1}(t) - x_i(t)\right\|, \eta \left\|x_j(t) - x_{N+1}(t)\right\|\right\}, \\ 0, & z > \min\left\{k, \eta \left\|x_{N+1}(t) - x_i(t)\right\|, \eta \left\|x_j(t) - x_{N+1}(t)\right\|\right\}, \\ \leq 0, & \text{otherwise.} \end{cases}$$

The unit vector connecting node i to node j is

$$n_{ij} = \frac{(x_j - x_i)}{\left\|x_j - x_i\right\|},$$

and by using that, one can formulate the functions $\bar{\psi}_{ij} : \mathbb{R}^+ \to \mathbb{R}^+$ as artificial potentials (edge tension) defined on the links of the network, which in the case of connectivity control and rendezvous objective take the following form:

$$\bar{\psi}_{ij} = \begin{cases} \frac{\|x_i - x_j\|_\sigma^2}{(k - \|x_i - x_j\|_\sigma)}, & \text{if } (v_i, v_j) \in \mathcal{E}_{\mathcal{G}}(t), \\ 0, & \text{otherwise,} \end{cases}$$

where

$$\frac{\partial \bar{\psi}_{ij}}{\partial x_i} = \begin{cases} \frac{2k - \|x_i - x_j\|_\sigma}{(k - \|x_i - x_j\|_\sigma)^2}(x_i - x_j), & \text{if } (v_i, v_j) \in \mathcal{E}_{\mathcal{G}}(t), \\ 0, & \text{otherwise,} \end{cases}$$

where the σ-norm is defined as before.

3.2 Performance Design

Specifically, every agent in the networked team for all $i \in \mathcal{N} := \{1, \ldots, N\}$ and the jammer want to maximize and minimize, respectively, the following criterion:

$$J = \Delta(s(T)) - \frac{1}{2} \int_{t_0}^{T} \left(\rho + \sum_{i=1}^{N} \left(\|s_i\|_{Q_{ii}}^2 + \|u_i\|_{R_{ii}}^2\right) + \|s_{N+1}\|_{Q_{N+1}}^2 - \|u_{N+1}\|_{R_{N+1}}^2\right) dt, \tag{29}$$

where $Q_{ii} \succeq 0$, $R_{ii} \succ 0$, $Q_{N+1} \succeq 0$, and $R_{N+1} \succ 0$ are symmetric matrices with appropriate dimensions, the final time T is free, and ρ is an additional parameter introduced as a relative weight between the quadratic energy and the elapsed time. Hence the optimal transit time $T^\star$ can be found by optimization of the cost (29). Note that $T^\star$ is the first time at which the states $s_1(t), s_2(t), \ldots, s_N(t), s_{N+1}(t)$ enter the closed set C_{stop} or $T = \infty$ if the states $s_1(t), s_2(t), \ldots, s_N(t), s_{N+1}(t)$ never enter C_{stop}. There is no constraint on the final state; thus the control objective is to make the final state sufficiently small. Because of the term $-\frac{\rho(T-t_0)}{2}$ arising from the integral cost, there is a concern to accomplish this in a long time. The terminal conditions are the states of the agents and the adversary in the target set C_{stop}. This is a performance index that allows for a trade-off between the optimal-time objective and a desire to keep the control inputs and states as small as possible. By selecting $\Delta(s(T))$ to be very large, one can guarantee that the optimal control will drive the final state $s_i(T)$ for all $i \in \mathcal{N}$ and $s_{N+1}(T)$ very close to zero to keep the cost large. Also, the tracking error $s_i \in \mathbb{R}^{n+1}$ for all $i \in \mathcal{N}$ is defined as

$$s_i = \sum_{j=1}^{N} a_{ij}(x) \begin{bmatrix} \bar{\psi}_{ij} \\ x_i - x_j \end{bmatrix}. \tag{30}$$

When the game starts at $t = 0$, the jammer connects to the closest agent. The signal $s_{N+1} \in \mathbb{R}^n$ describes the information the jammer gets from this node for all $j \in \mathcal{N} : j_c = \arg\min_j \left\| x_{N+1}(0) - x_j(0) \right\|$:

$$s_{N+1} = x_{N+1} - x_{jc} \tag{31}$$

The following definitions will give a formal fundamental notion of the strategies.

Definition 1. The feedback strategies for each agent in the networked team and the jammer are maps $\mu_i : \mathbb{R}^+ \times \mathbb{R}^m \to U_i$ for all $i \in \mathcal{N}$ and $\mu_{N+1} : \mathbb{R}^+ \times \mathbb{R}^m \to U_{N+1}$, respectively.

Another issue here is that, given arbitrary strategies $\mu_i : \mathbb{R}^+ \times \mathbb{R}^m \times \mathbb{R}^m \to U_i$ for all $i \in \mathcal{N}$ and $\mu_{N+1} : \mathbb{R}^+ \times \mathbb{R}^m \times \mathbb{R}^m \to U_{N+1}$ and initial positions and velocities given by x_{i0} for all $i \in \{1, 2, \ldots, N+1\}$, the systems (28) with those initial conditions do not necessarily have a solution. Thus one needs to restrict the analysis to admissible strategies.

Definition 2. The pair of $((\mu_1, \mu_2, \ldots, \mu_N), \mu_{N+1})$ strategies is admissible if:

1. for any $((\mu_1, \mu_2, \ldots, \mu_N), \mu_{N+1})$ of $((U_1, U_2, \ldots, U_N), U_{N+1})$ and any initial states x_{i0} for all $i \in \{1, 2, \ldots, N+1\}$, Eq. (28) has a unique solution.
2. all the Lebesgue measurable maps $\mu_i : [0, \infty) \to U_i$ for all $i \in \{1, 2, \ldots, N+1\}$ belong to U_i for all $i \in \{1, 2, \ldots, N+1\}$.

3.3 Error Dynamics

The error dynamics of the agents in the team (30) can be written as

$$\dot{s}_i = \sum_{j=1}^{N} a_{ij}(x) \begin{bmatrix} \dot{\bar{\psi}}_{ij} \\ F(x_i - x_j) + B_i u_i - B_j u_j \end{bmatrix} + \sum_{j=1}^{N} \dot{a}_{ij}(x) \begin{bmatrix} \bar{\psi}_{ij} \\ x_i - x_j \end{bmatrix}$$

or

$$\dot{s}_i = \begin{bmatrix} \sum_{j=1}^{N} a_{ij}(x)\dot{\bar{\psi}}_{ij} + \sum_{j=1}^{N} \dot{a}_{ij}(x)\bar{\psi}_{ij} \\ d_i(x)Fx_i - \sum_{j=1}^{N} a_{ij}(x)Fx_j + \sum_{j=1}^{N} \dot{a}_{ij}(x)(x_i - x_j) \end{bmatrix} + \begin{bmatrix} 0 \\ d_i(x)B_i \end{bmatrix} u_i - \begin{bmatrix} 0 \\ \sum_{j=1}^{N} a_{ij}(x)B_j \end{bmatrix} u_j,$$

which can be written in the compact form as follows:

$$\dot{s}_i = f(x) + \bar{B}_i u_i - \bar{B}_j u_j \tag{32}$$

where $f(x) := \begin{bmatrix} \sum_{j=1}^{N} a_{ij}(x)\dot{\bar{\psi}}_{ij} + \sum_{j=1}^{N} \dot{a}_{ij}(x)\bar{\psi}_{ij} \\ d_i(x)Fx_i - \sum_{j=1}^{N} a_{ij}(x)Fx_j + \sum_{j=1}^{N} \dot{a}_{ij}(x)(x_i - x_j) \end{bmatrix}$, $\bar{B}_i := \begin{bmatrix} 0 \\ d_i(x)B_i \end{bmatrix}$, $\bar{B}_j := \begin{bmatrix} 0 \\ \sum_{j=1}^{N} a_{ij}(x)B_j \end{bmatrix}$, $\dot{a}_{ij}(x) = \frac{\partial a_{ij}(x)}{\partial x}x$, and $\dot{\bar{\psi}}_{ij} = \frac{\partial \bar{\psi}_{ij}}{\partial x_i}x_i$.

Similarly, the error dynamics for the jammer (31) are given as

$$\dot{s}_{N+1} = Fs_{N+1} + B_{N+1}u_{N+1} - B_{jc}u_{jc}. \tag{33}$$

3.4 Zero-Sum Game Formulation

The problem can be formulated as a zero-sum game [7] between the agents in the network and the jammer. Hence one needs to derive optimal strategies for the game. The agents are *initially connected* and their objective is to remain connected and flock. The jammer, on the other hand, wants to jam the communication graph and disconnect the links. Hence the agents in the team want to maximize the time they are connected, such that flocking can be achieved, and the jammer wants to minimize this time. The game will terminate when the agents reach the set C_{stop}. We assume that the local clocks of the agents can read off the global time so all the agents are synchronized with it and also the agents can observe and also measure the cost.

The jammer does not reveal his position but knows the initial graph topology, and connects to his closest agent.

Let us explain now what it means when the agents reach the target C_{stop}. Given a trajectory $S : [0, \infty) \to \mathbb{R}^{(n+1)(N+1)}$, let

$$\Xi_{C_{\text{stop}}}(S) := \inf\{t \geq 0 | s(t) \in C_{\text{stop}}\}.$$

Now set $\Xi_{C_{\text{stop}}}(S) := \infty$ if $s(t) \notin C_{\text{stop}}$ for all $t \geq 0$. For the initial conditions $x_{i0} \notin C_{\text{stop}}$ for all $i \in \{1, 2, \ldots, N+1\}$, and the control inputs $u_i \in U_i$ for all $i \in \mathcal{N}$ and $u_{N+1} \in U_{N+1}$, one can define

$$J(s(t_0), u, u_{N+1}) := \Xi_{C_{\text{stop}}}(S).$$

The Hamiltonian of the system is given by (with abuse of the notation)

$$H(\cdot) = -\frac{1}{2}\left(\rho + \sum_{i=1}^{N}\left(\|s_i\|^2_{Q_{ii}} + \|u_i\|^2_{R_{ii}}\right) + \|s_{N+1}\|^2_{Q_{N+1}} - \|u_{N+1}\|^2_{R_{N+1}}\right)$$
$$+ \sum_{i=1}^{N} \Lambda_i^T \dot{s}_i + \Lambda_{N+1}^T \dot{s}_{N+1}, \tag{34}$$

where Λ_i for all $i \in \mathcal{N}$ and Λ_{N+1} are the costate (adjoint) variables.

Since the jammer is the minimizer and the agents are the maximizers, the Hamiltonian satisfies the following two conditions along the optimal trajectories:

$$(u^\star, u^\star_{N+1}) = \arg\max_{u} \min_{u_{N+1}} H\left(s, \frac{\partial V(s(t))}{\partial s}, u, u_{N+1}\right)$$

$$H^\star\left(s, \frac{\partial V^\star(s(t))}{\partial s}, u^\star, u^\star_{N+1}\right) = 0. \tag{35}$$

3.5 Hamilton-Jacobi-Isaacs Equation

It is important to interpret the control inputs as state-dependent strategies. Then the value function is given by

$$V(s(t_0)) = \Delta(s(T)) - \frac{1}{2}\int_{t_0}^{T}\left(\rho + \sum_{i=1}^{N}\left(\|s_i\|^2_{Q_{ii}}\right.\right.$$
$$\left.\left. + \|u_i\|^2_{R_{ii}}\right) + \|s_{N+1}\|^2_{Q_{N+1}} - \|u_{N+1}\|^2_{R_{N+1}}\right) d\tau. \tag{36}$$

The team-jammer objective is

$$V^{\star}(s(t_0)) = \max_{u_1 \in U_1, \ldots, u_N \in U_N} \min_{u_{N+1} \in U_{N+1}} \frac{1}{2} \int_{t_0}^{T} \Bigg(-\rho - \sum_{i=1}^{N} \left(\|s_i\|^2_{Q_{ii}} + \|u_i\|^2_{R_{ii}} \right) - \|s_{N+1}\|^2_{Q_{N+1}} + \|u_{N+1}\|^2_{R_{N+1}} \Bigg) d\tau \tag{37}$$

subject only to dynamical constraints (32) and (33) and on the target set C_{stop} being reached at time $T^{\star}$ all motion ceases. By using the Leibniz formula on Eq. (37), one can find the well-known Hamilton-Jacobi-Isaacs equation:

$$\begin{aligned}
\rho = &- \left(\sum_{i=1}^{N} \left(\|s_i\|^2_{Q_{ii}} + \|u_i\|^2_{R_{ii}} \right) + \|s_{N+1}\|^2_{Q_{N+1}} - \|u_{N+1}\|^2_{R_{N+1}} \right) \\
&+ 2 \sum_{i=1}^{N} \frac{\partial V^{\star}(s(t))}{\partial s_i}^{T} \\
&\left(\begin{bmatrix} \sum_{j=1}^{N} a_{ij}(x) \dot{\bar{\psi}}_{ij} + \sum_{j=1}^{N} \dot{a}_{ij}(x) \bar{\psi}_{ij} \\ d_i(x) F x_i - \sum_{j=1}^{N} a_{ij}(x) F x_j + \sum_{j=1}^{N} \dot{a}_{ij}(x)(x_i - x_j) \end{bmatrix} \right. \\
&\left. + \begin{bmatrix} 0 \\ d_i(x) B_i \end{bmatrix} u_i - \begin{bmatrix} 0 \\ \sum_{j=1}^{N} a_{ij}(x) B_j \end{bmatrix} u_j \right) \\
&+ 2 \frac{\partial V^{\star}(s(t))}{\partial s_{N+1}}^{T} \left(F s_{N+1} + B_{N+1} u_{N+1} - B_{jc} u_{jc} \right).
\end{aligned} \tag{38}$$

Eq. (38) is a partial differential equation that yields the value V for any given u_i for all $i \in \mathcal{N}$ and u_{N+1}. The boundary condition is $V(s(T)) = \Delta(s(T))$. Because of the highly nonlinear nature of Eq. (38), it is impossible to find a closed-form solution. In the subsequent section, we will use learning ideas to avoid that situation.

Moreover, Eq. (38) requires complete information on the system dynamics. To avoid this necessity, we will use an integral equivalent version [25] of Eq. (38) to write

$$V(s(t - T_s)) = V(s(t)) - \int_{t-T_s}^{t} \Bigg(\rho + \sum_{i=1}^{N} \left(\|s_i\|^2_{Q_{ii}} + \|u_i\|^2_{R_{ii}} \right) + \|s_{N+1}\|^2_{Q_{N+1}} - \|u_{N+1}\|^2_{R_{N+1}} \Bigg) d\tau, \tag{39}$$

given $\Delta(s(T))$ for any time $t \geq 0$ and time interval $T_s > 0$.

Now the closed-form solutions of the control and jam policies can be found by use of Bellman's optimality principle:

$$u_i^\star = \arg\max_{u_i} H(\cdot) = d_i(q) R_{ii}^{-1} \bar{B}_i^T \frac{\partial V^\star(s(t))}{\partial s_i} \tag{40}$$

$$u_{N+1}^\star = \arg\min_{u_{N+1}} H(\cdot) = -R_{N+1}^{-1} B_{N+1}^T \frac{\partial V^\star(s(t))}{\partial s_{N+1}}. \tag{41}$$

Recall the first-order necessary optimality conditions with $\Lambda_i = \frac{\partial V}{\partial s_i}$ for all $i \in \{1, 2, \ldots, N, N+1\}$. We know

$$H(s(t), \frac{\partial V^\star(s(t))}{\partial s(t)}, u^\star(t), u_{N+1}^\star(t)) = 0, \quad \forall t \geq 0. \tag{42}$$

The next theorem proves the existence of the saddle point and finds the value of the game.

Theorem 2. *Assume that the target set C_{stop} is closed. Suppose that there is a nonnegative map (belonging to class $\mathcal{KL}$) $\Phi(s(t)) \equiv (-V(s(t))) > 0 \in C^1$ over $\mathbb{R}^{(n+1)N+n}$, with $\Phi(s(t)) = 0$ on C_{stop} a smooth solution satisfying the Hamilton-Jacobi-Isaacs inequality given by*

$$H(s(t), \frac{\partial V^\star(s(t))}{\partial s(t)}, u^\star(t), u_{N+1}^\star(t)) \leq 0, \quad \forall s \in \mathbb{R}^{(n+1)N+n} \backslash C_{\text{stop}}, \tag{43}$$

with control policies $u_i^\star$ for all $i \in \mathcal{N}$, and $u_{N+1}^\star$ given in terms of this solution Φ. Then,

1. *As $T \to \infty$, systems (32), (33), (40), and (41) are asymptotically stable ($s_i \to 0$ for all $i \in \mathcal{N}$, and $s_{N+1} \to 0$), so the agents will achieve flocking and stay connected.*
2. *The policies $u_i^\star, i = 1, \ldots, N$, and $u_{N+1}^\star$ given by Eqs. (40) and (41), respectively, form a saddle point, and the corresponding value of the game is given by*

$$J^\star(s(t_0)) = -\frac{1}{2}\rho(T - t_0) + \Phi^\star(s(t_0)), \quad \forall s \in \mathbb{R}^{(n+1)N+n} \backslash C_{\text{stop}}. \tag{44}$$

Proof. The proof will appear elsewhere because of space limitations. □

3.6 Flocking with Learning Ideas in the Presence of an Intelligent Jammer

The Hamilton-Jacobi-Isaacs equation is, in general, impossible to solve for the cost $V(s(t))$. In this section we use ideas from approximation and learning theory to find an approximate local smooth optimal solution of the Hamilton-Jacobi-Isaacs equation.

Neural networks have been shown in [30] to be ideal approximators for uniformly approximating functions. The value function $V^{\star}(s(t))$ for $t \in [t_0, T]$ is smooth on a compact set Ω and hence can be constructed uniformly in t with use of a neural network with h basis sets on a compact set Ω as

$$V^{\star}(s(t)) = W_c^{\star T} \phi(s) + \epsilon, \tag{45}$$

where $W_c^{\star} \in \mathbb{R}^h$ are the ideal weights bounded as $\left\| W_c^{\star} \right\| \leq W_{\max}$, $\phi(s) \equiv \phi(s(t)) = [\phi_1(s)\ \phi_2(s)\ \ldots\ \phi_h(s)] : \mathbb{R}^{(n+1)N+n} \backslash C_{\text{stop}} \to \mathbb{R}^h$ is a bounded continuously differentiable basis function, and ϵ is the residual error bounded as $\epsilon \leq \epsilon_{\max}$. The basis sets are selected so that as $h \to \infty$, the functions $\phi(s)$ provide a complete independent basis for $V^{\star}(s(t))$.

Remark 1. According to the Weierstrass theorem, a real-valued continuous function defined on a bounded interval on $\mathbb{R}$ can be well approximated by use of the supremum norm if the number of basis functions goes to infinity.

Hence taking the partial derivative of Eq. (45) yields

$$\frac{\partial V^{\star}(s(t))}{\partial s} = \frac{\partial \phi(s)}{\partial s}^T W_c^{\star} + \frac{\partial \epsilon}{\partial s}.$$

It is straightforward to see that one needs an approximation of the value and its partial derivatives, and hence an approximation in Sobolev space $H^{1,2}(\Omega)$.

By use of Eqs. (40), (41), and (45) the optimal policies for the networked team and the jammer can be written as

$$u_i^{\star} = d_i(q) R_{ii}^{-1} \bar{B}_i^T \left(\frac{\partial \phi(s)}{\partial s_i}^T W_c^{\star} + \frac{\partial \epsilon}{\partial s_i} \right), \quad \forall i \in \mathcal{N}, \tag{46}$$

$$u_{N+1}^{\star} = -R_{N+1}^{-1} B_{N+1}^T \left(\frac{\partial \phi(s)}{\partial s_{N+1}}^T W_c^{\star} + \frac{\partial \epsilon}{\partial s_{N+1}} \right). \tag{47}$$

The signal (48) can be written in a compact form by use of Eq. (45) as

$$\pi(t) = -\int_{t-T_s}^{t} \left(\rho + \sum_{i=1}^{N} \left(\|s_i\|_{Q_{ii}}^2 + \left\| u_i^{\star} \right\|_{R_{ii}}^2 \right) + \|s_{N+1}\|_{Q_{N+1}}^2 - \left\| u_{N+1}^{\star} \right\|_{R_{N+1}}^2 \right) d\tau. \tag{48}$$

We also define the integral reinforcement signal as

$$W_c^\star(t)^T \Delta\phi(s) + \pi(t) = \epsilon_\pi, \tag{49}$$

where ϵ_π is a residual error and $\Delta\phi(s(t)) := \phi(s(t)) - \phi(s(t - T_s))$.

Since the ideal weights are unknown, the current approximation weights are given by $\hat{W}_c \in \mathbb{R}^h$, $\hat{W}_{ai} \in \mathbb{R}^h$, and $\hat{W}_{\text{jam}} \in \mathbb{R}^h$, and hence the outputs of the approximator of the value function (critic neural network) and the optimal policies for every agent and the jammer are given, respectively, by

$$\hat{V}(s(t)) = \hat{W}_c(t)^T \phi(s), \tag{50}$$

$$\hat{u}_i = \hat{W}_{ai}(t)^T \phi_{ai}(s_i(t)), \quad \forall i \in \mathcal{N}, \tag{51}$$

$$\hat{u}_{N+1} = \hat{W}_{\text{jam}}(t)^T \phi_{\text{jam}}(s_{N+1}(t)). \tag{52}$$

Toward making Eq. (49) hold with a zero error ϵ_π, our goal is to tune the weights $\hat{W}_c$ so that the following error is driven to zero:

$$e := \hat{W}_c(t)^T \Delta\phi(s(t)) + \hat{\pi}(t),$$

where $\hat{\pi}(t) := -\int_{t-T_s}^{t} \left(\rho + \sum_{i=1}^{N} \left(\|s_i\|_{Q_{ii}}^2 + \left\|\hat{u}_i\right\|_{R_{ii}}^2\right) + \|s_{N+1}\|_{Q_{N+1}}^2 - \left\|\hat{u}_{N+1}\right\|_{R_{N+1}}^2\right) d\tau$.

For that reason we will use a gradient descent law to select the critic weights $\hat{W}_c$ to minimize the squared error $E = \frac{1}{2}e^2$:

$$\dot{\hat{W}}_c = -\alpha \frac{\Delta\phi(s(t))}{(\Delta\phi(s(t))^T \Delta\phi(s(t)) + 1)^2} (\Delta\phi(s(t)) \hat{W}_c + \hat{\pi}), \tag{53}$$

where $\alpha \in \mathbb{R}^+$ determines the speed of convergence and the boundary condition, $V(s(T)) = \Delta(s(T))$.

The estimation error can be given as

$$\tilde{W}_c = W_c^\star - \hat{W}_c, \tag{54}$$

and the estimation error dynamics are

$$\dot{\tilde{W}}_c = -\dot{\hat{W}}_c. \tag{55}$$

The vector $\Delta\phi(s(t)))$ needs to be persistently exciting [27].

We define each agent's and the jammer's estimation error as

$$\tilde{W}_{ai} = W_{ai}^\star - \hat{W}_{ai} \tag{56}$$

$$\tilde{W}_{\text{jam}} = W_{\text{jam}}^\star - \hat{W}_{\text{jam}}, \tag{57}$$

where $W_{ai}^{\star}$ and $W_{\text{jam}}^{\star}$ are the optimal weights for each agent in the team and the jammer, respectively.

Now we have to define the error $e_{i\text{team}}$ for all $i \in \mathcal{N}$ for every agent in the team that we have to drive to zero. The error has the following form,

$$e_{i\text{team}} := \hat{W}_{ai}(t)^T \phi_{ai}(s_i(t)) + d_i(q) R_{ii}^{-1} \bar{B}_i^T \frac{\partial \phi(s)}{\partial s_i}^T \hat{W}_c.$$

To drive the error $e_{ii\text{team}}$ for every agent $i \in \mathcal{N}$, we have to select $\hat{W}_{ai}$ such that the following criterion is minimized:

$$E_{i\text{team}} = \frac{1}{2} e_{i\text{team}}^2. \tag{58}$$

Hence the tuning for the actor network in the team for all $i \in \mathcal{N}$ can be found by application of gradient descent to Eq. (58) as follows:

$$\dot{\hat{W}}_{ai} = -\alpha_{ai} \phi_{ai}(s_i(t))^T e_{i\text{team}}. \tag{59}$$

In a similar way we can derive the tuning for the jammer.

Now we have to define the error e_{jam} for the jammer that we have to drive to zero. The error has the following form:

$$e_{\text{jam}} := \hat{W}_{\text{jam}}(t)^T \phi_{\text{jam}}(s_{N+1}(t)) + R_{N+1}^{-1} B_{N+1}^T \frac{\partial \phi(s)}{\partial s_{N+1}}^T \hat{W}_c.$$

To drive the error e_{jam} for the jammer we have to select $\hat{W}_{ai}$ such that the following criterion is minimized:

$$E_{\text{jam}} = \frac{1}{2} e_{\text{jam}}^2. \tag{60}$$

Hence the tuning for the jammer actor network can be found by application of gradient descent to Eq. (60) as follows:

$$\dot{\hat{W}}_{\text{jam}} = -\alpha_{\text{jam}} \phi_{\text{jam}}(s_{N+1}(t))^T e_{\text{jam}}. \tag{61}$$

Theorem 3. *Assume that the target set C_{stop} is closed. Consider the dynamics for the networked team given by Eq. (32) and the dynamics for the jammer given by Eq. (33). Also consider the control input for every agent in the team given by Eq. (51) and the control input for the jammer given by Eq. (52) and that the signal $\Delta\phi(s(t))$ is persistently exciting. The tuning law for approximating the cost is given by Eq. (53), the control input tuning law is given by Eq. (59), and the jammer control input is given by Eq. (61). Then the solution $\tilde{Z} := [s(t)^T \ \tilde{W}_c(t)^T \ \tilde{W}_{a1}(t)^T \ \tilde{W}_{a2}(t)^T \ \ldots \ \tilde{W}_{aN}(t)^T \ \tilde{W}_{\text{jam}}(t)^T]^T$ is uniformly* *ultimately bounded for $t \in [t_0, T]$ and thus as $T \to \infty$, the agents flock and stay connected provided that the number of basis sets is large enough to guarantee that the residual error $\epsilon \to 0$.*

Proof. The proof will appear elsewhere because of space limitations. □

3.7 Simulation

This section illustrates the effectiveness of the proposed game algorithm with $\rho = 1$ and all user-defined matrices equal to the identity in the cost for a five-agent graph with radius $k = 6$, and the jammer will connect to the closest agent. The system dynamics are given in the form of Eq. (28) with a marginally stable $F = \begin{bmatrix} 0 & 1 \\ 0 & 0 \end{bmatrix}$, $B_1 = \begin{bmatrix} 0 \\ 1 \end{bmatrix}$, $B_2 = \begin{bmatrix} 0 \\ 2 \end{bmatrix}$, $B_4 = \begin{bmatrix} 0 \\ 1 \end{bmatrix}$, and $B_5 = \begin{bmatrix} 0 \\ 3 \end{bmatrix}$, and the input matrix for the jammer is $B_6 = \begin{bmatrix} 0 \\ 1 \end{bmatrix}$. All the agents are moving in a Euclidean space with random initial conditions. The gains in the tuning laws are selected as $\alpha = 50$, $\alpha_i = 1$, and $\alpha_{\text{jam}} = 1$ for all $i \in \mathcal{N}$. To ensure persistence of excitation, a probing signal $p(t) = \tanh t \left(\sin(2t) \cos t + 4 \sin t^2 + 10 \cos(5t) \sin(11t)\right)$ is added to the signals of the vector σ for the first 2 s of the simulation. The graph is generated randomly. The running time is $T = 300$ s and the jammer did not succeed in breaking a link. Fig. 3 shows the initial topology of the graph in blue and the jammer shown by green squares. The agents move to avoid jamming in a new formation (shown in pink in Fig. 3), and the trajectories followed by the agents are shown in red in Fig. 3.

4 CONCLUSION AND FUTURE WORK

We have derived two model-free game theory-based actor-critic algorithms for reaching consensus of multiagent systems with guaranteed performance and with fixed graph topologies when the dynamics are perturbed by persistent actuation attackers and jammers. Specifically, it has been shown that the proposed architectures reject the adversarial inputs so as to achieve consensus and flock. Each local controller optimizes its own cost by using only neighborhood information, solving in each case a Nash game. The optimizations are performed online and provide a consensus. To solve the problem of the curse of dimensionality and the infeasibility of partial differential equations, the algorithm proposed uses three approximators for every agent—one to approximate the cost, one to approximate the control input, and one to approximate the adversarial input—and thus three tuning

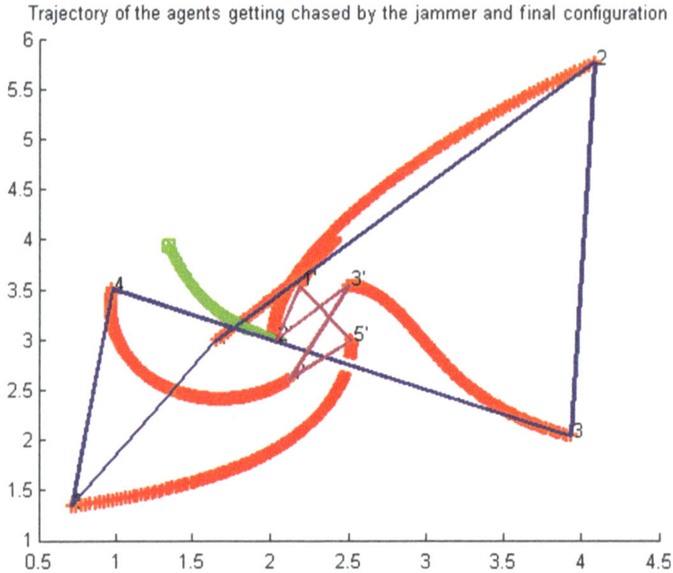

Fig. 3 The initial graph topology is illustrated in blue and the jammer is shown by green squares. The red lines show the trajectories of the agents to avoid the jammer until they reach a new formation, which is shown in pink.

laws are derived for every agent. Simulation results show the effectiveness of the proposed approaches.

So far we have identified and analyzed cyber attacks that involve actuation corruption and jamming attacks. We will study more sophisticated attacks such as the ones that use multiple points of attack or multiple methods. We will further look into coordinated attack possibilities where the attacker uses several attacks in a certain manner to induce more effective faults.

REFERENCES

[1] J.N. Tsitsiklis, Problems in decentralized decision making and computation (Ph.D. thesis), Massachusetts Institute of Technology, Boston, 1984.
[2] M. Mesbahi, M. Egerstedt, Graph Theoretic Methods for Multiagent Networks, Princeton University Press, Princeton, NJ, 2010.
[3] R. Olfati-Saber, J. Fax, R. Murray, Consensus and cooperation in networked multi-agent systems, Proc. IEEE 95 (1) (2007) 215–233.
[4] W. Ren, R. Beard, E. Atkins, A survey of consensus problems in multiagent coordination, in: Proceedings of the American Control Conference, 2005, pp. 1859–1864.
[5] Y. Shoham, K. Leyton-Brown, Multiagent Systems: Algorithmic, Game Theoretic, and Logical Foundations, Cambridge University Press, Cambridge, 2009.

[6] T. Alpcan, T. Basar, Network Security: A Decision and Game Theoretic Approach, Cambridge University Press, Cambridge, 2010.
[7] T. Basar, G.J. Olsder, Dynamic Noncooperative Game Theory, second ed., SIAM, Philadelphia, PA, 1999.
[8] J.P. Hespanha, H.J. Kim, S. Sastry, Multiple-agent probabilistic pursuit-evasion games, in: Proceedings of the 38th Conference on Decision and Control, vol. 3, 1999, pp. 2432–2437.
[9] R. Isaacs, Differential Games, Wiley, New York, NY, 1965.
[10] M. Wei, G. Chen, J.B. Cruz, L. Haynes, K. Pham, E. Blasch, Multi-pursuer multi-evader pursuit evasion games with jamming confrontation, J. Aerosp. Comput. Inf. Commun. 4 (3) (2007) 693–706.
[11] M.E. Harmon, L.C. Baird, A.H. Klopf, Reinforcement learning applied to a differential game, Adapt. Behav. 4 (1) (1995) 3–28.
[12] A. Cardenas, S. Amin, B. Sinopoli, A. Giani, A. Perrig, S. Sastry, Challenges for securing cyber physical systems, in: Workshop on Future Directions in Cyberphysical Systems Security, 2009.
[13] A. Cardenas, S. Amin, S.S. Sastry, Secure control: towards survivable cyber-physical systems, in: First International Workshop on Cyber-Physical Systems (WCPS2008), Beijing, China, 2008, pp. 495–500.
[14] F. Pasqualetti, A. Bicchi, F. Bullo, Consensus computation in unreliable networks: a system theoretic approach, IEEE Trans. Autom. Control 57 (1) (2012) 90–104.
[15] H.J. LeBlanc, X.D. Koutsoukos, Low complexity resilient consensus in networked multi-agent systems with adversaries, in: Proceedings of the 15th ACM international conference on Hybrid Systems: Computation and Control (HSCC '12), 2012, pp. 5–14.
[16] E. Semsar, K. Khorasani, Optimal control and game theoretic approaches to cooperative control of a team of multi-vehicle unmanned systems, in: Proceedings of the IEEE International Conference on Networking, Sensing and Control, 2007, pp. 628–633.
[17] E. Semsar-Kazerooni, K. Khorasani, An LMI approach to optimal consensus seeking in multi-agent systems, in: Proceedings of the American Control Conference, 2009, pp. 4519–4524.
[18] D. Bauso, L. Giarre, R. Pesenti, Mechanism design for optimal consensus problems, in: Proceedings of the Conference on Decision and Control, 2006, pp. 3381–3386.
[19] T. Yucelen, M. Egerstedt, Control of multiagent systems under persistent disturbances, in: Proceedings of the American Control Conference, 2012, pp. 5263–5269.
[20] M. Zhu, S. Martinez, Attack-resilient distributed formation control via online adaptation, in: Proc. CDC-ECE Conf., 2011, pp. 6624–6629.
[21] S. Gorman, Electricity grid in U.S. penetrated by spies, Wall Street J. (2009) A1, April 8, 2009.
[22] E.I. Verriest, F.L. Lewis, On linear quadratic minimum-time problem, IEEE Trans. Autom. Control 36 (7) (1991) 859–863.
[23] N. Elalami, N. Znaidi, On the discrete linear quadratic minimum-time problem, J. Franklin Inst. 335B (3) (1998) 525–532.
[24] R.S. Sutton, A.G. Barto, Reinforcement Learning: An Introduction, MIT Press, Cambridge, MA, 1998.
[25] D. Vrabie, K.G. Vamvoudakis, F.L. Lewis, Optimal adaptive control and differential games by reinforcement learning principles, in: Control Engineering Series, IET Press, London, 2012.
[26] K.G. Vamvoudakis, F.L. Lewis, G.R. Hudas, Multi-agent differential graphical games: online adaptive learning solution for synchronization with optimality, Automatica 48 (8) (2012) 1598–1611.

[27] P.A. Ioannou, B. Fidan, Adaptive control tutorial, in: Advances in Design and Control, SIAM, Philadelphia, PA, 2006.
[28] M. Ji, M. Egerstedt, Distributed coordination control of multi-agent systems while preserving connectedness, IEEE Trans. Robot. 23 (4) (2007) 693–703.
[29] S. Bhattacharya, T. Basar, Graph-theoretic approach for connectivity maintenance in dynamic networks in the presence of a jammer, in: Proceedings of the IEEE Conference on Decision and Control, 2010, pp. 3560–3565.
[30] E.W. Sandberg, Notes on uniform approximation of time-varying systems on finite time intervals, in: IEEE Transactions on Circuits and Systems-1: Fundamental Theory and Applications, vol. 45, 1998.
[31] K.G. Vamvoudakis, L.R. Garcia Carrillo, J.P. Hespanha, Learning consensus in adversarial environments, in: Proceedings of the SPIE Defense, Security and Sensing, Baltimore, MD, 2013.

SECTION 4

Networked Systems and Cooperative Control

CHAPTER ELEVEN

Adaptive Optimal Regulation of a Class of Uncertain Nonlinear Systems Using Event Sampled Neural Network Approximators

A. Sahoo*, S. Jagannathan†
*The DEI Group, Millersville, MD, United States
†Missouri University of Science and Technology, Rolla, MO, United States

1 INTRODUCTION

Growing interest in the networked control system (NCS) has given rise to an alternative control paradigm known as *event-triggered control* (ETC) [1–18] to reduce communication traffic and save computational load on the processors. Instead of transmission and execution of the controller in a traditional periodically sampled manner, the ETC approach provides a mechanism for deciding the sampling instants without compromising the desired performance. The analytically designed event-trigger or sampling condition allows the system error to grow before the transmission instant is decided without affecting the stability requirements of the system. This in turn reduces the communication and computation. Recently, various event-triggered control approaches have been presented in the literature [2–17], and different formulations have been introduced to analyze system stability. In general, the Lyapunov direct method is used to guarantee stability by designing an event-trigger or sampling condition.

Among earlier works, Tabuada [2] presented an event-triggered control scheme by assuming the input-to-state stability of the system with respect to measurement error. Further an event-trigger or sampling condition was developed to decide on the trigger or sampling instants at which to execute the controller with a desired closed-loop performance. A lower bound on the inter-event or inter-sample times is also guaranteed to avoid the accumulation point. The traditional ETC [2–4] was further extended to

Control of Complex Systems
http://dx.doi.org/10.1016/B978-0-12-805246-4.00011-2

a model-based scheme [5–8], which reduces the communication network traffic more effectively, demanding a higher level of computation. The ETC also finds application in large-scale and decentralized systems [9–11]. An extension to the ETC approach is the self-triggered control design [12–15], where the trigger instants are determined by the past state information and, hence, continuous monitoring of the current state is not required.

In previous work [2–7], known system dynamics were considered for the ETC design for both linear and nonlinear systems, with a few exceptions [8, 17]. Garcia and Antsaklis [8] considered known uncertainty for the system and developed a model-based event-triggered control scheme. Wang and Hovakimyan [17] proposed an L_1 adaptive control scheme where the nominal system dynamics are considered known, and uncertainties are compensated for by use of an adaptive term tuned with a projection-based tuning law. On the other hand, in our previous preliminary work [18], the requirement of complete knowledge of the system dynamics was relaxed by use of a neural network (NN)-based approximation of system dynamics, while a zero-order hold (ZOH) was used for the controller implementation.

In contrast, in this chapter we introduce the development of ETC of multi-input multi-output nonlinear continuous-time systems in affine form when the system and the controller are separated by an ideal communication network with no delays and packet losses. Further, an optimal control of such uncertain nonlinear continuous-time systems is also addressed. Instead of approximating the unknown nonlinear functions of the system dynamics by using two NNs [18], we approximate the controller by using a single linearly parameterized NN in the event-based sampling context under the assumption that the system states are measurable. An event-trigger condition based on the system state and estimated NN weight is designed to orchestrate the transmission of the state vector and the control input between the plant and the controller. Since the approximation is done with the event-sampled state vector, the event-trigger condition is made adaptive to obtain a trade-off between resource utilization and function approximation. In addition, the NN weights and the stabilizing control inputs are updated only at the event-trigger or sampling instants, which are aperiodic in nature, and are held until the next update. As a consequence, the proposed scheme reduces the overall computation when compared with the traditional NN schemes [19], where weights are updated periodically.

Optimal control [20, 21] of dynamic systems in continuous time is a challenging problem because of the difficulty involved in obtaining a closed-form solution to the Hamilton-Jacobi-Bellman (HJB) equation [21].

Adaptive dynamic programming (ADP) [20–32] techniques, on the other hand, are used to solve the optimal control of uncertain systems online by finding an approximated value function which becomes a solution to the HJB equation. Among the earlier work on ADP-based optimal control [20–32], an actor-critic NN-based framework [23] is used to generate an online yet approximate solution to the optimal control without the need of knowledge of the system dynamics. Later, online policy iteration schemes [23] are introduced to obtain the solution of the HJB equation and attain optimality. In addition, an alternative single NN-based ADP approach is presented in [29] for an affine nonlinear continuous-time system without use of an iterative technique. The NN weights are tuned online and periodically to achieve near optimality.

In practice, the system dynamics may not be known accurately for the traditional backward-in-time optimal control techniques to work. Therefore development of an ADP scheme in an event-triggered context is necessary and a challenging and open problem. Motivated by the above limitations, in this chapter we propose a novel ADP method [33] to solve the approximate optimal control of nonlinear continuous-time systems in an event-based sampling paradigm. After revisiting the NN approximation under this paradigm, we use NNs to identify the unknown system dynamics, and the value function which becomes the solution of the HJB equation. The optimal control policy is derived by use of both the approximated system dynamics and the value function.

In addition, to analyze the system stability and design the event-trigger or sampling condition, the nonlinear impulsive dynamical model of the closed-loop dynamics is considered. The well-developed Lyapunov approach for the nonlinear impulsive dynamical system [34] is used to study inter-event or inter-sample and sample time behavior, and is used to prove the local ultimate boundedness of the system state and the NN weight estimation errors. An NN-based control design for a traditional impulsive dynamical system was presented in [39]. In contrast to [39], in this chapter, we model the closed-loop event-triggered system as an impulsive dynamical system to analyze stability and performance. The main contributions of this chapter include (1) the design of an NN-based event-triggered stabilizing and optimal control policy of uncertain nonlinear continuous-time systems in affine form, (2) the design of an online approximate stabilizing and optimal controller in the event sampled context, (3) the development of aperiodic event-based NN weight update laws to reduce computation, (4) the design of novel adaptive event-trigger or sampling conditions for uncertain nonlinear dynamics to facilitate approximation and to maintain system stability and performance while reducing the transmission, and (5) the demonstration of closed-loop stability with the impulsive dynamical system formulation.

The remainder of this chapter is organized as follows. Section 2 discusses the preliminaries. Section 3 presents the state feedback design of the event-triggered control. The optimal regulator design is introduced in Section 4. The analytical results are verified with a numerical example in Section 5, and conclusions are presented in Section 6.

2 PRELIMINARIES

First, the notation used in this chapter is briefly introduced, followed by the stability notions. Subsequently, a brief background on traditional ETC is presented along with the problem formulation. Finally, the NN-based function approximation is revisited in the context of event-based sampling.

2.1 Notation

Let $\Re = (-\infty, \infty)$ be the set of real numbers, $\Re^+ = [0, \infty)$ the set of nonnegative real numbers, and $\Re^n$ the n-dimensional Euclidean space. For a vector, $x \in \Re^n$, $\|x\| \triangleq \sqrt{x^T x}$ is its 2-norm. For a matrix $A \in \Re^{n\times m}$, $\|A\| \triangleq \sqrt{\sum_{i=1}^{n} \sum_{j=1}^{m} |a_{ij}|^2}$ denotes the Frobenius norm. The transpose of A is denoted by $A^T \in \Re^{m\times n}$, and $\text{tr}\{A\}$ is the trace operator of a square matrix. Let $W = \begin{bmatrix} a & b \\ c & d \end{bmatrix}$ be a matrix, then $\mathbf{vec}(W) = [a \quad b \quad c \quad d]^T$ is the vectorization of the matrix W and $\mathbf{vec}(W)^T \mathbf{vec}(W) = \text{tr}\{W^T W\}$. For a square matrix $A \in \Re^{n\times n}$, we denote $\lambda_{\max}(A)$ and $\lambda_{\min}(A)$ as the maximum and minimum eigenvalues of A.

2.2 Stability Notion

Consider a nonlinear impulsive dynamical system represented as

$$\dot{\xi} = F_c(\xi), \quad \xi(0) = \xi_0, \quad \xi \in \mathcal{C} \subset \mathcal{D}, \quad \xi \notin \mathcal{Z}, \tag{1}$$

$$\Delta\xi = F_d(\xi), \quad \xi \in \mathcal{D}, \quad \xi \in \mathcal{Z}, \tag{2}$$

where $\xi \in \mathcal{D} \subset \Re^{n_\xi}$ is the state vector, $\mathcal{C} \subset \mathcal{D}$ and $\mathcal{Z} \subset \mathcal{D}$, which are, respectively, the flow and the jump sets, and $\mathcal{D}$ is an open set with

$0 \in \mathcal{D}$, $\Delta\xi = \xi(t^+) - \xi(t)$, where $\xi(t^+) = \lim_{\varepsilon\to 0} \xi\,(t+\varepsilon)$. The nonlinear functions $F_c\,(\xi) \in \Re^{n_\xi}$ and $F_d\,(\xi) \in \Re^{n_\xi}$ are, respectively, the continuous and reset dynamics of the impulsive dynamical system. Next, we state the definitions.

Definition 1 ([34]). The nonlinear impulsive dynamical system given by Eqs. (1) and (2) is locally bounded if there exists a $\gamma > 0$ such that for every $\delta \in (0,\gamma)$ there exists $\varepsilon = \varepsilon(\delta) > 0$ such that $\|\xi(0)\| < \delta$ implies $\|\xi\,(t)\| < \varepsilon,\ \ t \geq 0$.

Definition 2 ([34]). The nonlinear state-dependent impulsive dynamical system given by Eqs. (1) and (2) is ultimately bounded with bound ε if there exists $\gamma > 0$ such that for every $\delta \in (0,\gamma)$ there exists $T = T(\delta,\varepsilon) > 0$ such that $\|\xi(0)\| < \delta$ implies $\|\xi\,(t)\| < \varepsilon$, $t \geq T$ and is globally ultimately bounded with bound ε if, for every $\delta \in (0,\infty)$ there exists, $T = T(\delta,\varepsilon) > 0$ such that $\|\xi(0)\| < \delta$ implies $\|\xi(t)\| < \varepsilon,\ \ t \geq T$.

Definition 3 ([35]). A continuous function $\alpha : [0,a) \to \Re^+$ is said to belong to class $\mathcal{K}$ if it is strictly increasing and $\alpha\,(0) = 0$. It is said to belong to class K_∞ if $a = \infty$ and $\alpha\,(r) \to \infty$ as $r \to \infty$.

To prove the ultimate boundedness of the impulsive dynamical systems the following lemma will be used.

Lemma 1 ([34]). *Consider the impulsive dynamical system given by Eqs. (1) and (2). Assume that the jumps occur at distinct time instants and there exists a continuously differentiable function $V : \mathcal{D} \to \Re$ and class $\mathcal{K}$ functions $\alpha\,(\cdot)$ and $\beta\,(\cdot)$ such that*

$$\alpha\left(\|\xi\|\right) \leq V\left(\xi\right) \leq \beta\left(\|\xi\|\right),\ \ \xi \in \mathcal{D}, \tag{3}$$

$$\frac{\partial V\left(\xi\right)}{\partial \xi} F_c\left(\xi\right) < 0,\ \ \xi \in \mathcal{D},\ \ \xi \notin Z,\ \ \|\xi\| > \mu, \tag{4}$$

$$V\left(\xi + F_d\left(\xi\right)\right) - V\left(\xi\right) \leq 0,\ \ \xi \in \mathcal{D},\ \ \xi \in \mathcal{Z},\ \ \|\xi\| > \mu, \tag{5}$$

where $\mu > 0$ is such that $B_{\alpha^{-1}(\beta(\mu))} = \left\{\xi \in \Re^{n_\xi} : \|\xi\| < \alpha^{-1}\left(\beta\left(\mu\right)\right)\right\} \subset \mathcal{D}$ with $\eta > \beta\left(\mu\right)$. Further, assume $\theta \triangleq \sup_{\xi\in\bar{B}_\mu\cap Z} V\left(\xi + F_d\left(\xi\right)\right)$ exists. Then the nonlinear state-dependent impulsive dynamical system given by Eqs. (1) and (2) is ultimately bounded with bound $\Xi \triangleq \alpha^{-1}\left(\eta\right)$, where $\eta \triangleq \max\left\{\beta\left(\mu\right),\theta\right\}$. Furthermore, $\limsup_{t\to\infty} \|\xi(t)\| \leq \alpha^{-1}\left(\beta(\mu)\right)$.

In the next subsection, the problem formulation along with a brief background on the traditional event-triggered control will be introduced.

2.3 Background and Problem Formulation

Consider the multiinput-multioutput nonlinear uncertain continuous-time system represented in the affine form as

$$\dot{x} = f(x) + g(x)u, \quad x(0) = x_0, \tag{6}$$

where $x = [x_1 \quad x_2 \quad \cdots \quad x_n]^T \in \Re^{n_x}$ and $u \in \Re^{m_u}$ are the state and the input vectors of system (6), respectively. The nonlinear vector function, $f(x) \in \Re^{n_x}$, and the matrix function, $g(x) \in \Re^{n_x \times m_u}$, represent the internal dynamics and control coefficient function, respectively. The following assumption on system dynamics is needed to proceed.

Assumption 1. System (6) is controllable and input-to-state linearizable [35]. The internal dynamics $f(x)$ and control coefficient $g(x)$ are considered unknown, with the control coefficient matrix, $g(x)$, bounded above in a compact set for all $x \in \Omega_x \subset \Re^{n_x}$, satisfying $\|g(x)\| \le g_{\max}$, with $g_{\max} > 0$ being a known positive constant [19].

The input-to-state linearizable assumption is satisfied by a wide variety of practical systems, such as a robot manipulator and a mass damper system. For these classes of controllable nonlinear systems [35] in affine form with complete knowledge of the system dynamics, $f(x)$ and $g(x)$, it is demonstrated that there exists an ideal control input u_d for system (6) of the form

$$u_d = K(x) \tag{7}$$

that renders the closed-loop system asymptotically stable, where $K(x)$ is a function of the system state vector. The linear closed-loop dynamics can be represented by

$$\dot{x} = Ax, \tag{8}$$

where A is a Hurwitz matrix. By the converse Lyapunov theorem [35], an asymptotically stable system admits a Lyapunov function, $V(x) : \Omega_x \to \Re$, which satisfies the following inequalities:

$$\alpha_1(\|x\|) \le V(x) \le \alpha_2(\|x\|), \tag{9}$$

$$\dot{V}(x) \le -\alpha_3(\|x\|), \tag{10}$$

where α_1, α_2, and α_3 are class $\mathcal{K}$ functions. Moreover, for a standard quadratic Lyapunov function $V(x) = x^T Px$, for the closed-loop system (8), the class $\mathcal{K}$ functions are expressed as $\alpha_1(\|x\|) = \lambda_{\min}(P)\|x\|^2$, $\alpha_2(\|x\|) = \lambda_{\max}(P)\|x\|^2$, and $\alpha_3(\|x\|) = \lambda_{\min}(Q)\|x\|^2$. The matrices

$P \in \Re^{n_x \times n_x}$ and $Q \in \Re^{n_x \times n_x}$ are symmetric and positive definite and satisfy the Lyapunov equation given by

$$A^T P + PA = -Q. \tag{11}$$

In the case of a traditional NCS, the state vector x and the control input u are transmitted with a fixed sampling interval T_s. On the other hand, in an event sampled context, the system state vector is sampled and transmitted at the event-based sampling instants only. Let $\{\tau_k\}$ for $k = 1, 2, \ldots$ be a monotonically increasing sequence of time instants with $\tau_0 = 0$ such that $\tau_{k+1} > \tau_k$ and $\tau_k \to \infty$ as $k \to \infty$ represent when the events are triggered or the event-trigger or sampling instants and the system states, $x(\tau_k)$, and control inputs, $u(\tau_k)$, are transmitted. The event sampled/transmitted state and the control input vectors are held, respectively, at the controller and the plant by the ZOHs. It is important to note that τ_k is a function of system state x and the last transmitted system state $\breve{x} = x(\tau_k)$, $\tau_k < t \leq \tau_{k+1}$, and is aperiodic in nature. We define the event-trigger error, $e_s \in \Re^{n_x}$, as

$$e_s = x - \breve{x}, \quad \tau_k < t \leq \tau_{k+1}. \tag{12}$$

The trigger/sampling instant, τ_k, is determined by the event-trigger condition that consists of the event-trigger error (12) and a state-dependent threshold. Once the event-trigger error exceeds the threshold (time instant, $t = \tau_k$), the sensor and trigger mechanism initiates the transmission of the current state vector x. The last held event-based state vector $\breve{x}$ jumps to the new value (ie, $\breve{x}^+ = x$) for $t = \tau_k$ and is held for $\tau_k < t \leq \tau_{k+1}$, where $\breve{x}^+ = \breve{x}\left(\tau_k^+\right)$ and τ_k^+ is the time instant just after τ_k. The event-trigger error is reset to zero for the next event to occur. It is given by

$$e_s^+ = 0 \quad \text{for } t = \tau_k. \tag{13}$$

Since the system dynamics $f(x)$ and $g(x)$ are considered unknown, the controller given by Eq. (7) cannot be realized. Further, in the event-based sampling and transmission context, the intermittent availability of the system state vector at the controller precludes the traditional NN-based approximation with a periodic update of the NN weights. Therefore the NN function approximation property is revisited under the event-based sampling and transmission.

2.4 Function Approximation

By the universal approximation property of NNs, any continuous function $f(x)$ can be approximated over a compact set for all $x \in \Omega_x \subset \Re^{n_x}$ up to a desired level of accuracy ε_f by the selection of suitable activation functions and an adequate number of hidden layer neurons. Alternatively, there exists an unknown target weight matrix W such that $f(x)$ in a compact set can be written as

$$f(x) = W^T\varphi(V^Tx) + \varepsilon_f(x), \tag{14}$$

where $W \in \Re^{l\times b}$ and $V \in \Re^{a\times l}$ represent the target NN weight matrix for the output and input layers, respectively, and are defined as

$$(W, V) = \arg \min_{(W,\ V)} \left[\sup_{x\in\Omega_x} \left\| W^T\varphi(V^Tx) - f(x) \right\| \right]. \tag{15}$$

The activation function $\varphi(.) : \Re^a \to \Re^l$ is a hyperbolic tangent activation function and is given by $\varphi(.) = e^{2\bar{x}} - 1/e^{2\bar{x}} + 1$, with $\bar{x} = V^Tx$. The term $\varepsilon_f(x) \in \Re^{n_x}$ is the traditional reconstruction error and the constants l, a, and b are the number of neurons in the hidden layer, input layer, and output layer of the NN, respectively.

In this chapter we will consider the linearly parameterized [36] NNs, as shown in Fig. 1A, to approximate the unknown function as in Eq. (14), where the output layer weights $W \in \Re^{l\times b}$ are updated and the input layer weight matrix $V \in \Re^{a\times l}$ is initialized at random and held. This linearly parameterized NN is also known as a *random vector functional link network* [36]. The activation function $\varphi(V^Tx)$ forms a basis for the unknown function and the universal approximation property is retained [36]. The output layer activation functions, $\varphi_o(.)$ are selected to be purely linear.

With intermittent event-based transmission of the system state vector, x, the universal NN approximation property can be extended to achieve the

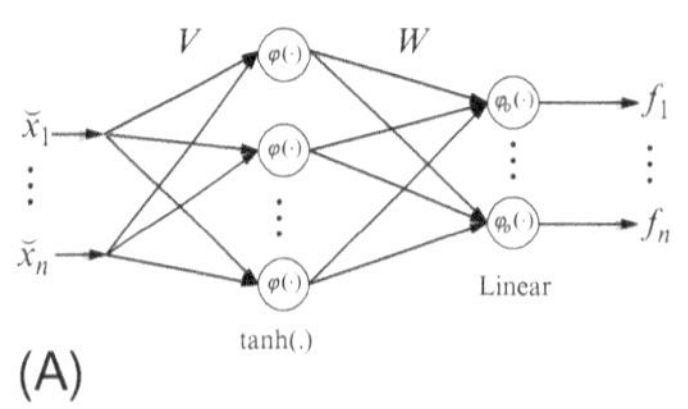

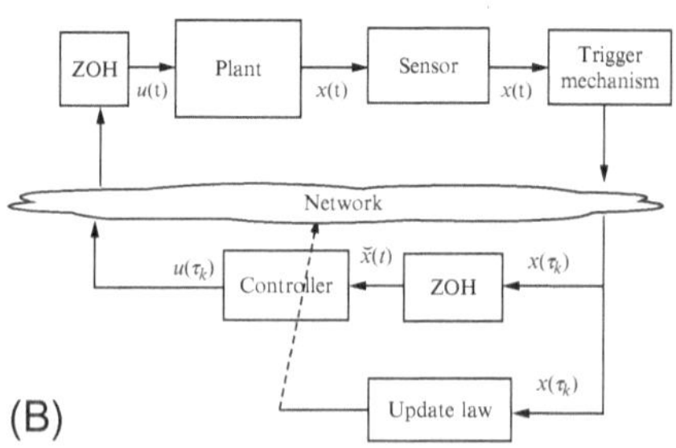

Fig. 1 (A) Neural network structure with an event-based activation function. (B) Structure of the adaptive state feedback event-triggered control system. *ZOH*, zero-order hold.

desired level of accuracy by proper design of an event-trigger condition. The trigger condition will generate the required number of events for the availability of the system state vector for approximation. The theorem introduced next extends the NN approximation property for event-based sampling.

Theorem 1 ([18]). *Let $\vec{f}(x) : \Omega_x \rightarrow \Re^{n_x}$ be a smooth and uniformly continuous function in a compact set for all $x \in \Omega_x \subset \Re^{n_x}$. Then there exists a single-layer NN with a sufficient number of neurons such that the function $\vec{f}(x)$ can be approximated with constant weights and an event sampled activation function such that*

$$\vec{f}(x) = W^T \varphi(\breve{x}) + \varepsilon_e(\breve{x}, e_s), \tag{16}$$

where $W \in \Re^{l \times n_x}$ is the target NN weight matrix, with l being the number of hidden-layer neurons, $\varphi(\breve{x})$ is the bounded event sampled activation function, and $\varepsilon_e(\breve{x}, e_s)$ is the event sampled NN reconstruction error, with $\breve{x}$ representing the last event sampled state held at the ZOH.

Remark 1. From the proof of Theorem 1, the event sampled reconstruction error $\varepsilon_e(\breve{x}, e_s) = W^T \left[\varphi(\vartheta(\breve{x}, e_s)) - \varphi(\breve{x})\right] + \varepsilon_{\vec{f}}(\vartheta(\breve{x}, e_s))$, where $\vartheta(\breve{x}, e_s) = \breve{x} + e_s$ is a function of the traditional reconstruction error $\varepsilon_{\vec{f}}(\cdot)$ and the event-trigger error e_s as in Eq. (12). A small event-based reconstruction error $\varepsilon_e(\breve{x}, e_s)$ can be observed if the frequency of event-based samples is increased. This requires a suitable event-trigger condition to obtain both approximation accuracy and a reduction in computation. A small event-based reconstruction error means a higher number of events, which results in more computations and transmissions. Hence a trade-off exists between reconstruction error and transmission.

The adaptive ETC scheme development is introduced next.

3 ADAPTIVE EVENT-TRIGGERED STATE FEEDBACK CONTROL

In this section, a state feedback design of the NN-based adaptive ETC is proposed.

3.1 Structure

The structure of the proposed adaptive ETC scheme with a communication network between the plant and the controller is depicted in Fig. 1B.

Further, for simplicity, the following assumption regarding the network is considered.

Assumption 2. The communication network between the plant and the controller is ideal [3, 17]—that is, the networked induced delays, including the computational delay, and the packet losses are not present.

In the proposed scheme a smart sensor and a trigger mechanism are included at the plant to decide on the event-trigger or sampling instants by evaluation of the event-trigger condition continuously. The state vector is transmitted first when the trigger condition is satisfied and then the controller is updated and transmitted to the plant. The ZOHs are used to hold the last transmitted state and control input, respectively, at the controller and the plant until the next transmission is received.

Since the system dynamics are considered unknown, the control input is approximated by use of an NN with event sampled inputs. Further, the NN weights are updated at every trigger instant aperiodically and are held during the inter-event or inter-sample durations. To achieve the desired approximation accuracy, an adaptive event-trigger condition is designed to generate the required number of events during the learning phase. Thus the event-trigger condition becomes a function of the NN weight estimates and the system state vector, whereas in the traditional ETC design, it is a function of the system state only [2, 3]. Therefore to evaluate the event-trigger condition locally, without transmission of the estimated NN weights, the NN weights are updated at both the trigger mechanism and the controller in synchrony. This increases the computation but because of the event-based aperiodic update at both places the overall computation reduces. The event-triggered controller design is presented next.

3.2 Controller Design

In this subsection the approximation-based event-triggered controller design is presented. By the universal approximation property of the NNs, the ideal control input (7) is written as

$$u_d = W_u^T \varphi_u(\bar{x}) + \varepsilon_u(x), \tag{17}$$

where $W_u \in \Re^{l_u \times m_u}$ is the output layer unknown ideal NN weight matrix and $\varphi_u(\bar{x}) \in \Re^{l_u}$ is the tangent hyperbolic activation function, with $\bar{x} = V_u^T x$. The function $\varepsilon_u(x) \in \Re^{m_u}$ is the traditional NN reconstruction error and $V_u \in \Re^{n_x \times l_u}$ is the input layer weight matrix, where l_u, n_x, and m_u are the number of neurons in the hidden layer, input layer, and output layer of the NN, respectively. Before presenting the approximation-based controller design, we introduce the following standard assumptions for the NN.

Assumption 3 ([19]). The target weights, W_u , the activation function, $\varphi_u(.)$, and the reconstruction error, $\varepsilon_u(.)$, of the NN are upper bounded in a compact set such that $\| W_u\| \leq W_{u,\max}$, $\|\varphi_u(.)\| \leq \varphi_{u,\max}$, and $\|\varepsilon_u(.)\| \leq \varepsilon_{u,\max}$, where $W_{u,\max}$, $\varphi_{u,\max}$, and $\varepsilon_{u,\max}$ are positive constants.

Assumption 4. The NN activation function, $\varphi_u(\bar{x})$, is considered to be Lipschitz continuous in a compact set for all $x \in \Omega_x \subset \Re^{n_x}$. Then for every $x \in \Omega_x \subset \Re^{n_x}$ there exists a constant $L_{\varphi_u} > 0$ such that $\left\|\varphi_u(\bar{x}) - \varphi_u\left(\bar{\breve{x}}\right)\right\| \leq L_{\varphi_u} \|x - \breve{x}\|$ is satisfied.

In the event-triggered context, the actual controller uses the event-based state vector $\breve{x}$ held at the ZOH. Hence, by Theorem 1, the actual event-based control input is represented as

$$u = \hat{W}_u^T \varphi_u(\bar{\breve{x}}), \quad \tau_k < t \leq \tau_{k+1}, \tag{18}$$

where $\hat{W}_u \in \Re^{l_u \times m_u}$ is the estimated NN weight matrix and $\varphi_u\left(\bar{\breve{x}}\right) \in \Re^{l_u}$ is the event sampled NN activation function, where $\bar{\breve{x}} = V_u^T \breve{x}$ is the scaled input to the NN. Since the last held state, $\breve{x}$ and the NN weights are updated at the event-trigger instants, $t = \tau_k$, the control input is also updated at the trigger instants, and is then transmitted to the plant and held by the ZOH until the next update is received.

Further, as proposed, the estimated NN weights, $\hat{W}_u \in \Re^{l_u \times m_u}$, are held during inter-event durations $\tau_k < t \leq \tau_{k+1}$ and are updated at the trigger instants or referred to as jumps at $t = \tau_k$. Therefore the NN update law during inter-event durations is defined as

$$\dot{\hat{W}}_u = 0, \quad \tau_k < t \leq \tau_{k+1}. \tag{19}$$

Further, at the event-trigger instants, the update law is selected as

$$\hat{W}_u^+ = \hat{W}_u - \frac{\alpha_u}{c + \|e_s\|^2} \varphi_u(\bar{x})\, e_s^T L - \kappa \hat{W}_u, \quad t = \tau_k, \tag{20}$$

where $\hat{W}_u^+ \in \Re^{l_u \times m_u}$ is the updated NN weight estimate just after the trigger instant, $\alpha_u > 0$ is the NN learning rate, $c > 0$ is a positive constant, $L \in \Re^{n_x \times m_u}$ is a design matrix to match the dimension, and $\kappa > 0$ is a positive constant serving the same role as the sigma modification [37] in the traditional adaptive control. The update law (20) uses traditional activation $\varphi_u(\bar{x})$ since the system state vector, x, is available for the update at the trigger instants. Next we define the NN weight estimation error as $\tilde{W}_u = W_u - \hat{W}_u$. The weight estimation error dynamics during the flow, by use of Eq. (19), can be written as

$$\dot{\tilde{W}}_u = \dot{W}_u - \dot{\hat{W}}_u = 0 \quad \tau_k < t \le \tau_{k+1}, \tag{21}$$

while for the jump instant, $t = \tau_k$, the NN weight estimation error dynamics derived from Eq. (20) becomes

$$\tilde{W}_u^+ = W_u - \hat{W}_u^+ = \tilde{W}_u + \alpha_u \chi_s \varphi_u(\bar{x})\, e_s^T L + \kappa \hat{W}_u, \quad t = \tau_k, \tag{22}$$

with $\chi_s = 1/\left(c + \|e_s\|^2\right)$. In the next subsection we will formulate the closed-loop dynamics of the adaptive ETC system as a nonlinear impulsive dynamical system to analyze the system behavior for both inter-event or inter-sample and sample time instants.

3.3 Closed-Loop System: Impulsive Dynamical Model

As per the proposed scheme the last transmitted state and the NN weights are updated at the trigger instants only. Hence the closed-loop event-triggered system behaves as an impulsive dynamical system. Assuming the event instants are distinct, ie, there exists a nonzero lower bound on the inter-event times, $\delta\tau_k = \tau_{k+1} - \tau_k > 0$, which is proven in [18], the closed-loop dynamics can be formulated in two steps. The first step toward the impulsive system modeling is to formulate the flow dynamics. The closed-loop system dynamics during the flow interval for $t \in (\tau_k, \tau_{k+1}]$ can be derived by use of both Eqs. (6) and (18), and is represented as

$$\dot{x} = f(x) + g(x)\hat{W}_u^T \varphi_u(\breve{x}), \quad t \in (\tau_k, \tau_{k+1}]. \tag{23}$$

Adding and subtracting the ideal control input u_d yields

$$\dot{x} = f(x) + g(x)\left(\hat{W}_u^T \varphi_u(\breve{x}) + u_d - u_d\right), \quad t \in (\tau_k, \tau_{k+1}]. \tag{24}$$

If we recall the NN approximation of the ideal controller (18) and the ideal closed-loop dynamics (8), Eq. (24) becomes

$$\dot{x} = Ax + g(x)\left(\hat{W}_u^T \varphi_u\left(\breve{x}\right) - W_u^T \varphi_u(\bar{x}) - \varepsilon_u(x)\right), \quad t \in (\tau_k, \tau_{k+1}]. \tag{25}$$

From the definition $W_u = \hat{W}_u + \tilde{W}_u$, the closed-loop dynamics (25) can be written as

$$\dot{x} = Ax - g(x)\left(\tilde{W}_u^T \varphi_u(\bar{x}) + \varepsilon_u(x)\right) + g(x)\left(\hat{W}_u^T \varphi_u\left(\bar{\breve{x}}\right) - \hat{W}_u^T \varphi_u(\bar{x})\right),$$
$$t \in (\tau_k, \tau_{k+1}]. \tag{26}$$

Similarly, the dynamics of the last transmitted state vector, $\breve{x}$, held by the ZOH, during the flow interval becomes

$$\dot{\breve{x}} = 0, \quad t \in (\tau_k, \tau_{k+1}]. \tag{27}$$

Further, the flow dynamics of the NN weight estimation error is given by Eq. (21).

In the second and final step, it remains only to formulate the reset dynamics to complete the impulsive modeling of the event-triggered system. This consists of the jumps in the system state and can be represented as

$$\breve{x}^+ = x, \quad t = \tau_k, \tag{28}$$

the last transmitted state held by the ZOH,

$$\breve{x}^+ = x, \quad t = \tau_k. \tag{29}$$

and the NN weight estimation error dynamics (22). From Eqs. (28), (29), and (22), the reset dynamics for the system are given by

$$\Delta x = x^+ - x = 0, \quad t = \tau_k, \tag{30}$$

$$\Delta \breve{x} = \breve{x}^+ - \breve{x} = x - \breve{x} = e_s, \quad t = \tau_k, \tag{31}$$

and

$$\Delta \tilde{W}_u = \tilde{W}_u^+ - \tilde{W}_u = \alpha_u \chi_s \varphi_u(\bar{x})\, e_s^T L + \kappa \hat{W}_u, \quad t = \tau_k. \tag{32}$$

To formulate the impulsive dynamical system, we consider $\xi_s = \begin{bmatrix} x^T & \breve{x}^T & \mathbf{vec}(\tilde{W}_u)^T \end{bmatrix} \in \Re^{n_{\xi_s}}$ as the augmented states, where $\mathbf{vec}\left(\tilde{W}_u\right) \in \Re^{l_u m_u}$ is the vector form of the NN weight estimation error matrix and $n_{\xi_s} = n_x + n_x + l_u m_u$. Now we combine Eqs. (26), (27), and (21) to obtain the flow dynamics as

$$\dot{\xi}_s = F_c^s(\xi_s), \quad \xi_s \in C \subset D_s, \quad \xi_s \notin Z_s. \tag{33}$$

Next we combine Eqs. (30)–(32) to get the reset dynamics as

$$\Delta \xi_s = F_d^s(\xi_s), \quad \xi_s \in D_s, \quad \xi_s \in Z_s \tag{34}$$

for the nonlinear impulsive dynamical system, where the nonlinear functions $F_c^s(\xi_s)$ and $F_d^s(\xi_s)$ are defined as $F_c^s(\xi_s) = \begin{bmatrix} H(\xi_s) \\ 0 \\ 0 \end{bmatrix}$ and

$F_d^s(\xi_s) = \begin{bmatrix} 0 \\ e_s \\ \mathbf{vec}\left(\alpha_u \chi_s \varphi_u(\bar{x}) e_s^T L + \kappa \hat{W}_u\right) \end{bmatrix}$, with $H(\xi_s) = Ax - g(x)$ $\left(\tilde{W}_u^T \varphi_u(\bar{x}) + \varepsilon_u(x)\right) + g(x)\left(\hat{W}_u^T \varphi_u\left(\bar{\bar{x}}\right) - \hat{W}_u^T \varphi_u(\bar{x})\right)$ and $\Delta\xi_s = \xi_s^+ - \xi_s$. The set $D_s \subset \Re^{n_\xi}$ is an open set with $0 \in D_s$. The flow set $C_s \subset D_s$ is defined as $C_s = \{\xi_s \in D_s : \|e_s\| \leq \sigma_s \|x\|\}$ and the jump set $Z_s \subset D_s$ is defined as $Z_s = \{\xi_s \in D : \|e_s\| > \sigma_s \|x\|\}$, where $\sigma_s \|x\|$ is the event-trigger threshold to be designed next.

3.4 Stability Analysis

In this section the stability results of the closed-loop system are established. Before we proceed, the following lemma for the boundedness of the NN weight estimation error both during the flow and during the jump instants is necessary.

Lemma 2. *[Boundedness of the NN weight estimation error [18]] Consider the nonlinear continuous-time system* (6) *and the controller* (18) *expressed as a nonlinear impulsive dynamical system given by Eqs. (33) and* (34). *Let Assumptions 1–4 be satisfied and the initial NN weights,* $\hat{W}_u(0)$, *be nonzero and initialized in the compact set* Ω_{W_u}. *Under the assumption that a nonzero positive lower bound on the inter-event times,* $\delta\tau_k = \tau_{k+1} - \tau_k > 0$, $k \in \aleph$ *exists, there exist positive constants* $\alpha_u > 0$, $0 < \kappa < 1/2$, $\bar{T}$ *and* T *such that the weight estimation error,* $\tilde{W}_u$, *is bounded during the flow period and ultimately bounded for all* $\tau_k > \bar{T}$ *or, alternatively* $t > T$ *when the NN weights are updated by use of Eqs. (19) and* (20).

Next we introduce the event-trigger or sampling condition given by

$$D(\|e_s\|) > \sigma_s \|x\|, \tag{35}$$

where

$$\sigma_s = \Gamma_s q_{\min} / 4 g_{\max} L_{\varphi_u} \left\| \hat{W}_u \right\| \|P\| \tag{36}$$

is the threshold coefficient, where $\left\| \hat{W}_u \right\| > \kappa_1$, $0 < \Gamma_s < 1$ and L_{φ_u} is the Lipschitz constant for the activation functions, $q_{\min}$ is the minimum eigenvalue of Q, and P is a symmetric positive definite matrix, with P and Q satisfying Eq. (11), and $D(.)$ is a dead-zone operator defined as

$$D(\|e_s\|) = \begin{cases} \|e_s\|, & \text{if } \|x\| > B^x_{s,\max}, \\ 0, & \text{otherwise}, \end{cases} \tag{37}$$

where $B^x_{s,\max}$ is the bound for the system state vector x. The system state vector is transmitted to the controller and the updated control input is transmitted to the plant when the event-trigger condition (35) is satisfied. Further, an event is triggered when the NN weight estimate $\left\|\hat{W}_u\right\| \le \kappa_1$ irrespective of the trigger condition (35), where κ_1 is a small positive constant. This ensures the adaptive threshold is well defined.

Next our main result on the local ultimate boundedness of the closed-loop impulsive dynamical system is introduced by use of the adaptive event-trigger condition (35) with the help of the Lyapunov approach [34].

Theorem 2 (Closed-loop stability). *Consider the nonlinear system* (6), *the control input* (18), *and NN update laws* (19) *and* (20), *expressed as an impulsive dynamical system given by Eqs. (33) and* (34). *Let Assumptions 1–4 hold. Assume there exists a nonzero positive lower bound on the inter-event times given by* $\delta\tau_k = \tau_{k+1} - \tau_k > 0$, $k \in \aleph$ *and the initial NN weights,* $\hat{W}_u(0)$, *are nonzero and initialized in the compact set* Ω_{W_u}. *Then the closed-loop system state vector* ξ_s *for any initial condition* $\xi_s(0) \in \mathcal{D}_s \subset \Re^{n_\xi}$ *is locally ultimately bounded with a bound* $\|\xi_s\| \le \Xi$ *provided the events are triggered when condition* (35) *and* $\left\|\hat{W}_u\right\| \le \kappa$ *are satisfied. Further, the ultimate bound is given by*

$$\Xi = \sqrt{\eta/\lambda_{\min}(\bar{P})}, \tag{38}$$

where $\bar{P} = \operatorname{diag}\{P, P, I\}$ *is a positive definite matrix, where* I *is the identity matrix with appropriate dimension, and* $\eta = \max\left\{\lambda_{\max}(\bar{P})\mu_s^2, \theta\right\}$, *where* $\theta = \sup_{\xi_s \in \bar{B}_{\mu_s} \cap Z_s} V_s\left(\xi_s + F_d^s(\xi_s)\right)$ *and* $\mu_s = \max\left\{B^x_{s,\max}, B^{\breve{x}}_{s,\max}, B^{\tilde{W}}_{s,\max}\right\}$, *where* $B^x_{s,\max}$, $B^{\breve{x}}_{s,\max}$, *and* $B^{\tilde{W}}_{s,\max}$ *are the bounds for the system state,* x, *the last transmitted state,* $\breve{x}$, *and the NN weight estimation error,* $\tilde{W}_u$, *respectively.*

Remark 2. The threshold coefficient σ_s of the event-trigger condition (1.35) is a function of the norm of NN weight estimates $\left\|\hat{W}_u\right\|$ and, hence, adaptive in nature. Since the weights are updated only at the trigger instant, $\left\|\hat{W}_u\right\|$ is piecewise constant and jumps at the trigger instant $t = \tau_k$, according to the update law (20). This implies that σ_s is also a piecewise constant function and changes at the trigger instant. This variation in σ_s implicitly depends on the NN weight estimation error, $\tilde{W}_u$, which generates the required number of triggers for the NN approximation of the control input during the learning phase. Once the NN weight matrix, $\hat{W}_u$, has converged close to the unknown constant target weight matrix, W_u, the weight estimates, $\hat{W}_u$ become steady; in turn, σ_s becomes a constant like the traditional event-triggered control with known knowledge of the system dynamics [2, 5–16].

Remark 3. The dead-zone operator $D(\cdot)$ is used to stop the unnecessary triggering of events due to the NN reconstruction error once the state vector had reached and stayed within the ultimately bounded region. This implies that for an event to trigger, the following three conditions need to be satisfied:

1. The system state vector is outside the bound—that is, $\|x\| > B^x_{s,\max}$.
2. The event-trigger condition (35) is satisfied—that is, $\|e_s\| > \sigma_s \|x\|$.
3. The NN weight estimates $\left\| \hat{W}_u \right\| \leq \kappa$.

Remark 4. The assumption on the nonzero positive lower bound on inter-event times in Theorem 2 is given in [18]. In addition, an explicit formula for analysis of the lower bound on the inter-event times when the system state vector $\|x\| > B^x_{s,\max}$ to avoid the accumulation point is also derived.

Remark 5. From the proof [18], the system state vector, x, and the NN weight estimation error, $\tilde{W}_u$, remain locally ultimately bounded for all $\tau_k > \bar{T}$ or alternatively for all $t > T$, where the time T depends on $\bar{T}$. This implies that the control input and the event-trigger error are also locally ultimately bounded. Consequently, all the closed-loop system parameters remain ultimately bounded for all time $t > T$.

4 APPROXIMATE OPTIMAL CONTROLLER DESIGN

4.1 Background

Consider the controllable nonlinear continuous-time systems in affine form represented as

$$\dot{x}(t) = f(x(t)) + g(x(t))u(t), x(0) = x_0, \tag{39}$$

where $x \in \Re^n$ and $u \in \Re^m$ represent the state and control input vectors, respectively. The unknown functions $f(x) \in \Re^n$ and $g(x) \in \Re^{n\times m}$ represent the nonlinear system dynamics satisfying $f(0) = 0$ with the following assumption.

Assumption 5 ([29]). The nonlinear system is controllable and observable. The nonlinear matrix function $g(x)$ for all $x \in \Omega_x$ satisfies $g_m \leq \|g(x)\| \leq g_M$, where g_M and g_m are known positive constants and Ω_x is a compact set. Consider the value function

$$V(t) = \int_t^{\infty} r(x(\tau), u(\tau))d\tau \tag{40}$$

to be optimized, where $r(x, u) = Q(x) + u^T Ru$ is the cost-to-go function, where $Q(x) \in \Re$ and $R \in \Re^{m\times m}$ represent a positive definite quadratic function and matrix, respectively, to penalize the system state vector and the control input. The initial control input u_0 must be admissible to keep the infinite-horizon value function (40) finite.

The Hamiltonian for the cost function (40) can be given by $H(x, u) = Q(x) + u^T Ru + \nabla_x V^T[f(x) + g(x)u]$, where $\nabla_x V = \partial V/\partial x$ is the gradient with respect to x. The optimal control policy $u^\star(x)$ that minimizes the value function (40) can be obtained by use of the stationary condition as

$$u^\star(x) = -(1/2)R^{-1}g^T(x)\left(\partial V^\star/\partial x\right), \tag{41}$$

where $V^\star \in \Re$ is the optimal value function. Then with substitution of the optimal control input into the Hamiltonian, the HJB equation becomes

$$\begin{aligned} H^\star(x, u^\star) = Q(x) + \nabla_x V^\star(x)f(x) - \\ (1/4)\nabla_x V^{\star T}(x)g(x)R^{-1}g^T(x)\nabla_x V^\star(x) = 0, \end{aligned} \tag{42}$$

where $\nabla_x V^\star(x) = \partial V^\star/\partial x$. It is extremely difficult to obtain an analytical solution to the HJB equation (42). Therefore ADP techniques [37] are used to generate an approximate solution in a forward-in-time manner by use of the periodically sampled state vector. The problem for the event sampled ADP is formulated next.

4.2 Problem Statement

In this section we will formulate the event sampled approximate optimal control problem by highlighting the challenges involved in the design with respect to approximation and stability.

For completeness and notational convenience we redefine the sampling instants as $\{t_k\}_{k=1}^{\infty}$, with $t_0 = 0$ is the initial sampling instant. The sampled state, $x(t_k)$, is released to the controller, and the last sampled state at the controller, denoted by $\breve{x}(t)$, is updated as

$$\breve{x}(t^+) = x(t_k), \quad t = t_k, \ \forall k = 1, 2, \ldots. \tag{43}$$

Then it is held at the controller until the next update and is given by

$$\breve{x}(t) = x(t_k), \quad t_k < t \leq t_{k+1}, \ \forall k = 1, 2, \ldots. \tag{44}$$

Similarly, the event-trigger error is rewritten as

$$e_{\mathrm{ET}}(t) = x(t) - \breve{x}(t), \quad t_k < t \leq t_{k+1}, \ \forall k = 1, 2, \ldots, \tag{45}$$

where $e_{\mathrm{ET}}(t)$ is the event-trigger error. Thus $e_{\mathrm{ET}}^{+} = 0$, $t = t_k$, for all $k = 1, 2, \ldots$.

For optimal policy generation using ADP in an event sampled framework, the value function and the system dynamics need to be approximated with the intermittently available system event sampled state vector. The optimal value function obtained with the NN-based approximation with the event sampled state vector (1.44) can be written as

$$V^{\star}(x) = W_V^T \phi(\breve{x}) + \varepsilon_{e,V}(\breve{x}, e_{\mathrm{ET}}), \quad t_k < t \leq t_{k+1}, \tag{46}$$

where $W_V \in \Re^{l_V}$ is the unknown constant target NN weights, $\phi(\breve{x}) \in \Re^{l_V}$ is the event sampled activation function, and $\varepsilon_{e,V}(\breve{x}, e_{\mathrm{ET}}) = W_V^T(\phi(x) - \phi(\breve{x})) + \varepsilon_V(\breve{x} + e_{\mathrm{ET}})$ is the event sampled reconstruction error, where $\varepsilon_V\left(\breve{x} + e_{\mathrm{ET}}\right) = \varepsilon_V(x) \in \Re$ is the traditional reconstruction error. The HJB equation (42) with event sampled approximation of the value function (46) can be expressed as

$$\begin{aligned} H^{\star}(x, u^{\star}) = {} & Q(x) + \left(W_V^T \nabla_x \phi(\breve{x}) + \nabla_x \varepsilon_{e,V}(\breve{x}, e_{\mathrm{ET}})\right) f(x) \\ & - (1/4)\left(W_V^T \nabla_x \phi(\breve{x}) + \nabla_x \varepsilon_{e,V}(\breve{x}, e_{\mathrm{ET}})\right) D(x) \\ & \times \left(\nabla_x \phi^T(\breve{x}) W_V + \nabla_x \varepsilon_{e,V}(\breve{x}, e_{\mathrm{ET}})\right), \end{aligned} \tag{47}$$

where $D(x) = g(x) R^{-1} g^T(x)$, $\nabla_x \phi(\breve{x}) = \partial \phi(\breve{x}) / \partial x$, and $\nabla_x \varepsilon_{e,V}(\breve{x}, e_{\mathrm{ET}}) = \partial \varepsilon_{e,V}(\breve{x}, e_{\mathrm{ET}}) / \partial x$. It is clear from Eq. (47) that the HJB equation is also a function of the event-trigger error e_{ET}. In other words, the performance is governed by the event-trigger or sampling condition design.

Thus the event sampled optimal control problem can be defined more precisely as follows: (1) Approximate the unknown system dynamics $f(x)$ and $g(x)$ and the value function V in an event sampled context with the desired level of accuracy; (2) design the event-trigger or sampling condition not only to reduce computation but also to minimize approximation error. (3) guarantee a positive lower bound on the inter-event or inter-sample time. A solution along with the detailed design procedure for the event sampled near optimal control design is presented in the next section.

The design is done with two NNs with the event sampled state vector. One NN is used as an identifier to approximate the unknown system dynamics and the second one is used to approximate the solution of the HJB equation, which is the value function. To reduce the computation and ensure accuracy of the approximation, we propose an adaptive event-trigger or sampling condition as a function of the event-trigger error, the estimated NN weights, and the system state vector (defined later). The system state is sent to the controller at the event-trigger or sampling instants and is used to tune the NN weights. The weights are held during the inter-sample times and hence the tuning scheme becomes aperiodic.

Remark 6. To evaluate the proposed adaptive event-trigger or sampling condition at the trigger mechanism, in case of an NCS [31], will require transmission of the NN weight estimates from the controller. To mitigate this additional transmission cost, mirror identifier and value function approximator NNs are used at the trigger mechanism to estimate the NN weights locally. Both the actual and the mirror NNs operate in synchronism at the event sampled instants. Thus this design can be considered as an event sampled NCS with negligible delays and packet losses. The detailed design procedure of the NN identifier and the controller is presented along with their weight tuning laws in the next subsections.

4.3 Identifier Design

Knowledge of the system dynamics is needed for the computation of the optimal control policy (41). To relax this requirement, an event sampled NN-based identifier design is presented in this subsection. By use of the event-based approximation (16), the nonlinear continuous-time system in Eq. (39) can be represented as

$$\dot{x} = f(x) + g(x)u = W_I^T \sigma_I(\breve{x})\bar{u} + \varepsilon_{e,I}, \tag{48}$$

where $W_I = [W_f^T \; W_g^T]^T \in \Re^{(l_f + ml_g)\times n}$, where $W_f \in \Re^{l_f \times n}$ and $W_g \in \Re^{ml_g \times n}$ are the unknown target NN weight matrices, $\sigma_I(\breve{x}) = \begin{bmatrix} \sigma_f(\breve{x}) & 0 \\ 0 & \sigma_g(\breve{x}) \end{bmatrix}$, where $\sigma_f(\breve{x}) \in \Re^{l_f}$ and $\sigma_g(\breve{x}) \in \Re^{ml_g \times m}$ are the event sampled activation functions, the error $\varepsilon_{e,I} =$

$\left[\varepsilon_{e,f}(\breve{x}, e_{\mathrm{ET}}) \quad \varepsilon_{e,g}(\breve{x}, e_{\mathrm{ET}})\right]\bar{u}$ is the event sampled reconstruction error, where $\varepsilon_{e,f}\left(\breve{x}, e_{\mathrm{ET}}\right) = W_f^T(\sigma_f(x) - \sigma_f(\breve{x})) + \varepsilon_f(\breve{x} + e_{\mathrm{ET}})$ and $\varepsilon_{e,g}\left(\breve{x}, e_{\mathrm{ET}}\right) = W_g^T\left(\sigma_g(x) - \sigma_g(\breve{x})\right) + \varepsilon_g\left(\breve{x} + e_{\mathrm{ET}}\right)$, and $\bar{u} = [1 \quad u^T]^T$. The subscripts f and g denotes parameters corresponding to the functions $f(.)$ and $g(.)$, respectively. The event sampled reconstruction error $\varepsilon_{e,I}$ can also be written as $\varepsilon_{e,I} = W_I^T\tilde{\sigma}_I(x, \breve{x})\bar{u} + \varepsilon_I$, where $\tilde{\sigma}_I(x, \breve{x}) = \sigma_I(x) - \sigma_I(\breve{x})$ is the activation function error and $\varepsilon_I = [\varepsilon_f \quad \varepsilon_g u]$ is the augmented traditional reconstruction error. The constants l_f and l_g denote the number of neurons in the NNs. The following assumption holds for the NN [36].

Assumption 6. The target weight vector, W_I, the activation function, $\sigma_I(.)$, and the augmented reconstruction error $\varepsilon_I(.)$ are bounded above in a compact set such that $\|W_I\| \leq W_{I,M}$, $\|\sigma_I(.)\| \leq \sigma_{I,M}$, and $\|\varepsilon_I(.)\| \leq \varepsilon_{I,M}$ are satisfied, where $W_{I,M}$, $\sigma_{I,M}$, and $\varepsilon_{I,M}$ are positive constants. Further, it is assumed that the activation function $\sigma_I(x)$ is Lipschitz continuous in the compact set for all $x \in \Omega_x$ and satisfies $\left\|\sigma_I(x) - \sigma_I(\breve{x})\right\| \leq C_{\sigma_I}\left\|x - \breve{x}\right\|$, where $C_{\sigma_I} > 0$ is a constant.

Since the system state vector is available only at event sampled instants, the event-based identifier dynamics are defined as

$$\dot{\hat{x}} = \bar{A}(\hat{x} - \breve{x}) + \hat{f}(\breve{x}) + \hat{g}(\breve{x})u, \quad t_k < t \leq t_{k+1}, \tag{49}$$

where $\hat{x} \in \Re^n$ is the identifier estimated state vector and $\bar{A}$ is a user-defined Hurwitz matrix and satisfies the Lyapunov equation $\bar{A}^T\bar{P} + \bar{P}\bar{A} = -\Pi$, where $\bar{P}$ and Π are positive definite matrices. The matrix $\bar{A}$ ensures the stability of the identifier. The functions $\hat{f}(\breve{x}) \in \Re^n$ and $\hat{g}(\breve{x}) \in \Re^{n\times m}$ are the estimated system dynamics. By use of the event sampled NN approximation for the system as in (50), the estimated value of the identifier dynamics is represented as

$$\dot{\hat{x}} = \bar{A}(\hat{x} - \breve{x}) + \hat{W}_I^T(t)\sigma_I(\breve{x})\bar{u}, t_k < t \leq t_{k+1}, \tag{50}$$

where $\hat{W}_I(t)$ is the estimated NN weight matrix and $\sigma_I(\breve{x})$ is the event sampled the activation function. If we define $e_I = x - \hat{x}$ as the identification error, the identification error dynamics, from Eqs. (48) and (50), can be expressed as

$$\dot{e}_I = \bar{A}e_I - \bar{A}e_{\mathrm{ET}} + \tilde{W}_I^T\sigma_I(\breve{x})\bar{u} + \hat{W}_I^T\tilde{\sigma}_I(x, \breve{x})\bar{u} + \varepsilon_I, \tag{51}$$

where $\tilde{W}_I = W_I - \hat{W}_I$ is the identifier NN weight estimation error. Since e_I can be computed only at the event sampled instants with the available current sampled state at the identifier, the NN identifier weight matrices are tuned at the event sampled instants only. This can be considered as a jump in the identifier NN weights, which is given by

$$\hat{W}_I^+ = \hat{W}_I + \left(\alpha_I \sigma_I(x)\bar{u}e_I^T/(c + \|\bar{u}\|^2\|e_I\|^2)\right) - \alpha_I \hat{W}_I, t = t_k, \tag{52}$$

where $\alpha_I > 0$ denotes the learning rate and $c > 0$ is a positive constant. During the inter-event or inter-sample times referred to as *flow duration* (ie, $t_k < t \leq t_{k+1}$), the weights are held at the previously tuned values. Therefore the tuning law for the flow duration is given by

$$\dot{\hat{W}}_I = 0, \quad t_k < t \leq t_{k+1}. \tag{53}$$

From Eqs. (52) and 53, it is clear that the NN weights are tuned aperiodically, saving computation when compared with traditional NN control [29, 36]. To ensure the ultimate boundedness of the closed-loop system parameters, the approximated control coefficient function in Eq. (49) is held once it becomes less than equal to the lower bound. It can be expressed as

$$\left\|\hat{g}(.)\right\| = \begin{cases} g_{\min}, & \text{if } \left\|\hat{g}(x)\right\| \leq g_{\min} \\ \left\|\hat{g}(.)\right\|, & \text{otherwise}. \end{cases} \tag{54}$$

The identifier NN weight estimation error dynamics from Eqs. (52) and (53) at both jump and flow duration can be expressed as

$$\tilde{W}_I^+ = \tilde{W}_I - \left(\alpha_I \sigma_I(x)\bar{u}e_I^T/(c + \|\bar{u}\|^2\|e_I\|^2)\right) + \alpha_I \hat{W}_I, \ t = t_k, \tag{55}$$

$$\dot{\tilde{W}}_I = \dot{W}_I - \dot{\hat{W}}_I = 0, \quad t_k < t < t_{k+1}. \tag{56}$$

We will use the identifier dynamics to design an event sampled near optimal controller in the next subsection.

4.4 Controller Design

In this section the solution to the HJB equation, essentially the value function, is approximated by use of a second NN with an event sampled state vector. The approximated value function is used to obtain the near optimal control input. Consider the event sampled approximation of the optimal value function in Eq. (46). The following assumptions hold.

Assumption 7. The target NN weights, activation functions, and the traditional reconstruction errors of the value function approximation NN are bounded above satisfying $\|W_V\| \leq W_{V,M}$, $\|\phi(.)\| \leq \phi_M$, and $|\varepsilon_V(.)| \leq \varepsilon_{V,M}$, where $W_{V,M}$, ϕ_M, and $\varepsilon_{V,M}$ are positive constants. It is further assumed that the gradients of the activation function and the reconstruction error are bounded by a positive constant, ie, $\nabla_x \phi(.) = \partial \phi(.)/\partial x \le \phi'_M$ and $\|\nabla_x \varepsilon_V(.)\| = \partial \varepsilon_V(.)/\partial x \le \varepsilon'_{V,M}$.

Assumption 8. The activation function and its gradient are Lipschitz continuous in a compact set such that for $x \in \Omega_x$ there exist positive constants $C_\phi > 0$ and $C_{\nabla\phi} > 0$ satisfying $\left\|\phi(x) - \phi\left(\breve{x}\right)\right\| \le C_\phi \left\|x - \breve{x}\right\|$ and $\left\|\nabla_x \phi(x) - \nabla_x \phi\left(\breve{x}\right)\right\| \le C_{\nabla\phi} \left\|x - \breve{x}\right\|$.

The optimal control policy (41) in terms of the event sampled NN approximation of the value function becomes

$$u^\star = -(1/2)R^{-1}g^T(x)\left(\nabla_x \phi^T(\breve{x})W_V + \nabla_x \varepsilon_{e,V}(\breve{x}, e_{\mathrm{ET}})\right). \tag{57}$$

Since

$$\hat{V}\left(\breve{x}\right) = \hat{W}_V^T \phi(\breve{x}), \quad t_k < t \le t_{k+1}, \tag{58}$$

the actual control policy by use of the estimated value function (58) and the identifier dynamics is given by

$$\begin{aligned} u(x) &= -(1/2)R^{-1}\hat{g}^T(\breve{x})\nabla_x \phi^T(\breve{x})\hat{W}_V \\ &= -(1/2)R^{-1}\left(\hat{W}_g^T \sigma_g(\breve{x})\right)^T \nabla_x \phi^T(\breve{x})\hat{W}_V, \quad t_k < t \le t_{k+1}. \end{aligned} \tag{59}$$

Now with the estimated value function (58) and approximated system dynamics (49), the error introduced in the HJB equation (47), referred to as the *temporal difference* or *HJB equation error*, can be expressed as

$$\hat{H}(\breve{x}, u) = Q(\breve{x}) + u^T R u + \nabla_x \hat{V}^T(\breve{x})[\hat{f}(\breve{x}) + g(\breve{x})u] \tag{60}$$

for $t_k < t \le t_{k+1}$, where $\nabla_x \hat{V}(\breve{x}) = \partial \hat{V}(\breve{x})/\partial x$. If we substitute the actual control policy (59) in Eq. (60), the HJB equation error can further be expressed as

$$\begin{aligned} \hat{H}(\breve{x}, \hat{W}_V) = {} & Q(\breve{x}) + \hat{W}_V^T \nabla_x \phi(\breve{x})\hat{f}(\breve{x}) \\ & - (1/4)\hat{W}_V^T \nabla_x \phi(\breve{x}) D(\breve{x}) \nabla_x^T \phi(\breve{x})\hat{W}_V, \\ & t_k < t \le t_{k+1}, \end{aligned} \tag{61}$$

where $\hat{D}(\breve{x}) = \hat{g}(\breve{x})R^{-1}\hat{g}^T(\breve{x})$.

Similarly to the identifier NN, the value function NN weight is updated at the event sampled instants with the updated HJB equation error. The HJB equation error (61) with the event sampled state at the sampled instants with $\breve{x}^+ = x$, $t = t_k$ can be written as

$$\begin{aligned} & \hat{H}^+(x, \hat{W}_V) \\ & = Q(x) + \hat{W}_V^T \nabla_x \phi(x)\hat{f}(x) - (1/4)\hat{W}_V^T \nabla_x \phi(x) D(x) \nabla_x^T \phi(x)\hat{W}_V, t = t_k. \end{aligned} \tag{62}$$

The NN tuning law at the event sampled instants is selected as

$$\hat{W}_V^+ = \hat{W}_V - \left(\alpha_V \hat{\omega} \hat{H}^{+T}(x, \hat{W}_V)/(1 + \hat{\omega}^T \hat{\omega})^2\right), \quad t = t_k \tag{63}$$

where $\alpha_V > 0$ is the NN learning gain parameter and

$$\hat{\omega} = \nabla_x \phi(x) \hat{f}(x) - (1/2) \nabla_x \phi(x) \hat{D}(x) \nabla_x^T \phi(x) \hat{W}_V. \tag{64}$$

During the inter-event or inter-sample times or flow period, the tuning law for the value function NN is given as

$$\dot{\hat{W}}_V = 0, \quad t_k < t \le t_{k+1}. \tag{65}$$

We define the value function NN weight estimation error $\tilde{W}_V = W_V - \hat{W}_V$. The NN weight estimation error $\tilde{W}_V$ dynamics by use of Eqs. (61) and (63) can be expressed as

$$\tilde{W}_V^+ = \tilde{W}_V + \left(\alpha_V \hat{\omega} \hat{H}^{+T}(x, \hat{W}_V)/(1 + \hat{\omega}^T \hat{\omega})^2\right), \quad t = t_k, \tag{66}$$

$$\dot{\tilde{W}}_V = 0, \quad t_k < t \le t_{k+1}, \tag{67}$$

The HJB equation error in terms of the value function NN weight estimation error, with use of Eqs. (47) and (61), can be expressed as

$$\begin{aligned}\hat{H}(\breve{x}, \hat{W}_V) = {} & Q(\breve{x}) - Q(x) + \hat{W}_V^T \nabla_x \phi(\breve{x}) \hat{f}(\breve{x}) - W_V^T \nabla_x \phi(x) f(x) \\ & - (1/4) \hat{W}_V^T \nabla_x \phi(\breve{x}) D(\breve{x}) \nabla_x^T \phi(\breve{x}) \hat{W}_V \\ & - (1/4) W_V^T \nabla_x \phi(x) D(x) \nabla_x \phi^T(x) W_V - \varepsilon_H, \quad t_k < t \le t_{k+1},\end{aligned} \tag{68}$$

where $\varepsilon_H = \nabla_x^T \varepsilon_V(x) \left(f(x) + g(x) u^\star\right) - (1/4) \nabla_x^T \varepsilon_V(x) D(x) \nabla_x \varepsilon_V(x)$. It is routine to check that $\|\varepsilon_H\| \le \varepsilon_{H,M}$, where $\varepsilon_{H,M}$ is a positive constant.

Similarly, the HJB equation error at the event sampled instants can be computed from Eq. (68) as

$$\begin{aligned}\hat{H}^+(x, \hat{W}_V) = {} & -\tilde{W}_V^T \nabla_x \phi(x) \hat{f}(x) - W_V^T \nabla_x \phi(x) \tilde{f}(x) \\ & + (1/2) \tilde{W}_V^T \nabla_x \phi(x) D(x) \nabla_x^T \phi(x) \hat{W}_V \\ & + (1/4) \tilde{W}_V^T \nabla_x \phi(x) D(x) \nabla_x^T \phi(x) \tilde{W}_V \\ & + (1/4) W_V^T \nabla_x \phi(x) D(x) \nabla_x \phi^T(x) W_V - \varepsilon_H, \quad t = t_k.\end{aligned} \tag{69}$$

In the next section the closed-loop event sampled system is formulated as a nonlinear impulsive dynamical system to analyze the system stability.

We design an adaptive event-trigger or sampling condition and present theoretical results.

4.5 Impulsive Dynamical System

Now consider the augmented state vector $\zeta = [x^T \; e_I^T \; \breve{x}^T \; \mathbf{vec}(\tilde{W}_I)^T \; \tilde{W}_V^T]^T$. The dynamics during the flow $t_k < t \leq t_{k+1}$ and the jump instants $t = t_k$ are computed as follows.

4.5.1 *Flow Dynamics*

The closed-loop system dynamics during the flow $t_k < t \leq t_{k+1}$ can be represented by use of Eq. (39) and the control policy (59) as

$$\dot{x} = f(x) + g(x)(-(1/2)\, R^{-1}\hat{g}^T(\breve{x})\nabla_x\phi^T(\breve{x})\,\hat{W}_V). \tag{70}$$

Adding and subtracting $g(x)u^\star$ in Eq. (70) and with some simple mathematical operations, we can write the closed-loop dynamics during the flow as

$$\begin{aligned}\dot{x} &= f(x) + g(x)u^\star + (1/2)\, g(x)R^{-1}\tilde{g}^T(x)\nabla_x\phi^T(x)\, W_V \\ &\quad + (1/2)\, g(x)R^{-1}\hat{g}^T(x)\nabla_x\phi^T(x)\,\tilde{W}_V + (1/2)\, g(x)R^{-1}\hat{g}^T(\breve{x}) \\ &\quad \times \left(\nabla_x\phi^T(x) - \nabla_x\phi^T(\breve{x})\right)\hat{W}_V + (1/2)\, g(x)R^{-1}g^T(x)\nabla_x\varepsilon_V(x), \\ &\quad t_k < t \leq t_{k+1}.\end{aligned} \tag{71}$$

The flow dynamics for the identification error e_I is same as in Eq. (51). The last held state, $\breve{x}$, during the flow period remains constant. Thus the dynamics can be represented as

$$\dot{\breve{x}} = 0, \quad t_k < t < t_{k+1}. \tag{72}$$

Finally, the dynamics of the NN weight estimation errors $\mathbf{vec}(\tilde{W}_I)$ and $\tilde{W}_V$ during the flow are as in Eqs. (56) and (67) represented in vector form.

Combining Eqs. (51), (56), (67), (71), and (72), we can write the flow dynamics of the impulsive dynamical system as

$$\dot{\zeta} = F_c(\zeta), \;\; \zeta \in \Omega_\zeta, \;\; \zeta \in C, \;\; \zeta \notin Z, \tag{73}$$

where $F_c(\zeta) = [F_c^{s}{}^T \; F_c^{e_I T} \; 0^T \; 0^T \; 0^T]^T$, where $F_c^s = f(x) + g(x)u^\star + (1/2)\, g(x)R^{-1}\tilde{g}^T(x)\nabla_x\phi^T(x)\, W_V + (1/2)\, g(x)R^{-1}\hat{g}^T(x)$, $F_c^{e_I} = Ae_I - Ae_{\mathrm{ET}} + \tilde{W}_I^T\sigma_I(\breve{x})\bar{u} + \hat{W}_I^T\tilde{\sigma}_I(x,\breve{x})\bar{u} + \varepsilon_I$, and the 0^T are null vectors of appropriate dimensions.

4.5.2 Jump Dynamics

The jump dynamics of the system state vector and the identification error are given by

$$x^+ = x, \quad t = t_k \tag{74}$$

$$e_I^+ = x^+ - \hat{x}^+ = x - \hat{x} = e_I, \quad t = t_k. \tag{75}$$

The jump dynamics of the last held state, $\breve{x}$, is given by

$$\breve{x}^+ = x, \quad t = t_k. \tag{76}$$

Further, the jumps in the NN weight estimation errors are given by Eqs. (55) and (66). If we define the first difference $\Delta\zeta = \zeta^+ - \zeta$ and use Eqs. (55), (66), (74), (75), and (76), the difference equation for the reset/jump dynamics can be written as

$$\Delta\zeta = F_d(\zeta), \ \zeta \in \Omega_\zeta, \ \zeta \in Z, \tag{77}$$

where $F_d(\zeta) = [0^T \ 0^T \ e_{\text{ET}}^T \ \mathbf{vec}(\Delta\tilde{W}_I)^T \ (\Delta\tilde{W}_V)^T]^T$, where $\Delta\tilde{W}_I = -(\alpha_I/(c + \|\bar{u}\|^2\|e_I\|^2))\sigma_I(x)\bar{u}e_I^T + \alpha_I(W_I - \tilde{W}_I)$ and $\Delta\tilde{W}_V = \alpha_V\hat{\omega}\hat{H}^+ (x, \hat{W}_V)/(1 + \hat{\omega}^T\hat{\omega})^2$. The set $C \subset \Omega_\zeta$ is the flow set, $Z \subset \Omega_\zeta$ is the jump set, and $\Omega_\zeta \subset \Re^{n_\zeta}$ an open set with $0 \in \Omega_\zeta$, where $n_\zeta = n+n+n+l_I+l_V$, where $l_I = (l_f + ml_g)n$. The flow and jump set are decided by the event-trigger or sampling condition introduced next.

Consider the event-trigger error in Eq. (45). We define the event-trigger or sampling condition given by

$$D(\|e_{\text{ET}}\|) > \sigma_{\text{ETC}}\left(x, \left\|\hat{W}_V\right\|, \left\|\hat{W}_I\right\|\right), \tag{78}$$

where $\sigma_{\text{ETC}}\left(x, \left\|\hat{W}_V\right\|, \left\|\hat{W}_I\right\|\right) = \min\left\{\Gamma Q(x)/\eta_1, \sqrt{2\Gamma Q(x)/\eta_2}\right\}$ is the threshold, where $\left\|\hat{W}_V\right\| > \kappa_2$, $\left\|\hat{W}_I\right\| > \kappa_3$, $\eta_1 = g_M\lambda_{\max}(R^{-1})\sigma_{I,M}$, $\phi'_M\left\|\hat{W}_V\right\|\left\|\hat{W}_I\right\|C_{\nabla\phi}$, and $\eta_2 = C_{\nabla\phi}g_M\lambda_{\max}(R^{-1})\sigma_{I,M}\phi'_M\left\|\hat{W}_I\right\| + (32/\Pi_{\min})\left(C_{\sigma_I}\left\|\hat{W}_I\right\|^2\|\bar{u}\|^2\left\|\bar{P}\right\|^2 + \left\|\bar{A}\bar{P}\right\|^2\right)$. The constants $0 < \Gamma < 1$, $C_{\nabla\phi}$, and C_{σ_I} are the Lipschitz constants and $\Pi_{\min} = \lambda_{\min}(\Pi)$. The dead-zone operator is defined as

$$D(\cdot) = \begin{cases} \cdot, & \|x\| > B_{ub}^x, \\ 0, & \end{cases} \tag{79}$$

where B_{ub}^x is the desired ultimate bound for the closed-loop system. The event sampled instants are decided when the condition (78) is satisfied.

Further, to ensure the estimated NN weights $\left\| \hat{W}_V \right\|$ and $\left\| \hat{W}_I \right\|$ in Eq. (78) are nonzero while the event sampling condition is being executed (ie, to ensure the threshold is well defined), an event is triggered once the NN weight estimates $\left\| \hat{W}_V \right\| \leq \kappa_2$ or $\left\| \hat{W}_I \right\| \leq \kappa_3$ irrespective of the threshold value, where κ_2 and κ_3 are small positive constants.

Remark 7. The main advantage of the adaptive event-trigger or sampling condition (78) is that the threshold $\sigma_{\text{ETC}}(\cdot)$ is tuned with NN weight estimates in addition to the system state. Therefore the system states will be sampled on the basis of the NN weight estimation errors and the system performance, ensuring the accuracy of approximation and stability. Once the NN weights have converged close to their target values and have become constant, the threshold becomes similar to the thresholds used in traditional event sampling [4, 38].

Remark 8. The dead-zone operator $D(\cdot)$ prevents unnecessary triggering of events due to the NN reconstruction error once the system state is inside the ultimate bound. To show the locally ultimately boundedness of the closed-loop event sampled system, we will use the extension of the Lyapunov direct method of stability from [34] for the nonlinear impulsive dynamical systems. Before we state the main results, the following technical results are necessary.

Lemma 3. *Consider the definition of $\hat{\omega}$ given in Eq. (64). For any positive number $N > 0$, the following inequality holds:*

$$\begin{aligned}
&-\tilde{W}_V^T\hat{\omega}\hat{\omega}^T\tilde{W}_V/(1+\hat{\omega}^T\hat{\omega})^2 \leq \left(1/N(1+\hat{\omega}^T\hat{\omega})^2\right)\\
&\quad \tilde{W}_V^T\left((1/2)\nabla_x\phi(x)\hat{D}(x)\nabla_x^T\phi(x)W_V - \nabla_x\phi(x)\hat{f}(x)\right)\\
&\times\left((1/2)\nabla_x\phi(x)\hat{D}(x)\nabla_x^T\phi(x)W_V - \nabla_x\phi(x)\hat{f}(x)\right)^T\tilde{W}_V\\
&-\left(1/(4(N+1)(1+\hat{\omega}^T\hat{\omega})^2)\right)(\tilde{W}_V^T\nabla_x\phi(x)\hat{D}(x)\nabla_x^T\phi(x)\tilde{W}_V)\\
&\times(\tilde{W}_V^T\nabla_x\phi(x)\hat{D}(x)\nabla_x^T\phi(x)\tilde{W}_V)^T.
\end{aligned} \tag{80}$$

We will use the above results to show the locally ultimately boundedness of the NN weight estimation errors in the following lemma.

Lemma 4. *Consider the nonlinear continuous-time system* (39) *along with the NN-based identifier* (50) *and the value function approximator* (58) *with the event sampled state vector. Let Assumptions 1–4 hold and let the initial identifier and value function NN weights, $\hat{W}_I(0)$ and $\hat{W}_V(0)$, respectively, be* *initialized with nonzero finite values in the compact sets* Ω_{W_I} *and* Ω_{W_V}. *Suppose there exists a positive minimum inter-event or inter-sample time between two consecutive event sampling instants,* $\tau_{\min} = \min\{t_{k+1} - t_k\} > 0$, *and the value function activation function* $\phi(x)$ *and its gradient* $\nabla_x \phi(x)$ *satisfy the persistency of excitation condition. Then the weight estimation errors are locally ultimately bounded for all event sampling instants provided the NN weights are tuned by use of Eqs. (52) and* (53), *and the learning gains satisfy* $0 < \alpha_I < 1/3$, $0 < \alpha_V < \min\{1/40, 2N(1-N)/2(N+1)(16+N)\}$, *where* $0 < N < 1$.

Theorem 3. *[[33]] Consider the nonlinear continuous-time system* (39), *identifier* (50), *and the value function approximator* (58) *represented as an impulsive dynamical system given by Eqs. (73) and* (77). *Let* u_0 *be an initial stabilizing control policy for Eq. (39). Let Assumptions 1–4 hold, and assume there exists a minimum inter-sample time* $\tau_{\min} > 0$. *Suppose the value function activation function* $\phi(x)$ *and its gradient* $\nabla_x \phi(x)$ *satisfy the persistency of excitation condition and the system states are transmitted at the violation of the event sampling condition* (78). *Let the initial identifier and value function NN initial weights,* $\hat{W}_I(0)$ *and* $\hat{W}_V(0)$, *respectively, be nonzero and bounded in compact sets* Ω_{W_I} *and* Ω_{W_V}, *and updated according to Eqs. (63) and* (65). *Then the closed-loop impulsive dynamical system is locally ultimately bounded for all event sampling instants* $t_k > \bar{T}$ *or* $t > T$ *for* $T > \bar{T}$. *Further, the estimated value function satisfies* $\left\| V^\star - \hat{V} \right\| \leq B_V$, *where* B_V *is a small positive constant provided the design parameters are selected as in Lemma 4.*

The assumption on the inter-event or inter-sample time in Theorem 3 is relaxed by the guarantee of the existence of a positive minimum intersample time in this section. The proposed structure of the event sampled approximate optimal design is illustrated in Fig. 2A.

4.6 Minimum Inter-sample Time

The following theorem guarantees the existence of the nonzero positive minimum inter-sample time $\tau_{\min} = \min_{k \in \aleph}\{t_{k+1} - t_k\}$.

Theorem 4. *[[33]] Consider the continuous-time system* (39) *represented as impulsive dynamical systems* (73) *and* (77) *along with the event-trigger or sampling condition* (78). *Then the minimum inter-sample time implicitly defined by Eq. (78) is lower bounded by a nonzero positive constant and is given by*

$$\tau_{\min} \geq \min_{k \in \aleph} \left\{ \left(1/K^\star\right) \ln\left(1 + \left(K^\star/M_k\right) \sigma_{\mathrm{ETC},\min}\right) \right\} > 0, \qquad (81)$$

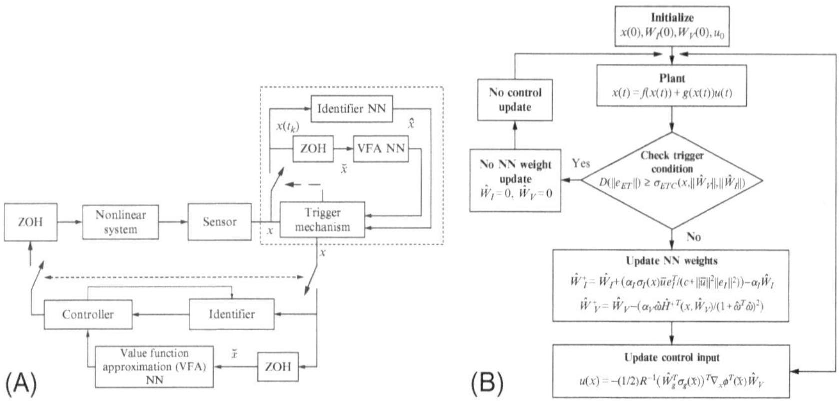

Fig. 2 (A) Near optimal event sampled control system. (B) Flowchart of the event sampled adaptive dynamic programming scheme. *NN*, neural network; *VFA*, value function approximation; *ZOH*, zero-order hold.

where $\sigma_{\text{ETC,min}}$ *is the minimum threshold coefficient value,* $K^\star > 0$ *is a constant, and* $M_k =$

$$g_M \lambda_{\max}(R^{-1}) \sigma_{I,M} \phi'_M \left\| \hat{W}_{I,k} \right\| \left\| \hat{W}_{V,k} \right\| + (1/2) g_M \lambda_{\max}(R^{-1}) \phi'_M W_{V,M} (\sigma_{I,M} \left\| \tilde{W}_{I,k} \right\| + \bar{\varepsilon}_I) + (1/2) g_M \times \lambda_{\max}(R^{-1}) \sigma_{I,M} \phi'_M \left\| \hat{W}_{I,k} \right\| \left\| \tilde{W}_{V,k} \right\| + (1/2) g_M^2 \lambda_{\max}(R^{-1}) \varepsilon'_{V,M},$$

where the subscript k represents the kth inter-sample time.

Remark 9. The constant $K^\star$ satisfies the inequality $\left\| f(x) + g(x) u^\star \right\| \leq K^\star \|x\|$. This inequality holds [38] since the optimal control input is stabilizing.

Remark 10. It is clear from Eq. (81) that the lower bound on inter-sample times depends on M_k or alternatively on the NN weight estimation errors $\tilde{W}_V$, and $\tilde{W}_I$ by the definitions $\hat{W}_I = W - \tilde{W}_I$ and $\hat{W}_V = W - \tilde{W}_V$. During the initial learning phase of the NN the term M_k in Eq. (81) may become large for certain initial values $\hat{W}_V(0)$ and $\hat{W}_I(0)$, which may lead to shorter inter-sample times. Hence a proper initialization of the NN weights is necessary during the learning phase. In addition, with an update of the NN weights, the convergence of the NN weight estimation errors will increase the inter-sample times, leading to fewer sampled events and less resource utilization. The flowchart in Fig. 2B illustrates the implementation of the event sampled ADP scheme designed in this section.

5 SIMULATIONS

In this section we validate the theoretical design in Sections 3 and 4 by using a numerical example. To show the effectiveness of the controller design in terms of saving of communication and computational resources, a practical industrial example of a two-link robot manipulator is provided.

A two-link robot manipulator is considered whose dynamics are given by

$$\dot{x} = f(x) + g(x)u, \tag{82}$$

where
$$f(x) = \begin{bmatrix} x_3 \\ x_4 \\ \dfrac{\begin{pmatrix} -(2x_3x_4 + x_4^2 - x_3^2 - x_3^2\cos x_2)\sin x_2 \\ +20\cos x_1 - 10\cos(x_1+x_2)\cos x_2 \end{pmatrix}}{\cos^2 x_2 - 2} \\ \dfrac{\begin{pmatrix} (2x_3x_4 + x_4^2 + 2x_3x_4\cos x_2 + x_4^2\cos x_2 + 3x_3^2 \\ +2x_3^2\cos x_2 + 20(\cos(x_1+x_2) - \cos x_1)\times \\ (1+\cos x_2) - 10\cos x_2\cos(x_1+x_2) \end{pmatrix}}{\cos^2 x_2 - 2} \end{bmatrix}$$

and
$$g(x) = \begin{bmatrix} 0 & 0 \\ 0 & 0 \\ \frac{1}{2-\cos^2 x_2} & \frac{-1-\cos x_2}{2-\cos^2 x_2} \\ \frac{-1-\cos x_2}{2-\cos^2 x_2} & \frac{3+2\cos x_2}{2-\cos^2 x_2} \end{bmatrix}.$$

5.1 Stabilizing Controller

The following parameters were selected for the simulation. The initial state vector is given by $x = \begin{bmatrix} \pi/3 & -\pi/10 & 0 & 0 \end{bmatrix}^T$, while the closed-loop matrix $A = \text{diag}\{-3, \ -4, \ -6, \ -8\}$ and the positive definite matrix was chosen as $Q = \text{diag}\{0.1, \ 0.1, \ 0.1, \ 0.1\}$. The learning gain was selected as $\alpha_u = 0.5$, $\Gamma_s = 0.9$, $\kappa = 0.0015$, $L \in \Re^{4\times 2}$with elements all 1, $g_{\max} = 3$, and $c = 1$. The bound for the system state vector was chosen as 0.001. The tangent hyperbolic activation function was used in the hidden layer of the NN with a randomly initialized fixed input weight V_u from the uniform distribution in the interval [0, 1]. The number of neurons in the hidden layer was selected as $l = 15$. The NN weight $\hat{W}_u(0)$ was initialized at random from the uniform distribution in the interval [0, 0.01]. The sampling time chosen for the simulation was 0.001 s.

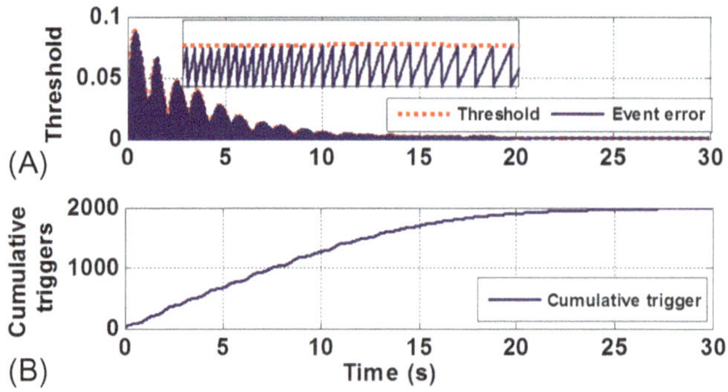

Fig. 3 Evolution of the event-trigger threshold, and the cumulative number of events.

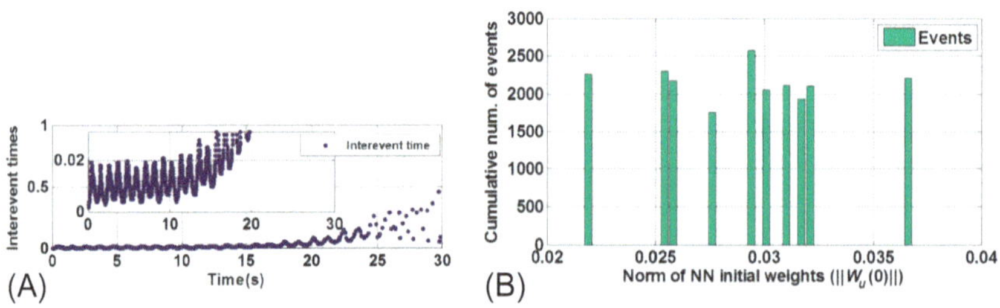

Fig. 4 (A) Existence of a nonzero positive lower bound on inter-event times and gradual increase with convergence of neural network (NN) weight estimates to the target. (B) Cumulative number of events with different NN initial weights.

The event-trigger threshold is shown in Fig. 3A along with the event-trigger error. The cumulative number of triggered events is a illustrated in Fig. 3B, which shows the state vector is transmitted only 2000 times, indicating the reduction in communication bandwidth usage when compared with a continuous transmission. Further, the lower bound on the interevent times is 0.002 s. In addition, the inter-event times increase with convergence of the NN weight estimates to the target as shown in Fig. 4A.

Further, from Theorem 4 the cumulative number of events depends on the initial NN weights. The histogram in Fig. 4B shows the relationship between the norm of the initial weights and the cumulative number of events. It is clear that the cumulative number of events varies with weight initialization.

Convergence of the system state and control input is shown in Fig. 5A, implying the event-based controller-regulated system states are close to zero.

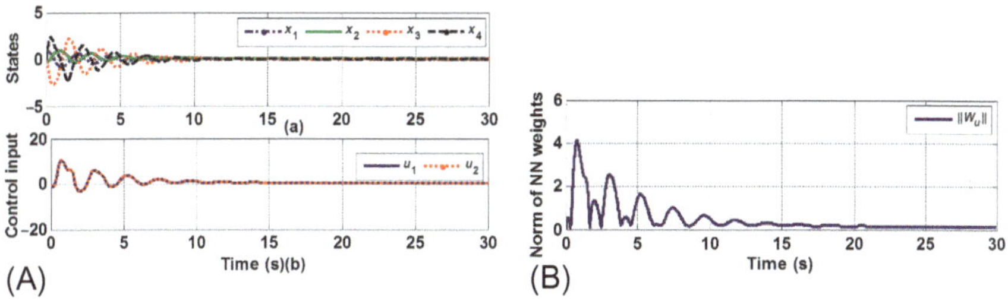

Fig. 5 (A) Convergence of system state vectors and control input. (B) Convergence of the neural network (NN) weight estimates.

Further, the convergence of the estimated NN weights to the target value with an aperiodic event-based update law is shown in Fig. 5B.

Finally, comparison results in terms of computation and the network traffic between a sampled data system with a fixed periodic sampling and the event-based sampling is presented in Table 1 and Fig. 6A, respectively. Table 1 gives the number of computations observed in terms of additions and multiplications that is needed to realize both methods. It is evident that with the event-based system compared with the sampled data approach, there is a 48% reduction in computation. Further, each

Table 1 Comparison of the computational load between a traditional periodically sampled system and an event sampled system

System		Traditional periodically sampled data system	Event-based nonperiodic sampled
Sampling instants		30,000	2000
Number of additions and multiplications at every sampling instant	NN update law at the controller	10	10
	Controller	3	3
	Update law at the trigger mechanism	0	10
	Trigger condition executed at every periodically sampled instant	0	6
Total number of computation		390,000	226,000

NN, neural network.

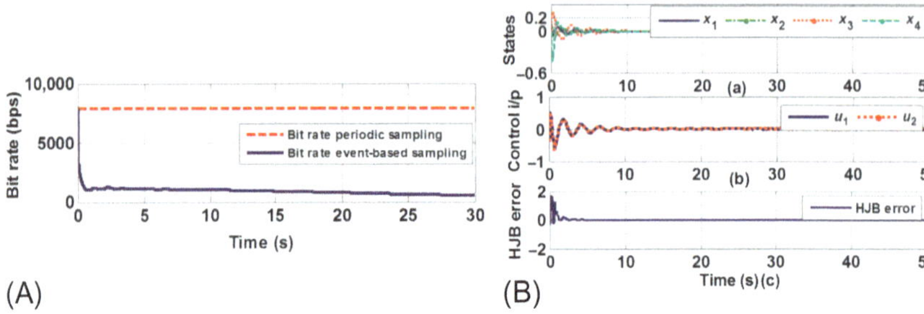

Fig. 6 (A) Comparison of data transfer rate between periodically sampled data and the event-sampled controller for a physical system with a network. (B) Convergence of the system state, the control input, and the Hamilton-Jacobi-Bellman (HJB) equation error.

packetized transmission is of eight-bit data through the ideal network. Fig. 6A shows a comparison between the data rate in both cases. It is clear that the data rate in the case of event-based sampling is lower, implying that the network bandwidth needed is less. This verifies the resourcefulness of the event-triggered control design.

5.2 Optimal Controller

The following simulation parameters were chosen for simulation, The initial system state vector was chosen as $\begin{bmatrix} \pi/6 & -\pi/6 & 0 & 0 \end{bmatrix}^T$. The cost function was selected as in Eq. (40) with $Q = I_{4\times4}$ and $R = I_{2\times2}$. The learning parameters are chosen as $\alpha_V = 0.005$ and $\alpha_I = 0.055$, and other design parameters were $g_M = 3$, $g_m = 1$, $\Gamma = 0.99$, $\bar{A} = -10I$, and $\bar{P} = I$, where I is the identify matrix. The basis function for approximating the value function is given by the expansion of $\phi(x) = [x_1^2; x_2^2; x_3^2; x_4^2; x_1x_2 \cdots; x_1^4; x_2^4; \cdots; x_1^3x_2; \cdots; x_1^2x_2x_3; \cdots; x_1x_2x_3x_4]$. The activation functions for the identifier $\sigma_I(.) = \text{diag}\{\tanh(.)\}\{\tanh(.)\}$. The numbers of hidden-layer neurons for the identifier and the value function NN are selected as 25 and 39, respectively. All the NN weight estimates are initialized at random from a uniform distribution in the interval $(0, 1)$. The ultimate bound for the system state is chosen as 0.001.

The performance of the event sampled control system is shown in Fig. 6B. The system state is regulated close to zero as shown in Fig. 6B along with the control input. The HJB equation error converges close to zero, confirming that a near optimal control policy is achieved with an event sampled implementation.

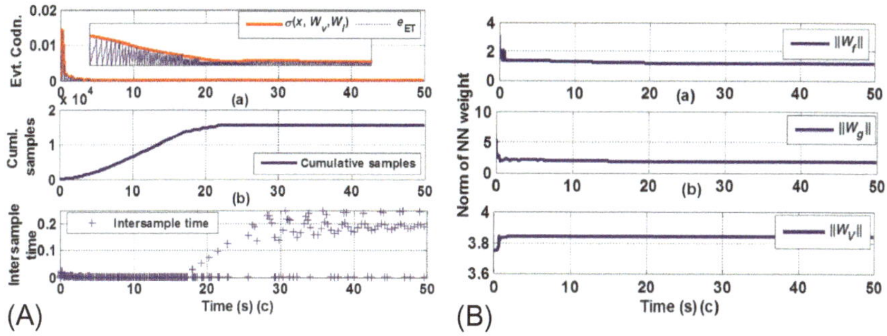

Fig. 7 (A) Evolution of the event sampling condition, cumulative number of event sampled instants, and inter-sample times. (B) Convergence of the norm of the neural network (NN) weight estimates.

The evolution of the event sampling threshold along with the event sampling error is shown in Fig. 7A. The cumulative number of event sampled instants is also shown along with the intersample times. From the cumulative number of event sampled instants, it is evident that fewer sampled instances occurred when compared with the traditional periodically sampled data system. The number of event sampled instants is 15,783 for a simulation duration of 50 s with a sampling interval of 0.001 s or 50,000 sampling instants for the traditional sampled data system.

It is further clear from Fig. 7A that the event sampling condition generated a large number of sampled instants in the initial NN online learning phase. This is due to the large weight estimation error and makes the NN learn the unknown system dynamics and the value function to achieve near optimality. Over time, as the NN approximates the system dynamics and the value function, the inter-sample times increase leading to a reduction in the number of sampled events. The convergence of all the NN weight estimates is shown in Fig. 7B.

6 CONCLUSIONS

We have presented an event-triggered stabilization of multi-input multi-output uncertain nonlinear continuous-time systems. The control input was directly approximated by use of an NN in the context of event-based transmission for stabilizing purposes. Subsequently, an optimal regulator was designed. A novel event-trigger condition was developed on the basis of the system state vector and the NN weight estimate to ensure the reduction in the transmission of the feedback control signal. The weights were updated in a nonperiodic manner at the trigger instants. The controller design guaranteed the desired performance while relaxing the need for system dynamics. Lyapunov analysis confirmed the closed-loop stability. Simulation results confirmed the validity of the control design and reduction in resource utilization.

ACKNOWLEDGMENTS

This work is supported in part by National Science Foundation grant ECCS #1406533 and in part by Intelligent Systems Center.

REFERENCES

[1] K.J. Aström, B.M. Bernhardsson, Comparison of periodic and event based sampling for first-order stochastic systems, in: Proceedings of the 14th IFAC World Congress, Beijing, China, July, 1999, pp. 301–306.
[2] P. Tabuada, Event-triggered real-time scheduling of stabilizing control tasks, IEEE Trans. Autom. Control 52 (2007) 1680–1685.
[3] M.C.F. Donkers, W.P. M.H. Heemels, Output-based event-triggered control with guaranteed L-gain and improved and decentralized event-triggering, IEEE Trans. Autom. Control 57 (2012) 1362–1376.
[4] R. Postoyan, A. Anta, D. Nesic, P. Tabuada, A unifying Lyapunov-based framework for event-triggered control of nonlinear systems, in: Proceedings of the 50th IEEE Conference on Decision and Control and European Control Conference, Orlando, FL, USA, December, 2011, pp. 2559–2564.
[5] W.P. M.H. Heemels, M.C.F. Donkers, Model-based periodic event-triggered control for linear systems, Automatica 49 (3) (2013) 698–711.
[6] J. Lunze, D. Lehmann, A state-feedback approach to event-based control, Automatica 46 (1) (2010) 211–215.
[7] C. Stocker, J. Lunze, Event-based control of nonlinear systems: an input-output linearization approach, in: Proceedings of the 50th IEEE Conference on Decision and Control and European Control Conference, Orlando, FL, USA, December, 2011, pp. 2541–2546.
[8] E. Garcia, P.J. Antsaklis, Model-based event-triggered control for systems with quantization and time-varying network delays, IEEE Trans. Autom. Control 58 (2013) 422–434.
[9] M. Mazo Jr., P. Tabuada, Decentralized event-triggered control over wireless sensor/actuator networks, IEEE Trans. Autom. Control 56 (2011) 2456–2461.
[10] X. Wang, M. Lemmon, Decentralized event-triggered broadcasts over networked control systems, in: M. Egerstedt, B. Mishra (Eds.), Hybrid Systems: Computation and Control, Springer, Berlin, Heidelberg, 2008.
[11] X. Wang, M. Lemmon, Event-triggering in distributed networked control systems, IEEE Trans. Autom. Control 56 (2011) 586–601.
[12] M. Mazo Jr., P. Tabuada, On event-triggered and self-triggered control over sensor/actuator networks, in: Proceedings of the 47th IEEE Conference on Decision and Control, Cancun, Mexico, December, 2008, pp. 435–440.
[13] A. Anta, P. Tabuada, Self-triggered stabilization of homogeneous control systems, in: Proceedings of the American Control Conference, Seattle, WA, USA, June, 2008, pp. 4129–4134.

[14] A. Anta, P. Tabuada, To sample or not to sample: self-triggered control for nonlinear systems, IEEE Trans. Autom. Control 55 (2010) 2030–2042.
[15] X. Wang, M. Lemmon, Self-triggered feedback control systems with finite-gain L2 stability, IEEE Trans. Autom. Control 45 (2009) 452–467.
[16] P. Tallapragada, N. Chopra, On event triggered tracking for nonlinear systems, IEEE Trans. Autom. Control 58 (2013) 2343–2348.
[17] X. Wang, N. Hovakimyan, L1 adaptive control of event-triggered networked systems, in: Proceedings of the American Control Conference, Baltimore, MD, USA, June, 2010, pp. 2458–2463.
[18] A. Sahoo, H. Xu, S. Jagannathan, Neural network-based adaptive event-triggered control of nonlinear continuous-time systems, in: Proceedings of the IEEE International Symposium on Intelligent Control Part of IEEE Multi-Conference on Systems and Control, Hyderabad, India, August, 2013, pp. 35–40.
[19] F. Lewis, S. Jagannathan, A. Yesildrek, Neural Network Control of Robot Manipulators and Nonlinear Systems, Taylor and Francis, London, UK, 1999.
[20] F.L. Lewis, V.L. Syrmos, Optimal Control, Wiley, Hoboken, NJ, 1995.
[21] D.P. Bertsekas, Dynamic Programming and Optimal Control, Athena Scientific, Belmont, MA, 2000.
[22] Z. Chen, S. Jagannathan, Generalized Hamilton-Jacobi-Bellman formulation based neural network control of affine nonlinear discrete time systems, IEEE Trans. Neural Netw. 19 (1) (2008) 90–106.
[23] J. Si, A.G. Barto, W.B. Powell, D. Wunsch, Handbook of Learning and Approximate Dynamics Programming, Wiley, New York, 2004.
[24] D. Vrabie, K. Vamvoudakis, F. Lewis, Adaptive optimal controllers based on generalized policy iteration in a continuous-time framework, in: Proceedings of the 17th IEEE Mediterranean Conference on Control and Automation, 2009, pp. 1402–1409.
[25] Y. Jiang, Z.P. Jiang, Global adaptive dynamic programming for continuous-time nonlinear systems, 2014, arXiv:1401.0020 [math.DS]. http://arxiv.org/pdf/1401.0020.pdf.
[26] D.V. Prokhorov, D. Wunsch, Adaptive critic designs, IEEE Trans. Neural Netw. 8 (5) (1997) 997–1007.
[27] D. Liu, Q. Wei, Finite-approximation-error based optimal control approach for discrete-time nonlinear systems, IEEE Trans. Cybern. 43 (2) (2013) 779–789.
[28] H. Zhang, Y. Luo, D. Liu, Neural-network-based near-optimal control for a class of discrete-time affine nonlinear systems with control constraints, IEEE Trans. Neural Netw. 20 (9) (2009) 1490–1503.
[29] T. Dierks, S. Jagannathan, Optimal control of affine nonlinear continuous-time systems, in: Proceedings of the American Control Conference, Baltimore, MD, USA, June, 2010, pp. 1568–1573.
[30] H. Xu, S. Jagannathan, F.L. Lewis, Stochastic optimal control of unknown linear networked control system in the presence of random delays and packet losses, Automatica 48 (6) (2012) 1017–1030.
[31] T. Dierks, S. Jagannathan, Online optimal control of affine nonlinear discrete-time system with unknown internal dynamics by using time-based policy update, IEEE Trans. Neural Netw. Learn. Syst. 23 (7) (2012) 1118–1129.
[32] H. Zhang, Q. Wei, Y. Luo, A novel infinite-time optimal tracking control scheme for a class of discrete-time nonlinear systems via the greedy HDP iteration algorithm, IEEE Trans. Syst. Man Cybern. B 38 (4) (2008) 937–942.
[33] A. Sahoo, H. Xu, S. Jagannathan, Event-triggered optimal control of nonlinear continuous-time systems in affine form by using neural networks, in: Proceedings of the 53rd IEEE conference on Decision and Control, Los Angeles, CA, USA, December, 2014, pp. 1227–1232.

[34] W.M. Haddad, V. Chellaboina, S.G. Nersesov, Impulsive and Hybrid Dynamical Systems: Stability, Dissipativity, and Control, Princeton University Press, Princeton, NJ, 2006.
[35] H.K. Khalil, Nonlinear Systems, Prentice Hall, Upper Saddle River, NJ, 2002.
[36] S. Jagannathan, Neural Network Control of Nonlinear Discrete-Time System, CRC Press, Boca Raton, FL, 2006.
[37] P. Ioannou, B. Fidan, Adaptive Control Tutorial, SIAM, Philadelphia, PA, 2006.
[38] Q. Zhao, H. Xu, T. Dierks, S. Jagannathan, Finite-horizon neural network-based optimal control design for affine nonlinear continuous-time systems, in: Proceeding of the International Joint Conference on Neural Networks, Dallas, TX, USA, August, 2013, pp. 1–6.
[39] T. Hayakawa, W. Haddad, Stable neural hybrid adaptive control for nonlinear uncertain impulsive dynamical systems, in: Proceedings of the 44th IEEE Conference on Decision and Control, and the European Control Conference, Seville, Spain, December, 2005, pp. 5510–5515.

CHAPTER TWELVE

Decentralized Cooperative Control in Degraded Communication Environments

D. Tolić*, I. Palunko†, A. Ivanović*, M. Car*, S. Bogdan*
*University of Zagreb, Zagreb, Croatia
†University of Dubrovnik, Dubrovnik, Croatia

1 INTRODUCTION

Recent years have witnessed increasing interest in *decentralized control* of multiagent systems (MASs) [1–9]. Decentralized control is characterized by *local* interactions between *neighbors*. In comparison with centralized control, decentralized control avoids a single point of failure, which in turn increases robustness of MASs, allows use of inexpensive and simple agents, and lowers the implementation cost. In addition, decentralized control scales better as the number of agents increases and is sometimes an intrinsic property of MASs. The problem of synchronizing agents' outputs is a typical problem solved in a decentralized fashion [2, 5]. The goal of output synchronization is to achieve a desired *collective behavior* of MASs. Examples of this are formation control, flocking, and consensus control.

Information exchange among neighbors is instrumental for coordination as discussed in all aforementioned references. However, degraded communication environments are commonly encountered in real-life applications. Namely, the exchanged information is often sampled, delayed, corrupted, and prone to communication channel dropouts. Furthermore, the agent models used to devise cooperative control laws might not be accurate or external disturbances might be present. Hence it is of interest to determine how robust a control law is with respect to realistic data exchange and to quantify this robustness. Herein the impact of external disturbances is quantified through $\mathcal{L}_p$-gains, while the influence of corrupted data is

Control of Complex Systems
http://dx.doi.org/10.1016/B978-0-12-805246-4.00012-4

quantified through the bias term in $\mathcal{L}_p$-stability with bias. As expected, different transmission rates and communication delays as well as different controller parameters yield different robustness levels. In addition, since many MASs possess a set of equilibria (rather than a sole equilibrium point), we employ the notions of $\mathcal{L}_p$-stability (with bias) and $\mathcal{L}_p$-detectability with respect to a set [10, 11].

The salient features of this chapter are threefold. First, our analysis applies to MASs with general heterogeneous continuous-time linear agents (not merely to single- or double-integrator dynamics) with exogenous disturbances, directed communication topologies, and output feedback. The settings and goals of [5, 7] appear to be the most similar to ours. However, Yu and Antsaklis [5] consider passive agents without external disturbances, which in turn implies that the number of inputs equals the number of outputs for all agents, and balanced topologies, while Liu and Jia [7] do not deal with transmission rates, directed graphs, or output feedback. Unlike Yu and Antsaklis [5], we do not impose specific requirements on the agent and controller dynamics per se nor on the underlying communication topology. Basically, when given local controllers do not yield $\mathcal{L}_p$-stability (with respect to a set) of the *nominal system* (defined precisely in Section 4.1), one can seek another topology or design alternative controllers. Second, when communication delays are greater than the sampling period, no other work takes into account the concurrent adverse effects of all aforementioned realistic data exchange phenomena [12]. For plant-controller settings, comprehensive theoretical and experimental studies of degraded communication environments are reported in [13, 14]. Nevertheless, these references impose zero-order hold (ZOH) sampling and do not consider noisy data or scheduling protocols [11, 12]. In addition, Kruszewski et al. [13] do not take into account disturbances. Third, the transmission intervals obtained are confirmed experimentally (see Fig. 1). It is worth mentioning that parts of this chapter were published in [15].

The remainder of the chapter is organized as follows. Section 2 provides the notation and terminology used herein. Section 3 states the robustness problem of linear MASs with respect to realistic communication. A method to solve this robustness problem is presented in Section 4 and experimentally verified in Section 5. Conclusions and future directions are given in Section 6. The proofs are provided in the Appendix.

Fig. 1 A snapshot of our experimental setup with three quadcopters.

2 PRELIMINARIES

2.1 Notation

We use $(x, y) := [x^\top \quad y^\top]^\top$. The dimension of a vector x is denoted n_x. Herein $\|\cdot\|$ refers to the Euclidean norm of a vector. If the argument of $\|\cdot\|$ is a matrix, then it denotes the induced matrix 2-norm. The kernel of a matrix A is denoted Ker(A). In addition, $|\cdot|$ denotes the set cardinality. An n-dimensional vector with all zero entries is denoted $\mathbf{0}_n$. Similarly, the $n \times m$ matrix with all zero entries is $\mathbf{0}_{n\times m}$.

We use

$$\|f[a,b]\|_{p,\mathcal{B}} := \left(\int_{[a,b]} \|f(s)\|_{\mathcal{B}}^p \mathrm{d}s \right)^{1/p}, \tag{1}$$

where $\|f(s)\|_{\mathcal{B}} := \inf_{b\in\mathcal{B}} \|f(s) - b\|$, to denote the $\mathcal{L}_p$-norm with respect to a set $\mathcal{B} \subset \mathbb{R}^n$ of a Lebesgue measurable function $f : \mathbb{R} \to \mathbb{R}^n$ restricted to the interval $[a, b] \subset \mathbb{R}$. Lastly, when $\mathcal{B} = \mathbf{0}_n$, we write $\|f[a, b]\|_p$ since this represents the standard $\mathcal{L}_p$-norm.

Left-hand and right-hand limits are denoted $x(t^-) = \lim_{t' \nearrow t} x(t')$ and $x(t^+) = \lim_{t' \searrow t} x(t')$, respectively. For a set $\mathcal{S} \subseteq \mathbb{R}^n$, let $PC([a, b], \mathcal{S}) = \{\phi : [a, b] \to \mathcal{S} \mid \phi(t) = \phi(t^+)$ for every $t \in [a, b)$, $\phi(t^-)$ exists in $\mathcal{S}$ for all $t \in (a, b]$, and $\phi(t^-) = \phi(t)$ for all but at most a finite number of points $t \in (a, b]\}$. For $\psi \in PC([a - d, a], \mathcal{S})$, where $d \geq 0$, the norm of ψ with respect to a set $\mathcal{B}$ is defined by $\|\psi\|_{d,\mathcal{B}} = \sup_{a-d\leq s\leq a} \|\psi(s)\|_{\mathcal{B}}$.

2.2 Graph Theory

A *directed graph*, or digraph, is a pair $\mathcal{G} = (\mathcal{V}, \mathcal{E})$, where $\mathcal{V} = \{v_1, \ldots, v_N\}$ is a nonempty set of *nodes* (or vertices) with unique ID numbers and $\mathcal{E} \subset \mathcal{V} \times \mathcal{V}$ is the set of the corresponding *edges*. When the edge (i,j) belongs to $\mathcal{E}$, this means that there is an information flow from node i to node j. We do not allow *self-loops* (ie, edges that connect a vertex to itself). The set of *neighbors* of the node v_i is $\mathcal{N}_i = \{j \in \mathcal{V} : (j, i) \in \mathcal{E}\}$, which is all nodes that the node v_i can obtain information from. A *directed tree* is a directed graph in which every node has exactly one parent except for one node. A *subgraph* $\mathcal{G}^s = (\mathcal{V}^s, \mathcal{E}^s)$ of $\mathcal{G}$ is a graph such that $\mathcal{V}^s \subseteq \mathcal{V}$ and $\mathcal{E}^s \subseteq \mathcal{E} \cap (\mathcal{V}^s \times \mathcal{V}^s)$. A *directed spanning tree* $\mathcal{G}^s$ of $\mathcal{G}$ is a subgraph of $\mathcal{G}$ such that $\mathcal{G}^s$ is a directed tree and $\mathcal{V}^s = \mathcal{V}$. A graph $\mathcal{G}$ contains a directed spanning tree if a directed spanning tree is a subgraph of $\mathcal{G}$.

Given a graph $\mathcal{G}$, the graph Laplacian matrix $L \in \mathbb{R}^{|\mathcal{V}| \times |\mathcal{V}|}$ is defined as

$$L = [l_{ij}], \quad l_{ij} = \begin{cases} -1, & j \in \mathcal{N}_i, \\ |\mathcal{N}_i|, & j = i, \\ 0, & \text{otherwise.} \end{cases}$$

For more details about algebraic graph theory, see [16].

2.3 Impulsive Delayed Systems

In this chapter we consider impulsive delayed systems

$$\Sigma \left\{ \begin{array}{ll} \chi(t^+) = h_\chi(\chi(t), \chi(t-d)) & t \in \mathcal{T}, \\ \left. \begin{array}{l} \dot{\chi}(t) = f_\chi(\chi(t), \chi(t-d), \omega) \\ y = \ell_\chi(\chi, \omega) \end{array} \right\} & \text{otherwise,} \end{array} \right. \tag{2}$$

where $\chi \in \mathbb{R}^{n_\chi}$ is the state, $\omega \in \mathbb{R}^{n_\omega}$ is the input, $y \in \mathbb{R}^{n_y}$ is the output, and $d \geq 0$ is the time delay. The functions f_χ and h_χ are regular enough to guarantee forward completeness of solutions which, given initial condition $\psi_\chi \in PC([t_0 - d, t_0], \mathbb{R}^{n_\chi})$ and initial time t_0, are given by right-continuous functions $t \mapsto \chi(t) \in PC([t_0 - d, \infty], \mathbb{R}^{n_\chi})$. Jumps of the state $\chi(t^+)$ occur at each $t \in \mathcal{T} := \{t_1, t_2, \ldots\}$, where $t_i < t_{i+1}$, $i \in \mathbb{N}_0$. For a comprehensive discussion regarding the solutions to Eq. (2) considered herein, see [17, Chapters 2 and 3]. Even though the considered solutions to Eq. (2) allow jumps at t_0, we exclude such jumps in favor of notational convenience.

In the following definitions we use the set

$$\mathcal{B}_y := \{y \in \mathbb{R}^{n_y} | \exists b \in \mathcal{B} \text{ such that } y = \ell_\chi(b, \mathbf{0}_{n_\omega})\}, \tag{3}$$

where $\mathcal{B} \subseteq \mathbb{R}^{n_x}$.

Definition 1 ($\mathcal{L}_p$-stability with bias b and with respect to a set). Let $p \in [1, \infty]$. The system Σ is $\mathcal{L}_p$-stable with respect to a set $\mathcal{B}$ and with bias $b(t) \equiv b \geq 0$ from ω to y with gain $\gamma \geq 0$ if there exists $K \geq 0$ such that for each $t_0 \in \mathbb{R}$ and each $\psi_\chi \in PC([t_0 - d, t_0], \mathbb{R}^{n_\chi})$ each solution to Σ from ψ_χ at $t = t_0$ satisfies $\|y[t_0, t]\|_{p,\mathcal{B}_y} \leq K\|\psi_\chi\|_{d,\mathcal{B}} + \gamma\|\omega[t_0, t]\|_p + \|b[t_0, t]\|_p$ for each $t \geq t_0$.

Definition 2 ($\mathcal{L}_p$-detectability with respect to a set). Let $p \in [1, \infty]$. The state χ of Σ is $\mathcal{L}_p$-detectable with respect to a set $\mathcal{B}$ from (y, ω) with gain $\gamma_d \geq 0$ if there exists $K_d \geq 0$ such that, for each $t_0 \in \mathbb{R}$ and each $\psi_\chi \in PC([t_0 - d, t_0], \mathbb{R}^{n_\chi})$ each solution to Σ from ψ_χ at $t = t_0$ satisfies $\|\chi[t_0, t]\|_{p,\mathcal{B}} \leq K_d\|\psi_\chi\|_{d,\mathcal{B}} + \gamma_d\|y[t_0, t]\|_{p,\mathcal{B}_y} + \gamma_d\|\omega[t_0, t]\|_p$ for each $t \geq t_0$.

Definition 1 is motivated by [18, 19] while Definition 2 is motivated by [18]. When $b = 0$, we say "$\mathcal{L}_p$-stability" instead of "$\mathcal{L}_p$-stability with bias 0."

3 PROBLEM STATEMENT

Consider N heterogeneous linear agents given by

$$\begin{aligned} \dot{\xi}_i &= A_i\xi_i + B_iu_i + \omega_i, \\ \zeta_i &= C_i\xi_i, \end{aligned} \tag{4}$$

where $\xi_i \in \mathbb{R}^{n_{\xi_i}}$ is the state, $u_i \in \mathbb{R}^{n_{u_i}}$ is the input, $\zeta_i \in \mathbb{R}^{n_\zeta}$ is the output of the ith agent, $i \in \{1, 2, \ldots, N\}$, and $\omega_i \in \mathbb{R}^{n_{\xi_i}}$ reflects modeling uncertainties and/or exogenous disturbances. The matrices A_i, B_i, and C_i are of appropriate dimensions. A common decentralized policy is

$$u_i = -K_i \sum_{j \in \mathcal{N}_i} (\zeta_i - \zeta_j), \tag{5}$$

where K_i is a $n_{u_i} \times n_\zeta$ matrix and $\mathcal{N}_i$ denotes the set of neighbors of the ith agent according to Section 2.2.

When real-life applications are considered, the controllers (5) are typically fed by estimates of signals ζ_i and ζ_j. Basically, Eq. (5) becomes

$$u_i = -K_i \sum_{j \in \mathcal{N}_i} (\hat{\zeta}_i - \hat{\zeta}_j), \tag{6}$$

where estimates $\hat{\zeta}_i \in ([t_0, \infty), \mathbb{R}^\zeta)$, $i \in \{1, \ldots, N\}$, are piecewise continuous functions with jump times contained in $\mathcal{T}$. The ZOH estimation strategy yields $\hat{\zeta}_i$ that are merely delayed, sampled (ie, piecewise constant), and noisy versions of the ζ_i. For simplicity, we assume that all communication links introduce the same propagation delay equal to some $d \geq 0$. In addition, we disregard message collisions and consider that all agents' outputs are received at time instants $t_i \in \mathcal{T}$ as indicated in Fig. 2. If one is concerned with packet collisions, the scheduling protocols among agents' outputs from [11, 12] readily complement the method presented herein.

Next we define the stack vectors $\xi := (\xi_1, \ldots, \xi_N)$, $\gamma := (\zeta_1, \ldots, \zeta_N)$, $\hat{\zeta} := (\hat{\zeta}_1, \ldots, \hat{\zeta}_N)$, and $\omega := (\omega_1, \ldots, \omega_N)$. The discrepancy between $\hat{\zeta}$ and the delayed signal ζ is captured by the error vector e—that is,

$$e(t) := \hat{\zeta}(t) - \zeta(t - d). \tag{7}$$

With sue of the Laplacian matrix L of the underlying communication graph $\mathcal{G}$, the closed-loop dynamic equation becomes

$$\dot{\xi}(t) = A^{\text{cl}}\xi(t) + A^{\text{cld}}\xi(t - d) + A^{\text{cle}}e(t) + \omega(t), \tag{8}$$

$$\zeta = C^{\text{cl}}\xi, \tag{9}$$

with

$$A^{\text{cl}} = \text{diag}(A_1, \ldots, A_N), \quad A^{\text{cld}} = [A^{\text{cld}}_{ij}], \quad A^{\text{cld}}_{ij} = -l_{ij}B_iK_iC_j,$$
$$A^{\text{cle}} = [A^{\text{cle}}_{ij}], \quad A^{\text{cle}}_{ij} = -l_{ij}B_iK_i, \quad C^{\text{cl}} = \text{diag}(C_1, \ldots, C_N),$$

where A^{cld}_{ij} and A^{cle}_{ij} are matrix blocks, while $\text{diag}(\cdot, \cdot, \ldots, \cdot)$ indicates a diagonal matrix. Our method allows the use of model-based estimators [12, 20]—that is,

$$\dot{\hat{\zeta}}(t) = E\hat{\zeta}(t - d), \tag{10}$$

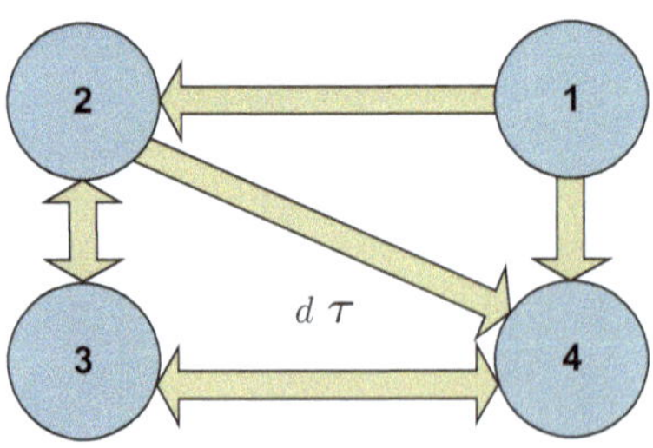

Fig. 2 Communication links among four agents characterized by the same propagation delays and sampling instants.

for each $t \in [t_0, \infty) \setminus \mathcal{T}$, where E is a block diagonal matrix. Notice that E is implemented locally because of the decentralized nature of the control problem. Basically, each agent runs an estimator regarding itself (eg, for $\hat{\zeta}_i$) and an estimator regarding each of its neighbors (ie, for each $\hat{\zeta}_j$, $j \in \mathcal{N}_i$). Notice that the estimates obtained are used in Eq. (6). For the estimates regarding each $\hat{\zeta}_i$, $i \in \{1, \ldots, N\}$, to match, the estimators are provided with the same initial conditions and the estimators' dynamics are identical and given by the respective block matrices of E. Using Eqs. (7), (9), and (10), we arrive at

$$\dot{e}(t) = \dot{\hat{\zeta}}(t) - \dot{\zeta}(t-d) = E\left(e(t-d) + C^{\mathrm{cl}}\xi(t-2d)\right) - C^{\mathrm{cl}}\dot{\xi}(t-d) \tag{11}$$

for each $t \in [t_0, \infty) \setminus \mathcal{T}$, while for each $t_i \in \mathcal{T}$ we have

$$e(t_i^+) = \nu(t_i), \tag{12}$$

where $\nu(t_i)$ models any discrepancy (eg, due to measurement noise and/or channel distortions) between the received values and their actual values at time $t_i - d$ (when the agents' outputs were sampled). We consider bounded ν—that is,

$$\sup_{t \in \mathbb{R}} \|\nu(t)\| = K_\nu \geq 0.$$

Let $\mathcal{B} := \mathrm{Ker}(A^{\mathrm{cl}} + A^{\mathrm{cld}})$ denote the set of equilibrium points of the MAS given by Eqs. (4)–(6) in the absence of realistic control system phenomena (eg, delays and exogenous disturbances as well as intermittent and distorted information). Since $\mathcal{B}$ specifies the desired MAS behavior (as the control law (5) is designed to steer the MAS toward $\mathcal{B}$), it is natural to analyze the robustness of the MAS given by Eqs. (4)–(6) with respect to $\mathcal{B}$. See [10] for a comprehensive discussion regarding stability with respect to sets.

Intuitively, the *nominal system* (defined explicitly in Section 4.1) pertaining to the MAS given by Eqs. (4)–(6) might be robustly stable with respect to $\mathcal{B}$ (in the $\mathcal{L}_p$-sense according to Eq. 15) only for some "small enough" values of d. We refer to such delays as *admissible delays*. Condition Eq. (15) typically entails the existence of a directed spanning tree [2], but one can envision (less challenging/appealing) MAS scenarios in which this directed spanning tree requirement is redundant. Furthermore, given an admissible delay d, the maximal transmission interval τ (ie, $t_{i+1} - t_i \leq \tau$ for each

$i \in \mathbb{N}_0$) that renders $\mathcal{L}_p$-stability of the MAS given by Eqs. (4)–(6) with respect to $\mathcal{B}$ is called the *maximally allowable transfer interval* (MATI) and is denoted $\overline{\tau}$. To exclude ill-disposed choices, such as $t_{i+1} - t_i = \frac{\tau}{i^2}$, that lead to Zeno behavior, we impose an arbitrary small lower bound on $t_{i+1} - t_i$—that is, $0 < \varepsilon \leq t_{i+1} - t_i$ for each $i \in \mathbb{N}_0$. The existence of such ε follows from Remark 3. We are now ready to state the main problem considered herein.

Problem 1. Given an admissible delay d existent in the MAS given by Eqs. (4)–(6) with estimator (10), determine the MATI $\overline{\tau}$ to update the control law (6) with novel neighbors' outputs such that the MAS of interest is $\mathcal{L}_p$-stable from ω to ξ with respect to $\mathcal{B}$ for some $p \in [1, \infty]$.

Remark 1. With use of our framework, it is straightforward to compute $\overline{\tau}$ resulting in $\mathcal{L}_p$-stability from ω to ξ with respect to $\mathcal{B}$ with a prespecified $\mathcal{L}_p$-gain for some $p \in [1, \infty]$. So as not to conceal the main points of this chapter, these details are omitted and the reader is referred to [12].

4 METHOD

To solve Problem 1, we bring together the concepts of $\mathcal{L}_p$-stability and $\mathcal{L}_p$-detectability with respect to a set (see [10] for more details) as well as $\mathcal{L}_p$-stability and $\mathcal{L}_p$-detectability for delayed (impulsive) systems [12].

4.1 Interconnecting the Nominal and Error System

The principal idea behind our method is to consider two interconnected systems and invoke the small-gain theorem [21]. Basically, we want to interconnect the dynamics given by Eq. (8) and the dynamics given by Eqs. (11) and (12). To that end, let us expand the right-hand side of Eq. (11):

$$\dot{e}(t) = E\left(e(t-d) + C^{\text{cl}}\xi(t-2d)\right) - C^{\text{cl}}\left(A^{\text{cle}}e(t-d) + A^{\text{cl}}\xi(t-d) + A^{\text{cld}}\xi(t-2d) + \omega(t-d)\right), \tag{13}$$

and select

$$\tilde{\zeta} := -C^{\text{cl}}A^{\text{cl}}\xi(t-d) + \left(EC^{\text{cl}} - C^{\text{cl}}A^{\text{cld}}\right)\xi(t-2d) - C^{\text{cl}}\omega(t-d) \tag{14}$$

to be the output of the *nominal system* Σ_n given by Eq. (8). Notice that the input to Σ_n is (e, ω). Now let us interconnect with Σ_n the *error system* Σ_e given by Eqs. (11) and (12), with the input and output being $\tilde{\zeta}$ and e, respectively. Notice that Σ_n is a delayed system and Σ_e is an impulsive delayed system from Section 2.3. According to Problem 1, the delay d is admissible, so the nominal system Σ_n is $\mathcal{L}_p$-stable with respect to $\mathcal{B}$ for some $p \in [1, \infty]$ and the corresponding $\mathcal{L}_p$-gain is denoted γ_n—that is, there exist $K_\mathrm{n}, \gamma_\mathrm{n} \geq 0$ such that

$$\|\tilde{\zeta}[t_0, t]\|_{p,\mathcal{B}_{\tilde{\zeta}}} \leq K_\mathrm{n}\|\psi_\xi\|_{d,\mathcal{B}} + \gamma_\mathrm{n}\|(e, \omega)[t_0, t]\|_p \tag{15}$$

for all $t \geq t_0$. To invoke the small-gain theorem and infer $\mathcal{L}_p$-stability, the transmission interval τ has to render the $\mathcal{L}_p$-gain γ_e of Σ_e such that $\gamma_\mathrm{n}\gamma_\mathrm{e} < 1$. Recall that the maximal such τ is in fact $\overline{\tau}$. When $\mathcal{B} = \mathbf{0}_{n_\xi}$, a procedure for attaining such a period τ is provided in [12, Sections IV.B and IV.C]. However, cases when $\mathcal{B} \neq \mathbf{0}_{n_\xi}$ are not covered in [12].

4.2 MASs with Nontrivial Sets $\mathcal{B}$

MASs with nontrivial sets $\mathcal{B}$ (ie, $\mathcal{B} \neq \mathbf{0}_{n_\xi}$) are designed to primarily seek an agreement, while the actual value of the agreement/consensus point depends on the agents' initial conditions. However, available tools for computing $\mathcal{L}_p$-stability and $\mathcal{L}_p$-detectability gains of delayed dynamics, such as the software program HINFN [22], cannot be used as they are. Nevertheless, Tolić [10, Section IV] delineates how to compute the respective gains in the quotient space $\mathbb{R}^{n_\xi} \setminus \mathcal{B}$, where the existing tools can be used. To exploit the method described in [10, Section IV], we impose the following two assumptions on the MAS given by Eqs. (4)–(6).

Assumption 1. The algebraic multiplicity of the zero eigenvalue of $A^{\mathrm{cl}} + A^{\mathrm{cld}}$, denoted $\mathcal{A}$, equals its geometric multiplicity, denoted $\mathcal{G}$.

From Assumption 1, we know that there exists a real Jordan form basis such that the last $\mathcal{A}$ rows and columns of the operator $A^{\mathrm{cl}} + A^{\mathrm{cld}}$ written in this basis contain all zero entries. For more details on this, see [10]. To streamline the exposition of this chapter, we consider such real Jordan form bases herein.

Assumption 2. There exists a real Jordan form basis of $A^{\mathrm{cl}} + A^{\mathrm{cld}}$ such that the last $\mathcal{A}$ rows and columns of operators A^{cl} and A^{cld} written in this basis contain all zero entries as well.

Remark 2. It is certainly of interest to investigate which class of $A^{\mathrm{cl}} + A^{\mathrm{cld}}$ yields Assumption 2. However, since about two dozen examples that we have considered so far readily fulfill Assumption 2, we have deferred this investigation for the future.

Assumption 3. The estimator E yields $\mathrm{Ker}\left(A^{\mathrm{cl}} + A^{\mathrm{cld}}\right) \subseteq \mathrm{Ker}\left(-C^{\mathrm{cl}}A^{\mathrm{cl}} + EC^{\mathrm{cl}} - C^{\mathrm{cl}}A^{\mathrm{cld}}\right)$.

Moreover, Tolić [10, Section IV] states that, given Assumptions 1 and 2, one can use the same procedure to infer $\mathcal{L}_p$-detectability with respect to $\mathcal{B}$ of the state ξ of Σ_{n} from $(e, \omega, \tilde{\zeta})$. Let γ_{d} denote the corresponding $\mathcal{L}_p$-detectability gain.

4.3 Computing Transmission Intervals τ

From [12, Lemma 1 and Theorem 1], we know that $\gamma_{\mathrm{e}} = \frac{2}{\lambda}\sqrt{M}$, $K_{\mathrm{e}} = 2\sqrt{M}\left(1 + \|E - C^{\mathrm{cl}}A^{\mathrm{cle}}\|\frac{2}{\lambda}\left(\mathrm{e}^{\frac{d\lambda}{2}} - 1\right)\right)\left(\frac{1}{p\lambda}\right)^{\frac{1}{p}}$, and bias $b = \frac{\tilde{K}_{\nu}\sqrt{M}}{\mathrm{e}^{\frac{\lambda\varepsilon}{2}} - 1}$ for any $p \in [1, \infty]$, where the constants $\lambda > 0$ and $M > 1$ satisfy the inequalities:

(I) $\tau\left(\lambda + r + \lambda_1 M\mathrm{e}^{-\lambda\tau}\right) < \ln M$,

(II) $\tau\left(\lambda + r + \frac{\lambda_1}{\lambda_2}\mathrm{e}^{\lambda d}\right) < -\ln\lambda_2$,

with $r > 0$ being an arbitrary constant, $\lambda_1 := \frac{\|E - C^{\mathrm{cl}}A^{\mathrm{cle}}\|^2}{r}$, and $\lambda_2 \in (0, 1)$. Lyapunov-Razumikhin techniques are employed to reach conditions (I) and (II). One obtains the expression for λ_1 by considering the try-once-discard protocol in [12, Theorem 2]. Namely, L in [12, Theorem 2] equals $\|E - C^{\mathrm{cl}}A^{\mathrm{cle}}\|$ according to Eq. (13). Altogether, we have reached the following result:

Theorem 1. *Assume that an admissible delay d for the MAS given by Eqs. (4)–(6) is given so that Eq. (15) holds for some $p \in [1, \infty]$. If the transmission interval τ satisfies conditions (I) and (II) for some $\lambda > 0$ and $M > 1$ such that $\frac{2}{\lambda}\sqrt{M}\gamma_{\mathrm{n}} < 1$, the interconnection of the nominal system Σ_{n} and the error system Σ_{e} is $\mathcal{L}_p$-stable from ω to $(\tilde{\zeta}, e)$ with respect to $(\mathcal{B}, \mathbf{0}_{n_e})$ (and with bias when $K_{\nu} \neq 0$).*

Corollary 2. *Assume that the conditions of Theorem 1 hold and that ξ is $\mathcal{L}_p$-detectable from $(e, \omega, \tilde{\zeta})$ with respect to $\mathcal{B}$. Then the MAS given by Eqs. (4)–(6) is $\mathcal{L}_p$-stable (with bias when $K_{\nu} \neq 0$) with respect to $(\mathcal{B}, \mathbf{0}_{n_e})$ from ω to (ξ, e).*

Section 5 delineates once more all the pieces put together so as to obtain Theorem 1 (ie, Corollary 2). However, the following remarks regarding the relevance of Theorem 1 to practical implementations are in order.

Remark 3. The left-hand sides of conditions (I) and (II) are nonnegative continuous functions of $\tau \geq 0$ and tend to ∞ as $\tau \to \infty$. Also, these left-hand sides equal zero for $\tau = 0$. Both sides of conditions (I) and (II) are continuous in λ, M, λ_1, λ_2, and d. Hence for every $\lambda > 0$, $\lambda_1 \geq 0$, $M > 1$, $\lambda_2 \in (0, 1)$, and $d \geq 0$ there exists $\tau > 0$ such that conditions (I) and (II) are satisfied. Finally, since $\frac{2}{\lambda}\sqrt{M}$ is continuous in λ and M, we infer that for every finite $\gamma_{\mathrm{n}} > 0$ there exists $\tau > 0$ such that $\frac{2}{\lambda}\sqrt{M}\gamma_{\mathrm{n}} < 1$. In other words, for each admissible d the unwanted Zeno behavior is avoided and the proposed method does not yield continuous feedback that might be impossible to implement. Notice that each τ yielding $\frac{2}{\lambda}\sqrt{M}\gamma_{\mathrm{n}} < 1$ is a candidate for $\overline{\tau}$. Depending on r, λ_2, λ, and M, the maximal such τ is in fact MATI $\overline{\tau}$.

Remark 4. Noisy measurements can be a consequence of quantization errors. According to [23], feedback control prone to quantization errors cannot yield closed-loop systems with linear $\mathcal{L}_p$-gains. Hence the bias term in the linear gain $\mathcal{L}_p$-stability with bias result of Theorem 1 cannot be removed without contradicting the points in [23]. Further investigations of quantized feedback are high on our future research agenda.

Remark 5. Let us consider the case of lossy communication channels. If there is an upper bound on the maximum number of successive dropouts, say, $N_{\mathrm{d}} \in \mathbb{N}$, simply use $\frac{\tau}{N_{\mathrm{d}}}$ as the transmission interval for Theorem 1 to hold.

5 EXPERIMENTAL VALIDATION

The agents in our experimental setup are commercially available AR.Drone Parrot quadcopter platforms [24]. To account for the specifics of our experimental settings (eg, baseline controller, custom battery supplies, latencies due to the OptiTrack motion caption system [25], and our ROS code [26]), we decided to perform system identification of these quadcopters on our own (rather than employing models or model parameters obtained by other researchers). As we are interested in coordinating the quadcopters in the plane, we identify the quadcopters' transfer functions in the x- and y-axes. The experimental data for identification are collected for each axis independently. Potential coupling among the x- and y-axes or slight model differences between quadcopters are taken into account by the ω_i in our MAS model given by Eqs. (4)–(6). With use of MATLAB System Identification Toolbox, for both axes the transfer functions obtained are

$$G_{x_i}(s) = \frac{X_i(s)}{U_{x_i}(s)} = \frac{K_p}{s(1+T_p s)} e^{-ds},$$
$$G_{y_i}(s) = \frac{Y_i(s)}{U_{y_i}(s)} = \frac{K_p}{s(1+T_p s)} e^{-ds}, \tag{16}$$

where $K_p = 5.2$, $T_p = 0.38$, and $d = 0.104$ s for all $i \in \{1, 2, 3, 4\}$. The inputs $U_{x_i}(s)$ and $U_{y_i}(s)$ are the Laplace transforms of the normalized pitch and roll angle of the i agent, respectively, and the outputs $X_i(s)$ and $Y_i(s)$ are the Laplace transforms of the x- and y-coordinates of the ith agent, respectively. This second-order transfer function with delay is a reasonable trade-off between the model complexity and accuracy. In addition, since our framework allows for modeling uncertainties through the ω_i, additional model accuracy is not essential. A fragment of comparison among the identified model and the actual model is provided in Fig. 3. To avoid notation ambiguity, we restrict our attention to $G_{x_i}(s)$ (ie, to the x-axis direction) from now on. The analysis with respect to the y-axis is the same.

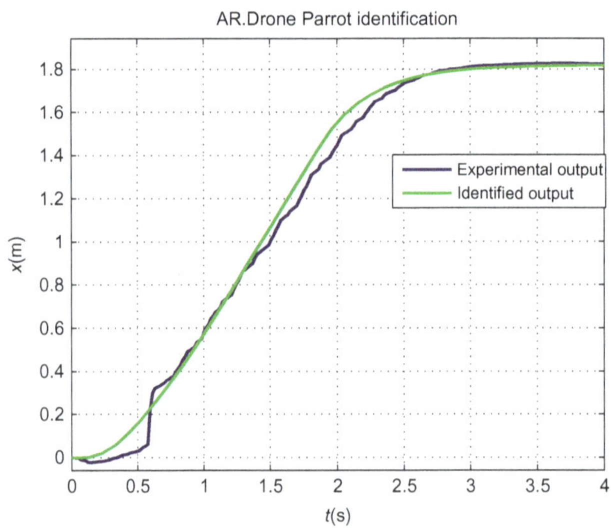

Fig. 3 Experimental verification of the identified transfer function for an AR.Drone Parrot quadcopter in the x-axis direction. These responses are obtained for the pulse input with duration of 1.75 s and amplitude of 0.2. This 0.2 is the normalized roll angle that is fed to the low-level controller and corresponds to the actual pitch angle of 9 degrees. Since our agents are homogeneous, we write x(m) instead of x_i(m).

Let us now write Eq. (16) in the form of the MAS given by Eqs. (4)–(6) with delays. Basically, the delay d from Eq. (16) becomes the communication link delay (see Fig. 2), while the remainder of the transfer function can be written in the state-space form (4) as

$$\dot{\xi}_i = \begin{bmatrix} 0 & 1 \\ 0 & -T_{\mathrm{p}} \end{bmatrix} \xi_i + \begin{bmatrix} 0 \\ K_{\mathrm{p}} \end{bmatrix} u_i + \omega_i.$$

Since the consensus controller (5) is a time-invariant static controller (ie, without memory/states on its own), the choice for d to represent the communication delay is justified. The first component of ξ_i corresponds to the position of the ith agent, while the second component corresponds to its velocity in the x-axis direction. The more intuitive notation for the first component of ξ_i is x_i, while for the second component of ξ_i it is v_i^x, which is the notation found in the plots. We weight the position and velocity in the output ζ_i as

$$\zeta_i = \begin{bmatrix} 0.05 & 0.025 \end{bmatrix} \xi_i$$

in order to put more emphasis (twice as much) on the position discrepancy among the agents because the velocity measurements are rather noisy (see Figs. 4 and 5). The selected graph topology is depicted in Fig. 2, and the graph Laplacian matrix is

$$L = \begin{bmatrix} 0 & 0 & 0 & 0 \\ -1 & 2 & -1 & 0 \\ 0 & -1 & 2 & -1 \\ -1 & -1 & -1 & 3 \end{bmatrix}.$$

As given by L, we consider one leader and three followers. Because of the space constraints of our laboratory, the leader is virtual (ie, generated on a computer). We include two sets of experiments in this chapter. The first set uses $K := K_1 = \cdots = K_4 = 2$, while the second set uses $K := K_1 = \cdots = K_4 = 4$.

For simplicity, we set $E = \mathbf{0}_{n_\zeta \times n_\zeta}$, which corresponds to the ZOH estimation strategy. Observe that this choice of E meets Assumption 3. In addition, one can readily verify that Assumption 1 holds for both $K = 2$ and $K = 4$ with $\mathcal{A} = \mathcal{G} = 1$. A real Jordan form basis of $A^{\mathrm{cl}} + A^{\mathrm{cld}}$ for $K = 2$ is given by the columns of the matrix

$$T_2 = \begin{bmatrix} 0 & 0 & 0 & 0 & 0 & 0 & -2.6316 & 1 \\ 0 & 0 & 0 & 0 & 0 & 0 & 1 & 0 \\ 0.3718 & 0.7091 & -0.8738 & 1.5873 & -0.3570 & 0.6601 & -2.6316 & 1 \\ -1 & 0 & 1 & 0 & 1 & 0 & 1 & 0 \\ -0.3718 & -0.7091 & -1.2357 & 2.2448 & 0.5049 & -0.9336 & -2.6316 & 1 \\ 1 & 0 & 1.4142 & 0 & -1.4142 & 0 & 1 & 0 \\ 0 & 0 & -0.8738 & 1.5873 & -0.3570 & 0.6601 & -2.6316 & 1 \\ 0 & 0 & 1 & 0 & 1 & 0 & 1 & 0 \end{bmatrix},$$

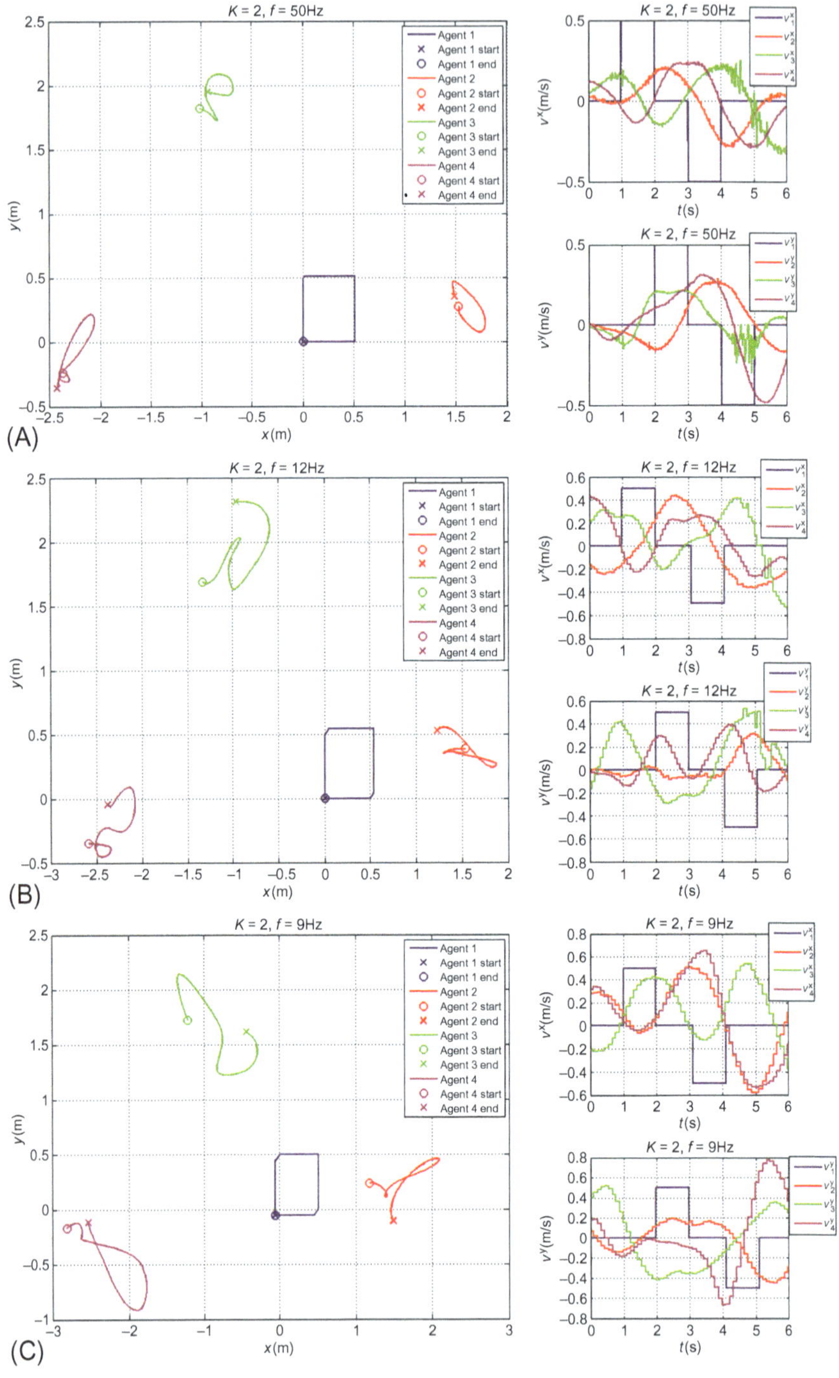

Fig. 4 Excerpts of the experimentally obtained data related to one square maneuver of the leader for $K = 2$. The computed marginal frequency is 12 Hz while the experimentally obtained one is 9 Hz. Each plot includes its description.

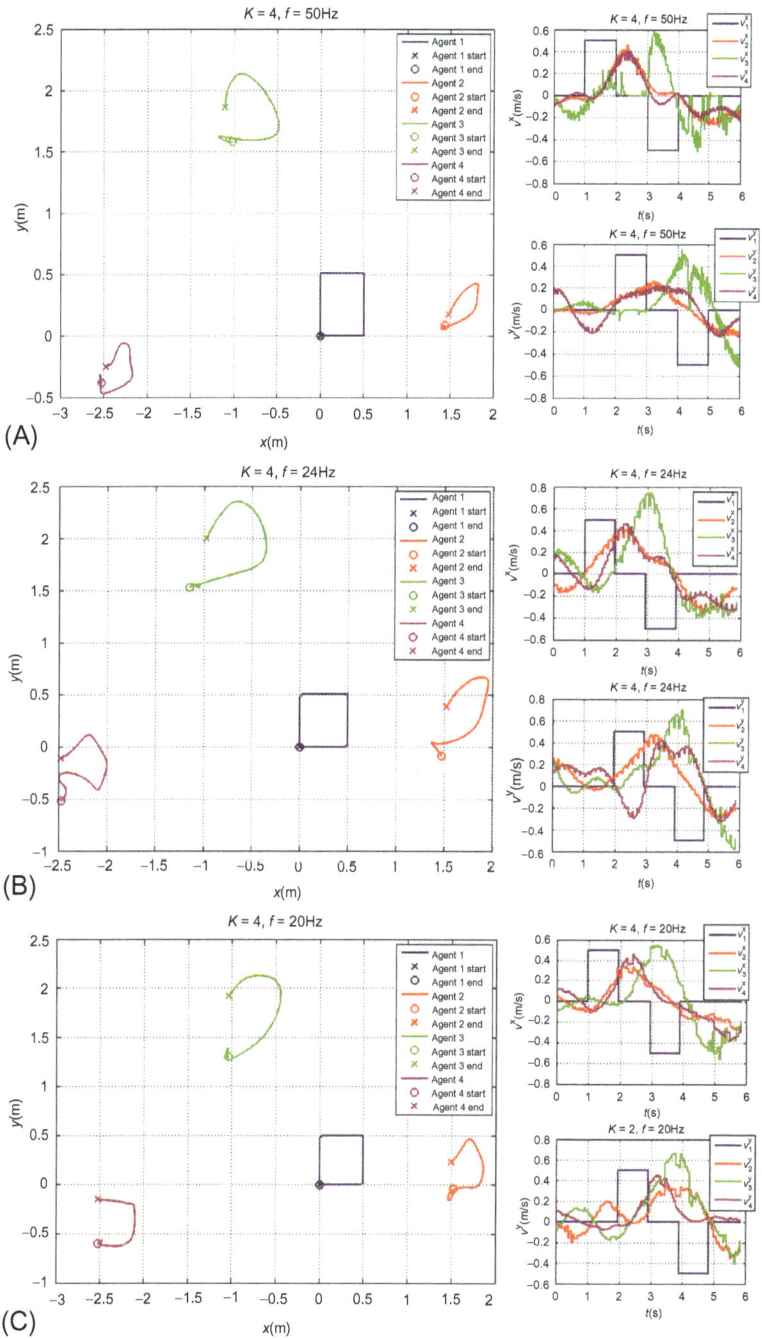

Fig. 5 Excerpts of the experimentally obtained data related to one square maneuver of the leader for $K = 4$. The computed marginal frequency is 24 Hz, while the experimentally obtained one is 20 Hz. Each plot includes its description.

while for $K = 4$ it is given by the columns of

$$T_4 = \begin{bmatrix} 0 & 0 & 0 & 0 & 0 & 0 & -2.6316 & 1 \\ 0 & 0 & 0 & 0 & 0 & 0 & 1 & 0 \\ 0.3109 & 0.4731 & -0.5619 & 1.1514 & -0.3035 & 0.4353 & -2.6316 & 1 \\ -1 & 0 & 1 & 0 & 1 & 0 & 1 & 0 \\ -0.3109 & -0.4731 & -0.7946 & 1.6283 & 0.4292 & -0.6156 & -2.6316 & 1 \\ 1 & 0 & 1.4142 & 0 & -1.4142 & 0 & 1 & 0 \\ 0 & 0 & -0.5619 & 1.1514 & -0.3035 & 0.4353 & -2.6316 & 1 \\ 0 & 0 & 1 & 0 & 1 & 0 & 1 & 0 \end{bmatrix}.$$

The last column of the above two matrices spans $\mathcal{B}$—that is,

$$\mathcal{B} := \rho(1, 0, 1, 0, 1, 0, 1, 0), \quad \rho \in \mathbb{R}.$$

Assumption 2 is readily verified as well. Apparently, the above $\mathcal{B}$ is not trivial as discussed in Section 4.2. Hence we need to work in the quotient space $\mathbb{R}^{n_\xi} \setminus \mathcal{B}$. To that end, we use the above real Jordan form bases and apply the procedure from [10, Section IV] to write the closed-loop dynamics (8) with the output $\tilde{\zeta}$, given by Eq. (14), in the reduced form. Basically, we introduce the substitution (ie, change of coordinates) $z = (T_K)^{-1}\xi$, where T_K is given above for the selected K, and prune the last component of z, obtaining z_r, so the reduced system for $K = 2$ becomes

$$\dot{z}_r(t) = \begin{bmatrix} -0.38 & 0 & 0 & 0 & 0 & 0 & 0 \\ -1.211 & 0 & 0 & 0 & 0 & 0 & 0 \\ 0 & 0 & -0.38 & 0 & 0 & 0 & 0 \\ 0 & 0 & 0.4208 & 0 & 0 & 0 & 0 \\ 0 & 0 & 0 & 0 & -0.38 & 0 & 0 \\ 0 & 0 & 0 & 0 & 1.3093 & 0 & 0 \\ 0 & 0 & 0 & 0 & 0 & 0 & -0.38 \end{bmatrix} z_r(t)$$

$$+ \begin{bmatrix} -0.2 & 1.1062 & 0 & 0 & 0 & 0 & 0 \\ 0.1049 & -0.58 & 0 & 0 & 0 & 0 & 0 \\ 0 & 0 & 0.1138 & -0.4835 & 0 & 0 & 0 \\ 0 & 0 & 0.0627 & -0.2662 & 0 & 0 & 0 \\ 0 & 0 & 0 & 0 & -0.2538 & -1.172 & 0 \\ 0 & 0 & 0 & 0 & -0.1373 & -0.6338 & 0 \\ 0 & 0 & 0 & 0 & 0 & 0 & 0 \end{bmatrix} z_r(t-d)$$

$$+ \begin{bmatrix} 0 & 31.2 & 0 & -31.2 & 0 & 0 & 0 & -1 & 0 & 0 & 0 & 1 \\ 0 & -16.3592 & 0 & 16.3592 & 0 & 0 & 1.4103 & 0.5243 & 0 & 0 & -1.4103 & -0.5243 \\ 5.2 & -2.1539 & -2.1539 & -0.8922 & 0 & -0.8536 & 0 & 0.3536 & 0 & 0.3536 & 0 & 0.1464 \\ 2.8624 & -1.1857 & -1.1857 & -0.4911 & -0.5377 & -0.4699 & 0.2227 & 0.1946 & 0.2227 & 0.1946 & 0.0923 & 0.0806 \\ 5.2000 & 12.5539 & 12.5539 & -30.3078 & 0 & -0.1464 & 0 & -0.3536 & 0 & -0.3536 & 0 & 0.8536 \\ 2.8122 & 6.7894 & 6.7894 & -16.3910 & -0.2218 & -0.0792 & -0.5356 & -0.1912 & -0.5356 & -0.1912 & 1.293 & 0.4616 \\ 0 & 0 & 0 & 0 & 0 & 1 & 0 & 0 & 0 & 0 & 0 & 0 \end{bmatrix} (e, \omega)(t),$$

where the output of the reduced system Σ_n is

$$\tilde{\zeta}_r(t) = \underbrace{\begin{bmatrix} 0 & 0 & 0 & 0 & 0 & 0 & -0.0405 \\ 0.0405 & 0 & -0.0405 & 0 & -0.0405 & 0 & -0.0405 \\ -0.0405 & 0 & -0.0573 & 0 & 0.0573 & 0 & -0.0405 \\ 0 & 0 & -0.0405 & 0 & -0.0405 & 0 & -0.0405 \end{bmatrix}}_{C_r} z_r(t)$$

$$
+\underbrace{\begin{bmatrix} 0 & 0 & 0 & 0 & 0 & 0 & 0 \\ -0.005 & 0.0277 & -0.0028 & 0.0121 & 0.0063 & 0.0293 & 0 \\ 0.005 & -0.0277 & -0.004 & 0.0171 & -0.009 & -0.0414 & 0 \\ 0 & 0 & -0.0028 & 0.0121 & 0.0063 & 0.0293 & 0 \end{bmatrix}}_{C_{\mathrm{r}}^{\mathrm{d}}} z_{\mathrm{r}}(t-d)
$$

$$
+\underbrace{\begin{bmatrix} 0 & 0 & 0 & 0 & -0.05 & -0.025 & 0 & 0 & 0 & 0 & 0 & 0 \\ 0 & 0 & 0 & 0 & 0 & 0 & -0.05 & -0.025 & 0 & 0 & 0 & 0 \\ 0 & 0 & 0 & 0 & 0 & 0 & 0 & 0 & -0.05 & -0.025 & 0 & 0 \\ 0 & 0 & 0 & 0 & 0 & 0 & 0 & 0 & 0 & 0 & -0.05 & -0.025 \end{bmatrix}}_{D_{\mathrm{r}}} (e,\omega)(t),
$$

delayed for an additional d time units (compare this with Eq. 14). The $\mathcal{L}_p$-gain corresponding to this additional delay d is unity [27]. Likewise, for $K = 4$ we obtain

$$
\dot{z}_{\mathrm{r}}(t) = \begin{bmatrix} -0.38 & 0 & 0 & 0 & 0 & 0 & 0 \\ -1.8639 & 0 & 0 & 0 & 0 & 0 & 0 \\ 0 & 0 & -0.38 & 0 & 0 & 0 & 0 \\ 0 & 0 & 0.6831 & 0 & 0 & 0 & 0 \\ 0 & 0 & 0 & 0 & -0.38 & 0 & 0 \\ 0 & 0 & 0 & 0 & 2.0322 & 0 & 0 \\ 0 & 0 & 0 & 0 & 0 & 0 & -0.38 \end{bmatrix} z_{\mathrm{r}}(t)
$$

$$
+\begin{bmatrix} -0.59 & 1.4762 & 0 & 0 & 0 & 0 & 0 \\ 0.3877 & -0.97 & 0 & 0 & 0 & 0 & 0 \\ 0 & 0 & 0.0377 & -0.7015 & 0 & 0 & 0 \\ 0 & 0 & 0.0184 & -0.3423 & 0 & 0 & 0 \\ 0 & 0 & 0 & 0 & -0.6977 & -1.5458 & 0 \\ 0 & 0 & 0 & 0 & -0.4864 & -1.0777 & 0 \\ 0 & 0 & 0 & 0 & 0 & 0 & 0 \end{bmatrix} z_{\mathrm{r}}(t-d)
$$

$$
+\begin{bmatrix} 0 & 62.4 & 0 & -62.4 & 0 & 0 & 0 & -1 & 0 & 0 & 0 & 1 \\ 0 & -41.0032 & 0 & 41.0032 & 0 & 0 & 2.1136 & 0.6571 & 0 & 0 & -2.1136 & -0.6571 \\ 10.4 & -4.3078 & -4.3078 & -1.7844 & 0 & -0.8536 & 0 & 0.3536 & 0 & 0.3536 & 0 & 0.1464 \\ 5.0751 & -2.1022 & -2.1022 & -0.8707 & -0.7413 & -0.4165 & 0.3071 & 0.1725 & 0.3071 & 0.1725 & 0.1272 & 0.0715 \\ 10.4 & 25.1078 & 25.1078 & -60.6156 & 0 & -0.1464 & 0 & -0.3536 & 0 & -0.3536 & 0 & 0.8536 \\ 7.2509 & 17.5051 & 17.5051 & -42.2611 & -0.3364 & -0.1021 & -0.8122 & -0.2465 & -0.8122 & -0.2465 & 1.9607 & 0.5951 \\ 0 & 0 & 0 & 0 & 0 & 1 & 0 & 0 & 0 & 0 & 0 & 0 \end{bmatrix} (e,\omega)(t),
$$

$$
\tilde{\zeta}_{\mathrm{r}}(t) = \underbrace{\begin{bmatrix} 0 & 0 & 0 & 0 & 0 & 0 & -0.0405 \\ 0.0405 & 0 & -0.0405 & 0 & -0.0405 & 0 & -0.0405 \\ -0.0405 & 0 & -0.0573 & 0 & 0.0573 & 0 & -0.0405 \\ 0 & 0 & -0.0405 & 0 & -0.0405 & 0 & -0.0405 \end{bmatrix}}_{C_{\mathrm{r}}} z_{\mathrm{r}}(t)
$$

$$
+\underbrace{\begin{bmatrix} 0 & 0 & 0 & 0 & 0 & 0 & 0 \\ -0.0147 & 0.0369 & -0.0009 & 0.0175 & 0.0174 & 0.0386 & 0 \\ 0.0147 & -0.0369 & -0.0013 & 0.0248 & -0.0247 & -0.0547 & 0 \\ 0 & 0 & -0.0009 & 0.0175 & 0.0174 & 0.0386 & 0 \end{bmatrix}}_{C_{\mathrm{r}}^{\mathrm{d}}} z_{\mathrm{r}}(t-d)
$$

$$
+\underbrace{\begin{bmatrix} 0 & 0 & 0 & 0 & -0.05 & -0.025 & 0 & 0 & 0 & 0 & 0 & 0 \\ 0 & 0 & 0 & 0 & 0 & 0 & -0.05 & -0.025 & 0 & 0 & 0 & 0 \\ 0 & 0 & 0 & 0 & 0 & 0 & 0 & 0 & -0.05 & -0.025 & 0 & 0 \\ 0 & 0 & 0 & 0 & 0 & 0 & 0 & 0 & 0 & 0 & -0.05 & -0.025 \end{bmatrix}}_{D_{\mathrm{r}}} (e,\omega)(t).
$$

We are now ready to apply the program HINFN [22] to compute γ_{n} and γ_{d} for $p = 2$. However, HINFN is not able to handle delayed states in the output $\tilde{\zeta}_{\mathrm{r}}(t)$. To accommodate $\tilde{\zeta}_{\mathrm{r}}(t)$ in HINFN, we first compute an estimate of the $\mathcal{L}_2$-gain, say, $\gamma_{\mathrm{n}1}$, from (e, ω) to $\tilde{\zeta}_{\mathrm{r}1}(t) := C_{\mathrm{r}} z_{\mathrm{r}}(t) + D_{\mathrm{r}}(e, \omega)(t)$, and than an estimate of the $\mathcal{L}_2$-gain, say, $\gamma_{\mathrm{n}2}$, from (e, ω) to $\tilde{\zeta}_{\mathrm{r}2}(t) := C_{\mathrm{r}}^{\mathrm{d}} z_{\mathrm{r}}(t)$. Afterward, we simply add together those two gains because

$$\begin{aligned}\|\tilde{\zeta}_{\mathrm{r}1}(t)[t_0, \infty) + \tilde{\zeta}_{\mathrm{r}2}(t-d)[t_0, \infty)\|_p &= \|\tilde{\zeta}_{\mathrm{r}1}(t)[t_0, \infty) + \tilde{\zeta}_{\mathrm{r}2}(t)[t_0 + d, \infty)\|_p \\ &\leq \|\tilde{\zeta}_{\mathrm{r}1}(t)[t_0, \infty)\|_p + \|\tilde{\zeta}_{\mathrm{r}2}(t)[t_0, \infty)\|_p \\ &\leq (\gamma_{\mathrm{n}1} + \gamma_{\mathrm{n}2})\|(e, \omega)[t_0, \infty)\|_p,\end{aligned}$$

where we used the fact that the $\mathcal{L}_p$-gain of the constant delay operator is less than unity (see [27]) and that $\tilde{\zeta}_{\mathrm{r}2}(t-d) = 0$ when $t-d < t_0$. Altogether, for $K = 2$ we obtain $\gamma_{\mathrm{n}} = 3.1588$ and $\gamma_{\mathrm{d}} = 62.0433$, while for $K = 4$ we obtain $\gamma_{\mathrm{n}} = 4.9444$ and $\gamma_{\mathrm{d}} = 66.5614$. Apparently, $d = 0.108$ s is an admissible delay. As far as the computation of γ_{e} is concerned, for $K = 2$ we have $\|C^{\mathrm{cl}} A^{\mathrm{cle}}\| = 0.9881$, while for $K = 4$ we have $\|C^{\mathrm{cl}} A^{\mathrm{cle}}\| = 1.9762$. Using Theorem 1, we obtain for $K = 2$ that $\overline{\tau} = 0.0835$ s when $r = 5$, $\lambda = 14.363$, $M = 5.168$, and $\lambda_2 = 0.08$. Additionally, for $K = 4$ we obtain $\overline{\tau} = 0.042$ s when $r = 8$, $\lambda = 17.2$, $M = 3$, and $\lambda_2 = 0.13$. In general, the ratio among two MATIs is not almost the inverse of the gain ratio. This is corroborated by the following: $K = 6$ results in $\overline{\tau} = 0.0262$ s, and $K = 8$ leads to $\overline{\tau} = 0.0176$ s. Apparently, an increase in K decreases $\overline{\tau}$.

Next we experimentally verify the computed MATIs $\overline{\tau} = 0.0835$ s and $\overline{\tau} = 0.042$ s for $K = 2$ and $K = 4$, respectively, which corresponds to 12 and 24 Hz, respectively. To test the claimed robustness of the experimental MAS (see Corollary 2) and to examine how different gains K and different sampling frequencies affect the system performance, we generate an exogenous disturbance (eg, wind) that moves the virtual agent with speed 0.5 m/s along a square of side length 0.5 m. This disturbance is generated in our computer so that it is repeatable while we are performing the experiments, which in turn yields a fair comparison. This disturbance was reproduced several times during each experimental trial. Also, we have introduced offsets among agent positions to avoid agent collisions. Of course, these offsets do not alter the MATIs. Experimental data are presented in Table 1 and Figs. 4 and 5. The accompanying video is available at https://www.youtube.com/watch?v=iBjIl-UVyM0.

Table 1 Comparison of the normalized $\|\xi[t_0, t_{\text{end}}]\|_{2,\mathcal{B}}$ for different gains K and sampling frequencies

		$\frac{\|\xi[t_0,t_{\text{end}}]\|_{2,\mathcal{B}}}{(t_{\text{end}}-t_0)}$
$K = 2$	50 Hz	0.054
	12 Hz	0.042
	9 Hz	0.078
$K = 4$	50 Hz	0.052
	24 Hz	0.057
	20 Hz	0.065

The consensus controller is turned on at t_0 and it is switched off at t_{end}. The difference $t_{\text{end}} - t_0$ is about 40 s for each experiment.

From Table 1 we infer that faster sampling improves the system performance (with the only exception being the case $K = 2, f = 12$), which is expected. The delay d is larger than the sampling intervals. Such delays are referred to as *large delays*. In addition, the computed MATIs appear to be rather accurate and useful for our experimental setup (see Figs. 4 and 5 and the accompanying video). Hence the experimental results obtained regarding the conservativeness of our method are encouraging. Nevertheless, further research efforts are needed in this regard.

6 CONCLUSION

In this chapter we investigated multiagent control in degraded communication environments. Our theoretical framework allows for sampled, delayed, and corrupted data as well as for general heterogeneous linear agents, output feedback, and directed (not necessarily balanced) topologies. We sought MATIs that yield a certain performance in terms of $\mathcal{L}_p$-gains. In addition, a salient feature of our framework is the consideration of so-called *large delays*. Essentially, we combined our previous results regarding $\mathcal{L}_p$-stability of delay systems and MASs under intermittent information, and experimentally verified the MATIs obtained. The experiments suggest that our method yields MATIs which are rather accurate and useful for practical purposes.

Future research avenues include nonuniform and time-varying delays as well as time-varying topologies. It is also of interest to further study the impact that different parameters (eg, different sampling frequencies, delays, controller gains, or noise levels) have on the control performance. Lastly, to further reduce communication efforts, we will consider the event- and self-triggered communication paradigms.

ACKNOWLEDGMENTS

This work is supported by the Air Force Research Laboratory under agreement number FA8655-13-1-3055. The US Government is authorized to reproduce and distribute reprints for Governmental purposes notwithstanding any copyright notation thereon. The views and conclusions contained herein are those of the authors and should not be interpreted as necessarily representing the official policies or endorsements, either expressed or implied, of the Air Force Research Laboratory or the US Government.

APPENDIX

6.1 Proof of Theorem 1

From the paragraph before Theorem 1, we have $\mathcal{L}_p$-stability from $\tilde{\zeta}$ to e with bias—that is,

$$\|e[t_0,t]\|_p \leq K_{\rm e}\|\psi_e\|_d + \gamma_{\rm e}\|\tilde{\zeta}[t_0,t]\|_p + \|b[t_0,t]\|_p, \tag{17}$$

for any $t \geq t_0$ and any $p \in [1,\infty]$. Expressions for $K_{\rm e}$, $\gamma_{\rm e}$, and b are provided immediately before Theorem 1.

Let us now infer $\mathcal{L}_p$-stability with bias from ω to $(\tilde{\zeta}, e)$ via the small-gain theorem [21]. Notice that Eq. (15) involves $\|\tilde{\zeta}[t_0,t]\|_{p,\mathcal{B}_{\tilde{\zeta}}}$, while Eq. (17) involves $\|\tilde{\zeta}[t_0,t]\|_p$. However, our choice of the output $\tilde{\zeta}$, stated in Eq. (14), yields $\mathcal{B}_{\tilde{\zeta}} = \mathbf{0}_{n_{\tilde{\zeta}}}$ because of Assumption 3. In other words, $\|\tilde{\zeta}[t_0,t]\|_p = \|\tilde{\zeta}[t_0,t]\|_{p,\mathcal{B}_{\tilde{\zeta}}}$. Next notice that $\|(\omega,e)[t_0,t]\|_p \leq \|\omega[t_0,t]\|_p + \|e[t_0,t]\|_p$. Combining Eq. (15) with Eq. (17) yields

$$\begin{aligned}\|e[t_0,t]\|_p \leq{}& \frac{K_{\rm e}}{1-\gamma_{\rm e}\gamma_{\rm n}}\|\psi_e\|_d + \frac{\gamma_{\rm e}K_{\rm n}}{1-\gamma_{\rm e}\gamma_{\rm n}}\|\psi_\xi\|_{d,\mathcal{B}}\\ &+ \frac{\gamma_{\rm e}\gamma_{\rm n}}{1-\gamma_{\rm e}\gamma_{\rm n}}\|\omega[t_0,t]\|_p + \frac{1}{1-\gamma_{\rm e}\gamma_{\rm n}}\|b[t_0,t]\|_p,\end{aligned} \tag{18}$$

$$\begin{aligned}\|\tilde{\zeta}[t_0,t]\|_p \leq{}& \frac{K_{\rm n}}{1-\gamma_{\rm e}\gamma_{\rm n}}\|\psi_\xi\|_{d,\mathcal{B}} + \frac{\gamma_{\rm n}K_{\rm e}}{1-\gamma_{\rm e}\gamma_{\rm n}}\|\psi_e\|_d\\ &+ \frac{\gamma_{\rm n}}{1-\gamma_{\rm e}\gamma_{\rm n}}\|\omega[t_0,t\|_p + \frac{\gamma_{\rm n}}{1-\gamma_{\rm e}\gamma_{\rm n}}\|b[t_0,t]\|_p.\end{aligned} \tag{19}$$

Using the Minkowski inequality [28, Chapter 6], we obtain

$$\|(\tilde{\zeta},e)[t_0,t]\|_p = \|(\tilde{\zeta},\mathbf{0}_{n_e})[t_0,t] + (\mathbf{0}_{n_{\tilde{\zeta}}},e)[t_0,t]\|_p \leq \|\tilde{\zeta}[t_0,t]\|_p + \|e[t_0,t]\|_p. \tag{20}$$

Combining Eqs. (18), (19), and (20), along with $\mathcal{B}_{\tilde{\zeta}} = \mathbf{0}_{n_{\tilde{\zeta}}}$, yields

$$\|(\tilde{\zeta}, e)[t_0, t]\|_{p,(\mathcal{B}_{\tilde{\zeta}},\mathbf{0}_{ne})} \leq K_1\|\psi_\xi\|_{d,\mathcal{B}} + K_2\|\psi_e\|_d + \gamma_1\|\omega[t_0, t]\|_p + \|b_1[t_0, t]\|_p \tag{21}$$

for all $t \geq t_0$, where $K_1 = \frac{\gamma_e K_n + K_n}{1-\gamma\gamma_e}$, $K_2 = \frac{K_e + \gamma_n K_e}{1-\gamma\gamma_e}$, $\gamma_1 = \frac{\gamma_n\gamma_e + \gamma_n}{1-\gamma\gamma_e}$, and $b_1 = \frac{1+\gamma_n}{1-\gamma_e\gamma_n} b$. We proceed by using the following expression:

$$\|\psi_\xi\|_{d,\mathcal{B}} + \|\psi_e\|_d \leq 2\max\{\|\psi_\xi\|_{d,\mathcal{B}}, \|\psi_e\|_d\} \leq 2\|(\psi_\xi, \psi_e)\|_{d,(\mathcal{B},\mathbf{0}_{ne})}. \tag{22}$$

Finally, putting together Eqs. (21) and (22), we obtain

$$|(\tilde{\zeta}, e)[t_0, t]\|_{p,(\mathcal{B}_{\tilde{\zeta}},\mathbf{0}_{ne})} \leq K_3\|(\psi_\xi, \psi_e)\|_{d,(\mathcal{B},\mathbf{0}_{ne})} + \gamma_1\|\omega[t_0, t]\|_p + \|b_1[t_0, t]\|_p \tag{23}$$

for all $t \geq t_0$, where $K_3 := 2\max\{K_1, K_2\}$. This proves the claim.

6.2 Proof of Corollary 2

The $\mathcal{L}_p$-detectability of ξ with respect to $\mathcal{B}$ from $(e, \omega, \tilde{\zeta})$ implies that there exist $K_d, \gamma_d \geq 0$ such that

$$\begin{aligned}\|\xi[t_0, t]\|_{p,\mathcal{B}} &\leq K_d\|\psi_\xi\|_{d,\mathcal{B}} + \gamma_d\|\tilde{\zeta}[t_0, t]\|_{p,\mathcal{B}_{\tilde{\zeta}}} + \gamma_d\|(e, \omega)[t_0, t]\|_p \\ &\leq K_d\|\psi_\xi\|_{d,\mathcal{B}} + \gamma_d\|\tilde{\zeta}[t_0, t]\|_{p,\mathcal{B}_{\tilde{\zeta}}} + \gamma_d\|e[t_0, t]\|_p + \gamma_d\|\omega[t_0, t]\|_p\end{aligned} \tag{24}$$

for all $t \geq t_0$. Plugging Eq. (15) into Eq. (24) leads to

$$\begin{aligned}\|\xi[t_0, t]\|_{p,\mathcal{B}} \leq\; & K_d\|\psi_\xi\|_{d,\mathcal{B}} + \gamma_d K_n\|\psi_\xi\|_{d,\mathcal{B}} + \gamma_d\gamma_n\|e[t, t_0]\|_p \\ & + \gamma_d\gamma_n\|\omega[t_0, t]\|_p + \gamma_d\|e[t_0, t]\|_p + \gamma_d\|\omega[t_0, t]\|_p \\ \leq\; & K_d\|\psi_\xi\|_{d,\mathcal{B}} + \gamma_d K_n\|\psi_\xi\|_{d,\mathcal{B}} + (\gamma_d\gamma_n + \gamma_d)\|e[t, t_0]\|_p \\ & + (\gamma_d\gamma_n + \gamma_d)\|\omega[t_0, t]\|_p\end{aligned}$$

for all $t \geq t_0$. Observe that (simply follow the proof for the classical Minkowski inequality [28, Chapter 6] and the fact that $\|\cdot\|_{\mathcal{B}}$ is a seminorm)

$$\begin{aligned}\|(\xi, e)[t_0, t]\|_{p,(\mathcal{B},\mathbf{0}_{ne})} &= \|(\xi, \mathbf{0}_{n_e})[t_0, t] + (\mathbf{0}_{n_\xi}, e)[t_0, t]\|_{p,(\mathcal{B},\mathbf{0}_{ne})} \\ &\leq \|(\xi, \mathbf{0}_{n_e})[t_0, t]\|_{p,(\mathcal{B},\mathbf{0}_{ne})} + \|(\mathbf{0}_{n_\xi}, e)[t_0, t]\|_{p,(\mathcal{B},\mathbf{0}_{ne})} \\ &= \|\xi[t_0, t]\|_{p,\mathcal{B}} + \|e[t_0, t]\|_p\end{aligned} \tag{25}$$

holds. In fact, one can think of Eq. (25) as a variant of the classical Minkowski inequality. Finally, we include Eq. (18) into Eq. (24) and add the inequality obtained to Eq. (18) which establishes $\mathcal{L}_p$-stability with respect to $(\mathcal{B}, \mathbf{0}_{n_e})$ with bias from ω to (ξ, e) following the same steps as in the proof of Theorem 1.

REFERENCES

[1] R. Olfati-Saber, R.M. Murray, Consensus problems in networks of agents with switching topology and time-delays, IEEE Trans. Autom. Control 49 (9) (2004) 1520–1533.

[2] W. Ren, R.W. Beard, Distributed Consensus in Multi-Vehicle Cooperative Control—Theory and Applications, in: Communications and Control Engineering, Springer, London, 2008.

[3] L. Fang, P.J. Antsaklis, On communication requirements for multi-agent consensus seeking, in: P. Antsaklis, P. Tabuada (Eds.), Networked Embedded Sensing and Control, Proc. of Workshop NESC05: University of Notre Dame, USA, October 17–18, 2005, Lecture Notes in Control and Information Sciences, vol. 331, Springer, Berlin, 2006, pp. 133–147.

[4] F. Xiao, L. Wang, Asynchronous consensus in continuous-time multi-agent systems with switching topology and time-varying delays, IEEE Trans. Autom. Control 53 (8) (2008) 1804–1816.

[5] H. Yu, P.J. Antsaklis, Output synchronization of multi-agent systems with event-driven communication: communication delay and signal quantization, Technical Report, Department of Electrical Engineering, University of Notre Dame, 2011.

[6] D.V. Dimarogonas, E. Frazzoli, K.H. Johansson, Distributed event-triggered control for multi-agent systems, IEEE Trans. Autom. Control 57 (5) (2012) 1291–1297.

[7] Y. Liu, Y. Jia, H_∞ consensus control for multi-agent systems with linear coupling dynamics and communication delays, Int. J. Syst. Sci. 43 (1) (2012) 50–62.

[8] M. Guinaldo, D.V. Dimarogonas, K.H. Johansson, J.S. Moreno, S. Dormido, Distributed event-based control strategies for interconnected linear systems, IET Control Theory Appl. 7 (6) (2013) 877–886.

[9] C.D. Persis, P. Frasca, Robust self-triggered coordination with ternary controllers, IEEE Trans. Autom. Control 58 (12) (2013) 3024–3038.

[10] D. Tolić, $\mathcal{L}_p$-stability with respect to sets applied towards self-triggered communication for single-integrator consensus, in: Proceedings of the IEEE Conference on Decision and Control, December, 2013, pp. 3409–3414.

[11] D. Tolić, V. Jeličić, V. Bilas, Resource management in cooperative multi-agent networks through self-triggering, IET Control Theory Appl. 9 (6) (2014) 915–928.

[12] D. Tolić, S. Hirche, Stabilizing transmission intervals and delays for nonlinear networked control systems: the large delay case, in: Proceedings of the IEEE Conference on Decision and Control, 2014, pp. 1203–1208.

[13] A. Kruszewski, W.J. Jiang, E. Fridman, J.P. Richard, A. Toguyeni, A switched system approach to exponential stabilization through communication network, IEEE Trans. Autom. Control 20 (4) (2012) 887–900.

[14] B. Zhang, A. Kruszewski, J.P. Richard, A novel control design for delayed teleoperation based on delay-scheduled Lyapunov-Krasovskii functionals, Int. J. Control 87 (8) (2014) 1694–1706.

[15] D. Tolić, I. Palunko, A. Ivanović, M. Car, S. Bogdan, Multi-agent control in degraded communication environments, in: European Control Conference, July, 2015, pp. 404–409.

[16] C. Godsil, G. Royle, Algebraic Graph Theory, in: Graduate Texts in Mathematics, Springer, New York, 2001.

[17] G.H. Ballinger, Qualitative theory of impulsive delay differential equations (Ph.D. thesis), University of Waterloo, Ontario, Canada, 1999.

[18] D. Nešić, A.R. Teel, Input-output stability properties of Networked Control Systems, IEEE Trans. Autom. Control 49 (10) (2004) 1650–1667.

[19] Z.P. Jiang, A.R. Teel, L. Praly, Small-gain theorem for ISS systems and applications, Math. Control Signals Syst. 7 (2) (1994) 95–120.

[20] T. Estrada, P.J. Antsaklis, Model-based control with intermittent feedback: bridging the gap between continuous and instantaneous feedback, Int. J. Control 83 (12) (2010) 2588–2605.
[21] H. Khalil, Nonlinear Systems, third ed., Prentice Hall, Englewood Cliffs, NJ, 2002.
[22] W. Michiels, S. Gumussoy, Characterization and computation of H-infinity norms of time-delay systems, IEEE Trans. Autom. Control 31 (4) (2010) 2093–2115.
[23] N. Martins, Finite gain l_p stability requires analog control, Syst. Control Lett. 55 (11) (2006) 949–954.
[24] S. Piskorski, N. Brulez, AR.Drone Developer Guide, Parrot, Technical Report, 2011.
[25] https://www.naturalpoint.com/optitrack/, 2014.
[26] M. Quigley, K. Conley, B.P. Gerkey, J. Faust, T. Foote, J. Leibs, R. Wheeler, A.Y. Ng, ROS: an open-source Robot Operating System, in: ICRA Workshop on Open Source Software, 2009.
[27] C.-Y. Kao, A. Rantzer, Stability analysis of systems with uncertain time-varying delays, Automatica 43 (6) (2007) 959–970.
[28] R.G. Bartle, The Elements of Integration and Lebesgue Measure, first ed., Wiley-Interscience, New York, 1995.

CHAPTER THIRTEEN

Multiagent Layered Formation Control Based on Rigid Graph Theory

S. Ramazani*, R.R. Selmic*, M. de Queiroz†
*Louisiana Tech University, Ruston, LA, United States
†Louisiana State University, Baton Rouge, LA, United States

1 INTRODUCTION

In recent years the formation control of multiagent systems has attracted significant attention. A multiagent system is a team of interacting agents that together perform a complex task [1]. The concept of multiagent formation can be observed in nature in many forms—for example, flock of birds, school of fish, and swarm of bees [2, 3]. The collective behavior of these biological systems is distributed and decentralized as each individual member of the group has its own local sensing and control mechanism without global knowledge or planning [1]. This multiagent system behavior is also demonstrated in engineering applications by teams of autonomous (aerial, ground, or underwater) vehicles that are deployed to perform surveillance, reconnaissance, and search of an area [1, 4].

The use of multiple agents for execution of a mission has many advantages over the use of a single agent, such as reduction in the time taken to accomplish a given task, robustness when one or more agents fail, and compensation for sensor uncertainty using the merging of overlapping information from the cooperating agents (therefore increasing the efficiency of the mission) [5].

Formation is a geometric shape, and maintaining this shape implies that the formation at one instant of time is congruent to the formation at another instant of time. Graphs have been used as a tool to model multiagent formations. In a graph that corresponds to a formation structure, each agent corresponds to a vertex, and for each pair of agents i and j, there must be

Control of Complex Systems
http://dx.doi.org/10.1016/B978-0-12-805246-4.00013-6

an edge (i, j) if the distance between agents i and j is to be maintained at all times [6]. In the solving of the formation control problem, the concept of graph rigidity is used to determine the number of interagent distances that are necessary to ensure a formation. In this case the rigidity matrix is important for the stability analysis of the formation control. Some previous work that used rigid graph theory to address the multiagent formation problem can be found in [1, 3, 7–11].

The concept of layered sensing was first introduced by the US Air Force Research Laboratory. It refers to the appropriate sensor or combination of sensors/platforms, infrastructure, and exploitation capabilities that generates situation awareness and directly supports "tailored effects" [12]. In this work the concepts of layered sensing and formation control are combined to address the multiagent layered formation control problem. The ability to have multiple agents, whether on the ground or airborne (coplanar), autonomously perform searching, detecting, and tracking tasks while one or more agents (coordinator/supervisor) move in an upper layer above the multiagent group is the main objective of this work. In addition to autonomously performing a task, all the agents must acquire a formation structure. For this approach the scenario of formation control of multiagent systems for layered sensing applications is considered by use of decentralized control laws where the agents acquire a desired formation shape.

An example of the scenario described above is shown in Fig. 1 and includes unmanned ground vehicles and micro aerial vehicles that serve as coordinators/supervisors. The supervising agents can be with or without sensors for the ground supporting mission.

This work specifically addresses the *layered formation control* of n planar agents that belong to a plane Q, modeled by an undirected graph and r coordinating/supervising agents that are moving outside the plane Q and belong to a plane W, as shown in Fig. 1. Triangulation of the layered formation is proposed to obtain a rigid framework in three-dimensional space.

As a special case the layered formation control with one agent in plane W (ie, $r = 1$) is also addressed (see Fig. 2). *The concept of coning [13, 14] is exploited in this case to create a three-dimensional formation that retains the rigidity of the framework for the planar agents in plane* Q.

In the solving of the formation control problem, the desired formations are achieved by direct control of interagent distances with the use of

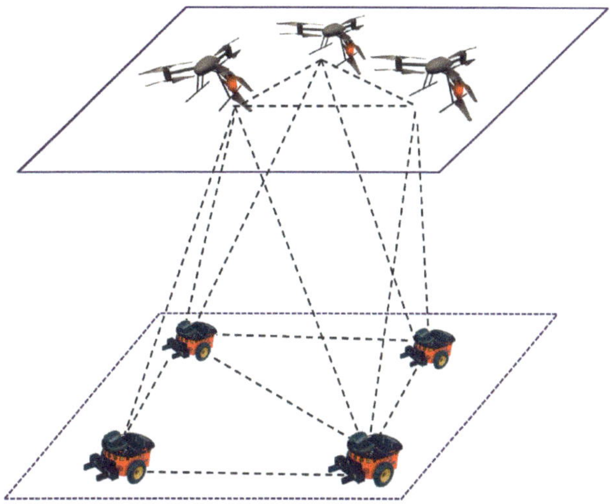

Fig. 1 Formation control concept under the layered sensing framework.

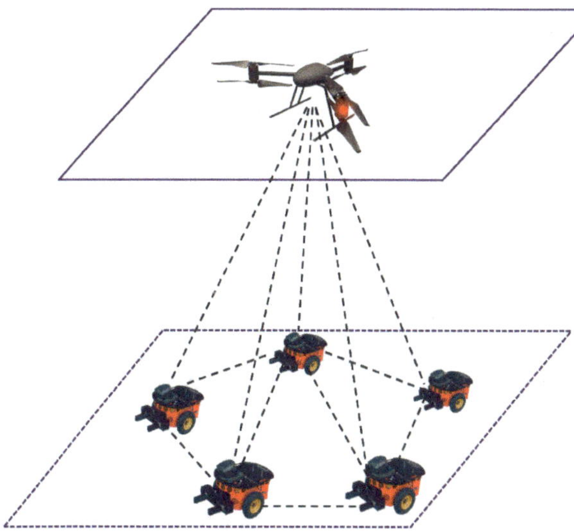

Fig. 2 Formation control concept under the coning framework.

single-integrator and double-integrator models for the agents' motions. Previous work [15] addressed the formation acquisition of n agents in two dimensions. The method in this work builds on the results presented in [15] and extends the control law developed so that it can be applied to three-dimensional layered graphs.

2 PRELIMINARIES

The formation shape of a multiagent system is represented by an undirected graph $G = (V, E)$, where $V = 1, 2, \ldots, n$ is the vertex set of this graph that represents the agents and E is the edge set that represents the communication links between the agents. The number of vertices and edges of G are denoted by $|V|$ and $|E|$, respectively.

A framework is a realization of a graph at given points in $\mathbb{R}^d$, where $d \in \{2, 3\}$. A d-dimensional framework F is a pair (G, p), where $p = (p_1, \ldots, p_n) \in \mathbb{R}^{dn}$, where $p_i \in \mathbb{R}^d$ is the coordinate of vertex i [15]. Given an arbitrary ordering of the edges of G, an edge function $\Phi_G : \mathbb{R}^{dn} \rightarrow \mathbb{R}^{|E|}$ associated with (G, p) is given by

$$\Phi_G(p) = (\ldots, \|p_i - p_j\|^2, \ldots), \quad (i,j) \in E, \tag{1}$$

where $\|\cdot\|$ denotes the Euclidean norm. The rigidity matrix $R(p) : \mathbb{R}^{dn} \rightarrow \mathbb{R}^{|E| \times dn}$ of (G, p) is defined as

$$R(p) = \frac{1}{2}\frac{\partial \Phi_G(p)}{\partial p}. \tag{2}$$

Two frameworks (G, p) and (G, q) are equivalent if $\Phi_G(p) = \Phi_G(q)$ and are congruent if $\|p_i - p_j\| = \|q_i - q_j\|$ for all $i, j \in V$ [15, 16]. In the case where two frameworks are equivalent but not congruent, then they are flip ambiguous [2]. The notion of flip ambiguity is illustrated in Fig. 3.

Rigid graph theory [2, 17, 18] plays a crucial role in analyzing the multiagent formation shape and describing the information architecture of the system. To determine the rigidity of a formation, both the number of the edges and their distribution among the graph vertices matter. Laman's

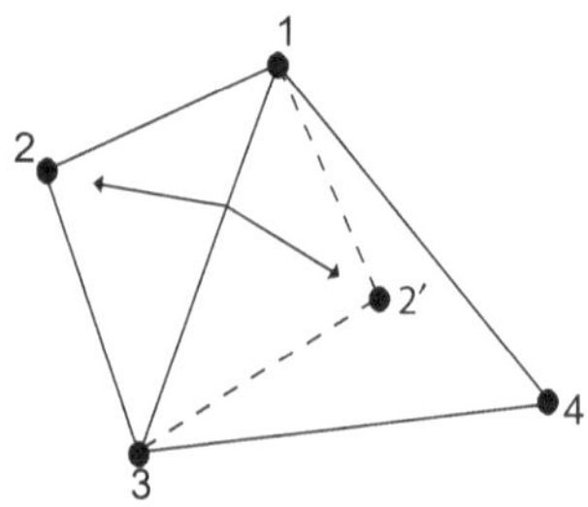

Fig. 3 A noncongruent framework that depicts flip ambiguity (vertex 2 can be flipped over the edge (1,3) to the symmetric position 2′).

theorem is known to be a key result that is used in solving the rigidity-based formation control problems in two dimensions.

Theorem 1 ([19]). *A graph* $G = (V, E)$ *modeling a formation in two dimensions is rigid if and only if there exists a subgraph* $G' = (V, E')$, $E' \subset E$ *with* $|E'| = 2|V| - 3$ *such that for any* $V' \subset V$, *the associated induced subgraph* $G'' = (V', E'')$ *of* G', *with* $E'' \subset E'$, *satisfies* $|E''| \leq 2|V'| - 3$.

Fig. 4 shows an example of flexible and rigid formations in two dimensions. In the flexible formation, part of the framework can be deformed by a smooth motion, while the distance between the agents that are connected with edges remains unchanged. In the rigid formation the only smooth motions of the framework correspond to translation or rotation; therefore the formation cannot be deformed.

A graph is *minimally rigid* if it is rigid and if no single edge can be removed from the graph without this causing the graph to lose its rigidity [20]. A graph $G = (V, E)$ is minimally rigid in two or three dimensions if and only if $|E| = 2|V| - 3$ or $|E| = 3|V| - 6$, respectively [2, 21].

A graph is *infinitesimally rigid* if it cannot flex or deform by even a very small amount. A framework (G, p) where $n > d$ and p is generic (the affine span of p is all of R^d) is infinitesimally rigid if and only if $\text{rank}[R(p)] = dn - \binom{d+1}{2}$. Therefore (G, p) is flexible if $|E| < dn - \binom{d+1}{2}$ [17, 22, 23]. A rigid framework is not always infinitesimally rigid. Fig. 5 shows an example of rigid, infinitesimally rigid, and infinitesimally and minimally rigid frameworks in two dimensions. The

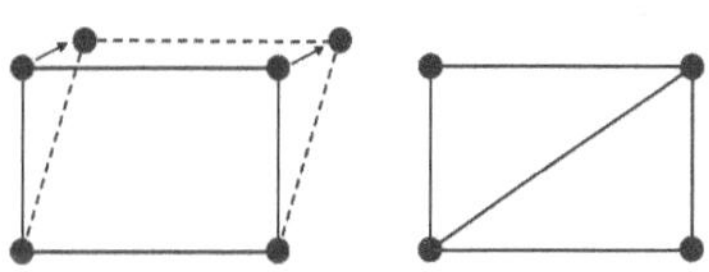

Fig. 4 Flexible (*left*) and rigid (*right*) formations in two dimensions.

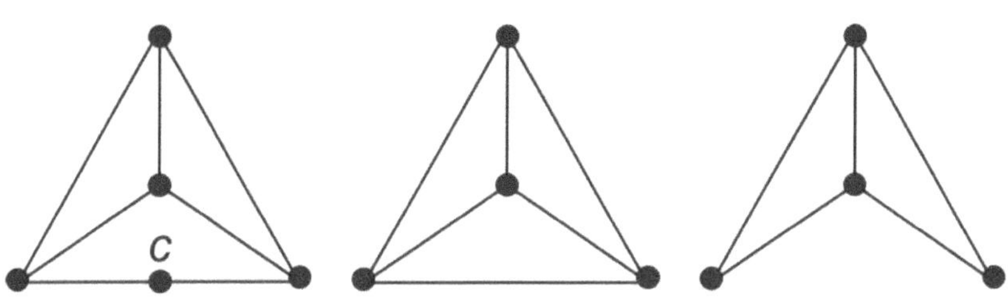

Fig. 5 A rigid but not infinitesimally and minimally rigid framework (*left*), an infinitesimally but not minimally rigid framework (*middle*), and an infinitesimally and minimally rigid framework (*right*).

framework on the left in Fig. 5 is not infinitesimally rigid, since it will have an infinitesimal deformation if force is applied to vertex c.

Lemma 1. *If the framework* $F = (G, p)$ *is infinitesimally and minimally rigid in three dimensions, then* $R(p)R^T(p)$ *is invertible.*

Proof. We know that if F is infinitesimally and minimally rigid in three dimensions, then $\text{rank}[R(p)] = 3|V| - 6$ and $|E| = 3|V| - 6$. Therefore $R(p)$ has full row rank. Since $R^T(p)$ has full column rank and $\text{rank}[R(p)] = \text{rank}[R(p)R^T(p)]$, $R(p)R^T(p) \in \mathbb{R}^{|E|\times|E|}$ is invertible. □

The rigidity matrix $R(p)$ for the framework F is constructed with an arbitrary ordering of vertices and edges and has $3|V|$ columns and $|E|$ rows. The rows of $R(p)$ correspond to the edges of G. If there is an edge between vertices i and j, then the entries $x_i - x_j$, $y_i - y_j$, $z_i - z_j$, $x_j - x_i$, $y_j - y_i$, and $z_j - z_i$ will fill columns $3i - 2$, $3i - 1$, $3i$, $3j - 2$, $3j - 1$, and $3j$ of $R(p)$, respectively. The entries of the other columns will be zero. For example, for the graph in Fig. 6, the rigidity matrix is

$$R(p) = \begin{bmatrix} x_1-x_2 & y_1-y_2 & z_1-z_2 & x_2-x_1 & y_2-y_1 & z_2-z_1 & 0 & 0 & 0 & 0 & 0 & 0 \\ 0 & 0 & 0 & x_2-x_3 & y_2-y_3 & z_2-z_3 & x_3-x_2 & y_3-y_2 & z_3-z_2 & 0 & 0 & 0 \\ x_1-x_3 & y_1-y_3 & z_1-z_3 & 0 & 0 & 0 & x_3-x_1 & y_3-y_1 & z_3-z_1 & 0 & 0 & 0 \\ x_1-x_4 & y_1-y_4 & z_1-z_4 & 0 & 0 & 0 & 0 & 0 & 0 & x_4-x_1 & y_4-y_1 & z_4-z_1 \\ 0 & 0 & 0 & x_2-x_4 & y_2-y_4 & z_2-z_4 & 0 & 0 & 0 & x_4-x_2 & y_4-y_2 & z_4-z_2 \\ 0 & 0 & 0 & 0 & 0 & 0 & x_3-x_4 & y_3-y_4 & z_3-z_4 & x_4-x_3 & y_4-y_3 & z_4-z_3 \end{bmatrix}.$$

A surface in $\mathbb{R}^3$ is triangulated if it is covered with a collection of triangles and if any two triangles intersect, their intersection is a common edge or vertex [24]. A graph $G = (V, E)$ is polyhedral if and only if there exists $p = (p_1, \ldots, p_n) \in \mathbb{R}^{3|V|}$ such that $p_i \neq p_j$ for $i \neq j$ and $(i, j) \in E$ are the edges of a convex polyhedron in $\mathbb{R}^3$ [25]. A triangulated convex polyhedral surface with vertices only in the natural edges is infinitesimally

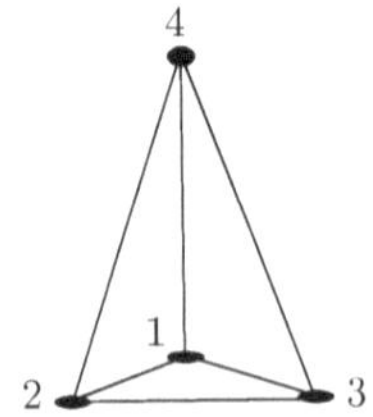

Fig. 6 A rigid formation in three dimensions.

rigid. The natural edges are one-dimensional intersections of a support plane with the convex polyhedral surface [26].

Coning is a technique in rigidity which takes frameworks in $\mathbb{R}^N$ to frameworks in $\mathbb{R}^{N+1}$, where $N \in \{1, 2, 3, \ldots\}$, while preserving first-order rigidity (or infinitesimal rigidity) [13, 14]. In this chapter the concept of coning is used for the case when there is only one coordinating agent. A graph $G = (V, E)$ is coned by addition of a new vertex c, and addition of edges from this vertex to all the original vertices in G. This will create the cone graph $G \star \{c\}$ with $V(G \star \{c\}) = V(G) \cup \{c\}$, where the cone vertex c is distinct from the vertices of G. If p_c is a configuration for $G \star \{c\}$ and $H \cong \mathbb{R}^N$ is a hyperplane, then we denote p_H to be the projection of p_c from the cone vertex into H. We call p_c and p_H a projection pair of configurations.

A general version of the infinitesimal rigidity coning theorem is presented in [14] and its proof can be found in [13]. The following lemma is directed toward the cone framework (one coordinator in the framework) of interest in this work.

Lemma 2 ([13]). *Let G be a graph that represents a group of n agents in $\mathbb{R}^2$, $G \star \{c\}$ be the cone graph with cone vertex c representing the coordinating agent, and $p^\star$ and p_H be a projection pair of configurations. Then $(G \star \{c\}, p^\star)$ is infinitesimally rigid in $\mathbb{R}^3$ if and only if (G, p_H) is infinitesimally rigid in $\mathbb{R}^2$.*

3 PROBLEM FORMULATION

In this work it is assumed that agents are equipped with sensors such as ultrasonic or infrared-based relative positioning sensors [27] that allow them to measure the distance and direction between selected agents.

3.1 Nonplanar Multiagent Layered Formation Control With a Single-Integrator Model

Consider a system of n agents in the plane Q and r coordinating agents in plane W outside the plane Q, modeled by the single integrator [1, 8, 9, 28]

$$\dot{p}_{l_i} = u_{l_i}, \quad i = 1, \ldots, n, n+1, \ldots, n+r, \tag{3}$$

where $p_{l_i} = (x_{l_i}, y_{l_i}, 0) \in \mathbb{R}^3$ for $i = 1, \ldots, n$ is the location of the ith agent in local coordinates in plane Q that is defined by $z = 0$. The location of the coordinators is defined by $p_{l_i} = (x_{l_i}, y_{l_i}, z_{l_i}) \in \mathbb{R}^3$ for $i = n+1, \ldots, n+r$, where z_{l_i} is the distance of the agents from plane Q. In general, there is no requirement for the planes Q and W to be parallel. The control input for the ith agent in plane Q and the coordinating agents is $u_{l_i} \in \mathbb{R}^3$ for $i = 1, \ldots, n, n+1, \ldots, n+r$.

Consider a formation in $\mathbb{R}^3$ that can be triangulated for the layered system mentioned above described by the framework $F_l^\star = (G_l^\star, p_l^\star)$, where $G_l^\star = (V_l, E_l)$ and $p_l^\star = (p_{l_1}^\star, \ldots, p_{l_n}^\star, p_{l_{n+1}}^\star, \ldots, p_{l_{n+r}}^\star)$. Let $F_l^\star$ represent the desired layered formation of the system of $n + r$ agents. Given the actual formation $F_l(t) = (G_l^\star, p_l(t))$, where $p_l = (p_{l_1}, \ldots, p_{l_n}, p_{l_{n+1}}, \ldots, p_{l_{n+r}})$, and assuming that at $t = 0$, $\|p_{l_i}(0) - p_{l_j}(0)\| \neq d_{l_{ij}}$ for $(i,j) \in E_l$, where $d_{l_{ij}} = \|p_{l_i}^\star - p_{l_j}^\star\| > 0$ is the constant desired distance between agents i and j in the layered formation, the control objective is to design the control input u_{l_i} such that

$$\|p_{l_i}(t) - p_{l_j}(t)\| \to d_{l_{ij}} \quad as \quad t \to \infty, \qquad (i,j) \in E_l. \tag{4}$$

3.2 Nonplanar Multiagent Layered Formation Control With a Double-Integrator Model

In this section the problem of nonplanar multiagent layered formation control is formulated with the double-integrator model. The use of a double-integrator model is more practical for a multiagent formation. In this case the formation reaches a state of balance when the agents are not scattered and their velocities match [29].

Consider the system mentioned in Section 3.1 modeled by the double integrator [15, 30]

$$\begin{cases} \dot{p}_{l_i} = v_{l_i} \\ \dot{v}_{l_i} = u_{l_i}, \quad i = 1, \ldots, n, n+1, \ldots, n+r, \end{cases} \tag{5}$$

where $v_{l_i} \in \mathbb{R}^3$ for $i = 1, \ldots, n, n+1, \ldots .n+r$ is the velocity and u_{l_i} is the acceleration-level control input for the ith agent. The control objective is to design the control input u_{l_i} such that Eq. (4) is satisfied.

4 CONTROL ALGORITHMS

We extend the results from [15] to solve the formation acquisition and stabilization problem in three dimensions where the coordinators are operating outside the plane defined by other agents. The approach is presented for both the single-integrator model and the double-integrator model.

4.1 Single-Integrator Model

Define the relative position of two agents in the layered formation as

$$\tilde{p}_{l_{ij}} = p_{l_i} - p_{l_j}, \quad (i,j) \in E_l. \tag{6}$$

The distance error for the group of $n + r$ agents and the corresponding distance error dynamics are given by

$$e_{l_{ij}} = \|\tilde{p}_{l_{ij}}\| - d_{l_{ij}}, \tag{7}$$

$$\begin{aligned} \dot{e}_{l_{ij}} &= \frac{d}{dt}\sqrt{\tilde{p}_{l_{ij}}^T \tilde{p}_{l_{ij}}} = (\tilde{p}_{l_{ij}}^T \tilde{p}_{l_{ij}})^{-\frac{1}{2}} \tilde{p}_{l_{ij}}^T (u_{l_i} - u_{l_j}) \\ &= \frac{\tilde{p}_{l_{ij}}^T (u_{l_i} - u_{l_j})}{e_{l_{ij}} + d_{l_{ij}}}. \end{aligned} \tag{8}$$

For control algorithm development and stability analysis, consider the potential function [8, 9, 15]

$$M_{ij} = \frac{1}{4} e_{l_{ij}}^2 (e_{l_{ij}} + 2d_{l_{ij}})^2, \tag{9}$$

where M_{ij} is positive definite and radially unbounded in $e_{l_{ij}}$. Define

$$M(e_l) = \sum_{(i,j)\in E_l} M_{ij}(e_{l_{ij}}), \tag{10}$$

where $M_{ij}(e_{l_{ij}})$ was given in Eq. (9). The time derivative of Eq. (10) is then given by

$$\dot{M} = \sum_{(i,j)\in E_l} \frac{\partial M_{ij}}{\partial e_{l_{ij}}} \frac{\tilde{p}_{l_{ij}}^T (u_{l_i} - u_{l_j})}{e_{l_{ij}} + d_{l_{ij}}}. \tag{11}$$

As shown in [15], it follows from Eqs. (1), (2), (7), and (8) that Eq. (11) can be expressed as

$$\dot{M} = \beta^T(e_l) R(p_l) u_l, \tag{12}$$

where $u_l = (u_{l_1}, \ldots, u_{l_n}, u_{l_{n+1}}, \ldots, u_{l_{n+r}}) \in \mathbb{R}^{3(n+r)}$, $e_l = (\ldots, e_{l_{ij}}, \ldots) \in \mathbb{R}^{|E_l|}$, and

$$\begin{aligned} \beta(e_l) &= \left(\ldots, \frac{\frac{\partial M_{ij}}{\partial e_{l_{ij}}}}{e_{l_{ij}} + d_{l_{ij}}}, \ldots \right) \\ &= (\ldots, e_{l_{ij}}(e_{l_{ij}} + 2d_{l_{ij}}), \ldots), \quad (i,j) \in E_l. \end{aligned} \tag{13}$$

The terms in e_l and $\beta(e_l)$ are ordered the same way as in Eq. (1).

The following theorem relates the triangulation of the layered framework in three-dimensional space and the infinitesimal and minimal rigidity of the formation.

Theorem 2 ([26]). *Let $F_l^\star = (G_l^\star, p_l^\star)$, which consists of n agents in the plane Q and r coordinators in the plane W, be the desired framework for a layered formation in $\mathbb{R}^3$. If the surface of $F_l^\star$ is arbitrarily triangulated, then the layered formation $F_l^\star$ is infinitesimally and minimally rigid in $\mathbb{R}^3$.*

The proof for infinitesimal rigidity of the framework $F_l^\star$ in Theorem 2 is presented in [26, 31]. Such a framework has $|E_l| = 3|V_l| - 6$ edges and therefore is minimally rigid in three dimensions [2, 21].

Corollary 1 ([15]). *Define the function*

$$\Omega(F_l, F_l^\star) = \sum_{(i,j)\in E_l} (\|p_{l_i} - p_{l_j}\| - \|p_{l_i}^\star - p_{l_j}^\star\|)^2. \tag{14}$$

If $F_l^\star$ is infinitesimally rigid and $\Omega(F_l, F_l^\star) \leq \varepsilon$, where ε is a sufficiently small positive constant, then F_l is also infinitesimally rigid.

The following theorem shows that the control law [15]

$$u_l = -k_c R^T(p_l)\beta(e_l), \tag{15}$$

where k_c is a positive control gain, achieves the local asymptotic stability of the desired layered formation $F_l^\star = (G_l^\star, p_l^\star)$. It is clear from Eq. (1) that the rigidity matrix function R is dependent on $\tilde{p}_l$, but in this work the argument of this function will be given as p_l.

Theorem 3 ([32]). *Given a group of n planar agents and r coordinators with a layered formation $F_l(t) = (G_l^\star, p_l(t))$ in $\mathbb{R}^3$, let δ be a small positive constant and $\tilde{p}_l(0)$, where $\tilde{p}_l = (\ldots, \tilde{p}_{ij}, \ldots) \in \mathbb{R}^{3|E_l|}$ is the initial condition. If $\tilde{p}_l(0) \in \mathcal{S}_1$, where*

$$\mathcal{S}_1 = \left\{ \tilde{p}_l \in \mathbb{R}^{3|E_l|} \middle| \left| \|\tilde{p}_{l_{ij}}\|^2 - d_{l_{ij}}^2 \right| \leq \frac{2\sqrt{\delta}}{\sqrt{|E_l|}}, (i,j) \in E_l \right\}, \tag{16}$$

then $F_l(t)$ is infinitesimally and minimally rigid, and the control law (15) *achieves the local asymptotic stability of the desired formation $F_l^\star$, ensuring $\|p_{l_i}(t) - p_{l_j}(t)\| \to d_{l_{ij}}$ as $t \to \infty$ for all $(i,j) \in E_l$.*

The following lemma uses the absolute initial and desired positions of the agents and shows that Eq. (15) achieves the local asymptotic stability of the desired formation $F_l^\star$. Therefore Lemma 3 establishes the conditions that prevent convergence to ambiguous formations. The proof of this lemma is presented in [33].

Lemma 3. *Let δ be a small positive constant and $p_l(0) \in \mathbb{R}^{3(n+r)}$ be the initial condition. If $p_l(0) \in \mathcal{S}_2$, where*

$$\mathcal{S}_2 = \left\{ p_l \in \mathbb{R}^{3(n+r)} \middle| \max_{i \in V_l} \|p_{l_i} - p^{\star}_{l_i}\| + D_l \le \sqrt[4]{\frac{\delta}{4|E_l|}}, \right\} \quad (17)$$

where $D_l = \max(d_{l_{ij}})$, then the control law (15) achieves the local asymptotic stability of the desired formation $F^{\star}_l$ and ensures $\|p_{l_i}(t) - p_{l_j}(t)\| \to d_{l_{ij}}$ as $t \to \infty$ for all $(i,j) \in E_l$.

Remark 1. Lemma 3 provides sufficient conditions for the formation convergence to the desired formation. *An upper bound of the norm of the initial position error for each agent is given by Eq. (17).* Conservatism of the bound (17) depends on uniformity of distances $d_{l_{ij}}$ (ie, more uniform distances will cause a less conservative bound).

Remark 2. The system of n agents in the plane Q and a single coordinating agent outside the plane Q is considered a special case for nonplanar multiagent layered formation control with single-integrator [33] and double-integrator models. In this problem the coordinating agent can be a central controller or a satellite. The triangulation method can be applied for the scenario where there is only one coordinator in the layered formation. In this case triangulation is referred to as coning, which allows the retention of the infinitesimal and minimal rigidity of the n planar agents in three-dimensional space.

The control algorithms in this section are applicable to a cone framework. In such a framework the coning method (Lemma 2) ensures that the layered formation $F^{\star}_l(t) = (G^{\star}_l, p^{\star}_l(t))$ is infinitesimally rigid in $\mathbb{R}^3$ [33]. Since the cone framework is triangulated and has $|E_l| = 3|V_l| - 6$ edges, $F^{\star}_l$ is also minimally rigid in three dimensions.

4.2 Double-Integrator Model

Define the relative position of two agents as in Eq. (6). The distance error for the group of n agents and r coordinators is defined in Eq. (7), and the distance error dynamics are given by

$$\dot{e}_{l_{ij}} = \frac{\tilde{p}^T_{l_{ij}}(v_{l_i} - v_{l_j})}{e_{l_{ij}} + d_{l_{ij}}}, \quad (18)$$

where Eq. (5) was used. Consider the potential function in Eq. (9).

Let

$$M_1(e_l) = \sum_{(i,j)\in E_l} M_{ij}(e_{l_{ij}}). \tag{19}$$

The time derivative of Eq. (19) along Eq. (18) is given by

$$\dot{M}_1 = \sum_{(i,j)\in E_l} \frac{\partial M_{ij}}{\partial e_{l_{ij}}} \frac{\tilde{p}_{l_{ij}}^T (v_{l_i} - v_{l_j})}{e_{l_{ij}} + d_{l_{ij}}}. \tag{20}$$

It follows from Eqs. (1), (2), and (7) that Eq. (20) can be written as

$$\dot{M}_1 = \beta^T(e_l)R(p_l)v_l, \tag{21}$$

where $v_l = (v_{l_1}, \ldots, v_{l_n}, v_{l_{n+1}}, \ldots, v_{l_{n+r}}) \in \mathbb{R}^{3(n+r)}$.

With use of the backstepping method [15, 34], consider the variable

$$s_l = v_l - f_l, \tag{22}$$

where $s_l = (s_{l_1}, \ldots, s_{l_n}, s_{l_{n+1}}, \ldots, s_{l_{n+r}}) \in \mathbb{R}^{3(n+r)}$, and $f_l = (f_{l_1}, \ldots, f_{l_n}, f_{l_{n+1}}, \ldots, f_{l_{n+r}}) \in \mathbb{R}^{3(n+r)}$ is a virtual velocity input for the layered formation that is given by

$$f_l = -k_v R^T(p_l)\beta(e_l), \tag{23}$$

where k_v is a positive constant.

Define the total potential function

$$M_2(e_l, s_l) = M_1(e_l) + \frac{1}{2}s_l^T s_l. \tag{24}$$

Taking the time derivative of Eq. (24) and using Eqs. (5), (21), and (22) yields

$$\begin{aligned} \dot{M}_2 &= \beta^T(e_l)R(p_l)v_l + s_l^T \dot{s}_l \\ &= \beta^T(e_l)R(p_l)f_l + s_l^T[u_l + R^T(p_l)\beta(e_l) - \dot{f}_l], \end{aligned} \tag{25}$$

where $\dot{f}_l = -k_v(\dot{R}^T(p_l)\beta(e_l) + R^T(p_l)\dot{\beta}(e_l))$, and $u_l = (u_{l_1}, \ldots, u_{l_n}, u_{l_{n+1}}, \ldots, u_{l_{n+r}}) \in \mathbb{R}^{3(n+r)}$ is the control input for the layered formation.

The following theorem shows that the control law [15]

$$u_l = -k_a s_l + \dot{f}_l - R^T(p_l)\beta(e_l), \tag{26}$$

where k_a is a positive control gain, achieves the asymptotic stability of the desired layered formation $F_l^\star = (G_l^\star, p_l^\star)$.

Theorem 4 ([35]). *Given a group of n planar agents and r coordinators with a layered formation* $F_l(t) = (G_l^\star, p_l(t))$ *in* $\mathbb{R}^3$, *let* δ *be a positive constant and* $\tilde{p}_l(0) \in \mathbb{R}^{3|E_l|}$ *and* $v_l(0) \in \mathbb{R}^{3(n+r)}$ *be the initial conditions. If* $(\tilde{p}_l(0), v_l(0)) \in \mathcal{T}_1$, *where*

$$\mathcal{T}_1 = \left\{ (\tilde{p}_l, v_l) \in \mathbb{R}^{3|E_l|} \times \mathbb{R}^{3(n+r)} \middle| \frac{1}{2}|E_l| \max_{(i,j)\in E_l} (\|\tilde{p}_{l_{ij}}\|^2 - d_{l_{ij}}^2)^2 \right.$$
$$\left. +|V_l| \max_{i\in V_l}(\|v_{l_i}\| + k_v|\Psi_i| \max_{j\in\Psi_i} \|\tilde{p}_{l_{ij}}\|^3)^2 \le 2\delta \right\}, \tag{27}$$

then $F_l(t)$ *is infinitesimally and minimally rigid, and the control law* (26) *achieves the asymptotic stability of the desired formation* $F_l^\star$ *and ensures* $\|p_{l_i}(t) - p_{l_j}(t)\| \to d_{l_{ij}}$ *as* $t \to \infty$ *for all* $(i,j) \in E_l$.

The condition (27) gives a set of allowed initial conditions $(\tilde{p}_l(0), v_l(0)) \in \mathcal{T}_1$ that will ensure the system stability.

Remark 3. The infinitesimal rigidity characteristic of the layered formation framework F_l ensures collision avoidance between all the connected agents i and j at any time $t > 0$.

The following lemma uses the absolute initial and desired positions of the agents and shows that Eq. (26) achieves the local asymptotic stability of the desired formation $F_l^\star$. Therefore Lemma 4 establishes the conditions that prevent convergence to ambiguous formations. The proof of this lemma can be found in [35].

Lemma 4. *Let* δ *be a small positive constant and* $p_l(0) \in \mathbb{R}^{3(n+r)}$ *and* $v_l(0) \in \mathbb{R}^{3(n+r)}$ *be the initial conditions. Given* $\Delta(G_l^\star)$ *is the maximum degree of the graph* $G_l^\star$, *if* $(p_l(0), v_l(0)) \in \mathcal{T}_2$, *where*

$$\mathcal{T}_2 = \left\{ (p_l, v_l) \in \mathbb{R}^{3(n+r)} \times \mathbb{R}^{3(n+r)} \middle| 8|E_l| \max_{i\in V_l} \|p_{l_i} - p_{l_i}^\star\|^2 (\|p_{l_i} - p_{l_i}^\star\| + D_l)^2 \right.$$
$$\left. +|V_l|(\max_{i\in V_l} \|v_{l_i}\| + \Delta(G_l^\star)k_v(2\max_{i\in V_l} \|p_{l_i} - p_{l_i}^\star\| + D_l)^3)^2 \le 2\delta \right\}, \tag{28}$$

then $F_l(t)$ *is infinitesimally and minimally rigid, and the control law* (26) *achieves the asymptotic stability of the desired formation* $F_l^\star$ *and ensures* $\|p_{l_i}(t) - p_{l_j}(t)\| \to d_{l_{ij}}$ *as* $t \to \infty$ *for all* $(i,j) \in E_l$.

Remark 4. Condition (28) provides a balanced trade-off between large initial velocity and large initial position errors. For example, for zero initial velocities, the condition sets an upper bound on initial position errors that ensures the system stability.

5 SIMULATION RESULTS

A set of simulations were performed to demonstrate the performance of the proposed method for the solving of the layered formation control problem.

5.1 Single-Integrator Model

In this section the simulation results for two scenarios with different numbers of agents with layered formations are presented to demonstrate the performance of the control law in Eq. (15) for a *three-dimensional case* (coordinators are outside the plane where other agents operate). The following initial condition was chosen for the agents:

$$p_{l_i}(0) = p_{l_i}^{\star} + \alpha I_1, \quad i = 1, \ldots, n, n+1, \ldots, n+r, \tag{29}$$

where α is a uniform random number on the interval (0,1) and I_1 generates a 3×1 unit vector of uniformly distributed random values on the interval $(0, 2\pi)$. In the simulations the value of k_c was set to 1 in Eq. (15).

The first simulation was conducted with six planar agents and one coordinating agent. The desired formation of the agents that was chosen for this simulation is shown in Fig. 7, where vertices 1–6 represent the six planar agents and vertex 7 represents the coordinator. A layered framework, in the shape of a cone that is infinitesimally rigid, was chosen as the desired framework. The edges of the desired framework in Fig. 7 are indexed by their vertices (eg, edge 13 connects vertices 1 and 3). With use of this edge

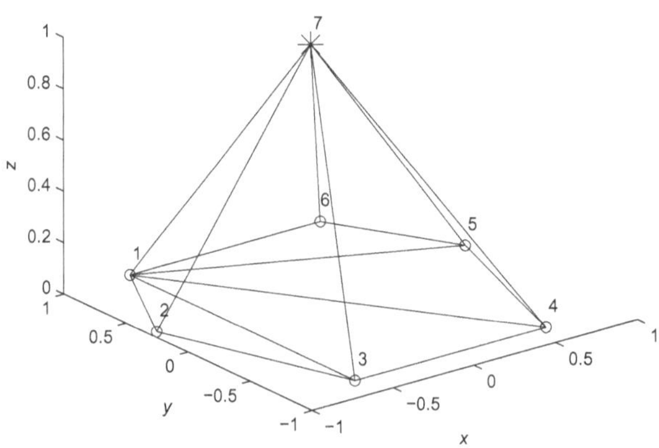

Fig. 7 Desired formation for six planar agents (*circles*) and one coordinator (*star*).

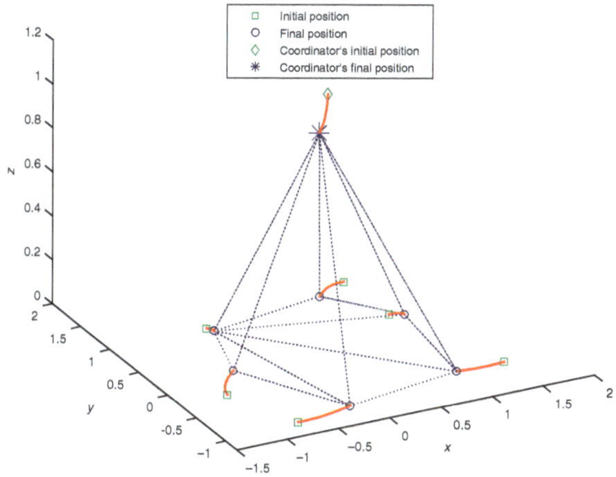

Fig. 8 Agents' trajectories $p_{l_i}(t), i = 1, \ldots, 7$ with the single integrator model (*solid line*) and the desired formation (*dotted line*).

notation, the desired distance between each pair of vertices was set with use of the Euclidean distance. For instance, in Fig. 7, the distance between vertices 1 and 2 is $d_{l_{12}} = \|p^{\star}_{l_1} - p^{\star}_{l_2}\|$.

Fig. 8 shows the trajectories for the six planar agents and the coordinator as they move from their initial position to the final position to form the desired layered formation. A sample of the distance errors $e_{l_{ij}}(t)$ for this simulation is shown in Fig. 9. It can be seen that $e_{l_{ij}}(t) \to 0$ at approximately

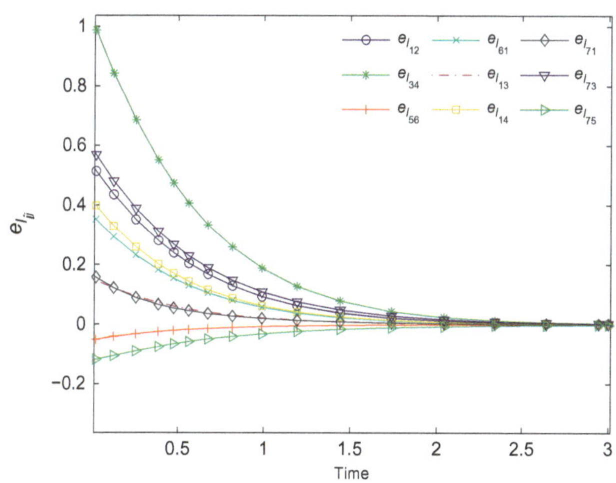

Fig. 9 Sample of distance errors for six planar agents and the coordinator.

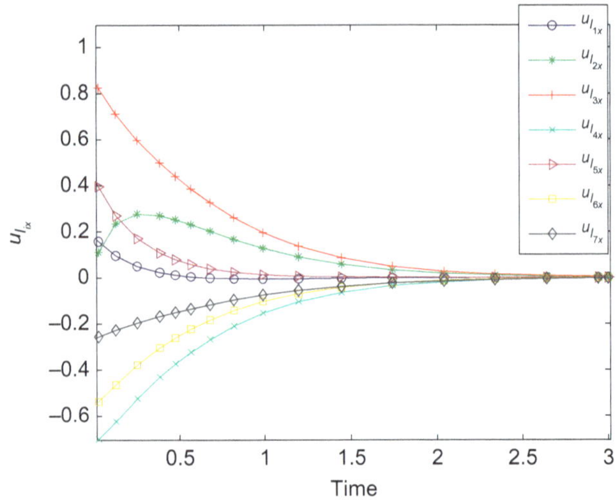

Fig. 10 Control inputs $u_{l_{ix}}(t), i = 1, \ldots, 7$, in the x-direction.

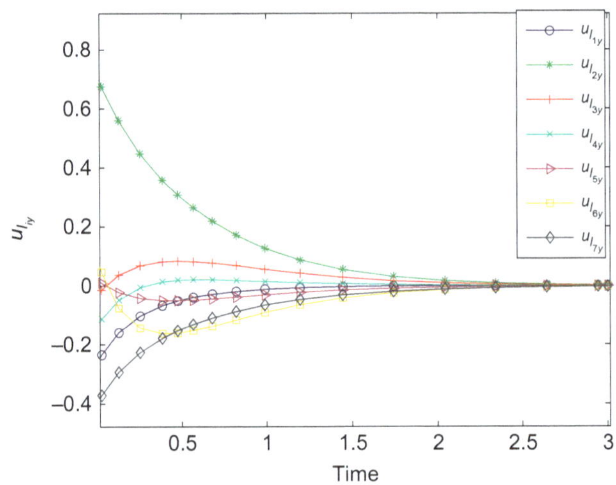

Fig. 11 Control inputs $u_{l_{iy}}(t), i = 1, \ldots, 7$, in the y-direction.

2.5 s. The control input $u_{l_i}(t)$ for $i = 1, \ldots, 7$ in the x- and y-directions is shown in Figs. 10 and 11, respectively.

An additional simulation was performed with four planar agents and four coordinators. In Fig. 12 the trajectories of the agents as they form

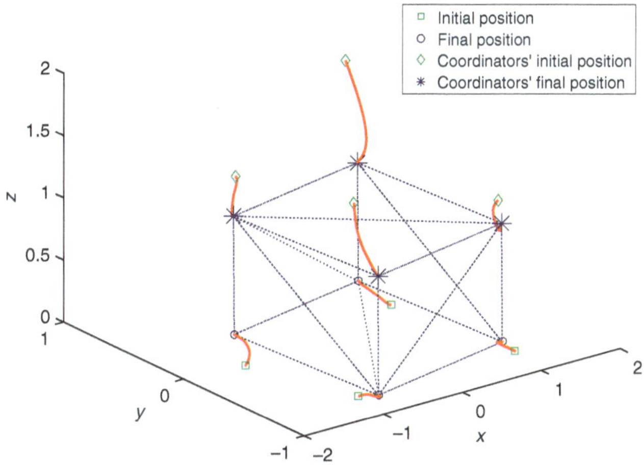

Fig. 12 Agents' trajectories $p_{l_i}(t), i = 1, \ldots, 8$, with the single-integrator model (*solid line*) and the desired formation (*dotted line*).

the desired formation are shown. A sample of the distance errors for this formation is presented in Fig. 13.

5.2 Double-Integrator Model

The simulations described in this section were performed for two scenarios and the results presented show the performance of the proposed control law (26). The initial conditions that were chosen for the agents were Eq. (29) and

$$v_{l_i}(0) = \alpha - 0.5 I_2, \quad i = 1, \ldots, n, n+1, \ldots, n+r, \tag{30}$$

where I_2 generates a 3×1 unit vector of uniformly distributed random values on the interval $(0, 1)$. The control gains k_a and k_v were set to 1 in Eq. (26).

In the first simulation there are five planar agents and three coordinators. The desired formation and trajectories of the agents as they move from the initial position to the final position are shown in Figs. 14 and 15, respectively. Fig. 16 demonstrates a sample of the distance errors and shows that $e_{l_{ij}}(t) \to 0$ at approximately 6 s. The control input $u_{l_i}(t)$ for $i = 6, 7, 8$ in the z-direction is presented in Fig. 17.

The second simulation was performed with two planar agents and four coordinators. Fig. 18 shows the trajectories of all the agents and Fig. 19 presents a sample of the distance errors.

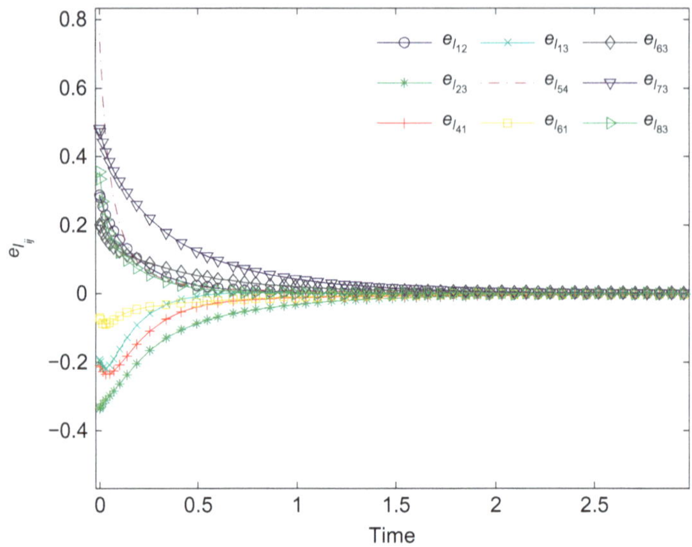

Fig. 13 Sample of distance errors for four planar agents and four coordinators.

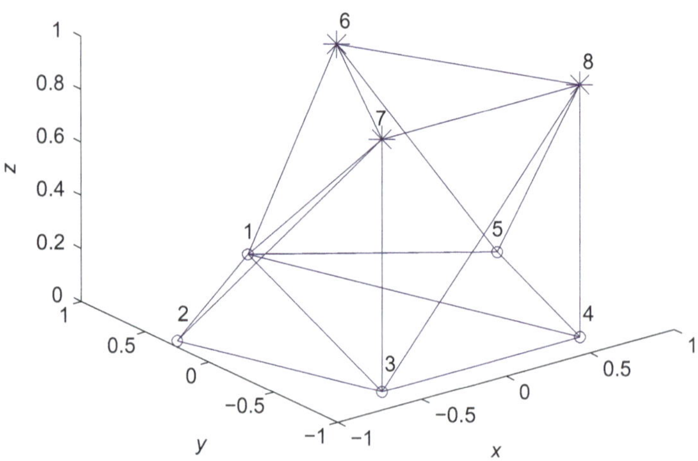

Fig. 14 Desired formation for five planar agents and three coordinators with the double-integrator model.

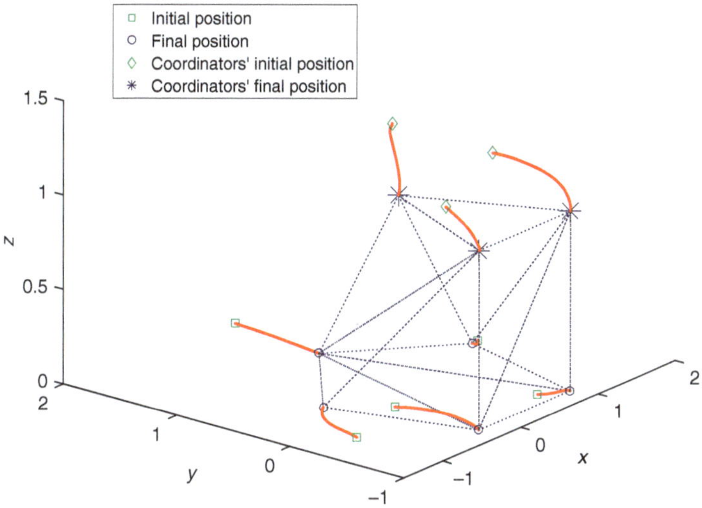

Fig. 15 Agents' trajectories $p_{l_i}(t)$, $i = 1, \ldots, 8$, with the double-integrator model (*solid line*) and the desired formation (*dotted line*).

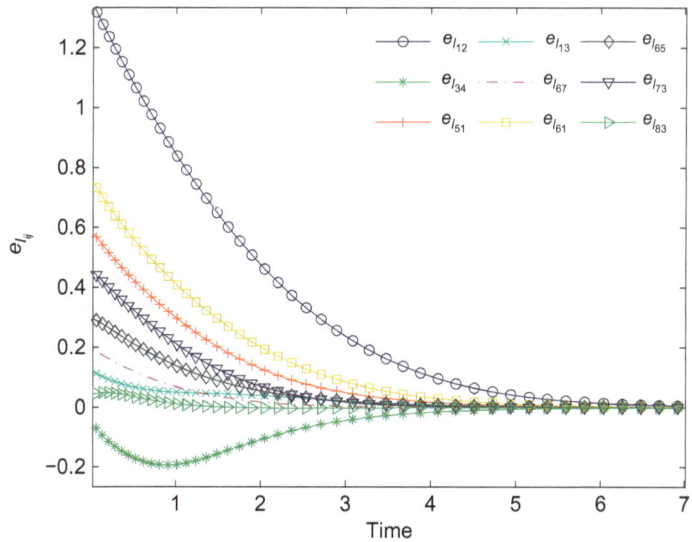

Fig. 16 Sample of distance errors for five planar agents and three coordinators.

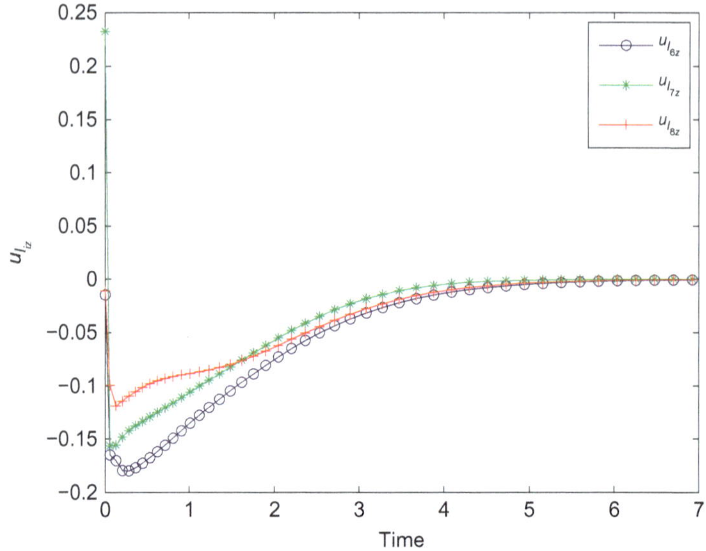

Fig. 17 Control inputs in the *z*-direction for the coordinators.

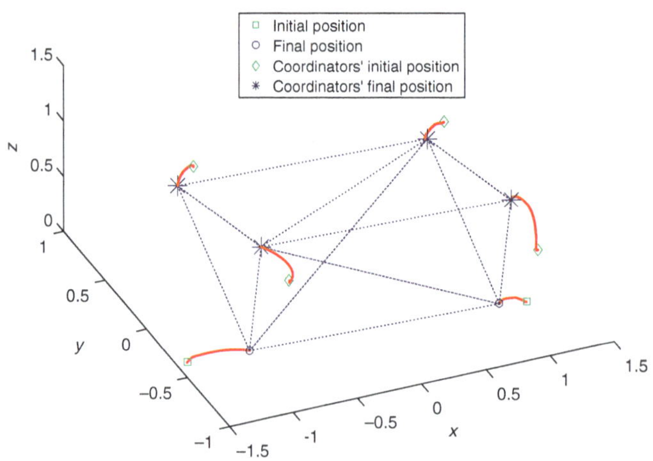

Fig. 18 Agents' trajectories $p_{l_i}(t)$, $i = 1, \ldots, 6$, with the double integrator model (*solid line*) and the desired formation (*dotted line*).

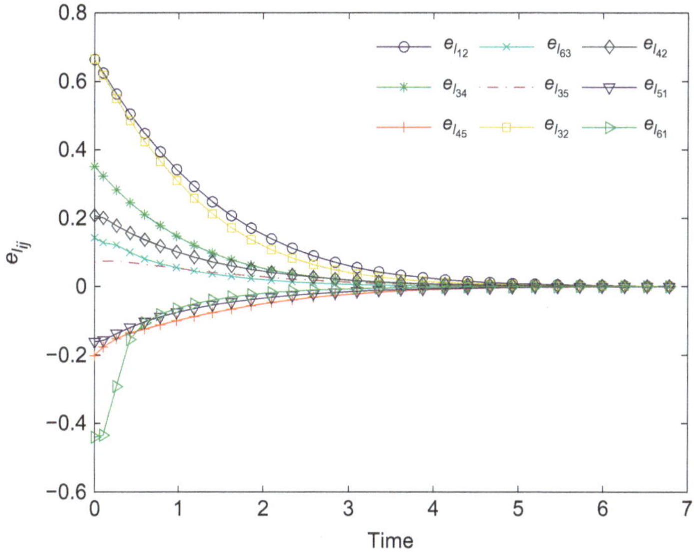

Fig. 19 Sample of distance errors for two planar agents and four coordinators.

6 CONCLUSION

In this work we considered the problem of formation acquisition for a group of n agents that belong to a plane Q and r coordinators that move in a plane W outside the plane Q. Graph coning and triangulation concepts were used to create a three-dimensional framework that retains the infinitesimal and minimal rigidity characteristics of the framework of n planar agents.

By use of the proposed approach, the interagent distances were stabilized to acquire a predefined shape in three dimensions. Sufficient conditions were provided for the initial conditions that guarantee convergence of the layered formation to the desired framework. The simulation results showed that the proposed control method, using single- and double-integrator models, yields asymptotic stability of the desired formation.

REFERENCES

[1] X. Cai, M. de Queiroz, Multi-agent formation maintenance and target tracking, in: Proceedings of American Control Conference (ACC), 2013, pp. 2521–2526.
[2] B.D.O. Anderson, C. Yu, B. Fidan, J. Hendrickx, Rigid graph control architectures for autonomous formations, IEEE Control Syst. 28 (6) (2008) 48–63.
[3] B.D. Anderson, C. Yu, S. Dasgupta, A.S. Morse, Control of a three-coleader formation in the plane, Syst. Control Lett. 56 (9–10) (2007) 573–578.

[4] D. Pack, P. DeLima, G. Toussaint, G. York, Cooperative control of UAVs for localization of intermittently emitting mobile targets, IEEE Trans. Syst. Man Cybern. B: Cybern. 39 (4) (2009) 959–970.
[5] W. Burgard, F. Schneider, Collaborative exploration of unknown environments with teams of mobile robots, in: Proceedings of Dagstuhl Seminar on Plan-based Control of Robotic Agents, Springer Verlag, Berlin, 2002, pp. 52–70.
[6] C. Yu, J.M. Hendrickx, B. Fidan, B.D. Anderson, V.D. Blondel, Three and higher dimensional autonomous formations: rigidity, persistence and structural persistence, Automatica 43 (3) (2007) 387–402.
[7] X. Cai, M. de Queiroz, Adaptive rigidity-based formation control for multirobotic vehicles with dynamics, IEEE Trans. Control Syst. Technol. 23 (1) (2015) 389–396.
[8] L. Krick, M. Broucke, B. Francis, Stabilization of infinitesimally rigid formations of multi-robot networks, in: Proceedings of 47th IEEE Conference on Decision and Control (CDC), 2008, pp. 477–482.
[9] K.-K. Oh, H.-S. Ahn, Formation control of mobile agents based on inter-agent distance dynamics, Automatica 47 (10) (2011) 2306–2312.
[10] K.-K. Oh, H.-S. Ahn, Distance-based undirected formations of single-integrator and double-integrator modeled agents in n-dimensional space, Int. J. Robust Nonlinear Control (12) (2014) 1809–1820.
[11] K.-K. Oh, H.-S. Ahn, Distance-based control of cycle-free persistent formations, in: Proceedings of IEEE Multi-Conference on Systems and Control, September, 2011, pp. 816–821.
[12] M. Bryant, P. Johnson, B.M. Kent, M. Nowak, S. Rogers, Layered sensing, White Paper Version 6, 1 May 2008, Air Force Research Laboratory, 2008.
[13] W. Whiteley, Cones, infinity and 1-story buildings, Struct. Topol. (8) (1983) 53–70.
[14] R. Connelly, W. Whiteley, Global rigidity: the effect of coning, Discrete Comput. Geom. 43 (4) (2010) 717–735.
[15] X. Cai, M. de Queiroz, Rigidity-based stabilization of multi-agent formations, ASME J. Dyn. Syst. Meas. Control 136 (1) (2014).
[16] B. Jackson, T. Jordán, Connected rigidity matroids and unique realizations of graphs, J. Combin. Theory B 94 (1) (2005) 1–29.
[17] L. Asimow, B. Roth, The rigidity of graphs, II, J. Math. Anal. Appl. 68 (1) (1979) 171–190.
[18] T.-S. Tay, W. Whiteley, Generating Isostatic Frameworks, Struct. Topol. 11 (1985) 21–69.
[19] G. Laman, On graphs and rigidity of plane skeletal structures, J. Eng. Math. 4 (4) (1970) 331–340.
[20] C. Yu, B.D.O. Anderson, S. Dasgupta, B. Fidan, Control of minimally persistent formations in the plane, SIAM J. Control Optim. 48 (1) (2009) 206–233.
[21] W. Whiteley, Some matroids from discrete applied geometry, Contemp. Math. 197 (1996) 171–312.
[22] H. Maehara, Geometry of frameworks, Yokohama Math. J. 47 (1999) 41–65 (special issue).
[23] R. Connelly, Generic global rigidity, Discrete Comput. Geom. 33 (4) (2005) 549–563.
[24] R. Connelly, The rigidity of polyhedral surfaces, Math. Mag. 52 (5) (1979) 275–283.
[25] L. Asimow, B. Roth, The rigidity of graphs, Trans. Am. Math. Soc. 245 (1978) 279–289.
[26] R. Connelly, The rigidity of certain cabled frameworks and the second-order rigidity of arbitrarily triangulated convex surfaces, Adv. Math. 37 (3) (1980) 272–299.
[27] J. Roberts, T. Stirling, J.-C. Zufferey, D. Floreano, 3-D relative positioning sensor for indoor flying robots, Auton. Robot. 33 (1–2) (2012) 5–20.
[28] F. Dorfler, B. Francis, Geometric analysis of the formation problem for autonomous robots, IEEE Trans. Autom. Control 55 (10) (2010) 2379–2384.

[29] D. Jin, L. Gao, Stability analysis of a double integrator swarm model related to position and velocity, Trans. Inst. Meas. Control 30 (3–4) (2008) 275–293.
[30] R. Olfati-Saber, P. Jalalkamali, Coupled distributed estimation and control for mobile sensor networks, IEEE Trans. Autom. Control 57 (10) (2012) 2609–2614.
[31] H. Gluck, Almost all simply connected closed surfaces are rigid, Geom. Topol. 438 (1975) 225–239.
[32] S. Ramazani, Unmanned Vehicles Formation Control in 3D Space and Cooperative Search, Ramazani dissertation, Louisiana Tech University, 2016.
[33] S. Ramazani, R. Selmic, M. de Queiroz, Non-planar multi-agent formation control using coning graphs, in: Proceedings of IEEE International Conference on Systems, Man and Cybernetics (SMC), 2014, pp. 3091–3096.
[34] Y.-Y. Chen, Y.-P. Tian, A backstepping design for directed formation control of three-coleader agents in the plane, Int. J. Robust Nonlinear Control 19 (7) (2009) 729–745.
[35] S. Ramazani, R.R. Selmic, M. de Queiroz, Stabilization of non-planar multi-agent layered formations with double integrator model, in: 2015 IEEE Conference on Control Applications (CCA), September, 2015, pp. 1386–1391.

CHAPTER FOURTEEN

Certainty Equivalence, Separation Principle, and Cooperative Output Regulation of Multiagent Systems by the Distributed Observer Approach

J. Huang
The Chinese University of Hong Kong, Shatin, Hong Kong

1 INTRODUCTION

The cooperative output regulation problem by the distributed observer approach was first studied for linear multiagent systems subject to static communication topology in [1], and then for linear multiagent systems subject to dynamic communication topology in [2]. The problem is interesting because its formulation includes the leader-following consensus, synchronization, or formation as special cases. In contrast with the output regulation problem of a single linear system [3–5], the information of the exogenous signal may not be available for every subsystem because of communication constraints. Thus information sharing, or, what is the same, cooperation among different subsystems is essential in the design of the control law. We call a control law that satisfies the communication constraints a *distributed control law*. The core of the approach in [1, 2] is the employment of the so-called distributed observer, which provides the estimation of the leader's signal to each follower so that a distributed controller can be synthesized on the basis of a purely decentralized controller and the distributed observer. Such an approach to designing a distributed controller is known as the certainty equivalence principle.

Since the publication of [1, 2], several variants and extensions of [1, 2] have been proposed [6–10]. The objectives of this chapter are threefold.

Control of Complex Systems
http://dx.doi.org/10.1016/B978-0-12-805246-4.00014-8

First, we will present a more general formulation of the cooperative output regulation problem for linear multiagent systems that includes some existing versions of the cooperative output regulation problem as special cases. Second, we will describe a more general distributed observer. Third, we will simplify the proof of the main results by more explicitly using the separation principle and the certainty equivalence principle.

The cooperative output regulation problem by the distributed observer approach can also be generalized to some nonlinear systems, such as multiple Euler-Lagrange systems [11, 12] and multiple rigid-body systems [13]. Moreover, the distributed observer approach can also be applied to such problems as the leader-following flocking/rendezvous with connectivity preservation [14, 15].

The cooperative output regulation problem of multiagent systems can also be handled by the distributed internal model approach [16, 17]. This approach has an additional advantage that it can tolerate perturbations of the plant parameters, and it does not need to solve the regulator equations. A combined distributed internal model and distributed observer approach was proposed in [7, 18]. Nevertheless, in this chapter we will focus only on the distributed observer approach.

Notation

$\otimes$ denotes the Kronecker product of matrices. $\sigma(A)$ denotes the spectrum of a square matrix A. For matrices $x_i \in \mathbb{R}^{n_i \times p}$, $i = 1, \ldots, m$, $\mathrm{col}(x_1, \ldots, x_\mathrm{m}) = [x_1^T, \ldots, x_\mathrm{m}^T]^T$. We use $\sigma(t)$ to denote a piecewise constant switching signal $\sigma : [0, +\infty) \to \mathcal{P} = \{1, 2, \ldots, \rho\}$ for some positive integer ρ, where $\mathcal{P}$ is the switching index set. We assume the switching instants $t_0 = 0, t_1, t_2, \ldots$ of σ satisfy $t_{k+1} - t_k \geq \tau > 0$ for some constant τ and for any $k \geq 0$, where τ is the dwell time. I_n denotes the identity matrix of dimension n by n.

2 LINEAR OUTPUT REGULATION

In this section we review the linear output regulation problem for the class of linear time-invariant systems as follows:

$$\begin{aligned}
\dot{x}(t) &= Ax(t) + Bu(t) + Ev(t), \quad x(0) = x_0, \quad t \geq 0, \\
y_\mathrm{m}(t) &= C_\mathrm{m}x(t) + D_\mathrm{m}u(t) + F_\mathrm{m}v(t), \\
e(t) &= Cx(t) + Du(t) + Fv(t),
\end{aligned} \tag{1}$$

where $x \in \mathbb{R}^n$, $y_m \in \mathbb{R}^{p_m}$, $e \in \mathbb{R}^p$, and $u \in \mathbb{R}^m$ are the state, measurement output, error output, and input of the plant, and $v \in \mathbb{R}^q$ is the exogenous signal generated by an exosystem of the form

$$\dot{v}(t) = Sv(t), \quad v(0) = v_0, \ t \geq 0, \tag{2}$$

where $S \in \mathbb{R}^{q \times q}$ is some constant matrix.

Typically, the tracking error e is the difference between the system output y and the reference input r (ie, $e = y - r$, where $y = Cx + Du + F_1 v$ for some matrix F_1 and $r = F_2 v$ for some matrix F_2). Thus, typically, we have $F = F_1 - F_2$ in Eq. (1). The tracking error e is assumed to be measurable but it may not be the only measurable variable available for feedback control. Using the measurement output feedback control allows us to solve the output regulation problem for some systems which cannot be solved by the error output feedback control.

For convenience, we put the plant (1) and the exosystem (2) together into the following form:

$$\begin{aligned} \dot{x} &= Ax + Bu + Ev, \\ \dot{v} &= Sv, \\ y_m &= C_m x + D_m u + F_m v, \\ e &= Cx + Du + Fv, \end{aligned} \tag{3}$$

and we call Eq. (3) a composite system with $\text{col}(x, v)$ as the composite state.

In general, some components of the exogenous signal v, say, the reference inputs, are measurable and some other components of the exogenous signal v, say, the unknown external disturbances, are not measurable. We denote the unmeasured and measured components of v by $v_u \in \mathbb{R}^{q_u}$ and $v_m \in \mathbb{R}^{q_m}$, respectively, where $0 \leq q_u, q_m \leq q$, with $q_u + q_m = q$. Without loss of generality, we can assume v_u and v_m are generated, respectively, by the following systems:

$$\dot{v}_u = S_u v_u, \quad \dot{v}_m = S_m v_m, \tag{4}$$

for some constant matrices S_u and S_m. The expressions in Eq. (4) are still in the form of the expressions in Eq. (2) with $v = \text{col}(v_u, v_m)$ and $S = \text{block diag}[S_u, S_m]$.

To emphasize that v may contain both measurable and unmeasurable components, we can rewrite the plant (1) as

$$\begin{aligned} \dot{x}(t) &= Ax(t) + Bu(t) + E_u v_u(t) + E_m v_m(t), \quad x(0) = x_0, \ t \geq 0, \\ y_m(t) &= C_m x(t) + D_m u(t) + F_{mu} v_u(t) + F_{mm} v_m(t), \\ e(t) &= Cx(t) + Du(t) + F_u v_u(t) + F_m v_m(t), \end{aligned} \tag{5}$$

where $E = [E_u \ E_m]$, $F_m = [F_{mu} \ F_{mm}]$, and $F = [F_u \ F_m]$.

We will consider the following so-called measurement output feedback control law:

$$\begin{aligned} u &= K_z z + K_y y_m, \\ \dot{z} &= \mathcal{G}_1 z + \mathcal{G}_2 y_m, \end{aligned} \tag{6}$$

where $z \in \mathbb{R}^{n_z}$, with n_z to be specified later, and where $K_z \in \mathbb{R}^{m\times n_z}$, $K_y \in \mathbb{R}^{m\times p_m}$, $\mathcal{G}_1 \in \mathbb{R}^{n_z\times n_z}$, and $\mathcal{G}_2 \in \mathbb{R}^{n_z\times p_m}$ are constant matrices.

This control law contains the following four types of control laws as special cases:

1. Full information when $y_m = \mathrm{col}(x, v)$ and $n_z = 0$:

$$u = K_1 x + K_2 v, \tag{7}$$

where $K_1 \in \mathbb{R}^{m\times n}$ and $K_2 \in \mathbb{R}^{m\times q}$ are constant matrices.

2. (Strictly proper) measurement output feedback when $K_y = 0$:

$$\begin{aligned} u &= K_z z, \\ \dot{z} &= \mathcal{G}_1 z + \mathcal{G}_2 y_m. \end{aligned} \tag{8}$$

3. Error output feedback when $y_m = e$ and $K_y = 0$:

$$\begin{aligned} u &= K_z z, \\ \dot{z} &= \mathcal{G}_1 z + \mathcal{G}_2 e. \end{aligned} \tag{9}$$

4. Combined error feedback and feedforward when $y_m = \mathrm{col}(e, v)$:

$$\begin{aligned} u &= K_z z + K_v v, \\ \dot{z} &= \mathcal{G}_1 z + \mathcal{G}_e e + \mathcal{G}_v v, \end{aligned} \tag{10}$$

where $K_y = [0_{m\times p}, K_v]$ and $\mathcal{G}_2 = [\mathcal{G}_e, \mathcal{G}_v]$.

Needless to say the control law (6) contains cases other than the control laws (7)–(10).

From the second equation of the plant (1) and the first equation of the control law (6), the control input u satisfies

$$u = K_z z + K_y(C_m x + D_m u + F_m v),$$

or

$$(I_m - K_y D_m)u = K_y C_m x + K_z z + K_y F_m v. \tag{11}$$

Therefore the control law (6) is well defined if and only if $I_m - K_y D_m$ is nonsingular. It can be easily verified that all four control laws (7)–(10) satisfy

$K_y D_m = 0$. Thus, in what follows, we will assume $K_y D_m = 0$ though it suffices to assume $I_m - K_y D_m$ is nonsingular. As a result, we have

$$u = K_y C_m x + K_z z + K_y F_m v. \tag{12}$$

Under the assumption that $K_y D_m = 0$, the control law (6) can be expressed as follows:

$$\begin{aligned} u &= K_y C_m x + K_z z + K_y F_m v, \\ \dot{z} &= (\mathcal{G}_2(C_m + D_m K_y C_m))x + (\mathcal{G}_1 + \mathcal{G}_2 D_m K_z)z + \mathcal{G}_2(F_m + D_m K_y F_m)v. \end{aligned} \tag{13}$$

Thus the closed-loop system composed of the plant (1) and the control law (6) can be expressed as follows:

$$\begin{aligned} \dot{x}_c &= A_c x_c + B_c v, \\ e &= C_c x_c + D_c v, \end{aligned} \tag{14}$$

where $x_c = \text{col}(x, z)$,

$$A_c = \begin{bmatrix} A + BK_y C_m & BK_z \\ \mathcal{G}_2(C_m + D_m K_y C_m) & \mathcal{G}_1 + \mathcal{G}_2 D_m K_z \end{bmatrix}, \quad B_c = \begin{bmatrix} E + BK_y F_m \\ \mathcal{G}_2(F_m + D_m K_y F_m) \end{bmatrix},$$
$$C_c = [C + DK_y C_m \quad DK_z], \quad D_c = F + DK_y F_m.$$

In particular, for the full information control law (7), we have $y_m = \text{col}(x, v)$ and $n_z = 0$. Thus, $x_c = x$, $K_y C_m = K_1$, and $K_y F_m = K_2$. As a result, we have

$$\begin{aligned} A_c &= A + BK_1, \quad B_c = E + BK_2, \\ C_c &= C + DK_1, \quad D_c = F + DK_2. \end{aligned}$$

We now describe the linear output regulation problem as follows.

Problem 1. Given the plant (1) and the exosystem (2), find the control law of the form (6) such that the closed-loop system has the following properties:

Property 1. *The matrix A_c is Hurwitz (ie, all the eigenvalues of A_c have negative real parts).*

Property 2. *For any $x_c(0)$ and $v(0)$, $\lim_{t \to \infty} e(t) = 0$.*

At the outset, we list some standard assumptions as follows:

Assumption 1. S has no eigenvalues with negative real parts.

Assumption 2. The pair (A, B) is stabilizable.

Assumption 3. The pair $\left([C_m \quad F_{mu}], \begin{bmatrix} A & E_u \\ 0 & S_u \end{bmatrix} \right)$ is detectable.

Assumption 4. The linear equations

$$\begin{aligned} XS &= AX + BU + E, \\ 0 &= CX + DU + F \end{aligned} \tag{15}$$

admit a solution pair (X, U).

Remark 1. Assumption 2 is made so that Property 1 (ie, the exponential stability of A_c) can be achieved by the state feedback control. Assumption 3 together with Assumption 2 is used to render the exponential stability of A_c by the measurement output feedback control. Assumption 1 is made only for convenience and loses no generality. In fact, if Assumption 1 is violated, then, without loss of generality, we can assume $S =$ block diag $[S_1, S_2]$, where S_1 satisfies Assumption 1, and all the eigenvalues of S_2 have negative real parts. Thus if a control law of the form (6) solves Problem 1 with the exosystem being given by $\dot{v}_1 = S_1 v_1$, then the same control law solves Problem 1 with the original exosystem $\dot{v} = Sv$. This is because Property 1 is guaranteed by Assumption 2 and/or Assumption 3, and, as long as the closed-loop system has Property 1, Property 2 will not be affected by exogenous signals that exponentially decay to zero.

Remark 2. Eqs. (15) are known as the regulator equations [4]. It will be shown in Theorem 2 and Remark 6 that under Assumptions 1–3, Problem 1 is solvable by a control law of the form (6) only if the regulator equations are solvable. Moreover, if Problem 1 is solvable at all, the trajectory of the closed-loop system starting from any initial condition is such that

$$\lim_{t\to\infty} (x(t) - Xv(t)) = 0 \quad \text{and} \quad \lim_{t\to\infty} (u(t) - Uv(t)) = 0. \tag{16}$$

Therefore Xv and Uv are the *steady-state* state and the *steady-state* input of the closed-loop system at which the tracking error e is identically zero. Thus the steady-state behavior of the closed-loop system is completely characterized by the solution of the regulator equations.

Remark 3. By Theorem 1.9 in [19], for any matrices E and F the regulator equations (15) are solvable if and only if

$$\text{rank} \begin{bmatrix} A - \lambda I & B \\ C & D \end{bmatrix} = n + p, \quad \forall\ \lambda \in \sigma(S). \tag{17}$$

Nevertheless, for a particular pair of (E, F), the regulator equations may still have a solution even if Eq. (17) fails.

3 SOLVABILITY OF THE LINEAR OUTPUT REGULATION PROBLEM

In this section we will study the solvability of Problem 1. Let us first present the following lemma on the closed-loop system.

Lemma 1. *Suppose, under the control law* (6), *closed-loop system* (14) *satisfies Property 1 (ie, A_c is Hurwitz). Then closed-loop system* (14) *also satisfies Property 2—that is, $\lim_{t\to\infty} e(t) = 0$—if there exists a matrix X_c that satisfies the following matrix equations:*

$$\begin{aligned} X_c S &= A_c X_c + B_c, \\ 0 &= C_c X_c + D_c. \end{aligned} \tag{18}$$

Moreover, under the additional assumption given by Assumption 1, closed-loop system (14) *also satisfies Property 2 only if there exists a unique matrix X_c that satisfies Eq. (18).*

Remark 4. The proof is the same as that of Lemma 1.4 in [19] and is omitted. Here we note only that if X_c satisfies Eq. (18), then the variable $\bar{x} = x_c - X_c v$ satisfies

$$\begin{aligned} \dot{\bar{x}} &= A_c \bar{x}, \\ e &= C_c \bar{x}. \end{aligned}$$

Since A_c is Hurwitz, $\lim_{t\to\infty} \bar{x}(t) = 0$, and hence, $\lim_{t\to\infty} e(t) = 0$. The solvability of the first equation in Eq. (18) is guaranteed as long as the eigenvalues of A_c do not coincide with those of S. Thus Assumption 1 is not necessary for the sufficient part of Lemma 1. It suffices to require that the eigenvalues of A_c do not coincide with those of S.

Lemma 2. *Under Assumption 1, suppose there exists a control law of the form* (6) *such that Property 1 holds. Then Property 2 also holds if and only if there exist matrices X and U that satisfy the following regulator equations:*

$$\begin{aligned} XS &= AX + BU + E, \\ 0 &= CX + DU + F. \end{aligned} \tag{19}$$

The proof is similar to that of Lemma 1.13 in [19] and is thus omitted.

Now let us first consider the full information case where the control law (7) is defined by two constant matrices K_1 and K_2. The two matrices K_1 and K_2 will be called the *feedback gain* and the *feedforward gain*, respectively.

Theorem 1. *Under Assumption 2, let the feedback gain K_1 be such that $(A+BK_1)$ is Hurwitz. Then Problem 1 is solvable by the full information control law (7) if Assumption 4 holds and the feedforward gain K_2 is given by*

$$K_2 = U - K_1 X. \tag{20}$$

Proof. Under Assumption 2, there exists K_1 such that $A_c = A + BK_1$ is Hurwitz. Thus under the control law (7), Property 1 is satisfied. Under Assumption 4, let $\bar{x} = x - Xv$ and K_2 be given by Eq. (20). Then we have

$$\begin{aligned} \dot{\bar{x}} &= (A+BK_1)\bar{x}, \\ e &= (C+DK_1)\bar{x}. \end{aligned} \tag{21}$$

Since $(A+BK_1)$ is Hurwitz, $\bar{x}(t)$ and hence $e(t)$ will approach zero as t tends to infinity. Thus Property 2 is also satisfied. □

Remark 5. By Lemma 2, Assumption 4 is also necessary for the solvability of Problem 1 by the full information control law (7) if Assumption 1 also holds.

We now turn to the construction of the measurement output feedback control law (6). Since we already know how to synthesize the full information control law, which takes the plant state x and the exosystem state v as its inputs, we seek to synthesize a measurement output feedback control law by estimating the state x and the unmeasurable exogenous signal v_u. To this end, we lump the state x and the unmeasured exogenous signals v_u together to obtain the following system:

$$\begin{aligned} \begin{bmatrix} \dot{x} \\ \dot{v}_u \end{bmatrix} &= \begin{bmatrix} A & E_u \\ 0 & S_u \end{bmatrix} \begin{bmatrix} x \\ v_u \end{bmatrix} + \begin{bmatrix} B \\ 0 \end{bmatrix} u + \begin{bmatrix} E_m \\ 0 \end{bmatrix} v_m, \\ y_m &= [C_m \ F_{mu}] \begin{bmatrix} x \\ v_u \end{bmatrix} + D_m u + F_{mm} v_m. \end{aligned} \tag{22}$$

Use of the well-known Luenberger observer theory suggests the following observer-based control law:

$$\begin{aligned} u &= [K_1 \ \ K_{2u}] z + K_{2m} v_m, \\ \dot{z} &= \begin{bmatrix} A & E_u \\ 0 & S_u \end{bmatrix} z + \begin{bmatrix} B \\ 0 \end{bmatrix} u + \begin{bmatrix} E_m \\ 0 \end{bmatrix} v_m + L(y_m - [C_m \ \ F_{mu}] z - D_m u - F_{mm} v_m), \end{aligned} \tag{23}$$

where $K_{2u} \in \mathbb{R}^{m \times q_u}$, $K_{2m} \in \mathbb{R}^{m \times q_m}$, and $L \in \mathbb{R}^{(n+q_u) \times p_m}$ is an observer gain matrix.

The control law (23) can be expressed in the following form:

$$\begin{aligned} u &= K_z z + K_{2m} v_m, \\ \dot{z} &= \mathcal{G}_1 z + \mathcal{G}_{21} y_m + \mathcal{G}_{22} v_m, \end{aligned} \tag{24}$$

where

$$K_z = [K_1 \quad K_{2u}],$$
$$\mathcal{G}_1 = \begin{bmatrix} A & E_u \\ 0 & S_u \end{bmatrix} + \begin{bmatrix} B \\ 0 \end{bmatrix} K_z - L(\begin{bmatrix} C_m & F_{mu} \end{bmatrix} + D_m K_z),$$
$$\mathcal{G}_{21} = L, \quad \mathcal{G}_{22} = \begin{bmatrix} B \\ 0 \end{bmatrix} K_{2m} + \begin{bmatrix} E_m \\ 0 \end{bmatrix} - L(F_{mm} + D_m K_{2m}).$$

Since v_m is measurable, there exists a matrix C_v such that $v_m = C_v y_m$. Thus the control law (24) can be further expressed in the standard form (6) with $K_y = K_{2m} C_v$ and $\mathcal{G}_2 = \mathcal{G}_{21} + \mathcal{G}_{22} C_v$.

Theorem 2. *Under Assumptions 2 and 3, Problem 1 is solvable by the measurement output feedback control law* (6) *if Assumption 4 holds.*

Proof. First note that by Assumption 2 there exists a state feedback gain K_1 such that $(A+BK_1)$ is Hurwitz and by Assumption 3 there exist matrices L_1 and L_2 such that

$$A_L = \begin{bmatrix} A & E_u \\ 0 & S_u \end{bmatrix} - \begin{bmatrix} L_1 \\ L_2 \end{bmatrix} \begin{bmatrix} C_m & F_{mu} \end{bmatrix} = \begin{bmatrix} A - L_1 C_m & E_u - L_1 F_{mu} \\ -L_2 C_m & S_u - L_2 F_{mu} \end{bmatrix}$$

is Hurwitz. Let (X, U) satisfy the regulator equations, $L = \mathrm{col}(L_1, L_2)$, and $K_2 = U - K_1 X$, and partition K_2 as $K_2 = [K_{2u}, K_{2m}]$. Let $\bar{x} = (x - Xv)$, $\bar{u} = (u - Uv)$, and $z_e = \begin{bmatrix} x \\ v_u \end{bmatrix} - z$. Then it can be verified that

$$\begin{aligned} \bar{u} &= [K_1, K_{2u}] \begin{bmatrix} x \\ v_u \end{bmatrix} - K_z z_e + K_{2m} v_m - Uv \\ &= -K_z z_e + K_1 x + K_2 v - (K_2 + K_1 X) v \\ &= -K_z z_e + K_1 \bar{x}. \end{aligned}$$

In terms of $\bar{x}$ and z_e, the closed-loop system is given by

$$\begin{aligned} \dot{\bar{x}} &= A\bar{x} + B\bar{u} = (A + BK_1)\bar{x} - BK_z z_e, \\ \dot{z}_e &= A_L z_e. \end{aligned} \tag{25}$$

Let A_c be the closed-loop system matrix. Then $\sigma(A_c) = \sigma(A + BK_1) \cup \sigma(A_L)$. Thus Property 1 is satisfied. To show $\lim_{t\to\infty} e(t) = 0$, first note that Eq. (25) implies that $\lim_{t\to\infty} \bar{x}(t) = 0$ and $\lim_{t\to\infty} z_e(t) = 0$. Then note that $e = Cx + Du + Fv = C(x - Xv) + D(u - Uv) + (CX + DU + F)v = C(x - Xv) + D(u - Uv) = (C + DK_1)\bar{x} - DK_z z_e$. □

Remark 6. By Lemma 2, Assumption 4 is also necessary for the solvability of Problem 1 by a measurement output feedback control law of the form (6) if Assumption 1 also holds.

Specializing Eq. (24) to the two special cases with $v = v_{\mathrm{u}}$ and $v = v_{\mathrm{m}}$, respectively, gives the following two corollaries of Theorem 2.

Corollary 1. *Under Assumptions 2–4 with $v_{\mathrm{u}} = v$, Problem 1 is solvable by the following observer-based control law:*

$$\begin{aligned} u &= [K_1 \;\; K_2]z, \\ \dot{z} &= \begin{bmatrix} A & E \\ 0 & S \end{bmatrix} z + \begin{bmatrix} B \\ 0 \end{bmatrix} u + L(y_{\mathrm{m}} - [C_{\mathrm{m}} \;\; F_{\mathrm{m}}]z - D_{\mathrm{m}}u), \end{aligned} \tag{26}$$

where L is an observer gain matrix of dimension $(n+q)$ by p_{m}.

Corollary 2. *Under Assumptions 2—4 with $v_{\mathrm{m}} = v$, Problem 1 is solvable by the following observer-based control law:*

$$\begin{aligned} u &= K_1 z + K_2 v, \\ \dot{z} &= Az + Bu + Ev + L(y_{\mathrm{m}} - C_{\mathrm{m}}z - F_{\mathrm{m}}v - D_{\mathrm{m}}u), \end{aligned} \tag{27}$$

where L is an observer gain matrix of dimension n by p_{m}.

4 LINEAR MULTIAGENT SYSTEMS AND A DISTRIBUTED OBSERVER

In this section we turn to the cooperative output regulation problem for a group of linear systems as follows:

$$\begin{aligned} \dot{x}_i &= A_i x_i + B_i u_i + E_i v, \\ y_{\mathrm{m}i} &= C_{\mathrm{m}i} x_i + D_{\mathrm{m}i} u_i + F_{\mathrm{m}i} v, \\ e_i &= C_i x_i + D_i u_i + F_i v, \quad i = 1, \ldots, N, \end{aligned} \tag{28}$$

where $x_i \in \mathbb{R}^{n_i}$, $y_{\mathrm{m}i} \in \mathbb{R}^{p_{\mathrm{m}i}}$, $e_i \in \mathbb{R}^{p_i}$, and $u_i \in \mathbb{R}^{m_i}$ are the state, measurement output, error output, and input of the ith subsystem, and $v \in \mathbb{R}^q$ is the exogenous signal generated by a so-called exosystem of the form

$$\dot{v} = Sv, \quad y_{\mathrm{m}0} = C_0 v, \tag{29}$$

where $y_{\mathrm{m}0} \in \mathbb{R}^{p_0}$ is the output of the exosystem.

Like the special case with $N = 1$, the exogenous signal v may also contain both unmeasured components $v_{\mathrm{u}} \in \mathbb{R}^{q_{\mathrm{u}}}$ and measured components $v_{\mathrm{m}} \in \mathbb{R}^{q_{\mathrm{m}}}$, where $0 \leq q_{\mathrm{u}}, q_{\mathrm{m}} \leq q$, with $q_{\mathrm{u}} + q_{\mathrm{m}} = q$. Then, as in Eq. (1), we can assume v_{u} and v_{m} are generated by Eq. (4). Correspondingly, we assume $C_0 = [0_{p_0 \times q_u}, C_{m0}]$ for some matrix $C_{m0} \in \mathbb{R}^{p_0 \times q_m}$. As a result, the plant (28) can be further written as follows:

$$\begin{aligned} \dot{x}_i &= A_i x_i + B_i u_i + E_{iu} v_u + E_{im} v_m, \\ y_{mi} &= C_{mi} x_i + D_{mi} u_i + F_{miu} v_u + F_{mim} v_m, \\ e_i &= C_i x_i + D_i u_i + F_{iu} v_u + F_{im} v_m, \quad i = 1, \ldots, N, \end{aligned} \tag{30}$$

where $E_i = [E_{iu}\ E_{im}]$, $F_{mi} = [F_{miu}\ F_{mim}]$, and $F_i = [F_{iu}\ F_{im}]$.

Various assumptions are as follows.

Assumption 5. S has no eigenvalues with negative real parts.

Assumption 6. For $i = 1, \ldots, N$, the pairs (A_i, B_i) are stabilizable.

Assumption 7. For $i = 1, \ldots, N$, the pairs $\left([C_{mi}\ \ F_{miu}], \begin{bmatrix} A_i & E_{iu} \\ 0 & S_u \end{bmatrix}\right)$ are detectable.

Assumption 8. The linear matrix equations

$$\begin{aligned} X_i S &= A_i X_i + B_i U_i + E_i, \\ 0 &= C_i X_i + D_i U_i + F_i, \end{aligned} \quad i = 1, \ldots, N \tag{31}$$

have solution pairs (X_i, U_i).

Remark 7. The system (28) is still in the form of Eq. (1) with $x = \text{col}(x_1, \ldots, x_N)$, $u = \text{col}(u_1, \ldots, u_N)$, $y_m = \text{col}(y_{m1}, \ldots, y_{mN})$, and $e = \text{col}(e_1, \ldots, e_N)$. Thus if the state v of the exosystem can be used by the control u_i of each follower, then by Theorem 1, under Assumptions 6–8, the output regulation problem of the system (28) and the exosystem (29) can be solved by the following full information control law:

$$u_i = K_{1i} x_i + K_{2i} v, \quad i = 1, \ldots, N, \tag{32}$$

where K_{1i} are such that $A_i + B_i K_{1i}$ are Hurwitz, and $K_{2i} = U_i - K_{1i} X_i$.

Under the additional assumption given by Assumption 7, there exist $L_i \in \mathbb{R}^{(n_i + q_u) \times p_{mi}}$ such that

$$\begin{bmatrix} A_i & E_{iu} \\ 0 & S_u \end{bmatrix} - L_i \begin{bmatrix} C_{mi} & F_{miu} \end{bmatrix}$$

are Hurwitz. Partition K_{2i} as $K_{2i} = [K_{2iu}, K_{2im}]$ with $K_{2iu} \in \mathbb{R}^{m_i \times q_u}$ and $K_{2im} \in \mathbb{R}^{m_i \times q_m}$. Then by Theorem 2, under Assumptions 6–8, the output regulation problem of the system (28) and the exosystem (29) can be solved by the following measurement output feedback control law:

$$\begin{aligned} u_i &= [K_{1i}\ \ K_{2iu}] z_i + K_{2im} v_m, \quad i = 1, \ldots, N, \\ \dot{z}_i &= \begin{bmatrix} A_i & E_{iu} \\ 0 & S_u \end{bmatrix} z_i + \begin{bmatrix} B_i \\ 0 \end{bmatrix} u_i + \begin{bmatrix} E_{im} \\ 0 \end{bmatrix} v_m \\ &\quad + L_i (y_{mi} - [C_{mi}\ \ F_{miu}] z_i - D_{mi} u_i - F_{mim} v_m). \end{aligned} \tag{33}$$

Nevertheless, in practice, the communication among different subsystems of Eq. (28) is subject to some constraints due to, say, the physical distance among these subsystems. Thus the exogenous signal v or the measurable exogenous signal v_m may not be available for the control u_i of all the followers. Since, typically, $e_i = y_i - y_0$, the tracking error e_i may not be available for the control u_i of all the followers. To describe the communication constraints among various subsystems, we view system (28) and system (29) as a multiagent system with Eq. (29) as the leader and the N subsystems of Eq. (28) as the followers. Let $\bar{\mathcal{G}}_{\sigma(t)} = (\bar{\mathcal{V}}, \bar{\mathcal{E}}_{\sigma(t)})$,[1] where $\bar{\mathcal{V}} = \{0, 1, \ldots, N\}$ and $\bar{\mathcal{E}}_{\sigma(t)} \subseteq \bar{\mathcal{V}} \times \bar{\mathcal{V}}$ for all $t \geq 0$ is a switching graph, where the node 0 is associated with the leader system (29) and the node i, $i = 1, \ldots, N$, is associated with the ith subsystem of system (28). For $i = 1, \ldots, N$, $j = 0, 1, \ldots, N$, $(j, i) \in \bar{\mathcal{E}}_{\sigma(t)}$ if and only if u_i can use y_{mj} for control at time instant t. Let $\bar{\mathcal{N}}_i(t) = \{j \mid (j, i) \in \bar{\mathcal{E}}_{\sigma(t)}\}$ denote the neighbor set of agent i at time instant t.

The case where the network topology is static can be viewed as a special case of switching network topology when the switching index set contains only one element. We will use the simplified notation $\bar{\mathcal{G}}$ to denote a static graph.

We will consider the following class of control laws:

$$\begin{aligned} u_i &= f_i(x_i, \xi_i), \quad i = 1, \ldots, N, \\ \dot{\xi}_i &= g_i(\xi_i, y_{mi}, \xi_j, y_{mj}, j \in \bar{\mathcal{N}}_i(t)), \end{aligned} \tag{34}$$

where both f_i and g_i are linear in their argument, and g_i is time varying if the graph $\bar{\mathcal{G}}_{\sigma(t)}$ is. It can be seen that, at each time $t \geq 0$, for any $i = 1, \ldots, N$, u_i can make use of y_{m0} if and only if the leader is a neighbor of the subsystem i. Such a control law is called a *distributed control law*. If, for $i = 1, \ldots, N$, f_i is independent of x_i, then the control law is called a *distributed measurement output feedback control law*. If, for $i = 1, \ldots, N$, $\bar{\mathcal{N}}_i(t) = \{0\}$ for all $t \geq 0$, then the control law (34) is called a *purely decentralized control law*. In particular, Eqs. (32) and (33) are called the *purely decentralized full information control law* and the *purely decentralized measurement output feedback control law*.

We now describe our problem as follows.

Problem 2. Given systems (28) and (29) and a switching graph $\bar{\mathcal{G}}_{\sigma(t)}$, find a distributed control law of the form (34) such that the closed-loop system has the following properties:

[1] See the Appendix for a summary of graph.

Property 3. *The origin of the closed-loop system with v being set to zero is asymptotically stable.*

Property 4. *For any initial condition $x_i(0)$, $\xi_i(0)$, $i = 1, \ldots, N$, and $v(0)$, the solution of the closed-loop system is such that*

$$\lim_{t\to\infty} e_i(t) = 0, \quad i = 1, \ldots, N. \tag{35}$$

Clearly, the solvability of the above problem depends not only on the dynamics of systems (28) and (29), but also on the property of the graph $\bar{\mathcal{G}}_{\sigma(t)}$. A typical assumption on the graph $\bar{\mathcal{G}}_{\sigma(t)}$ is as follows.

Assumption 9. There exists a subsequence $\{i_k\}$ of $\{i : i = 0, 1, \ldots\}$ with $t_{i_{k+1}} - t_{i_k} < \nu$ for some positive ν such that every node $i = 1, \ldots, N$ is reachable from the node 0 in the union graph $\bigcup_{j=i_k}^{i_{k+1}-1} \bar{\mathcal{G}}_{\sigma(t_j)}$.

Remark 8. Assumption 9 is similar to what was proposed in [2, 20, 21], and will be called the jointly connected condition in the sequel. Since under Assumption 9 the graph $\bar{\mathcal{G}}_{\sigma(t)}$ can be disconnected for all $t \geq 0$, it is perhaps the least stringent condition on the graph as opposed to some other conditions such as every time connected or frequently connected. In particular, Assumption 9 is satisfied if Assumption 2 in [8] is. Thus the main result in [8] is essentially included in [2] even though the approach in [8] appears somewhat different from that in [2].

The static graph is a special case of the switching graph when $\rho = 1$. For this special case, Assumption 9 reduces to the following.

Assumption 10. Every node $i = 1, \ldots, N$ is reachable from the node 0 in the static graph $\bar{\mathcal{G}}$.

Remark 9. Let $\mathcal{G}_{\sigma(t)} = (\mathcal{V}, \mathcal{E}_{\sigma(t)})$ denote the subgraph of $\bar{\mathcal{G}}_{\sigma(t)}$, where $\mathcal{V} = \{1, \ldots, N\}$ and $\mathcal{E}_{\sigma(t)} \subseteq \mathcal{V} \times \mathcal{V}$ is obtained from $\bar{\mathcal{E}}_{\sigma(t)}$ by removal of all edges between the node 0 and the nodes in $\mathcal{V}$. Let $\bar{\mathcal{A}}_{\sigma(t)} = [a_{ij}(t)]_{i,j=0}^{N}$ denote the weighted adjacency matrix of $\bar{\mathcal{G}}_{\sigma(t)}$, let $\mathcal{L}_{\sigma(t)}$ be the Laplacian matrix of $\mathcal{G}_{\sigma(t)}$ and let $\Delta_{\sigma(t)} = \text{diag}(a_{10}(t), \ldots, a_{N0}(t))$. Then it is shown in Remark 14 in [22] that, under Assumption 9, the matrix $H_{\sigma(t)} = \mathcal{L}_{\sigma(t)} + \Delta_{\sigma(t)}$ has the property that all the eigenvalues of the matrix $\sum_{j=i_k}^{i_{k+1}-1} H_{\sigma(j)}$ have positive real parts. Furthermore, if the graph $\mathcal{G}_{\sigma(t)}$ is undirected, then the matrix $\sum_{j=i_k}^{i_{k+1}-1} H_{\sigma(j)}$ is positive definite. In particular, under Assumption 10, the constant matrix $-H$ is Hurwitz.

5 SOME STABILITY RESULTS

As pointed out in Section 1, our approach is based on the employment of the distributed observer. To introduce the distributed observer.

Let us first consider the stability property for the following class of switched linear systems:

$$\dot{x}(t) = \left(I_N \otimes A - \mu F_{\sigma(t)} \otimes (BK)\right) x(t), \quad \sigma(t) \in \mathcal{P}, \tag{36}$$

where $A \in \mathbb{R}^{n\times n}$, $B \in \mathbb{R}^{n\times m}$, and $F_{\sigma(t)} \in \mathbb{R}^{n\times n}$ are given, and $\mu > 0$ and $K \in \mathbb{R}^{m\times n}$ are to be designed.

Assumption 11. There exists a subsequence $\{i_k\}$ of $\{i : i = 0, 1, \ldots\}$ with $t_{i_{k+1}} - t_{i_k} \leq \nu$ for some positive ν such that all the eigenvalues of the matrix $\sum_{j=i_k}^{i_{k+1}-1} F_{\sigma(j)}$ have positive real parts.

The stability property of the system of the form (36) has been extensively studied in the literature. We summarize the main results in two lemmas corresponding to the switching network and the static network, respectively, as follows.

Lemma 3. *Under Assumption 11, suppose the pair (A, B) is controllable. Then:*

(i) *If A is marginally stable—that is, there exists a unique positive definite matrix P such that $PA + A^TP \leq 0$—and $F_{\sigma(t)}$ is symmetric, then, with $\mu = 1$, and $K = B^TP$, Eq. (36) is asymptotically stable.*

(ii) *If $B = I_n$ and A has no eigenvalues with positive real parts, then, with any $\mu > 0$ and $K = I_n$, Eq. (36) is asymptotically stable.*

Remark 10. The stability property of the system of the form (36) was first studied in [22]. Part (i) of Lemma 3 was established in Theorem 1 in [22]. Part (ii) of Lemma 3 was established in Lemma 2 in [2]. As a corollary of Lemma 2 in [2], under Assumption 11, for any $\mu > 0$, the system

$$\dot{x} = -\mu F_{\sigma(t)} x \tag{37}$$

is asymptotically stable. As a special case of this result, when $\mathcal{P} = \{1\}$, the matrix $F_{\sigma(t)}$ is constant [23], and will be denoted by F. For this special case, the result of Lemma 3 can be strengthened to the following.

Lemma 4. *Under Assumption 11 with $\mathcal{P} = \{1\}$, suppose the pair (A, B) is stabilizable. Then:*

(i) *Let P be the unique positive definite matrix satisfying $PA + A^TP - PBB^TP + I_n \leq 0$ and let $\mu \geq \delta^{-1}$, where $\delta = \min\{\mathrm{Re}(\lambda_i(F)\}$. Then, with $K = B^TP$, Eq. (36) is asymptotically stable.*

(ii) *If $B = I_n$, then for any A, Eq. (36) is asymptotically stable with $K = I_n$ and sufficiently large μ.*

Remark 11. The proof of part (i) of Lemma 4 can be extracted from the proof of Theorem 2 in [16]. In fact, under Assumption 11 with $\mathcal{P} = \{1\}$, all the eigenvalues of F have a positive real part. Let T be such that $TFT^{-1} = J$ is in the Jordan form of F. Denote the eigenvalues of F by $\lambda_1, \ldots, \lambda_N$. Then $(I_N \otimes A) - \mu(F \otimes BK) = (T^{-1} \otimes I_n)((I_N \otimes A) - \mu(J \otimes BK))(T \otimes I_n)$. From the block triangular structure of J, we know that the eigenvalues of $(I_N \otimes A) - \mu(F \otimes BK)$ coincide with those of $A - \mu\lambda_i BK$, $i = 1, \ldots, N$. Since the pair (A, B) is controllable, by Lemma 1 in [24], the algebraic Riccati equation

$$A^T P + PA - PBB^T P + I_n = 0 \tag{38}$$

admits a unique positive definite solution P. Moreover, for any $\mu \geq \delta^{-1}$, where $\delta = \min\{\text{Re}(\lambda_i(F)\}$, the gain matrix $K = B^T P$ is such that $A - \mu\lambda_i BK$ and hence $(I_N \otimes A) - \mu(F \otimes BK)$ are Hurwitz.

Part (ii) of Lemma 4 was established in Theorem 1 in [1]. It is also a direct result of the fact that the eigenvalues of the matrix $(I_N \otimes A) - \mu(F \otimes I_n)$ are

$$\{\lambda_i(A) - \mu\lambda_j(F) : i = 1, \ldots, n, j = 1, \ldots, N\},$$

where $\lambda_i(A)$ and $\lambda_j(F)$ are the eigenvalues of A and F, respectively. Thus the matrix $(I_N \otimes A) - \mu(F \otimes I_n)$ is Hurwitz for sufficiently large μ, and is Hurwitz for any positive μ if A does not have any eigenvalue with a positive real part.

6 SOLVABILITY OF THE COOPERATIVE LINEAR OUTPUT REGULATION PROBLEM

Given systems (28) and (29) and the switching graph $\bar{\mathcal{G}}_{\sigma(t)}$ whose weighted adjacency matrix is denoted by $\bar{\mathcal{A}}_{\sigma(t)} = [a_{ij}(t)]_{i,j=0}^N$, we call the following compensator

$$\dot{\eta}_i = S_{\rm m}\eta_i + \mu L_0 \left(\sum_{j \in \bar{\mathcal{N}}_i(t)} a_{ij}(t) C_{\rm m0}(\eta_j - \eta_i) \right), \quad i = 1, \ldots, N, \tag{39}$$

where $\eta_0 = v_{\rm m}$, $\mu > 0$ and $L_0 \in \mathbb{R}^{q_{\rm m} \times p_0}$ are two design parameters, a *distributed observer candidate* for $v_{\rm m}$, and call it a *distributed observer* for $v_{\rm m}$ if for any $v_{\rm m}(0)$ and $\eta_i(0)$, $i = 1, \ldots, N$,

$$\lim_{t \to \infty} (\eta_i(t) - v_{\rm m}(t)) = 0, \quad i = 1, \ldots, N. \tag{40}$$

Whether or not Eq. (39) is a distributed observer of v_{m} depends on both the pair $(C_{\mathrm{m}0}, S_{\mathrm{m}})$ and the property of the graph.

Let $\tilde{\eta}_i = (\eta_i - v_{\mathrm{m}})$ and $\tilde{\eta} = \mathrm{col}(\tilde{\eta}_1, \ldots, \tilde{\eta}_N)$. Then system (39) can be expressed in the following compact form:

$$\dot{\tilde{\eta}} = \big((I_N \otimes S_{\mathrm{m}}) - \mu(H_{\sigma(t)} \otimes L_0 C_{\mathrm{m}0})\big)\, \tilde{\eta}. \tag{41}$$

Thus system (39) is a distributed observer of v_{m} if and only if system (41) is asymptotically stable.

Remark 12. Since $((I_N \otimes S_{\mathrm{m}}) - \mu(H_{\sigma(t)} \otimes L_0 C_{\mathrm{m}0}))^T = (I_N \otimes S_{\mathrm{m}}^{\mathrm{T}}) - \mu(H_{\sigma(t)}^{\mathrm{T}} \otimes C_{\mathrm{m}0}^{\mathrm{T}} L_0^T)$, system (41) is asymptotically stable if and only if a system of the form (36) with $A = S_{\mathrm{m}}^{\mathrm{T}}$, $F_{\sigma(t)} = H_{\sigma(t)}^T$, $B = C_{\mathrm{m}0}^{\mathrm{T}}$, and $K = L_0^T$ is asymptotically stable. Moreover, by Remark 9, under Assumption 9, all the eigenvalues of the matrix $\sum_{j=i_k}^{i_{k+1}-1} H_{\sigma(j)}$ have positive real parts. Furthermore, if the graph $\mathcal{G}_{\sigma(t)}$ is undirected, then the matrix $\sum_{j=i_k}^{i_{k+1}-1} H_{\sigma(j)}$ is positive definite.

Corresponding to the two purely decentralized control laws (32) and (33), we can synthesize two types of distributed control laws as follows:

1. *Distributed full information control law*:

$$\begin{aligned} u_i &= K_{1i} x_i + K_{2i} \eta_i, \quad i = 1, \ldots, N, \\ \dot{\eta}_i &= S \eta_i + \mu L_0 \left(\sum_{j \in \bar{\mathcal{N}}_i(t)} a_{ij}(t) C_0 (\eta_j - \eta_i) \right), \end{aligned} \tag{42}$$

where $K_{1i} \in \mathbb{R}^{m_i \times n_i}$ are such that $A_i + B_i K_{1i}$ are Hurwitz, $K_{2i} = U_i - K_{1i} X_i$, μ is some positive constant, and $\eta_0 = v = v_{\mathrm{m}}$.

2. *Distributed measurement output feedback control law*:

$$\begin{aligned} u_i &= [K_{1i} \;\; K_{2i\mathrm{u}}] z_i + K_{2i\mathrm{m}} \eta_i, \quad i = 1, \ldots, N, \\ \dot{z}_i &= \begin{bmatrix} A_i & E_{i\mathrm{u}} \\ 0 & S_{\mathrm{u}} \end{bmatrix} z_i + \begin{bmatrix} B_i \\ 0 \end{bmatrix} u_i + \begin{bmatrix} E_{i\mathrm{m}} \\ 0 \end{bmatrix} \eta_i \\ &\quad + L_i (y_{\mathrm{m}i} - [C_{\mathrm{m}i} \;\; F_{\mathrm{m}i\mathrm{u}}] z_i - D_{\mathrm{m}i} u_i - F_{\mathrm{m}i\mathrm{m}} \eta_i), \\ \dot{\eta}_i &= S_{\mathrm{m}} \eta_i + \mu L_0 \left(\sum_{j \in \bar{\mathcal{N}}_i(t)} a_{ij}(t) C_{\mathrm{m}0} (\eta_j - \eta_i) \right), \end{aligned} \tag{43}$$

where $L_i \in \mathbb{R}^{(n_i + q_{\mathrm{u}}) \times p_{\mathrm{m}i}}$ are such that

$$A_{Li} = \begin{bmatrix} A_i & E_{i\mathrm{u}} \\ 0 & S_{\mathrm{u}} \end{bmatrix} - L_i \begin{bmatrix} C_{\mathrm{m}i} & F_{\mathrm{m}i\mathrm{u}} \end{bmatrix}$$

are Hurwitz and $\eta_0 = v_{\mathrm{m}}$.

Remark 13. Control law (42) contains the distributed state feedback control law in [2] as a special case by letting $L_0 = C_0 = I_q$, or what is the same, $y_{m0} = v$, and control law (43) contains the distributed measurement output feedback control law in [2] as a special case by letting $v_m = v$ and $L_0 = C_{m0} = I_q$.

Remark 14. Both of the control laws (42) and (43) are synthesized on the basis of the certainty equivalence principle in the sense that they are obtained from the purely decentralized control laws (32) and (33) by replacement of v in Eq. (32) and v_m in Eq. (33) with η_i, where η_i is generated by a distributed observer of the form (39).

In [1, 2], the solvability of the cooperative output regulation problem was established by means of Lemma 1, which incurred tedious matrix manipulation. In what follows, we will further simplify the proof of the solvability of the problem by means of the following lemmas.

Lemma 5. *Consider the linear time-invariant system*

$$\dot{x} = Ax + Bu, \ t \geq 0, \tag{44}$$

where $x \in \mathbb{R}^n$, $A \in \mathbb{R}^{n \times n}$ is Hurwitz, and $u \in \mathbb{R}^m$ is piecewise continuous in t and $\lim_{t\to\infty} u(t) = 0$. Then for any initial condition $x(0)$, $\lim_{t\to\infty} x(t) = 0$.

Proof. The conclusion follows directly from the fact that system (44) is input-to-state stable with the input u decaying to the origin asymptotically (Lemma 10.4.4 in [25]). A more elementary self-contained proof can be given as follows. For any $x(0)$, let

$$x(T) = e^{AT}x(0) + \int_0^T e^{A(T-\tau)}Bu(\tau)d\tau, \ T \geq 0, \tag{45}$$

and

$$x(t) = e^{A(t-T)}x(T) + \int_T^t e^{A(t-\tau)}Bu(\tau)d\tau, \ t \geq T. \tag{46}$$

It suffices to show $\lim_{t\to\infty} x(t) = 0$. Since A is Hurwitz, we have $||e^{A(t-T)}|| \leq ke^{-\lambda(t-T)}$ for some $k > 0$ and $\lambda > 0$. Thus for any $T \geq 0$, $\lim_{t\to\infty} ||e^{A(t-T)}x(T)|| = 0$. We need to show only that, for sufficiently large T, $\lim_{t\to\infty} || \int_T^t e^{A(t-\tau)}Bu(\tau)d\tau|| = 0$. In fact, for any $t \geq T$,

$$\begin{aligned}\left|\left|\int_T^t e^{A(t-\tau)}Bu(\tau)d\tau\right|\right| &\leq \int_T^t ke^{-\lambda(t-\tau)}||B||\,||u(\tau)||d\tau \\ &\leq \frac{k||B||}{\lambda} \sup_{T\leq\tau\leq t} ||u(\tau)||(1 - e^{-\lambda(t-T)}) \\ &\leq \frac{k||B||}{\lambda} \sup_{T\leq\tau\leq t} ||u(\tau)||. \end{aligned} \tag{47}$$

Since $\lim_{t\to\infty} u(t) = 0$, for any $\varepsilon > 0$ there exists $T > 0$ such that for any $t \geq T$, $||u(t)|| \leq \frac{\lambda}{k||B||}\varepsilon$. Thus Eq. (47) implies

$$\left|\left|\int_T^t e^{A(\tau - t_0)} Bu(\tau)d\tau\right|\right| \leq \varepsilon, \quad t \geq T. \tag{48}$$

Thus, $\lim_{t\to\infty} x(t) = 0$. □

Lemma 6. *Under Assumption 1, suppose a control law of the form (6) solves the output regulation problem of system (3). Let* $\delta_u : [0, \infty) \to \mathbb{R}^m$ *and* $\delta_z : [0, \infty) \to \mathbb{R}^{n_z}$ *be two piecewise continuous functions such that* $\lim_{t\to\infty} \delta_u(t) = 0$, *and* $\lim_{t\to\infty} \delta_z(t) = 0$. *Then the control law*

$$\begin{aligned} u &= K_z z + K_y y_m + \delta_u(t), \\ \dot{z} &= \mathcal{G}_1 z + \mathcal{G}_2 y_m + \delta_z(t) \end{aligned} \tag{49}$$

is such that $\lim_{t\to\infty} e(t) = 0$.

Proof. Denote the closed-loop system composed of Eqs. (3) and (6) by Eq. (14). Then A_c is Hurwitz. By Lemma 1 there exists a unique matrix X_c that satisfies Eq. (18). Let $\bar{x}_c(t) = x_c(t) - X_c v(t)$. Then the closed-loop system composed of Eqs. (3) and (49) satisfies

$$\begin{aligned} \dot{\bar{x}}_c &= A_c \bar{x}_c + B_u \delta_u(t) + B_z \delta_z(t), \\ e &= C_c \bar{x}_c + D\delta_u(t), \end{aligned}$$

where $B_u = \text{col}(B, \mathcal{G}_2 D_m)$ and $B_z = \text{col}(0_{n \times n_z}, I_{n_z})$.

By Lemma 5 we have $\lim_{t\to\infty} \bar{x}_c(t) = 0$, and thus $\lim_{t\to\infty} e(t) = 0$. □

From the proof of Lemma 6 we can immediately obtain the following result.

Corollary 3. *Under Assumption 1, suppose a control law of the form (6) solves Problem 1. Let* $\mathcal{S}$ *be a piecewise continuous bounded square matrix defined over* $[0, \infty)$ *such that* $\dot{\eta} = \mathcal{S}(t)\eta$ *is asymptotically stable, and let* K_u *and* K_z *be any constant matrices. Then under the control law*

$$\begin{aligned} u &= K_z z + K_y y_m + K_u \eta, \\ \dot{z} &= \mathcal{G}_1 z + \mathcal{G}_2 y_m + K_z \eta, \\ \dot{\eta} &= \mathcal{S}(t)\eta, \end{aligned} \tag{50}$$

the closed-loop system also satisfies the two properties in Problem 1.

We now consider the solvability of Problem 2.

Lemma 7. *Suppose the distributed observer (41) is asymptotically stable. Then:*

(i) *Under Assumptions 5, 6, and 8, Problem 2 is solved by the distributed full information control law* (42).

(ii) *Under the additional assumption given by Assumption 7, Problem 2 is solved by the distributed measurement output feedback control law* (43).

Proof.

(i) Let $K_{1i} \in \mathbb{R}^{m_i \times n_i}$ be such that $A_i + B_i K_{1i}$ are Hurwitz and $K_{2i} = U_i - K_{1i} X_i$, $i = 1, \ldots, N$. Then, by Remark 7, the purely decentralized full information control law (32) solves Problem 2. The control law (42) can be expressed in the following form:

$$\begin{aligned} u_i &= K_{1i} x_i + K_{2i} v + K_{2i} \tilde{\eta}_i, \qquad i = 1, \ldots, N, \\ \dot{\tilde{\eta}} &= \big((I_N \otimes S) - \mu (H_{\sigma(t)} \otimes L_0 C_0) \big) \tilde{\eta}, \end{aligned} \tag{51}$$

where $\lim_{t \to \infty} \tilde{\eta}(t) = 0$. By Corollary 3, the proof is complete.

(ii) By Remark 7, under the additional assumption given by Assumption 7, Problem 2 can be solved by a control law of the form (33). Now denote control law (33) by $u_i = k_i(z_i, v_{\rm m}), \dot{z}_{\rm i} = g_i(z_i, k_i(z_i, v_{\rm m}), y_{{\rm m}i}, v_{\rm m})$, $i = 1, \ldots, N$, and control law (43) by

$$\begin{aligned} u_i &= k_i(z_i, \eta_i), \qquad i = 1, \ldots, N, \\ \dot{z}_i &= g_i(z_i, k_i(z_i, \eta_i), y_{{\rm m}i}, \eta_i), \\ \dot{\tilde{\eta}} &= \big((I_N \otimes S_{\rm m}) - \mu (H_{\sigma(t)} \otimes L_0 C_{\rm m0}) \big) \tilde{\eta}. \end{aligned} \tag{52}$$

Then it is easy to verify that

$$k_i(z_i, \eta_i) = k_i(z_i, \tilde{\eta}_i + v_{\rm m}) = k_i(z_i, v_{\rm m}) + K_{2i{\rm m}} \tilde{\eta}_i \tag{53}$$

and

$$\begin{aligned} & g_i(z_i, k_i(z_i, \eta_i), y_{{\rm m}i}, \eta_i) \\ &= g_i(z_i, k_i(z_i, v_{\rm m}) + K_{2i{\rm m}} \tilde{\eta}_i, y_{{\rm m}i}, \tilde{\eta}_i + v_{\rm m}) \\ &= g_i(z_i, k_i(z_i, v_{\rm m}), y_{{\rm m}i}, v_{\rm m}) + \Gamma_i \tilde{\eta}_i, \end{aligned} \tag{54}$$

where

$$\Gamma_i = \begin{bmatrix} B_i \\ 0 \end{bmatrix} K_{2i{\rm m}} + \begin{bmatrix} E_{i{\rm m}} \\ 0 \end{bmatrix} - L_i (F_{{\rm m}i{\rm m}} + D_{{\rm m}i} K_{2i{\rm m}}).$$

Since $\lim_{t \to \infty} \tilde{\eta}_i(t) = 0$ for $i = 1, \ldots, N$, the proof follows from Corollary 3.

□

Remark 15. Lemma 7 is reminiscent of the well-known separation principle for the design of the Luenberger observer-based output feedback control law. It is worth noting that the closed-loop system is a time-varying system when the graph is a switching graph.

Combining Lemma 7 with Lemmas 3 and 4, respectively, leads to the following two theorems.

Theorem 3. *Suppose Assumption 9 holds. Then:*

(i) *If the graph $\bar{\mathcal{G}}_{\sigma(t)}$ is undirected, the pair (C_0, S) is observable, and S is marginally stable. Then under Assumptions 6 and 8, Problem 2 is solved by the distributed full information control law* (42) *with $\mu = 1$ and $L_0 = PC_0^T$, where P is the unique positive definite solution of the inequality $PS^T + SP \leq 0$, and under the additional assumption given by Assumption 7, Problem 2 is solved by the distributed measurement output feedback control law* (43) *with $\mu = 1$ and $L_0 = PC_{\mathrm{m0}}^{\mathrm{T}}$, where P is the unique positive definite solution of the inequality $PS_{\mathrm{m}}^{\mathrm{T}} + S_{\mathrm{m}}P \leq 0$.*

(ii) *If $y_{\mathrm{m0}} = v$ and none of the eigenvalues of S have a positive real part, then under Assumptions 6 and 8, Problem 2 is solved by the distributed full information control law* (42) *for any $\mu > 0$ and $L_0 = I_q$.*

(iii) *If $y_{\mathrm{m0}} = v_{\mathrm{m}}$ and none of the eigenvalues of S_m have a positive real part, then under Assumptions 6–8, Problem 2 is solved by the distributed dynamic measurement output feedback control law* (43) *for any $\mu > 0$ and $L_0 = I_{q_{\mathrm{m}}}$.*

Theorem 4. *Suppose Assumption 10 holds. Let $\delta = \min\{Re(\lambda_i(H)\}$. Then:*

(i) *If the pair (C_0, S) is detectable, then under Assumptions 5, 6, and 8, Problem 2 is solved by the distributed full information control law* (42) *with $\mu \geq \delta^{-1}$ and $L_0 = PC_0^T$, where P is the unique positive definite solution of the inequality $PS^T + SP - PC_0^T C_0 P + I_q \leq 0$, and under the additional assumption given by Assumption 7, Problem 2 is solved by the distributed measurement output feedback control law* (43) *with $\mu \geq \delta^{-1}$ and $L_0 = PC_{\mathrm{m0}}^{\mathrm{T}}$, where P is the unique positive definite solution of the inequality $PS_{\mathrm{m}}^{\mathrm{T}} + S_{\mathrm{m}}P - PC_{\mathrm{m0}}^{\mathrm{T}} C_{\mathrm{m0}}P + I_{q_{\mathrm{m}}} \leq 0$.*

(ii) *If $y_{\mathrm{m0}} = v$, then under Assumptions 5, 6, and 8, Problem 2 is solved by the distributed full information control law* (42) *for sufficiently large μ and $L_0 = I_q$.*

(iii) *If $y_{\mathrm{m0}} = v_{\mathrm{m}}$, then under Assumptions 5–8, Problem 2 is solved by the distributed measurement output feedback control law* (43) *for sufficiently large μ and $L_0 = I_{q_{\mathrm{m}}}$.*

Remark 16. In parts (ii) and (iii) of Theorem 4, if none of the eigenvalues of S or S_{m} have a positive real part, then μ can be any positive real number.

Remark 17. For the special case where $v = v_{\rm m}$, control law (43) reduces to the following special form:

$$\begin{aligned} u_i &= K_{1i} z_i + K_{2i} \eta_i, \quad i = 1, \ldots, N, \\ \dot{z}_i &= A_i z_i + B_i u_i + E_i \eta_i + L_i (y_{mi} - C_{mi} z_i - D_{mi} u_i - F_{mi} \eta_i), \\ \dot{\eta}_i &= S \eta_i + \mu L_0 \left(\sum_{j \in \bar{\mathcal{N}}_i(t)} a_{ij}(t) C_0 (\eta_j - \eta_i) \right), \end{aligned} \tag{55}$$

where $L_i \in \mathbb{R}^{n_i \times p_{mi}}$ are such that $(A_i - L_i C_{mi})$ are Hurwitz.

On the other hand, for the special case where $v = v_{\rm u}$, there is no measurable leader's signal $v_{\rm m}$ to estimate, and control law (43) reduces to the following special form:

$$\begin{aligned} u_i &= [K_{1i} \;\; K_{2i}] z_i, \quad i = 1, \ldots, N, \\ \dot{z}_i &= \begin{bmatrix} A_i & E_i \\ 0 & S \end{bmatrix} z_i + \begin{bmatrix} B_i \\ 0 \end{bmatrix} u_i + L_i (y_{mi} - [C_{mi} \;\; F_{mi}] z_i - D_{mi} u_i). \end{aligned} \tag{56}$$

For this special case, the distributed observer is not needed and the control law is a purely decentralized one.

7 SOME VARIANTS AND EXTENSIONS

In this section we will make some remarks on some variants and extensions of the problem studied in this chapter.

7.1 Multiple Leaders and Containment Control

The containment control problem involves multiple leaders and the asymptotic tracking of the output of followers to a convex hull of the state variables of the multiple leaders [26]. The problem formulation in Section 4 also includes the containment control problem as a special case by appropriately interpreting the leader system and the tracking error e_i. In fact, suppose there are multiple leaders of the following form:

$$\dot{v}_i = S_i v_i, \quad i = 1, \ldots, l, \tag{57}$$

where, for $i = 1, \ldots, l$, $v_i \in \mathbb{R}^{q_0}$ for some positive integer q_0, and l is some integer greater than 1. Let $\text{Co} = \{\sum_{i=1}^{l} \alpha_i v_i, \alpha_i \geq 0, \sum_{i=1}^{l} \alpha_i = 1\}$, where Co is the convex hull of the points $v_1, \ldots, v_l$. Let $v = \text{col}(v_1, \ldots, v_l)$

Now define, for $i = 1, \ldots, N$, the output of subsystem i as $y_i = C_i x_i + D_i u_i + F_{1i} v$ for some matrix F_{1i}. Denote the reference input of each follower by $r = F_2 v$, where $F_2 = (\alpha_1, \ldots, \alpha_l) \otimes I_{q_0}$. Let $e_i = y_i - r$. Then e_i is in the form given in Eq. (28) with $F_i = F_{1i} - F_2$. Finally, let $S =$ block diag$[S_1, \ldots, S_l]$ and $C_0 = F_2$. Then the multiple leader systems (57) can be expressed in the standard form (29).

It can be seen that the objective of making the tracking error e_i approach the origin asymptotically implies the asymptotic convergence of the output of all follower subsystems to the convex hull Co.

7.2 Local Exogenous Signals Versus Global Exogenous Signals

Another variant of the group of systems given by Eq. (28) is given as follows:

$$\begin{aligned} \dot{x}_i &= A_i x_i + B_i u_i + E_i v_i, \\ y_{\mathrm{m}i} &= C_{\mathrm{m}i} x_i + D_{\mathrm{m}i} u_i + F_{\mathrm{m}i} v_i, \\ e_i &= C_i x_i + D_i u_i + F_i v_i, \quad i = 1, \ldots, N, \end{aligned} \tag{58}$$

where

$$\dot{v}_i = S_i v_i, \quad i = 1, \ldots, N, \tag{59}$$

for some constant matrices S_i. This formulation is actually contained in Eq. (28) if we define $v = \mathrm{col}(v_1, \ldots, v_N)$, and $S =$ block diag$[S_1, \ldots, S_N]$ and redefine the matrices E_i, $F_{\mathrm{m}i}$, and F_i, $i = 1, \ldots, N$.

7.3 Synchronized Reference Generator and the Output Synchronization

Given maps $\xi_i : [0, \infty) \to \mathbb{R}^p$ for $i = 1, \ldots, N$ and a map $\bar{\xi} : [0, \infty) \to \mathbb{R}^p$, the elements of the set $\{\xi_i(\cdot) : i = 1, \ldots, N\}$ are said to synchronize to $\bar{\xi}(\cdot)$ if $\lim_{t \to \infty}(\xi_i(t) - \bar{\xi}(t)) = 0$ for all $i = 1, \ldots, N$ and are said to synchronize if they synchronize to some $\bar{\xi}(\cdot)$ [24].

Consider the following dynamic compensator:

$$\dot{\eta}_i = S\eta_i + \mu L_0 \left(\sum_{j \in \mathcal{N}_i(t)} a_{ij}(t) C_0 (\eta_j - \eta_i) \right), \quad i = 1, \ldots, N, \tag{60}$$

where $S \in \mathbb{R}^{q \times q}$ and $C_0 \in \mathbb{R}^{p_0 \times q}$ are some given constant matrices, $\mathcal{N}_i(t)$ denotes the neighbor set of node i in the graph $\mathcal{G}_{\sigma(t)}$, and $\mu > 0$ and $L_0 \in \mathbb{R}^{q \times p_0}$ are to be designed.

The compensator (60) can be obtained from Eq. (39) by replacement of $\bar{\mathcal{N}}_i(t)$ by $\mathcal{N}_i(t)$.

We assume the graph $\mathcal{G}_{\sigma(t)}$ satisfies the following assumption.

Assumption 12. There exists a subsequence $\{i_k\}$ of $\{i : i = 0, 1, \ldots\}$ with $t_{i_{k+1}} - t_{i_k} \leq \nu$ for some positive ν such that the union graph $\bigcup_{j=i_k}^{i_{k+1}-1} \mathcal{G}_{\sigma(t_j)}$ is connected.

Then we have the following result.

Theorem 5. *Under Assumption 12:*

(i) *Suppose the graph $\mathcal{G}_{\sigma(t)}$ is undirected, S is marginally stable, and the pair (C_0, S) is observable. Then, with $\mu = 1$ and $L_0 = PC_0^T$, where the matrix P is the unique positive definite matrix such that $PS^T + SP \leq 0$, for any initial condition $\eta_i(0)$, $i = 1, \ldots, N$, the solution of Eq. (60) is such that*

$$\lim_{t\to\infty} \left(\eta_i(t) - e^{St} \frac{\sum_{j=1}^{N} \eta_j(0)}{N} \right) = 0 \tag{61}$$

exponentially.

(ii) *If $C_0 = I_q$ and none of the eigenvalues of S have a positive real part, then with $L_0 = I_q$ and any $\mu > 0$, and any initial condition $\eta_i(0)$, $i = 1, \ldots, N$, there exists some $\bar{\eta}_0 \in \mathbb{R}^q$ determined by $\eta_j(0)$, $j = 1, \ldots, N$, such that*

$$\lim_{t\to\infty} \left(\eta_i(t) - e^{St}\bar{\eta}_0 \right) = 0 \tag{62}$$

exponentially.

Remark 18. The proof of part (i) of Theorem 5 can be extracted from the proof of Theorem 2 in [22]. Part (ii) of Theorem 5 was studied in Lemma 1 in [27], which is in turn based on the result in [28].

Remark 19. If the graph $\mathcal{G}_{\sigma(t)}$ is static and connected and the pair (C_0, S) is detectable, then Theorem 5 can be strengthened as follows. Let $\lambda_2(\mathcal{L})$ denote the smallest positive eigenvalue of the Laplacian matrix $\mathcal{L}$ of the graph $\mathcal{G}$ and let P be the unique positive definite matrix such that $SP + PS^T - PC_0^T C_0 P + I_q \leq 0$. Then for $\mu \geq \max\{1, \frac{1}{\lambda_2(\mathcal{L})}\}$ and $L_0 = PC_0^T$, the solution of Eq. (60) is such that

$$\lim_{t\to\infty} \left(\eta_i(t) - e^{St} \sum_{j=1}^{N} r_j \eta_j(0) \right) = 0, \tag{63}$$

where $r = \mathrm{col}(r_1, \ldots, r_N) \in \mathbb{R}^N$ is the unit vector such that $r^T \mathcal{L} = 0$. This special case of Theorem 5 is the direct result of Lemma 1 in [24].

Remark 20. The special case with $\mu = 1$, $C_0 = I_q$, and $L_0 = I_q$ of the dynamic compensator (60) was proposed in [29], and was called a synchronized reference generator. Its main difference from the distributed observer (39) is that it does not contain a feedforward term $a_{i0}(v - \eta_i)$. If we define a virtual leader $\dot{v} = Sv$ and let $\tilde{\eta}_i = (\eta_i - v)$ and $\tilde{\eta} = \text{col}(\tilde{\eta}_1, \ldots, \tilde{\eta}_N)$, then system (60) can be expressed in the following compact form:

$$\dot{\tilde{\eta}} = \left(I_N \otimes S - \mu(L_{\sigma(t)} \otimes L_0 C_0)\right) \tilde{\eta}. \tag{64}$$

By Theorem 5, the compensator is not a distributed observer of the virtual leader $\dot{v} = Sv$ because, for $i = 1, \ldots, N$, the convergence of η_i to v happens only if $\eta_i(0)$ and $v(0)$ satisfy some equality. As a result, control law (42) with the distributed observer replaced by the synchronized reference generator (60) will not solve the output regulation problem of Eq. (28) with the exosystem being $\dot{v} = Sv$, $y_{m0} = C_0 v$. Nevertheless, by the same technique as used in the proof of Lemma 7, it is possible to show that this control law can still make the output y_i, $i = 1, \ldots, N$, of Eq. (28) synchronize to a signal of the form $e^{St}\bar{\eta}_0$ for some $\bar{\eta}_0$ determined by $\eta_i(0)$, $i = 1, \ldots, N$.

7.4 Discrete Distributed Observer

To introduce our problem, let $\mathbb{Z}^+$ denote the set of nonnegative integers and let $\sigma_d : \mathbb{Z}^+ \to \mathcal{P}$ with $\mathcal{P} = \{1, 2, \ldots, \rho\}$ be a piecewise constant switching signal in the sense that there exists a subsequence t_i of $\mathbb{Z}^+$, called *switching instants*, such that $\sigma_d(t) = p$ for some $p \in \mathcal{P}$, where $t_i \leq t < t_{i+1}$ and $t \in \mathbb{Z}^+$.

Consider the discrete-time counterpart of the linear system (36) of the following form:

$$x(t+1) = \left(I_N \otimes A - \mu F_{\sigma_d(t)} \otimes (BK)\right) x(t), \quad t = 0, 1, \ldots, \infty, \tag{65}$$

where $A \in \mathbb{R}^{n\times n}$, $B \in \mathbb{R}^{n\times m}$, and $F_{\sigma_d(t)} \in \mathbb{R}^{n\times n}$ is a piecewise switching matrix, and where $\mu > 0$ and $K \in \mathbb{R}^{m\times n}$ are to be designed.

Assumption 13. **(i)** The switching times satisfy $t_{i+1} - t_i \geq \tau$ for some positive integer $\tau > 1$ for all t_i.

(ii) The matrix $F_{\sigma_d(t)}$ is symmetric for all $t \in \mathbb{Z}^+$, and there exists a subsequence $\{i_k\}$ of $\mathbb{Z}^+$ with $t_{i_{k+1}} - t_{i_k} \leq \nu$ for some positive ν such that all the eigenvalues of the matrix $\sum_{j=i_k}^{i_{k+1}-1} F_{\sigma_d(j)}$ have positive real parts.

Let $\bar{A}$ be the real Jordan form of A, let P be the nonsingular matrix such that $A = P^{-1}\bar{A}P$, $\bar{B} = P^{-1}B$, and $\bar{K} = KP^{-1}$, and let $\bar{x}(t) = (I_N \otimes P)x(t)$. Then

$$\bar{x}(t+1) = \left(I_N \otimes \bar{A} - \mu F_{\sigma_d(t)} \otimes (\bar{B}\bar{K})\right) \bar{x}(t), \quad t = 0, 1, \ldots, \infty. \tag{66}$$

We have the following result:

Lemma 8. *Under Assumption 13, suppose all the eigenvalues of A are semisimple with modulus 1 and the pair (A, B) is controllable. Then with $\bar{K} = \bar{B}^T\bar{A}$ or $K = \bar{K}P = B^TP^TPA$, system (65) or system (66) is asymptotically stable for all μ satisfying*

$$0 < \mu \leq \min_{p\in\mathcal{P}} \left\{ \frac{1}{||F_p \otimes (\bar{A}^T\bar{B}\bar{B}^T\bar{A})||} \right\}. \tag{67}$$

Remark 21. Lemma 8 is taken from Lemma 3.1 in [30]. As pointed out in Remark 2.2 in [30], the assumption that all the eigenvalues of A are semisimple with modulus 1 can be relaxed to the assumption that A is marginally stable (ie, all the eigenvalues of A are inside the unit circle) and those eigenvalues of A with modulus 1 are semisimple. Also it can be proved that condition (67) can be relaxed to the following:

$$0 < \mu < \min_{p\in\mathcal{P}} \left\{ \frac{2}{||F_p \otimes (\bar{A}^T\bar{B}\bar{B}^T\bar{A})||} \right\}. \tag{68}$$

Now consider a discrete-time linear multiagent system of the following form:

$$\begin{aligned} x_i(t+1) &= A_i x_i(t) + B_i u_i(t) + E_i v(t), \\ y_{mi}(t) &= C_{mi} x_i(t) + D_{mi} u_i(t) + F_{mi} v(t), \\ e_i(t) &= C_i x_i(t) + D_i u_i(t) + F_i v(t), \quad i = 1, \ldots, N, \ t = 0, 1, \ldots, \infty, \end{aligned} \tag{69}$$

where $x_i \in \mathbb{R}^{n_i}$, $y_{mi} \in \mathbb{R}^{p_{mi}}$, $e_i \in \mathbb{R}^{p_i}$, and $u_i \in \mathbb{R}^{m_i}$ are the state, measurement output, error output, and input of the ith subsystem, and $v \in \mathbb{R}^q$ is the exogenous signal generated by a discrete-time exosystem of the form

$$v(t+1) = Sv(t), \quad y_{m0}(t) = C_0 v(t), \ t = 0, 1, \ldots, \infty, \tag{70}$$

where $S \in \mathbb{R}^{q\times q}$ is marginally stable. Let $\bar{\mathcal{G}}_{\sigma_d(t)}$ be a switching graph associated with Eqs. (69) and (70) whose weighted adjacency matrix is denoted by $\bar{\mathcal{A}}_{\sigma_d(t)} = [a_{ij}(t)]_{i,j=0}^N$.

Define the following dynamic compensator

$$\eta_i(t+1) = S\eta_i(t) + \mu L_0 \left(\sum_{j\in\bar{\mathcal{N}}_i(t)} a_{ij}(t) C_0(\eta_j(t) - \eta_i(t)) \right), \quad i = 1, \ldots, N, \tag{71}$$

where $\eta_0 = v$, and the scaler $\mu > 0$ and the matrix $L_0 \in \mathbb{R}^{q\times p_0}$ are to be designed. System (71) is called a *discrete distributed observer candidate* for v and is called a *discrete distributed observer* for v if for any $v(0)$ and $\eta_i(0)$,

$$\lim_{t\to\infty}(\eta_i(t)-v(t))=0,\quad i=1,\ldots,N. \tag{72}$$

Let $\tilde{\eta}_i=(\eta_i-v)$ and $\tilde{\eta}=\mathrm{col}(\tilde{\eta}_1,\ldots,\tilde{\eta}_N)$. Then system (71) can be expressed in the following compact form:

$$\tilde{\eta}(t+1)=\left((I_N\otimes S)-\mu(H_{\sigma_d(t)}\otimes L_0C_0)\right)\tilde{\eta}(t),\quad t=0,1,\ldots,\infty. \tag{73}$$

Since $\left((I_N\otimes S)-\mu(H_{\sigma_d(t)}\otimes L_0C_0)\right)^T=\left((I_N\otimes S^T)-\mu(H^T_{\sigma_d(t)}\otimes C_0^TL_0^T)\right)$, it is not difficult to deduce the following lemma from Lemma 8.

Lemma 9. *Under Assumption 13, suppose all the eigenvalues of S are semisimple with modulus 1 and the pair (C_0,S) is observable. Then with $L_0=SP^TPC_0^T$, system (73) is asymptotically stable for all μ satisfying*

$$0<\mu<\min_{p\in\mathcal{P}}\left\{\frac{2}{||F_p\otimes(\bar{S}^T\bar{C}_0^T\bar{C}_0\bar{S}^T)||}\right\}, \tag{74}$$

where $\bar{S}$ is the real Jordan form of S, P is the nonsingular matrix such that $S^T=P^{-1}\bar{S}^TP$, and $\bar{C}_0^T=P^{-1}C_0^T$.

From this lemma, the discrete counterparts of Theorems 3 and 4 can be obtained.

7.5 Adaptive Distributed Observer

A drawback of the distributed observer (39) is that the matrix S or S_m is used by the controller of every follower. A more realistic controller should allow only those followers that are the children of the leader to know the matrix S or S_m. In [12], assuming $y_{m0}=v$ and $S_m=S$, an adaptive distributed observer was proposed as follows:

$$\begin{aligned}\dot{S}_i&=\mu_1\sum_{j=0}^{N}a_{ij}(t)(S_j-S_i),\quad i=1,\cdots,N,\\ \dot{\eta}_i&=S_i\eta_i+\mu_2\sum_{j=0}^{N}a_{ij}(t)(\eta_j-\eta_i),\end{aligned} \tag{75}$$

where $S_i\in R^{q\times q}$, $\eta_i\in R^q$, $S_0=S$, $\eta_0=v$, and $\mu_1,\mu_2>0$.

Moreover, the following result was established in [12]:

Lemma 10. *Consider system (75). Under Assumption 9, suppose all the eigenvalues of the matrix S are semisimple with zero-real parts. Then for any $\mu_1,\mu_2>0$ and for any initial condition $S_i(0)$, $\eta_i(0)$ and $v(0)$, for $i=1,\ldots,N$,*

$S_i(t)$ *and* $\eta_i(t)$ *exist and are bounded for all* $t \geq 0$, *and*

$$\lim_{t\to\infty}(S_i(t) - S) = 0, \quad \lim_{t\to\infty}(\eta_i(t) - v(t)) = 0. \tag{76}$$

Remark 22. Lemma 10 holds for any S if the graph is static and connected.

8 CONCLUDING REMARKS

We have presented a unified framework for handling the cooperative output regulation problem of multiagent systems using the distributed observer approach. The main result not only contains various versions of the cooperative output regulation problem for linear multiagent systems in the literature as special cases, but also presents a more general distributed observer. We have also simplified the proof of the main result by more explicitly using the separation principle and the certainty equivalence principle. In summary, we conclude that as long as a distributed observer exists, the cooperative output regulation problem of multiagent systems is solvable if and only if the classical output regulation problem of each subsystem is solvable in the classical way as summarized in Section 3.

ACKNOWLEDGMENTS

This work has been supported in part by the Research Grants Council of the Hong Kong Special Administrative Region under grant no. 412813 and in part by the National Natural Science Foundation of China under grant no. 61174049.

APPENDIX: GRAPH

A digraph $\mathcal{G} = (\mathcal{V}, \mathcal{E})$ consists of a finite set of nodes $\mathcal{V} = \{1, \ldots, N\}$ and an edge set $\mathcal{E} = \{(i,j), i,j \in \mathcal{V}, i \neq j\}$. An edge from node i to node j is denoted by (i,j). Node i is called the *father* of node j and node j is called the *child* of node i. Node i is also called the *neighbor* of node j. If the digraph $\mathcal{G}$ contains a sequence of edges of the form $\{(i_1, i_2), (i_2, i_3), \ldots, (i_k, i_{k+1})\}$, then the set $\{(i_1, i_2), (i_2, i_3), \ldots, (i_k, i_{k+1})\}$ is called a *directed path* of $\mathcal{G}$ from i_1 to i_{k+1}, and node i_{k+1} is said to be reachable from node i_1. A digraph is said to be connected if it has a node from which there exists a directed path to every other node. The edge (i,j) is called *undirected* if $(i,j) \in \mathcal{E}$ implies $(j,i) \in \mathcal{E}$. The digraph is called *undirected* if every edge in $\mathcal{E}$ is undirected. A graph $\mathcal{G}_s = (\mathcal{V}_s, \mathcal{E}_s)$ is called a *subgraph* of $\mathcal{G} = (\mathcal{V}, \mathcal{E})$ if $\mathcal{V}_s \subseteq \mathcal{V}$ and $\mathcal{E}_s \subseteq \mathcal{E} \cap (\mathcal{V}_s \times \mathcal{V}_s)$. Given a set of r graphs $\mathcal{G}_i = (\mathcal{V}, \mathcal{E}_i)$,

$i = 1, \ldots, r$, the graph $\mathcal{G} = (\mathcal{V}, \mathcal{E})$ with $\mathcal{E} = \bigcup_{i=1}^{r} \mathcal{E}_i$ is called the *union of graphs* $\mathcal{G}_i$ and is denoted by $\mathcal{G} = \bigcup_{i=1}^{r} \mathcal{G}_i$. The weighted adjacency matrix $\mathcal{A} = [a_{ij}]_{i,j=1}^{N}$ of $\mathcal{G}$ is defined as $a_{ii} = 0$, for $i \neq j$, $a_{ij} > 0 \Leftrightarrow (j, i) \in \mathcal{E}$, and $a_{ij} = a_{ji}$ if the edge (j, i) is undirected. The Laplacian of $\mathcal{G}$ is defined as $\mathcal{L} = [l_{ij}]_{i,j=1}^{N}$, where $l_{ii} = \sum_{j=1}^{N} a_{ij}$, $l_{ij} = -a_{ij}$ for $i \neq j$. To define a switching graph, let $\mathcal{P} = \{1, 2, \ldots, \rho\}$ for some positive integer ρ. We call a time function $\sigma : [0, +\infty) \to \mathcal{P} = \{1, 2, \ldots, \rho\}$ a *piecewise constant switching signal* if there exists a sequence $t_0 = 0 < t_1 < t_2, \ldots$ satisfying $\lim_{i\to\infty} t_i = \infty$ such that for any $k \geq 0$ for all $t \in [t_{\mathrm{k}}, t_{k+1})$, $\sigma(t) = i$ for some $i \in \mathcal{P}$. $\mathcal{P}$ is called the *switching index set*. Given a piecewise constant switching signal $\sigma : [0, +\infty) \to \mathcal{P} = \{1, 2, \ldots, \rho\}$ and a set of ρ graphs $\mathcal{G}_i = (\mathcal{V}, \mathcal{E}_i)$, $i = 1, \ldots, \rho$, with the corresponding weighted adjacency matrices being denoted by $\mathcal{A}_i$, $i = 1, \ldots, \rho$, we call a time-varying graph $\mathcal{G}_{\sigma(t)} = (\mathcal{V}, \mathcal{E}_{\sigma(t)})$ a *switching graph* with the weighted adjacency matrix $\mathcal{A}_{\sigma(t)}$ if for any $k \geq 0$ for all $t \in [t_{\mathrm{k}}, t_{\mathrm{k}+1})$, $\mathcal{A}_{\sigma(t)} = \mathcal{A}_i$ for some $i \in \mathcal{P}$.

REFERENCES

[1] Y. Su, J. Huang, Cooperative output regulation of linear multi-agent systems, IEEE Trans. Autom. Control 57 (4) (2012) 1062–1066.

[2] Y. Su, J. Huang, Cooperative output regulation with application to multi-agent consensus under switching network, IEEE Trans. Syst. Man Cybern. B: Cybern. 42 (3) (2012) 864–875.

[3] E.J. Davison, The robust control of a servomechanism problem for linear time-invariant multivariable systems, IEEE Trans. Autom. Control 21 (1) (1976) 25–34.

[4] B.A. Francis, The linear multivariable regulator problem, SIAM J. Control Optim. 15 (3) (1977) 486–505.

[5] B.A. Francis, W.M. Wonham, The internal model principle of control theory, Automatica 12 (5) (1976) 457–465.

[6] S. Li, G. Feng, X. Guan, X. Luo, J. Wang, Distributed adaptive pinning control for cooperative linear output regulation of multi-agent systems, in: Proceedings of the 32nd Chinese Control Conference, Xi'an, China, 2013, pp. 6885–6890.

[7] H. Liu, C.D. Persis, M. Cao, Robust decentralized output regulation with single or multiple reference signals for uncertain heterogeneous systems, Int. J. Robust Nonlinear Control 25 (9) (2015) 1399–1422.

[8] Z. Meng, T. Yang, D.V. Dimarogonas, K. Johansson, Coordinated output regulation of heterogeneous linear systems under switching topologies, Automatica 53 (2015) 362–368.

[9] G.S. Seyboth, W. Ren, F. Allgower, Cooperative Control of Linear Multi-Agent Systems via Distributed Output Regulation and Transient Synchronization 2014, Available via DIALOG. http://arxiv.org/abs/1406.0085 .

[10] Y. Su, J. Huang, Cooperative output regulation of linear multi-agent systems by output feedback, Syst. Control Lett. 61 (12) (2012) 1248–1253.

[11] H. Cai, J. Huang, Leader-following consensus of multiple uncertain Euler-Lagrange systems under switching network topology, Int. J. Gen. Syst. 43 (3–4) (2014) 294–304.

[12] H. Cai, J. Huang, The leader-following consensus for multiple uncertain Euler-Lagrange systems with an adaptive distributed observer, IEEE Trans. Autom. Control, (DOI:10.1109/TAC.2015.2504728).
[13] H. Cai, J. Huang, The leader-following consensus control of multiple rigid spacecraft systems, Automatica 50 (4) (2014) 1109–1115.
[14] Y. Dong, J. Huang, A leader-following rendezvous problem of double integrator multi-agent systems, Automatica 49 (5) (2013) 1386–1391.
[15] Y. Dong, J. Huang, Leader-following connectivity preservation rendezvous of multiple double integrator systems based on position measurement only, IEEE Trans. Autom. Control 59 (9) (2014) 2598–2603.
[16] Y. Su, Y. Hong, J. Huang, A general result on the robust cooperative output regulation for linear uncertain multi-agent systems, IEEE Trans. Autom. Control 58 (5) (2013) 1275–1279.
[17] X. Wang, Y. Hong, J. Huang, Z. Jiang, A distributed control approach to a robust output regulation problem for multi-agent linear systems, IEEE Trans. Autom. Control 55 (12) (2010) 2891–2895.
[18] W. Liu, J. Huang, Cooperative robust output regulation of linear minimum-phase multi-agent systems under switching network, in: Proceedings of the 10th Asian Control Conference, Kota Kinabalu, Malaysia, 2015.
[19] J. Huang, Nonlinear Output Regulation: Theory and Applications, SIAM, Philadelphia, PA, 2004.
[20] A. Jadbabaie, J. Lin, A.S. Morse, Coordination of groups of mobile agents using nearest neighbor rules, IEEE Trans. Autom. Control 48 (6) (2003) 988–1001.
[21] W. Ni, D. Cheng, Leader-following consensus of multi-agent systems under fixed and switching topologies, Syst. Control Lett. 59 (3) (2010) 209–217.
[22] Y. Su, J. Huang, Stability of a class of linear switching systems with applications to two consensus problems, IEEE Trans. Automat. Control 57 (6) (2012) 1420–1430.
[23] J. Hu, Y. Hong, Leader-following coordination of multi-agent systems with coupling time delays, Physica A: Stat. Mech. Appl. 374 (2) (2007) 853–863.
[24] S.E. Tuna, LQR-based coupling gain for synchronization of linear systems, 2008, available via DIALOG. http://arxiv.org/abs/0801.3390 .
[25] A. Isidori, Nonlinear Control Systems II, Springer, Berlin, 1999.
[26] J. Mei, W. Ren, G. Ma, Distributed containment control for Lagrangian networks with parametric uncertainties under a directed graph, Automatica 48 (4) (2012) 653–659.
[27] L. Scardovi, R. Sepulchre, Synchronization in networks of identical linear systems, Automatica 45 (11) (2009) 2557–2562.
[28] L. Moreau, Stability of continuous-time distributed consensus algorithms, in: Proceedings of the 43rd IEEE Conference on Decision and Control, Atlantis, Paradise Island, Bahamas, 2004, pp. 3998–4003.
[29] P. Wieland, R. Sepulchre, A.F., An internal model principle is necessary and sufficient for linear output synchronization, Automatica 47 (5) (2011) 1068–1074.
[30] Y. Su, J. Huang, Two consensus problems for discrete-time multi-agent systems with switching topology, Automatica 48 (9) (2012) 1988–1997.

CHAPTER FIFTEEN

Cooperative Learning for Robust Connectivity in Multirobot Heterogeneous Networks

P.J. Cruz, R. Fierro, C.T. Abdallah
The University of New Mexico, Albuquerque, NM, United States

1 INTRODUCTION

In the last decade we have witnessed numerous improvements in the field of cooperative robotics. For example, technology has pushed the limits further, providing microscale high-performance processors and advanced high-efficiency sensors and actuators. At the same time, novel methods and strategies to control and coordinate teams of robots have been developed, including reinforcement learning (RL) techniques. As a result, commercially available robotic platforms are more capable now of providing essential support to human teams in a variety of civilian and military missions, such as urban search and rescue, environmental surveillance, and target localization. The accomplishment of these missions places special requirements on the robotic agents that cannot easily be fulfilled by a team of homogeneous robots. Furthermore, the use of robots with different dynamics and/or capabilities has the potential to meet complex task requirements. For example, unmanned aerial vehicles (UAVs) can monitor an area faster but with low resolution, while unmanned ground vehicles (UGVs) can sense a limited area but do so with much more accuracy. Thus these platforms are complementary, and by coordinating them, one can mitigate their weaknesses.

Coordinated control refers to the ability of a team of robots to work together to accomplish a task [1]. Typical coordination tasks for a multirobot network include sensor coverage [2], area exploration [3], and flocking control [4], to mention a few. In these applications a multirobot team is generally deployed in an environment populated with obstacles. Thus the

Control of Complex Systems
http://dx.doi.org/10.1016/B978-0-12-805246-4.00015-X

robotic agents experience uncertainty in communication, navigation, and sensing. For example, the robots have to move toward a region of interest while avoiding collisions with obstacles and other robots. At the same time, the objects in the environment can create phenomena such as shadow effects and secondary reflections which degrade the performance of the interagent wireless communication. Consequently, the coordinated control of the multirobot network should incorporate wireless communication constraints.

The objective of maintaining connectivity together with additional requirements such as collision avoidance of a multiagent system has been extensively studied in the literature. Fink et al. [1] demonstrated that a team of networked robots can maintain end-to-end connectivity in complex environments while they move to accomplish their preassigned task. Decentralized controllers based on navigation functions for a group of robots were developed in [5] to satisfy individual sensing goals while some neighborhood connectivity relationships are maintained. In [6], communication range and line of sight are used as motion constraints for a swarm of point robots which goes from an initial to a final configuration in a cluttered environment. Rooker and Birk [7] introduced a multirobot exploration algorithm that uses a utility function built with the constraints of wireless networking being taken into account. Monitoring the communication link quality and the construction of a signal strength map are strategies described in [8] for a good link quality maintenance in the deployment of a mobile robot network.

The use of heterogeneous robots to enhance the communication capabilities of the whole multirobot network has attracted a lot of attention because it brings new research challenges compared with its homogeneous counterpart. Indeed, the introduction of robots with different kinematics, sensing, and communication resources increases the team robustness to operate in highly dynamic environments. Cortez et al. [2] described a mobile communication relay to a network of sensors and derived connectivity constraints among the network members. Moreover, these constraints are used to maximize the feasible motion sets of the sensing agents. In this work, the network is assumed to move in a free-obstacle environment and its agents are considered as point robots. In [9], a team of UGVs performs a collaborative task, while a team of UAVs is positioned in a configuration such that they optimize the communication link quality to support the team of UGVs, but Gil et al. [9] assume that the UGVs are static to guarantee the connectivity of the UAV-UGV network. By use of a heterogeneous robotic system, a search/pursuit scenario is implemented in [10], where a control algorithm guarantees a certain level of signal-to-interference plus noise ratio among the members of the system. Even though the field of view of the sensors in the network is considered, the geometry of the agents is ignored. Communication maintenance for a group of heterogeneous robots is enforced in [11] by a passivity-based decentralized strategy. This approach allows the creation/deletion of communication links at any time as long as global connectivity is preserved, but the strategy is not tested in the case that the network has a main goal such as sensing on top of maintaining connectivity.

Motivated by these applications, we are interested in exploiting the heterogeneity of a multiagent network made of aerial and ground robots. We envision a scenario where a group of UAVs (eg, quadrotors [12]) interacts with a group of ground robots (eg, OctoRoACHes [13]) such that they extend the situational awareness of human task forces in a disaster area with collapsed structures. A sketch of this scenario is shown in Fig. 1A. This multirobot heterogeneous network can provide operational capabilities to first responders that would otherwise be costly, impossible, or dangerous to achieve. Therefore we seek to develop strategies to maintain network connectivity at all times while different regions of interest are sensed. We assume that the ground agents are better equipped for sensing purposes, while the aerial vehicles act as mobile wireless communication routers. We use potential field methods to coordinate the ground mobile sensors, so they move toward the regions of interest, avoiding obstacle and interagent collisions. For the case of the aerial routers, we employ a cooperative RL

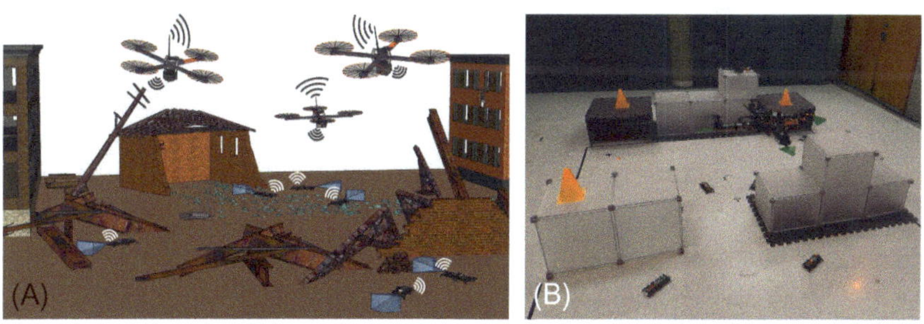

Fig. 1 (A) A group of aerial relays cooperating with a group of mobile ground sensors to explore a disaster area with collapsed structures. (B) A heterogeneous robotic network made by a quadrotor equipped with four directional antennas, one in each arm, and three OctoRoACHes moving around a cluttered environment.

technique to dynamically relocate them in such a way that the whole multirobot network stays connected. In particular, our algorithm seeks to increase the received signal strength (RSS) among the relays and the sensors. RSS is a reasonable proxy for link quality, so by improving the RSS, one can enhance connectivity in networked robotic teams [14–16].

The rest of the chapter is organized as follows. Section 2 describes the heterogeneous robotic team considered in this work. Also in this section, the motion dynamics assumed for these agents are presented. The concepts related to the connectivity of the heterogeneous robotic network are explained in Section 3. For example, we introduce the definitions of RSS, antenna diversity, and the connectivity matrix. In Section 4 we state the problem addressed herein. The control methods for the mobile sensors and for the flying relays are detailed in Sections 5 and 6, respectively. The proposed control algorithms are tested in simulation and their results are discussed in Section 7. Section 8 gives our concluding remarks.

2 HETEROGENEOUS ROBOTIC TEAM

The term *heterogeneous* is used in robotics to describe a group of agents with differences in their hardware structure (*hardware-based heterogeneity*) and/or their mission objective (*objective-based heterogeneity*) [2]. Hardware heterogeneity includes, for example, different agent dynamics and distinct sensing capabilities and communication ranges. With respect to objective-based heterogeneity, the overall mission goal can be divided into multiple subobjectives which are assigned to each agent. Thus the particular abilities of each member of the network are employed to accomplish effectively the subgoals, which in turn results in successful completion of the global objective. Fig. 1B shows an example of a robotic network that has hardware- and objective-based heterogeneity. In Fig. 1B a robotic team formed by three crawler robots (OctoRoACHes) and one quadrotor is deployed in a cluttered environment. The aerial vehicle has four antennas, one in each arm, to improve the network communication range while the OctoRoACHes explore the environment.

In this work we focus on a set of robots $\mathcal{A}$ made of two types of agents: aerial vehicles acting mainly as communication relays and ground robots better equipped for sensing. Thus $\mathcal{A}$ is a network of robots that shows hardware- and objective-based heterogeneity. We define $\mathcal{S} = \{s_1, s_2, \ldots, s_N\}$ as the set of N mobile sensors and $\mathcal{R} = \{r_1, r_2, \ldots, r_M\}$ as the set of M aerial relays. Notice that $\mathcal{A} = \mathcal{S} \cup \mathcal{R}$ and $\mathcal{S} \cap \mathcal{R} = \emptyset$.

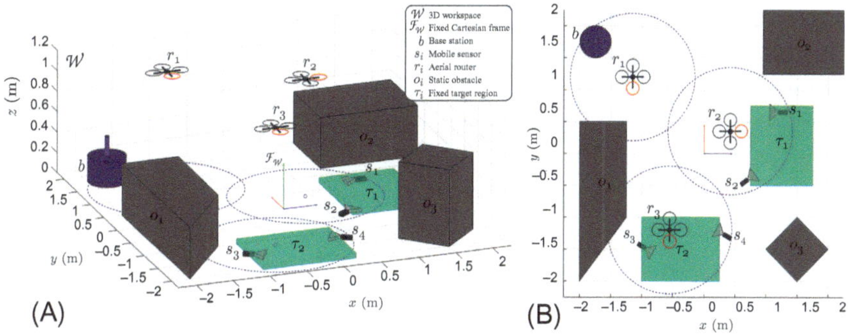

Fig. 2 Workspace where the heterogeneous robotic network operates. (A) Three-dimensional view and (B) two-dimensional view. The key in (A) facilitates the reference of the elements that are part of the workspace $\mathcal{W}$.

We denote $\mathcal{I}_\mathcal{S}$ and $\mathcal{I}_\mathcal{R}$ as the index sets of $\mathcal{S}$ and $\mathcal{R}$, respectively. We assume that the set of robots $\mathcal{A}$ operates in $\mathcal{W} \subset \mathbb{R}^3$, a compact subset of the three-dimensional Euclidean space. Inside $\mathcal{W}$ there is a fixed base station b, whose position is denoted as $\mathbf{q}_b$. Also in $\mathcal{W}$ there are L convex obstacles grouped in the set $\mathcal{O} = \{o_1, o_2, \ldots, o_L\}$ and K target regions that form the set $\mathcal{T} = \{\tau_1, \tau_2, \ldots, \tau_K\}$ such that $\mathcal{O} \cap \mathcal{T} = \emptyset$. We denote $\mathcal{I}_\mathcal{O}$ and $\mathcal{I}_\mathcal{T}$ as the index sets of $\mathcal{O}$ and $\mathcal{S}$, respectively. Embedded in $\mathcal{W}$ there is a Cartesian frame $\mathcal{F}_\mathcal{W}$ that helps to describe the position and orientation of the robots, objects, and targets in the workspace $\mathcal{W}$. Fig. 2 depicts the environment created for simulation purposes. The key in Fig. 2A facilitates the reference of the elements inside $\mathcal{W}$. In the next section we introduce the models for the mobile sensors and the aerial routers.

2.1 Motion Dynamics

For every i in the index set $\mathcal{I}_\mathcal{S}$, we define the state vector of the ith mobile sensor as $\mathbf{q}_{s_i} = [x_{s_i} \quad y_{s_i} \quad v_{s_i} \quad \theta_{s_i}]^T$. Here,$(x_{s_i}, y_{s_i})$ is the xy position of the ith mobile sensor and θ_{s_i} is its heading angle. We assume that the equations of motion of the ith mobile sensor are given by the unicycle model

$$\begin{aligned} \dot{x}_{s_i} &= v_{s_i} \cos\theta_{s_i}, \\ \dot{y}_{s_i} &= v_{s_i} \sin\theta_{s_i}, \\ \dot{v}_{s_i} &= a_{s_i}, \\ \dot{\theta}_{s_i} &= \omega_{s_i}, \end{aligned} \tag{1}$$

where a_{s_i} and ω_{s_i} are the linear acceleration and the angular velocity, respectively. We define $\mathbf{u}_{s_i} = [a_{s_i} \quad \omega_{s_i}]^T$ as the control input of the ith mobile sensor.

For every j in $\mathcal{I}_{\mathcal{R}}$, the state vector of the jth aerial router is given by $\mathbf{q}_{r_j} = [x_{r_j} \quad y_{r_j} \quad z_{r_j}]^T$. This state specifies the position of the router with respect to $\mathcal{F}_{\mathcal{W}}$. The motion dynamics for this case are the ones given by the holonomic model

$$\ddot{\mathbf{q}}_{r_j} = \mathbf{u}_{r_j}, \tag{2}$$

where $\mathbf{u}_{r_j} \in \mathbb{R}^3$ is the acceleration control input.

In the next section we detail our assumptions with respect to the communication capabilities of the heterogeneous robotic team.

3 NETWORK CONNECTIVITY

We consider as communication nodes the base station b, the set of sensors $\mathcal{S}$, and the set of relays $\mathcal{R}$. Thus we define $\mathcal{M} = \{b\} \cup \mathcal{S} \cup \mathcal{R}$ as the set of communication nodes. The cardinality of this set is $|\mathcal{M}| = N+M+1$. Let $\mathcal{I}_{\mathcal{M}}$ be the index set of $\mathcal{M}$. Then for all $i \in \mathcal{I}_{\mathcal{M}}$, the set of neighbors in the vicinity of the communication node $n_i \in \mathcal{M}$ is given by

$$\mathcal{N}_{n_i} = \left\{ n_j \in \mathcal{M} \mid d(n_i, n_j) \leq \varrho_{n_i} \right\}. \tag{3}$$

Here $d(n_i, n_j)$ is the distance between the nodes n_i and n_j, and ϱ_{n_i} is the maximum range of node n_i.

3.1 Received Signal Strength

Assuming the ith node acts as a receiver and the jth node acts as a transmitter, we adopt a log-distance path-loss model [17] to estimate the power loss between the transmitter and the receiver. For this model, the noise-free received power at the ith node from the jth node is given by

$$p_{ij} = p_{\mathrm{TX}_j} \kappa \left[\frac{d_o}{d(n_i, n_j)} \right]^{\alpha}, \tag{4}$$

where p_{TX_j} is the power transmitted by the jth node, d_o is a reference close-in distance for the antenna far field, α is the path-loss exponent, and κ is a unitless constant that depends on the antenna characteristics and the frequency of operation. We assume that p_{TX_j} is the same for all nodes, so $p_{\mathrm{TX}_j} = p_{\mathrm{TX}}$. The log-distance model is a simplified piecewise model with two segments: free-space propagation is assumed up to distance d_o and non-free-space propagation is assumed for distances greater than d_o. Typical values for d_o are 1–10 m indoors and 10–100 m outdoors. The value of the path-loss exponent α depends on the propagation environment. It ranges from 2 for free-space propagation to 6 for heavy cluttered environments [17]. The constant κ can be set to the free-space path gain at distance d_o given by

$$\kappa = G_{n_i} G_{n_j} \left(\frac{\lambda}{4\pi d_o} \right)^2 . \tag{5}$$

Here G_{n_i} and G_{n_j} are the antenna gains for the nodes n_i and n_j, respectively, and $\lambda = c/f$ is the wavelength of the transmitted signal (c is the speed of light and f is the communication frequency). We assume that the base station and the mobile sensors are equipped with unidirectional antennas. Thus the antenna gain for these cases is 1. On the other hand, we consider that the aerial relays have directional antennas, which means that the antenna gain takes on a value of 1 or less.

Every received-transmitter link is affected by obstructions and clutter, so the effective received power at the ith node from the jth node can be expressed as

$$P_{ij} = p_{ij}\, \zeta_{ij}, \tag{6}$$

with $\zeta_{ij} = 10^{\frac{\chi_{ij}}{10}}$ being the model of a log-normal shadowing noise [17]. Each χ_{ij} is assumed to have a Gaussian distribution with zero mean and variance σ_{ij}^2. Furthermore, we assume independent and homogeneous log-normal shadowing across all nodes; then $\sigma_{ij}^2 = \sigma^2$. The total RSS at the ith node, RSS_{n_i}, represents an aggregate of the power emanating from all the transmitters in its vicinity. This parameter measured in decibels can be expressed as [18]

$$\mathrm{RSS}_{n_i} = 10 \log_{10} \sum_{j \in \mathcal{I}_{\mathcal{N}_{n_i}}} P_{ij}, \tag{7}$$

where $\mathcal{I}_{\mathcal{N}_{n_i}}$ is the index set of $\mathcal{N}_{n_i}$. From Eq. (6), we also define the RSS at the ith node with respect to the jth node as

$$\mathrm{RSS}_{n_i,n_j} = 10 \log_{10} \left(p_{ij}\, \zeta_{ij} \right) . \tag{8}$$

3.2 Antenna Diversity

For the case of the aerial relays, we consider that each of them has four directional antennas. Each antenna is mounted at the end of each arm (see Fig. 3). The use of multiple antennas, known as *antenna diversity*,

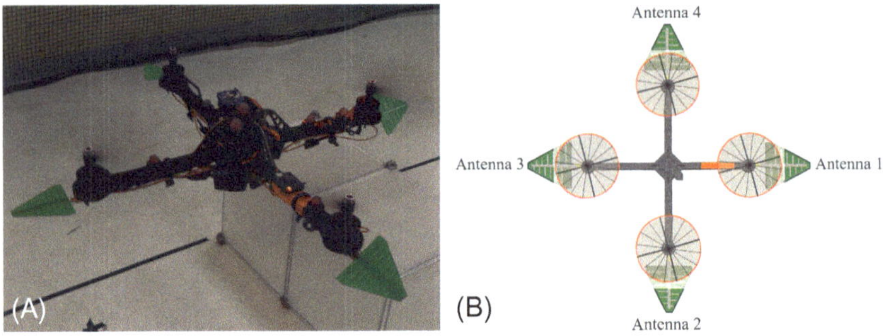

Fig. 3 (A) An aerial relay equipped with four directional antennas. (B) The numbers assigned to each antenna.

improves wireless links and is especially effective at mitigating multipath. Antenna diversity also has the advantage that while one antenna is in a deep fade, it is probable that another antenna has a strong enough signal [17, 19]. We assume directional antennas since they radiate greater power in a specific direction as opposed to omnidirectional antennas, which radiate power uniformly. This characteristic allows the quadrotor to sense the unknown RF environment and find a direction to move, improving the communication performance.

We assume the incoming signal is received by all four antennas but only the signal taken from the antenna with the highest RSS is used. Therefore four RSS values have to be calculated for every aerial relay $r_j \in \mathcal{R}$. These values can be found by application of Eqs. (4)–(7) considering the distance with respect to each antenna as well as the individual antenna gain. Let $\text{RSS}_{r_j}^{(k)}$ be the RSS of the kth antenna at the jth aerial relay. Notice that $k = 1, \ldots, 4$. Then the RSS for the jth relay is

$$\text{RSS}_{r_j} = \max\left(\text{RSS}_{r_j}^{(k)}\right). \tag{9}$$

3.3 Connectivity Matrix

To describe the connectivity of the network formed by the nodes in $\mathcal{M}$, we construct a graph $\mathcal{G} = (\mathcal{V}, \mathcal{E})$. This graph consists of the set of vertices $\mathcal{V} = \mathcal{M}$ and the set of edges $\mathcal{E}$ such that for all $n_i, n_j \in \mathcal{M}$

$$(n_i, n_j) \in \mathcal{E} \quad \text{iff } i \neq j \text{ and } \text{RSS}_{n_i,n_j} \geq \upsilon_{ij}. \tag{10}$$

Here v_{ij} is the RSS threshold needed to maintain a minimal link quality between the nodes n_i and n_j. We consider that $v_{ij} = v$ for all nodes. We also assume that the links between all nodes are symmetric (ie, $\text{RSS}_{n_i,n_j} = \text{RSS}_{n_j,n_i}$). From this assumption and by condition (10), the graph $\mathcal{G}$ is undirected and has no self-loops. For this type of graph, $\mathcal{G}$ is connected if and only if there exists a path between any two vertices [20, 21]. Hence we define an $|\mathcal{M}| \times |\mathcal{M}|$ matrix $\mathbf{C} = [C_{ij}]$ such that

$$C_{ij} = \begin{cases} 1 & \text{if } (n_i, n_j) \in \mathcal{E}, \\ 0 & \text{otherwise.} \end{cases}$$

Thus C_{ij} is nonzero just when RSS_{n_i,n_j} is sufficiently high to establish a link between n_i and n_j. For this reason we call $\mathbf{C}$ the *connectivity matrix*. Let $\mathbf{L}$ be the Laplacian matrix associated with the graph $\mathcal{G}$ defined by

$$\mathbf{L} = \text{diag}(\delta_i) - \mathbf{C}, \tag{11}$$

with $\delta_i = \sum_{j=1}^{|\mathcal{M}|} C_{ij}$. From graph theory [20, 21] we know that

$$\mathcal{G} \text{ is connected iff } \text{rank}(\mathbf{L}) = |\mathcal{M}| - 1 = N + M.$$

Now that we have introduced our assumptions, we state the specific problem addressed in this chapter next.

4 DEPLOYMENT MAINTAINING NETWORK CONNECTIVITY

Given a heterogeneous robotic team $\mathcal{A}$ formed by a set of mobile sensors $\mathcal{S}$ with dynamics given by Eq. (1) and a set of aerial relays $\mathcal{R}$ with dynamics given by Eq. (2), our goal is to design controllers for these two sets such that the sensors get measurements of the target regions specified in the set $\mathcal{T}$, while the aerial relays maintain network connectivity between the mobile sensors and the base station b.

We consider that every mobile sensor in $\mathcal{S}$ has been assigned a target region in $\mathcal{T}$ whose location is known a priori. We know from Section 2 that the workspace $\mathcal{W}$ is populated with L static obstacles. Therefore the controller for each mobile sensor is required to guide it to get measurements of its preassigned target region avoiding collisions with obstacles and other mobile sensors. Section 5 provides the details of this controller designed based on potential functions [22–24].

For the case of the aerial relays, we assume they fly at a safe height above the obstacles at all times. Thus the aerial relays do not require obstacle avoidance, but they have to keep a safe distance among themselves. While the goal of the mobile sensors is to get measurements of the target regions, the flying relays have to keep the connectivity of the whole network, particularly between the base station and the mobile sensors. The proposed controller for the relays (see Section 6) uses a cooperative learning technique [4] to maintain the RSS between the relays and the sensors above a threshold so as to have a good link quality.

5 MOBILE SENSOR CONTROLLER

The control objectives for the case of the mobile sensors are (i) getting measurements of the target region, (ii) avoiding obstacles, and (iii) avoiding collisions with other sensors. Thus we use potential field methods having an attractive potential function to move toward the assigned target region and a repulsive potential function to avoid collisions.

5.1 Potential Field Method

This motion planning method controls the movement of a robot on the basis of the gradient of a potential function [22]. Potential field methods were originally developed as an online collision avoidance approach applicable when a robot does not have a model of the obstacles a priori but senses them during motion execution. Roughly speaking, a potential field method acts as a fastest descent optimization procedure, so it may get stuck at a local minimum other than the goal configuration. However, various techniques [23, 24] have been demonstrated to be valid for escaping from local minima. In our simulations we use random walk [22] when an agent gets stuck at a local minimum. This is the method most commonly used to avoid this problem.

Let U_{s_i} be the total potential function for the mobile sensor $s_i \in \mathcal{S}$. This function is given by

$$U_{s_i} = U_{\text{att}}(\mathbf{q}_{s_i}) + U_{\text{rep}}(\mathbf{q}_{s_i}), \tag{12}$$

where $U_{\text{att}}(\mathbf{q}_{s_i})$ is the attractive potential defined on the basis of the position of s_i and the position of its assigned target region, and $U_{\text{rep}}(\mathbf{q}_{s_i})$ is the repulsive potential generated considering the obstacles and the location of the other mobile sensors. Indeed, the attractive potential is defined as

$$U_{\text{att}}(\mathbf{q}_{s_i}) = \begin{cases} \frac{1}{2}\beta_{\text{att}} d^2(\mathbf{q}_{s_i}, \tau_{s_i}) & \text{if } d(\mathbf{q}_{s_i}, \tau_{s_i}) \geq \varrho_{\text{sen}}, \\ 0 & \text{otherwise.} \end{cases} \tag{13}$$

Here $d(\mathbf{q}_{s_i}, \tau_{s_i})$ is the distance between s_i and its assigned target region τ_{s_i}, ϱ_{sen} is the minimum range to get measurements of a target region, and β_{att} is a scaling factor. The repulsive potential is given by

$$U_{\text{rep}}(\mathbf{q}_{s_i}) = U_{\text{s}}(\mathbf{q}_{s_i}) + U_{\text{o}}(\mathbf{q}_{s_i}), \tag{14}$$

where $U_{\text{s}}(\mathbf{q}_{s_i})$ is used to avoid intersensor collisions and $U_{\text{O}}(\mathbf{q}_{s_i})$ is used to prevent obstacle collisions. These functions are defined as

$$U_{\text{s}}(\mathbf{q}_{s_i}) = \sum_{j \neq i, j=1}^{N} U(\mathbf{q}_{s_i}, \mathbf{q}_{s_j}), \text{ with } U(\mathbf{q}_{s_i}, \mathbf{q}_{s_j}) = \begin{cases} \frac{1}{2}\beta_{\text{s}} \left[\frac{1}{d(\mathbf{q}_{s_i}, \mathbf{q}_{s_j})} - \frac{1}{\varrho_{\text{s}}}\right]^2 & \text{if } d(\mathbf{q}_{s_i}, \mathbf{q}_{s_j}) \leq \varrho_{\text{s}}, \\ 0 & \text{otherwise} \end{cases} \tag{15}$$

and

$$U_o(\mathbf{q}_{s_i}) = \sum_{j=1}^{L} U(\mathbf{q}_{s_i}, o_j), \text{ with } U(\mathbf{q}_{s_i}, o_j) = \begin{cases} \frac{1}{2}\beta_{\text{O}} \left[\frac{1}{d(\mathbf{q}_{s_i}, o_j)} - \frac{1}{\varrho_{\text{O}}}\right]^2 & \text{if } d(\mathbf{q}_{s_i}, o_j) \leq \varrho_{\text{O}}, \\ 0 & \text{otherwise} \end{cases}, \tag{16}$$

respectively. In Eqs. (15) and (16), $d(\mathbf{q}_{s_i}, \mathbf{q}_{s_j})$ is the distance between the ith and the jth mobile sensors, $d(\mathbf{q}_{s_i}, o_j)$ is the distance between the ith mobile sensor and the jth obstacle, ϱ_{s} is the clearance distance to avoid intersensor collision, ϱ_{O} is the influence distance of the obstacles, and β_{s} and β_{O} are scaling factors. We assume that $0 < \varrho_{\text{s}} < \varrho_{\text{O}}$.

From Section 2 we have that the control vector for the mobile sensor s_i is $\mathbf{u}_{s_i} = [a_{s_i} \quad \omega_i]^T$. Using the gradient of the total potential function U_{s_i} given by

$$\nabla U_{s_i} = \left[\frac{\partial U_{s_i}}{\partial x_{s_i}} \quad \frac{\partial U_{s_i}}{\partial y_{s_i}} \quad \frac{\partial U_{s_i}}{\partial \theta_{s_i}}\right]^T,$$

we define the control inputs for the ith mobile sensor as

$$\begin{aligned} a_{s_i} &= \left[\cos\theta_{s_i} \quad \sin\theta_{s_i} \quad 0\right] \nabla U_{s_i} - k_0 v_{s_i}, \quad \text{and} \\ a\omega_{s_i} &= k_1 \left[\text{arctan2}\left(\frac{\partial U_{s_i}}{\partial x_{s_i}}, \frac{\partial U_{s_i}}{\partial y_{s_i}}\right) - \theta_{s_i}\right], \end{aligned} \tag{17}$$

where k_0 and k_1 are control constants.

6 AERIAL RELAY CONTROLLER

The main goal of the flying relays is to maintain the connectivity of the network, specially between the base station and the mobile sensors. We define $\mathcal{D}_{r_j} \subset \{b\} \cup \mathcal{R}$ as the set of neighbors for which the jth relay has to keep a relative distance μ_r. This set for each relay is defined a priori. We assume that one of the relays has the base station as a neighbor and that all the relays have at least one neighbor. Also, we consider that if r_j is a neighbor of r_k, then r_k is also a neighbor of r_j. From our definition of $\mathcal{D}_{r_j}$, $\mathcal{D}_{r_j} \cap \mathcal{S} = \emptyset$ for every j. We do not consider that a relay should maintain a sensor in its vicinity since the sensors explore around $\mathcal{W}$, moving from the vicinity of one relay to another. However, we use the RSS among relays and the sensors in their vicinity to maintain network connectivity.

We define the acceleration control input for the jth aerial router as

$$\mathbf{u}_{r_j} = k_2 \sum_{i \in \mathcal{I}_{\mathcal{D}_{r_j}}} \left(\ell_{ij} - \mu_r\right) \mathbf{e}_{ij} - k_3 \dot{\mathbf{q}}_{r_j} - \mathrm{k}_4 \left(\mathbf{q}_{r_j} - \mathbf{q}_{\mathrm{ref}}\right), \tag{18}$$

where $\mathcal{I}_{\mathcal{D}_{r_j}}$ is the index set of $\mathcal{D}_{r_j}$. The first two terms in Eq. (18) create a virtual spring [10] among r_j and its neighbors, so the relative distance μ_r can be maintained. For these terms, k_2 and k_3 are the spring constant and the damping coefficient, respectively. Also, we define for every ith neighbor of r_j in $\mathcal{D}_{r_j}$ the length and the direction of the force of the virtual spring as

$$\ell_{ij} = \|\mathbf{q}_i - \mathbf{q}_{r_j}\| \quad \text{and} \quad \mathbf{e}_{ij} = \frac{\mathbf{q}_i - \mathbf{q}_{r_j}}{\|\mathbf{q}_i - \mathbf{q}_{r_j}\|},$$

respectively. Here $\mathbf{q}_i$ is the position with respect to $\mathcal{F}_{\mathcal{W}}$ of the ith neighbor of r_j.

The last term in Eq. (18) helps to move the aerial relay r_j such that the RSS of r_j is improved, taking into account the RSS with respect to the mobile sensors in its vicinity. From Section 3.2 and using Eq. (7), we redefine the RSS for the jth mobile relay considering just the mobile sensors in its vicinity. Thus we get

$$\mathrm{RSS}_{r_j}^{(k)} = 10 \log_{10} \sum_{i, n_i \in \mathcal{N}_{r_j} \cap \mathcal{S}} P_{ij}, \tag{19}$$

where $k = 1, 2, 3, 4$ is the number of the antenna. Then redefining Eq. (9), we obtain

$$[\mathrm{RSS}_{r_j}, A_{r_j}] = \max\left(\mathrm{RSS}_{r_j}^{(k)}\right), \quad \text{with } k = 1, 2, 3, 4. \tag{20}$$

Here the max function returns the maximum, not only RSS_{r_j} (the maximum RSS between all four antennas) but also the number of the antenna with this RSS A_{r_j}. Using RSS_{r_j} and A_{r_j}, we select on the basis of a cooperative RL technique the reference position $\mathbf{q}_{\text{ref}}$ required by the last term in Eq. (18).

6.1 Cooperative *Q*-Learning

Recently, RL has grown as an effective approach for control applications [25] and in particular to allow multirobot systems to learn cooperation [4, 26, 27]. *Reinforcement learning* (RL) refers to an actor (eg, a robotic agent) which interacts with its environment and modifies its control policy on the basis of a reward received in response to its actions. Thus RL algorithms seek to find the optical control policy that maximizes the future reward. In RL the actor knows only its previous and current states and the reward of how good the previous action was. Let x^t and u^t be the agent state and the action at time t. Then the action u^t produces a new state x^{t+1} by interacting with the environment and a reward ρ^{t+1} is observed. Applying RL, the agent then selects a new action based on x^t, x^{t+1}, and ρ^{t+1}. Typically, RL algorithms must balance the need to collect informative data (by *exploring* novel action choices) with the need to control the process (by *exploiting* the currently available knowledge) [28–30]. Exploration is when the agent selects random actions while learning regardless of the current state. In this way the agent can explore the entire state-action space to avoid the problem of learning local maxima. In contrast, exploitation is the use of the agent's knowledge in selecting actions. The agents choose the action that will produce the maximum reward given the current state. The trade-off between exploration and exploitation is challenging. For example, use of mostly exploration can cause the agent to take a long time to learn. Usually at the beginning of the learning, exploration is mainly used. Then the exploration diminishes over time and exploitation takes over. One approach to transition between exploration and exploitation is the ε-greedy technique [28]. We use this technique in our algorithm for implementing the exploration-exploitation trade-off.

Q-learning is a model-free value iteration RL algorithm which uses a state-action value function known as a Q-function [28–30]. This function represents the expected utility of taking a given action in a given state and following the optimal policy thereafter. In fact, Q-learning starts from an arbitrary Q-function and updates it without requiring a model and using instead observed state transitions and rewards. Let x_j^t, u_j^t, and ρ_j^t be the state, action, and reward of the jth aerial relay r_j at time t, respectively. Then we have the following models for the state, action, and reward.

The state. For every r_j, the state is defined by the number of the antenna whose RSS is the highest. From Eq. (20) we have that $x_j = A_{r_j}$, and so $x_j \in \{1, 2, 3, 4\}$. Thus the total number of states per aerial relay is four.

The action. We assume five possible actions: north, south, east, west, and stay, which are associated with the unitary vectors

$$\mathbf{e}_1 = [1 \quad 0 \quad 0]^T, \quad \mathbf{e}_2 = -\mathbf{e}_1, \quad \mathbf{e}_3 = [0 \quad 1 \quad 0]^T, \quad \mathbf{e}_4 = -\mathbf{e}_3,$$
$$\text{and} \quad \mathbf{e}_5 = \mathbf{0},$$

respectively. Here, $\mathbf{0} \in \mathbb{R}^3$ is the zero vector. Once an action has been chosen, we use its associated vector to find the reference position $\mathbf{q}_{\text{ref}}$ required by Eq. (18). For example, let $u_j =$ east. Then

$$\mathbf{q}_{\text{ref}} = \mathbf{q}_{r_j} + \Delta L \mathbf{e}_3,$$

where ΔL is a reference distance. Since the total number of states is four and the number of actions is five, we have 20 state-action pairs per aerial relay.

The reward. This is calculated on the basis of RSS_{r_j} obtained by application of Eq. (20). Since our purpose is to maintain RSS_{r_j} above the threshold υ, we define the reward function as follows:

$$\rho_j = \begin{cases} 100 & \text{if } \text{RSS}_{r_j} > \upsilon, \\ 0 & \text{otherwise.} \end{cases} \tag{21}$$

Each aerial relay r_j updates an individual Q-function $\bar{Q}_j(x_j, u_j)$ as follows:

$$\bar{Q}_j^{t+1}(x_j, u_j) = Q_j^t(x_j, u_j) + \eta \left[\rho_j^{t+1} + \gamma \max_{\hat{u}_j} Q_j^t(x_j^{t+1}, \hat{u}_j) - Q_j^t(x_j, u_j) \right], \tag{22}$$

where $\eta \in (0, 1]$ is the learning rate and $\gamma \in (0, 1]$ is the discount factor. However, our intention is to allow each aerial relay to aggregate information on its relay neighbors via their individual Q-functions. Let $\mathcal{B}_{r_j} = \mathcal{D}_{r_j} \setminus \{b\}$ be the set of the relay neighbors of relay r_j. Then each r_j updates the total Q-function $Q_j(x_j, u_j)$ following the equation for the cooperative Q-learning technique described in [4]:

$$Q_j^{t+1}(x_j, u_j) = w\bar{Q}_j^t(x_j, u_j) + (1 - w) \sum_{i \in \mathcal{I}_{\mathcal{B}_{r_j}}} \bar{Q}_i^t(x_i, u_i), \tag{23}$$

where $\mathcal{I}_{\mathcal{B}_{r_j}}$ is the index set of $\mathcal{B}_{r_j}$ and $w \in [0, 1]$ is the trust weight. If $w = 1$, the Q-function is updated with trust just of the knowledge of the relay itself (independent learning). On the other hand, if $w = 0$, the aerial relay updates its Q-function only on the basis of the Q-functions of its relay neighbors. When $0 < w < 1$, the learning is performed in a cooperative fashion. The convergence of the cooperative Q-learning algorithm is analyzed in [4]. Q^t will converge if $\mu \in (0, 1)$ and $\gamma \in (0, \frac{1}{n})$, with n being the number of agents performing the cooperative learning. For a complete discussion, the reader is referred to [4].

Once the Q-function has been updated, we use the ε-greedy technique to balance exploration with exploitation [28, 30]. This method selects actions according to

$$u_j^t = \begin{cases} \arg\max_{\hat{u}} Q_j^t(x_j, \hat{u}_j) & \text{with probability } 1 - \varepsilon_t \text{ (exploit)}, \\ \text{uniformly random action} & \text{with probability } \varepsilon_t \text{ (explore)}. \end{cases} \tag{24}$$

Here $\varepsilon_t \in (0, 1]$ is the exploration probability at time step t. A typical value for the exploration probability is $\varepsilon_t = \frac{1}{t}$. Algorithm 15.1 summarizes the cooperative Q-learning method for every aerial relay.

Algorithm 15.1 Cooperative *Q*-Learning for All $r_j \in \mathcal{R}$

Require: learning rate η, discount factor γ, initial Q-function Q_j^0 (eg, $Q_j^0 \leftarrow 0$), initial state x_j^0, trust weight w

1: **for** every time step $t = 0, 1, 2, \ldots$ **do**

2: $u_j^t = \begin{cases} \arg\max_{\hat{u}} Q_j^t(x_j, \hat{u}_j) & \text{with probability } 1 - \frac{1}{t+1} \\ \text{uniformly random action in \{north, south, east, west, stay\}} & \text{with probability } \frac{1}{t+1} \end{cases}$

3: Apply u_j^t

4: Find next state x_j^{t+1} and the reward ρ_j^{t+1}

5: $\bar{Q}_j^{t+1}(x_j, u_j) \leftarrow Q_j^t(x_j, u_j) + \eta\left[\rho_j^{t+1} + \gamma \max_{\hat{u}_j} Q_j^t(x_j^{t+1}, \hat{u}_j) - Q_j^t(x_j, u_j)\right]$

6: $Q_j^{t+1}(x_j, u_j) \leftarrow w\bar{Q}_j^t(x_j, u_j) + (1 - w)\sum_{i \in \mathcal{I}_{\mathcal{B}_{r_j}}} \bar{Q}_i^t(x_i, u_i)$

7: **end for**

7 SIMULATION RESULTS

In this section we present simulation results of the proposed method for a network of mobile sensors (Section 5) and aerial relays (Section 6). For the case of the cooperative Q-learning method (Algorithm 15.1), we compare independent learning ($w = 1$) with cooperative learning (assuming $w = 0.5$) in terms of the network connectivity. From our discussion in Section 3, we know that the network formed by the base station b, the set of N mobile sensors $\mathcal{S}$, and the set of M serial relays $\mathcal{R}$ is connected if the rank of the Laplacian matrix $\mathbf{L}$ given by Eq. (11) is equal to $N + M$. Therefore the relative connectivity index [4] can be defined as

$$c = \frac{\text{rank}(\mathbf{L})}{N + M}. \tag{25}$$

The network is connected if and only if $c = 1$. Otherwise it is broken, and as c tends to zero, the network has more and more disconnected components [21]. We compute this index during our simulation tests.

We consider a heterogeneous team of robots made by four mobile sensors (ie, $N = 4$) and three aerial relays (ie, $M = 3$). The workspace $\mathcal{W}$ is of size $4 \times 4 \times 1.5$ m^3 populated with four convex obstacles and two target regions. Fig. 4 illustrates this workspace, where the obstacles are

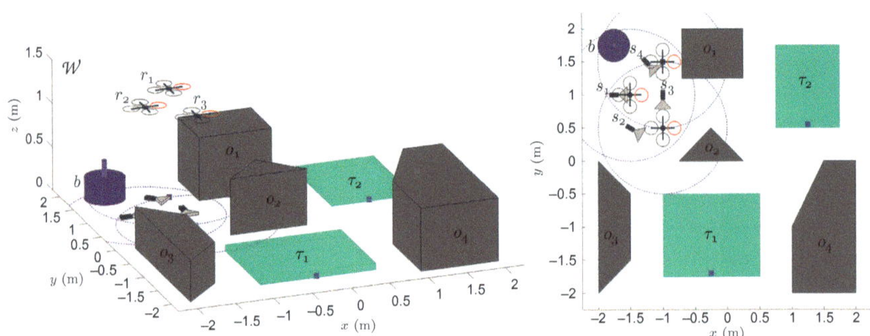

(A) (B)

Fig. 4 (A) Three-dimensional view and (B) two-dimensional view of the workspace employed for the simulation tests. The workspace $\mathcal{W}$ has a base station b, four obstacles $\mathcal{O} = \{o_1, o_2, o_3, o_4\}$, and two target regions $\mathcal{T} = \{\tau_1, \tau_2\}$. The heterogeneous robotic network is formed by four mobile sensors, $\mathcal{S} = \{s_1, s_2, s_3, s_4\}$, and three aerial relays, $\mathcal{R} = \{r_1, r_2, r_3\}$. The blue squares inside τ_1 and τ_2 denote the reference points used for the initial deployment for the aerial relays.

in gray and the target regions are in green. The base station is represented in blue and its position is $\mathbf{q}_{\text{base}} = [-1.75 \quad 1.75 \quad 0]^T$ m. Also, the initial positions of the ground sensors and the aerial relays are shown. The blue dashed circles denote the range for each flying relay. We assume that for all the communication nodes the range is 1 m (ie, for all $n_i \in \mathcal{M}, \varrho_{n_i} = 1$ m. For the set of relays $\mathcal{R} = \{r_1, r_2, r_3\}$, we have

$$\mathcal{D}_{r_1} = \{b, r_2, r_3\}, \quad \mathcal{D}_{r_2} = \{r_3\}, \quad \text{and} \quad \mathcal{D}_{r_3} = \{r_2\}.$$

Thus r_1 has to keep a relative distance with the base station and with the other sensors, while r_2 and r_3 have to do the same with r_1. The time step is 0.005 s and we run the simulation for 30 s. The rest of the simulation parameters are summarized in Table 1.

At the beginning of the simulation, we deploy just the aerial routers such that r_2 gets closer to the first target τ_1, and r_3 gets closer to the second target τ_2. To achieve this, we use the control law (18), but $\mathbf{q}_{\text{ref}}$ is fixed to a position inside their respective target regions. The blue squares within τ_1 and τ_2 represent the reference positions. For the case of r_1 we also use Eq. (18) but without considering the third term (ie, $k_4 = 0$). Therefore we do not use the learning algorithm during this initial deployment. Fig. 5 shows a series of snapshots of the deployment of the aerial relays. Since we take into consideration damping effects in the controller (18), the mesh of virtual springs between the flying relays has behavior similar to that of a real spring system, in which dissipative forces act against the movement of the springs, ensuring the velocity eventually reaches zero. This behavior guarantees that the network of aerial relays eventually reaches a rest state [10], which indicates the end of the initial deployment. Indeed, the initial deployment maneuver finishes (ie, the flying relays reach the rest state) around 4 s after it started. As can be seen in Fig. 5, the relays move such that they maintain the relative distance with respect to their predefined

Table 1 Simulation parameters

Network connectivity	$p_{\text{TX}} = 60\,\text{mW}$, $\alpha = 2.5$, $d_o = 0.5\,\text{m}$, $f = 2.4\,\text{GHz}$, $\sigma = 0.25$, $\upsilon = -70\,\text{dB}$
Mobile sensor controller	$\varrho_{\text{sens}} = 0.1$ m, $\varrho_{\text{s}} = 0.2$ m, $\varrho_{\text{o}} = 0.3$ m, $\beta_{\text{sens}} = 0.6$, $\beta_{\text{s}} = \beta_{\text{o}} = 0.4$, $k_0 = 0.75$, $k_1 = 1.25$
Aerial relay controller	$\mu_r = 1.5$ m, $\eta = 0.2$, $\gamma = 0.25$, $\Delta L = 0.75$ m, $k_2 = 5.75$, $k_3 = 3.75$, $k_4 = 2.5$

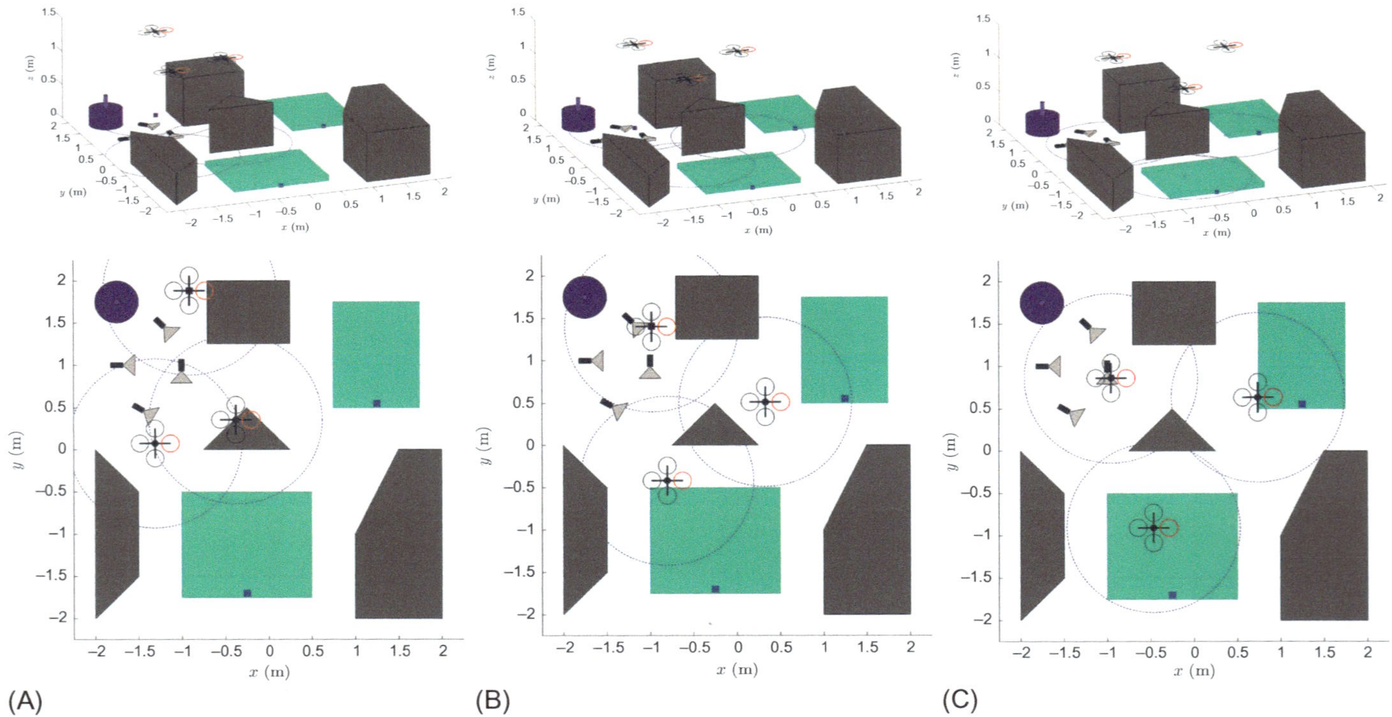

Fig. 5 Snapshots of the initial deployment of the aerial routers. Time elapsed since the start of the simulation (A) 1 s, (B) 2 s, and (C) 4 s. The three-dimensional views are at the top and the corresponding two-dimensional views are at the bottom. The maneuver finishes around 4 s after the start of the simulation.

neighbors closer to μ_r while they go toward their preassigned reference positions in the target regions, specially for the case of r_2 and r_3. Fig. 5C shows the rest position that the aerial relays achieve.

Once the deployment of the aerial relays is complete, the mobile sensors are commanded to start moving around $\mathcal{W}$ such that they get measurements of the target regions τ_1 and τ_2 by applying the control law (17). Also, the aerial relays start using the learning algorithm together with their controller (18) to enhance the network connectivity. Two of the mobile sensors are assigned to τ_1 and the others are assigned to τ_2. Once a sensor gets to its predefined target region, it keeps moving around within the region. The flying relays do not use the learning algorithm (Algorithm 15.1) at every time step. Instead, Algorithm 15.1 is called every 0.25 s. We do this to allow the aerial relays to have enough time to move to the reference position $\mathbf{q}_{\text{ref}}$ computed on the basis of this algorithm. Fig. 6 shows different time instants of the simulation for the case of independent learning (ie, $w = 1$). For each relay we indicate which antenna has the maximum RSS by changing to red the circle at the end of its corresponding arm.

From Fig. 6, we see the mobile sensors move toward their preassigned targets, evading obstacles and collisions with other sensors. At the same time, the aerial relays select the antenna with the maximum RSS and change its position according to the reference position computed on the basis of Algorithm 15.1. In Fig. 6B one of the mobile sensors assigned to τ_1 is out of the range of the relays. Indeed, it is not inside any of the blue dashed circles associated with each flying relay. This causes the network to be disconnected, making the relative connectivity index, defined by Eq. (25), be less than 1. Furthermore, the evaluation of this index during the simulation experiments is shown in Fig. 7A for the case of independent learning and in Fig. 7B for cooperative learning. As one can see in Fig. 7A and B, the network connectivity is lost more times when the independent learning method is applied. For example, two mobile sensors are out of the communication range of the relays in Fig. 6C, decreasing the RSS and then affecting the relative connectivity index c. Even though c is not always 1 for cooperative learning (meaning that the network is always connected), we see from our tests that in the worst-case two of the mobile sensors are disconnected for a short time. This is not the case for independent learning, where at some time instant three of the four sensors were disconnected.

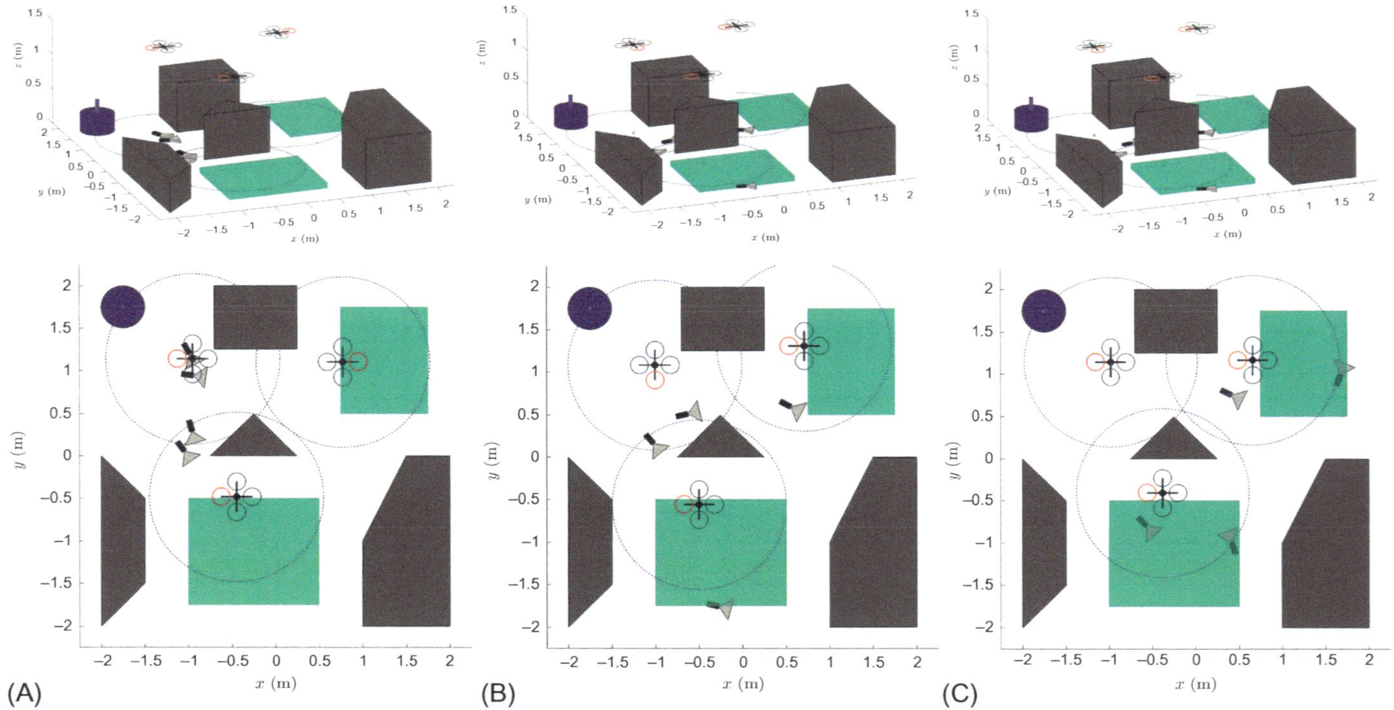

Fig. 6 Snapshots of the simulation experiment under the assumption of independent learning (ie, $w = 1$). Time elapsed since the start of the simulation (A) 6 s, (B) 10 s, and (C) 20 s. The three-dimensional views are at the top and the corresponding two-dimensional view are at the bottom. In (B) and (C) the network is disconnected since one or more mobile sensors are out of the range of the mobile routers.

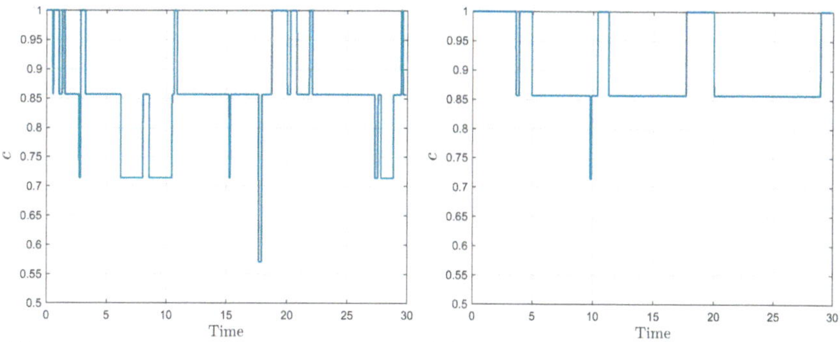

Fig. 7 Relative connectivity index *c* given by (25): (A) for the case of independent learning (ie, $w = 1$) and (B) for the case of cooperative learning (ie, $0 < w < 1$). We set $w = 0.5$ to simulate the latter case.

8 CONCLUSIONS

We took advantage of the different dynamics, resources, and capabilities of a heterogeneous robotic network made of ground and aerial agents to explore a cluttered environment. Exploiting heterogeneity allows us to effectively use the agents' specialized abilities and therefore accomplish subgoals more efficiently, which in turn results in successful completion of the overall mission. By use of potential field methods, the ground mobile sensors were coordinated to sense regions of interest in the environment, avoiding collisions. At the same time, we combined antenna diversity and cooperative Q-learning to maintain a certain RSS level between the sensors and the aerial communication relays. We analyzed the effectiveness of our proposed method in a simulation scenario where we considered for the relays the cases of independent and cooperative learning. Our results showed that the network achieved a better relative connectivity index when the learning was performed in a cooperative fashion.

Since the connectivity of the network is not maintained all the time even when the cooperative method is employed, our future work will consist in expanding the learning method to the mobile sensors. Thus they will be able to react when their RSS drops under a minimum communication level. In our proposed method, just the aerial relays are sensing and maintaining the RSS between themselves and the sensors. We believe that adding the consideration of the minimum RSS communication level to the mobile sensor controller will help us get better results for the whole network connectivity.

ACKNOWLEDGMENTS

This work was supported in part by the Army Research Lab Micro Autonomous Systems and Technology Collaborative Alliance (ARL MAST-CTA #W911NF-08-2-0004). We would like to thank the Ecuadorian scholarship program administrated by the Secretaría de Educación Superior, Ciencia, Tecnología e Innovación (SENESCYT) for providing part of the financial support for P. J. Cruz.

REFERENCES

[1] J. Fink, A. Ribeiro, V. Kumar, Robust control of mobility and communications in autonomous robot teams, IEEE Access 1 (2013) 290–309.

[2] A. Cortez, R. Fierro, J. Wood, Connectivity maintenance of a heterogeneous sensor network, in: A. Martinoli, F. Mondada, N. Correll, G. Mermoud, M. Egerstedt, M.A. Hsieh, L.E. Parker, K. Støy (Eds.), Distributed Autonomous Robotic Systems, Springer Tracts in Advanced Robotics, vol. 83, Springer, Berlin, Heidelberg, 2013, pp. 33–46.

[3] P. Cruz, R. Fierro, W. Lu, S. Ferrari, Maintaining robust connectivity in heterogeneous robotic networks, in: Proc. SPIE, vol. 8741, Unmanned Systems Technology XV, 2013, pp. 87410N–87410N-15.

[4] H.M. La, R. Lim, W. Sheng, Multirobot cooperative learning for predator avoidance, IEEE Trans. Control Syst. Technol. 23 (1) (2015) 52–63.

[5] G.A.S. Pereira, A.K. Das, V. Kumar, M.F.M. Campos, Decentralized motion planning for multiple robots subject to sensing and communication constraints, in: Proceedings of the Second MultiRobot Systems Workshop, Kluwer Academic Press, Dordrecht, 2003, pp. 267–278.

[6] J. Esposito, T. Dunbar, Maintaining wireless connectivity constraints for swarms in the presence of obstacles, in: Proceedings IEEE International Conference on Robotics and Automation, ICRA, 2006, pp. 946–951.

[7] M.N. Rooker, A. Birk, Multi-robot exploration under the constraints of wireless networking, Control Eng. Pract. 15 (4) (2007) 435–445.

[8] M.A. Hsieh, A. Cowley, V. Kumar, C.J. Taylor, Maintaining network connectivity and performance in robot teams, J. Field Rob. 25 (1–2) (2008) 111–131.

[9] S. Gil, M. Schwager, B.J. Julian, D. Rus, Optimizing communication in air-ground robot networks using decentralized control, in: Proceedings IEEE International Conference on Robotics and Automation, ICRA, 2010, pp. 1964–1971.

[10] N. Bezzo, M.S. Anderson, R. Fierro, J. Wood, A real world coordination framework for connected heterogeneous robotic systems, in: M. Ani Hsieh, G. Chirikjian (Eds.), Distributed Autonomous Robotic Systems, Springer Tracts in Advanced Robotics, vol. 104, Springer, Berlin, Heidelberg, 2014, pp. 75–89.

[11] P. Robuffo Giordano, A. Franchi, C. Secchi, H.H. Bülthoff, A passivity-based decentralized strategy for generalized connectivity maintenance, Int. J. Robot. Res. 32 (3) (2013) 299–323.

[12] R. Mahony, V. Kumar, P. Corke, Multirotor aerial vehicles: modeling, estimation, and control of quadrotor, IEEE Robot. Autom. Mag. 19 (3) (2012) 20–32.

[13] A. Pullin, N. Kohut, D. Zarrouk, R. Fearing, Dynamic turning of 13 cm robot comparing tail and differential drive, in: Proceedings IEEE International Conference on Robotics and Automation, ICRA, 2012, pp. 5086–5093.

[14] Y. Mostofi, M. Malmirchegini, A. Ghaffarkhah, Estimation of communication signal strength in robotic networks, in: Proceedings IEEE International Conference on Robotics and Automation, ICRA, 2010, pp. 1946–1951.

[15] J. Twigg, J. Fink, P. Yu, B. Sadler, RSS gradient-assisted frontier exploration and radio source localization, in: Proceedings IEEE International Conference on Robotics and Automation, ICRA, 2012, pp. 889–895.
[16] G. Tuna, V.C. Gungor, K. Gulez, An autonomous wireless sensor network deployment system using mobile robots for human existence detection in case of disasters, Ad Hoc Netw. 13 (Part A) (2014) 54–68, (1) Special Issue: Wireless Technologies for Humanitarian Relief & (2) Special Issue: Models and Algorithms for Wireless Mesh Networks.
[17] A. Goldsmith, Wireless Communications, Cambridge University Press, Cambridge, 2005.
[18] J. Nelson, J. Almodovar, M. Gupta, W. Mortensen, Estimating multiple transmitter locations from power measurements at multiple receivers, in: IEEE International Conference on Acoustics, Speech and Signal Processing, ICASSP, 2009, pp. 2761–2764.
[19] B. Griffin, R. Fierro, I. Palunko, An autonomous communications relay in GPS-denied environments via antenna diversity, J. Def. Model. Simul. Appl. Methodol. Technol. 9 (1) (2012) 33–44.
[20] G. Agnarsson, R. Greenlaw, Graph Theory: Modeling, Applications, and Algorithms, Prentice-Hall, Inc., Upper Saddle River, NJ, USA, 2006.
[21] C. Godsil, G. Royle, Algebraic Graph Theory, in: Graduate Texts in Mathematics, Springer, New York, 2001.
[22] J. Latombe, Robot Motion Planning, Kluwer Academic Publishers, Dordrecht, 1991.
[23] L. Chengqing, V. Ang, H. Krishnan, S.Y. Lim, Virtual obstacle concept for local-minimum-recovery in potential-field based navigation, in: IEEE International Conference on Robotics and Automation, ICRA, vol. 2, 2000, pp. 983–988.
[24] S. Ge, Y. Cui, New potential functions for mobile robot path planning, IEEE Trans. Robot. Autom. 16 (5) (2000) 615–620.
[25] F. Lewis, D. Liu (Eds.), Reinforcement Learning and Approximate Dynamic Programming for Feedback Control, IEEE Press Series on Computational Intelligence, Wiley, New York, 2012.
[26] L. Buşoniu, R. Babuška, B. De Schutter, A comprehensive survey of multiagent reinforcement learning, IEEE Trans. Syst. Man Cybern. C: Appl. Rev. 38 (2) (2008) 156–172.
[27] H. Guo, Y. Meng, Distributed reinforcement learning for coordinate multi-robot foraging, J. Intell. Robot. Syst. 60 (3–4) (2010) 531–551.
[28] R.S. Sutton, A.G. Barto, Reinforcement Learning: An Introduction, first ed., MIT Press, Cambridge, MA, USA, 1998.
[29] D.P. Bertsekas, Dynamic Programming and Optimal Control, vol. I and vol. II, third ed., Athena Scientific, Belmont, MA, 2005 (vol. I) and 2012 (vol. II).
[30] L. Buşoniu, R. Babuška, B.D. Schutter, D. Ernst, Reinforcement Learning and Dynamic Programming Using Function Approximators, first ed., CRC Press, Inc., Boca Raton, FL, USA, 2010.

CHAPTER SIXTEEN

Discrete-Time Flocking of Wheeled Vehicles With a Large Communication Delay Through a Potential Function Approach[a]

H. Hu*, S.Y. Yoon†, Z. Lin*
*University of Virginia, Charlottesville, VA, United States
†University of New Hampshire, Durham, NH, United States

1 INTRODUCTION

Collective motion of agents, such as flocking of birds and schooling of fishes, in a large community with a common objective is easy to observe in nature. This cooperative behavior has been studied extensively by scientists and engineers for its potential in applications such as coordination of robots, cooperation of unmanned air vehicles, and traffic control.

Basic coordination problems that have been widely studied include flocking, consensus, formation control, and swarm tracking. The objective of formation control is to design a controller that brings all agents to their desired relative positions, which are specified by a predefined formation geometry. A formation control law was proposed for linear systems with communication delay in [1], and the same problem was studied for a class of nonlinear agents in [2, 3]. Flocking control of agents with linear dynamics was explored in [4]. A potential function-based control law for a class of nonholonomic agents was proposed in [5] and the flocking behavior was theoretically verified. The same system of nonholonomic agents was considered in [6] but communication delays were also accounted for.

[a] This work was supported in part by the National Science Foundation under grants CMMI-1129752 and CMMI-1462171, and in part by the US Air Force Office of Scientific Research under grant FA9550-12-0163.

Control of Complex Systems
http://dx.doi.org/10.1016/B978-0-12-805246-4.00016-1

Artificial potential functions are commonly used to dictate the interaction between the agents in collective motion problems. Differently from formation control methods where the agents' positions relative to a common leader are predetermined, coordinated control based on a potential function does not require a virtual leader and is robust to disruption in the agent ordering. The potential function method mentioned above has been developed in the continuous-time setting. As observed in [7, 8], applications of continuous-time controllers through direct discretization could be very restrictive because of the requirement for small sampling periods. In this work we consider a discrete-time nonholonomic model of wheeled vehicles and develop a potential function-based discrete-time low-gain controller for flocking control in the presence of arbitrarily large delays in the communication shared between the agents.

The remainder of this chapter is organized as follows. After defining the problem in Section 2, we introduce some preliminary results in Section 3. A distributed control law for each vehicle and the stability analysis of the closed-loop system are presented in Section 4. In Section 5 the effectiveness of the proposed control laws is verified by numerical simulation. Finally, conclusions are drawn in Section 6.

2 PROBLEM FORMULATION

We consider a multiagent system consisting of N nonholonomic wheeled vehicles. For $i = 1, 2, \ldots, N$ the continuous-time kinematic model of agent i is given by

$$\begin{aligned} \dot{x}_i(t) &= v_i(t)\cos\theta_i(t), \\ \dot{y}_i(t) &= v_i(t)\sin\theta_i(t), \\ \dot{\theta}_i(t) &= \omega_i(t). \end{aligned} \tag{1}$$

The position of agent i is determined by the vector $r_i(t) = [x_i(t)\ y_i(t)]^{\mathrm{T}}$, and the orientation is given by $\theta_i(t)$. There are two control inputs for each agent: the linear velocity $v_i(t)$ and the angular velocity $\omega_i(t)$. For agents i and j, the relative position is denoted as $r_{ij}(t) = r_i(t) - r_j(t)$. The heading error between agent i and agent j is defined as $e_{ij}(t) = \theta_i(t) - \theta_j(t)$. The dynamics of our wheeled vehicle is illustrated in Fig. 1.

The wheeled vehicle model is discretized through a zero-order hold with the sampling period T. The discrete-time model of system (1) is obtained as follows:

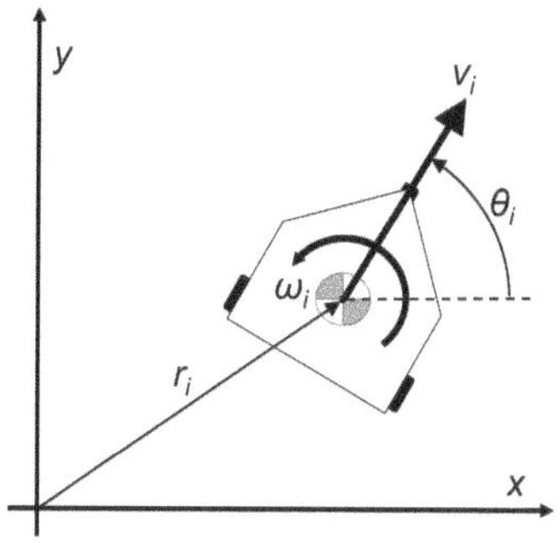

Fig. 1 Dynamics of agent *i*.

$$\begin{aligned} x_i(k+1) &= x_i(k) + Tv_i(k)\cos\theta_i(k), \\ y_i(k+1) &= y_i(k) + Tv_i(k)\sin\theta_i(k), \\ \theta_i(k+1) &= \theta_i(k) + T\omega_i(k), \end{aligned} \tag{2}$$

where the discrete states and inputs are defined as $x(k) = x(kT)$, $y(k) = y(kT)$, $\theta(k) = \theta(kT)$, $v_i(k) = v(kT)$, and $\omega(k) = \omega(kT)$, respectively.

We assume that all agents have their states and inputs updated simultaneously, and so have the information on the relative position and heading error of their neighboring agents. We also assume that the magnitude of the communication delay between agents is a multiple of the sampling period (ie, τT, where $\tau \in \mathbb{N}$). In the following sections, we propose for each vehicle a potential function-based low-gain control law which creates the flocking behavior of the multiagent system.

3 PRELIMINARIES

3.1 Graph Theory

The coordinated control of multiagent systems, including flocking, requires the exchange of information among the agents. In this work we represent the communication network among the agents by an undirected graph.

Definition 1. An undirected graph $\mathcal{G} = \{\mathcal{V}, \mathcal{E}\}$ consists of a nonempty set of nodes $\mathcal{V} = \{v_1, v_2, \ldots, v_N\}$, each representing an agent, and a set of edges $\mathcal{E} \subset \mathcal{V} \times \mathcal{V}$. The unordered pair $(v_i, v_j) \in \mathcal{E}$ if and only if there exists a bidirectional communication link between node v_i and node v_j. Let $\mathcal{N}_i$ denote the index set of the neighboring nodes of v_i—that is,

$$\mathcal{N}_i = \{1 \leq j \leq N : (v_i, v_j) \in \mathcal{E}\}.$$

Definition 2. The topology of a graph $\mathcal{G}$ can be described by an associated adjacency matrix $A_{\text{adj}} = [a_{ij}] \in \mathbb{R}^{N\times N}$, which is defined as

$$a_{ij} = a_{ji} = \begin{cases} 1, & \text{if } (v_i, v_j) \in \mathcal{E}, \\ 0, & \text{otherwise.} \end{cases}$$

We define the Laplacian matrix of $\mathcal{G}$ as $L = D - A_{\text{adj}}$, where $D = \text{diag}\{d_1, d_2, \ldots, d_N\}$ and $d_i = \sum_{j=1}^{N} a_{ij}, i = 1, 2, \ldots, N$. For an undirected graph, $L = L^{\text{T}}$.

Without loss of generality, we assume that the communication network among agents does not allow self-loops. In other words, $(v_i, v_i) \notin \mathcal{E}$, and therefore $a_{ii} = 0$ for $i = 1, 2, \ldots, N$ throughout this chapter.

Lemma 1. *[9]. An undirected graph is connected if and only if zero is a simple eigenvalue of the Laplacian matrix L with* $\bar{\mathbf{1}}_N = [1, 1, \ldots, 1]^T \in \mathbb{R}^N$ *being the only corresponding eigenvector.*

3.2 Artificial Potential Functions

Suppose that there are attractive/repulsive forces between each pair of neighboring agents. With the help of the paired forces, neighboring agents maintain a desired distance. As in physics, these artificial forces are determined by the artificial potential functions. We refer the reader to [10] for more details about the design of artificial potential functions.

A flocking control law was proposed in [6] on the basis of the following potential function:

$$V_{ij}(\|r_{ij}\|) = \rho^2 \left(\frac{1}{\left\| \frac{1}{\rho} r_{ij} \right\|^2} + \log\left(\left\| \frac{1}{\rho} r_{ij} \right\|^2 \right) \right), \tag{3}$$

which represents the potential between agent i and j as plotted in Fig. 2.

It is observed in Fig. 2 that the potential function (3) has its minimum value at $\|r_{ij}\| = \rho$, where ρ is defined as the desired distance between agents. In addition, the value of the function approaches infinity as $\|r_{ij}\|$ goes to 0. This feature guarantees collision avoidance between neighboring agents, which will be explained and demonstrated in later sections. We define the total potential of an agent i as

$$V_i = \sum_{j \in \mathcal{N}_i} V_{ij}(\|r_{ij}\|).$$

It was proved in [6] that for an arbitrarily large communication delay τ, the continuous-time potential function-based low-gain control law

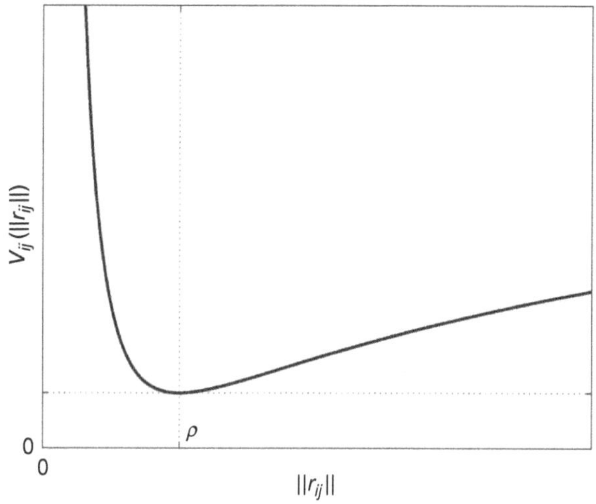

Fig. 2 Artificial potential function.

$$v_i(t) = -\varepsilon[\cos\theta_i(t)\ \sin\theta_i(t)]\nabla_{r_i} V_i(t-\tau),$$
$$\omega_i(t) = -\varepsilon\kappa \sum_{j\in\mathcal{N}_i} e_{ij}(t-\tau) - \varepsilon[-\sin\theta_i(t)\ \cos\theta_i(t)]\nabla_{r_i} V_i(t-\tau)$$

steers the continuous-time system (1) into a formation corresponding to a minimum of $\sum_{i=0}^{N} V_i$ and a common orientation, as long as the value of the low-gain parameter ε is sufficiently small.

4 MAIN RESULTS

We make the following assumptions regarding the communication network and the initial conditions.

Assumption 1. The communication topology of the system (2) is described by an undirected and connected graph $\mathcal{G} = \{\mathcal{V}, \mathcal{E}\}$, where $\mathcal{V}$ is the indexed set of agents in the system and $\mathcal{E}$ represents the communication links among the agents.

Assumption 2. For an arbitrary positive constant c, let the initial conditions of system (2) for $k = -\tau, -\tau+1, \ldots, 0$ be inside the bounded set

$$\Omega = \left\{(r_{ij}, e_{ij}) : \sum_{i=0}^{N} V_i \leq c,\ -2\pi < e_{ij} < 2\pi\right\},$$

where the bound on e_{ij} comes from the condition that $-\pi < \theta_i \leq \pi$.

We propose the following discrete-time potential function-based low-gain control law for each agent i:

$$\begin{aligned}
v_i(k) &= -\varepsilon\,[\cos\theta_i(k)\ \ \sin\theta_i(k)]\,\nabla_{r_i} V_i(k-\tau), \\
\tilde{\omega}_i(k) &= -\varepsilon\kappa \sum_{j\in\mathcal{N}_i} e_{ij}(k-\tau) - \varepsilon\,[-\sin\theta_i(k)\ \ \cos\theta_i(k)]\,\nabla_{r_i} V_i(k-\tau), \\
\omega_i(k) &= \operatorname{sign}(\tilde{\omega}_i(k)) \min\left\{|\tilde{\omega}_i(k)|, \frac{\pi}{2T}\right\}.
\end{aligned} \tag{4}$$

The following lemmas will be needed to develop our main results.

Lemma 2. *[11]. Consider a set of real vectors a_i, $i = 1, 2, \ldots, m$. Then*

$$\left(\sum_{i=1}^{m} a_i\right)^T \left(\sum_{i=1}^{m} a_i\right) \le m \sum_{i=1}^{m} a_i^T a_i.$$

Lemma 3. *Define a new discrete-time function in terms of any given discrete-time function $f(k)$ as follows:*

$$V(k) = \sum_{s_1=0}^{\tau} \sum_{s_2=k-s_1}^{k-1} f(s_2),$$

where $\tau \in \mathbb{N}$ is a given constant. Then the variation of $V(k)$ is given by

$$V(k+1) - V(k) = \tau f(k) - \sum_{s_2=k-\tau}^{k-1} f(s_2).$$

Proof. Reordering the summation indices, we can rewrite $V(k)$ as

$$V(k) = \sum_{s_2=k-\tau}^{k-1} \sum_{s_1=k-s_2}^{\tau} f(s_2). \tag{5}$$

Since the function f in Eq. (5) does not depend on s_1, Eq. (5) can be simplified as

$$\begin{aligned}
V(k) &= \sum_{s_2=k-\tau}^{k-1} (\tau - k + s_2 + 1) f(s_2), \\
&= (\tau - k + 1) \sum_{s_2=k-\tau}^{k-1} f(s_2) + \sum_{s_2=k-\tau}^{k-1} s_2 f(s_2).
\end{aligned}$$

Then the variation of $V(k)$ is computed as

$$\begin{aligned} V(k+1)-V(k) &= (\tau-k)\sum_{s_2=k-\tau+1}^{k} f(s_2) - (\tau-k+1)\sum_{s_2=k-\tau}^{k-1} f(s_2) \\ &\quad + \sum_{s_2=k-\tau+1}^{k} s_2 f(s_2) - \sum_{s_2=k-\tau}^{k-1} s_2 f(s_2), \\ &= (\tau-k)(f(k)-f(k-\tau)) - \sum_{s_2=k-\tau}^{k-1} f(s_2) \\ &\quad + kf(k)-(k-\tau)f(k-\tau), \\ &= \tau f(k) - \sum_{s_2=k-\tau}^{k-1} f(s_2). \end{aligned} \tag{6}$$

□

The next lemma demonstrates for a sufficiently small value of the low-gain parameter ε that system (2) achieves position aggregation under the control laws (4), even in the presence of an arbitrarily large communication delay τ.

Lemma 4. *Consider multiagent system* (2) *with communication delay* τ. *If Assumptions 1 and 2 are both satisfied, there exists an* $\bar{\varepsilon}_1 > 0$ *such that for any* $0 < \varepsilon < \bar{\varepsilon}_1$ *the low-gain control laws* (4) *drive the system into a formation corresponding to a local minimum of* $\sum_{i=0}^{N} V_i$.

Proof. By a Taylor series expansion we have

$$V_{ij}(k+1) = V_{ij}(k) + \nabla_{r_{ij}}^{\mathrm{T}} V_{ij}(k)(r_{ij}(k+1) - r_{ij}(k)) + R_{ij}(k), \tag{7}$$

where $R_{ij}(k)$ is the Lagrange remainder.

By Assumption 2 the initial states of the system are bounded in the set Ω, and the second- and higher-order derivatives of V_{ij} are also bounded. Thus there always exists an $\alpha > 0$ that satisfies

$$R_{ij}(k) \le \alpha (r_{ij}(k+1) - r_{ij}(k))^{\mathrm{T}} (r_{ij}(k+1) - r_{ij}(k)) \tag{8}$$

for all i and j.

Consider the potential function

$$V_{r,1}(k) = \sum_{i=1}^{N} \sum_{j\in\mathcal{N}_i} V_{ij}(k). \tag{9}$$

Combining Eqs. (7) and (8) yields

$$\begin{aligned}\Delta V_{r,1}(k) &= V_{r,1}(k+1) - V_{r,1}(k),\\ &\leq \sum_{i=1}^{N}\sum_{j\in\mathcal{N}_i} \nabla_{r_{ij}}^{\mathrm{T}} V_{ij}(k)(r_{ij}(k+1) - r_{ij}(k))\\ &\quad + \alpha\sum_{i=1}^{N}\sum_{j\in\mathcal{N}_i}(r_{ij}(k+1) - r_{ij}(k))^{\mathrm{T}}(r_{ij}(k+1) - r_{ij}(k)).\end{aligned} \tag{10}$$

In view of Eqs. (2) and (4), we have

$$r_{ij}(k+1) = r_{ij}(k) - \varepsilon T\Phi_i(k)\nabla_{r_i} V_i(k-\tau) + \varepsilon T\Phi_j(k)\nabla_{r_j} V_j(k-\tau), \tag{11}$$

where $\Phi_i(k)$ is a positive semidefinite matrix and is defined as

$$\Phi_i(k) = \begin{bmatrix}\cos\theta_i(k)\\ \sin\theta_i(k)\end{bmatrix}\begin{bmatrix}\cos\theta_i(k) & \sin\theta_i(k)\end{bmatrix}.$$

For simplicity, we introduce the following notation:

$$\nabla_{r_i} V_{ij}(k) = 2r_{ij}(k)\Pi_{ij}(k), \tag{12}$$

where

$$\Pi_{ij}(k) = \frac{\rho^2}{\|r_{ij}(k)\|^2} - \frac{\rho^4}{\|r_{ij}(k)\|^4}. \tag{13}$$

Substituting Eqs. (11) and (12) into Eq. (10) yields

$$\begin{aligned}\Delta V_{r,1}(k) &\leq \sum_{i=1}^{N}\sum_{j\in\mathcal{N}_i} 2r_{ij}^{\mathrm{T}}(k)\Pi_{ij}(k)\\ &\times\Bigg(-2\varepsilon T\Phi_i(k)\sum_{l\in\mathcal{N}_i} r_{il}(k-\tau)\Pi_{il}(k-\tau) + 2\varepsilon T\Phi_j(k)\\ &\times\sum_{l\in\mathcal{N}_j} r_{jl}(k-\tau)\Pi_{jl}(k-\tau)\Bigg) + \alpha\sum_{i=1}^{N}\sum_{j\in\mathcal{N}_i} 2\varepsilon T\\ &\times\Bigg(-\Phi_i(k)\sum_{l\in\mathcal{N}_i} r_{il}(k-\tau)\Pi_{il}(k-\tau) + \Phi_j(k)\sum_{l\in\mathcal{N}_j} r_{jl}(k-\tau)\Pi_{jl}(k-\tau)\Bigg)^{\mathrm{T}}\end{aligned}$$

$$\times 2\varepsilon T\left(-\Phi_i(k)\sum_{l\in\mathcal{N}_i} r_{il}(k-\tau)\Pi_{il}(k-\tau)\right.$$
$$\left.+\Phi_j(k)\sum_{l\in\mathcal{N}_j} r_{jl}(k-\tau)\Pi_{jl}(k-\tau)\right).$$

Applying Lemma 2 to the second term on the right-hand side of the above inequality and reordering summation indices gives

$$\begin{aligned}\Delta V_{r,1}(k) \le &-8\varepsilon T\sum_{i=1}^{N}\left(\sum_{j\in\mathcal{N}_i} r_{ij}(k)\Pi_{ij}(k)\right)^{\mathrm{T}}\Phi_i(k)\left(\sum_{j\in\mathcal{N}_i} r_{ij}(k-\tau)\Pi_{ij}(k-\tau)\right)\\ &+16\alpha\varepsilon^2T^2 d_{\max}\sum_{i=1}^{N}\left(\sum_{j\in\mathcal{N}_i} r_{ij}(k-\tau)\Pi_{ij}(k-\tau)\right)^{\mathrm{T}}\Phi_i^2(k)\\ &\times\left(\sum_{j\in\mathcal{N}_i} r_{ij}(k-\tau)\Pi_{ij}(k-\tau)\right),\end{aligned} \tag{14}$$

where $d_{\max} = \max_{i=1,2,\ldots,N}\{d_i\}$ and d_i is defined in Definition 2. By the mean value theorem we have

$$r_{ij}(k)\Pi_{ij}(k) = r_{ij}(k-\tau)\Pi_{ij}(k-\tau) + \Lambda(\hat{r}_{ij})(r_{ij}(k) - r_{ij}(k-\tau)), \tag{15}$$

where $\hat{r}_{ij} = r_{ij}(k-\tau) + c_{ij}(r_{ij}(k) - r_{ij}(k-\tau))$, for some $c_{ij} \in (0,1)$, and

$$\Lambda(\hat{r}_{ij}) = \left(\frac{\rho^2}{\|\hat{r}_{ij}\|^2} - \frac{\rho^4}{\|\hat{r}_{ij}\|^4}\right)I + 2\left(\frac{2\rho^4}{\|\hat{r}_{ij}\|^6} - \frac{\rho^2}{\|\hat{r}_{ij}\|^4}\right)\hat{r}_{ij}\hat{r}_{ij}^{\mathrm{T}}.$$

We define $\lambda_{ij}(k) = \Lambda(\hat{r}_{ij})(r_{ij}(k+1) - r_{ij}(k))$, and then Eq. (15) can be rewritten as

$$r_{ij}(k)\Pi_{ij}(k) = r_{ij}(k-\tau)\Pi_{ij}(k-\tau) + \sum_{\sigma=k-\tau}^{k-1}\lambda_{ij}(\sigma). \tag{16}$$

Substituting Eq. (16) into Eq. (14), we obtain

$$\begin{aligned}\Delta V_{r,1}(k) \le &-8\varepsilon T\sum_{i=1}^{N}\left(\sum_{j\in\mathcal{N}_i} r_{ij}(k-\tau)\Pi_{ij}(k-\tau)\right)^{\mathrm{T}}\Phi_i(k)\\ &\times\left(\sum_{j\in\mathcal{N}_i} r_{ij}(k-\tau)\Pi_{ij}(k-\tau)\right)\end{aligned}$$

$$
\begin{aligned}
&- 8\varepsilon T \sum_{i=1}^{N} \left(\sum_{j\in\mathcal{N}_i} \sum_{\sigma=k-\tau}^{k-1} \lambda_{ij}(\sigma) \right)^{\mathrm{T}} \Phi_i(k) \left(\sum_{j\in\mathcal{N}_i} r_{ij}(k-\tau)\Pi_{ij}(k-\tau) \right) \\
&+ 16\alpha\varepsilon^2 T^2 d_{\max} \sum_{i=1}^{N} \left(\sum_{j\in\mathcal{N}_i} r_{ij}(k-\tau)\Pi_{ij}(k-\tau) \right)^{\mathrm{T}} \Phi_i^2(k) \\
&\times \left(\sum_{j\in\mathcal{N}_i} r_{ij}(k-\tau)\Pi_{ij}(k-\tau) \right).
\end{aligned}
$$

Applying Lemma 2 to the second term on the right-hand side of the above inequality gives

$$
\begin{aligned}
\Delta V_{r,1}(k) \leq &-8\varepsilon T \sum_{i=1}^{N} \left(\sum_{j\in\mathcal{N}_i} r_{ij}(k-\tau)\Pi_{ij}(k-\tau) \right)^{\mathrm{T}} \Phi_i(k) \\
&\times \left(\sum_{j\in\mathcal{N}_i} r_{ij}(k-\tau)\Pi_{ij}(k-\tau) \right) \\
&+ 16\varepsilon^2 T^2 \sum_{i=1}^{N} \left(\sum_{j\in\mathcal{N}_i} r_{ij}(k-\tau)\Pi_{ij}(k-\tau) \right)^{\mathrm{T}} \Phi_i^2(k) \\
&\times \left(\sum_{j\in\mathcal{N}_i} r_{ij}(k-\tau)\Pi_{ij}(k-\tau) \right) \\
&+ 16\alpha\varepsilon^2 T^2 d_{\max} \sum_{i=1}^{N} \left(\sum_{j\in\mathcal{N}_i} r_{ij}(k-\tau)\Pi_{ij}(k-\tau) \right)^{\mathrm{T}} \Phi_i^2(k) \\
&\times \left(\sum_{j\in\mathcal{N}_i} r_{ij}(k-\tau)\Pi_{ij}(k-\tau) \right) \\
&+ \sum_{i=1}^{N} \left(\sum_{\sigma=k-\tau}^{k-1} \sum_{j\in\mathcal{N}_i} \lambda_{ij}(\sigma) \right)^{\mathrm{T}} \left(\sum_{\sigma=k-\tau}^{k-1} \sum_{j\in\mathcal{N}_i} \lambda_{ij}(\sigma) \right).
\end{aligned}
$$

Lemma 2 can be applied again to the last term on the right-hand side of the above inequality to give

$$\begin{aligned}\Delta V_{r,1}(k) \le & -8\varepsilon T\sum_{i=1}^{N}\left(\sum_{j\in\mathcal{N}_i} r_{ij}(k-\tau)\Pi_{ij}(k-\tau)\right)^{\mathrm{T}} \\ & \times\left(\Phi_i(k)-2\alpha\varepsilon T d_{\max}\Phi_i^2(k)-2\varepsilon T\Phi_i^2(k)\right) \\ & \times\left(\sum_{j\in\mathcal{N}_i} r_{ij}(k-\tau)\Pi_{ij}(k-\tau)\right)+\tau\sum_{i=1}^{N}\sum_{\sigma=k-\tau}^{k-1}\left(\sum_{j\in\mathcal{N}_i}\lambda_{ij}(\sigma)\right)^{\mathrm{T}} \\ & \times\left(\sum_{j\in\mathcal{N}_i}\lambda_{ij}(\sigma)\right).\end{aligned} \tag{17}$$

Consider a second potential function

$$V_{r,2}(k)=\tau\sum_{i=1}^{N}\sum_{s_1=0}^{\tau}\sum_{s_2=k-s_1}^{k-1}\left(\sum_{j\in\mathcal{N}_i}\lambda_{ij}(s_2)\right)^{\mathrm{T}}\left(\sum_{j\in\mathcal{N}_i}\lambda_{ij}(s_2)\right).$$

In view of Lemma 3, the variation of $V_{r,2}(k)$ is computed as

$$\begin{aligned}\Delta V_{r,2}(k) &= V_{r,2}(k+1)-V_{r,2}(k), \\ &= \tau^2\sum_{i=1}^{N}\left(\sum_{j\in\mathcal{N}_i}\lambda_{ij}(k)\right)^{\mathrm{T}}\left(\sum_{j\in\mathcal{N}_i}\lambda_{ij}(k)\right) \\ &\quad -\tau\sum_{i=1}^{N}\sum_{\sigma=k-\tau}^{k-1}\left(\sum_{j\in\mathcal{N}_i}\lambda_{ij}(\sigma)\right)^{\mathrm{T}}\left(\sum_{j\in\mathcal{N}_i}\lambda_{ij}(\sigma)\right).\end{aligned} \tag{18}$$

Finally, we define a proper Lyapunov function $V_r(k) = V_{r,1}(k)+V_{r,2}(k)$, the variation of which can be easily determined by combination of Eqs. (17) and (18) as follows:

$$\begin{aligned}\Delta V_r(k) \le & -8\varepsilon T\sum_{i=1}^{N}\left(\sum_{j\in\mathcal{N}_i} r_{ij}(k-\tau)\Pi_{ij}(k-\tau)\right)^{\mathrm{T}} \\ & \times\left(\Phi_i(k)-2\alpha\varepsilon T d_{\max}\Phi_i^2(k)-2\varepsilon T\Phi_i^2(k)\right) \\ & \times\left(\sum_{j\in\mathcal{N}_i} r_{ij}(k-\tau)\Pi_{ij}(k-\tau)\right)+\tau^2\sum_{i=1}^{N}\left(\sum_{j\in\mathcal{N}_i}\lambda_{ij}(k)\right)^{\mathrm{T}}\left(\sum_{j\in\mathcal{N}_i}\lambda_{ij}(k)\right).\end{aligned}$$

In view of Lemma 2, the above inequality implies that

$$\Delta V_r(k) \leq -8\varepsilon T \sum_{i=1}^{N} \left(\sum_{j\in\mathcal{N}_i} r_{ij}(k-\tau)\Pi_{ij}(k-\tau) \right)^{\mathrm{T}} \times \left(\Phi_i(k) - 2\alpha\varepsilon T d_{\max} \Phi_i^2(k) - 2\varepsilon T \Phi_i^2(k) \right) \times \left(\sum_{j\in\mathcal{N}_i} r_{ij}(k-\tau)\Pi_{ij}(k-\tau) \right) + \tau^2 d_{\max} \sum_{i=1}^{N} \sum_{j\in\mathcal{N}_i} \lambda_{ij}^{\mathrm{T}}(k)\lambda_{ij}(k). \tag{19}$$

With the aid of Eq. (11), $\lambda_{ij}(k)$ can be rewritten as

$$\lambda_{ij}(k) = 2\varepsilon T \Lambda(\hat{r}_{ij}) \left(-\Phi_i(k) \sum_{l\in\mathcal{N}_i} r_{il}(k-\tau)\Pi_{il}(k-\tau) + \Phi_j(k) \sum_{l\in\mathcal{N}_j} r_{jl}(k-\tau)\Pi_{jl}(k-\tau) \right),$$

and Lemma 2 can be used to demonstrate that

$$\lambda_{ij}^{\mathrm{T}}(k)\lambda_{ij}(k) \leq 8\varepsilon^2 T^2 \left(\sum_{l\in\mathcal{N}_i} r_{il}(k-\tau)\Pi_{il}(k-\tau) \right)^{\mathrm{T}} \Phi_i(k)\Lambda^2(\hat{r}_{ij})\Phi_i(k) \times \left(\sum_{l\in\mathcal{N}_i} r_{il}(k-\tau)\Pi_{il}(k-\tau) \right) + 8\varepsilon^2 T^2 \left(\sum_{l\in\mathcal{N}_j} r_{jl}(k-\tau)\Pi_{jl}(k-\tau) \right)^{\mathrm{T}} \Phi_j(k)\Lambda^2(\hat{r}_{ij})\Phi_j(k) \times \left(\sum_{l\in\mathcal{N}_j} r_{jl}(k-\tau)\Pi_{jl}(k-\tau) \right). \tag{20}$$

By Assumption 2 there exists some constant $\eta > 0$ such that $\eta I \geq \Lambda^2(\hat{r}_{ij})$ for any i and $j \neq i$. After summing both sides of Eq. (20) over i and j, we deduce that

$$\sum_{i=1}^{N}\sum_{j\in\mathcal{N}_i}\lambda_{ij}^{\mathrm{T}}(k)\lambda_{ij}(k)\le 16\varepsilon^2T^2\eta d_{\max}\sum_{i=1}^{N}\left(\sum_{j\in\mathcal{N}_i}r_{ij}(k-\tau)\Pi_{ij}(k-\tau)\right)^{\mathrm{T}}\Phi_i^2(k)$$
$$\times\left(\sum_{j\in\mathcal{N}_i}r_{ij}(k-\tau)\Pi_{ij}(k-\tau)\right). \tag{21}$$

Substituting Eq. (21) into Eq. (19) together with the fact that $\Phi_i^2(k) = \Phi_i(k)$ is positive semidefinite, we obtain

$$\Delta V_r(k) \le -8\varepsilon T\left(1-2\alpha\varepsilon T d_{\max}-2\varepsilon T-2\varepsilon T\eta\tau^2 d_{\max}^2\right)$$
$$\times\sum_{i=1}^{N}\left(\sum_{j\in\mathcal{N}_i}r_{ij}(k-\tau)\Pi_{ij}(k-\tau)\right)^{\mathrm{T}}\Phi_i(k)$$
$$\times\left(\sum_{j\in\mathcal{N}_i}r_{ij}(k-\tau)\Pi_{ij}(k-\tau)\right). \tag{22}$$

We define the following constant:

$$\bar{\varepsilon}_1 = \frac{1}{2T(1+\alpha d_{\max}+\eta\tau^2 d_{\max}^2)}. \tag{23}$$

Then it follows that $\Delta V_r(k) \le 0$ for any $0 < \varepsilon < \bar{\varepsilon}_1$.

From Eq. (22) we can easily see that $\Delta V_r(k) \equiv 0$ if and only if

$$\left[\cos\theta_i(k)\ \ \sin\theta_i(k)\right]\sum_{j\in\mathcal{N}_i}r_{ij}(k-\tau)\Pi_{ij}(k-\tau)\equiv 0, \tag{24}$$

for all agents i and all $k \ge 0$, and hence the following is true as well:

$$\left[\cos\theta_i(k+1)\ \ \sin\theta_i(k+1)\right]\sum_{j\in\mathcal{N}_i}r_{ij}(k+1-\tau)\Pi_{ij}(k+1-\tau)\equiv 0. \tag{25}$$

Eq. (24) indicates that $v_i(k) = 0$ for all agents i. Any complete trajectory of system (2) satisfying Eq. (24) also possesses the property that $r_{ij}(k+1) = r_{ij}(k)$. Therefore we have

$$\sum_{j\in\mathcal{N}_i}r_{ij}(k+1-\tau)\Pi_{ij}(k+1-\tau)=\sum_{j\in\mathcal{N}_i}r_{ij}(k-\tau)\Pi_{ij}(k-\tau),$$

which, together with Eq. (25), implies that

$$\left[\cos\theta_i(k+1)\ \ \sin\theta_i(k+1)\right]\sum_{j\in\mathcal{N}_i}r_{ij}(k-\tau)\Pi_{ij}(k-\tau)=0, \tag{26}$$

for any agent i. Suppose that $\sum_{j \in \mathcal{N}_i} r_{ij} \Pi_{ij} \neq 0$ for all agents i. Then, after comparing Eqs. (24) and (26), we find that

$$\theta_i(k+1) - \theta_i(k) = m\pi,$$

where m could be any integer. According to the proposed control laws (4), it is true that

$$|\theta_i(k+1) - \theta_i(k)| \leq \frac{\pi}{2}, \quad i = 1, 2, \ldots, N.$$

Consequently, we have $\theta_i(k+1) = \theta_i(k)$ and $\omega_i(k) = 0$ for all agents i.

Finally, following arguments similar to those in the proof of Lemma 1 in [6], we conclude that $\omega_i(k) = 0$ if and only if

$$\sum_{j \in \mathcal{N}_i} r_{ij} \Pi_{ij} = 0 \tag{27}$$

for all agents i. Therefore by LaSalle's invariance principle, system (2) is driven into a formation that satisfies Eq. (27). In the meantime, it can be easily verified from Eq. (12) that any equilibrium satisfying Eq. (27) corresponds to a minimum of $\sum_{i=0}^{N} V_i$. This completes the proof. □

Remark 1. In the above proof we derived a sufficient condition on the low-gain parameter ε for position aggregation in the presence of communication delay τ. For the delay-free cases where $\tau = 0$, the first term in Eq. (14) becomes negative semidefinite and the second term is positive semidefinite. Then without further manipulations we obtain that $\Delta V_{r,1} \leq 0$ when $\varepsilon < \frac{1}{2\alpha T d_{\max}}$.

Lemma 4 proves that system (2) converges to a desired formation under the distributed low-gain control laws (4). In the next two lemmas, it will be shown that heading alignment is also guaranteed.

Lemma 5. *Consider the system*

$$z(k+1) = z(k) + u_1(k),$$

where $z \in \mathbb{R}$ is the state variable and $u_1 \in \mathbb{R}$ is the control input. Let $V(z(k)) = z^2(k)$. Suppose that there exists some feedback control law $u_1(z(k))$ under which

$$\Delta V(k) = V(z(k+1)) - V(z(k)) < 0, \quad z(k) \neq 0.$$

Then,

$$\Delta V(k) < 0, \quad z(k) \neq 0,$$

if we replace $u_1(z(k))$ with $u_2(z(k)) = \operatorname{sign}(u_1(z(k))) \min\{|u_1(z(k))|, \beta\}$, where β is a positive constant.

Proof. Consider a simple Lyapunov function

$$V(k) = z^2(k).$$

Given that the closed-loop system is asymptotically stable if driven by $u_1(k)$, we know that

$$\begin{aligned}\Delta V(k) &= z^2(k+1) - z^2(k),\\ &= 2z(k)u_1(k) + u_1^2(k) < 0, \quad z(k) \neq 0.\end{aligned} \tag{28}$$

The saturated control input $u_2(k)$, $u_2(k) \neq 0$, can be rewritten as

$$u_2(k) = au_1(k),$$
$$a = \frac{\min(|u_1(k)|, \beta)}{|u_1(k)|},$$

and $a \in (0, 1]$. If the system is steered by $u_2(k)$ instead, the variation of $V(k)$ becomes

$$\begin{aligned}\Delta V(k) &= 2z(k)u_2(k) + u_2^2(k),\\ &= 2az(k)u_1(k) + a^2u_1^2(k),\\ &\leq a(2z(k)u_1(k) + u_1^2(k)).\end{aligned}$$

In view of Eq. (28), we have

$$\Delta V(k) < 0, \quad z(k) \neq 0.$$

□

Lemma 6. *Consider multiagent system* (2) *with communication delay* τ. *If Assumptions 1 and 2 are both satisfied, there exists an* $\bar{\varepsilon}_2 > 0$ *such that for any* $0 < \varepsilon < \bar{\varepsilon}_2$ *the low-gain control laws* (4) *align the orientations of all agents as the formation convergence is achieved asymptotically.*

Proof. Consider a positive definite function

$$V_{e,1}(k) = \sum_{i=1}^{N} \sum_{j \in \mathcal{N}_i} e_{ij}^2(k). \tag{29}$$

In view of Lemma 5, it is equivalent to prove that heading alignment will be achieved if the system (2) is steered by $\tilde{\omega}_i$ instead, where $\tilde{\omega}_i$ is defined in Eq. (4). In other words, we consider the following dynamics of orientation errors:

$$\begin{aligned}e_{ij}(k+1) = e_{ij}(k) + \varepsilon T \Bigg(&-\kappa \sum_{l \in \mathcal{N}_i} e_{il}(k-\tau) - \Psi_i(k)\nabla_{r_i} V_i(k-\tau)\\ &+\kappa \sum_{l \in \mathcal{N}_j} e_{jl}(k-\tau) + \Psi_j(k)\nabla_{r_j} V_j(k-\tau) \Bigg),\end{aligned} \tag{30}$$

where to simplify the notation we have we have introduced the definition

$$\Psi_i(k) = \begin{bmatrix} -\sin\theta_i(k) & \cos\theta_i(k) \end{bmatrix}.$$

Let $e_{ij}(k+1) = (e_{ij}(k+1) - e_{ij}(k)) + e_{ij}(k)$. After simple manipulations we obtain the following equality:

$$e_{ij}^2(k+1) = e_{ij}^2(k) + 2e_{ij}(k)(e_{ij}(k+1) - e_{ij}(k)) + (e_{ij}(k+1) - e_{ij}(k))^2. \quad (31)$$

Combining Eqs. (30) and (31) yields

$$\begin{aligned} \Delta V_{e,1}(k) &= V_{e,1}(k+1) - V_{e,1}(k), \\ &= 2\varepsilon T \sum_{i=1}^{N} \sum_{j\in\mathcal{N}_i} e_{ij}(k) \left(-\kappa \sum_{l\in\mathcal{N}_i} e_{il}(k-\tau) - \Psi_i(k)\nabla_{r_i} V_i(k-\tau) \right. \\ &\quad \left. + \kappa \sum_{l\in\mathcal{N}_j} e_{jl}(k-\tau) + \Psi_j(k)\nabla_{r_j} V_j(k-\tau) \right) + \sum_{i=1}^{N} \sum_{j\in\mathcal{N}_i} \Delta e_{ij}^2(k), \end{aligned} \quad (32)$$

where $\Delta e_{ij}(k) = e_{ij}(k+1) - e_{ij}(k)$. By reordering the summation indices and collecting similar terms, we can simplify Eq. (32) as

$$\begin{aligned} \Delta V_{e,1}(k) = &-4\varepsilon\kappa T \sum_{i=1}^{N} \sum_{j\in\mathcal{N}_i} e_{ij}(k) \sum_{l\in\mathcal{N}_i} e_{il}(k-\tau) \\ &- 4\varepsilon T \sum_{i=1}^{N} \sum_{j\in\mathcal{N}_i} e_{ij}(k)\Psi_i(k)\nabla_{r_i} V_i(k-\tau) \\ &+ \sum_{i=1}^{N} \sum_{j\in\mathcal{N}_i} \Delta e_{ij}^2(k). \end{aligned} \quad (33)$$

We express $e_{ij}(k)$ in terms of $e_{ij}(k-\tau)$ and Δe_{ij} as follows:

$$e_{ij}(k) = e_{ij}(k-\tau) + \sum_{\sigma=k-\tau}^{k-1} \Delta e_{ij}(\sigma).$$

Substituting the above expression into Eq. (33) results in

$$\begin{aligned} \Delta V_{e,1}(k) = &-4\varepsilon\kappa T \sum_{i=1}^{N} \left(\sum_{j\in\mathcal{N}_i} e_{ij}(k-\tau) \right)^2 \\ &- 4\varepsilon\kappa T \sum_{i=1}^{N} \sum_{\sigma=k-\tau}^{k-1} \sum_{j\in\mathcal{N}_i} \Delta e_{ij}(\sigma) \sum_{l\in\mathcal{N}_i} e_{il}(k-\tau) \end{aligned}$$

$$- 4\varepsilon T \sum_{i=1}^{N} \sum_{j \in \mathcal{N}_i} e_{ij}(k-\tau)\Psi_i(k)\nabla_{r_i} V_i(k-\tau)$$

$$- 4\varepsilon T\sum_{i=1}^{N} \sum_{\sigma=k-\tau}^{k-1} \sum_{j\in\mathcal{N}_i} \Delta e_{ij}(\sigma)\Psi_i(k)\nabla_{r_i} V_i(k-\tau)+\sum_{i=1}^{N}\sum_{j\in\mathcal{N}_i} \Delta e_{ij}^2(k). \tag{34}$$

Recalling that $2ab \leq a^2 + b^2$ for any $a, b \in \mathbb{R}$ and applying it to the nonquadratic terms in Eq. (34), we obtain

$$\Delta V_{e,1}(k) \leq \left(-4\varepsilon\kappa T + 5\varepsilon^2\kappa^2 T^2\right) \sum_{i=1}^{N} \left(\sum_{j\in\mathcal{N}_i} e_{ij}(k-\tau) \right)^2$$
$$+ \left(\frac{4}{\kappa^2} + 4\varepsilon^2 T^2\right) \sum_{i=1}^{N} (\Psi_i(k)\nabla_{r_i} V_i(k-\tau))^2$$
$$+ 2\sum_{i=1}^{N} \left(\sum_{\sigma=k-\tau}^{k-1} \sum_{j\in\mathcal{N}_i} \Delta e_{ij}(\sigma) \right)^2 + \sum_{i=1}^{N} \sum_{j\in\mathcal{N}_i} \Delta e_{ij}^2(k).$$

In view of Lemma 2, it follows from the above inequality that

$$\Delta V_{e,1}(k) \leq (-4\varepsilon\kappa T + 5\varepsilon^2\kappa^2 T^2) \sum_{i=1}^{N} \left(\sum_{j\in\mathcal{N}_i} e_{ij}(k-\tau) \right)^2$$
$$+ \left(\frac{4}{\kappa^2} + 4\varepsilon^2 T^2\right) \sum_{i=1}^{N} (\Psi_i(k)\nabla_{r_i} V_i(k-\tau))^2$$
$$+ 2\tau \sum_{i=1}^{N} \sum_{\sigma=k-\tau}^{k-1} \left(\sum_{j\in\mathcal{N}_i} \Delta e_{ij}(\sigma) \right)^2 + \sum_{i=1}^{N} \sum_{j\in\mathcal{N}_i} \Delta e_{ij}^2(k). \tag{35}$$

We define a second positive definite function

$$V_{e,2}(k) = 2\tau \sum_{i=1}^{N} \sum_{s_1=0}^{\tau} \sum_{s_2=k-s_1}^{k-1} \left(\sum_{j\in\mathcal{N}_i} \Delta e_{ij}(s_2) \right)^2.$$

By Lemma 3 the variation of $V_{e,2}(k)$ is computed as

$$\Delta V_{e,2}(k) = V_{e,2}(k+1) - V_{e,2}(k)$$
$$= 2\tau^2 \sum_{i=1}^{N} \left(\sum_{j\in\mathcal{N}_i} \Delta e_{ij}(k) \right)^2 - 2\tau \sum_{i=1}^{N} \sum_{\sigma=k-\tau}^{k-1} \left(\sum_{j\in\mathcal{N}_i} \Delta e_{ij}(\sigma) \right)^2. \quad (36)$$

Finally, consider the Lyapunov function $V_e(k) = V_{e,1}(k) + V_{e,2}(k)$. We can determine its variation by combining Eqs. (35) and (36) as follows:

$$\Delta V_e(k) \le (-4\varepsilon\kappa T + 5\varepsilon^2\kappa^2T^2) \sum_{i=1}^{N} \left(\sum_{j\in\mathcal{N}_i} e_{ij}(k-\tau) \right)^2 + \left(\frac{4}{\kappa^2} + 4\varepsilon^2 T^2 \right)$$
$$\times \sum_{i=1}^{N} (\Psi_i(k)\nabla_{r_i} V_i(k-\tau))^2 + 2\tau^2 \sum_{i=1}^{N} \left(\sum_{j\in\mathcal{N}_i} \Delta e_{ij}(k) \right)^2$$
$$+ \sum_{i=1}^{N} \sum_{j\in\mathcal{N}_i} \Delta e_{ij}^2(k). \quad (37)$$

By Lemma 2 the last two terms in Eq. (37) can be simplified and combined:

$$\Delta V_e(k) \le (-4\varepsilon\kappa T + 5\varepsilon^2\kappa^2T^2) \sum_{i=1}^{N} \left(\sum_{j\in\mathcal{N}_i} e_{ij}(k-\tau) \right)^2$$
$$+ \left(\frac{4}{\kappa^2} + 4\varepsilon^2 T^2 \right) \sum_{i=1}^{N} (\Psi_i(k)\nabla_{r_i} V_i(k-\tau))^2$$
$$+ \left(2\tau^2 d_{\max} + 1\right) \sum_{i=1}^{N} \sum_{j\in\mathcal{N}_i} \Delta e_{ij}^2(k). \quad (38)$$

Applying Lemma 2 to the following definition of $\Delta e_{ij}(k)$,

$$\Delta e_{ij}(k) = \varepsilon T \left(-\kappa \sum_{l\in\mathcal{N}_i} e_{il}(k-\tau) - \Psi_i(k)\nabla_{r_i} V_i(k-\tau) + \kappa \sum_{l\in\mathcal{N}_j} e_{jl}(k-\tau) \right.$$
$$\left. + \Psi_j(k)\nabla_{r_j} V_j(k-\tau) \right),$$

we have

$$\Delta e_{ij}^2(k) \le 4\varepsilon^2\kappa^2T^2\left(\sum_{l\in\mathcal{N}_i} e_{il}(k-\tau)\right)^2 + 4\varepsilon^2\kappa^2T^2\left(\sum_{l\in\mathcal{N}_j} e_{jl}(k-\tau)\right)^2$$
$$+ 4\varepsilon^2T^2\left(\Psi_i(k)\nabla_{r_i}V_i(k-\tau)\right)^2 + 4\varepsilon^2T^2\left(\Psi_j(k)\nabla_{r_j}V_j(k-\tau)\right)^2. \tag{39}$$

Summing both sides of Eq. (39) over i and j and reordering the summation indices, we have the resulting inequality as

$$\sum_{i=1}^{N}\sum_{j\in\mathcal{N}_i}\Delta e_{ij}^2(k) \le 8\varepsilon^2\kappa^2T^2d_{\max}\sum_{i=1}^{N}\left(\sum_{j\in\mathcal{N}_i} e_{ij}(k-\tau)\right)^2$$
$$+ 8\varepsilon^2T^2d_{\max}\sum_{i=1}^{N}\left(\Psi_i(k)\nabla_{r_i}V_i(k-\tau)\right)^2. \tag{40}$$

Substituting Eqs. (40) into (38) yields

$$\Delta V_e(k) \le (-4\varepsilon\kappa T + (8d_{\max}+5)\varepsilon^2\kappa^2T^2 + 16\varepsilon^2\kappa^2T^2\tau^2d_{\max}^2)$$
$$\times\sum_{i=1}^{N}\left(\sum_{j\in\mathcal{N}_i} e_{ij}(k-\tau)\right)^2 + \left(\frac{4}{\kappa^2} + (8d_{\max}+4)\varepsilon^2T^2 + 16\varepsilon^2T^2\tau^2d_{\max}^2\right)$$
$$\times\sum_{i=1}^{N}(\Psi_i(k)\nabla_{r_i}V_i(k-\tau))^2. \tag{41}$$

It is proved in Lemma 4 that $\nabla_{r_i}V_i(k-\tau)$ asymptotically approaches 0 as $k\to\infty$ for all agents i if $\varepsilon < \bar{\varepsilon}_1$. This indicates that for any given $\gamma > 0$ there exists some $K_1 > 0$ such that for $k \ge K_1$,

$$\Delta V_e(k) \le \left(-4\varepsilon\kappa T + (8d_{\max}+5)\varepsilon^2\kappa^2T^2 + 16\varepsilon^2\kappa^2T^2\tau^2d_{\max}^2\right)$$
$$\times\sum_{i=1}^{N}\left(\sum_{j\in\mathcal{N}_i} e_{ij}(k-\tau)\right)^2 + \gamma^2. \tag{42}$$

We define the following positive constant:

$$\bar{\varepsilon}_2 = \frac{4}{\kappa T(16\tau^2d_{\max}^2 + 8d_{\max} + 5)}.$$

In Eq. (42) we easily observe that for any $0 < \varepsilon < \min\{\bar{\varepsilon}_1, \bar{\varepsilon}_2\}$, $\Delta V_e(k) < 0$ when

$$\left|\sum_{j\in\mathcal{N}_i} e_{ij}(k-\tau)\right| > \frac{\gamma}{\sqrt{4\varepsilon\kappa T - (8d_{\max}+5)\varepsilon^2\kappa^2T^2 - 16\varepsilon^2\kappa^2T^2\tau^2d_{\max}^2}},$$

for any agent i. Since the dominator in the above expression is a constant and γ can be arbitrarily small, this implies that $\sum_{j\in\mathcal{N}_i} e_{ij}(k-\tau) \to 0$ as $k \to \infty$ for all agents i. Consequently, $L\theta \to 0$ as $k \to \infty$, where $\theta = [\theta_1, \theta_2, \ldots, \theta_N]^{\mathrm{T}}$ and L is the Laplacian matrix defined in Definition 2. Under the assumption that the communication network is undirected and connected, by Lemma 1 the only nontrivial solution of $L\theta = 0$ is $\theta = [b, b, \ldots, b]^{\mathrm{T}}$ for some constant b dependent on the initial conditions. Hence orientation alignment is guaranteed. □

Flocking behavior includes position aggregation and heading alignment. Combining Lemmas 4 and 6 gives the following theorem, which demonstrates the performance of the proposed control laws (4).

Theorem 1. *Consider the multivehicle system* (2) *with a communication delay* $\tau \in \mathbb{N}$. *If Assumptions 1 and 2 both hold, then for a sufficiently small* ε, *the distributed control laws* (4) *steer the system to a formation corresponding to a minimum of* $\sum_{i=1}^{N} V_i$ *and a common orientation.*

Proof. Properly choose some positive $\varepsilon < \min\{\bar{\varepsilon}_1, \bar{\varepsilon}_2\}$, where $\bar{\varepsilon}_1$ and $\bar{\varepsilon}_2$ are specified in Lemmas 4 and 6, respectively. By Lemma 4 the positions of all agents gradually converge to a formation corresponding to a minimum of $\sum_{i=1}^{N} V_i$ and $\nabla_{r_i} V_i \to 0$ as $k \to \infty$ for $i = 1, 2, \ldots, N$. Then by Lemma 6 the orientations of all agents converge asymptotically to a common value as the desired formation is approached. □

5 SIMULATION

In this section the performance of the proposed control laws is demonstrated through numerical simulation. Throughout this section we consider five agents with sampling period $T = 1$ s and the initial positions $[-2.9, 2.8]^{\mathrm{T}}$, $[-2.5, -0.5]^{\mathrm{T}}$, $[0.2, -2.8]^{\mathrm{T}}$, $[2.4, -2]^{\mathrm{T}}$, and $[2.5, 1.5]^{\mathrm{T}}$. Their initial orientations are -2.25 rad, -0.49 rad, 2.61 rad, 1.84 rad and 2.88 rad, respectively. The parameters in the control laws are chosen as $\rho = 1$, $\kappa = 0.1$, and $\varepsilon = 0.015$. The communication delay is $\tau = 2$.

5.1 Complete Graph

We first consider the case where the communication graph is complete, which means that each agent is able to access the state of all other agents. Fig. 3 shows the trajectories and aligned headings of the agents, where circles indicate the initial positions and the final positions are denoted by crosses. Fig. 4 shows the evolution of $V_r(k)$ and $V_e(k)$. The flocking behavior is verified in Figs. 3 and 4. Agents are steered into a formation corresponding to the predefined optimal average distance ρ. Heading errors also converge to zero as expected.

In addition, collisions are avoided in this example. The potential of the entire system must remain bounded according to the derivation in the preceding section. This together with the fact that any pair of agents are directly connected implies that $\|r_{ij}\| \nrightarrow 0$ for any agent i and j. Therefore collision avoidance is guaranteed.

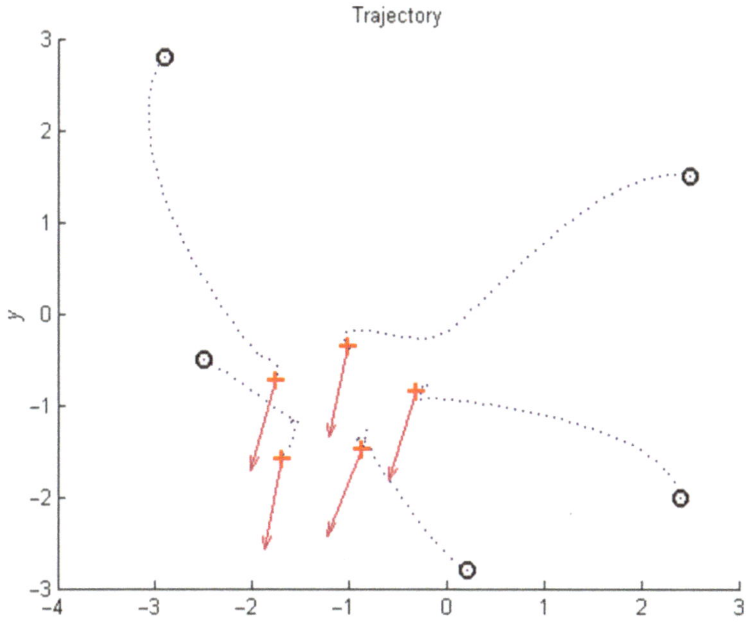

Fig. 3 Trajectories of the closed-loop system under a complete graph.

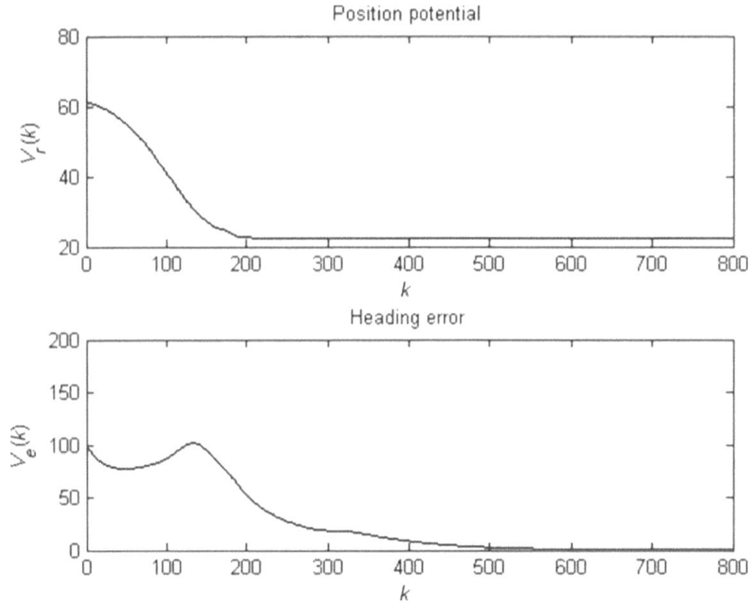

Fig. 4 Evolution of $V_r(k)$ and $V_e(k)$ under a complete graph.

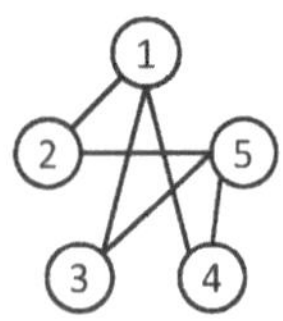

Fig. 5 A connected incomplete graph.

5.2 Tree Graph

Consider a connected, but not complete, communication graph as illustrated in Fig. 5. The trajectory of each agent is plotted in Fig. 6, in which we can see that both position aggregation and orientation alignment are achieved. Fig. 7 shows that the total potential of the five-agent system converges toward its minimum and the heading error approaches zero as the desired formation is being formed.

However, unlike in the previous example, where the graph is complete, we observe that two agents collide in Fig. 6. Since these two agents are not directly connected, they are not able to collect the position information on each other and their relative position is not accounted for in the potential function. Hence collision avoidance cannot be guaranteed.

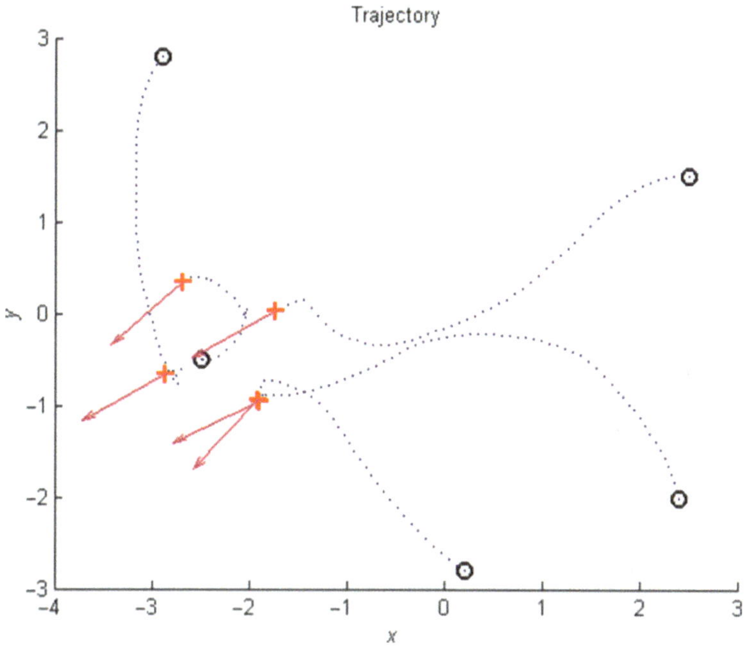

Fig. 6 Trajectories of the closed-loop system under a connected, but not complete, graph.

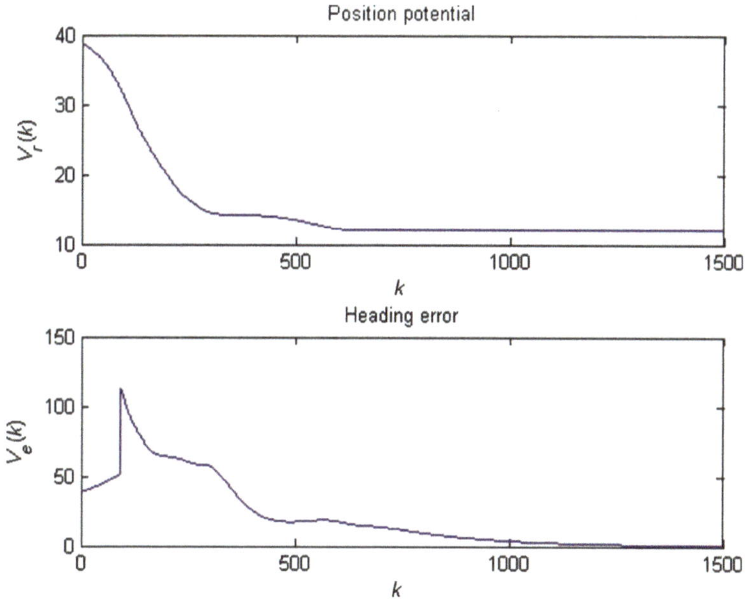

Fig. 7 Evolution of $V_r(k)$ and $V_e(k)$ under a connected, but not complete, graph.

6 CONCLUSIONS

We have studied the flocking control problem for a group of nonholonomic vehicles with discrete-time dynamics and in the presence of an arbitrarily large communication delay. A potential function-based low-gain control law was designed for each agent. Through the Lyapunov function approach we proved that the closed-loop multiagent system possesses the desired flocking behavior, including position aggregation and orientation alignment. Because of the characteristics of the potential function chosen, collision avoidance is guaranteed among neighboring agents. In addition, the effectiveness of the controller was demonstrated by numerical examples.

REFERENCES

[1] U. Munz, A. Papachristodoulou, F. Allgower, Delay-dependent rendezvous and flocking of large scale multi-agent systems with communication delays, in: Proceedings of the 47th IEEE Conference on Decision and Control, Cancun, Mexico, 2008, pp. 2038–2043.
[2] W. Dong, J.A. Farrell, Decentralized cooperative control of multiple nonholonomic systems, in: Proceedings of the 46th IEEE Conference on Decision and Control, New Orleans, LA, 2007, pp. 1486–1491.
[3] W. Dong, Flocking of multiple mobile robots based on backstepping, IEEE Trans. Syst. Man. Cybern. B: Cybern. 41 (2011) 414–424.
[4] N.E. Leonard, E. Fiorelli, Virtual leaders, artificial potentials and coordinated control of groups, in: Proceedings of the 40th IEEE Conference on Decision and Control, Orlando, FL, 2001, pp. 2968–2973.
[5] H. Tanner, A. Jadbabaie, G.J. Pappas, Coordination of multiple autonomous vehicles, in: IEEE Mediterranean Conference on Control and Automation, Rhodes, Greece, 2003, pp. 869–876.
[6] H. Hu, S.Y. Yoon, Z. Lin, Coordinated control of wheeled vehicles in the presence of a large communication delay through a potential functional approach, IEEE Trans. Intell. Transp. Syst. 15 (2014) 2261–2272.
[7] P.A. Nino-Suarez, E. Aranda-Bricaire, M. Velasco-Villa, Discrete-time sliding mode path-tracking control for a wheeled mobile robot, in: Proceedings of the 45th IEEE Conference on Decision and Control, San Diego, CA, 2006, pp. 3052–3057.
[8] B. Dumitrscu, A. Filipescu, A. Radaschin, A. Filipescu, E. Minca, Discrete-time sliding mode control of wheeled mobile robots, in: Proceeding of the 8th Asian Control Conference, Kaohsiung, Taiwan, 2011, pp. 771–776.
[9] W. Ren, R. Beard, Distributed Consensus in Multi-vehicle Cooperative Control: Theory and Applications, Springer, Berlin, 2008.
[10] V. Gazi, K.M. Passino, Swarm Stability and Optimization, Springer, Berlin, 2011.
[11] I.S. Gradshteyn, I.M. Ryzhik, Tables of Integrals, Series and Products, Academic Press, New York, 2007.

CHAPTER SEVENTEEN

Cooperative Control and Networked Operation of Passivity-Short Systems

R. Harvey, Z. Qu
University of Central Florida, Orlando, FL, United States

1 INTRODUCTION

Recent designs of cooperative systems involve decentralized and distributed controls that communicate locally rather than globally. Designs for physical systems with underlying communication networks and cooperative controls have been developed by many different techniques. Some techniques include the use of passivity as a design tool [1], Lyapunov methods involving passivity [2], and graph theory methods [3]. Systems with a shortage of passivity lie in a broad category of dissipative systems which include non-minimum-phase systems and systems with relative degree greater than 1, whereas passive systems include systems of low relative degree and minimum phase as shown in [4] and [10]. The utilization of passivity indices allows controls for switched systems to be designed as shown in [8]. Passive systems may be connected under certain conditions, such as implementation of negative feedback of two passive systems, or connection of passive systems in parallel. These properties are valuable for networking passive systems; however, if a system is not passive, the same properties of connection will not hold. Differences in passive systems and passivity-short systems and their respective properties will be investigated in terms of stability and input-output analysis in Sections 2 and 3. Some variations of nonpassive systems, specifically systems that have a shortage of passivity, or passivity-short systems, possess desirable connectivity properties that may be used in a networked design of such systems. A specific subclass of passivity-short systems—namely, input feedforward passive systems—are the most desirable type of passivity-short systems for networked operation

Control of Complex Systems
http://dx.doi.org/10.1016/B978-0-12-805246-4.00017-3

because of their input-output relationships, as shown in Section 4. Methods for designing controls for input feedforward passive systems are the subject of Section 5. It is shown in Section 6 that there are various ways to interconnect passivity-short systems, and in Section 7 it is shown that by use of the characteristics described in Section 4 and combination of the design method in Section 5 with network-level controls, the overall system can achieve consensus, and by design, becomes a plug-and-play system.

2 PASSIVE SYSTEMS

Let us consider a nonlinear dynamic system in the generic form

$$\dot{z} = F(z, u), \quad y = H(z), \tag{1}$$

where $z \in \Re^n$ is the state, $u \in \Re^m$ is the input, and $y \in \Re^p$ is the output. Qualitative properties such as Lyapunov stability, asymptotic stability, and input-output stability (eg, finite L_2 gain) are at the core of nonlinear analysis and control design.

The concept of dissipativity provides a general framework to study various qualitative properties of nonlinear systems, and it is defined below as introduced in [5].

Definition 1. System (1) is said to be *dissipative* with respect to a storage function $V(z)$ and a supply rate $\Phi(z, u)$ if the following inequality holds:

$$V(z(t)) - V(z(0)) \leq -\int_0^t \Phi(z(\tau), u(\tau))d\tau. \tag{2}$$

To establish inequality (2), both the storage function and the supply rate function can be chosen. A commonly used class of supply rate functions have the following quadratic form: if $p = m$ (ie, the system is square in its input-output relationship),

$$\Phi(z, u) = \eta(z) - u^{\mathrm{T}}y + \alpha\varphi(u) + \beta\phi(y), \tag{3}$$

where $\eta(\cdot)$, $\varphi(\cdot)$, and $\phi(\cdot)$ are positive semidefinite functions, and α and β are constant (design) parameters. If V is also differentiable, we have

$$\dot{V} \leq -\eta(z) + u^{\mathrm{T}}y - \alpha\varphi(u) - \beta\phi(y). \tag{4}$$

The most common choices of $\eta(\cdot)$, $\varphi(\cdot)$, and $\phi(\cdot)$ are the quadratic functions

$$\eta(z) = \|z\|^2, \quad \phi(y) = \|y\|^2, \quad \varphi(u) = \|u\|^2.$$

A special and very useful case of dissipativity that may arise is the concept of passivity. In essence, if the energy stored in a system is bounded from above by the energy supplied to the system, then the system is said to be *passive*.

Definition 2. System (1) is said to be *passive* if its storage function $V(z)$ is positive semidefinite and its supply rate $\Phi(z, u)$ is of the form of Eq. (3) with $\alpha, \beta \geq 0$.

Passive systems can be further classified into several types: a passive system is *lossless* if

$$\dot{V} = u^{\mathrm{T}} y,$$

input strictly passive if $\varphi(\cdot)$ is positive definite and

$$\dot{V} \leq u^{\mathrm{T}} y - \alpha\varphi(u) \quad \text{for} \quad \alpha > 0,$$

output strictly passive if $\phi(\cdot)$ is positive definite and

$$\dot{V} \leq u^{\mathrm{T}} y - \beta\phi(y) \quad \text{for} \quad \beta > 0,$$

or *strictly passive* if $\eta(\cdot)$ is positive definite and

$$\dot{V} \leq u^{\mathrm{T}} y - \eta(z).$$

Passivity has direct implications on system stability if the storage function $V(z)$ is positive definite. The relationship between passivity and stability was thoroughly investigated in [4]. The following properties summarize the results in [4, pp. 242–243].

Property 1. *The origin of system* (1) *is Lyapunov stable if it is passive with a positive definite storage function* $V(x)$.

Property 2. *The origin of system* (1) *is asymptotically stable if it is strictly passive.*

Property 3. *If system* (1) *is output strictly passive and zero-state observable,*[1] *then the system is asymptotically stable at the origin.*

When the dynamics of system (1) are unavailable, input-output mapping becomes a useful way of quantifying input-output properties of the system. The following definition, adopted from [4], is one of the commonly used input-output stability concepts.

Definition 3. System (1) is finite-gain L_2 stable if

$$\|y\|_{L_2} \leq \kappa \|u\|_{L_2} + \rho,$$

[1] Zero-state observability refers to the property that no solution other than the trivial solution of $z(t) \equiv 0$ satisfies $\dot{z} = F(z, 0) \equiv 0$.

where κ is a nonnegative constant referred to as the L_2 *gain*, ρ is also a nonnegative constant called the *bias term*, and $\|\cdot\|_{L_2}$ denotes the functional 2-norm.

Passivity lends itself naturally to quantification of the relationship between input and output. The following two properties of passive systems are well known, and their proofs can be found in [4].

Property 4. *If system* (1) *is output strictly passive in the form of*

$$\dot{V} \leq u^{\mathrm{T}} y - \beta y^{\mathrm{T}} y,$$

it is finite-gain L_2 stable with its L_2 gain less than or equal to $1/\beta$.

Property 5. *Suppose that system* (1) *is passive. Then under self-feedback control $u = -\epsilon y + v$, it is finite-gain L_2 stable with its L_2 gain less than or equal to $1/\epsilon$.*

Obviously, not all dynamic systems may be passive; in fact, most systems are not passive. It is well known that the entirety of classes of non-minimum-phase systems and minimum-phase systems of relative degree 2 or higher are not passive. This is due to the following definition coupled with the positive-real lemma [4].

Definition 4. A $p \times p$ rational transfer function matrix $G(s)$ is called *positive real* if poles of all elements of $G(s)$ are in $\mathrm{Re}[s] \leq 0$ and if the matrix $G(j\omega) + G^{\mathrm{T}}(-j\omega)$ is positive semidefinite for all real ω, where $j\omega$ is not a pole of $G(s)$, and if any imaginary pole $j\omega$ of $G(s)$ and the residue matrix $\lim_{s \to j\omega}(s - j\omega)G(s)$ is positive semidefinite and Hermitian.

This definition leads us to the following lemma.

Lemma 1. *A linear time-invariant minimal realization with transfer function $G(s)$, as defined above, is passive if and only if $G(s)$ is positive real.*

Hence nonpassive systems need to be studied so that their analysis and control design can be performed systematically.

3 PASSIVITY-SHORT SYSTEMS

In terms of the storage function and the supply rate function, system (1) becomes nonpassive or *passivity-short* if the system falls into one of the following three cases [11]:

1. The system is *state passivity short* if for some indefinite or negative definite function $\eta(x)$,

$$\dot{V} \leq u^{\mathrm{T}} y - \eta(x). \tag{5}$$

2. The system is *output passivity short* if for some positive definite function $\phi(\cdot)$ and some $\beta < 0$,

$$\dot{V} \le u^{\mathrm{T}} y - \beta\phi(y). \tag{6}$$

3. The system is *input passivity short* if for some positive definite function $\varphi(\cdot)$ and some $\alpha < 0$,

$$\dot{V} \le u^{\mathrm{T}} y - \alpha\varphi(u). \tag{7}$$

In the case of Eq. (5), the system stability property may be altered if the system is controllable and a full-state feedback control can be found. Nonetheless, if the system is of high order but the control design is limited to output feedback controls, it is unrealistic to expect that the system stability can be achieved in general.

The remaining two cases of passivity-short systems do have solutions to stability analysis and control design. In the case of Eq. (6), the system is called *output feedback passive* since under the output feedback control of $u = u' + \beta\phi'(y)$ (provided that $\phi(y) = y^{\mathrm{T}}\phi'(y)$, with $\phi'(y)$ being well defined), the system becomes passive (with respect to the newly introduced input u'). As a result, the results on passive systems can be readily applied to the class of output feedback passive systems. In the case of Eq. (7), the system is called *input feedforward passive* since, with the definition of a fictitious output $y' = y - \alpha\varphi'(u)$ (provided that $\varphi(u) = u^{\mathrm{T}}\varphi'(u)$, with $\varphi'(u)$ being well defined), the system becomes passive (with respect to the newly introduced output y'). While this appears to be simple, detailed analysis and control design need to be done with respect to the original input-output pair, which is the subject of the next section.

4 INPUT FEEDFORWARD PASSIVE SYSTEMS

Let us consider an input feedforward passive system whose storage function V is differentiable and whose supply rate function is quadratic:

$$\dot{V} \le u^{\mathrm{T}} y + \frac{\epsilon}{2} u^{\mathrm{T}} u = u^{\mathrm{T}} y', \tag{8}$$

where ϵ is a nonnegative constant called the *impact coefficient* and

$$y' \stackrel{\triangle}{=} y + \frac{\epsilon}{2} u.$$

It is apparent that passive systems are included as the special case of $\epsilon = 0$. The subsequent analysis is for the case of $\epsilon > 0$. Similar analysis can be applied to systems of the general form of Eq. (7).

The following property is the direct result of Eq. (8).

Property 6. *If system* (1) *is input feedforward passive and satisfies Eq. (7), it is passive with respect to input-output pair* $\{u,\ y'\}$.

Consider the class of linear systems

$$\dot{z} = Az + Bu, \quad y = Cz; \quad \text{or} \quad G(s) = C(sI - A)^{-1}B. \tag{9}$$

It follows from the positive-real lemma that $\mathrm{Re}[G(j\omega)] \geq 0$ holds for a scalar passive system—that is, the Nyquist plot of the system frequency response

$$\frac{Y(j\omega)}{U(j\omega)} = G(j\omega)$$

must lie within the first and fourth quadrants, as illustrated in Fig. 1A. Combining the positive-real lemma and Property 6 yields the following result (whose proof is included in the Appendix).

Property 7. *(1) System* (9) *is input feedforward passive if it is Lyapunov stable with all the poles in the left half plane except for at most one pole at the origin. (2) System* (9) *can be made (under a dynamic output-feedback control) input feedforward passive if it is Lyapunov stable with imaginary poles and is both controllable and observable.*

Case (1) of Property 7, which is illustrated in Fig. 1B, tells us that all linear systems in use (hence at least Lyapunov stable) are input feedforward passive. In other words, input feedforward passive systems include minimum-phase systems, non-minimum-phase systems, high-relative-degree systems, and their combinations, which are illustrated by the following examples.

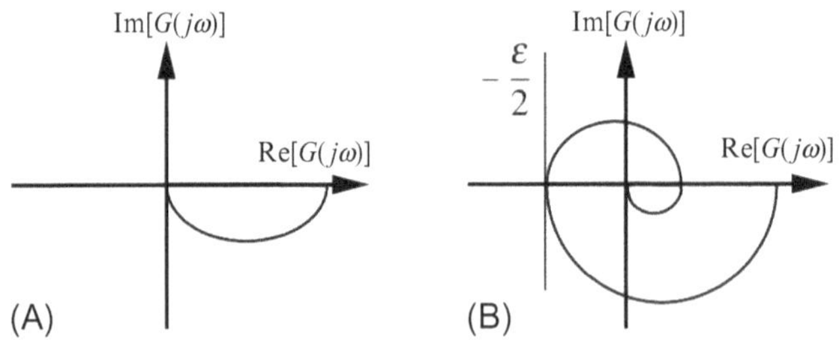

Fig. 1 Nyquist plots: (A) passive linear systems; (B) input feedforward passive linear systems.

Examples. Given a transfer function $G(s) = \frac{(s-1)}{(s+1)(s+2)}$, the corresponding Nyquist plot is presented below.

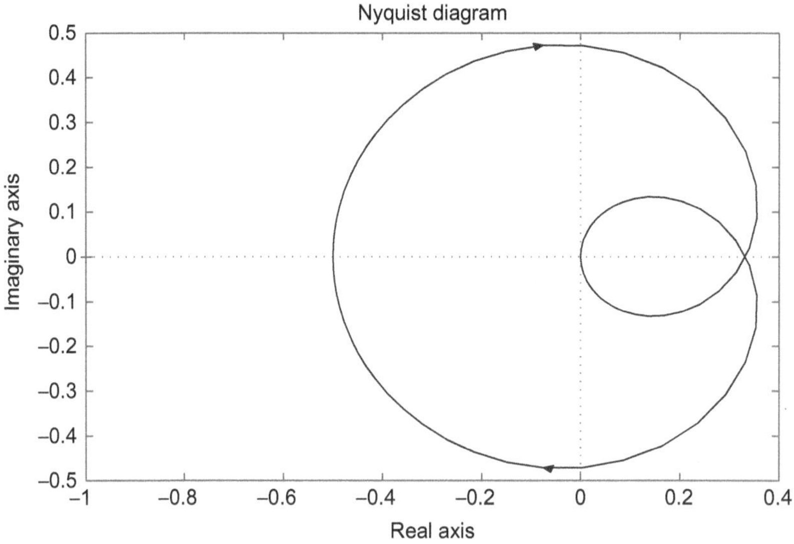

For the transfer function $G(s) = \frac{1}{(s+1)(s+2)}$, the corresponding Nyquist plot is presented below.

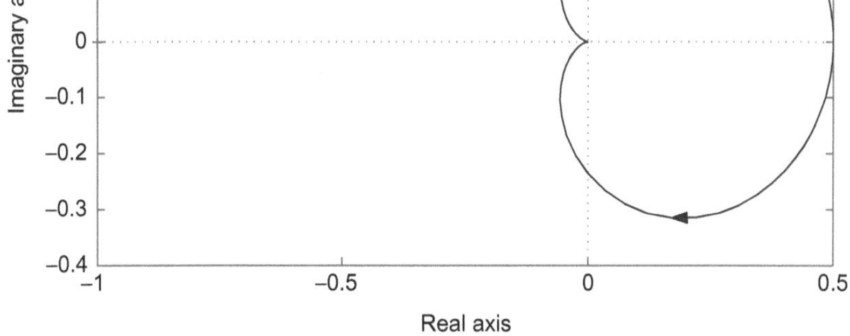

Case (2) can be illustrated by the following example: Given a transfer function in the form of $G(s) = (s + a)/(s^2 + 1)$ and attempting to satisfy conditions in Definition 4, we see that

$$G(j\omega) = \frac{j\omega + a}{-\omega^2 + 1},$$

which is not positive real. Furthermore,

$$\lim_{s\to j}(s - j)\frac{a + j}{(s + j)(s - j)} = \lim_{s\to j}\frac{a + j}{s + j} = \frac{1}{2}(1 - aj) \tag{10}$$

is neither positive semidefinite nor Hermitian for any $a > 0$. If $a = 0$, the system is input feedforward passive. In summary, a system with imaginary poles may not be input feedforward passive. Nonetheless, it is shown in the proof of Property 7 that systems of this type may become input feedforward passive under an observer-based control.

For nonlinear input feedforward passive systems, the following properties are available, and their proofs are included in the Appendix.

Property 8. *If system* (1) *is input feedforward passive in the form of Eq. (8) and with a positive definite storage function $V(x)$, the origin is Lyapunov stable.*

Property 9. *If system* (1) *is input feedforward passive in the form of Eq. (8), it is finite-gain-of-1 L_2 stable under the output feedback control $u = -y/\epsilon + v$.*

Property 10. *If system* (1) *is input feedforward passive in the form of Eq. (8) and if the autonomous system $\dot{z} = F(z, -ky)$ and $y = h(z)$ is zero-state observable, it is asymptotically stable under the output feedback control $u = -ky$ with $0 < k < 2/\epsilon$.*

5 CONTROL DESIGNS

By use of the properties presented in the previous sections, control designs may be developed to make any given controllable system input feedforward passive. Let us consider a set of affine systems: For $i = 1, \ldots, n_s$ for some positive integer n_s,

$$\dot{z}_i = F_i(z_i) + G_i(z_i)v_i, \quad y_i = H_i(z_i), \tag{11}$$

where $z_i \in \Re^{n_i}$ is the state of the ith system, and $v_i, y_i \in \Re^m$ are the input and output of the ith system, respectively. In the case that system (11) is linear and Lyapunov stable, no individual control design is required since it is clear from Property 7 that system (11) is already input feedforward passive. If system (11) is linear and controllable but not Lyapunov stable, a simple pole placement control (eg, an observer-based output feedback control) can be applied to achieve input feedforward passivity. If system (11) is nonlinear, minimum phase but unstable, a control can be designed with use of the feedback linearization technique. If system (11) is nonlinear and controllable but nonminimum phase and unstable, no systematic approach exists for control design; nonetheless, controls can be (and have been) found for practical problems.

For networked operation of passivity-short systems, a network-level control needs to be designed. As shown in [6], it is advantageous to pursue the networked control design in three steps. The first step is to make each of the systems input feedforward passive, as described above. The second step is to find diffeomorphic[2] transformations $z_i = Z_i(w_i)$ and $u_i = U_i(v_i, z_i)$ such that the individual systems are mapped into the following canonical form. The third step is to separately design a cooperative control at the network level, which is the topic of Section 7.

Definition 5. The canonical form for cooperative control of individual passivity-short systems is

$$\begin{bmatrix} \dot{w}_{i1} \\ \dot{w}_{i2} \end{bmatrix} = \begin{bmatrix} F_{w_{i1}}(w_{i1}, w_{i2}) \\ 0 \end{bmatrix} + \begin{bmatrix} G_{w_{i1}}(w_{i1}, w_{i2}) \\ I \end{bmatrix} u_i, \tag{12}$$

$$y_i = H_{w_i}(w_{i1}, w_{i2}) + \beta_i w_{i2}, \tag{13}$$

where $H_{w_i}(0, w_{i2}) = 0$, $\beta_i \neq 0$, the reduced order system of $\dot{w}_{i1} = F_{w_{i1}}(w_{i1}, w_{i2}) + G_{w_{i1}}(w_{i1}, w_{i2})u_i$ is globally asymptotically stable, and $H_{w_i}(w_{i1}, w_{i2}) \in L_2$ if w_{i2} is uniformly bounded and if $u_i \in L_2$.

The following two examples illustrate the design process and the above canonical form. The first example is an non-minimum-phase linear system, while the second example is a nonlinear feedback linearizable system.

Example 1. Consider the linear non-minimum-phase system

$$\dot{z}_1 = z_2, \quad \dot{z}_2 = v, \quad y = -z_1 + z_2,$$

where $y(s)/v(s) = (s-1)/s^2$. Under the control law $v = -2z_2 - u$, the closed-loop system becomes

$$\dot{z} = \begin{bmatrix} 0 & 1 \\ 0 & -2 \end{bmatrix} z + \begin{bmatrix} 0 \\ -1 \end{bmatrix} u, \quad y = [-1 \quad 1]\, z.$$

[2] A diffeomorphism is a differentiable mapping that is bijective, and its inverse is also differentiable.

Then transforming the system using $w_1 = z_2$ and $w_2 = -(2z_1 + z_2)$ results in the system

$$\dot{w} = \begin{bmatrix} -2 & 0 \\ 0 & 0 \end{bmatrix} w + \begin{bmatrix} -1 \\ 1 \end{bmatrix} u, \quad y_i = \begin{bmatrix} \frac{3}{2} & \frac{1}{2} \end{bmatrix} w,$$

which is Lyapunov stable, and proves to be an input feedforward passive system.

Example 2. Consider the nonlinear system

$$\begin{aligned} \dot{z}_1 &= z_1^3 + z_2, \\ \dot{z}_2 &= -(2k + 3z_1^2)(z_1^3 + z_2) + u, \\ y &= z_1. \end{aligned}$$

If the state transformation of $w_1 = z_1^3 + z_2,\ w_2 = 2kz_1 + (z_1^3 + z_2)$ is used, the system is transformed into

$$\begin{aligned} \dot{w}_1 &= -2kw_1 + u, \\ \dot{w}_2 &= u, \\ y &= -\frac{1}{2k}w_1 + \frac{1}{2k}w_2, \end{aligned}$$

where w_1 has L_2 stability.

It is obvious that system (12) is Lyapunov stable, and the pole at the origin is intentionally introduced so that heterogeneous nonlinear systems can reach consensus. It is shown in [6] that system (12) is input feedforward passive. Networked operation of these passivity-short systems will be discussed in the subsequent sections.

6 INTERCONNECTION OF PASSIVITY-SHORT SYSTEMS

In this section interconnections of passive systems and passivity-short systems will be investigated. Specifically, methods of interconnecting input feedforward passive systems will be studied. By setting the impact coefficients to zero, one can readily apply the results to passive systems as well as output feedback passive systems.

6.1 Parallel Connection

A parallel connection of two passivity-short systems, as shown in Fig. 2, is passivity short since

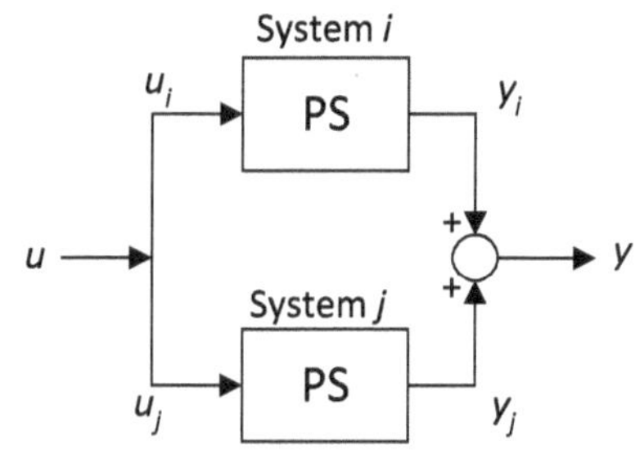

Fig. 2 Parallel connection of passivity-short (PS) systems.

$$
\begin{aligned}
\dot{V}_i + \dot{V}_j &\leq \left[u_i^{\mathrm{T}} y_i + \frac{\epsilon_i}{2}\|u_i\|^2\right] + \left[u_j^{\mathrm{T}} y_j + \frac{\epsilon_j}{2}\|u_j\|^2\right] \\
&= u^{\mathrm{T}} y + \frac{\epsilon_i + \epsilon_j}{2}\|u\|^2,
\end{aligned}
$$

where $u = u_i = u_j$ and $y = y_i + y_j$. If ϵ_i and ϵ_j are both zero, this reduces to the known result that parallel connection of passive systems is also passive.

6.2 Series Connection

Serial connection of two passivity-short systems with individual negative feedback, as shown in Fig. 3, is passivity short. To demonstrate this, consider the ith system with $u_i = v_i - k_i y_i$, where $k_i > 0$ if $\epsilon_i = 0$ and $k_i \in (0, 1/\epsilon_i]$ if $\epsilon_i > 0$. It follows from $2ab \leq a^2 + b^2$ that

$$
\begin{aligned}
\dot{V}_i &\leq u_i^{\mathrm{T}} y_i + \frac{\epsilon_i}{2}\|u_i\|^2 \\
&= -\frac{\epsilon_i}{2}k_i^2\|y_i\|^2 + (1 - \epsilon_i k_i) v_i^{\mathrm{T}} y_i - k_i(1 - \epsilon_i k_i)\|y_i\|^2 + \frac{\epsilon_i}{2}\|v_i\|^2 \\
&\leq -\frac{\epsilon_i}{2}k_i^2\|y_i\|^2 - \frac{k_i}{2}(1 - \epsilon_i k_i)\|y_i\|^2 + \frac{1}{2k_i}(1 - \epsilon_i k_i)\|v_i\|^2 + \frac{\epsilon_i}{2}\|v_i\|^2 \\
&= -\frac{k_i}{2}\|y_i\|^2 + \frac{1}{2k_i}\|v_i\|^2.
\end{aligned} \tag{14}
$$

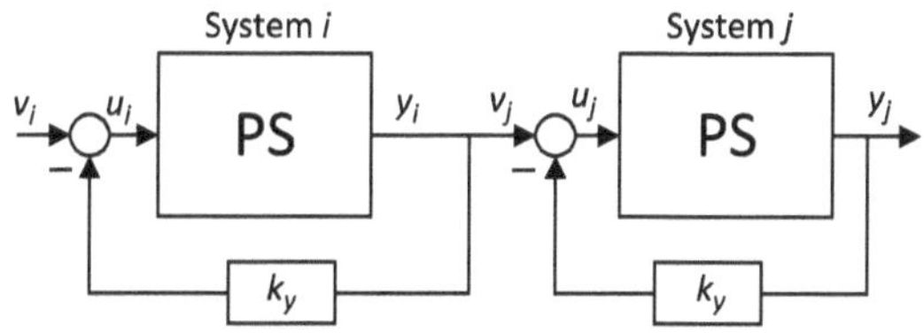

Fig. 3 Series connection of passivity-short (PS) systems.

Similarly, the jth system under $u_j = v_j - k_j y_j$ has the property that

$$\dot{V}_j \leq -\frac{k_j}{2}\|y_j\|^2 + \frac{1}{2k_j}\|v_j\|^2. \tag{15}$$

Combining Eqs. (14) and (15) and recalling that $y = y_j$, $v_j = y_i$, and $v_i = u$, we have

$$\frac{1}{k_i k_j^2}\dot{V}_i + \frac{1}{k_j}\dot{V}_j \leq \frac{1}{2k_i^2 k_j^2}\|u\|^2 - \frac{1}{2}\|y\|^2 \tag{16}$$

$$\leq u^{\mathrm{T}} y + \frac{1}{2}\left(\frac{1}{k_i^2 k_j^2} + 1\right)\|u\|^2. \tag{17}$$

Inequality (17) shows that the passivity-short property is invariant under serial connection (of passivity-short systems which individually have self negative feedback). It is worth noting from Eqs. (14)–(17) that input feedforward systems under negative self-feedback have finite L_2 gains, that a serial connection of finite L_2 gain systems has a finite L_2 gain, and that a system of a finite L_2 gain is input feedforward passive (or passivity short). In these inequalities, ϵ_i or ϵ_j is not present; hence series connection of passive systems may not be (and mostly like is not) passive, but the resulting system is always passivity short (specifically, input feedforward passive, the focus of our study).

6.3 Positive Feedback Connection

A positive feedback connection of two passivity-short systems with negative self-feedbacks, as illustrated in Fig. 4, is passivity short. To demonstrate this result, we prescribe the interaction using the control laws

$$u_i = k_y(y_j - y_i) + v_i, \quad u_j = k_y(y_i - y_j) + v_j,$$

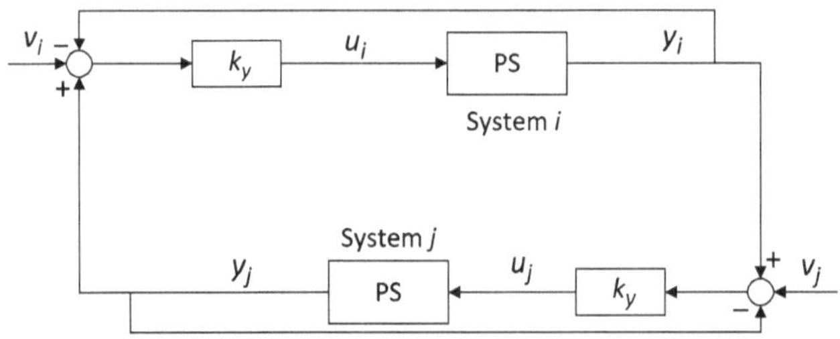

Fig. 4 Positive feedback connection of passivity-short (PS) systems.

where positive gain k_y is to be chosen. It follows that

$$\begin{aligned}\dot{V}_i + \dot{V}_j &\le u_i^{\mathrm{T}} y_i + u_j^{\mathrm{T}} y_j + \frac{\epsilon_i}{2}\|u_i\|^2 + \frac{\epsilon_j}{2}\|u_j\|^2 \\ &= v_i^{\mathrm{T}} y_i + v_j^{\mathrm{T}} y_j + \frac{\epsilon_i}{2}\|v_i\|^2 + \frac{\epsilon_j}{2}\|v_j\|^2 + \epsilon_i k_y (y_j - y_i) v_i + \epsilon_j k_y (y_i - y_j) v_j \\ &\quad + k_y (y_j - y_i)^{\mathrm{T}} y_i + k_y (y_i - y_j)^{\mathrm{T}} y_j + \frac{\epsilon_i}{2} k_y^2 (y_j - y_i)^2 + \frac{\epsilon_j}{2} k_y^2 (y_i - y_j)^2 \qquad (18) \\ &\le v^{\mathrm{T}} y + \epsilon \|v\|^2 - k_y \left[1 - (\epsilon_i + \epsilon_j) k_y \|y_i - y_j\|^2\right], \qquad (19)\end{aligned}$$

where $v = [v_i^{\mathrm{T}}\ v_j^{\mathrm{T}}]^{\mathrm{T}}$, $y = [y_i^{\mathrm{T}}\ y_j^{\mathrm{T}}]^{\mathrm{T}}$, and $\epsilon = \max\{\epsilon_i,\ \epsilon_j\}$. Inequality (19) implies that the closed-loop system is passivity short for all $k_y \in \left(0,\ 1/(\epsilon_i + \epsilon_j)\right)$. On the other hand, setting $v_i = v_j = 0$ in Eq. (18) yields

$$\dot{V}_i + \dot{V}_j \le -k_y \left[1 - \frac{\epsilon_i + \epsilon_j}{2} k_y \|y_i - y_j\|^2\right], \qquad (20)$$

which means that output consensus is reached for any $k_y \in \left(0,\ 2/(\epsilon_i + \epsilon_j)\right)$. This fact will be extended later to networked operation of passivity-short systems under arbitrary topology. If $\epsilon_i = \epsilon_j = 0$, the individual systems and the closed-loop system are all passive, and the closed-loop stability and output consensus can be ensured for any $k_y > 0$.

6.4 Negative Feedback Connections

In this section three negative feedback configurations are proposed to interconnect passivity-short systems. The first configuration involves a negative feedback loop in which two passivity-short systems have individual input feedforwards, as illustrated in Fig. 5. Specifically, considering

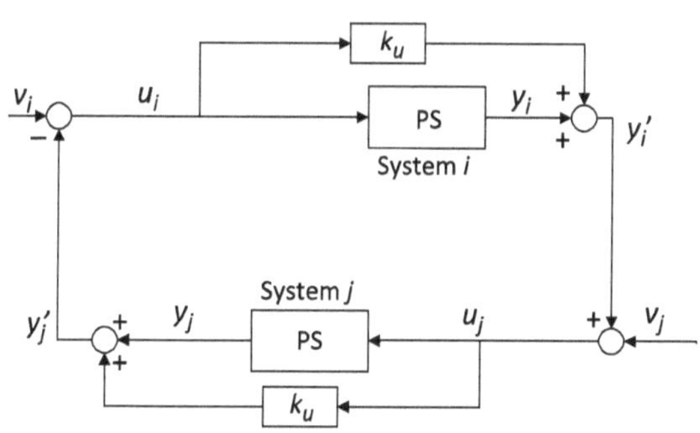

Fig. 5 Negative feedback connection of passivity-short (PS) systems: individual input feedforward.

$$y_i' = y_i + k_u u_i, \quad y_j' = y_j + k_u u_j, \quad u_i = v_i - y_j', \quad u_j = v_j + y_i', \tag{21}$$

where $k_u \geq \max\{\epsilon_i, \epsilon_j\}/2$, we have

$$\begin{aligned}\dot{V}_i + \dot{V}_j &\leq \left[u_i^{\mathrm{T}} y_i + \frac{\epsilon_i}{2}\|u_i\|^2\right] + \left[u_j^{\mathrm{T}} y_j + \frac{\epsilon_j}{2}\|u_j\|^2\right] \\ &\leq u_i^{\mathrm{T}} y_i' + u_j^{\mathrm{T}} y_j' \\ &= v_i^{\mathrm{T}} y_i' + v_j^{\mathrm{T}} y_j' \\ &= v^{\mathrm{T}} y',\end{aligned}$$

where $v = [v_i^{\mathrm{T}} \ v_j^{\mathrm{T}}]^{\mathrm{T}}$ and $y' = [(y_i')^{\mathrm{T}} \ (y_j')^{\mathrm{T}}]^{\mathrm{T}}$. The above inequality shows that the closed-loop system is passive with respect to the augmented output y'. It follows from Eq. (21) that

$$y' = \frac{1}{1+k_u^2}\begin{bmatrix} 1 & -k_u \\ k_u & 1 \end{bmatrix}(y + k_u v).$$

If both systems are passive ($\epsilon_i = \epsilon_j = 0$), we can choose $k_u = 0$, and the above result reduces the standard one that a negative feedback connection of two passive system is also passive [4].

In the second negative feedback configuration, two passivity-short systems with negative self-feedbacks are connected by means of the standard negative feedback loop. As illustrated in Fig. 6, the interaction can be expressed by the control laws

$$u_i = k_y(-y_j - y_i) + v_i, \quad u_j = k_y(y_i - y_j) + v_j,$$

where positive gain k_y is to be chosen. It follows that

$$\begin{aligned}\dot{V}_i + \dot{V}_j &\leq u_i^{\mathrm{T}} y_i + u_j^{\mathrm{T}} y_j + \frac{\epsilon_i}{2}\|u_i\|^2 + \frac{\epsilon_j}{2}\|u_j\|^2 \\ &= v_i^{\mathrm{T}} y_i + v_j^{\mathrm{T}} y_j + \frac{\epsilon_i}{2}\|v_i\|^2 + \frac{\epsilon_j}{2}\|v_j\|^2 + \epsilon_i k_y(-y_j - y_i)v_i + \epsilon_j k_y(y_i - y_j)v_j\end{aligned}$$

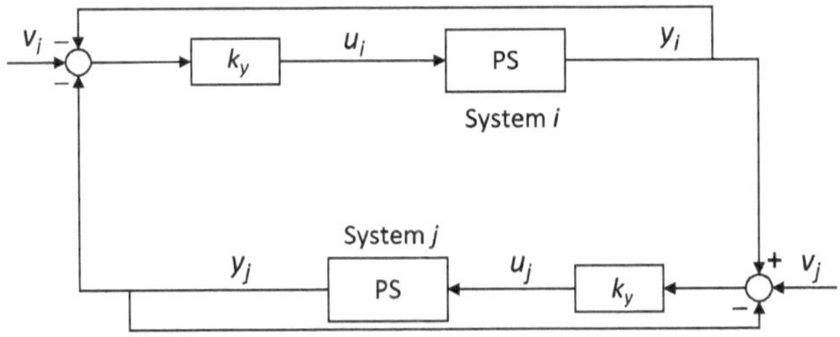

Fig. 6 Negative feedback connection of passivity-short (PS) systems: individual output feedback.

$$+ k_y(-y_j - y_i)^{\mathrm{T}} y_i + k_y(y_i - y_j)^{\mathrm{T}} y_j + \frac{\epsilon_i}{2} k_y^2 (-y_j - y_i)^2 + \frac{\epsilon_j}{2} k_y^2 (y_i - y_j)^2$$
$$\leq v^{\mathrm{T}} y + \frac{3\epsilon_i}{2} \|u_i\|^2 + \frac{3\epsilon_j}{2} \|u_j\|^2 - k_y(1 - 2\epsilon_i k_y) \|y_i\|^2$$
$$- k_y(1 - 2\epsilon_j k_y) \|y_j\|^2, \tag{22}$$

where $v = [v_i^{\mathrm{T}}\ v_j^{\mathrm{T}}]^{\mathrm{T}}$ and $y = [y_i^{\mathrm{T}}\ y_j^{\mathrm{T}}]^{\mathrm{T}}$. Inequality (22) indicates that for all $k_y \in \left(0,\ 1/\max\{2\epsilon_i,\ 2\epsilon_j\}\right)$, the closed-loop system is passivity short and has a finite L_2 gain. If $\epsilon_i = \epsilon_j = 0$, the individual systems and the closed-loop system are all passive, and the closed-loop stability can be ensured for any $k_y > 0$.

The third negative feedback configuration is the combination of the first two configurations—that is, in the standard feedback loop, two passivity-short systems are interconnected but their individual connections are different: one with output feedback and the other with input feedforward. As shown in Fig. 7, the interconnection is prescribed by

$$y_i' = y_i + k_u u_i, \quad u_i = v_i - y_j - k_{yi} y_i', \quad u_j = v_j - k_{yj} y_j + y_i',$$

where $k_u \geq \epsilon_i/2$.

$$\dot{V}_i + \dot{V}_j \leq u_i^{\mathrm{T}} y_i + u_j^{\mathrm{T}} y_j + \frac{\epsilon_i}{2} \|u_i\|^2 + \frac{\epsilon_j}{2} \|u_j\|^2$$
$$\leq u_i^{\mathrm{T}} y_i' + u_j^{\mathrm{T}} y_j + \frac{\epsilon_j}{2} \|u_j\|^2$$
$$= v_i^{\mathrm{T}} y_i' - y_j^{\mathrm{T}} y_i' - k_{yi} \|y_i'\|^2 + v_j^{\mathrm{T}} y_j - k_{yj} \|y_j\|^2 + (y_i')^T y_j + \frac{\epsilon_j}{2} \|v_j\|^2$$

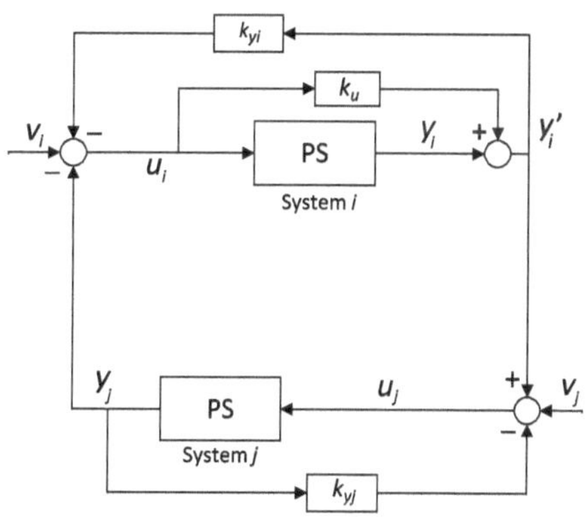

Fig. 7 Negative feedback connection of passivity-short (PS) systems: individual output feedback and individual input feedforward.

$$
\begin{aligned}
&+\frac{\epsilon_j}{2}\|y_i'\|^2+\frac{\epsilon_j}{2}k_{yj}^2\|y_j\|^2-\epsilon_j k_{yj}(y_j^{\mathrm{T}}v_j+y_j^{\mathrm{T}}y_i')+\epsilon_j v_j^{\mathrm{T}}y_i'\\
&\leq v^{\mathrm{T}}y+\frac{3\epsilon_j}{2}\|v_j\|^2-k_{yj}\left(1-\frac{3\epsilon_j}{2}k_{yj}\right)\|y_j\|^2-\left(k_{yi}-\frac{3\epsilon_j}{2}\right)\|y_i'\|^2,
\end{aligned}
\tag{23}
$$

where $v=[v_i^{\mathrm{T}}\ v_j^{\mathrm{T}}]^{\mathrm{T}}$ and $y=[(y_i')^{\mathrm{T}}\ y_j^{\mathrm{T}}]^{\mathrm{T}}$. Inequality (23) illustrates the passivity-short property (with respect to y, which involves the augmented output y_i') provided that $k_{yj}\in[0,2/(3\epsilon_j)]$ and $k_{yi}\geq 3\epsilon_j/2$. It follows that if $\epsilon_i=\epsilon_j=0$, the individual systems and the overall system reduce to passive systems.

Among the three negative feedback configurations, the second one (with individual negative feedbacks) is most appropriate for applications since the input feedforward passivity and L_2 properties hold for the original input-output pairs (while the other two involve augmented outputs).

7 COOPERATIVE CONTROL OF PASSIVITY-SHORT SYSTEMS

Summarizing all the discussions in the previous three sections, we know that the concept of input feedforward passivity broadly admits all linear Lyapunov stable systems as well as classes of nonlinear Lyapunov stable systems and that the input feedforward passivity property is invariant under all interconnections: parallel, series, positive feedback loop, and negative feedback loop. Other properties (such as passivity or output feedback passivity) possess some but not all of these invariance characteristics. In addition, consensus arises naturally from positive feedback interactions of input feedforward passive systems. If the impact coefficients are designed to be less than or equal to a given threshold of $\bar{\epsilon}$, the choice of feedback gains can be chosen (to be small and based on $1/\bar{\epsilon}$), any number of input feedforward systems may be interconnected, some or all of these connections could be intermittent, and the aforementioned passivity-short property (ie, input feedforward passive) and other relevant properties (such as L_2 stability) are maintained, which means plug-and-play capabilities of nonlinear systems.

Consider a set of heterogeneous systems in the form of Eq. (11). The problem of cooperative control for heterogeneous systems is to design network-enabled controls for the systems in Eq. (11) while taking into account the shared local communication network. In this case the communication network is best represented by a digraph composed of a set of n_s nodes and a set of directed edges, denoted by Υ. The information structure for the communication network can be characterized by the binary matrix function $S(t)$, which may be time varying [12]. $S(t)$ is defined as

$$S(t) = \begin{bmatrix} 1 & S_{12} & \ldots & S_{1n_s} \\ S_{21} & \ddots & \ddots & S_{2n_s} \\ \vdots & \ddots & \ddots & \vdots \\ S_{n_s1} & \ldots & S_{n_sn_s-1} & 1 \end{bmatrix}, \tag{24}$$

where $S_{ii} \equiv 1$ and $S_{ij} = 1$ if $\{j \rightarrow i\} \in \Upsilon$ (the directed edge) or $S_{ij}(t) = 0$ otherwise.

The network topology is specified by $S(t)$, from which a Laplacian L may be derived as $L = D - S$, where $D = \text{diag}\{S\mathbf{1}\}$. Here matrix D is the diagonal of the in-degrees of the digraph.

The overall control design for networked passivity-short systems consists of two layers, which may be referred to as the *high-level* and *low-level controls.* The low-level controls were presented in the previous section, and each control ensures that the systems are passivity short. Each control layer may be synthesized separately such that the high-level control ensures the system can achieve a global consensus. The high-level, or network-level, cooperative control is chosen as

$$u_i = k_{y_i} \sum_{i=1}^{n} S_{ij}(y_j - y_i), \tag{25}$$

where S_{ij} denotes the communication matrix for the cooperative system and $k_{y_i} > 0$.

The *impact equivalence principle* presented in [7] shows that it is possible to achieve nontrivial output consensus with networked passivity-short systems. The following theorem restates the impact equivalence principle for these systems.

Theorem 1. *For cooperative control* (25), *the gain* k_{y_i} *may be constrained as*

$$0 < k_{y_i} < \frac{\lambda'(\text{diag}\{\gamma_i\})L + L^{\text{T}}\text{diag}\{\gamma_i\})}{\bar{\epsilon}\lambda_{\max}(L^{\text{T}}\text{diag}\{\gamma_i\}L)},$$

in which L is a Laplacian of the directed network, $\bar{\epsilon} > 0$, λ' *denotes the smallest nonzero eigenvalue,* $\lambda_{\max}$ *denotes the largest eigenvalue, the values of* γ_i *are the elements of the eigenvector according to* $\gamma^{\text{T}}L = 0$, *and the network topology is strongly connected. Then a nontrivial output consensus is guaranteed.*

The high-level control does not require explicit information about the individual systems; however, the relationship between the high-level control and the low-level control is dependent on the value of $\bar{\epsilon}$.

It follows from Theorem 1 that from the Lyapunov equation

$$P_h K_h L + L^{\mathrm{T}} K_h P_h - \bar{\epsilon} L^{\mathrm{T}} K_h P_h K_h L - Q_h = 0, \tag{26}$$

where $K_h = \mathrm{diag}\{k_{y_i}\} \in \Re^{n_s \times n_s}$, $P_h \in \Re^{n_s \times n_s}$ is a positive definite solution (given that S is irreducible), and $Q_h \in \Re^{n_s \times n_s}$ is a positive semidefinite solution with rank $(n_s - 1)$, such that as the gain k_{y_i} gets smaller, the solution Q_h is positive semidefinite for all values of k_{y_i}. This is proven in [7] by the choosing of $P_h = \mathrm{diag}\{\gamma_i / k_{y_i}\}$ in Eq. (26), and by the defining of a positive semidefinite matrix W_h as

$$W_h = \mathrm{diag}\{\gamma_i\} L + L^{\mathrm{T}} \mathrm{diag}\{\gamma_i\},$$

such that

$$Q_h = W_h - \bar{\epsilon} L^{\mathrm{T}} \mathrm{diag}\{k_{y_i} \gamma_i\} L$$

and $\mathbf{1}^{\mathrm{T}} Q_h \mathbf{1} = 0$. It can be seen that W_h dominates as the gains k_{y_i} become smaller, thus ensuring that Q_h is positive semidefinite.

8 CONCLUSION

Since most (closed-loop individually-controlled) physical systems are not passive, a general class of passivity-short systems (specifically, input feed-forward passive systems) was studied. This class of passivity-short systems includes all Lyapunov stable linear systems as well as numerous nonlinear systems. It was shown that with individual output negative feedback, this class of systems can be interconnected either in parallel or in series or in a positive feedback loop or a negative feedback loop while maintaining the same passivity-short property (to yield L_2 stability and, under zero state observability, asymptotic stability). Furthermore, this class of passivity-short systems can be interconnected by means of distributed positive feedback interconnections (enabled by local communication networks) to achieve output consensus. The so-called impact equivalence principle enables the separation of control design for individual systems and the network-level cooperative control design, and it was shown that the overall system becomes a plug-and-play system as long as the individual systems are designed to be passivity short and have their impact coefficients bounded from above. Applications of networking passivity-short systems and their cooperative control may include sensor networks, autonomous vehicle systems [9], and smart grids.

ACKNOWLEDGMENT

Youngjun Joo provided useful discussions in the development of Property 7.

APPENDIX

Proof of Property 7. For case (1), it follows from partial fraction expansion that every entry in the scalar transfer function $G(s)$ is a weighted sum of the following terms:

$$\frac{1}{s}, \quad \frac{1}{s+\sigma_i}, \quad \frac{a_i s + b_i}{[(s+\sigma_i)^2 + \omega_i^2]^{l_i}},$$

where $\sigma_i > 0$ for all i, and l_i assumes finite positive integer values, and $G(s)$ is positive real excluding poles on the imaginary axis. Hence it follows from Fig. 1B that real parts of the entries of $G(j\omega)$ are uniformly bounded from below for all $\omega \in \Re$.

In general, for a transfer function matrix $G(j\omega)$, there exists a constant $\epsilon \geq 0$ such that

$$G(j\omega) + G^{\mathrm{T}}(-j\omega) + \frac{\epsilon}{2} I$$

is positive definite for all $\omega \in \Re$.

For case (2), if $G(s)$ is controllable and observable, we may implement an observer-based control to achieve an input feedforward passive system model as follows.

In general, given a system of the form of system (9), an observer for this system is presented as

$$\dot{\hat{z}} = A\hat{z} + Bu + L(y - \hat{y}) \quad \hat{y} = C\hat{z}, \tag{27}$$

where $\hat{z}$ and $\hat{y}$ signify the estimated state and output, respectively. The error term can be defined as $e = z - \hat{z}$, and it follows that $\dot{e} = (A - LC)e$. Then eigenvalues may be chosen to place poles and calculate the corresponding observer gain matrix L with Ackermann's formula or the Bass-Gura formula. An output feedback control may be designed in the form of $u = -K\hat{z} + u'$, where the feedback gain matrix K can be used to place the poles of $(A - BK)$ to shape the characteristics of the closed-loop state equation

$$\dot{z} = (A - BK)z + BK(z - \hat{z}) + Bu',$$

where u' is the higher-level control term.

Proof of Property 8. It follows that V is the Lyapunov function and that if $u = 0$, $\dot{V} \leq 0$.

Proof of Property 9. Under the control $u = -y/\epsilon + v$, the system is finite-gain L_2 stable since

$$\begin{aligned}\dot{V} &\leq y^{\mathrm{T}}\left(-\frac{y}{\epsilon} + v\right) + \frac{\epsilon}{2}\left(-\frac{y}{\epsilon} + v\right)^{\mathrm{T}}\left(-\frac{y}{\epsilon} + v\right)\\ &= \frac{\epsilon}{2}v^{\mathrm{T}}v - \frac{\epsilon}{2}y^{\mathrm{T}}y.\end{aligned}$$

Proof of Property 9. Under the control $u = -ky$, inequality (8) becomes

$$\dot{V} \leq -k\left(1 - \frac{\epsilon}{2}k\right)y^{\mathrm{T}}y \quad \Longrightarrow \quad \lim_{t\to\infty} y = 0.$$

Asymptotic stability can be concluded from zero state observability.

REFERENCES

[1] M. Arcak, Passivity as a design tool for group coordination, IEEE Trans. Autom. Control 52 (8) (2007) 1380–1390.

[2] N. Chopra, M.W. Spong, Passivity-based control of multi-agent systems, in: Advances in Robot Control, Springer, Berlin, 2006, pp. 107–134.

[3] A. Jadbabaie, J. Lin, A.S. Morse, Coordination of groups of mobile autonomous agents using nearest neighbor rules, IEEE Trans. Autom. Control 48 (6) (2003) 988–1001.

[4] H.K. Khalil, Nonlinear Systems, third ed., Prentice Hall, Englewood Cliffs, NJ, 2003.

[5] J.C. Willems, Dissipative dynamical systems. Part I: General theory, Arch. Ration. Mech. Anal. 45 (5) (1972) 321–351.

[6] Z. Qu, M.A. Simaan, Modularized design for cooperative control and plug-and-play operation of networked heterogeneous systems, Automatica 50 (9) (2014) 2405–2414.

[7] Z. Qu, An impact equivalence principle of separating control designs for networked heterogeneous affine systems, in: Estimation and Control of Networked Systems, vol. 3, 2012, pp. 210–215.

[8] M.J. McCourt, P.J. Antsaklis, Control design for switched systems using passivity indices, in: American Control Conference (ACC), 2010, IEEE, 2010, pp. 2499–2504.

[9] Z. Qu, Cooperative Control of Dynamical Systems: Applications to Autonomous Vehicles, Springer Science & Business Media, Berlin, 2009.

[10] C.I. Byrnes, A. Isidori, J.C. Willems, Passivity, feedback equivalence, and the global stabilization of minimum phase nonlinear systems, IEEE Trans. Autom. Control 36 (11) (1991) 1228–1240.

[11] R. Sepulchre, M. Jankovic, P. Kokotovic, Constructive Nonlinear Control, Springer, New York, 1997.

[12] H. Zhang, F.L. Lewis, Z. Qu, Lyapunov, adaptive, and optimal design techniques for cooperative systems on directed communication graphs, IEEE Trans. Ind. Electron. 59 (7) (2012) 3026–3041.

CHAPTER EIGHTEEN

Synchronizing Region Approach for Identical Linear Time-Invariant Agents[a]

K.H. Movric, M. Sebek
Czech Technical University in Prague (CVUT), Czech Republic

1 INTRODUCTION

The last 2 decades have witnessed increasing interest in multiagent networked cooperative systems, inspired by the natural occurrence of flocking and formation forming. In nature, individual creatures act only according to locally available, nearest-neighbor information [1]; however, the group as a whole succeeds in accomplishing global goals (eg, predator evasion in a heard or school, or conserving energy in avian formations). These concepts and decision schemes are applied in engineering for control of spacecraft formations, unmanned aerial vehicles, mobile robots, distributed sensor networks, etc. [2]. The canonical control problem there is *consensus* or *synchronization* of all agents' states. Early work with networked cooperative systems in continuous-time and discrete-time is presented in [3–7]. These articles generally refer to consensus without a leader, where the final synchronized state depends on precise initial conditions of all agents. By adding a command generator that is pinned to a group of other agents, one can obtain synchronization to a command trajectory for all initial conditions using a virtual leader [5], which is also termed *pinning control* [8, 9]. We name this here the *cooperative tracker problem*. Necessary and sufficient conditions for synchronization are given by the *master stability function* and the related concept of the *synchronizing region* in [9–11]. The synchronizing region method crystallized over time as a versatile and robust tool for analysis and design of cooperative control systems. For continuous-time systems it guarantees synchronization [9, 12, 13] using

[a] This publication is supported by the Czech Granting Agency GAČR grant 16-25493Y.

http://dx.doi.org/10.1016/B978-0-12-805246-4.00018-5

optimal state-feedback derived from the continuous-time algebraic Riccati equation. It is shown that feedback gain at each node yields an unbounded right-half-plane synchronizing region in the complex plane. For discrete-time systems the guaranteed synchronizing region is inherently bounded, which leads to stricter synchronizability conditions. There the discrete-time algebraic Riccati equation serves, in some sense, to maximize the synchronizing region [14, 15].

In the multiagent setting, the problems of cooperative observer convergence and synchronization are related by a duality concept for distributed systems on directed graphs [16]. It is also shown that cooperative control design and cooperative observer design can both be approached by the decoupling of the effects of graph structure from the feedback gain design via algebraic Riccati equation-based procedure. Sufficient conditions are given guaranteeing observer convergence.

Lacking the full state feedback, the static output-feedback (OPFB) cooperative synchronization control design is possible under certain stipulations. Usually one treats OPFB control problems by observers (ie, dynamic OPFB), thus increasing the order of dynamics, but static OPFB, when feasible, offers a simpler solution. The synchronizing region estimate for distributed static OPFB control of continuous-time agents is described and is found to be a conical sector in the complex plane. This is to be contrasted with the infinite right-half-plane synchronizing region of the full state feedback [17].

With practical means of distributed control (eg, wireless communication, dedicated buses, networked control), delays in information signals from neighboring agents are unavoidable [4]. In realistic applications one is interested in the effects these delays have on cooperative stability (ie, the robustness of cooperative stability to delays). A generally well-behaved system, robust to disturbances, may be sensitive to feedback delays, which can lead to poor performance and even result in loss of stability. Guaranteeing robustness of cooperative stability to delays is therefore important for all practical implementations of multiagent distributed control. Some early articles on cooperative consensus (eg, [4]) address the problem of delayed signals from the neighbors. For discrete-time synchronization, delays are treated with use of stochastic matrices [18]. For continuous-time, a uniform delay in one-dimensional single-integrator agents on undirected graphs was introduced in [4, 19, 20]. More recent work [19] extends the work in [4] to directed graphs, introducing the delay margin and a responsible eigenvalue concept.

One way to extend both early [4] and more recent [19] results on delays to general continuous-time identical linear time-invariant agents on directed graphs is offered by the theory of retarded functional differential equations (RFDEs), [20]. An essential question is that of single-agent stability. In a search for simple and flexible stability conditions, the delay-dependent criteria based on the Razumikhin theorem are a good choice. Here we are concerned with identical linear time-invariant agents, all having the same (*uniform*), constant control delay. The simplifying assumption of uniform delays is standard in the literature [4, 19, 20]. A static cooperative synchronization control design is presented that is based on the estimated delay-dependent synchronizing region. Stricter conditions for the stronger property of exponential cooperative stability with a prescribed convergence rate are also given. These results extend the delay-free conditions for convergence in [13] to agents with a uniform control delay.

The synchronizing region concept, robust by the merit of its definition, also offers a means to guarantee robustness of multiagent system cooperative stability to external disturbances and uncertainties in single-agent systems. For external disturbances one applies the H_∞ approach [21], leading to an $\mathcal{L}_2$ bound of the synchronization errors with respect to acting disturbances. For single-agent uncertainties, one assumes that agents remain identical. Nonidentical (ie, *heterogeneous*) agents require methods other than those offered by the synchronizing region, which are not considered here. If the single-agent uncertainties are considered closely, it is found that they produce different effects depending on which part of the single-agent system they occur in. The combined effect is found to be qualitatively similar to that occurring with static OPFB. Greater magnitudes of uncertainties in single-agent systems ostensibly lead to stricter conditions on graph Laplacian matrix eigenvalues.

The outline of this survey is as follows: Section 2 presents graph theory preliminaries and notational conventions. Section 3 introduces the dynamics of the systems considered, defines the control goals, and specifies the forms of cooperative signals used. Section 4 forms the main part of the survey. First it defines synchronizing regions, presenting necessary and sufficient conditions for cooperative stability. Then it provides Lyapunov estimates of those synchronizing regions, yielding sufficient conditions for cooperative stability and providing the basis for the corresponding design schemes. Section 5 briefly addresses robustness of multiagent systems to external disturbances and single-agent uncertainties by a synchronizing region method, and Section 6 presents the conclusions.

2 GRAPH PROPERTIES AND NOTATION

Consider a graph $\mathcal{G} = (\mathcal{V}, \mathcal{E})$ with a nonempty finite set of N nodes $\mathcal{V} = \{v_1, \cdots, v_N\}$ and a set of edges $\mathcal{E} \subseteq \mathcal{V} \times \mathcal{V}$. It is assumed that the graph is simple (ie, there are no repeated edges or self-loops). Directed graphs are considered, and information propagates through the graph along the edges. Two nodes v_j and v_k connected by an edge $(v_j, v_k) \in \mathcal{E}$ are termed the *parent node* (v_k) and the *child node* (v_j)—that is, the edge leaves the parent node and connects into the child node. We denote the adjacency matrix as $E = [e_{ij}]$, with $e_{ij} > 0$ if $(v_i, v_j) \in \mathcal{E}$ and $e_{ij} = 0$ otherwise. Diagonal elements satisfy $e_{ii} = 0$. The set of neighbors of node v_i is $\mathcal{N}_i = \{v_j : (v_i, v_j) \in \mathcal{E}\}$—that is, the set of nodes with edges connecting into v_i. We define the in-degree matrix as a diagonal matrix, $D = \text{diag}(d_1 \ldots d_N)$, with $d_i = \sum_j e_{ij}$, the (weighted) in-degree of node i. We define the graph Laplacian matrix as $L = D - E$, which has all row sums equal to 0.

A *directed path* is a sequence of edges joining two nodes. A graph is said to be *strongly connected* if any two nodes can be joined by a directed path. A node is termed *isolated* if it has no incoming edges. Hence in strongly connected graphs there are no isolated nodes. A *directed tree* is a subgraph containing a single node v_0, isolated in that subgraph, such that all other nodes in the subgraph except v_0 have only one parent and are joined to v_0 by a directed path. Node v_0 is called a *root node*. A graph is said to contain a directed *spanning tree* if there exists a directed tree containing every node in $\mathcal{V}$. The Laplacian matrix L has a simple zero eigenvalue if and only if its directed graph contains a spanning tree.

For a matrix $Q = Q^T \geq 0$, a square root, $Q^{1/2}$, is a matrix satisfying $(Q^{1/2})^T Q^{1/2} = Q$. We denote its matrix transpose as $Q^{T/2}$. Complex conjugation of a scalar $\sigma \in \mathbb{C}$ is denoted by an overbar: $\bar{\sigma} \in \mathbb{C}$. The abbreviations *a.s.* and *e.s.* stand for asymptotic and exponential stability, respectively.

3 SYSTEM DYNAMICS AND CONTROL GOAL

This section introduces the single-agent dynamics considered in this survey. Control goals are defined and the cooperative signals used are specified. Given a graph $\mathcal{G}(\mathcal{V}, \mathcal{E})$, endow each of its N nodes with a state vector $x_i \in \mathbb{R}^n$ and a control input $u_i \in \mathbb{R}^m$. Consider at each node the dynamics (ie, state evolution) given by a continuous-time differential equation,

$$\dot{x}_i(t) = Ax_i(t) + Bu_i(t), \tag{1}$$

or by a discrete-time difference equation,

$$x_i(k+1) = Ax_i(k) + Bu_i(k). \tag{2}$$

With control signal delays the single-agent dynamics in continuous-time is instead of Eq. (1) described by a single-delay RFDE:

$$\dot{x}_i(t) = Ax_i(t) + Bu_i(t-\tau). \tag{3}$$

The treatment of control delays in discrete-time is considerably simpler, with the dynamics described by a variant of difference equation (2) where $u(k)$ is replaced by $u(k-l)$, $l > 0$ [18].

Let the output relations with $y_i \in \mathbb{R}^p$ be given as

$$y_i(t) = Cx_i(t), \quad y_i(k) = Cx_i(k). \tag{4}$$

Assume that (A, B) is stabilizable, (A, C) is detectable, B has full column rank m, and C has full row rank, p; hence there are no redundant control inputs or redundant measurement outputs. Consider also a leader node, control node, or command generator, following the autonomous dynamics in continuous-time,

$$\dot{x}_0(t) = Ax_0(t), \tag{5}$$

or in discrete-time,

$$x_0(k+1) = Ax_0(k), \tag{6}$$

with $x_0 \in \mathbb{R}^n$. For instance, if $n = 2$ and matrix A has pure imaginary eigenvalues, then the reference trajectory is harmonically oscillating. Let the leader have the same output relation as Eq. (4), $y_0 = Cx_0 \in \mathbb{R}^p$.

Then for the above one has the following:

Control goal: The *cooperative tracker* or *synchronization problem* is to select the control signals u_i for agent i such that the states at all nodes asymptotically synchronize to the state of the control node—that is, $\lim_{t\to\infty} \|x_i(t) - x_0(t)\| = 0$ for all i in continuous-time or $\lim_{k\to\infty} \|x_i(k) - x_0(k)\| = 0$ for all i in discrete-time. These requirements must be fulfilled for all initial conditions $x_i(0)$.

3.1 Continuous-Time and Discrete-Time Cooperative Tracker

For the continuous-time multiagent systems (1) and (5), define the local neighborhood tracking errors

$$e_i = \sum_{j\in N_i} e_{ij}(x_j - x_i) + g_i(x_0 - x_i), \tag{7}$$

where the pinning gain $g_i \geq 0$ is nonzero if node v_i can sense the state of the control node (5). The intent is that only a small percentage of nodes have $g_i > 0$, yet all nodes should synchronize through distributed communication to the trajectory of the control node using local neighbor control protocols [8]. It is assumed that at least one pinning gain is nonzero. The local neighborhood tracking error represents the information available to agent i for control purposes. Choose the input of agent i as the weighted local control protocol

$$u_i = cKe_i, \tag{8}$$

where $c \in \mathbb{R}^+$ is a coupling gain to be detailed later. Then the closed-loop dynamics of individual agents (1) are

$$\dot{x}_i(t) = Ax_i(t) + cBKe_i(t). \tag{9}$$

For the discrete-time multiagent systems (2) and (6), the same local neighborhood tracking errors (7) are used for synchronization control; however, they are weighted differently [15]. Choose the input of agent i as the weighted local control protocol

$$u_i = c(1 + d_i + g_i)^{-1}Ke_i, \tag{10}$$

where $c \in \mathbb{R}^+$ is a coupling gain to be detailed later. Then the closed-loop dynamics of individual agents (2) are

$$x_i(k+1) = Ax_i(k) + c(1 + d_i + g_i)^{-1}BKe_i(k). \tag{11}$$

The multiagent systems (9) and (11) in total state-space form read

$$\dot{x}(t) = (I_N \otimes A - c(L+G) \otimes BK)(x(t) - \bar{x}_0(t)), \tag{12}$$

$$x(k+1) = (I_N \otimes A - c(I+D+G)^{-1}(L+G) \otimes BK)(x(k) - \bar{x}_0(k)), \tag{13}$$

where $x = \begin{bmatrix} x_1^T & \cdots & x_N^T \end{bmatrix}^T \in \mathbb{R}^{Nn}$, $\bar{x}_0 = \begin{bmatrix} x_0^T & \cdots & x_0^T \end{bmatrix}^T \in \mathbb{R}^{Nn}$, and $G = \text{diag}\left(g_1, \ldots, g_N\right)$ is the diagonal matrix of pinning gains. Introducing the synchronization error,

$$\delta_i = x_i - x_0, \tag{14}$$

one obtains the closed-loop dynamics in total state-space form as

$$\dot{\delta}(t) = (I_N \otimes A - c(L+G) \otimes BK)\delta(t) \tag{15}$$

in continuous-time and as

$$\delta(k+1) = (I_N \otimes A - c(I+D+G)^{-1}(L+G) \otimes BK)\delta(k) \tag{16}$$

in discrete-time. Asymptotic stability of closed-loop synchronization error systems (15) and (16) guarantees cooperative synchronization. Hence the continuous-time stability and discrete-time stability of system matrices

$$I_N \otimes A - c(L+G) \otimes BK, \tag{17}$$

$$I_N \otimes A - c(I+D+G)^{-1}(L+G) \otimes BK \tag{18}$$

determine the synchronization of multiagent systems (9) and (11).

3.2 Discrete-Time Cooperative Observer

With the availability of only outputs for feedback control, one can proceed with the design of output dynamic controllers. The first step in that direction is to design cooperative observers with $\hat{x}_i(k)$ estimating states of agents $x_i(k)$ given measurements of their outputs $y_j(k)$. Only the discrete-time systems are presented here, but the same considerations and design methods are applicable mutatis mutandis to the continuous-time case [13].

Given an estimated output $\hat{y}_i = C\hat{x}_i$ at node i, define the *local output error*

$$\tilde{y}_i = y_i - \hat{y}_i. \tag{19}$$

To construct an observer that takes into account information from the neighbors of node i, define the *local neighborhood output disagreement*

$$\varepsilon_i = \sum_j e_{ij}(\tilde{y}_j - \tilde{y}_i) + g_i(\tilde{y}_0 - \tilde{y}_i), \tag{20}$$

where the observer pinning gains are $g_i \geq 0$, with $g_i > 0$ for only a small percentage of nodes that have direct measurements of the output error $\tilde{y}_0$ of the control node. With the total output $y = \begin{bmatrix} y_1^T & \cdots & y_N^T \end{bmatrix}^T \in \mathbb{R}^{pN}$, the total output disagreement error $\varepsilon(k) \in \mathbb{R}^{pN}$ is

$$\varepsilon(k) = -(L+G) \otimes I_p\tilde{y}(k) + (L+G) \otimes I_p \underline{1}\tilde{y}_0(k). \tag{21}$$

Throughout this subsection it is assumed that $x_0 = \hat{x}_0 \Rightarrow \tilde{y}_0 \equiv 0$ (ie, the control node's state is accurately known). Exemption from that assumption would require a separate observer for the leader's states that fulfills the above assumption asymptotically. We define the *distributed observer* dynamics

$$\hat{x}_i(k+1) = A\hat{x}_i(k) + Bu_i(k) - c(1+d_i+g_i)^{-1}F\varepsilon_i(k), \tag{22}$$

where F is an observer gain matrix and c a coupling gain to be detailed later. The total distributed observer dynamics then takes the form

$$\hat{x}(k+1) = I_N \otimes A\hat{x}(k) + I_N \otimes Bu(k) + c(I+D+G)^{-1}(L+G) \otimes F\tilde{y}(k)$$

$$\hat{x}(k+1) = \left[I_N \otimes A - c\,(I+D+G)^{-1}\,(L+G) \otimes FC\right]\hat{x}(k)$$
$$+I_N \otimes Bu(k) + c\,(I+D+G)^{-1}\,(L+G) \otimes Fy(k).$$

We define the ith *state observer error*

$$\eta_i(k) = x_i(k) - \hat{x}_i(k),$$

where $x_i(k)$ follows dynamics (2). The total state-space observer error $\eta = \left[\eta_1^T \ \ \ldots \ \ \eta_N^T\right]^T \in \mathbb{R}^{nN}$ follows the discrete-time dynamics

$$\eta(k+1) = \left[I_N \otimes A - c(I+D+G)^{-1}(L+G) \otimes FC\right]\eta(k). \tag{23}$$

The total observer system matrix

$$I_N \otimes A - c(I+D+G)^{-1}(L+G) \otimes FC \tag{24}$$

determines the observer estimate $\hat{x}_i$ dynamics with respect to the true state x_i. The goal is asymptotic observer convergence for all distributed observers (ie, $\eta \to 0$). The pinning terms in the local neighborhood output disagreement mean pinning with zero. Comparing this with the cooperative tracker setup, we require a brief explanation. Pinning with zero here serves the purpose of making a fixed zero reference, which is needed for convergence of all observers' estimates, directly available to some particular observers. The remaining observers rely on relative measurements only, thus receiving information on the fixed zero reference in a distributed way.

3.3 Static Distributed Output-Feedback Cooperative Tracker

Let the multiagent systems be given as Eqs. (1), (4), and (5). We define the *local neighborhood output error*

$$e_{yi} = \sum_j e_{ij}(y_j - y_i) + g_i(y_0 - y_i), \tag{25}$$

where $g_i > 0$ for only a few nodes having direct measurements of the leader's output $y_0(t)$. The feedback control signal for agent i is chosen as a linear distributed static OPFB:

$$u_i = cKe_{yi}, \tag{26}$$

where $c > 0$ is the coupling gain and K is the local OPFB gain matrix. Standard cooperative control approaches assume distributed control in terms of state neighborhood error (7), or construct dynamic controllers in terms of output disagreement (20). By contrast, in this subsection we study distributed static OPFB of the form (26). In total state-space form, the local neighborhood output error (25) reads

$$e_y = -(L + G) \otimes C\delta, \tag{27}$$

where $e_y = \begin{bmatrix} e_{y1}^T & \cdots & e_{yN}^T \end{bmatrix}^T \in \mathbb{R}^{Np}$. We define as before (ie, by stacking single-agent vectors) all other total quantities. The total state-space form of the distributed OPFB (26) is then

$$u = -c(L + G) \otimes KC\delta. \tag{28}$$

Note that Eq. (27) reflects the restrictions on the information available for distributed static OPFB control. Matrix C specifies the local single-agent restrictions due to OPFB, while $L + G$ specifies the information restrictions on a coarser scale due to limited communication. Both local and global restrictions appear in the Kronecker product (27) in a symmetric manner.

Feedback (26) gives closed-loop single-agent systems (1) as

$$\dot{x}_i = Ax_i + cBKCe_i, \tag{29}$$

which yields the multiagent system dynamics in total state-space form:

$$\dot{x} = (I_N \otimes A)x - c(L + G) \otimes BKC\delta. \tag{30}$$

The total dynamics of synchronization errors follows readily as

$$\dot{\delta} = (I_N \otimes A - c(L + G) \otimes BKC)\delta. \tag{31}$$

Hence cooperative stability under distributed OPFB is determined by asymptotic stability of the system matrix in Eq. (31),

$$I_N \otimes A - c(L + G) \otimes BKC. \tag{32}$$

3.4 Multiagent Systems with Uniform Control Delays

This subsection describes the single agents and the total multiagent system dynamics under uniform time delays in the control signal [20]. Let the single-agent dynamics be given as a continuous-time linear time-invariant system with a fixed control signal delay $\tau \geq 0$,

$$\dot{x}_i(t) = Ax_i(t) + Bu_i(t - \tau), \tag{33}$$

where $x_i \in \mathbb{R}^n$ and $u_i \in \mathbb{R}^m$ for all i are the states and inputs. Let there be an autonomous leader given by Eq. (5). The delay $\tau \geq 0$ is taken to be constant and the same for all agents (ie, uniform). Though this need not be true in more realistic models, having different delays for different agents makes them, strictly speaking, not identical as dynamical systems. The assumption on uniform delays is a standard simplifying assumption [4, 20], and it is crucial for the following development.

With the local neighborhood error (7), let the control signal for agent i be calculated as a linear feedback of the local neighborhood error (8), where K is the local linear feedback gain matrix and $c > 0$ is the coupling gain, to be detailed later. This gives the closed-loop single-agent dynamics

$$\dot{x}_i(t) = Ax_i(t) + cBKe_i(t - \tau). \tag{34}$$

Introducing the synchronization error (14), one has in total state-space form

$$\dot{x}(t) = (I_N \otimes A)x(t) - c(L + G) \otimes BK\delta(t - \tau), \tag{35}$$

where $x = \begin{bmatrix} x_1^T & \cdots & x_N^T \end{bmatrix}^T \in \mathbb{R}^{Nn}$ and $\delta = \begin{bmatrix} \delta_1^T & \cdots & \delta_N^T \end{bmatrix}^T \in \mathbb{R}^{Nn}$. The total synchronization error system is then given as

$$\dot{\delta}(t) = (I_N \otimes A)\delta(t) - c(L + G) \otimes BK\delta(t - \tau). \tag{36}$$

Eq. (36) is a linear RFDE of the form

$$\dot{\delta}(t) = \bar{A}\delta(t) + \bar{A}_d\delta(t - \tau), \tag{37}$$

whose asymptotic stability guarantees synchronization.

4 SUFFICIENT CONDITIONS FOR SYNCHRONIZATION

This section first presents definitions of synchronizing regions for the systems presented in Section 3, and the pertaining necessary and sufficient conditions for cooperative synchronization [10, 13, 15, 16]. Then it provides Lyapunov estimates of these synchronizing regions, yielding sufficient conditions for cooperative stability. Those sufficient conditions in turn lead to design procedures.

4.1 Synchronizing Regions and Cooperative Stability Conditions

Two lemmas are given below that provide a simplification crucial in designing local feedback gains K and provide the basis for the definitions of synchronizing regions. Lemma 1 is presented without a proof, as proofs of its parts are readily found in the literature. The proof is, however, provided for Lemma 2 as it is not as straightforward as that of Lemma 1 and not as readily available.

Lemma 1 ([9, 13, 15, 16]). *The structured matrix $I_N \otimes A - c\Gamma \otimes B$ is Hurwitz if and only if all the matrices $A - c\lambda_j B$ are Hurwitz, where λ_j are the eigenvalues of Γ.*

Hence the relevant matrices in Eqs. (17), (18), (24), and (32) are Hurwitz if and only if:

1. $A - c\lambda_j BK$ *is continuous-time asymptotically stable for all* λ_j *eigenvalues of* $L + G$.
2. $A - c\lambda_j BK$ *is discrete-time asymptotically stable for all* λ_j *eigenvalues of* $(I + D + G)^{-1}(L + G)$.
3. $A - c\lambda_j FC$ *is discrete-time asymptotically stable for all* λ_j *eigenvalues of* $(I + D + G)^{-1}(L + G)$,
4. $A - c\lambda_j BKC$ *is continuous-time asymptotically stable for all* λ_j *eigenvalues of* $L + G$.

Lemma 1 applies to continuous-time and discrete-time cooperative state-feedback control (conditions 1 and 2), discrete-time cooperative observer design (condition 3) and continuous-time static OPFB control design (condition 4). Lemma 2 provides an analogous result for distributed state feedback in the case of uniform control delays.

Lemma 2 ([20]). *The trivial solution of system (36), $\delta(t) \equiv 0$, is asymptotically stable if and only if the trivial solutions of systems*

$$\dot{y}(t) = Ay(t) - c\lambda_j BKy(t - \tau), \tag{38}$$

each having an order of a single-agent system, where λ_j are the eigenvalues of the graph matrix $(L + G)$, are asymptotically stable for all j.

Proof. The stability of Eq. (38) is determined by the transcendental transfer function matrix:

$$\left[sI_{Nn} - (I_N \otimes A) + c(L + G) \otimes BKe^{-s\tau}\right]^{-1}. \tag{39}$$

The frequency-domain expression (39) needs to be asymptotically stable, a property depending on the roots of the corresponding characteristic quasipolynomial:

$$\Delta(s) = \det\left[sI_{Nn} - (I_N \otimes A) + c(L + G) \otimes BKe^{-s\tau}\right]. \tag{40}$$

Knowing that linear transformations do not change the determinant of a matrix, one can use a coordinate transformation $T \otimes I_n$ in Eq. (40), where $T^{-1}(L + G)T = \Lambda$ is an upper triangular matrix, thereby simplifying the problem considerably. One obtains

$$\det\left[sI_{Nn} - (I_N \otimes A) + c(L + G) \otimes BKe^{-s\tau}\right]$$

$$
\begin{aligned}
&= \det\left[(T \otimes I_n)^{-1}(sI_{Nn} - (I_N \otimes A) + c(L+G) \otimes BKe^{-s\tau})(T \otimes I_n)\right] \\
&= \det\left[sI_{Nn} - I_N \otimes A + c\Lambda \otimes BKe^{-s\tau}\right] \qquad (41) \\
&= \prod_j \det\left[sI_n - A + c\lambda_j BKe^{-s\tau}\right],
\end{aligned}
$$

where λ_j are the eigenvalues of Λ, equaling the eigenvalues of the graph matrix $(L+G)$. Therefore each factor in the product (41), $\det(sI_n - A + c\lambda_j BKe^{-s\tau})$, must be asymptotically stable. This requirement is equivalent to the asymptotic stability of the trivial solution for a set of linear RFDEs:

$$\dot{y}(t) = Ay(t) - c\lambda_j BKy(t-\tau). \qquad (42)$$

Systems (42) are in the form of Eq. (37) but have the order of a single-agent dynamics (ie, n). □

The proof of Lemma 2 uses a detour in the frequency domain but the result pertains to the time domain, and as such it will be used. This is in contrast with many treatments of delays in the literature using frequency-domain stability criteria (eg, the Nyquist criterion) and is more along the lines of methods guaranteeing a synchronizing region. In fact, the conclusions of Lemma 1 for continuous-time cases can be considered as special cases of Lemma 2 with zero delay. Note that Lemmas 1 and 2 provide necessary and sufficient conditions for cooperative stability. Lemmas 1 and 2 both rely on appropriate state transformations reducing the cooperative stability of high-order total dynamics to asymptotic stability of a set of lower-order systems. This reduced problem is then addressed by the synchronizing region approach.

The following definitions, motivated by Lemmas 1 and 2, are required for subsequent development.

Definition 1. Given matrices (A, B), the *synchronizing region* for a matrix pencil

$$A - \sigma B \qquad (43)$$

is a set given by $S_c = \left\{\sigma \in \mathbb{C} : A - \sigma B \textit{ is asymptotically stable}\right\}$.

Definition 2. Given a complex RDFE (38), the *delay-dependent synchronizing region* is a subset of the complex plane $\mathbb{C}$ depending on the delay τ:

$$S_c(\tau) = \left\{\sigma \in \mathbb{C} : Ay(t) + \sigma A_d y(t-\tau) \textit{ is asymptotically stable}\right\}.$$

Remark 1. The appropriate notions of stability used in Definitions 1 and 2 depend on the particular system considered. Definitions for continuous-time and discrete-time synchronizing regions applicable here are given in [13, 15–17]. Furthermore, the actual matrix B in matrix pencil (43) one has control over depends on whether one has full state feedback or only OPFB and is different in controller and observer designs. Hence according to Lemmas 1 and 2, the conditions for cooperative stability in each case reduce to having all the relevant graph matrix eigenvalues scaled by an appropriate value of the coupling gain $c > 0$ into the corresponding synchronizing region. The goal is then to design the local feedback gain matrices K and F in some way that they maximize the synchronizing region, thus facilitating cooperative stabilization.

To assess a synchronizing region, one resorts to its guaranteed estimates. Sufficient conditions guaranteeing synchronizing region estimates for matrix pencils (43) are based on Lyapunov methods and evaluation of the following single-agent Lyapunov quadratic forms:

$$(A - \sigma B)^{\dagger} P + P(A - \sigma B) \tag{44}$$

in continuous-time [12, 13, 17], and

$$(A - \sigma B)^{\dagger} P(A - \sigma B) - P \tag{45}$$

in discrete-time [15]. The presence of time delays requires a similar albeit more involved approach using Lyapunov-Razumikhin functions instead of Lyapunov functions. The following theorems provide sufficient conditions and design schemes guaranteeing synchronization for a wide class of communication graphs. Some details of the Lyapunov-Razumikhin approach are sketched only briefly, the interested reader being referred to [20] for a more detailed exposition.

4.2 Discrete-Time Cooperative Tracker Distributed Control Design

The following two theorems present design methods for distributed control gains K that guarantee discrete-time multiagent system synchronization.

Theorem 1 ([15] H_∞ inequality design for synchronization). *Given systems* (2) *and* (6)*, with protocol* (10)*, assume that the interaction graph contains a spanning tree with at least one pinning gain that connects into a root node. Then the eigenvalues of the graph matrix* $(I + D + G)^{-1}(L + G)$*,* Λ_i*, satisfy* $Re\,\Lambda_i > 0$ *for all* i*. If there exists an* $\omega \in \mathbb{R}$ *such that*

$$\delta(\omega) := \max_{j\in\{1,\ldots,N\}} |1 - \omega \Lambda_j| < \delta_c, \tag{46}$$

where δ_c is obtained by

$$\delta_c = \sup_{\delta>0} \left\{\delta | \exists P > 0 \text{ subject to } P > A^T PA - (1-\delta^2)A^T PB(B^T PB)^{-1} B^T PA\right\}, \quad (47)$$

then there exists a $P > 0$ solving

$$P > A^T PA - (1-\delta^2)A^T PB(B^T PB)^{-1} B^T PA, \quad (48)$$

with $\delta = \delta(\omega)$. Moreover, the control gain $K = (B^T PB)^{-1} B^T PA$ and the coupling gain $c = \omega$ guarantee synchronization.

Proof. Denote $\delta_j = 1 - \omega \Lambda_j$. It is found that

$$\begin{aligned} P - (A - \omega\Lambda_j BK)^{\dagger} P(A - \omega\Lambda_j BK) &= P - A^T PA + (1 - |\delta_j|^2)A^T PB(B^T PB)^{-1} B^T PA \\ &\geq P - A^T PA + (1-\delta^2)A^T PB(B^T PB)^{-1} B^T PA > 0, \end{aligned}$$

where the dagger denotes the complex conjugate transpose. Thus it follows from the single-agent discrete-time Lyapunov inequality that $\rho(A - \omega\Lambda_j BK) < 1$, which is hence discrete-time stable. The rest of the proof follows from Lemma 1 [15]. □

Remark 2. For the single-input case—that is, $\text{rank}(B) = 1$—the sufficient condition given in Theorem 1 is also necessary [15]. The importance of Theorem 1 is in showing that cooperative feedback design based on the inequality allows the separation of the single-agent feedback gain design from the detailed properties of the communication graph topology, as long as the graph contains a spanning tree with a nonzero pinning gain into a root node. Specifically, it reveals that if Riccati-based design is used for the feedback K at each node, then synchronization is guaranteed for a class of communication graphs whose graph matrix eigenvalues satisfy condition (46). This condition is appealing because it allows a disentanglement of the properties of individual agents' feedback gains, as reflected by δ_c, and the graph topology, described by Λ_k, relating these through an inequality.

Theorem 2 ([15] H_2 design for synchronization). *Given systems* (2) *and* (6), *with protocol* (10), *assume that the interaction graph contains a spanning tree with at least one pinning gain nonzero that connects into the root node. Then the eigenvalues of the graph matrix $(I+D+G)^{-1}(L+G)$, Λ_i, satisfy Re $\Lambda_i > 0$ for all i. Let $P > 0$ be a solution of the single-agent discrete-time Riccati-like equation*

$$A^T PA - P + Q - A^T PB(B^T PB)^{-1} B^T PA = 0 \quad (49)$$

for some prescribed $Q = Q^T > 0$. Define

$$r := \left[\sigma_{\max}(Q^{-T/2} A^T PB(B^T PB)^{-1} B^T PAQ^{-1/2})\right]^{-1/2}. \quad (50)$$

Then protocol (10) *guarantees synchronization of multiagent systems* (11) *for some* K *if there exists a covering circle* $C(c_0, r_0)$ *of the graph matrix eigenvalues* Λ_k, $k = 1 \ldots N$ *such that*

$$\frac{r_0}{c_0} < r. \tag{51}$$

Moreover, if condition (51) *holds, then the feedback gain matrix*

$$K = (B^T PB)^{-1} B^T PA \tag{52}$$

and the coupling gain

$$c = \frac{1}{c_0} \tag{53}$$

guarantee state synchronization.

Proof. The synchronizing region for the choice of K given in Theorem 2, Eq. (52) contains the open circle $C(1, r)$. This is seen by our choosing the state-feedback gain matrix as Eq. (52), with $P > 0$ solving Eq. (49). One has

$$A^T PA - P + Q - K^T \left(B^T PB\right) K = 0.$$

From this equation, by completing the square, one obtains

$$A^T PA - P + Q - K^T B^T PBK = (A - BK)^T P(A - BK) - P + Q = 0.$$

Therefore given the single-agent quadratic Lyapunov function $V(y) = y^T Py$, $y \in \mathbb{R}^n$, the choice of feedback gain stabilizes the system (A, B). A synchronizing region estimate can be found from the condition

$$(A - \sigma BK)^\dagger P(A - \sigma BK) = (A - \mathrm{Re}\sigma BK)^T P(A - \mathrm{Re}\sigma BK) + \mathrm{Im}^2\sigma (BK)^T PBK < P,$$

where the dagger denotes the complex conjugate transpose (Hermitian adjoint), $\sigma \in \mathbb{C}$. Therefore from the above one has

$$\begin{aligned}
&(A - \sigma BK)^\dagger P(A - \sigma BK) - P \\
&= A^T PA - P - 2\mathrm{Re}\sigma K^T (B^T PB)K + \mathrm{Re}^2\sigma (BK)^T PBK + \mathrm{Im}^2\sigma (BK)^T PBK \\
&= A^T PA - P - (1 - |\sigma - 1|^2)(BK)^T PBK = -Q + K^T (B^T PB)K \\
&\quad - (1 - |\sigma - 1|^2)(BK)^T PBK \\
&= -Q + |\sigma - 1|^2 K^T B^T PBK.
\end{aligned}$$

This implies stability if

$$Q - |\sigma - 1|^2 K^T B^T PBK > 0,$$

which gives a simple bounded complex region, more precisely a part of the complex gain margin region

$$Q > |\sigma - 1|^2 K^T B^T PBK \Rightarrow 1 > |\sigma - 1|^2 \sigma_{\max}(Q^{-T/2}(BK)^T P(BK)Q^{-1/2}).$$

This is an open circle $C(1, r)$ specified by

$$|\sigma - 1|^2 < \frac{1}{\sigma_{\max}(Q^{-T/2}K^T B^T PBKQ^{-1/2})}.$$

Furthermore, expressing K as the H_2 Riccati equation state-feedback completes this part of the proof. Now, given a circle $C(1, r)$, centered at 1 of radius r, contained in the synchronizing region S of matrix pencil $A - \sigma BK$, and the properties of dilation (homothety), assuming there exists a directed spanning tree in the graph with a nonzero pinning gain into a root node, it follows that synchronization is guaranteed if all eigenvalues Λ_k are contained in a circle $C(c_0, r_0)$ similar with respect to homothety to a circle concentric with and contained within $C(1, r)$.

The center of the covering circle c_0 can be taken on the real axis because of symmetry, and the radius equals $r_0 = \max_k |\Lambda_k - c_0|$. Taking these as given, one should have

$$\frac{r_0}{c_0} < \frac{r}{1}. \tag{54}$$

If this equation is satisfied, then our choosing $c = 1/c_0$ maps with homothety the covering circle of all eigenvalues $C(c_0, r_0)$ into a circle $C(1, r_0/c_0)$ concentric with and for $r > r_0/c_0$ contained in the interior of the circle $C(1, r)$. □

If there exists a solution $P > 0$ to Eq. (49), B must have full column rank. Assuming B has full column rank, there exists a positive definite solution P to Eq. (49) only if (A, B) is stabilizable.

Remark 3 (Comparison of H_∞ and H_2 designs for synchronization). By comparing Eq. (50) with Eq. (46), we see that a similar role is played by δ_c in Theorem 1 and r in Theorem 2. The latter can be explicitly computed with Eq. (50), but it depends on the selected Q. On the other hand, δ_c is found by numerical linear matrix inequality techniques [15]. Further, in Theorem 1, $\omega \in \mathbb{R}$ plays a role analogous to $1/c_0$ in Theorem 2. Theorem 1 relies on analysis based on the circle $C(1, \delta_c)$, whereas Theorem 2 uses $C(1, r)$. Both circles are contained in the synchronizing region of the respective designed feedbacks.

More importantly, Theorem 2 gives synchronization conditions in terms of the radius r easily computed as Eq. (50) in terms of an H_2 Riccati equation solution. Computation of the radius δ_c used in Theorem 1 must generally be done *via* a linear matrix inequality. However, the condition in Theorem 1 is milder than that in Theorem 2—that is, $C(1, r)$ is generally contained in $C(1, \delta_c)$, so systems that fail to meet the condition in Theorem 2 may yet be found synchronizable when tested according to the condition in Theorem 1.

4.3 Discrete-Time Cooperative Observer Design

Having addressed the problem of cooperative stabilization in Section 4.2, we now consider the dual problem of cooperative estimation. Cooperative estimators and observers find use, among other applications, in dynamic cooperative controllers designed to guarantee state synchronization [16]. The following theorem is based on an observer equation and provides a result dual to that of Theorem 2.

Theorem 3 ([16] Design of distributed observer gains). *Given systems* (22), *assume the interaction graph contains a spanning tree with at least one pinning gain that connects into the root node. Let $P > 0$ be a solution of the single-agent discrete-time observer equation*

$$APA^T - P + Q - APC^T(CPC^T)^{-1}CPA^T = 0, \tag{55}$$

where $Q = Q^T > 0$. Choose the observer gain matrix as

$$F = APC^T(CPC^T)^{-1}. \tag{56}$$

Define

$$r_{\text{obs}} := \left[\sigma_{\max}(Q^{-T/2}APC^T(CPC^T)^{-1}CPA^TQ^{-1/2})\right]^{-1/2}. \tag{57}$$

Then the observer errors (23) *are asymptotically stable if there exists a covering circle $C(c_0, r_0)$ of the graph matrix $(I + D + G)^{-1}(L + G)$ eigenvalues Λ_k, $k = 1 \dots N$, such that*

$$\frac{r_0}{c_0} < r_{\text{obs}}. \tag{58}$$

If Eq. (58) holds, then the coupling gain

$$c = \frac{1}{c_0} \tag{59}$$

makes the observer error dynamics (23) *stable.*

The proof follows similarly to that of Theorem 2 [16].

4.4 Continuous-Time Cooperative Tracker Distributed Static Output-Feedback Design

Under special conditions on single-agent systems it is possible to achieve cooperative stabilization by static distributed OPFB. Considerations lead to distributed design of static OPFBs, thereby avoiding the need for dynamic cooperative controllers.

Definition 3. Given matrices (A, B, C) and the static OPFB gain K, the *synchronizing region for static OPFB* is a set given by $S_y = \{\sigma \in \mathbb{C} : A - \sigma BKC \text{ is asymptotically stable}\}$.

Theorem 4 ([17]). *Let the multiagent system be given by Eqs. (1),* (4), *and* (5). *Let the graph have a spanning tree with at least one pinning gain connecting to a root node. Choose the distributed static OPFB control* (26), *where the OPFB gain K satisfies the single-agent output-type Riccati equation*

$$A^T P + PA + Q - PBR^{-1}B^T P + M^T R^{-1} M = 0 \tag{60}$$

for some $Q = Q^T > 0$, $R = R^T > 0$, and $R = R^T > 0$, together with

$$KC = R^{-1}(B^T P + M). \tag{61}$$

In addition, let the inequality

$$C^T K^T RKC - C^T K^T M - M^T KC \geq 0 \tag{62}$$

be satisfied. Then the guaranteed synchronizing region for static OPFB is a conical sector in the complex plane.

Proof. Introduce the following abbreviations:

$$a := \lambda_{\min>0}(Q^{-T/2} C^T K^T RKCQ^{-1/2}), \tag{63}$$

$$b := \lambda_{\min>0}(Q^{-T/2}(C^T K^T RKC - C^T K^T M - M^T KC)Q^{-1/2}), \tag{64}$$

$$f := \lambda_{\max}(jQ^{-T/2}(M^T KC - C^T K^T M)Q^{-1/2}), \tag{65}$$

where $\lambda_{\min>0}$ denotes the smallest eigenvalue on the complement of the kernel.

Take the single-agent quadratic Lyapunov function $V(x) = x^\dagger Px$, $x \in \mathbb{C}^n$, for the complex matrix pencil $A - \sigma BKC$. The dagger denotes the Hermitian adjoint. The positive definite real matrix, $P = P^T$, is chosen as a solution of the output-type Riccati equation (60). The time derivative of this Lyapunov function is determined by the quadratic form of a Hermitean matrix:

$$(A - \sigma BKC)^\dagger P + P(A - \sigma BKC) = A^T P + PA - \bar{\sigma} C^T K^T B^T P - \sigma PBKC,$$

which, by completion of the squares, reads

$$A^TP + PA + (\sigma KC - R^{-1}(B^TP + M))^{\dagger}R(\sigma KC - R^{-1}(B^TP + M)) \\ - |\sigma|^2\, C^TK^TRKC + \bar{\sigma}\, C^TK^TM + \sigma M^TKC - C^TK^TRKC.$$

The choice of the OPFB satisfying Eq. (61) makes this expression equal to

$$A^TP + PA + |\sigma - 1|^2\, C^TK^TRKC - |\sigma|^2\, C^TK^TRKC + \bar{\sigma}\, C^TK^TM \\ + \sigma M^TKC - C^TK^TRKC.$$

Using the single-agent output-type Riccati equation (60), one obtains

$$= -Q - (B^TP + M)^TR^{-1}(B^TP + M) - M^TR^{-1}M + PBR^{-1}B^TP \\ + (1 - 2\mathrm{Re}\sigma)C^TK^TRKC + \bar{\sigma}\, C^TK^TM + \sigma M^TKC.$$

The sufficient condition guaranteeing asymptotic stability is then

$$-Q - (B^TP + M)^TR^{-1}(B^TP + M) - M^TR^{-1}M + PBR^{-1}B^TP \\ + (1 - 2\mathrm{Re}\sigma)C^TK^TRKC + \bar{\sigma}\, C^TK^TM + \sigma M^TKC < 0,$$

allowing assessment of the synchronizing region for static OPFB in $\mathbb{C}$. This expression is equivalent to

$$-Q - |\sigma|^2\, C^TK^TRKC + |\sigma - 1|^2\, C^TK^TRKC \\ + (\bar{\sigma} - 1)C^TK^TM + (\sigma - 1)M^TKC < 0. \tag{66}$$

With $|\sigma - 1|^2 - |\sigma|^2 = (1 - 2\mathrm{Re}\sigma)$, Eq. (66) becomes

$$-Q - (2\mathrm{Re}\sigma - 1)C^TK^TRKC + (\mathrm{Re}\sigma - 1)(C^TK^TM + M^TKC) \\ + j\mathrm{Im}\sigma(M^TKC - C^TK^TM) < 0.$$

After straightforward manipulations, condition (66) reads

$$-I - \mathrm{Re}\sigma(Q^{-T/2}C^TK^TRKCQ^{-1/2}) \\ - (\mathrm{Re}\sigma - 1)Q^{-T/2}(C^TK^TRKC - C^TK^TM - M^TKC)Q^{-1/2} \\ + \mathrm{Im}\sigma(jQ^{-T/2}(M^TKC - C^TK^TM)Q^{-1/2}) < 0. \tag{67}$$

Inequality Eq. (67) is satisfied for $\mathrm{Re}\sigma \geq 1$ if

$$-1 - \mathrm{Re}\sigma\lambda_{\min>0}(Q^{-T/2}C^TK^TRKCQ^{-1/2}) \\ - (\mathrm{Re}\sigma - 1)\lambda_{\min>0}(Q^{-T/2}(C^TK^TRKC - C^TK^TM - M^TKC)Q^{-1/2}) \\ + |\mathrm{Im}\sigma|\,\lambda_{\max}(jQ^{-T/2}(M^TKC - C^TK^TM)Q^{-T/2}) < 0. \tag{68}$$

Even though the last matrix term in Eq. (67) is antisymmetric, it cannot be disregarded for the evaluation of the quadratic form. This is because

j multiplying it makes it Hermitian, and one has complex vector spaces here. Namely, the proof of Lemma 1 uses a coordinate transformation that is generally complex. This matrix has a special property that its eigenvalues appear as $\pm\lambda \in \mathbb{R}$, which is seen by complex conjugation of the eigenvalue-eigenvector relation. Hence the absolute value of $\text{Im}\sigma$ is needed in Eq. (68) compared with Eq. (67).

With the abbreviations and shorthand notation introduced, $v := \text{Re}\sigma, w := \text{Im}\sigma$, a guaranteed synchronizing region for OPFB is described as $v \geq 1$ and

$$|w| < [1 - b + (a + b)v]/f, \tag{69}$$

which describes a conical sector in the complex plane. This completes the proof. □

Corollary 1. *If $M = 0$, one recovers the synchronizing region characteristic of the full state feedback [12, 13], described by the matrix inequality*

$$-Q + (1 - 2\text{Re}\sigma)C^T K^T RKC < 0.$$

However, given the prescribed output measurement matrix C in Eq. (27), the choice $M = 0$ generally cannot be made [17]. That is, there may exist $M \neq 0$ such that Eq. (61) holds—that is, $K^*C = R^{-1}(B^T P + M)$—yet $K^*C = R^{-1}B^T P$ fails to hold.

As in all cases hitherto presented with a guaranteed OPFB synchronizing region, one obtains cooperative stability by scaling all the graph matrix eigenvalues into that region by an appropriate choice of a coupling gain $c > 0$. With the synchronizing region having the form of a conical sector, one finds a restriction on the argument of the complex eigenvalues of the graph matrix. As a special case of OPFB, with $C = I_n$ and $M = 0$, the proof of Theorem 4 reproduces the well-known continuous-time full state feedback unbounded right-half-plane synchronizing region [13].[2] The cooperative stability conditions presented here are only sufficient, being based on Lyapunov estimates of guaranteed synchronizing regions, while Lemmas 1 and 2 give necessary and sufficient conditions.

[2] It is mainly for this reason that the conventional continuous-time full state distributed feedback cooperative tracker is not treated separately to parallel its discrete-time counterpart in Section 4.2.

4.5 Delayed Control Cooperative Tracker Design

This subsection presents an application of the synchronizing region method to asymptotic and exponential cooperative stability of multiagent systems having a uniform time delay in control signals. Full state distributed feedback is considered for simplicity. Investigating such systems is the first step in tackling the effects of time delays on cooperative stability. Although in realistic systems the assumption of uniform delays seems restrictive, it is nonetheless frequently adopted for simplicity and feasibility of analysis.

4.5.1 Asymptotic Delay-Dependent Cooperative Stability

As variants of Lyapunov stability analysis applicable to time-delay systems are not considered to be generally familiar, the basic results on Lyapunov-Razumikhin stability conditions are given in the following theorem. Those results are then specialized to linear RFDEs and quadratic Lyapunov-Razumikhin functions, which are subsequently applied to complex RFDE (38) to guarantee a delay-dependent synchronizing region estimate.

Theorem 5 (Lyapunov-Razumikhin stability theorem [20]). *Let α, β, and γ be class $\mathcal{K}$ functions, and let $p(s) > s$ for all $s > 0$ be a scalar continuous nondecreasing function. If there exists a continuous function $V(x) : \mathbb{R}^n \to \mathbb{R}$ such that for some $k > 0$, for all $x_t \in \mathcal{C}$ satisfying $V(x_t(\vartheta)) \leq k$ for all $\vartheta \in [t-\tau, t]$, one has $\alpha(\|x\|) \leq V(x) \leq \beta(\|x\|)$,*

$$\frac{d}{dt}V(x(t)) \leq -\gamma(\|x(t)\|) \text{ if } V(x(\vartheta)) < p(V(x(t)), \quad \forall\vartheta \in [t-\tau, t],$$

then the trivial solution $x(t) \equiv 0$ of the RFDE considered is uniformly asymptotically stable. Furthermore, the set in $\mathcal{C}$ where $V(x_t(\vartheta)) \leq k$ for all $\vartheta \in [t-\tau, t]$ is an invariant subset in the region of attraction in the space $\mathcal{C}$.

On the basis of Theorem 5, sufficient conditions for asymptotic stability depending on delays are derived for the special case of a linear single-delay complex RFDE having the form (37) or (38). The Lyapunov-Razumikhin functions appearing in Theorem 5 are modified to accommodate the fact that the system is complex. State vectors x are allowed to be complex and a dagger denotes the Hermitian adjoint (ie, transposition and complex conjugation). Real $P = P^T > 0$ then give real-valued single-agent quadratic Lyapunov-Razumikhin functions, $V(x) = x^{\dagger}Px > 0$, $x \in \mathbb{C}^n$.

Theorem 6 (Lyapunov-Razumikhin delay-dependent result for the complex equation [20]). *Let there exist real positive definite symmetric matrices $P, P_1, P_2 > 0$, and a constant $\tau_0 > 0$ so that the following conditions are satisfied:*

$$P_1^{-1} < P, P_2^{-1} < P, \tag{70}$$

$$(A+\sigma A_d)^{\dagger}P + P(A+\sigma A_d) + \tau_0 |\sigma|^2 PA_dAP_1A^TA_d^TP$$
$$+ \tau_0 |\sigma|^4 PA_d^2P_2(A_d^2)^TP + 2\tau_0 P < 0. \tag{71}$$

Then, for given $\sigma \in \mathbb{C}$, the trivial solution, $x(t) \equiv 0$, of the equation

$$\dot{x}(t) = Ax(t) + \sigma A_d x(t-\tau) \tag{72}$$

is uniformly asymptotically stable for $0 \le \tau \le \tau_0$.

The Lyapunov-Razumikhin delay-dependent sufficient stability conditions of Theorem 6 in turn provide a guaranteed estimate of the delay-dependent synchronizing region when applied to single-agent complex RFDE (72).

Theorem 7 ([20]). *Let (A, B) be stabilizable. Choose the local feedback gain as*

$$K = R^{-1}B^TP, \tag{73}$$

with $P > 0$ a real symmetric solution of the single-agent algebraic matrix equation

$$A^TP + PA + Q - PBR^{-1}B^TP + \tau_0(PA_dAP_1A^TA_d^TP + PA_d^2P_2(A_d^2)^TP + 2P) = 0. \tag{74}$$

where $A_d = -BK = -BR^{-1}B^TP$ and $Q > 0$, and where $P_1 > 0$ and $P_2 > 0$ are symmetric matrices chosen to satisfy

$$P_1^{-1} < P, P_2^{-1} < P.$$

Then the single-agent closed-loop system, given by $A + A_d = A - BK$, is asymptotically stable with the delay margin τ_0, and the guaranteed synchronizing region estimate for delay margin τ is determined by the inequalities

$$\mathrm{Re}\sigma > 1/2, \tag{75}$$

$$(1-2\mathrm{Re}\sigma)Q^{-T/2}PBR^{-1}B^TPQ^{-1/2} + Q^{-T/2}((\tau|\sigma|^2 - \tau_0)PA_dAP_1A^TA_d^TP$$
$$+ (\tau|\sigma|^4 - \tau_0)PA_d^2P_2(A_d^2)^TP + 2(\tau - \tau_0)P)Q^{-1/2} < I. \tag{76}$$

Proof. If the real positive definite symmetric solution of Eq. (71) exists, with the feedback gain (70), the closed-loop system matrix $A + A_d = A - BR^{-1}B^TP$ satisfies

$$(A - BR^{-1}B^TP)^TP + P(A - BR^{-1}B^TP) + \tau_0(PA_dAP_1A^TA_d^TP + PA_d^2P_2(A_d^2)^TP + 2P)$$
$$= -Q - PBR^{-1}B^TP < 0,$$

guaranteeing stability of the closed-loop single-agent system ($\sigma = 1$) with the delay margin τ_0. The synchronizing region for the delay margin τ, generally different from τ_0, follows as

$$\begin{aligned}
&(A - \sigma BR^{-1}B^T P)^{\dagger}P + P(A - \sigma BR^{-1}B^T P) + \tau(|\sigma|^2 PA_dAP_1A^TA_d^TP \\
&\quad + |\sigma|^4 PA_d^2P_2(A_d^2)^TP + 2P) \\
&= A^TP + PA - 2\mathrm{Re}\sigma(PBR^{-1}B^TP) + \tau(|\sigma|^2 PA_dAP_1A^TA_d^TP \\
&\quad + |\sigma|^4 PA_d^2P_2(A_d^2)^TP + 2P) \\
&= -Q + (1 - 2\mathrm{Re}\sigma)PBR^{-1}B^TP - \tau_0(PA_dAP_1A^TA_d^TP \\
&\quad + PA_d^2P_2(A_d^2)^TP + 2P) + \tau(|\sigma|^2 PA_dAP_1A^TA_d^TP \\
&\quad + |\sigma|^4 PA_d^2P_2(A_d^2)^TP + 2P) \\
&= -Q + (1 - 2\mathrm{Re}\sigma)PBR^{-1}B^TP + ((\tau|\sigma|^2 - \tau_0)PA_dAP_1A^TA_d^TP \\
&\quad + (\tau|\sigma|^4 - \tau_0)PA_d^2P_2(A_d^2)^TP + 2(\tau - \tau_0)P).
\end{aligned}$$

For stability the above matrix expression needs to be negative definite, which is equivalent to

$$\begin{aligned}
&-I + (1 - 2\mathrm{Re}\sigma)Q^{-T/2}PBR^{-1}B^TPQ^{-1/2} + Q^{-T/2}((\tau|\sigma|^2 - \tau_0) \\
&\quad PA_dAP_1A^TA_d^TP + (\tau|\sigma|^4 - \tau_0)PA_d^2P_2(A_d^2)^TP \\
&+ 2(\tau - \tau_0)P)Q^{-1/2} < 0.
\end{aligned}$$

Condition (74) guarantees single-agent stability with delay margin τ_0 when delays τ are disregarded. This concludes the proof. □

Note that a real positive definite solution to Eq. (74) surely exists under the conditions of Theorem 7 for the delay margin τ_0 sufficiently small, as it equals the solution of the Algebraic Riccati Equation with the Q matrix appropriately redefined by subtraction of the resulting term multiplied by τ_0. For sufficiently small τ_0, the Q matrix thus redefined is still positive definite.

Corollary 2 ([20]). *Let (A, B) be stabilizable. Choose the local feedback gain as*

$$K = R^{-1}B^TP, \tag{77}$$

with $P > 0$ a real symmetric solution of the single-agent algebraic equation

$$A^TP + PA + Q - PBR^{-1}B^TP = 0, \tag{78}$$

where $Q = Q^T > 0$. Choose symmetric matrices $P_1 > 0$ and $P_2 > 0$ satisfying $P_1^{-1} < P, P_2^{-1} < P$. Then the guaranteed estimate of the delay-dependent synchronizing region $S_c(\tau)$ is determined by the inequalities

$$\mathrm{Re}\sigma > 1/2, \tag{79}$$

$$(1-2\mathrm{Re}\sigma)Q^{-T/2}PBR^{-1}B^TPQ^{-1/2} + \tau Q^{-T/2}\left[|\sigma|^2 PBKAP_1A^TK^TB^TP + |\sigma|^4 P(BK)^2P_2(K^TB^T)^2P + 2P\right]Q^{-1/2} < I. \tag{80}$$

Building on a guaranteed delay-dependent synchronizing region estimate of Theorem 7 and Corollary 2, the following result provides a design prescription for state synchronization under delayed distributed state feedback.

Theorem 8 ([20]). *Let the graph contain the spanning tree with pinning to a root node. Choose the local feedback matrix as the gain of Corollary 2. Then the multiagent system* (35) *synchronizes for the choice of*

$$c \geq \frac{1}{2\min \mathrm{Re}\lambda(L+G)} \tag{81}$$

if the delay margin τ *satisfies the condition*

$$(1-2\mathrm{Re}c\lambda_j)Q^{-T/2}PBR^{-1}B^TPQ^{-1/2} + \tau Q^{-T/2}\left[\left|c\lambda_j\right|^2 PBKAP_1A^TK^TB^TP + \left|c\lambda_j\right|^4 P(BK)^2P_2(K^TB^T)^2P + 2P\right]Q^{-1/2} < I \tag{82}$$

for all $c\lambda_j$.

4.5.2 Exponential Delay-Dependent Cooperative Stability

Conditions for exponential cooperative stability are presented here as a strengthening of the asymptotic cooperative stability conditions presented in Section 4.5.1.

Definition 4. Given a complex RFDE (38), the *delay-dependent synchronizing region for exponential stability* with a prescribed convergence rate $a \in \mathbb{R}^+$ is a subset of the complex plane $\mathbb{C}$ depending on the delay τ,

$$S_c(\tau, a) = \left\{\sigma \in \mathbb{C} : Ay(t) + \sigma A_d y(t-\tau) \text{ is exponentially stabile}\right\}, \tag{83}$$

with exponential stability characterized by the prescribed convergence rate a.

A sufficient condition for asymptotic stability with an exponential convergence rate for RFDEs based on Razumikhin stability functions is detailed in the following theorem [20].

Theorem 9 (Lyapunov-Razumikhin exponential stability result). *Let the conditions of Theorem 5 be satisfied with* $\lambda \|x\|^p = \alpha(\|x\|)$ *for some* $\lambda > 0$, $p \in \mathbb{R}$, $p \geq 1$, $\gamma > 0$, *and*

$$\frac{d}{dt}V(x(t)) \leq -\gamma V(x(t)) \text{ if } V(x(\vartheta)) < V(x(t))e^{\gamma\tau}, \quad \forall \vartheta \in [t-\tau, t]. \tag{84}$$

Then the trivial solution $x(t) \equiv 0$ of the RFDE considered is exponentially stable with convergence rate γ/p.

Linear delay differential systems—for example, Eq. (37)—have a spectrum-determined growth property. Therefore in the case of exponential stability, the convergence rate is determined by the spectral abscissa. Theorem 9 applied to Eq. (37) provides a Lyapunov-Razumikhin bound on the spectral abscissa. The following result guarantees exponential stability with a prescribed convergence rate dependent on delays for the complex RFDE, [20]. It is considered a stricter version of Theorem 6.

Theorem 10 (Lyapunov-Razumikhin delay-dependent exponential stability result for the complex equation [20]). *Let there exist real positive definite symmetric matrices $P, P_1, P_2 > 0$ and a constant $\tau_0 > 0$ so that the following conditions are satisfied:*

$$\begin{aligned} &P_1^{-1} < P, P_2^{-1} < P, \\ &\quad (A+\sigma A_d)^\dagger P + P(A+\sigma A_d) + \tau_0 |\sigma|^2 PA_dAP_1A^TA_d^TP \\ &\quad + \tau_0 |\sigma|^4 PA_d^2P_2(A_d^2)^TP + 2\tau_0 e^{2\gamma\tau_0}P < -\gamma P, \end{aligned} \tag{85}$$

where $\gamma \in \mathbb{R}^+$. Then, for given $\sigma \in \mathbb{C}$, the trivial solution, $y(t) \equiv 0$, of the equation

$$\dot{y}(t) = Ay(t) + \sigma A_d y(t-\tau)$$

is exponentially stable for $0 \leq \tau \leq \tau_0$, with convergence rate $\gamma/2$.

Proof. The proof follows from Theorems 9 and 6. With the quadratic Lyapunov-Razumikhin function, one has $p = 2$ in Theorem 9. Provisions (84) of Theorem 9 on the interval $[t-2\tau, t]$; $V(x(\vartheta)) < V(x(t))e^{2\gamma\tau}$ for all $\vartheta \in [t-2\tau, t]$ yield the inequality

$$\begin{aligned} \frac{d}{dt}V(x(t)) \leq x^\dagger(t)\big[&(A+\sigma A_d)^\dagger P + P(A+\sigma A_d) + \tau |\sigma|^2 PA_dAP_1A^TA_d^TP \\ &+\tau |\sigma|^4 PA_d^2P_2(A_d^2)^TP + 2\tau e^{2\gamma\tau}P\big]x(t). \end{aligned}$$

By condition (85) of Theorem 10, for $0 \leq \tau \leq \tau_0$,

$$\frac{d}{dt}V(x(t)) \leq -\gamma V(x(t)),$$

thus, according to Theorem 9, completing the proof. □

Exponential stability of the transformed model with a prescribed convergence rate, used to derive the stability condition (85), is sufficient to guarantee the same for the original system [20]. Exponential stability with a prescribed convergence rate is stronger than asymptotic stability; hence a synchronizing region satisfying the prescribed exponential convergence rate is contained in the asymptotic stability synchronizing region for the same time delay,

$$S_c(\tau, a) \subseteq S_c(\tau).$$

Because of exponential terms, the exponential synchronizing region estimate Eq. (85) strongly depends on the values of the delay τ and the prescribed convergence rate a. As Theorem 10 shows, the inclusion property with respect to decreasing delays holds naturally for prescribed convergence exponential stability synchronizing region estimates Eq. (85). Furthermore, the synchronizing region for greater convergence rates a is necessarily contained in that for smaller ones,

$$S_c(\tau, a_1) \subseteq S_c(\tau, a_2) \ \text{ for } a_1 \geq a_2.$$

These considerations reflect the robustness of exponential stability with respect to delays. The design method of Theorem 8 is equally applicable to a synchronizing region with a prescribed exponential convergence rate. One needs to consider only the single-agent algebraic matrix equation

$$\begin{aligned} &A^T P + PA + Q - PBR^{-1}B^T P + \tau_0(PA_d A P_1 A^T A_d^T P + PA_d^2 P_2 \\ &\qquad (A_d^2)^T P + 2\mathrm{e}^{2\gamma\tau_0} P) = -\gamma P \end{aligned}$$

guaranteeing exponential stability for the single-agent system, instead of asymptotic stability guaranteed by Eq. (74). Then an estimate of the synchronizing region for exponential stability with the same convergence rate a but possibly different delays τ follows similarly as in Theorem 7 from the single-agent matrix inequality

$$\begin{aligned} &- Q + (1 - 2\mathrm{Re}\sigma)PBR^{-1}B^T P + (\tau\, |\sigma|^2 - \tau_0)PA_d A P_1 A^T A_d^T P \\ &+ (\tau\, |\sigma|^4 - \tau_0)PA_d^2 P_2 (A_d^2)^T P + 2(\tau \mathrm{e}^{2\gamma\tau} - \tau_0 \mathrm{e}^{2\gamma\tau_0})P < 0. \end{aligned}$$

For the delay-free case, $\tau = 0 = \tau_0$, the above results naturally reduce to the results presented in the preceding subsections (ie, those not considering time delays), leading to unbounded synchronizing regions.

If all the diagonal blocks in Eq. (41)—that is, all the systems in Eq. (42)—are exponentially stable with a prescribed convergence rate, as guaranteed by the delay-dependent synchronizing region, then the original system (36) shares the same convergence property. Therefore synchronization is guaranteed with a prescribed exponential convergence rate if all the graph matrix eigenvalues are scaled into the pertaining synchronizing region by an appropriate coupling gain.

5 ROBUST SYNCHRONIZING REGION FOR DISTURBANCES AND UNCERTAIN SYSTEMS

Design of cooperative controls robust to external disturbances and single-agent uncertainties can also be approached by a synchronizing region method. This section presents H_∞-synchronizing regions for robustness to disturbances and synchronizing regions for uncertain agents.

5.1 H_∞-Synchronizing Regions

Reference [21] introduced the concept of an H_∞-synchronizing region for agents

$$\dot{x}_i = Ax_i + Bu_i + D\omega_i$$

acted on by external disturbances ω_i. The regulated outputs are $z_i = N^{-1}C\sum_{j=1}^{N}(x_i - x_j)$ and those signals are related to the total synchronization error. Graph topology is assumed undirected and the distributed state feedback (8) is used. An appropriate state transformation of the total multiagent system dynamics leads to consideration of systems of the form

$$\begin{aligned}\dot{\hat{x}}_i &= (A - c\lambda_i BK)\hat{x}_i + D\hat{\omega}_i,\\ \hat{z}_i &= C\hat{x}_i,\end{aligned}$$

which are required to be asymptotically stable in the absence of disturbances, and to have an H_∞ bound on output $\hat{z}_i$ with respect to disturbance $\hat{\omega}_i$ less than γ for all relevant eigenvalues of the graph matrix, λ_i. This motivates our interest in the system

$$\begin{aligned}\dot{\zeta}_i &= (A - \sigma BK)\zeta_i + D\omega_i,\\ z_i &= C\zeta_i,\end{aligned}$$

which defines the H_∞-synchronizing region, $\sigma \in \mathbb{R}$, for undirected graphs. Then an H_∞ inequality

$$A^TP + PA - \delta PBB^TP + \gamma^{-2}PDD^TP + C^TC < 0$$

for any $\delta > 0$ is used to provide a guaranteed H_∞-synchronizing region estimate and to subsequently construct the single-agent feedback gain:

$$K = \frac{1}{2} B^T P.$$

This feedback gain in distributed state feedback (8) guarantees an $\mathcal{L}_2$ bound less than γ for all regulated outputs z_i with respect to disturbances if the coupling gain $c > 0$ is chosen such that all relevant graph matrix eigenvalues λ_i satisfy $c\lambda_i \geq \delta$. For more details on H_∞-synchronizing regions, the interested reader is referred to [21].

5.2 Synchronizing Regions for Uncertain Agents

For single-agent uncertainties, assume bounded yet otherwise unknown uncertainties in the system matrix A, denoted by δA, and in the input-to-state coupling matrix B, denoted likewise by δB. For clarity, only distributed state feedback is presented here. Consider the guaranteed synchronizing region estimate for an uncertain matrix pencil $(A + \delta A, B + \delta B)$ with feedback gain K:

$$A + \delta A - \sigma(B + \delta B)K.$$

The single-agent Lyapunov stability condition is constructed as

$$(A + \delta A)^T P + P(A + \delta A) - \sigma P(B + \delta B)K - \bar{\sigma} K^T (B + \delta B)^T P < 0.$$

Choosing the state-feedback gain as designed for a certain part of single-agent system (A, B),

$$K = R^{-1} B^T P,$$

where $P > 0$ is a real symmetric solution of the single-agent algebraic Riccati equation

$$A^T P + PA - PBR^{-1} B^T P = -Q,$$

leads to the stability requirement

$$\begin{aligned} &A^T P + PA + \delta A^T P + P\delta A - \sigma PBR^{-1} B^T P - \sigma P\delta BR^{-1} B^T P - \bar{\sigma} PBR^{-1} \\ &\qquad B^T P - \bar{\sigma} PBR^{-1} \delta B^T P < 0 \\ &\qquad \Rightarrow A^T P + PA - 2\mathrm{Re}\sigma PBR^{-1} B^T P + \delta A^T P + P\delta A - \sigma P\delta BR^{-1} \\ &\qquad\quad B^T P - \bar{\sigma} PBR^{-1} \delta B^T P \\ &\qquad = -Q + (1 - 2\mathrm{Re}\sigma) PBR^{-1} B^T P + (\delta A^T P + P\delta A) \end{aligned}$$

$$
\begin{aligned}
&- (\sigma P\delta BR^{-1}B^TP + \bar{\sigma}PBR^{-1}\delta B^TP)\\
&= -Q + (1 - 2\mathrm{Re}\sigma)PBR^{-1}B^TP + (\delta A^TP + P\delta A)\\
&- \mathrm{Re}\sigma(P\delta BR^{-1}B^TP + PBR^{-1}\delta B^TP) - j\mathrm{Im}\sigma(P\delta BR^{-1}B^TP\\
&- PBR^{-1}\delta B^TP) < 0.
\end{aligned}
$$

It is seen from the above expression that uncertainties δA contribute terms independent of σ which can be robustly taken care of by sufficiently large $\mathrm{Re}\sigma$. For a fixed solution P, a magnitude bound on δA gives a constant bounded indefinite term. Hence, because of δA uncertainties, the shape of a guaranteed synchronizing region remains qualitatively the same. However, uncertainties δB contribute terms depending on σ, with $\mathrm{Re}\sigma$ and $\mathrm{Im}\sigma$ multiplying Hermitian matrices. Comparing this matrix expression with that for constant OPFB, Eq. (67), leads one to conclude that the guaranteed robust synchronizing region has a conical form; the uncertainties in B change the shape of the guaranteed synchronizing region qualitatively as well, (c.f. proof of Theorem 4). So the greater the magnitude of uncertainties δB, the narrower the guaranteed synchronizing region estimate in the complex plane.

It is worth emphasizing here yet again that this is only a sufficient and not a necessary stability condition; hence one operates with guaranteed synchronizing region estimates rather than exact synchronizing regions.

6 CONCLUSIONS

We have reviewed some methods of cooperative stability analysis and control design for various identical-agent multiagent system state-synchronization problems. Cooperative state feedback, observer design, static OPFB, and delayed state feedback were addressed. The common tread is the synchronizing region approach: separating the detailed graph topology from single-agent dynamics and thus guaranteeing robustness and flexibility. This versatile approach crystallized in the past few years into a useful analysis and design tool. It was shown in this survey, for the first time to the knowledge of the authors, that this method is also applicable to uncertain single-agent systems. Single-agent uncertainties are found to restrict the graph topologies that allow robust synchronization in terms of magnitudes of uncertainties.

REFERENCES

[1] T. Vicsek, A. Czirok, E. Ben-Jacob, I. Cohen, O. Schochet, Novel type of phase transitions in a system of self-driven particles, Phys. Rev. Lett. 75 (1995) 1226–1229.

[2] R. Olfati-Saber, J.A. Fax, R. Murray, Consensus and cooperation in networked multi-agent systems (invited paper), Proc. IEEE 95 (1) (2007) 215–233.

[3] J. Tsitsiklis, Problems in decentralized decision making and computation (Ph.D. dissertation), Dept. Elect. Eng. and Comput. Sci., MIT, Cambridge, MA, 1984.

[4] R. Olfati-Saber, R.M. Murray, Consensus problems in networks of agents with switching topology and time-delays, IEEE Trans. Autom. Control 49 (9) (2004) 1520–1533.

[5] A. Jadbabaie, J. Lin, A. Morse, Coordination of groups of mobile autonomous agents using nearest neighbour rules, IEEE Trans. Autom. Control 48 (6) (2003) 988–1001.

[6] J.A. Fax, R.M. Murray, Information flow and cooperative control of vehicle formations, IEEE Trans. Autom. Control 49 (9) (2004) 1465–1476.

[7] W. Ren, R.W. Beard, Consensus seeking in multiagent systems under dynamically changing interaction topologies, IEEE Trans. Autom. Control 50 (2005) 655–661.

[8] X.F. Wang, G. Chen, Pinning control of scale free dynamical networks, Physica A 310 (2002) 521–531.

[9] Z. Li, Z. Duan, G. Chen, L. Huang, Consensus of multiagent systems and synchronization of complex networks: a unified viewpoint, IEEE Trans. Circuits Syst. I: Reg. Pap. 57 (1) (2010) 213–224.

[10] L.M. Pecora, T.L. Carroll, Master stability functions for synchronized coupled systems, Phys. Rev. Lett. 80 (10) (1998) 2109–2112.

[11] Z. Duan, G. Chen, L. Huang, Disconnected synchronized regions of complex dynamical networks, IEEE Trans. Autom. Control 54 (4) (2009) 845–849.

[12] S.E. Tuna, LQR-based coupling gain for synchronization of linear systems, 2008, arXiv:0801.3390v1 [math.OC].

[13] H. Zhang, F.L. Lewis, Optimal design for synchronization of cooperative systems: state feedback, observer and output feedback, IEEE Trans. Autom. Control 56 (8) (2011) 1948–1953.

[14] N. Elia, S.K. Mitter, Stabilization of linear systems with limited information, IEEE Trans. Autom. Control 46 (9) (2001) 1384–1400.

[15] K. Hengster-Movric, K. You, F.L. Lewis, L. Xie, Synchronization of discrete-time multi-agent systems on graphs using Riccati design, Automatica 49 (2) (2012) 414–423.

[16] K. Hengster-Movric, F.L. Lewis, Cooperative observers and regulators for discrete-time multiagent systems, Int. J. Robust Nonlinear Control 23 (14) (2012) 1545–1562.

[17] K. Hengster-Movric, L. Lewis, Distributed static output-feedback control for state synchronization in networks of identical LTI agents, Automatica 53 (2015) 282–290.

[18] H.G. Tanner, D.K. Christodoulakis, State synchronization in local-interaction networks is robust with respect to time delays, in: 44th IEEE Conference on Decision and Control. 2005 and 2005 European Control Conference. CDC-ECC '05, December 12–15, 2005, pp. 4945–4950, doi:10.1109/CDC.2005.1582945.

[19] R. Sipahi, W. Qiao, Responsible eigenvalue concept for the stability of a class of single-delay consensus dynamics with fixed topology, IET Control Theory Appl. 5 (4) (2011) 622–629.

[20] K. Hengster-Movric, F. Lewis, M. Šebek, T. Vyhlidal, Cooperative synchronization control for agents with control delays: a synchronizing region approach, J. Frankl. Inst. 352 (5) (2015) 2002.

[21] Z. Li, Z. Duan, G. Chen, H_∞ consensus regions of multi-agent systems, in: Proceedings of the 29th Chinese Control Conference, Beijing, China, July 29–31, 2010.

SECTION 5

Applications

CHAPTER NINETEEN

The Stereographic Product of Positive-Real Functions is Positive Real[a]

R.W. Newcomb
University of Maryland, College Park, MD, United States

1 INTRODUCTION: THE o OPERATION

Recently, a study of stereographic trigonometry has led to the introduction of the o operation, which has allowed simplified expressions of trigonometric functions of trigonometric functions [1, p. 46]. In observing this operation, one sees that it may have some use in circuit analysis and especially in synthesis of positive-real functions. As positive-real functions are also of interest to control system design [2] since they describe passive systems which are stable and relatively insensitive to changes [3].

Specifically, the o operation combines a and b via the formula

$$aob = \frac{a+b}{1+ab} = c, \tag{1a}$$

from which we also see that

$$b = \frac{c-a}{1-ac} = (-a)\,\mathrm{o}c = (-a)\,\mathrm{o}(aob). \tag{1b}$$

In our case, a and b will be functions of the complex frequency variable s.

The key thing to note is that if $a(s)$ and $b(s)$ are positive-real functions, then aob is also positive real. Some other things to note are that o is a commutative operator (ie, $aob = boa$ and $ao0 = a$) so 0 is the identity

[a] In honor of Frank Lewis on his 65th birthday with fond remembrance of his hospitality during my visits to Fort Worth.

Control of Complex Systems
http://dx.doi.org/10.1016/B978-0-12-805246-4.00019-7

in an algebraic system based on o. Of interest also is $a\text{o}\infty = 1/a$ and $a\text{o}(-a) = 0$.

2 CIRCUITS OF INTEREST

In the case where $a = y_1$ and $b = y_2$ are admittances, then $y_1 \text{o} y_2$ is an impedance z as seen by the numerator being in mhos and the denominator being in mhos squared. Expanding

$$z = y_1 \text{o} y_2 = \frac{y_1 + y_2}{1 + y_1 y_2} = 1/\left(\frac{1}{y_1 + y_2} + \frac{1}{\frac{1}{y_1}} + \frac{1}{\frac{1}{y_2}}\right) \tag{2}$$

is seen to describe the series parallel connection in Fig. 1.

From Eq. (2) we see that the 1 in the denominator is a normalized conductance g squared and that if y_1 and y_2 describe a passive circuit then so does z since the passive connection of passive circuits is still a passive circuit. This means that if y_1 and y_2 are rational and positive real, then so is z, since a rational positive-real function can be synthesized by a passive circuit [4, p. 352]. The result that z is positive real is true even for nonrational positive-real functions as can be seen by algebraic manipulation of $y_i(s) = \text{Re}[y_i(s)] + j\text{Im}[y_i(s)]$. At this point we see that under the o operation the set of positive-real functions is a closed set, with the set forming a semigroup, in which case o should be useful for synthesis. In the next two sections we consider the case using the Richards function. That is, given a positive-real $y = 1/z$ we wish to find positive-real y_1 and y_2, simpler than y, where by simpler is meant of lower degree, so that a construction of y_1 and y_2 gives a construction of y. In essence we will choose y_1 capacitive, $y_1 = Cs$ with $C > 0$, and form y_2 via a Richards function of y.

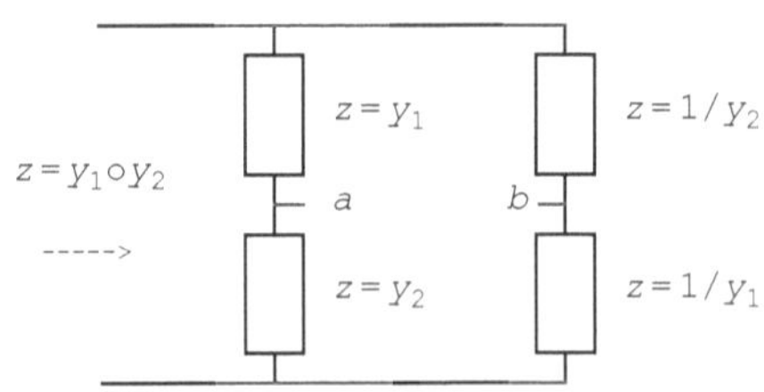

Fig. 1 Interpretation of the o operation. As the bridge is balanced, the cross products being 1, points *a* and *b* can be connected.

3 POSITIVE-REAL AND RICHARDS FUNCTIONS

Of importance for this chapter is another form of visualization of the o operation. For this consider a 2-port described by its admittance matrix, $Y = [y_{ij}]$, loaded in an admittance y_2 as shown in Fig. 2.

The input admittance is calculated from $I = YV$ via $I_2 = -y_2 V_2 = y_{21} V_1 + y_{22} V_2$, so $V_2 = -[1/(y_{21} + y_{22})] V_1$, giving $I_1 = y V_1 = [y_{11} - y_{12} y_{21}/(y_2 + y_{22})] V_1$, to yield, with determinant $\Delta = y_{11} y_{22} - y_{12} y_{21}$,

$$y = \frac{\Delta + y_{11} y_2}{y_{22} + y_2} = 1/z \quad \Rightarrow \quad y_2 = \frac{\Delta - y_{22} y}{y - y_{11}}. \qquad \text{(3a,b)}$$

Next to note is the connection in Fig. 3, where a gyrator-y_1 circuit forms the coupling circuit having the admittance matrix

$$y = \begin{bmatrix} y_1 & g - y_1 \\ -g - y_1 & y_1 \end{bmatrix}, \qquad (4)$$

where g is the gyration conductance.

In this case $\Delta = g^2$, so given y, then by Eq. (3b),

$$y_2 = \frac{g^2 - y y_1}{y - y_1} = g \left(\frac{1 - \left(\frac{y}{g}\right)\left(\frac{y_1}{g}\right)}{\left(\frac{y}{g}\right) - \left(\frac{y_1}{g}\right)} \right) \quad \Rightarrow \quad \frac{y_2}{g} = \frac{1}{\left(\frac{y}{g}\right) \circ \left(-\frac{y_1}{g}\right)}. \qquad \text{(5a,b)}$$

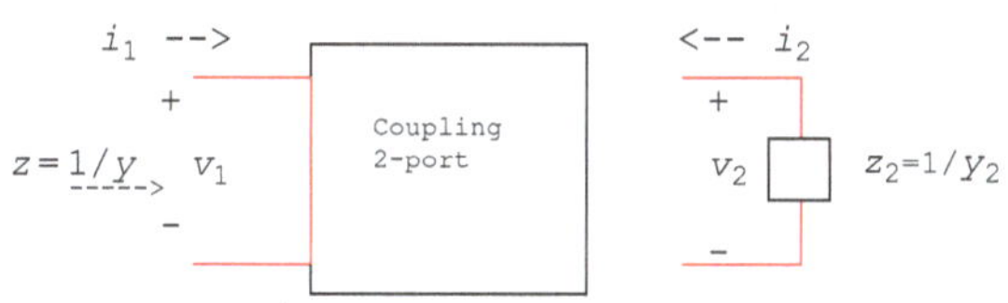

Fig. 2 Loaded 2-port.

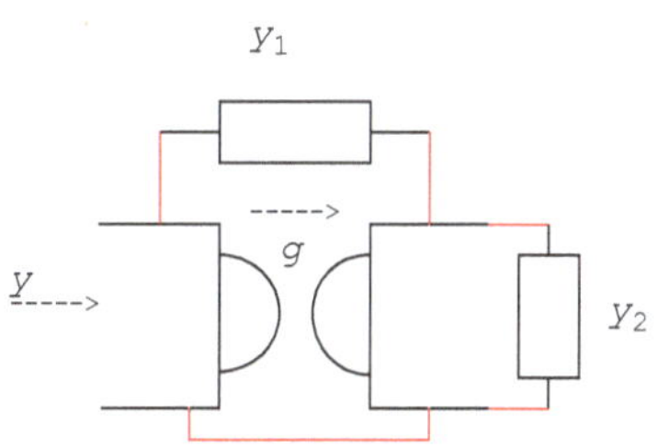

Fig. 3 A most useful circuit giving Eq. (5a,b).

At this point we see that if we determine a normalization g and a bridging admittance y_1, then we can synthesize y by finding y_2. This is possible via the Richards function [4, p. 361], which has the advantage of decreasing the degree if a normalization $y(k)$ is chosen via an appropriate k. For this, k is a complex constant and y is a positive real-function. Then

$$y_R(s) = y(k)\frac{ky(k) - sy(s)}{ky(s) - sy(k)} \quad \Rightarrow \quad \frac{y_R(s)}{y(k)} = \frac{1 - \left(\frac{y(s)}{y(k)}\right)\left(\frac{s}{k}\right)}{\left(\frac{y(s)}{y(k)}\right) - \left(\frac{s}{k}\right)}. \tag{6a,b}$$

4 SYNTHESIS

Comparing Eq. (6a,b) with Eq. (5a,b) shows that we can identify

$$g = y(k), \quad \frac{y_1(s)}{y(k)} = \frac{s}{k}, \quad y_2(s) = y_R(s) = \frac{y(k)}{\left(\frac{y(s)}{y(k)}\right) \circ \left(-\frac{s}{k}\right)}. \tag{7a,b,c}$$

This means the circuit in Fig. 4 can be extracted by use of the o operation, which is really a Richards section extraction.

The key then is the choice of k. Richards [5, p. 779] proved that if k is real, then $y_R(s)$ is positive real if $y(s)$ is. Further, if k is chosen as a zero of the even part of $y(s)$, then if $y(s)$ is positive real, there is a reduction in degree by 1, so $y_2(s)$ is simpler and the process can be repeated until we obtain zero degree.

The even part of $y(s)$ is defined by

$$2Ev[y(s)] = y(s) + y(-s). \tag{8}$$

Zeros of the even part mirror themselves in the origin, so they need not be real but if $y(s) = -y(-s)$, which it is in the lossless case, then any k is a zero of the even part. In summary, in the lossless case choose any $k > 0$, for which automatically

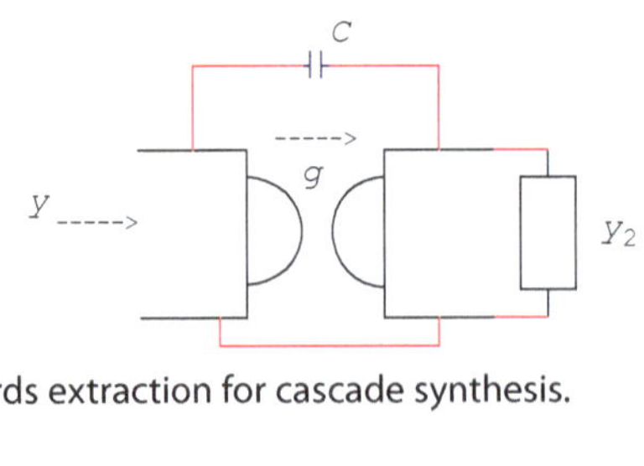

Fig. 4 Circuit for Richards extraction for cascade synthesis.

$$\gamma(k) + \gamma(-k) = 0 \tag{9a}$$

and then, with $\gamma_1(s) = Cs$,

$$g = \gamma(k), \quad C = \gamma(k)/k, \quad \gamma_2(s) = g/[(\gamma(s)/g)o(-Cs/g)]. \tag{9b,c,d}$$

In the nonlossless case, k need not be real, in which case a first extraction at a zero of the even part followed by one at the conjugate k, $k^\star$, will lead to a reduction of degree by 2 and lead to a positive real result, even though complex components result but disappear after appropriate combinations [6, p. 132]; for details, see [7]. Here we will limit ourselves to two dual examples for the lossless case.

5 EXAMPLES

1. Let it be desired to synthesize

$$\gamma(s) = \frac{s(s^2+3)}{s^2+1}. \tag{10a}$$

If we choose $k = k_1 = 1$, then

$$\gamma(k) = 2, \quad g_1 = \gamma(1) = 2, \quad C_1 = \gamma(1)/1 = 2, \tag{10b,c,d,e}$$

giving

$$\begin{aligned}\gamma_2(s) &= \frac{\gamma(k)}{\left(\frac{\gamma(s)}{\gamma(k)}\right) o\left(-\frac{s}{k}\right)} = \frac{2\left[1 - \left(\frac{s(s^2+3)}{2(s^2+1)}\right)\left(\frac{s}{1}\right)\right]}{\left(\frac{s(s^2+3)}{2(s^2+1)}\right) - \left(\frac{s}{1}\right)} \\ &= 2\frac{s^4+s^2-2}{s^3-s} = 2\frac{s^2+2}{s},\end{aligned} \tag{10f}$$

in which we note that $s-k$ and $s+k$ both cancel to lower the degree from 3 to 2.

If we repeat this procedure on γ_2, choosing $k_2 = 2$, then

$$\gamma_2(2) = 6, \quad g_2 = \gamma_2(2) = 6, \quad C_2 = \gamma_2(2)/2 = 3, \tag{10g,h,j,k}$$

giving

$$\gamma_3(s) = \frac{\gamma_2(2)}{\left[\left[\frac{\gamma_2(s)}{\gamma_2(2)}\right] 0\left(-\frac{s}{2}\right)\right]} = 6\frac{(s^2-4)}{\left(\frac{1}{s}\right)(s^2-4)} = 6s = C_3 s, \tag{10l}$$

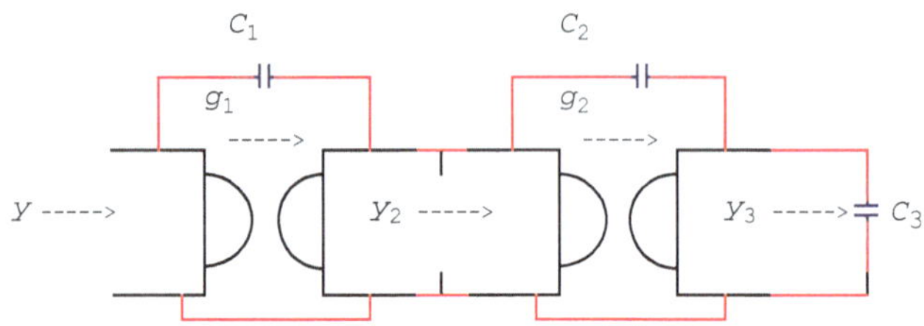

Fig. 5 Cascade synthesis of a degree 3 $y(s)$.

Again $s^2 - 4$ is canceled and the degree decreases by 1. The synthesis uses two cascaded stages of the gyrator-C 2-port loaded in the $y_3(s) = 6s$ capacitor. The final circuit is shown in Fig. 5, where it is seen that the circuit uses the minimum number of dynamical elements (ie, degree 3 capacitors). Note also that the load capacitor can also be formulated as a gyrator-C section loaded in a short circuit by the use of

$$y_4 = \frac{1}{\left(\frac{6s}{6k}\right) \circ \left(-\frac{s}{k}\right)} = \infty. \tag{10m}$$

2. Consider the dual

$$y(s) = \frac{(s^2+1)}{s(s^2+3)}, \tag{11a}$$

with the same k's as in the first example. Then $k = k_1 = 1$, with

$$y(k) = 1/2, \quad g_1 = y(1) = 1/2, \quad C_1 = y(1)/1 = 1/2, \tag{11b,c,d,e}$$

giving

$$y_2(s) = \frac{y(k)}{\left(\frac{y(s)}{y(k)}\right) \circ \left(-\frac{s}{k}\right)} = \frac{\left(\frac{1}{2}\right)\left[1 - \left(\frac{2(s^2+1)}{s(s^2+3)}\right)\left(\frac{s}{1}\right)\right]}{\left(\frac{2(s^2+1)}{s(s^2+3)}\right) - s} = \frac{1}{2}\frac{s}{s^2+2}, \tag{11f}$$

in which we note that $s - k = s - 1$ and $s + k$ both cancel to lower the degree from 3 to 2.

If we repeat the procedure on y_2, choosing $k_2 = 2$, then

$$y_2(2) = 1/6, \quad g_2 = y_2(2) = 1/6, \quad C_2 = y_2(2)/2 = 1/12, \tag{11g,h,j,k}$$

giving

$$y_3(s) = \frac{y_2(2)}{\frac{y_2(s)}{y_2(2)}} \circ \left(-\frac{s}{2}\right) = \frac{1}{6}\frac{(s^2-4)}{(s)(s^2-4)} = \frac{1}{6s}. \tag{11l}$$

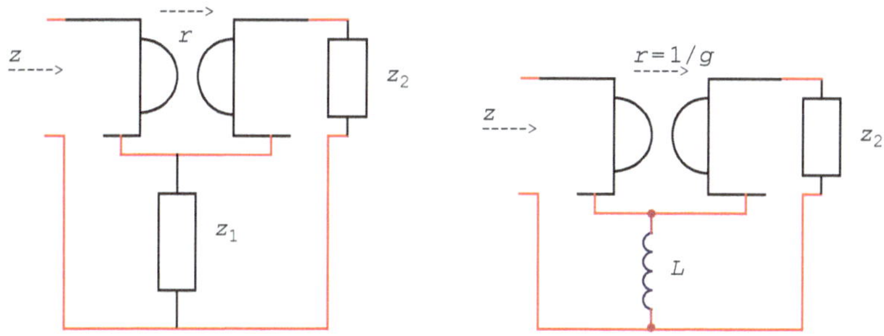

Fig. 6 Circuits useful for a dual treatment.

Again $s^2 - 4$ cancels and the degree decreases by 1. The synthesis again uses two cascaded stages of the gyrator-C 2-port but now loaded in the $y_3(s) = 1/(6s)$ inductor. Fig. 5 still holds but with different element values and the right capacitor changed to an inductor (which can be made by a gyrator-loaded capacitor). Alternatively the realization could have proceeded with use of impedances and the gyrator-C sections replaced by the gyrator-L sections in Fig. 6.

In the lossless positive-real case, by concatenation of o operations on successive normalizations, the method gives an alternative to the well-known Foster synthesis, via partial-fraction expansions, and Cauer synthesis, via continued fraction expansions [4, p. 342], [6, p. 64].

6 CONCLUSIONS

The o operation is performed via a Richards extraction, in which case it gives nice notation for the synthesis of lossless positive-real functions. Here we limited ourselves to the circuit most useful for admittance extractions, probably the most important when used with transistor circuits since transistors are customarily designed on a voltage-controlled basis. But its dual, shown in Fig. 6, is sometimes useful for impedance extractions.

Since z resulted when o was used on y's, it may have been better to use a reciprocal of o, but since the mathematics was set up to mirror that used in [1], the definition in [1] was adhered to. Note also that $0o0 = 0$, whereas should the reciprocal, $ao'b = 1/(aob)$, have been used, then the result $0o'0 = \infty$ is inconvenient. With the definition used, the set of positive-real functions under o forms a commutative semigroup, something which may be of interest for the development of abstract theories of synthesis. However, the normalization needed to use the gyrator sections somewhat gets in the way, so a better definition may be via a g-dependent operation, o_g:

$$a\left(o_g\right)b = \frac{a+b}{g^2+ab}. \tag{12}$$

An alternative is to precede the capacitor section by a gyrator of $g^2 = y(k)$ section with no capacitor via $y_2 = g^2/y$ (for which $y_1 = 0$) and which, however, uses an extra gyrator.

Extension to matrices where there is a Richards type of extraction available (especially in the symmetric case [6, p. 290]) looks possible. And it is of interest to see the relationship on the (bounded real) scattering matrix (reflection coefficient in the single input-single output case), $s_y(s) = (1 - y(s))/(1 + y(s))$. By algebraic manipulation

$$s_{y_1 o y_2} = -s_{y_1}s_{y_2}, \tag{13}$$

In which case the circuit for $y_1 o y_2$ is realized via the dual (due to the -1) of a circulator (realizing the product) connected to the circuits for y_1 and y_2, something of practical importance for microwave circuit design and readily generalized to matrices.

REFERENCES

[1] M. Hardy, Stereographic trigonometric identities, Am. Math. Mon. 122 (1) (2015) 43–47.
[2] S. Wang, J.H. Chow, Coprime factors, strictly positive real functions, LMI and low-order controller design for SISO systems, in: Proceedings of the 37th IEEE Conference on Decision & Control, Tampa, Florida, December, 1998, pp. 2738–2744.
[3] Positive Real Transfer Functions & Connection With Lyapunov Stability, www.egr.msu.edu/~khalil/Nonlinearsystems/Sample/Lect_15.pdf.
[4] B. Peikari, Fundamentals of Network Analysis and Synthesis, Robert E. Krieger Publishing Co., Malabar, FL, 1982.
[5] P. Richards, A special class of functions with positive real part in half-plane, Duke Math. J. 14 (3) (1947) 777–786.
[6] D. Hazony, Elements of Network Synthesis, Reinhold, New York, 1963.
[7] Richards function for cascade, www.ee.umd.edu/~newcomb/courses/fall2007/610/complex_Richards.pdf.

CHAPTER TWENTY

Collective Target Tracking Mean Field Control for Markovian Jump-Driven Models of Electric Water Heating Loads

A.C. Kizilkale[*,†], R.P. Malhamé[*,†]
[*]Polytechnique Montreal, Montreal, QC, Canada
[†]Research Group in Decision Analysis (GERAD), Montreal, QC, Canada

1 INTRODUCTION

In this work the potential of energy storage in dispersed devices naturally present in electric water heaters is employed as a tool to mitigate renewable generation variability and reduce peak load. The envisioned control architecture is hybrid: (i) centralized in terms of target trajectory generation for homogeneous groups of energy storage-capable electric devices, so as to preserve overall optimality characteristics; (ii) decentralized at the implementation level so as to locally enforce safety and comfort constraints, as well as to minimize communication requirements. More specifically, we mention the following implementation principles, and argue that a class of decentralized control schemes based on a so-called mean field game setup [2–5] can actually meet all the requirements:

1. *Each controller has to be situated locally.* A completely centralized control architecture micromanaging every individual device to be controlled requires a significant bandwidth as well as very large computational power. Moreover, in the event of a loss of communication, users might face difficulties in the sense of comfort and safety. When the controller is situated locally, these worries are void since the controller can locally enforce comfort and safety constraints. Even if the communication with the central authority is lost, it is able to maintain the safety and comfort requirements of the user.

Control of Complex Systems
http://dx.doi.org/10.1016/B978-0-12-805246-4.00020-3

2. *Data exchange should be kept to a minimum both with the central authority and among users.* The control architecture must be able to operate with a highly reduced volume of data exchange since, given the very large number of agents involved, a requirement of constant flows of information between agents and the central authority, as well as among agents themselves, would create scalability problems. In this context, the mean field game-based controllers turn the large numbers issue from a hurdle into an asset by invoking the predictability stemming from the law of large numbers.
3. *User disturbance should be kept at a minimum.* Since the centralized authority is solely interested in the aggregate consumption, individual trajectories do not need to necessarily follow the targets set by the authority. On the contrary, in fact in the case of a population of controlled electric space heaters, air conditioners, or electric water heaters, for example, it is desirable to shape the mean temperature of the population with the least disturbance, ideally without customers even noticing the effects of the control actions on their comfort level. Also, it is important to maintain some measure of fairness among the users when it comes to sharing the control effort. We will show that the recent developments in [6] allow the flexibility to address these issues.

We introduced *collective target tracking mean field control* [6], where the presence of large numbers of space-heating electric devices is employed to develop a decentralized mean field control-based approach to the problem of these devices following a desired mean trajectory. The proposed solution deviates from the classical formulation, which would have each element track the desired mean temperature, thus introducing unnecessary control actions. The solution made possible by mean field theory enforces collective mean temperature tracking while leaving individual devices freer to remain, if possible, closer to their initially desired zone. In this context the mean field effect is mediated by quadratic cost function parameters under the form of an integral error, as compared with currently prevailing mean field control formulations (eg, [5]), where the mean field effect is concentrated on shaping a common signal to be tracked by all states.

The model in [6] applies to space heaters where temperature evolves according to linear dynamics subject to random processes of heat addition and heat losses stemming from human activity within the dwelling. The random heat processes are modeled as Brownian motion. This diffusion model is not sufficient to characterize the dynamics of a water heater where the system is mostly isolated and dynamical parameter changes occur intermittently with respect to the extraction of water from the water heater tank. An alternative model that captures the essence of water heater dynamics is a Markovian jump process where dynamical parameters randomly switch between states [7]. In this work, we extend the model introduced in [6] to the system of populations where the dynamical parameters are time varying and characterized by a sequence of independent Markov chains with identical generators.

Statistical mechanics-inspired models of large aggregates of energy storage-associated loads, particularly space heating and cooling loads, to be controlled within peak shaving and valley filling load management programs by direct control were presented in [8], whereas the modeling method was applied to electric water heaters in [9].

Use of dispersed storage for accommodating renewable sources is a growing area of research. An aggregate model for a large number of pool pumps using mean field limits was developed in [10–12]. The storage potential for HVAC systems is studied in [13]. Dispersed energy storage for frequency regulation in the presence of wind energy is investigated in [14, 15]. Callaway [14] uses the aggregate load modeling framework in [8] and extends it for improved transient analysis. Dynamic pricing for controlling the load of aggregates of large commercial buildings is analyzed in [16], and domestic heating systems are employed as heat buffers in [17]. A decentralized charging control strategy for large populations of plug-in electric vehicles using the mean field method is presented in [18]. Both centralized, model predictive and distributed control with a randomized dispatch strategy are applied in [19] to manage a large number of thermostat-based appliances.

We denote the set of nonnegative real numbers by $\mathbb{R}_+$. The norm $\|\cdot\|$ denotes the 2-norm of vectors and matrices, and $\|x\|_Q^2 \triangleq x^\top Qx$. $\mathbf{C}_b[0, T] := \{x : x \in \mathbf{C}, \sup_{t\geq 0}\|x_t\| < \infty\}$ denotes the family of all bounded continuous functions, and for any $x \in \mathbf{C}_b$, $\|\cdot\|_\infty$ denotes the supremum norm: $\|x\|_\infty := \sup_{t\geq 0}\|x_t\|$. $X^\top$ denotes the transpose of a matrix X, and $\lambda_{\max}$ and $\lambda_{\max}$ denote the maximum and minimum eigenvalues, respectively of symmetric matrices.

In Section 2 the mean field game theory of the Markovian jump linear-quadratic Gaussian multiagent mean state-dependent target tracking problem is reviewed. Then in Section 3 we present the linearly controlled water heater dynamics together with a quadratic cost function which differs from typical mean field game cost structures in that the mean field effect is propagated not through a target mean field-dependent trajectory but rather through the weights in the quadratic cost expression. In Section 4 a classical linear quadratic solution is presented, and in Section 5 the collective target tracking Markovian jump mean field solution associated with a fixed point equation system is presented. The existence and uniqueness of solutions to this fixed point system is established in Section 6 under appropriate technical conditions. Also it is shown that the fixed point solution constitutes an ϵ-Nash equilibrium for a finite population of agents. Lastly, in Section 7, we provide associated simulation results together with comparisons with results from a simplistic target tracking control formulation of the problem.

2 BACKGROUND ON MARKOVIAN JUMP LINEAR-QUADRATIC GAUSSIAN MEAN FIELD CONTROL (FOLLOWING [1])

To meet the three requirements previously described, the principles of the mean field control method [5] are employed. This framework is based on a decentralized scheme whereby each agent calculates its individual best response to the anticipated action profile of the population. The anticipation of the population response curve is possible because the numbers are sufficiently large that the law of large numbers applies. Under technical constraints together with the assumption of each agent's individual rationality, the approach looks for a Nash equilibrium as the number of users goes to infinity when agents implement their stabilizing best response actions. The anticipation of the action profile is done off-line and locally with statistical information obtained on the basis of a parsimonious population measurement scheme and available at the start of the control horizon. The control scheme is fully decentralized—that is, *no communication is required among individual controllers throughout the horizon.*

A large population of N stochastic dynamic agents is considered where agents are stochastically independent, but which will be cost coupled and such that the individual dynamics are defined by

$$dx_t^i = \left(A^{\theta_t^i} x_t^i + B^{\theta_t^i} u_t^i + c^{\theta_t^i}\right) dt + D^{\theta_t^i} dw_t^i, \quad t \geq 0, \tag{1}$$

$1 \leq i \leq N$, where for agent $\mathcal{A}_i$, $x^i \in \mathbb{R}^n$ is the state, $u^i \in \mathbb{R}^m$ is the control input, and $w^i \in \mathbb{R}^r$ is a standard Wiener process on a sufficiently large underlying probability space $(\Omega, \mathcal{F}, P)$ such that w^i is progressively measurable with respect to $\mathcal{F}^{w^i,\theta^i} := \{\mathcal{F}_t^{w^i} \times F_t^{\theta^i}; t \geq 0\}$.

$\{\theta_t^i, 1 \leq i \leq N\}$ is a sequence of independent continuous-time Markov chains taking values in $\Theta = \{1, 2, \ldots, p\}$ with the infinitesimal generator $\Lambda = \{\lambda_{ij}, i, j, = 1, \ldots, p\}$ progressively measurable with respect to $\mathcal{F}^{w^i, \theta^i}$. We denote the population average state by $x^{(N)} = (1/N) \sum_{i=1}^{N} x^i$.

The cost function for agent $\mathcal{A}_i$, $1 \leq i \leq N$, is given by

$$J_i^N(u^i, u^{-i}) = E \int_0^T \left[\|x_t^i - m_t^N\|_Q^2 + \|u_t^i\|_R^2 \right] dt + \|x_T^i - m_T^N\|_Q^2, \tag{2}$$

where $Q \geq 0$ and $R > 0$. The cost coupling is assumed to be in the form of an averaging function $m_t^N := m(x_t^{(N)} + \eta)$, $\eta \in \mathbb{R}^n$. The term u^i is the control input of the agent $\mathcal{A}_i$, and u^{-i} denotes the control inputs of the complementary set of agents $\mathcal{A}_{-i} = \{\mathcal{A}_j, j \neq i, 1 \leq j \leq N\}$.

For the basic mean field control problem, the following assumptions are adopted.

Assumption 1. The identically distributed disturbance processes $\{w^i, 1 \leq i \leq N\}$ are mutually independent and independent of mutually independent and identically distributed $\{\theta_t^i, 1 \leq i \leq N\}$ and $\{x_0^i, 1 \leq i \leq N\}$, where $\sup_{i \geq 1}[E\|x_0^i\|^2] < \infty$.

Assumption 2. The following coupled Riccati equations have a set of unique positive definite solutions Π^j, $1 \leq j \leq p$, throughout the control horizon $[0, T]$:

$$\begin{aligned} \frac{-d\Pi_t^j}{dt} &= \Pi_t^j A^j + A^{j\top} \Pi_t^j \\ &\quad - \Pi_t^j B^j R^{-1} B^{j\top} \Pi_t^j + \sum_{k=1}^{p} \lambda_{jk} \Pi_t^j + Q, \quad 1 \leq j \leq p, \end{aligned} \tag{3}$$

where $\Pi_T^j = Q$, $1 \leq j \leq p$.

Remark 1. Assumption 2, which often appears in the study on stabilization of Markovian jump systems, ensures the stability of the closed-loop system [20].

Remark 2. Under Assumption 2, if we further assume that the time-varying eigenvalues of the continuous Π_t^j are continuous functions of time, then given that $[0, T]$ is closed and bounded, it is possible to conclude that $\inf_{t \in [0,T]}[\lambda_{\min}(\Pi_t^j)] > 0$, $1 \leq j \leq p$.

Also $\sup_{t \in [0,T]}[\lambda_{\max}(\Pi_t^j)] < \infty$, $1 \leq j \leq p$. Thus in the rest of the chapter we will also assume continuity of the eigenvalues associated with the Π_t^j as functions of time. This augmented assumption will be referred to as Assumption 2$\star$.

For a given posited *mass tracking signal* $m^\star \in \mathbf{C}[0,T]$, the mass offset function s^j is generated by the coupled set of differential equations

$$-\frac{ds_t^j}{dt} = (A^j - B^j R^{-1} B^{j\top} \Pi_t^j)^\top s_t^j - Q m_t^\star + \Pi_t^j c^j + \sum_{k=1}^{p} \lambda_{jk} s_t^k, \quad 1 \le j \le p, \tag{4}$$

where $s_T^j = -Q m_T^\star$, $1 \le j \le p$. Then under Assumptions 1 and 2 (i) there exists a unique solution $s^\star$ in the class $\mathbf{C}[0,T]$ and (ii) the optimal tracking control law is given by

$$u_t^\circ = -\sum_{j=1}^{p} \mathbb{I}_{[\theta_t=j]} R^{-1} B^{j\top} (\Pi^j x_t + s_t^j), \quad t \ge 0. \tag{5}$$

Note that $m^\star$ is assumed to be a deterministic albeit unknown trajectory. This is implicitly assuming that the population of agents is growing without bounds, and that the law of large numbers sets in as the coupling between pairs of agents becomes weaker and weaker. For that $m^\star$ to be sustainable, it must be collectively replicated by the agents implementing their best responses to that signal. Thus the system given by Eqs. (4) and (5) must be complemented by a fixed point requirement leading to the mean field equation system in Definition 1.

Definition 1. (Markovian jump mean field equation system on $t \in [0,T]$)

$$\begin{aligned}
-\frac{ds_t^j}{dt} &= (A^j - B^j R^{-1} B^{j\top} \Pi_t^j)^\top s_t^j - Q m_t^\star + \Pi_t^j c^j \\
&\quad + \sum_{k=1}^{p} \lambda_{jk} s_t^k, \quad s_T^j = -Q m_T^\star, \quad 1 \le j \le p, \\
\frac{d\bar{x}_t^j}{dt} &= (A^j - B^j R^{-1} B^{j\top} \Pi_t^j)\bar{x}_t^j + \sum_{k=1}^{p} \lambda_{kj} \bar{x}_t^k + \zeta_t^j c^j \\
&\quad - \zeta_t^j B^j R^{-1} B^{j\top} s_t^j, \quad 1 \le j \le p, \\
\bar{x}_t &= \sum_{j=1}^{p} \bar{x}_t^j, \\
m_t^\star &= m(\bar{x}_t + \eta), \quad t \in [0,T], \\
\frac{d\zeta_t}{dt} &= \zeta_t \Lambda^\top,
\end{aligned} \tag{6}$$

where $\bar{x}_t^j := E(\bar{x}_t \mathbb{I}_{[\theta_t=j]})$ and

$$\zeta_t = [\zeta_t^1, \ldots, \zeta_t^p] \tag{7}$$

is the probability distribution of the Markov chain.

Before we present the main theorem in this section we introduce another assumption and we define a control set where the admissible controls are defined.

Assumption 3. The cost coupling is of the form: $m^N(\cdot) \triangleq m(x^{(N)} + \eta), \eta \in \mathbb{R}^n$, where the function $m(\cdot)$ is Lipschitz continuous on $\mathbb{R}^n$ with a Lipschitz constant $\gamma > 0$—that is, $\|m(x) - m(y)\| \leq \gamma \|x - y\|$ for all $x, y \in \mathbb{R}^n$.

Global observation control set $\mathcal{U}_g^N$: The set of control inputs $\mathcal{U}_g^N$ consists of all feedback controls adapted to $\mathcal{F}_t^N$, where $\mathcal{F}_t^N$ is the σ-field generated by the set $\{x_\tau^j, \theta_\tau^j; 0 \leq \tau \leq t, 1 \leq j \leq N\}$.

We now present the main theorem.

Theorem 1 (Markovian jump mean field stochastic control theorem [1]). *Under Assumptions 1–3 the Markovian jump mean field stochastic control law* (5) *generates a set of controls* $\mathcal{U}_{\mathrm{MF}}^N \triangleq \{(u^i)^0; 1 \leq i \leq N\}, 1 \leq N < \infty$, *with*

$$u_t^0 = -\sum_{j=1}^{p} \mathbb{I}_{[\theta_t=j]} R^{-1} B^{j\top} (\Pi^j x_t + s_t^j), \quad t \geq 0, \tag{8}$$

such that:

1. *The mean field equations* (6) *have a unique solution under the contraction hypothesis (A4) in [1].*
2. *All agent system trajectories* x^i, $1 \leq i \leq N$, *are second-order stable.*
3. $\{\mathcal{U}_{\mathrm{MF}}^N; 1 \leq N < \infty\}$ *yields an* ϵ*-Nash equilibrium for all* $\epsilon > 0$*—that is, for all* $\epsilon > 0$*, there exists* $N(\epsilon)$ *such that for all* $N \geq N(\epsilon)$

$$J_i^N\left((u^i)^0, (u^{-i})^0\right) - \epsilon \leq \inf_{u^i \in \mathcal{U}_g^N} J_i^N\left(u^i, (u^{-i})^0\right) \leq J_i^N\left((u^i)^0, (u^{-i})^0\right).$$

In essence Theorem 1 states that the Markovian jump mean field equation system produces a set of decentralized control policies for each agent, which collectively become arbitrarily close in performance to a Nash equilibrium in the space of feedback strategies, provided the number of agents increases sufficiently.

3 ELECTRIC WATER HEATER MODELS

The dynamics of the temperature of a water heater tank subject to thermal stratification can be modeled by our assuming that the tank consists

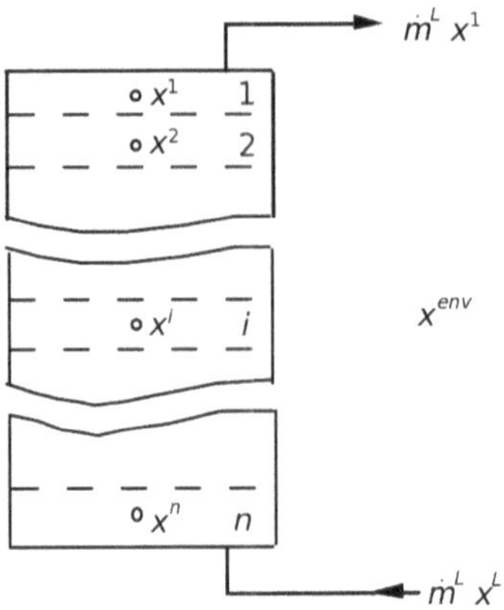

Fig. 1 Stratification in a water heater.

Table 1 Parameters for water heater dynamics

Parameter	Description
x_l	Temperature of the lth segment
u_l	Control action at the lth segment
$\dot{m}_L$	Fluid mass flow rate to the load
$\dot{Q}_l$	Rate of energy input by the heating element to the lth segment
x^{env}	Temperature of the environment
x^L	Temperature of the inlet fluid
M_l	Mass of the fluid in the lth segment
A_l	Lateral surface area of the lth segment
C^{pf}	Specific heat of the fluid
U	Loss coefficient between the tank and its environment

of n fully mixed equal volume segments as shown in Fig. 1. For the thermal dynamics we adopt the linear model given in [21] and written below with the nomenclature specified in Table 1.

$$\begin{aligned} M_l C^{pf} \frac{dx_{l,t}}{dt} &= UA_l(x^{\text{env}} - x_{l,t}) + \dot{m}_t^L C^{pf}(x_{(l+1),t} - x_{l,t}) + \dot{Q}_l u_{l,t}, \\ & t \geq 0, \quad l \neq n, \\ M_l C^{pf} \frac{dx_{l,t}}{dt} &= UA_l(x^{\text{env}} - x_{l,t}) + \dot{m}_t^L C^{pf}(x_t^L - x_{l,t}) + \dot{Q}_l u_{l,t}, \\ & t \geq 0, \quad l = n. \end{aligned} \tag{9}$$

In Eq. (9) $\dot{m}^L$ represents the amount of water being pulled from the tank per time unit. We will treat the water demand mode as a stochastic process denoted by θ_t, $t \geq 0$, which is a continuous-time Markov chain taking values in $\Theta = \{1, 2, \ldots, p\}$, where the modes represent different types of water events such as dishwashing, hand washing, and showering. In general, electric heating can be contributed only in the top and bottom segments.

The dynamics of a water tank therefore can be written in linear form as

$$\frac{dx_t}{dt} = A'^{\theta_t} x_t + Bu_t + c''^{\theta_t}, \quad t \geq 0.$$

Recalling implementation principle 3, we consider that users "naturally" would like their devices to stay at their initial temperatures (normally attained via thermostatic action and before the intervention of the power utility control center). Thus we have reformulated the control effort as the signal required to make them deviate from that initial temperature; more precisely, the expected global control effort to maintain the average water heater temperature at the initial value is considered *free* and will not be penalized by the cost function to be defined. The corresponding dynamical equation is written as

$$\frac{dx_t}{dt} = A^{\theta_t} x_t + B(u_t + u_t^{\text{free}}) + c'^{\theta_t}, \quad t \geq 0,$$

where

$$u_m^{\text{free}} = \sum_{l=1}^{n} UA_l (x_{l,0} - x^{\text{env}}) + E \sum_{j=1}^{p} \zeta^j \dot{m}_t^L(j) C^{pf} (x_{1,t} - x_t^L). \tag{10}$$

In Eq. (10) although the control input is assumed to be multidimensional, all the *free* power is accumulated on the bottom layer actuation component u_m. In the linear model adopted, heat propagates from the bottom upward only. As a result of that, the system is controllable if and only if there is a controller at the bottom stratification layer. Since Assumption 2 implicitly assumes controllability, it is assumed that all water heaters have actuation at their bottom layers.

Finally we write the dynamics as

$$\frac{dx_t}{dt} = A^{\theta_t} x_t + Bu_t + c^{\theta_t}, \quad t \geq 0. \tag{11}$$

4 CLASSICAL MARKOVIAN JUMP LINEAR-QUADRATIC TRACKING

The dynamics for a population of N water heaters is given as

$$\frac{dx_t^i}{dt} = A^{\theta_t^i} x_t^i + Bu_t^i + c^{\theta_t^i}, \quad t \geq 0, \quad 1 \leq i \leq N. \tag{12}$$

Following the results of a global optimization analysis, we consider a *building block* decentralized control problem whereby it is assumed that the central authority has determined that it is best that the mean temperature of a given population of water heater tracks some fixed target temperature signal y, thus creating, for example, a temporary relief in the global electric power demand of the system. In the Markovian jump linear-quadratic tracking formulation each agent's cost function is defined as

$$J_i^N(u^i, u^{-i}) = E \int_0^T \left[(Hx_t^i - y)^2 q + \|u_t^i\|_R^2 \right] dt + (Hx_T^i - y)^2 q, \tag{13}$$

where $H = [1/n, \ldots, 1/n]$.

The problem with this approach is that each agent minimizes its own cost function and tracks the same signal. Even though the central authority is only interested in aggregate behavior and in particular the *mean* temperature, this simplistic control approach causes *all agents* to track the target signal.

5 COLLECTIVE TARGET TRACKING MEAN FIELD MODEL

We employ the dynamics for the heaters given in Eq. (12). We also assume without loss of generality that the initial mean temperature of the population of the heaters is above the control center-dictated target y for that mean (power relief required). The cost function for agent $\mathcal{A}_i$, $1 \leq i \leq N$, is defined instead as follows:

$$\begin{aligned} J_i^N(u^i, u^{-i}) = E \int_0^T & \left[(Hx_t^i - z)^2 q_t^y + (Hx_t^i - Hx_0^i)^2 q^{x_0} + \|u_t^i\|_R^2 \right] dt \\ & + (Hx_T^i - z)^2 q_T^y + (Hx_T^i - Hx_0^i)^2 q^{x_0}, \end{aligned} \tag{14}$$

where $H = [1/n, \ldots, 1/n]$, and where z is a *direction* assigned to each agent in the population and the mean of each agent's layer temperatures deviation from this direction is penalized by the deviation penalty coefficient q_t^y, $t \in [0, T]$, which captures the mean field information and is calculated as the *integrated error signal*

$$q_t^y = \left| \lambda \int_0^t (Hx_t^{(N)} - y)dt \right|. \tag{15}$$

The justification for the above cost function is that for a system where $low < y < x_0^{(N)}$, by pointing individual agents toward what is considered as the minimum comfort mean temperature $z = low$, it dictates a global decrease in their individual temperatures. This pressure for decrease persists as long as the differential between the mean temperature and the mean target y is high. The role of the integral controller is to mechanically compute the *right* level of penalty coefficient q which, in a potential steady state as the length of the horizon T goes to infinity, should maintain the mean population temperature at y. When this happens, because of the second term on the cost function which penalizes each individual's mean temperature deviation from its initial value, and the initial condition-dependent definition of free input, individual agents reach each their individualized steady state (in general different from y and closer to their comfort zone than classical Markovian jump linear-quadratic tracking would dictate). To derive the limiting infinite population mean field equation system, and analogously to the more classical Markovian jump mean field linear-quadratic case in Section 2, we start this time assuming a given (although unknown) cost penalty trajectory $q^y \in \mathbf{C}_b[0, T]$ and a constant q^{x_0}. Given q_t^y, $t \geq 0$, individual agents $\mathcal{A}_i$, $1 \leq i \leq N$, solve a classical target tracking Markovian jump problem with time-varying cost coefficient with Riccati gain Π_t, $t \geq 0$, evolving as follows [22]:

$$\begin{aligned} -\frac{d\Pi_t^j}{dt} &= \Pi_t^j A^j + A^{j\top} \Pi_t^j \\ &\quad - \Pi_t^j BR^{-1}B^\top \Pi_t^j + \sum_{k=1}^{p} \lambda_{jk}\Pi_t^k + (q_t^y + q^{x_0})H^\top H, \end{aligned} \tag{16}$$

$1 \leq j \leq p$ for $t \in [0, T]$, where $\Pi_T^j = (q_T^y + q^{x_0})H^\top H$, $1 \leq j \leq p$. Moreover, for a given direction z, the Markov state-dependent offset function s_t^j is generated by the differential equation

$$\begin{aligned} -\frac{ds_{i,t}^j}{dt} &= (A^j - BR^{-1}B^\top \Pi_t^j)^\top s_{i,t}^j - q_t^y H^\top z \\ &\quad - q^{x_0} H^\top x_0^i + \Pi_t^j c_i^j + \sum_{k=1}^{p} \lambda_{jk} s_{i,t}^k, \quad 1 \leq j \leq p, \end{aligned} \tag{17}$$

where $s^j_{i,T} = -[q^y_T H^\top z + q^{x_0} H^\top x^i_0]$, $1 \le j \le p$, $1 \le i \le N$. Then the optimal tracking control law is given by

$$u^\circ_{i,t} = -\sum_{j=1}^{p} \mathbb{I}_{[\theta_{i,t}=j]} R^{-1} B^\top (\Pi^j_t x_{i,t} + s^j_{i,t}), \quad t \ge 0. \tag{18}$$

The calculation of the unknown q^y_t, $t \ge 0$, is obtained by our requiring that $q^y_t, t \ge 0$, must be such that the individual agents performing their corresponding optimal responses must collectively replicate q^y_t, $t \ge 0$, itself. This fixed point requirement leads to the last equation in Definition 2 for the collective target tracking Markovian jump mean field equation system.

Definition 2. *Collective target tracking Markovian jump mean field equation system on* $t \in [0, T]$, $1 \le j \le p$:

$$\begin{aligned}
-\frac{d\Pi^j_t}{dt} &= \Pi^j_t A^j + A^{j\top} \Pi^j_t - \Pi^j_t B R^{-1} B^\top \Pi^j_t \\
&\quad + \sum_{k=1}^{p} \lambda_{jk} \Pi^k_t + (q^y_t + q^{x_0}) H^\top H, \quad \Pi^j_T = (q^y_T + q^{x_0}) H^\top H, \quad 1 \le j \le p, \\
-\frac{ds^j_t}{dt} &= (A^j - BR^{-1}B^\top \Pi^j_t)^\top s^j_t - q^y_t H^\top z - q^{x_0} H^\top \bar{x}_0 \\
&\quad + \Pi^j_t c^j + \sum_{k=1}^{p} \lambda_{jk} s^k_t, \quad s^j_T = -[q^y_T H^\top z + q^{x_0} H^\top \bar{x}_0], \quad 1 \le j \le p, \\
\frac{d\bar{x}^j_t}{dt} &= (A^j - BR^{-1}B^\top \Pi^j_t)\bar{x}^j_t + \sum_{k=1}^{p} \lambda_{kj} \bar{x}^k_t + \zeta^j_t c^j \\
&\quad - \zeta^j_t B R^{-1} B^\top s^j_t, \quad 1 \le j \le p, \\
\bar{x}_t &= \sum_{j=1}^{p} \bar{x}^j_t, \\
q^y_t &= \left| \lambda \int_0^t (H\bar{x}_\tau - y) d\tau \right|, \\
\frac{d\zeta_t}{dt} &= \zeta_t \Lambda^\top.
\end{aligned} \tag{19}$$

One recalls:

- $\bar{x}^j_t = E(\bar{x}_t I_{[\theta_t=j]})$,
- $\Lambda = \{\lambda_{i'j'},\ i', j' = 1, \ldots, p\}$ is the infinitesimal generator of the Markov chain,
- $\zeta_t = [\zeta^1_t, \ldots, \zeta^p_t]$ is the probability distribution of the Markov chain.

The mean field equations for this model are significantly different from Eq. (6). Indeed system (6) is entirely amenable to analysis within a linear systems framework, while system (19) is fundamentally nonlinear (because of the form of q_t^{γ}, $t \geq 0$) and special arguments have to be developed for analysis of the existence and uniqueness of solutions.

6 FIXED POINT ANALYSIS AND ϵ-NASH THEOREM

In this section we study the conditions that guarantee the existence and uniqueness of a solution to the equation system given in Eq. (19).

First we characterize the collective target tracking Markovian jump mean field equation system as the operator $\mathcal{M} : \mathbf{C}_b[0, T] \to \mathbf{C}[0, T]$. From its definition, one can write $\mathcal{M}\bar{x} = \bar{x}$. In the remainder of this section we present conditions leading to guaranteed existence and uniqueness of solution to this system.

We state the properties of the underlying function space:

Definition 3. Define the set $\mathcal{G}$ endowed with the norm $\|\cdot\|_\infty$ as that including all functions such that $f \in \mathbf{C}_b[0, T]$, $f(0) = \bar{x}_0$, and $z \leq f(t) \leq \bar{x}_0$ for $t \in [0, T]$. Note that $\mathcal{G}$ is nonempty.

Lemma 1. *Under $\|\cdot\|_\infty$, $\mathcal{G}$ is convex and closed in $\mathbf{C}_b[0, T]$; therefore it is a Banach space.*

The proof is given in [23].

We state a final assumption and then present our fixed point existence theorem.

Assumption 4.

$$\frac{2b_1}{a_1 \min_j \sqrt{\inf_{0\leq t\leq T}[\lambda_{\min}(\Pi_t^j)]}} < 1,$$

where

$$a_1 = \min_j \frac{\inf_{0\leq t\leq T}\left[\lambda_{\min}(Q_t^j + \Pi_t^j B R^{-1} B^\top \Pi_t^j)\right]}{\sup_{0\leq t\leq T}\left[\lambda_{\sup}(\Pi_t^j)\right]},$$

$$b_1 = \max_j \sqrt{\sup_{0\leq t\leq T}\left[\lambda_{\max}(\Pi_t^j)\right]} \|BR^{-1}B^\top\|(\gamma_1 + \gamma_2 + \gamma_3 + \gamma_4),$$

$$\gamma_1 = (\max_j \kappa_3^j \lambda)\left[(\sqrt{p/n}q^{x_0}\bar{x}_0) + (\sqrt{p/n}\kappa_1 z)\right],$$

$$\gamma_2 = \lambda\sqrt{p/n}z\Psi(t, T)'T,$$

$$\gamma_3 = (\max_j \kappa_2^j \lambda)(\sqrt{p/n}q^{x_0}\bar{x}_0)(\sqrt{p/n}\kappa_1 z)(c_j \sup_{0 \le t \le T} [\lambda_{\max}(\Pi_t^j)]),$$

$$\gamma_4 = (\sqrt{p/n}z + \sqrt{n}\max_j \|c^j\| \kappa_2^j)\lambda \int_T^t \Psi(t,\tau)''\tau \, d\tau.$$

Theorem 2. *Under Assumptions 1–4 there exists a unique fixed point for the map $\mathcal{M} : \mathcal{G} \to \mathcal{G}$.*

The proof is given in the Appendix.

6.1 ϵ-Nash Theorem

Here we present the main theorem of the chapter. It provides an infinite population based mean field stochastic control law that achieves an ϵ-Nash equilibrium when applied within the finite population.

Theorem 3 (Collective target tracking Markovian jump mean field stochastic control theorem). *Under Assumptions 1–4 the collective target tracking Markovian jump mean field stochastic control law* (18) *generates a set of controls $\mathcal{U}_{\text{col}}^N \triangleq \{(u^i)^0; 1 \le i \le N\}$, $1 \le N < \infty$, with*

$$u_t^{\circ} = -\sum_{j=1}^{p} I_{[\theta_t = j]} R^{-1} B^{\top} (\Pi_t^j x_t + s_t^j), \quad t \ge 0,$$

such that:

1. *The collective target tracking mean field equations* (19) *have a unique solution.*
2. *All agent system trajectories x^i, $1 \le i \le N$, are second-order stable.*
3. *$\{\mathcal{U}_{\text{col}}^N; 1 \le N < \infty\}$ yields an ϵ-Nash equilibrium for all $\epsilon > 0$, where $\epsilon = O(1/\sqrt{N})$.*

The proof is given in [23].

7 SIMULATIONS

In the first simulation we consider a population of 200 identical water heaters with two stratification layers where the mean initial temperature in the population is 55°C. We test two scenarios where the central authority sets the target temperature to (i) 53°C and (ii) 57°C over a 6-h horizon, and provides the target temperature trajectory and initial population temperature mean to each controller.

For the simulations we use the dynamics given in Eq. (9), where $\dot{m}_t^L := K\theta_t$, $t \ge 0$, and $\theta_t \in \{0, 1\}$, $t \ge 0$, is a two-state continuous-time Markov chain with infinitesimal generator

Table 2 Parameter values for water heater dynamics

Parameter	Description
K	0.05 L/s
$\dot{Q}^i$	4500 J/s
x^{env}	25°C
x^L	15°C
M^i	139 L
A^i	1.715 m^3
C^{pf}	4190 J/(kg °C)
U	1.05 J/(s m^2 K)

$$L = \begin{bmatrix} -0.5 & 0.5 \\ 7 & -7 \end{bmatrix}.$$

These values correspond to an average of 288 L of water consumption in 24 h. The selected parameter values are provided in Table 2. We employ the cost function provided in Eq. (14), where $q^{x_0} = 8000$ and $R = 0.05 \times I_{2\times2}$. In searching for fixed point trajectories, we adopt a naive successive substitutions algorithm starting from a flat profile of the water heater mean temperature population throughout the control horizon.

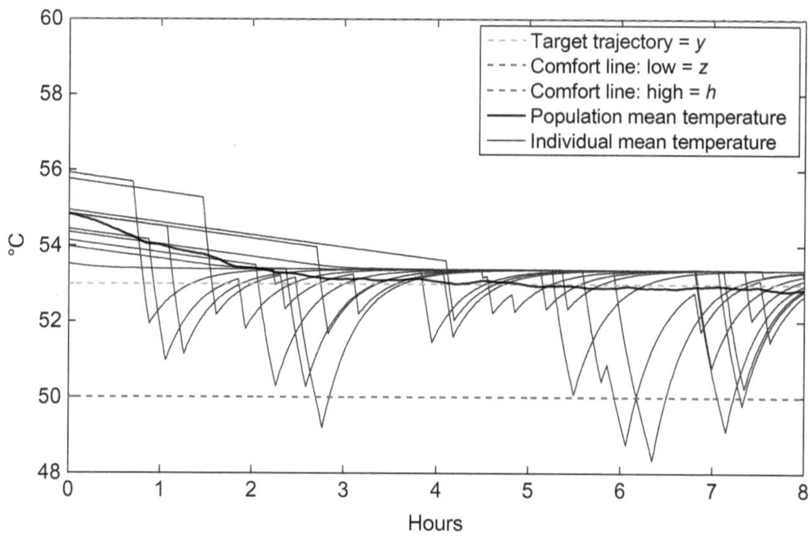

Fig. 2 Temperature trajectories where agents apply Markovian jump linear-quadratic tracking.

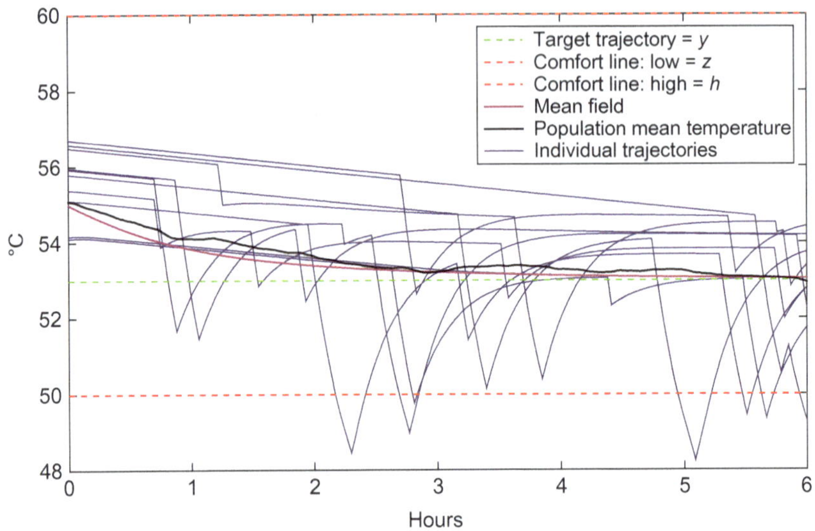

Fig. 3 Agents applying collective target tracking Markovian jump mean field control: all agents moving toward low comfort temperature.

Fig. 2 shows the temperature trajectories of all water heaters in the system where each agent in the population tracks the target desired mean temperature of 53°C using Markovian jump linear-quadratic control. In this implementation all agents unnecessarily track 53°C so as to ensure that the mean temperature tracks 53°C. In this control scheme, customer disturbance is maximized, thus violating implementation principle 3.

In the next experiment the central authority sets the target temperature to 53°C, and all agents are driven toward $l = 50$°C, the low comfort temperature as they sense the mean field pressure through the collective target tracking Markovian jump mean field control structure. The trajectories are presented for only 10 agents, and the calculated mean field signal is shown in Fig. 3. It can be seen that while the mean temperature settles at 53°C, individuals are disturbed much less than in the Markovian jump linear-quadratic tracking implementation.

The aggregate power consumption plot is provided in Fig. 4 for the system. The algorithm not only provides immense relief at the early stages of the horizon, it also provides a smooth transition to the steady-state power consumption profile without a delayed payback peak typical of direct control schemes of thermostats.

Next, the central authority sets the target temperature to 57°C, and all agents are assigned to track the upper safety temperature $h = 60$°C

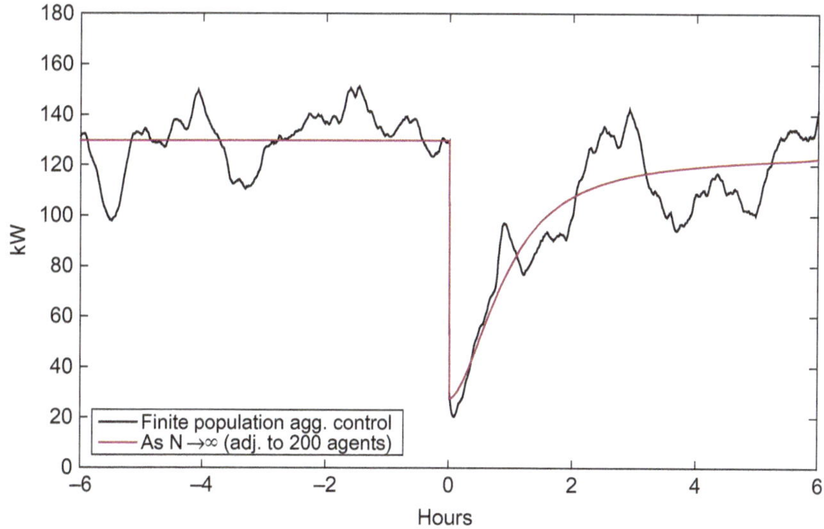

Fig. 4 Aggregate power consumption.

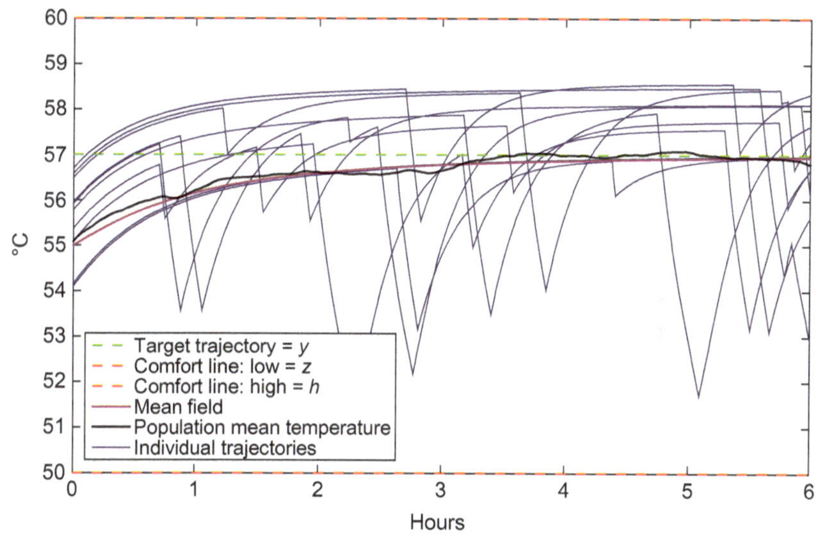

Fig. 5 Agents applying collective target tracking Markovian jump mean field control: all agents moving toward their upper safety temperature limit.

by applying collective target tracking Markovian jump mean field control. The trajectories are presented for only 10 agents, and the calculated mean field signal is shown in Fig. 5. It can be seen that the algorithm comfortably handles energy accumulation as well. The corresponding power consumption plot is given in Fig. 6.

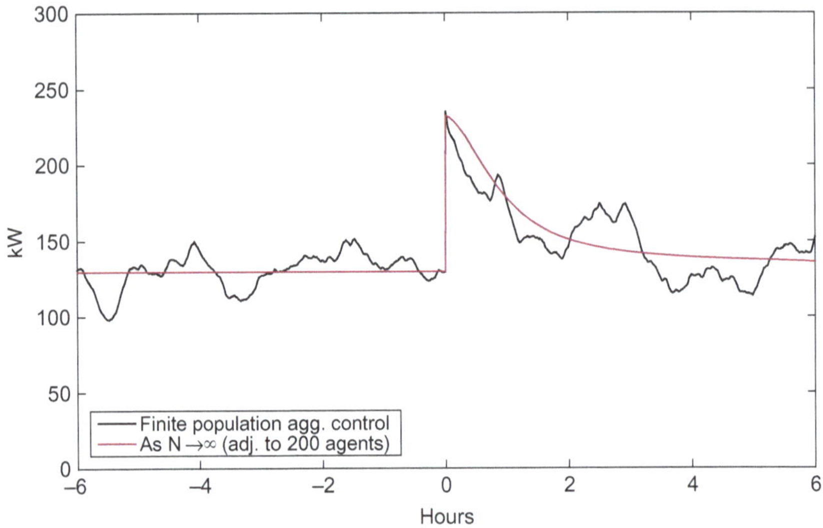

Fig. 6 Aggregate power consumption.

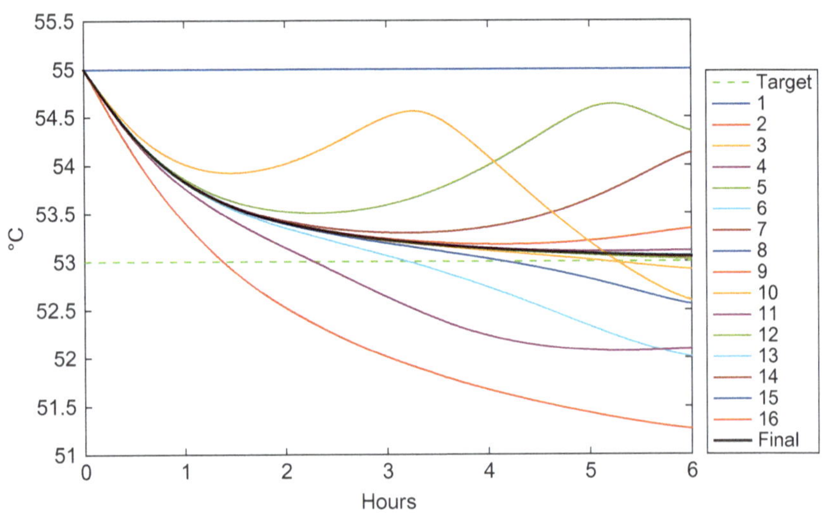

Fig. 7 Collective target tracking Markovian jump mean field iterations.

In Fig. 7 for the same simulation we plot the iterations of a collective target tracking Markovian jump mean field successive substitutions algorithm to identify the fixed point mean temperature trajectory until convergence occurs.

For the next experiment we separate the population into two groups, where the first group consists of the agents above 55°C initial temperature

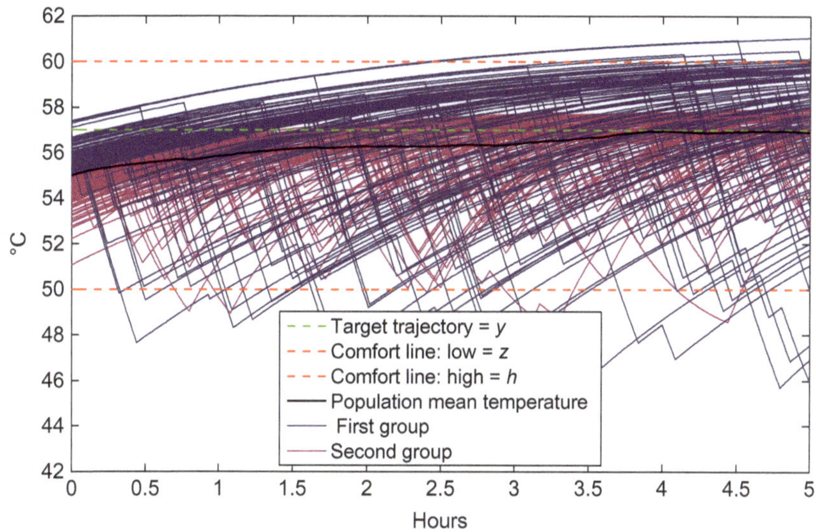

Fig. 8 Agents applying collective target tracking Markovian jump mean field control: different *R* for subpopulations.

and the second group consists of the ones below 55°C. Both groups are assigned to track 60°C. To achieve a certain level of fairness, the first group is assigned a larger control penalty coefficient R. Collective target tracking Markovian jump mean field control is applied to these groups, and the simulation result is provided in Fig. 8. It can be seen that the mean field-based integral error control scheme leads the mean temperature of the whole water heater population to 57°C while soliciting the agents with higher initial temperatures less intensely than those with lower initial temperatures. The comparison with the case of uniform R illustrates how one could shape the collective tracking response for greater fairness (Fig. 9).

For the next experiment we simulate a system of a further generalization where (i) the target is time varying (see Fig. 10) and (ii) the infinitesimal generator is time varying for 24 h. This plot is given to show the method's applicability to cases beyond the assumptions made in this chapter for the sake of analysis. In the plot, the mean temperature of the simulations does not match the mean field trajectory between 9 and 16 h because the mean field controller in this work is allowed to generate negative power, whereas while we were running the simulations physical constraints were enforced. The analysis of constrained control and state, and time varying targets are subjects of future research.

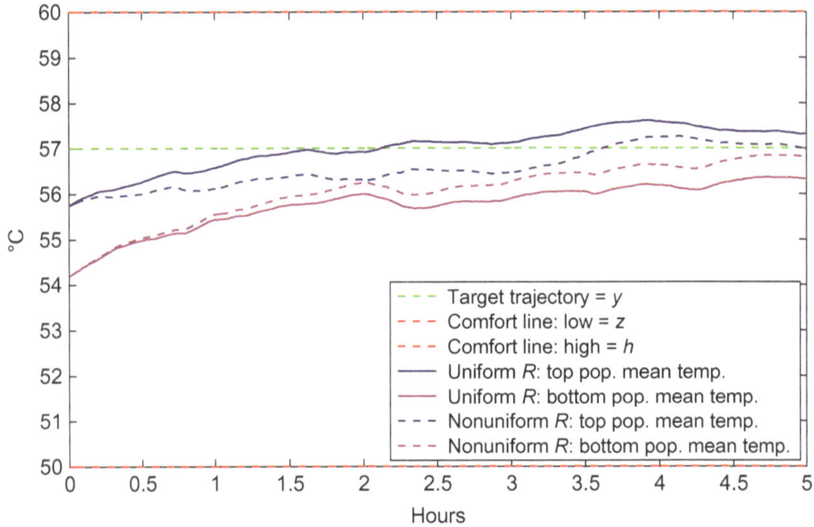

Fig. 9 Agents applying collective target tracking Markovian jump mean field control: uniform R versus nonuniform R.

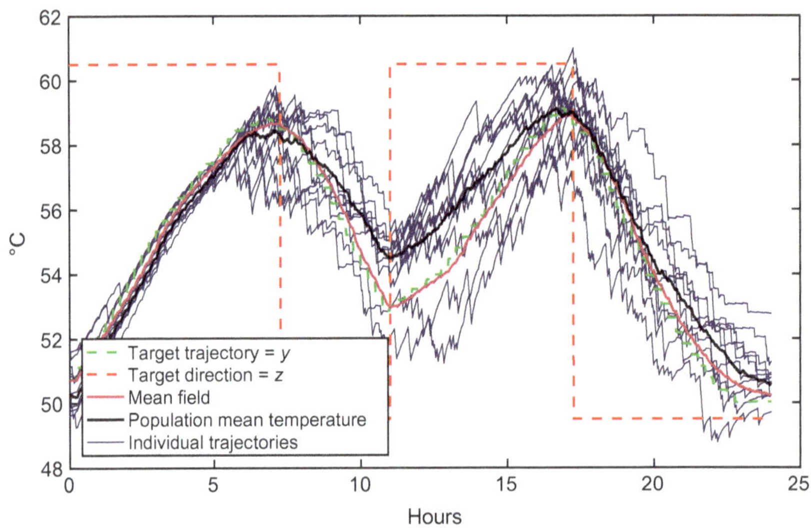

Fig. 10 Agents applying collective target tracking Markovian jump mean field control: time-varying target and time-varying Markov chain.

8 CONCLUSION

In this chapter we extended the *collective target tracking mean field control* method to the cases where the dynamics are modeled through Markovian jump processes. This formulation allows the characterization of the dynamics of a water heater where the water consumption event is modeled as a Markovian jump process. The presence of large numbers of electric devices associated with energy storage was employed to develop a decentralized mean field control-based approach to the problem of these devices following a desired mean trajectory. Given the statistics of the hot water consumption events and provided with the mean temperature target trajectory as well as the initial mean temperature in the controlled group, the devices generate their own control locally, and thus enforce their safety and comfort constraints locally as well.

In future work we will consider the impact of temperature comfort and safety constraints, as well as maximum and minimum bounds on the control action on the aggregate control performance. Finally, the case of periodic target trajectories will also be considered.

ACKNOWLEDGMENT

The authors gratefully acknowledge the support of Natural Resources Canada.

APPENDIX

Proof of Theorem 2

Repeating Eq. (17) here gives

$$-\frac{ds_t^j}{dt} = (A^j - BR^{-1}B^\top \Pi_t^j)^\top s_t^j - q_t^\gamma H^\top z - q^{x_0} H^\top \bar{x}_0 + \Pi_t^j c^j + \sum_{k=1}^{p} \lambda_{jk} s_t^k.$$

Solving with the boundary condition $s_T^j = -[H^\top q_T^\gamma z + H^\top q^{x_0} \bar{x}_0]$ gives

$$\begin{aligned} s_t^j &= (e_j^\top \otimes I)\Psi(t, T)[-1 \otimes H^\top q_T^\gamma z - 1 \otimes H^\top q^{x_0} \bar{x}_0] \\ &\quad + (e_j^\top \otimes I) \int_T^t \Psi(t, \tau) \left[1 \otimes H^\top q_\tau^\gamma z + 1 \otimes H^\top q^{x_0} \bar{x}_0 - \Pi_\tau c\right] d\tau, \end{aligned}$$

where

$$\frac{d\Psi(t,\tau)}{dt} = -[G^\top + \Lambda \otimes I]\Psi(t,\tau), \tag{20}$$

$G = \mathrm{diag}(G_1, \ldots, G_p)$, and $G_j = A_j - BR^{-1}B^\top \Pi^j$.

Now take $x' = x'_t,\ t \in [0,T]$, and $x'' = x''_t,\ t \in [0,T]$. The corresponding q' and q'' are obtained by use of Eq. (15), Π' and Π'' are obtained by use of Eq. (16), and Ψ' and Ψ'' are obtained by use of Eq. (20). Now we are ready to write

$$\begin{aligned}
\|s_t{}' - s_t{}''\| &= \Bigg\| \Psi(t,T)'(\cdot)' - \Psi(t,T)''(\cdot)'' \\
&\quad + \int_T^t \left[\Psi(t,\tau)'(\cdot)' - \Psi(t,\tau)''(\cdot)''\right] d\tau \Bigg\| \\
&\le \left\| \Psi(t,T)'(\cdot)' - \Psi(t,T)''(\cdot)' \right\| \\
&\quad + \left\| \Psi(t,T)''(\cdot)' - \Psi(t,T)''(\cdot)'' \right\| \\
&\quad + \left\| \int_T^t \left[\Psi(t,\tau)'(\cdot)' - \Psi(t,\tau)''(\cdot)'\right] d\tau \right\| \\
&\quad + \left\| \int_T^t \left[\Psi(t,\tau)''(\cdot)' - \Psi(t,\tau)''(\cdot)''\right] d\tau \right\| \\
&\triangleq \Delta_1 + \Delta_2 + \Delta_3 + \Delta_4.
\end{aligned}$$

We start with Δ_1:

$$\Delta_1 = \left\| (\Psi(t,T)' - \Psi(t,T)'')[-1 \otimes H^\top (q_T^y)'z - 1 \otimes H^\top q^{x_0}\bar{x}_0] \right\|.$$

Now $\|1 \otimes H^\top q^{x_0}\bar{x}_0\| \le \sqrt{p/n} q^{x_0}\bar{x}_0$ and $\|1 \otimes H^\top (q_\tau^y)'z\| \le \sqrt{p/n}\kappa_1 \tau z$, where κ_1 is given in Lemma A.1 in [23]

Before we examine $\left(\Psi(t,T)' - \Psi(t,T)''\right)$, we look into $\|\Pi_t(q') - \Pi_t(q'')\|$.

$$\|\Pi_t(q') - \Pi_t(q'')\| \le \max_j \|\Pi_t^j(q') - \Pi_t^j(q'')\|.$$

Employing Proposition 4.3 in [23], we get

$$\|\Pi_t(q') - \Pi_t(q'')\| \le \max_j \kappa_2^j \lambda t \|x' - x''\|_\infty.$$

Then, employing Lemma A.2 in [23] we obtain

$$\left\|\Psi(t,\tau)' - \Psi(t,\tau)''\right\| \tau \le \max_j \kappa_3^j \lambda \|x' - x''\|_\infty.$$

Therefore

$$\begin{aligned}
\Delta_1 &\le (\max_j \kappa_3^j \lambda) \left[(\sqrt{p/n} q^{x_0}\bar{x}_0) + (\sqrt{p/n}\kappa_1 z) \right] \|x' - x''\|_\infty \\
&\triangleq \gamma_1 \|x' - x''\|_\infty.
\end{aligned}$$

For Δ_2 we have

$$\Delta_2 = \left\| \Psi(t,T)' \left(-1 \otimes H^\top [(q_T^y)' - (q_T^y)'']z \right) \right\|.$$

We then have

$$\begin{aligned} \Delta_2 &\leq \lambda \|x' - x''\|_\infty \sqrt{p/n} z \Psi(t,T)' T \\ &\triangleq \gamma_2 \|x' - x''\|_\infty. \end{aligned}$$

For Δ_3 we have

$$\begin{aligned} \Delta_3 = \Bigg\| \int_T^t & \left(\Psi(t,\tau)' - \Psi(t,\tau)'' \right) \\ & \times \left[1 \otimes H^\top (q_\tau^y)' z + 1 \otimes H^\top q^{x_0} \bar{x}_0 - (\Pi_\tau)' c \right] d\tau \Bigg\|. \end{aligned}$$

Now $\|1 \otimes H^\top q^{x_0} \bar{x}_0\| \leq \sqrt{p/n} q^{x_0} \bar{x}_0$ and $\|1 \otimes H^\top (q_\tau^y)' z\| \leq \sqrt{p/n} \kappa_1 \tau z$, where κ_1 is given in Lemma A.1 in [23], whereas $(\Pi_\tau)' c \leq \max_j \|c^j\| \sup_{0 \leq t \leq T} [\lambda_{\max}(\Pi_t')]$.

Employing Lemma A.3 in [23], we obtain

$$\int_T^t \|\Psi(t,\tau)' - \Psi(t,\tau)''\| \tau \, d\tau \leq \max_j \kappa_2^j \lambda \|x' - x''\|_\infty.$$

Therefore

$$\begin{aligned} \Delta_3 &\leq (\max_j \kappa_2^j \lambda) \Bigg[(\sqrt{p/n} q^{x_0} \bar{x}_0) + (\sqrt{p/n} \kappa_1 z) \\ &\quad + (\max_j \|c_j\| \sup_{0 \leq t \leq T} [\lambda_{\max}(\Pi_t')]) \Bigg] \|x' - x''\|_\infty \\ &\triangleq \gamma_3 \|x' - x''\|_\infty. \end{aligned}$$

For Δ_4 we have

$$\begin{aligned} \Delta_4 &= \left| \int_T^t \Psi(t,\tau)'' \left[1 \otimes H^\top (q_\tau^y)' z - 1 \otimes H^\top (q_\tau^y)'' z + (\Pi_\tau)'' c - (\Pi_\tau)' c \right] d\tau \right| \\ &\leq \left| \int_T^t \Psi(t,\tau)'' \left[1 \otimes H^\top (q_\tau^y)' z - 1 \otimes H^\top (q_\tau^y)'' z \right] d\tau \right| \\ &\quad + \left| \int_T^t \Psi(t,\tau)'' \left[(\Pi_\tau)'' c - (\Pi_\tau)' c \right] d\tau \right| \\ &\triangleq \Delta_{41} + \Delta_{42}. \end{aligned}$$

We then have

$$\Delta_{41} \leq \lambda \|x' - x''\|_\infty \sqrt{p/n} z \int_T^t \Psi(t,\tau)'' \tau \, d\tau$$

and

$$\Delta_{42} \leq \sqrt{n} \max_j \|c^j\| \kappa_2^j \lambda \|x' - x'\|_\infty \int_T^t \Psi(t,\tau)'' \tau \, d\tau .$$

Hence

$$\begin{aligned}\Delta_4 &\leq (\sqrt{p/n} z + \sqrt{n} \max_j \|c^j\| \kappa_2^j) \lambda \|x' - x''\|_\infty \int_T^t \Psi(t,\tau)'' \tau \, d\tau \\ &\triangleq \gamma_2 \|x' - x''\|_\infty ,\end{aligned}$$

which implies

$$\|s_t{}' - s_t{}''\| \leq (\gamma_1 + \gamma_2 + \gamma_3 + \gamma_4) \|x' - x''\|_\infty .$$

Now let $\Delta x_t = x'(t) - x''(t)$. Then Eq. (11) together with Eq. (18) gives

$$d\Delta x_t = G^{\theta_t{}'} x_t' dt - G^{\theta_t{}''} x_t'' dt - BR^{-1} B^\top \Delta s(\theta_t, t) dt.$$

Define $V(j, x', x'') = (x' - x'')^\top \Pi^j (x' - x'')$, $1 \leq j \leq p$ and $\mathcal{A}$ to be the infinitesimal generator of $\{\theta_t, \Delta x_t\}$. Then

$$\begin{aligned}\mathcal{A}V(j, x', x'') &= \lim_{s \to 0} \frac{1}{s} \{E[V(\theta(t+s), x'(t+s), x''(t+s)) \\ &\quad |\theta_t = j, x_t' = x', x_t'' = x''] - V(j, x', x'')\} \\ &= \frac{\partial V}{\partial x'}(j, x', x'')(G^j x' - BR^{-1} B^\top s^{j'} - G^j x'' + BR^{-1} B^\top s^{j''}) \\ &\quad + \frac{\partial V}{\partial x''}(j, x', x'')(G^j x' - BR^{-1} B^\top s^{j'} - G^j x'' + BR^{-1} B^\top s^{j''}) \\ &\quad + \sum_k \lambda_{jk} V(k, x', x''),\end{aligned}$$

which following the steps in Theorem 3.4 in [1] leads to

$$\begin{aligned}\frac{dEV(\theta_t, x_t', x_t'')}{dt} &= E\mathcal{A}V(\theta_t, x_t', x_t'') \\ &\leq -a_1 EV(\theta_t, x_t', x_t'') + 2b_1 \|x' - x''\|_\infty \sqrt{EV(\theta_t, x_t', x_t'')},\end{aligned}$$

where

$$a_1 = \min_j \frac{\inf_{0\le t\le T}\left[\lambda_{\min}(Q_t^j + \Pi_t^j BR^{-1}B^\top \Pi_t^j)\right]}{\sup_{0\le t\le T}\left[\lambda_{\max}(\Pi_t^j)\right]},$$

$$b_1 = \max_j \sqrt{\sup_{0\le t\le T}\left[\lambda_{\max}(\Pi_t^j)\right]}\|BR^{-1}B^\top\|(\gamma_1+\gamma_2).$$

This implies

$$\begin{aligned} EV(\theta_t, x_t', x_t'') &\le \left\{\int_0^t \exp(-a_1/2(t-s))b_1\|x'-x''\|_\infty ds\right\}^2 \\ &\le \left(\frac{2b_1}{a_1}\|x'-x''\|_\infty\right)^2 \end{aligned}$$

and

$$E\|\Delta x_t\|^2 \le \frac{EV(\theta_t, x_t', x_t'')}{\min_j \inf_{0\le t\le T}[\lambda_{\min}(\Pi_t^j)]} \le \frac{\left(\frac{2b_1}{a_1}\|x'-x''\|_\infty\right)^2}{\min_j \inf_{0\le t\le T}[\lambda_{\min}(\Pi_t^j)]}.$$

Lastly we write

$$\|\mathcal{M}x' - \mathcal{M}x''\|_\infty = \sup_{t\in[0,T]} \|E[x_t' - x_t'']\|.$$

Applying the Cauchy-Schwarz inequality gives

$$\|\mathcal{M}x' - \mathcal{M}x''\|_\infty \le \sup_{t\in[0,T]} \left[E\|x_t' - x_t''\|^2\right]^{\frac{1}{2}} \le \frac{\frac{2b_1}{a_1}\|x'-x''\|_\infty}{\min_j \sqrt{\inf_{0\le t\le T}[\lambda_{\min}(\Pi_t^j)]}}.$$

REFERENCES

[1] B. Wang, J.-F. Zhang, Mean field games for large-population multiagent systems with Markov jump parameters, SIAM J. Control Optim. 50 (4) (2012) 2308–2334.

[2] J.M. Lasry, P.-L. Lions, Jeux à champ moyen. I—le cas stationnaire, C. R. Acad. Sci. Paris, Ser. I 343 (2006) 619–625.

[3] M. Huang, P.E. Caines, R.P. Malhamé, Individual and mass behaviour in large population stochastic wireless power control problems: centralized and Nash equilibrium solutions, in: 42nd IEEE Conference on Decision and Control, Maui, Hawaii, December, 2003, pp. 98–103.

[4] M. Huang, R.P. Malhamé, P.E. Caines, Large population stochastic dynamic games: closed loop McKean-Vlasov systems and the Nash certainty equivalence principle, Commun. Inf. Syst. 6 (2006) 221–252, special issue in honour of the 65th birthday of Tyrone Duncan.

[5] M. Huang, P.E. Caines, R.P. Malhamé, Large population cost-coupled LQG problems with non-uniform agents: individual-mass behaviour and decentralized ϵ-Nash equilibria, IEEE Trans. Autom. Control 52 (9) (2007) 1560–1571.
[6] A.C. Kizilkale, R.P. Malhamé, Mean field based control of power system dispersed energy storage devices for peak load relief, in: 52nd IEEE Conference on Decision and Control, December, 2013, pp. 4971–4976, invited paper.
[7] R.P. Malhamé, A jump-driven Markovian electric load model, Adv. Appl. Probab. 22 (1990) 564–586.
[8] R.P. Malhamé, C.-Y. Chong, Electric load model synthesis by diffusion approximation of a high-order hybrid-state stochastic system, IEEE Trans. Autom. Control 30 (9) (1985) 854–860.
[9] J.C. Laurent, R.P. Malhamé, A physically-based computer model of aggregate electric water heating loads, IEEE Trans. Power Syst. 9 (3) (1994) 1209–1217.
[10] S. Meyn, P. Barooah, A. Bušić, J. Ehren, Ancillary service to the grid from deferrable loads: the case for intelligent pool pumps in Florida, in: Proceedings of the 52nd IEEE Conf. on Decision and Control, December, 2013, pp. 6946–6953.
[11] Y. Chen, A. Bušić, S. Meyn, Individual risk in mean-field control models for decentralized control, with application to automated demand response, in: Proceedings of the 53rd IEEE Conf. on Decision and Control, December, 2014, pp. 6425–6432.
[12] Y. Chen, A. Bušić, S. Meyn, State estimation and mean field control with application to demand dispatch, 2015 [Online]. Available: http://arxiv.org/abs/1504.00088.
[13] Y. Lin, P. Barooah, S. Meyn, T. Middelkoop, Experimental evaluation of frequency regulation from commercial building HVAC systems, IEEE Trans. Smart Grid 6 (2015) 776–783.
[14] D. Callaway, Tapping the energy storage potential in electric loads to deliver load following and regulation, with application to wind energy, Energy Convers. Manag. 50 (2009) 1389–1400.
[15] J. Mathieu, S. Koch, D. Callaway, State estimation and control of electric loads to manage real-time energy imbalance, in: Power and Energy Society General Meeting (PES), IEEE, July, 2013, p. 1.
[16] J.L. Mathieu, D.S. Callaway, S. Kiliccote, Examining uncertainty in demand response baseline models and variability in automated response to dynamic pricing, in: 49th IEEE Conference on Decision and Control, 2010, pp. 4332–4339.
[17] F. Tahersima, J. Stoustrup, S. Afkhami, H. Rasmussen, Contribution of domestic heating systems to smart grid control, in: 50th IEEE Conference on Decision and Control, 2011, pp. 3677–3681.
[18] Z. Ma, D.S. Callaway, I.A. Hiskens, Decentralized charging control of large populations of plug-in electric vehicles, IEEE Trans. Control Syst. Technol. 21 (1) (2013) 67–78.
[19] L. Totu, J. Leth, R. Wisniewski, Control for large scale demand response of thermostatic loads, in: American Control Conference (ACC), June, 2013, pp. 5023–5028.
[20] Y. Ji, J. Chizeck, Controllability, stabilizability and continuous-time Markovian jump linear quadratic control, IEEE Trans. Autom. Control 35 (1990) 777–788.
[21] S. Klein, A design procedure for solar heating systems, Ph.D. dissertation, Department of Chemical Engineering, University of Wisconsin-Madison, 1976.
[22] W.M. Wonham, Random differential equations in control theory, in: Probabilistic Methods in Applied Mathematics, vol. 2, Academic, New York, 1971.
[23] A.C. Kizilkale, R.P. Malhamé, Mean field based control of power system dispersed energy storage devices for peak load relief, Technical Report, Groupe d'Études et de Recherche en Analyse des Décisions (GERAD), Canada, 2015.

CHAPTER TWENTY-ONE

Trajectory Planning Based on Collocation Methods for Adaptive Motion Control of Multiple Aerial and Ground Autonomous Vehicles

S. Vera*, F. Petric†, G. Heredia*, A. Ollero*, Z. Kovacic†
*University of Seville, Seville, Spain
†University of Zagreb, Zagreb, Croatia

1 INTRODUCTION

Interest in unmanned vehicles has greatly increased in the last decade. Given the ability to plan a trajectory on the basis of sensor input, unmanned vehicles have the capability to avoid obstacles or other vehicles, track targets, optimize certain performance characteristics, such as endurance, and generally adapt to a dynamic situation.

Initially this interest was focused on military uses, mainly in the use of unmanned aerial vehicles (UAVs) for different applications, but over time the number of civil applications has increased, with both UAVs and unmanned ground vehicles (UGVs). In recent years the technologies for autonomous aerial vehicles have experienced an important development, making research into aerial autonomous systems affordable for universities and research centers.

This chapter presents a centralized method of optimal trajectory planning for multiple unmanned vehicles. The method can be used with aerial or ground vehicles, as well as in mixed scenarios with aerial and ground vehicles. This method is based on pseudospectral (PS) collocation. PS collocation is a technique that solves numerically optimal control problems and which has gained popularity in recent years in several applications [1, 2].

Vehicle trajectory planning is formulated as an optimal control problem, including the differential equations of the vehicle dynamics and physical constraints, such as actuator saturations, avoiding collisions with the

Control of Complex Systems
http://dx.doi.org/10.1016/B978-0-12-805246-4.00021-5

environment or other vehicles, etc., and an optimization criterion that should be minimized—for example, the total distance traveled or the energy consumption. PS collocation methods try to solve these optimal control problems by approximating the states and control inputs using polynomials, and by enforcing the differential equations and constraints on specific instants of time called *collocation points*. Thus PS collocation methods effectively discretize the optimal control problem and convert it to an algebraic nonlinear programming problem, which can be efficiently solved by application of optimization tools such as the solver SNOPT [3].

This formulation of the trajectory planning problem allows a good representation of the real vehicle behavior in real applications, and thus the solution trajectories will be realistic and easily tracked by the autonomous vehicles in real time.

The study presented in this chapter covers behavior analysis of different PS collocation methods when they are used in vehicle trajectory planning. Although PS collocation methods have been formulated in the general case in the literature, it is important to analyze how certain aspects such as discretization or approximation of the solution, for example, may affect their performance for trajectory planning problems. The work presented in this chapter covers specific aspects such as accuracy, scalability (number of vehicles in a scenario), and computation time for implementation in real time. Finally, a new collocation algorithm designed especially for trajectory planning of multiple vehicles moving simultaneously in the same area is developed and presented, and comparative results with the previous methods are presented in multiple scenarios, validated in simulation and real experiments.

1.1 Trajectory Planning for Multiple Autonomous Vehicles

Efficient trajectory planning and conflict detection and resolution (CDR) methods have been extensively studied in the literature. This is an important aspect in coordinated missions with multiple UAVs. A detailed survey on the former is presented in [4], and a review of articles on the latter is presented in [5].

CDR methods in the literature can be coarsely classified as either deterministic or stochastic. The deterministic methods perform an exploration of the feasible solution set using search procedures usually directed by the local behavior of the optimization function. These methods include nonlinear programming (NLP) [6, 7], integer programming [8, 9], dynamic programming if the problem can be broken down into simpler subproblems [10], and collocation methods reducing the number of dimensions of the problem [11, 12]. The stochastic methods perform a random exploration of the optimization problem by using random variables [13]. This involves random objective functions or random constraints. These methods include random search methods [14, 15], graph search such as A⋆ and D⋆ [16], rapidly exploring random trees [17], particle swarm optimization [18], evolutionary computation methods [19, 20], multiobjective evolutionary computation methods [21], ant colony optimization methods [22], and simulated annealing [23].

Most CDR methods use three types of allowed maneuvers to solve the conflicts between aerial vehicles: changes of height, heading, or speed. The methods in [19, 20] are based on the use of genetic algorithms to solve conflicts by changing the heading. The main drawback of genetic algorithms is that the computation time is not predictable and the convergence to a solution is not ensured in a finite time interval. The optimal control method in [24] solves conflicts for two or three aircraft by changing the heading.

A stochastic method based on the Monte Carlo approach solves conflicts in air traffic control [25] but the computation time is again high. A hybrid system model that generates safe maneuvers is applied in [26]. A CDR method based on a mixed-integer linear program optimizes the total flight time by modifying the velocity or heading [9]. However, this method allows only one speed change for each aircraft.

The method proposed in [27] considers the problem in three dimensions using a mixed-integer linear program to solve the conflicts with only velocity changes by minimizing the total flight time. However, this approach requires a long computation time. The method described in [8] is also based on a mixed-integer linear program to solve pairwise conflicts by changing speeds for a large number of aircraft. Changes of speed to solve the conflicts are also considered in [28]. Three different collision detection and resolution methods based on speed planning are presented in [29].

Collocation methods [30] pose the trajectory generation problem as an optimal control problem which can be solved to generate optimal trajectories. They have several advantages over other methods for multivehicle collision-free trajectory generation. One advantage is that collocation methods can employ mathematical models of different detail for the vehicles (UAVs or UGVs). Furthermore, different models can be used for each vehicle, thus allowing for heterogeneous teams. The collocation method formulation can incorporate restrictions in the control inputs and the state variables, representing the physical characteristics of the autonomous vehicles and yielding solution trajectories that are very close to what the UAVs or UGVs can do. Collocation methods generate optimized trajectories with respect to a user-defined optimization index but the execution time of the algorithm is short, allowing fast recomputation of the trajectories if conditions change. The next sections describe the basics of the method and its formulation for the generation of trajectories for multiple UAVs/UGVs.

2 COLLOCATION METHODS

This section presents an overview of different collocation methods and an outline of the history and the origin of the theory of collocation methods. First an introduction to numerical methods is presented. Then the optimal control problem is introduced as well as its definition as a nonlinear problem. Finally a global classification of the different collocation methods and a full description of the theory of the most important algorithms are given.

2.1 Numerical Analysis and Spectral Methods

Differential equations appear in many areas of engineering and science as a fundamental way of modeling the physical behavior of systems. Although analytical methods for the solution of differential equation exist, they are only applicable to specific and simple cases. In most applications, differential equations have to be solved by numerical approximation methods that simplify or discretize the problem.

Numerical analysis for solving partial and ordinary differential equations constitutes a scientific discipline with well-known methods of finite differences and finite elements but it also includes a family of methods known as *spectral methods*.

Given a differential equation with boundary conditions, the idea of spectral methods is to try to approximate a solution ν of the differential equation by means of truncated series of basic functions:

$$\nu_N = \sum_{i=0}^{N} c_i \varphi_i, \tag{1}$$

where φ_i are the basis functions, N + 1 is the number of terms of the approximation, and c_i are the real coefficients, which are unknown. Thus two main questions arise: from which function class should φ_i be chosen; and how should the expansion coefficients c_i be determined.

The φ_i function class should meet several criteria: the approximations $\sum_{i=0}^{N} c_i\varphi_i$ of v must converge rapidly with increasing N; it should be easy to determine the coefficients of derivatives; and it should be fast to convert between the c_i coefficients and the value of v_N at some set of nodes.

For periodic problems the choice is usually $\varphi_i(x) = e^{ikx}$, which is the base of the well-known Fourier spectral method. If the function v is infinitely smooth and periodic together with all its derivatives, then the kth coefficient of the Fourier expansion decays faster than any inverse power of k. This property is known as *spectral accuracy*. In practice this never happens, but spectral accuracy is attainable also for nonperiodic but smooth functions. Another classic result of the approximation theory is that for analytical functions exponential (or spectral) decay of the coefficients can be obtained for basic functions of the Jacobi polynomials type, which are mutually orthogonal over the interval (−1,1). Chebyshev and Legendre polynomials are the most important special cases of Jacobi polynomials.

2.1.1 Pseudospectral Approximation

The spectral approximation defines a transformation from the physical space to the spectral space (like the Fourier coefficients in a Fourier transform). The coefficients c_i in the spectral approximation depend on all the values of the function v in the physical space and can be computed only by numerical integration. Since this cannot be performed exactly for arbitrary functions v, a set of approximate coefficients $\hat{c}_i$ can be obtained with use of an interpolating polynomial v_P of the function v defined by a finite set of interpolation points:

$$v_P = \sum_{i=0}^{N} \hat{c}_i\varphi_i. \tag{2}$$

The interpolating polynomial must be exactly equal to the function in the interpolation points:

$$v_P(x_k) = \sum_{i=0}^{N} \hat{c}_i\varphi_i(x_k) = v(x_k). \tag{3}$$

These points are usually called *collocation points*, and are the points at which the approximation is equal to the exact solution.

Now the question that arises is how to choose these collocation points x_k in a way that the residual (differences between the real function value and the approximation) at the remaining points is a minimum. The answer is that the best choice is what is known as *Gauss quadrature points*. A *quadrature rule* is just a method to approximate the integral of a function by a weighted sum of the values of the function at some specific points, called *quadrature points*. The Gauss quadrature rule generates an optimal approximation:

$$\int_{-1}^{1} f(x)dx \approx \sum_{i=0}^{N} w_i f(x_i), \tag{4}$$

where w_i are the *quadrature weights* and N is the number of points of the quadrature.

Three of the more extended Gaussian quadratures are known as *Gauss*, *Gauss-Radau*, and *Gauss-Lobatto*, respectively. The *Gauss quadrature points* are distributed on the interior of the interval but do not include the end points. On the other hand, *Gauss-Radau quadrature points* include one end point, and *Gauss-Lobatto quadrature points* include both end points of the interval. Although the three types of quadrature points are similar, the effects in the numerical approximation are relevant. A detailed explanation can be found in [31].

Gauss-type quadrature points are more concentrated on both end points of the interval to reduce the *Runge phenomenon*, the fact that the approximation error of a function by a polynomial is larger at the edges of the domain than in the interior. Then, although intuitively it may seem more adequate for a given problem to choose the collocation points evenly distributed in the domain, this is not possible because it would introduce substantial numerical errors, and the number of collocation points would also be increased significantly, greatly slowing down the computations. This is a very important fact that will be used in the derivation of the new S-adaptive PS collocation method in the next section.

It can be shown that spectral convergence is retained when the continuous transform (1) is replaced by the interpolating polynomial (2) if the interpolation (or collocation) points are the corresponding Gauss-type quadrature points. Still, the coefficients $\hat{c}_i$ have to be computed. In practice, however, the interpolation polynomials are sometimes written as a linear combination of *Lagrange interpolation polynomials* φ_i through the Gauss-type quadrature points:

$$v_P = \sum_{i=0}^{N} v_i \varphi_i. \tag{5}$$

The Lagrange polynomials are mathematically equivalent to the interpolating polynomials, and have the property that the coefficients are given by the value of the function in the interpolation $v_i = v(x_k)$. Furthermore, the derivative of the function can be computed by a matrix multiplication, which makes numerical approximation faster and is the reason for choosing Lagrange polynomials.

From a general perspective the fundamental interpretation of spectral methods was originally that the differential equation is transformed with the polynomial expansion in Eq. (1), indicating that c_i can be used as a representation of the solution to the differential equation. But this leads to limitations when it comes to differential equations with variable coefficients or nonlinear problems. Instead we use the spectral representation for only part of the solution procedure, while we use the function in the physical space for the other part. This leads to what are known as *pseudospectral (PS) methods*.

Different software implementations of PS collocation methods have been published, all of which focus on the resolution of optimal control problems. The most recent are the software package DIDO [32], a MATLAB implementation of the Legendre PS method; GPOPSII [33] a MATLAB implementation of a multiinterval Radau PS method; and PSOPT [34], an open source software package written in C++ that uses both Legendre and Chebyshev polynomials. Another software implementation, OTIS [35], was developed by Boeing for NASA. Other useful solvers include SOCS [36], DIRCOL [37], GESOP [38], ICLOCS [39], and ACADO [40].

2.2 Optimal Control

Optimal control is the process of finding the control and state law for a dynamic system over a period of time so that the performance of the system is optimal with respect to some criterion, such as control effort, tracking error, energy consumption, or amount of time taken to reach a target.

The formulation of an optimal control problem usually requires:

- a mathematical model of the system to be controlled;
- a specification of all boundary conditions on states, and constraints to be satisfied by states and controls;
- a specification of the performance index.

Because of the increasing complexity of optimal control applications, in the past 2 decades the subject of optimal control has transitioned from theory to computation.

In particular, computational optimal control has become a science in and of itself, resulting in a variety of numerical methods and their corresponding software implementations. The roots of these methods can be traced back to the work of Bernoulli and Euler [41]. The simplicity in this approach is based on a wide range of deep theoretical issues that lie at the intersection of approximation theory, control theory, and optimization. Examples of this progress can be found in [42–46].

Numerical methods for solving optimal control problems are divided into two major classes: indirect methods and direct methods. In an *indirect method*, the calculus of variations, is used to determine the first-order optimality conditions of the original optimal control problem. The indirect approach leads to a multiple-point boundary-value problem that is solved to determine candidate optimal trajectories called *extremals*. Each of the computed extremes is then examined to see if it is a local minimum, maximum, or saddle point. Of the locally optimizing solutions, the particular extreme with the lowest cost is chosen.

On the other hand, in a *direct method*, the state and/or control of the optimal control problem is discretized in some manner and the problem is transcribed to a nonlinear optimization problem or NLP problem. The NLP problem is then solved by use of well-known optimization techniques. Since a numerical solution of an NLP problem can be obtained much faster and more efficiently than a solution with the indirect method that satisfies end point and/or interior point conditions, direct methods have been widely adopted to solve optimal control problems in real applications.

2.2.1 General Formulation of the Optimal Control Problem

The objective of an optimal control problem in continuous time is to determine the state and control variables that optimize an objective function, subject to dynamic constraints and boundary conditions. In collocation methods the optimal control problem is given by the following standard form (known as the *Bolza form*): Determine the state (equivalently the *trajectory* or path), $x(t) \in \mathbb{R}^n$, the control input, $u(t) \in \mathbb{R}^m$, the initial time, $t_0 \in \mathbb{R}$, and the terminal time, $t_f \in \mathbb{R}$ (where $t \in [t_0, t_f]$ is the independent variable) that optimize the performance index:

$$J = \varphi\left(x\left(t_0\right), t_0, x\left(t_f\right), t_f\right) + \int_{t_0}^{t_f} g(x\left(t\right), u(t))dt \tag{6}$$

subject to the *dynamic constraints* (differential equations that describe the system dynamics)

$$\dot{x}\left(t\right) = \frac{dx}{dt} = f(x\left(t\right), u(t)), \tag{7}$$

inequality *path constraints* (continuous constraints of the state and control input during the whole trajectory $t \in [t_0, t_f]$)

$$C\left(x\left(t\right), u\left(t\right)\right) \leq 0, \tag{8}$$

and the *boundary conditions*

$$E\left(x\left(t_0\right), t_0, x\left(t_f\right), t_f\right) = 0, \tag{9}$$

where $f : \mathbb{R}^n \times \mathbb{R}^m \to \mathbb{R}^n$, $C : \mathbb{R}^n \times \mathbb{R}^m \to \mathbb{R}^s$, and $E : \mathbb{R}^n \times \mathbb{R} \times \mathbb{R}^n \times \mathbb{R} \to \mathbb{R}^q$. Furthermore, the function $\varphi : \mathbb{R}^n \times \mathbb{R} \times \mathbb{R}^n \times \mathbb{R} \to \mathbb{R}$ is the *Mayer cost function* and $g : \mathbb{R}^n \times \mathbb{R}^m \to \mathbb{R}$ is the *Lagrangian cost function*. The Mayer cost function focuses on the boundaries of the problem (essentially focusing on the optimization in the initial and final states of the problem). On the other hand, the Lagrangian cost function focuses on optimization throughout the entire process.

2.3 Nonlinear Optimization

A key ingredient to solving optimal control problems is the ability to solve nonlinear optimization problems or *NLP problems*. An NLP problem takes the following general mathematical form:

Determine the vector of decision variables $z \in \mathbb{R}^n$ that minimize the cost function $f(z)$ subject to the algebraic constraints:

$$g\left(z\right) = 0, \tag{10}$$

$$h(z) \leq 0,$$

where $g(z) \in \mathbb{R}^m$ and $h(z) \in \mathbb{R}^p$.

NLP problems are usually solved with use of gradient-based methods. The most commonly used gradient-based NLP solution methods are *sequential quadratic programming* and *interior-point* or *barrier* methods.

Extensive research has been done in the past 2 decades in sequential quadratic programming and interior-point methods, leading to extremely efficient and robust software for the numerical solution of NLPs. Examples of well-known software programs that use sequential quadratic programming methods include the dense NLP solver NPSOL [47] and the sparse NLP solvers SNOPT [3] and SPRNLP. Well-known sparse interior-point NLP solvers include BARNLP, LOQO, KNITRO, and IPOPT.

2.4 Collocation Methods

Collocation methods are based on transcription of a continuous-time problem into a finite-dimensional NLP problem, by enforcement of a suitable interpolating function through the state values, and maintenance of the state derivatives at the collocation points spanning one time interval.

Collocation methods are characterized by the numerical integration method and the discretization method. In the case of discretization, there are three important concepts that have to be differentiated: *intervals* and *subintervals*, *nodes*, and *collocation points*. The problem timeline is defined by one interval $[t_0, t_f]$, where t_0 is the initial time in the problem and t_f is the final time. This interval can be split into subintervals $[t_{k-1}, t_k]$, where k is the number of subintervals. Every interval or subinterval is splinted by nodes. The definition of nodes is very close to the discretization concept of the problem. The distribution of the nodes throughout the segment is different in every collocation method. For example, Fig. 1 shows the distribution of nodes in the case of the direct collocation method as well as in the case of the PS collocation method. It can be seen that in direct

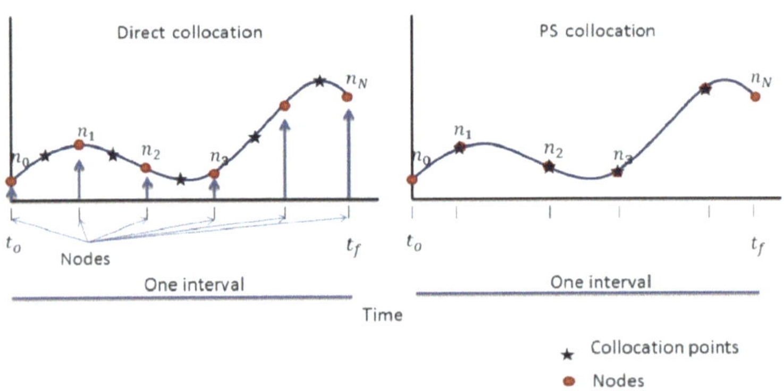

Fig. 1 General distribution of nodes and collocation points in collocation methods.

collocation, the distribution of nodes is equally spaced in the time interval $[t_0, t_f]$, while this does not happen in the PS method.

Collocation points are the points at which dynamic equations of motion are enforced through quadrature rules or interpolation. Depending on the numerical integration method, the collocation point is set at the midpoint between two nodes, which happens in the direct collocation method, or the collocation points are set at the same nodes, as seen in PS collocation method. Specifically in the PS method, collocation points are considered the subset of nodes $[n_1, n_2, \ldots, n_{N-1}]$, where N is the number of nodes. End point nodes n_0 and n_N of the segment are never used as collocation points.

Considering this specification, collocation methods can be divided into three groups: *direct collocation*, *PS collocation*, and *subinterval PS collocation* methods.

2.4.1 Pseudospectral Collocation Method

Recently, a great deal of research has been done in the class of Gaussian quadrature orthogonal collocation methods, commonly known as *pseudospectral (PS) collocation methods* [45, 48, 49]. Some improvements and advances are introduced in relation to the direct collocation method. PS collocation focuses on a global perspective of the problem, while direct collocation solves the problem piece by piece. In PS methods the entire problem is approximated by a quadrature rule, while in direct collocation the problem is solved by $(N - 1)$ Hermite polynomials, where N is the number of nodes. This change from local to global happens because it has been proven in numerical analysis that more sophisticated node selection methods are able to achieve significantly improved accuracy with many fewer nodes. The PS collocation method has a simple structure and it converges at an exponential rate [50–52], so it fulfils the inequality:

$$||f\,(\mathrm{t}) - I_N f\,(\mathrm{t})\,||_{L^2} \leq \frac{\mathrm{C}}{N^m}, \tag{11}$$

where $f\,(t)$ is a function, I_N is the polynomial interpolation of the function with N points, m is the smoothness of $f(t)$, and C is a constant independent of N. As $N \to \infty$, the polynomial interpolation at the Legendre-Gauss-Lobatto (LGL) nodes, for example, converges to $f(t)$ under the L^2 norm at a rate of $1/N^m$. If $f(t)$ is C^∞, then $m = \infty$. This implies that the polynomial interpolation at these nodes converges at a spectral rate (faster than any given polynomial rate).

With use of the PS collocation method, variables of the entire time interval are typically approximated by a Lagrange polynomial, where the support points of this polynomial are chosen on the basis of their association with a Gaussian quadrature. These methods are fully conditioned by the quadrature rule used in the approximation of the solution. This affects the distribution of the nodes as well as numerical integration of the cost function. The definitions of nodes, also named *quadrature nodes*, are given by the zeros of the derivative of an orthogonal Jacobi polynomial (mainly Legendre or Chebyshev polynomial). Thus, between nodes and interpolants, a very large family of PS methods can be used to solve optimal control problems. The most popular are the:

- Legendre-Gauss PS collocation method [49],
- Legendre-Gauss-Radau PS collocation method [53–55],
- Legendre-Gauss-Lobatto (LGL) PS collocation method [45],
- Chebyshev-Gauss-Lobatto PS collocation method [56].

A complete description of PS collocation methods is presented next. This outline describes in detail an LGL PS collocation method [45, 57–60], although this demonstration may be applicable to any PS method.

The problem defined in Eqs. (6)–(9) is translated from the time domain $t \in [t_0, t_f]$ to the normalized domain $\tau \in [-1, 1]$ by a related transformation:

$$\tau = \frac{2t - (t_0 + t_f)}{t_f - t_0}. \tag{12}$$

The inverse is given by

$$t = \frac{t_f - t_0}{2}\tau + \frac{t_0 - t_f}{2}, \tag{13}$$

and it is noted that

$$\frac{dt}{d\tau} = \frac{t_f - t_0}{2}. \tag{14}$$

The optimization problem is formulated in the Bolza form (Eqs. 6–9), with the control vector function $u(\tau)$, state vector $x(\tau)$, $t_0 \to -1$ and $t_f \to 1$.

Because of to the use of the LGL quadrature, the problem is discretized by an LGL distribution of nodes. Fig. 2 shows an example of the distribution of 16 nodes. LGL nodes are defined by domain $\tau \in [-1, 1]$ as $\tau_0 = -1 < \tau_1 < \tau_2 < \ldots < \tau_N = 1$, where $\tau_1, \tau_2, \ldots \tau_{N-1}$ are the roots of the derivate of the Nth order Legendre polynomial.

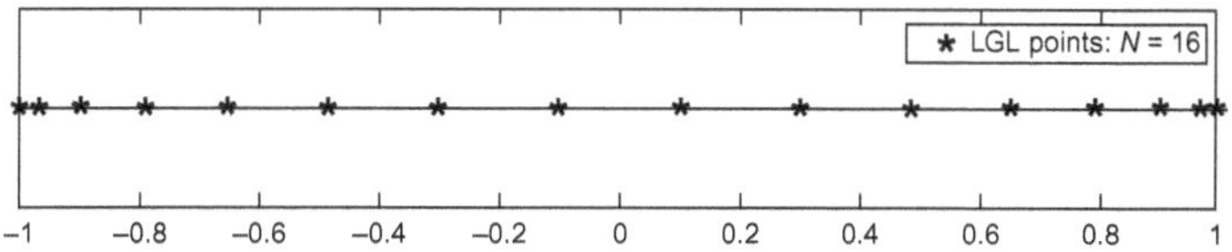

Fig. 2 The distribution of 16 Legendre-Gauss-Lobatto nodes.

The state and control vectors of the problem, $x(\tau)$ and $u(\tau)$, are approximated by $\boldsymbol{X}(\tau) \in \mathbb{R}^r$ and $\boldsymbol{U}(\tau) \in \mathbb{R}^r$. In the discretization the state variables at the nodes are approximated by column vectors:

$$x\left(\tau_j\right) \approx \boldsymbol{X}^j = [\boldsymbol{X}_1^j, \boldsymbol{X}_2^j, \ldots, \boldsymbol{X}_r^j]^T, \tag{15}$$

where r is the number or variable that composes the state vector, and $\tau_j = [\tau_0, \tau_1, \ldots, \tau_N]$ are the nodes. Similarly, $u\left(\tau_j\right) \approx \boldsymbol{U}^j$ is the approximation of the control inputs. Thus a discrete approximation of the function $x_i(\tau)$ across all nodes is the row vector

$$\boldsymbol{X}_i = [\boldsymbol{X}_i^0, \boldsymbol{X}_i^1, \ldots, \boldsymbol{X}_i^N]. \tag{16}$$

In the notation the discrete variables are denoted by bold uppercase letters. Subscript i is used to discriminate the variable in the state vector $i = 1, 2, \ldots, r$ and superscript j is used to discriminate the nodes $j = 0, 1, \ldots, N$. $\boldsymbol{X}_i^0$ and $\boldsymbol{X}_i^N$ are the variables i of the state vector at the initial and the final instant. Then the entire state vector can be represented as

$$\mathbf{X} = \begin{bmatrix} X_1^1 & \cdots & X_1^N \\ \vdots & \ddots & \vdots \\ X_r^1 & \cdots & X_r^N \end{bmatrix}. \tag{17}$$

A continuous approximation of $x(\tau)$ is defined by its polynomial in the Lagrange form, denoted $x_i^N(\tau)$:

$$x_i(\tau) \approx x_i^N(\tau) = \sum_{j=0}^{N} \boldsymbol{X}_i^j \phi_j(\tau), \tag{18}$$

$$u_i(\tau) \approx u_i^N(\tau) = \sum_{j=0}^{N} \boldsymbol{U}_i^j \phi_j(\tau),$$

where ϕ_j are the Lagrange basis polynomials

$$\phi_j = \frac{1}{N(N+1)L_N(\tau_j)} \frac{(\tau^2 - 1)\dot{L}_N(\tau)}{\tau - \tau_j} \tag{19}$$

and L_N is the Nth order Legendre polynomial function, because we are using the LGL method.

The derivative of $x_i^N(\tau)$ at the LGL node τ_j is easily computed by differentiation of the approximation of Eq. (18) with respect to τ:

$$\frac{dx_i^N(\tau)}{d\tau} = \sum_{j=0}^{N} \mathbf{X}_i^j \dot{\phi}_j(\tau), \tag{20}$$

where $\dot{\phi}_j(\tau_l)$ is an $(N+1) \times (N+1)$ matrix, also named a *differentiating matrix*, D, given by

$$D = D_{j,l} = \begin{cases} \frac{L_N(\tau_j)}{L_N(\tau_l)} \frac{1}{(\tau_j - \tau_l)} & j \neq l, \\ -\frac{N(N+1)}{4} & j = l = 0, \\ \frac{N(N+1)}{4} & j = l = N, \\ 0, & \text{otherwise.} \end{cases} \tag{21}$$

Finally, the discretization of the cost function is given by the LGL quadrature:

$$J(x, u) \approx \mathbf{J}(\mathbf{X}, \mathbf{U}) = \varphi\left(\mathbf{X}^0, \mathbf{X}^N\right) + \sum_{j=0}^{N} g(\mathbf{X}^j, \mathbf{U}^j) w_j, \tag{22}$$

where w_j are the LGL quadrature weights, defined by

$$w_j = \frac{2}{N(N+1)} \frac{1}{[L_N(\tau_j)]^2}. \tag{23}$$

In summary, the entire problem defined in Eqs. (6)–(9) is translated to a discrete problem where the goal is to find a state vector $\mathbf{X}^j \in \mathbb{X}$ and a control vector $\mathbf{U}^j \in \mathbb{U}$, with $j = 0, 1, \ldots, N$, that minimize the cost function

$$J(\mathbf{X}, \mathbf{U}) = \varphi\left(\mathbf{X}^0, \mathbf{X}^N\right) + \sum_{j=0}^{N} g(\mathbf{X}^j, \mathbf{U}^j) w_j \tag{24}$$

subject to the discrete dynamics

$$\sum_{j=0}^{N} D_{j,l} \mathbf{X}^j - f\left(\mathbf{X}^j, \mathbf{U}^j\right) = 0, \tag{25}$$

path constraints

$$C\left(\mathbf{X}^j, \mathbf{U}^j\right) \leqslant 0, \tag{26}$$

and the end points

$$E\left(\mathbf{X}^1, \mathbf{X}^N\right) = 0. \tag{27}$$

One achieves convergence of the PS method by increasing the degree of the Lagrange polynomial approximation (increasing the number of nodes).

2.4.2 Subinterval Pseudospectral Collocation Method

The subinterval PS method introduces a new concept, which is a hybrid between direct collocation and the PS method. In the direct method, one achieves convergence by increasing the number of subproblems. A similar concept is applied here. In this case the time interval $[t_0, t_f]$ is split into subintervals $[t_{k-1}, t_k]$, and every subinterval is solved by the classic PS method. Now, border equalities of the subinterval are added to PS method constraints because continuity must be maintained. Differences between various subinterval PS methods are given by how the problem is split, and by which PS method is used to solve every subinterval.

Subinterval PS methods are very recent, and only a few implementations of this technique can be found in the literature. One example of this method is *hp-adaptive collocation* [61]. This method was developed to significantly reduce the size of the finite-dimensional approximation problem, and thus improve the computational efficiency of solving the NLP problem. This method increases the order of the polynomial when the solution is smooth and the convergence rate is fast and also splits the problem into subintervals when the solution changes rapidly. Then the segmentation rate and the order of the polynomial interpolation could be different in every subinterval. hp-adaptive collocation is an iterative method, and the stop condition is given when a suitable error or a bounded iteration is achieved.

A complete description of the subinterval PS collocation method [62] is presented next. Subinterval methods are considered to reformulate the problem defined in Eqs. (6)–(9) by dividing it into K intervals. An interval k is defined to begin at t_{k-1} and end at t_k, where $t_0 < t_1 < t_2 < \cdots < t_k = t_f$. The time domain in each subinterval k, $[t_{k-1}, t_k]$, is changed to the normalized domain $\tau \in [-1, 1]$ by the transformation

$$\tau = \frac{2t - (t_{k-1} + t_k)}{t_k - t_{k-1}}. \tag{28}$$

An illustration of this concept is given by Fig. 3, in which the entire problem has been split into two subintervals.

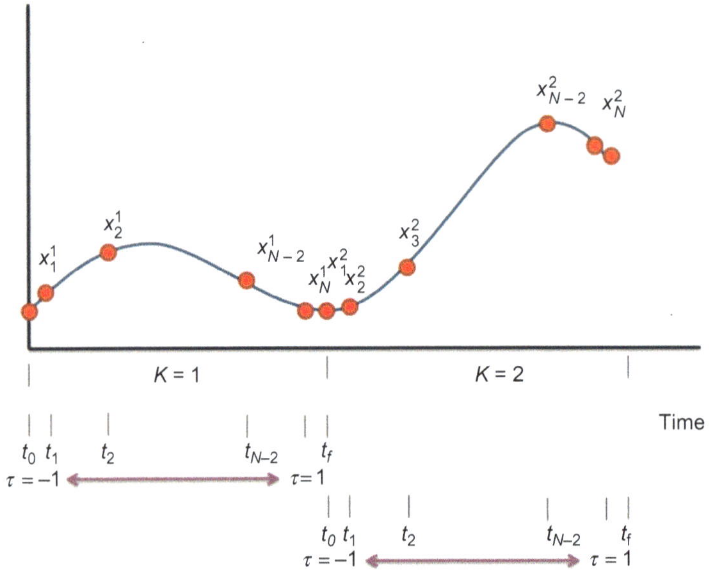

Fig. 3 Multiinterval pseudospectral collocation method scheme.

The optimization problem is described as follows: Let $x^{(k)}(\tau)$ and $u^{(k)}(\tau)$ be the state and control in the kth interval, and let the continuity in the state be enforced between intervals; therefore $x^{(k)}(+1) = x^{(k+1)}(-1)$ with $(k = 1, \ldots, K-1)$. Therefore the optimal control problem is expressed in K intervals and is defined by the minimization of the cost function

$$J = \varphi\left(x^{(1)}(-1), t_{k-1}, x^{(K)}(1), t_K\right) + \sum_{k=1}^{K} \frac{t_k - t_{k-1}}{2} \int_{-1}^{1} g(x^{(k)}(\tau), u^{(k)}(\tau))d\tau \tag{29}$$

subject to the dynamic constraints

$$\dot{x}(\tau) = \frac{dx^{(k)}(\tau)}{d\tau} = \frac{t_k - t_{k-1}}{2} f(x^{(k)}(\tau), u^{(k)}(\tau)) \quad (k = 1, \ldots, K) \tag{30}$$

inequality path constraints

$$\frac{t_k - t_{k-1}}{2} C\left(x^{(k)}(\tau), u^{(k)}(\tau)\right) \leq 0 \quad (k = 1, \ldots, K) \tag{31}$$

the boundary conditions

$$E\left(x^{(1)}(-1),\ t_0, x^{(K)}(1), t_K\right) = 0 \tag{32}$$

and the interior point constraints

$$x^{(k)}(+1) - x^{(k+1)}(-1) = 0 \quad (k = 1, \ldots, K-1) \tag{33}$$

Every kth segment is solved using classic PS collocation method. Again, the entire problem defined in Eqs. (6)–(9) is transformed to NLP for any given interval. Now the goal is to find the state vector $\boldsymbol{X}_j^{(k)} \in \mathbb{X}$ and a control vector $\boldsymbol{U}_j^{(k)} \in \mathbb{U}$ that minimize the cost function:

$$J \approx \varphi\left(\boldsymbol{X}_0^{(1)},\ t_0, \boldsymbol{X}_N^{(\mathbf{K})}, t_K\right) + \sum_{k=1}^{K} \sum_{j=1}^{N} \left(\frac{t_k - t_{k-1}}{2}\right) w_j^{(k)} g(\boldsymbol{X}_j^{(k)}, \boldsymbol{U}_j^{(k)}) \tag{34}$$

where $w_j^{(k)}$ are the LGL weights in interval k:

$$w_j = \frac{2}{N(N+1)} \frac{1}{[L_N(t_j)]^2} \tag{35}$$

subject to the discrete dynamics

$$\sum_{j=0}^{N} \boldsymbol{X}_j^{(k)} D_{ij}^{(k)} - \left(\frac{t_k - t_{k-1}}{2}\right) f(\boldsymbol{X}_j^{(k)}, \boldsymbol{U}_j^{(k)}) = 0, \tag{36}$$

the inequality path constraints

$$\frac{t_k - t_{k-1}}{2} \mathbf{C}^{(k)} \left(\boldsymbol{X}_j^{(k)}, \boldsymbol{U}_j^{(k)}\right) \leq 0, \tag{37}$$

and the end points

$$E\left(\boldsymbol{X}_1^1, \boldsymbol{X}_N^K\right) = 0. \tag{38}$$

The formulation presented here allows an arbitrary number of intervals K with arbitrary placement, and with an arbitrary number of nodes N.

Following this method's philosophy, a new method named the *S-adaptive collocation method* has been developed. This method focuses on an advanced method of segmentation of the problem, and then every subinterval is solved by a fixed Nth Lagrange polynomial. In this way the convergence of the solution is achieved by a low number of nodes, and consequently a significant improvement in the computation time compared with the hp-adaptive method is achieved. The S-adaptive method has been especially designed to solve trajectory planning problems, as shown in the following sections.

3 MULTIVEHICLE OPTIMAL TRAJECTORY PLANNING WITH COLLOCATION METHODS

This section focuses on solving the trajectory planning problem using collocation methods, and the advantages and drawbacks will be analyzed. Collocation methods have been largely used for trajectory planning of single aerospace vehicles. An example of multiaircraft trajectory planning was presented in [2], which deals with the problem of multiple aircraft landings to reduce the waiting time, improve airport efficiency, and minimize fuel consumption while maintaining safety. We will then present a new algorithm called the *S-adaptive PS method*, which proposes an advanced segmentation process in the PS collocation method.

3.1 Trajectory Planning Using Collocation Methods

In the case of collocation methods the trajectory planning problem is translated into an NLP problem by our defining the optimal criterion, constraints, and boundary conditions, and taking into account vehicle kinematics and dynamics. The state vector $x(t)$ is usually defined by the position (x, y, z) and the orientation (roll, pitch, yaw) of the vehicle in time. Sometimes the speed variable is added to define a full waypoint. The input or control vector $u(t)$ is defined by the controls of the vehicle (forces, acceleration, angles of heading, etc.). In the model, vehicles are considered as point masses, while the trajectory is defined by a set of waypoints in time. Fig. 4 shows a trajectory discretized by eight waypoints.

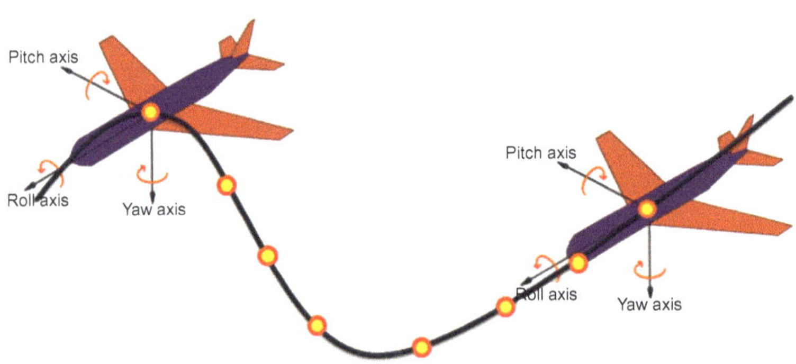

Fig. 4 Trajectory defined by waypoints.

Different optimal criteria can be considered:

- *Time:* minimum arrival time;
- *Distance:* minimum distance traveled;
- *Consumption:* minimum consumption of the vehicle;
- *Smooth trajectories:* smoothing control inputs.

Also a mixed criterion can be used when two of more criteria are weighted. Constraints are related to the limits of the problem variables: boundaries of the working area, speed, angles of heading of the vehicle, etc. State and control vectors have to be fully bounded with an upper and a lower limit. Boundary conditions are determined by the initial and final states of the problem. Typically the initial state is fully defined, while a final state is only at times fully defined, and sometimes some variables of the state are free.

3.1.1 Multivehicle Problem

When two or more vehicles are moving in the same scenario, new constraints must be added to the trajectory planning problem regarding collision avoidance.

In trajectory planning with collision avoidance, a safety distance has to be stabilized between every two vehicles. When this safety distance is violated, a collision resolution method is executed. Mainly, there are three collision avoidance techniques:

1. *Velocity planning:* This consists in the generation of a new speed profile with the vehicles involved. Changes in the orientation or altitude are not allowed. In most cases of trajectory planning, the trajectories are straight, and collision is avoided by the changing of the speed of vehicles. This technique is usually applied in ground and aerial vehicles.
2. *Changes of altitude:* In this case, when a collision is detected, the vehicles involved change the pitch angle and solve the problem, one passing over the other. Speed and orientation could change, but this is not always necessary. This technique is mainly used in aerial vehicles.
3. *Changes of orientation:* In this case a collision is avoided by the changing of the yaw angle. With the change in the orientation, a new trajectory is generated with the vehicles involved. This technique is applied in ground and aerial vehicles.

Combinations of some of these techniques can be made in a mixed policy when a vehicle with three degrees of freedom is used, but the computation time of the collision detection methods increases with the use of different avoidance techniques.

When the multivehicle trajectory planning problem is translated into an NLP problem, in addition to the constraints shown in the previous section,

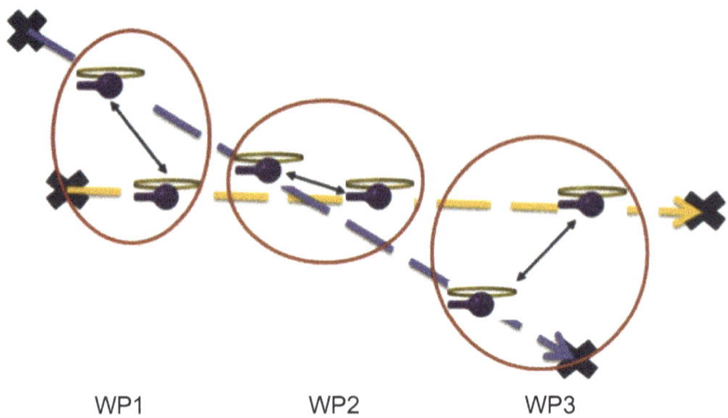

Fig. 5 Safety distance between vehicles.

new constraints are taken into account. In this case, path constraints are added to the safety distance constraints. The safety distance is modeled by the straight-line distance between the center of mass of two vehicles (see Fig. 5). All combinations of two vehicles are considered.

In case of a multivehicle problem the optimal criterion is extended to all vehicles. Also a weight policy can be established to set up a priority order between vehicles.

3.1.2 Advantages and Drawbacks of Collocation Methods With Respect to Trajectory Planning

Trajectory planning is an NP-hard problem, and the execution time of planning methods is usually long. Furthermore, when the number of vehicles involved increases, the computation time usually grows at an exponential rate (ie, scalability[1] of the optimal trajectory method is usually bad or very bad).

Collocation methods have recently received significant attention pertaining to complete space and orbit trajectories [63, 64]. The success of these methods is based on the discretization of the dynamics problem and constraints at some points, instead of the solving of the continuous problem. One drawback of collocation methods is that the dependence between the computation time and number of collocation points is very strong. Computation time increases as does the number of collocation points.

One main advantage of collocation methods is that they use the kinematic models of vehicles in order to deploy the trajectories, and

[1] Scalability is defined in this chapter as computation time dependence with the increase in the number of vehicles (ie, how the computation time increases as well as the number of vehicles in the same scenario).

then the trajectories more closely resemble reality. On the other hand, the discretization process of this method produces some drawbacks which must be controlled. Collocation methods only solve the problem into the collocation points (waypoints), and then trajectories are generated by some interpolation method. According to the number of collocation points used and the difficulty of the problem, the results could be insufficient to maintain the safety distance throughout the entire trajectory. That is, the collocation method ensures that the trajectory constraints will be met at the collocation points, but there is the possibility of collision when a vehicle is moving from one waypoint to another. Because of this uncertainty, the safety distances between vehicles have to be checked after a solution has been obtained and the number of collocation points must be readjusted if there is a collision.

Selection of the correct number of collocation points is not easy. This number of collocation points needs to be large enough to generate good solutions (ie, trajectories are safe after the interpolation) but not so large that the computation time would be very long (the computation time increases with the number of collocation points).

The advantages of collocation methods can be summarized as follows:

- Trajectories are very close to reality.
- Multiple optimal criteria can be used.
- Priorities between vehicles can be established.
- Short computation time (related to the number of points).
- Good scalability.

The drawbacks of collocation methods can be summarized as follows:

- The safety distance is not completely ensured: Trajectories are discretized and the safety distance between two points (waypoints) in the trajectory is not ensured.
- The quality of the solution depends on the number of collocation points.
- Selection of the correct number of collocation points is a difficult balance, in order to find a good solution and a good execution time.

3.2 S-Adaptive Pseudospectral Collocation Method

The S-adaptive PS method belongs to the family of subinterval PS collocation methods and exhibits some improvements with respect to other collocation methods. Firstly, the state vector and the control vector are based on the input and output of the control of one or more vehicles.

Secondly, multiple vehicles are considered and the distance between every two vehicles is checked at the nodes. Thirdly and most importantly, the safety distance is checked for the entire continuous trajectory and not only at the nodes. Fourthly, the dynamic model of vehicles is considered in the trajectory generation. Fifthly, the solution uses a low number of nodes, and the computation time is short.

The S-adaptive method considers splitting the time interval into segments and increasing the number of nodes within the segments only if this is necessary (when safety distance constraints are not met). For this a *verification process* which checks if a minimum distance is maintained between every two vehicles (in continuous time) is used. The instants of time at which the safety distance is violated are included in segments. In this way, after the segmentation process, we have temporal segments in which all vehicles are separated at least by the safety distance and temporal segments in which not all vehicles respect the safety distance.

The segmentation process also considers the dynamics of the vehicle. Sometimes the trajectory proposed in each segment varies greatly, and although the trajectory continuity is ensured between two segments, an abrupt transition could occur that may be unrealizable for some vehicles. To deal with such a problem, every segment is increased by the horizon time t_{h} in the limits of the segment. This parameter models the inertia of changing the trajectory from one segment to another. For example, t_{h} in a fixed-wing UAV will be greater than that in a rotary-wing UAV because the curvatures of the trajectories are bigger. Also, a rotary-wing UAV can decrease its speed and stop but a fixed-wing UAV is able to decrease its speed only until a minimum stall speed.

The S-adaptive method is an iterative method with a bounded number of iterations. In the first iteration, the entire problem is defined by the initial segment, which is $[t_0, t_{\mathrm{f}}]$. Then the safety distance between vehicles is checked in the entire segment by the verification process. If the solution is rejected, the segmentation process splits the problem into segments. One or more segments can be generated in this step. After this, a new solution (trajectory) will be searched for the segments which do not meet the safety constraints. New collocation points are added only in these segments. Collocation points of the segments that meet the constraints remain unchanged. At the end of the iteration, the number and the distribution of the collocation points has changed in the entire problem $[t_0, t_{\mathrm{f}}]$. Then this new solution returns to the verification process (in the second iteration). This method runs in iterations until a solution is obtained or the limit of iteration is reached. In the verification process, a small tolerance error (which is defined by the user) is allowed.

The following parameters are considered in the S-adaptive SP collocation method:

- *Horizon time* (t_h): Dynamic aspects of the vehicle, such as the inertia effect, are taken into account by this parameter of the vehicle. This parameter affects the segmentation process by adding a short period of time at the beginning and the end of every segment. Therefore the time interval will be increased by this parameter. The value of this parameter has to be set by some knowledge of the dynamics of the vehicle.
- *Number of collocation points per segment:* This parameter defines the number of collocation points used in every segment (ie, it defines the degree of the polynomial interpolant of the PS method). It always uses the same polynomial.
- *Limit of iteration:* This parameter defines the limit of the number of iterations that are allowed in the method. Because of the robustness of method, this is a bounded algorithm. Then the algorithm is not complete.
- *Error:* This parameter defines a small margin of error in the verification process (safety distance between vehicles). The dimension of this parameter is the same as that of the distance (meters).
- *Resolution step:* This parameter defines the interpolation step used in the verification process which simulates continuous time. This parameter will be explained in the next section.

Fig. 6 shows an example of the behavior of the S-adaptive PS method. The initial segment is defined by t_i and t_f, and eight collocation points are placed along the segment (black points) in the first iteration ($iter_1$). The verification process checks the safety distance between vehicles and returns an interval defined by the time t_{ci} and t_{cf}, where the constraints are not met. The horizon time, t_h, is added to define the final segment (red ellipse). In the second iteration ($iter_2$), points of the segment are replaced for eight collocation points following the PS method solution (red points). The remainder of collocation points stay as in the first iteration (black points). In the next iteration ($iter_3$), after the verification process, two segments which do not meet the constraints are detected (red ellipses). Again eight collocation points replace the collocation points in each segment (red points), and the rest of collocation points stay as in the previous iteration ($iter_2$). The verification process ensured that the safety distance constraints

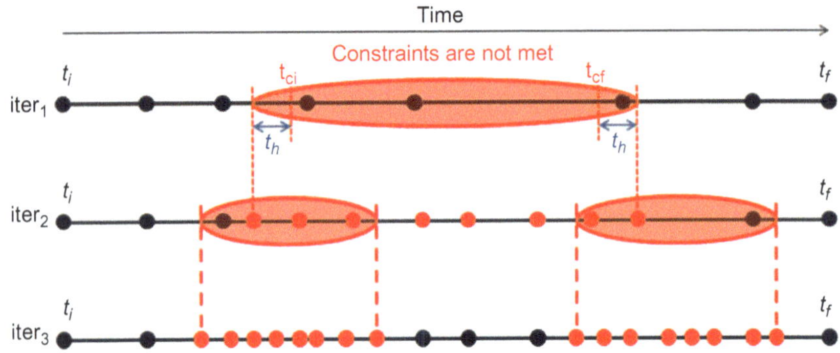

Fig. 6 Application of the S-adaptive pseudospectral method.

are met in the entire problem. Thus the algorithm finds a solution to the problem.

Every subinterval is solved by the PS collocation method, specifically, using the LGL PS collocation algorithm. The LGL distribution of points considers the boundary limits of the time interval problem $[t_0, t_f]$ (or $[-1, 1]$ in normalized time). This fact is very important to ensure the continuity of the trajectory. With use of this method there is an overlap in the boundaries of the subsegments, and the trajectories are continuous throughout a problem.

3.2.1 Verification Process

The verification process is a completely new step which has been introduced into the collocation methods. This method consists in the interpolation of the coordinates from the collocation points (waypoint position), taking into account the time; thus the speed in the trajectory is interpolated as well. When the trajectory is in continuous time, it is discretized by a small fixed step (defined as a resolution step). After this all distances between all vehicles in every discrete instant are checked.

Two parameters must be set in the verification process:

1. *Interpolation function:* This parameter is related to the type of trajectory that the vehicle realizes (ie, how the navigation module of the vehicle interpolates the waypoints). The most commonly used functions are a straight line, a third-order spline, and a third-order polynomial.
2. *Resolution step:* The fixed step depends on the problem. It can be used with any range from 1 to 0.01 s.

4 APPLICATIONS AND VALIDATION OF COLLOCATION METHODS IN MULTIVEHICLE TRAJECTORY PLANNING

4.1 Teams of Multiple Rotary-Wing Unmanned Aerial Vehicles

The real-time systems for vehicle guidance can be considered in a two-level module: a high-level module which generates the waypoints by trajectory planning (following some optimal criterion), and a low-level module that focuses on reaching every waypoint on a singular goal. In this section the design of a high-level trajectory planning method is presented.

A model of a quadrotor (see Fig. 7) is considered for rotary-wing vehicles. Dynamic analysis of a simplified model with six degrees of freedom, disregarding the Coriolis terms, is presented in [65] and omitted here.

This model considers the following inputs and outputs:

- Inputs: thrust force (T) and torques for the x-axis τ_{ϕ}, y-axis τ_{θ}, and z-axis τ_{ψ};
- Outputs: inertial position $(p_{\rm n}, p_{\rm e}, p_{\rm d})$, and orientation roll, pitch, and yaw angle (ϕ, θ, Ψ).

The parameters of the AscTec Hummingbird, given in Table 1, have been plugged in the model which is used for trajectory planning.

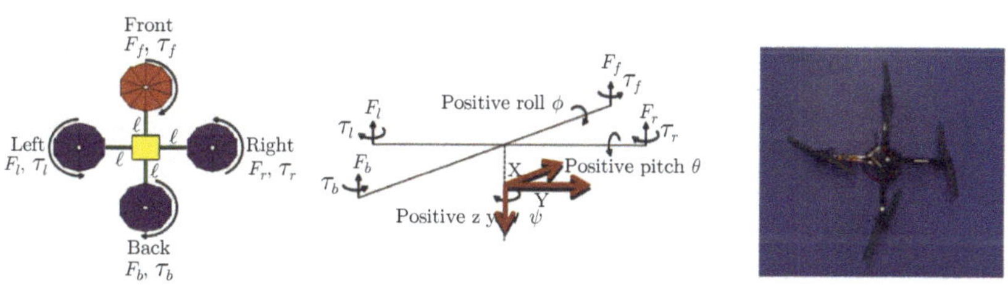

Fig. 7 Rotary-wing model (quadrotor) and the AscTec Hummingbird.

Table 1 Parameters of the Hummingbird unmanned aerial vehicle

Characteristics	Values
Minimum speed	0.1 m/s
Maximum speed	2 m/s
Cruise speed	0.65 m/s
Mass	200 g

4.1.1 Collision Avoidance

In this section, two studies will be presented: the first one analyzes the results of a scenario in which two different optimal criteria are used, and the second one focuses on the scalability and computation time. Afterward, a comparison between the S-adaptive method and the basic LGL PS method is presented in which important advantages of the S-adaptive method will be seen.

Two different combinations of trajectories in the same scenario (S1) are used. A full description of the scenario is given by Table 2. The waypoints are defined by coordinates (x, y) and the speed from that waypoint (v).

It is assumed that all UAV trajectories are known and the expected time of arrival and the safety distances must be maintained.

The goal is to find the conflict-free trajectories that minimize the optimal criteria J_d and J_a while the safety distance constraints are met in the entire problem.

4.1.2 Results of the S-Adaptive PS Method

Fig. 8A shows the trajectories of the solution for the scenario when cost function J_d is used. It can be seen that the trajectories obtained are straight because of the optimal criterion. This means that the collision avoidance is solved by the changing of the speed profile, also known as *velocity planning*. The number of collocation points used in the solution is very low: only 15 points are needed to solve the problem.

Fig. 8B shows the speed profile of the solution. From the profile it can be seen that the range of speeds is very large (from 0.3 to 1.3 m/s) and the changes of speed are abrupt.

Table 2 Specifications of scenario S1

Characteristics	**Values**
Area dimensions	15 × 15 m
Flight distance	5 m
ETA	9 s
Safety distance	1 m
Optimal criterion J_a	$\sum_{i=1}^{M}\left(\frac{\sum_{j=1}^{N}\lvert\theta_{i,j}\rvert}{N}+\frac{\sum_{j=1}^{N}\lvert\phi_{i,j}\rvert}{N}\right)$
Optimal criterion J_d	$\sum_{i=1}^{M}\sum_{j=1}^{N}\sqrt{(x_{i,j}-x_{i,j-1})^2+(y_{i,j}-y_{i,j-1})^2}$
CDR	Changes of speed and changes of heading

CDR, conflict detection and resolution; *ETA*, expected time of arrival.

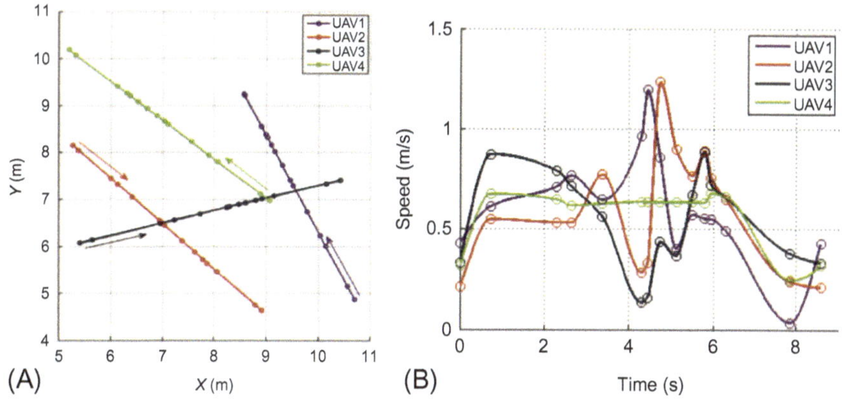

Fig. 8 Solution for scenario S1 with optimal criterion J_d: (A) trajectories; (B) speed profiles.

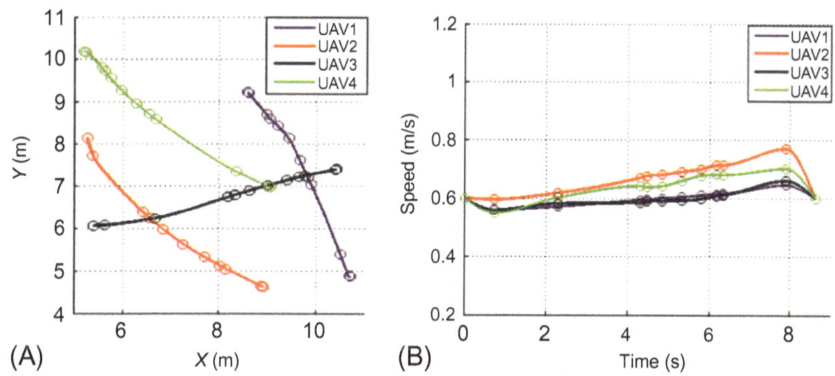

Fig. 9 Solution for scenario S1 with optimal criterion J_a: (A) trajectories; (b) speed profiles.

It can be seen that coordination between UAV 3 and UAVs 1 and 2 occurred to avoid a collision: UAV 3 decreases its speed and UAVs 1 and 2 increase their speeds. UAV 4 is mainly unaffected and maintains a nearly constant speed profile.

To perform a comparison, the resulting trajectories for the same scenario when the optimal criterion J_a is used are presented in Fig. 9A. It can be seen that the trajectories are no longer straight but are rather smooth curves. Additionally, the solution uses fewer collocation points than in the cases of J_d, and so the algorithm is faster.

Fig. 9B shows the speed profile of the solution. It can be seen that the speed profiles obtained with J_a and J_d are very different. In the case of J_a, the optimization focuses on the changes of the control inputs, and then smooth curvatures in the trajectories and the speed profiles are obtained. Additionally, this criterion rewards that use of a small speed range. The conclusion is that this criterion performs better and generates a more acceptable solution (because the abrupt changes of speed generated by the criterion J_d could make the small UAVs unstable).

4.1.3 Scalability of the S-Adaptive Method

In this section a scalability test of the S-adaptive method is reported. The aim is to test the performance of this new method when the number of vehicles increases in the same scenario. For this study, nine tests were performed: from 2 to 10 UAVs were considered in scenario S1. Fig. 10A shows the scenario when 10 UAVs are considered. Fifty scenarios were randomly generated in each of the nine cases, to obtain an average computation time.

Fig. 10B shows the execution time[2] and its standard deviation. As in previous studies, the computation time depends on the number of UAVs. As the number of UAVs considered in a scenario of the same size increases, the number of collisions increases, and the problem is more difficult to solve. In particular, cases with 9 or 10 UAVs are very difficult to solve, and the computation time is long. However, it can be seen in Fig. 10B that the rate of increase is not exponential as usually happens in NP-hard problems.

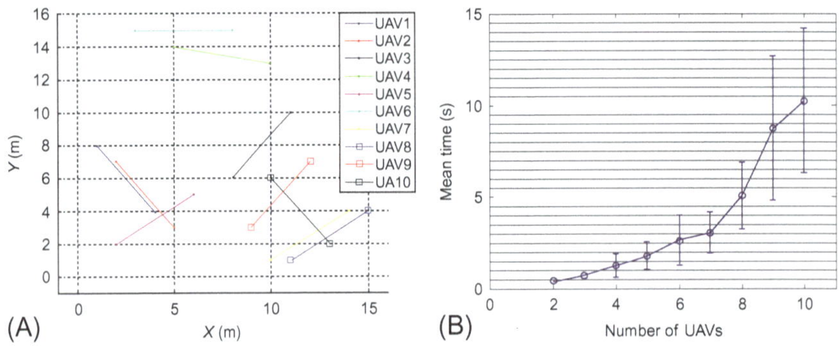

Fig. 10 Scalability of the S-adaptive method for scenario S1: (A) trajectories; (B) computation time.

[2] Computation time of the algorithm in MATLAB.

Then it can be concluded that the scalability of the S-adaptive method is satisfactory.

4.1.4 S-Adaptive Versus Legendre-Gauss-Lobatto Pseudospectral Method Comparison

This study focuses on the quality of the solution and the computation time needed to obtain the solution. In this case scenario S2 has been considered for comparison. Scenario S2 has the same specifications as scenario S1 (given in Table 2) but uses different start and end points for UAV trajectories. This is a simple scenario composed of four UAVs which have two conflicts (two by two). J_{d} is the optimal criterion considered in both cases. This criterion rewards straight-line trajectories.

The number of collocation points affects the quality of the solution and the computation time. Tests with a different number of points were performed and, finally, the LGL standard implementation was set with seven collocation points, providing a good balance between quality and computation time.

Fig. 11A shows the solution given by the LGL PS method. It can be observed that the trajectories obtained are straight lines, so the conflicts have been solved by velocity planning. At first, the trajectories obtained seem correct, but Fig. 11B shows the verification process of the solution. The verification process consists in the temporal interpolation of the collocation points and the safety distances checked between every two UAVs using a small fixed step. In Fig. 11B it can be seen that the safety distances between

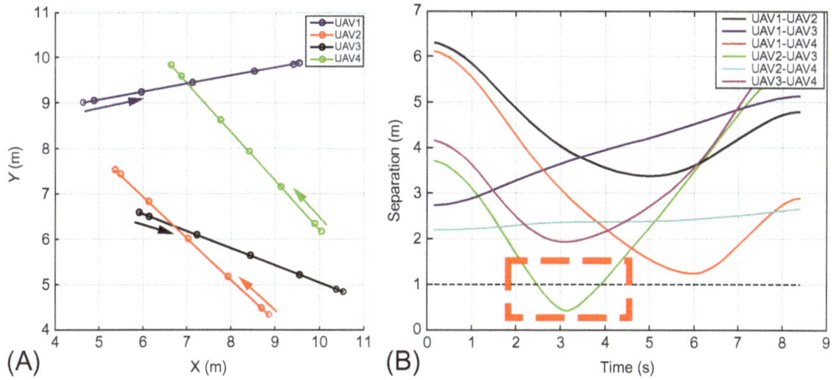

Fig. 11 Solution for scenario S2 with the Legendre-Gauss-Lobatto Pseudospectral method: (A) trajectories; (B) safety distances.

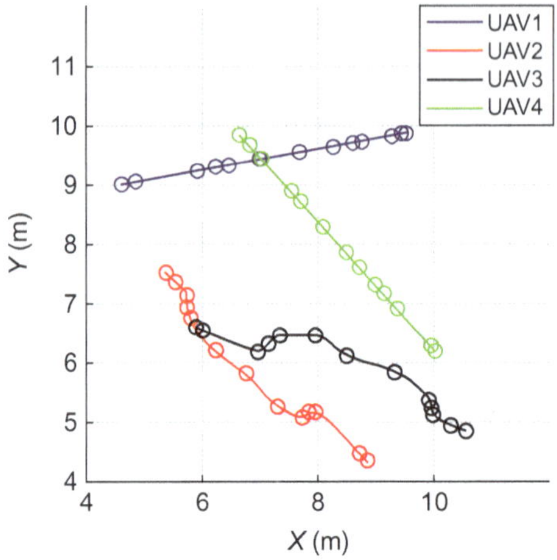

Fig. 12 Solution of scenario S2 with optimal criterion J_d: trajectories.

UAVs 2 and 3 have were during the period of time [2.5, 4] (green line), so the solution is not correct. The computation time needed for this solution was 1.326 s.

Several tests were performed in which the number of collocation points was increased for the LGL PS method. We obtained a valid solution by computing 30 collocation points; several hours was needed to obtain the solution. The same problem was solved by the S-adaptive method; the trajectories obtained are shown in Fig. 12.

In this case, one of the conflicts (between UAV 1 and UAV 4) was solved by velocity planning, while the other conflict was solved by changes in the heading. In this case the S-adaptive method needed only 13 collocation points to obtain a valid solution.

The computation time needed by the S-adaptive method was 1.30 s. Considering that the LGL method needs hours to solve the same problem successfully, it can be concluded that the S-adaptive method brings noticeable improvements.

4.2 Experimental Validation on Unmanned Aerial Vehicles

The experimental validation of the proposed algorithms with aerial vehicles was performed at the indoor test bed of the Center for Advanced Aerospace Technologies (CATEC) in Seville (Spain). This test bed is based on a rectangular prism of size $16 \times 15 \times 6\ m^3$. It uses a localization system based on 20 VICON cameras which is able to provide the position and altitude of each object with centimeter accuracy in real time. Up to 20 mobile objects can by tracked by the system. Wireless communication between the master computer and mobile vehicles is performed by the Zigbee network.

Up to eight UAV quadrotors like the one shown in Fig. 7 can be used in this test bed.

4.2.1 Collision Avoidance With the S-Adaptive Pseudospectral Method

Four experiments are reported in this section. The two variants of scenario S5 shown in Fig. 20 were executed, each with two different cost functions. The goal in this section is to validate the trajectories obtained as well as the comparison of the trajectories when two different cost functions are used in the same scenario.

The real scenario has a dimension of $8 \times 4\ m^2$ and the trajectories are 9 m long. Four UAVs are considered. The conflict resolution method is based on changes of speed and heading, and the safety distance considered in this case is 1.2 m. The minimum UAV speed is 0 m/s and the maximum speed is 2.0 m/s. Two different optimal criteria are considered: J_d is the minimum distance traveled and J_a produces smooth trajectories.

Real trajectories for scenario S1 are shown in Fig. 13, which shows the trajectories when the criterion J_d is considered. It can be seen that the trajectories are almost straight lines because of the optimal criterion. In a three-dimensional view it can be seen that all UAVs fly at the same flight level (there is a small millimeter variation because of the altitude control of the quadrotor), so the conflict resolution is solved by velocity planning in this case.

To check the accuracy of the trajectories, the safety distances are also presented, and it can be seen that all distances are greater than 1.2 m, so there are no collisions between vehicles.

In the second experiment, the same method was executed with the same scenario (S1) but with a difference in the optimality criterion considered, J_a in this case (minimum changes in the UAV inputs controls), and the safety distance was 0.8 m. Fig. 14 shows the UAV real trajectories and safety distances. Changes of speed and heading for each UAV have occurred in this scenario, but all the distances between the UAVs are greater than the safety distance of 0.8 m, ensuring there are no collisions.

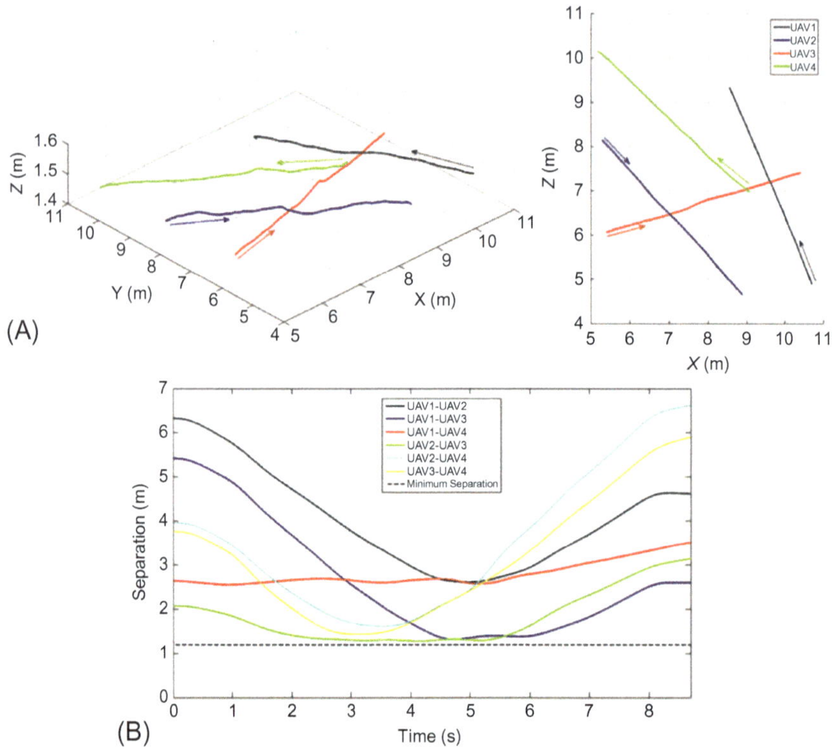

Fig. 13 Scenario S1 experimental results with optimal criterion J_d: (A) trajectories; (B) safety distances.

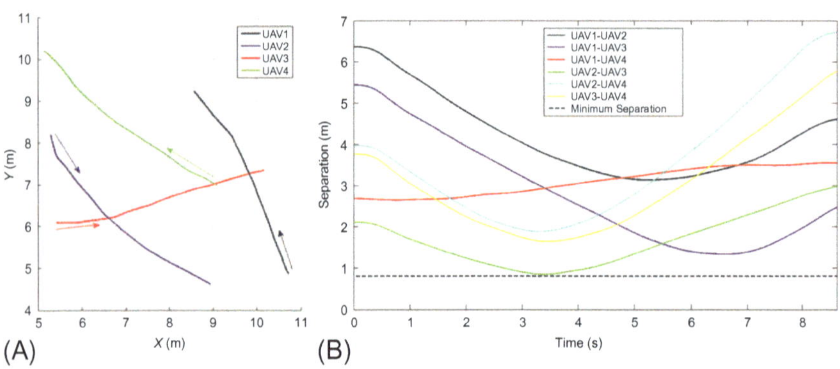

Fig. 14 Scenario S1 experimental results with optimal criterion J_a: (A) trajectories; (B) safety distances.

Fig. 15 shows three pictures at different points in time for the experiment performed at the multiple UAV test bed at the Center for Advanced Aerospace Technologies.

Fig. 16 shows the real trajectories and the safety distances between UAVs for scenario S2 when criterion J_d is considered. The effects of the segmentation process of the S-adaptive PS method are shown. It can be seen, for example, in the trajectory of UAV 1 that only part of the trajectory has been changed in order to avoid the collision. The remainder of the trajectory follows a straight line. This effect is due to the optimality criterion. Something similar happens with UAVs and UAV 3.

To check the validity of the trajectories, the safety distances are also presented, showing that all distances are greater than 1 m (the safety distance considered in this experiment).

Fig. 15 Snapshots of the multiple unmanned aerial vehicle test bed during the experiment for scenario S2. The initial state of the experiment is presented at the top, and two moments of the navigation are presented below.

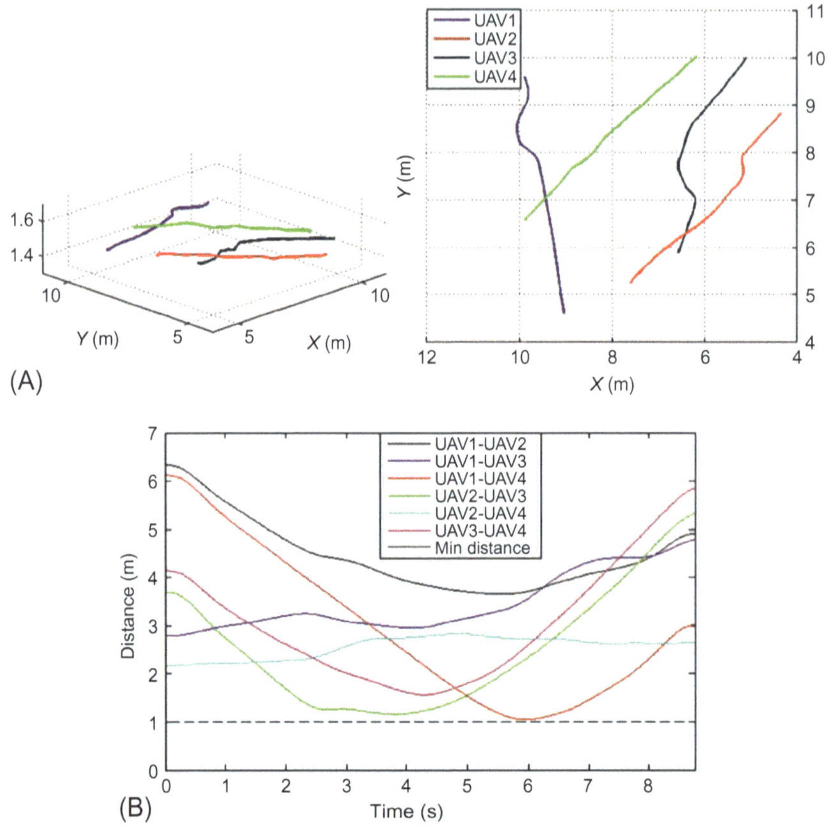

Fig. 16 Scenario S2 experimental results with optimal criterion J_d: (A) trajectories; (B) safety distances.

The second experiment was executed in the same scenario (S2) with optimal criterion J_a. Fig. 17 shows the UAV real trajectories. Four smooth trajectories are obtained in this case, and safety distances are also maintained.

4.3 Teams of Multiple Unmanned Ground Vehicles

A trajectory planning application with UGVs is presented in this section. Particularly an automated warehousing as shown in Fig. 18 is considered. In this scenario a set of autonomous forklifts is used to transport pallets between storage areas, and for palletization stations and delivery trucks, all the while sharing their workspace with human-operated vehicles and human workers.

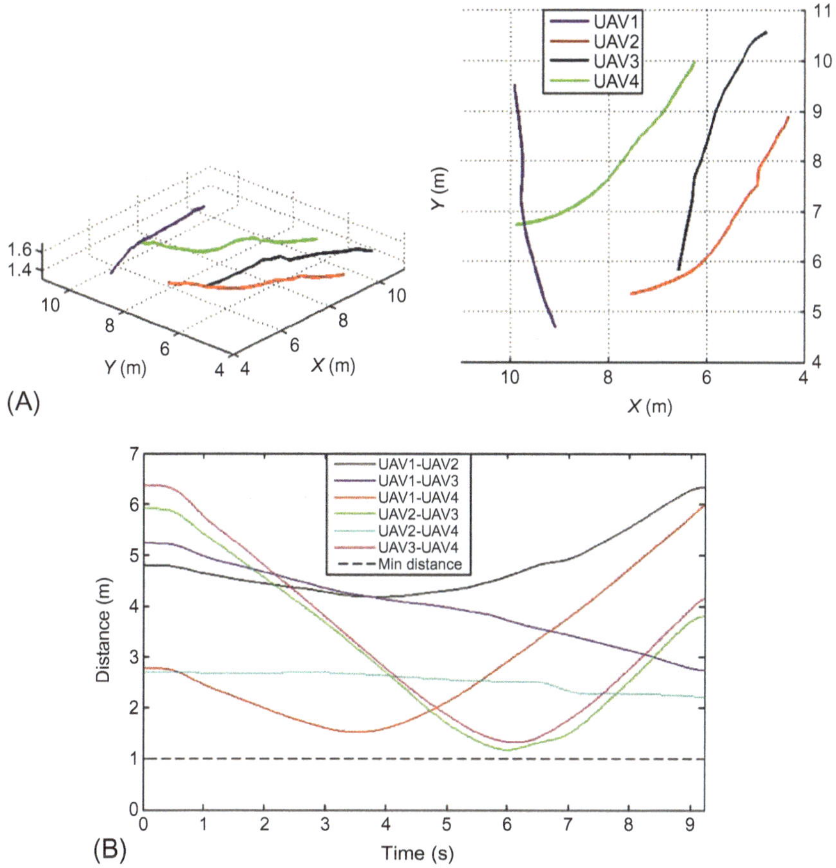

Fig. 17 Scenario S2 experimental results with optimal criterion J_a: (A) trajectories; (B) safety distances.

The fixed obstacles in trajectory planning scenarios solved by collocation method are presented in this section as a novelty. The resolution of optimal trajectory planning problems with fixed obstacles assumes a substantial increase in the difficulty of finding a solution. The number of constraints increases with the number of obstacles considered.

Three scenarios are presented for this application. The first scenario represents the case in which several UGVs have to move in the same limited area without colliding. The second and third scenarios represent a scenario similar to the first one but with one and three obstacles, respectively, in the limited area.

The simulations reported in this section were performed according to the model of a differential-drive [66] UGV. These vehicles have two main

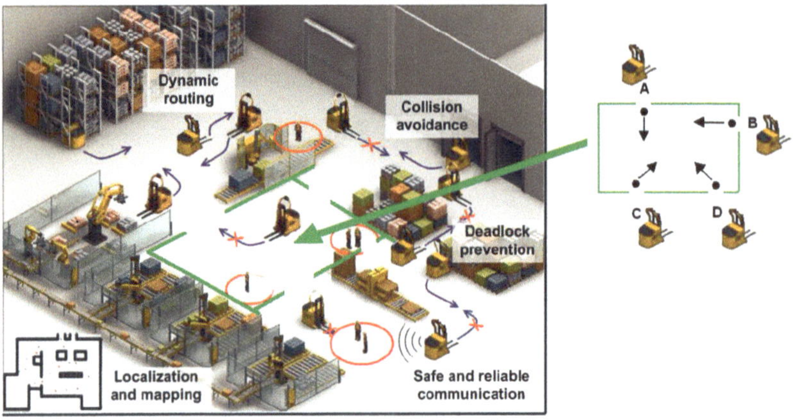

Fig. 18 Context of scenarios S3, S4, and S5: automated warehousing.

Fig. 19 Pioneer 3-DX unmanned ground vehicle and differential-drive model.

wheels, each being attached to a motor. A third wheel (not visible in Fig. 19) is placed at the rear to passively roll along while preventing the robot from falling over.

This model considers the following inputs and outputs:

- Inputs: front right (R) and left (L) wheel angular speed (ω_R, ω_L);
- Outputs: position (x, y) and heading angle (yaw) (Ψ).

The kinematic equations of the model are given by

$$\begin{aligned} \dot{x} &= \frac{r(\omega_R + \omega_L)}{2} \cos \Psi, \\ \dot{y} &= \frac{r(\omega_R + \omega_L)}{2} \sin \Psi, \\ \dot{\Psi} &= \frac{r(\omega_R - \omega_L)}{L}, \end{aligned} \tag{39}$$

where L is the distance between wheels and r is the radius of the wheel.

Table 3 Parameters of the Pioneer 3-DX unmanned ground vehicle

Characteristics	Values
Minimum speed	0 m/s
Maximum speed	1.2 m/s
Cruise speed	0.5 m/s
Maximum wheel angular speed	12.30 rad/s
Horizon time	0.01 s

This model was completed with the parameters of the Pioneer 3-DX robot which are given in Table 3.

4.3.1 Collision Avoidance Without Obstacles

Fig. 18 shows the real scenario of a warehousing problem. Case study scenarios S3, S4, and S5 are extracted as shown in Fig. 20 and involve four UGVs initially located in each corner of the scenario. Waypoints are defined by two-dimensional coordinates (x, y) and speed from that waypoint (v). The characteristics of scenario S3 are given in Table 4.

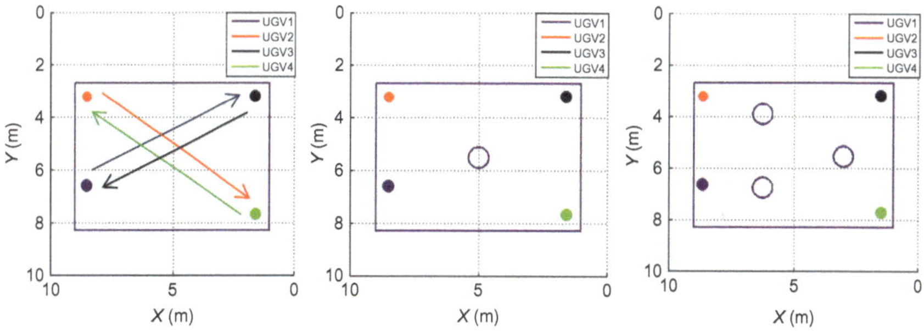

Fig. 20 Scenarios S3, S4, and S5.

Table 4 Specifications of scenario S3

Characteristics	Values
Navigation area dimensions	8 × 5 m
Distance of trajectory	9 m
Safety distance	1 m
CDR	Changes of speed and changes of heading

CDR, conflict detection and resolution.

The optimal criterion considered in this case is the minimum distance. The goal is to find the conflict-free trajectories that minimize the optimal criterion and meet the safety distance constraints of the problem.

4.3.2 Results of the S-Adaptive PS Method

The trajectory planning results for scenario S3 are given in Fig. 21. Fig. 21A presents the trajectories given by the S-adaptive method. The solution comprises by 14 waypoints. It can be seen how UGVs move to the center in this scenario, they move in a circle to avoid a collision in the center of the scenario. When the collision is avoided, each UGV moves to its goal point.

Most of the collocation points are set in the center of the scenario, where the collision is avoided. This is because the S-adaptive method increases the number of collocation points only when this is necessary. Fig. 21B shows the speed profiles of the solution. In this case the profiles are similar.

4.3.3 Collision Avoidance With Fixed Obstacles

As far as can be ascertained, this is the first instance that fixed obstacles have been considered in collocation methods to solve a trajectory planning problem with ground vehicles. Two scenarios are presented in this section related to the collision avoidance application. The problem is more difficult to solve because the safety distance has to be maintained between every pair of UGVs and between each UGV and the obstacles.

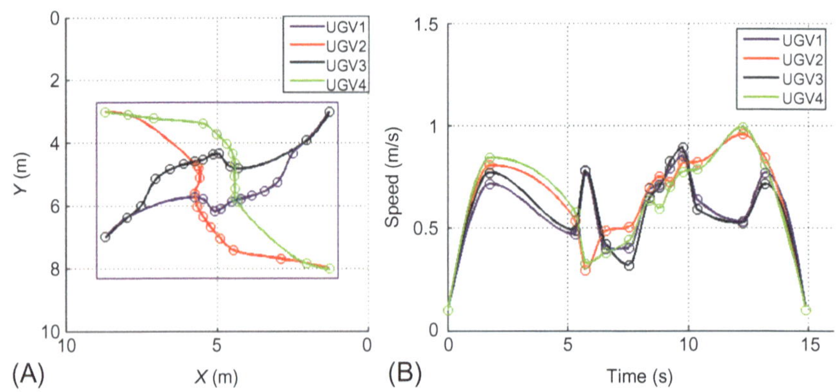

Fig. 21 Solution for scenario S3 with four unmanned ground vehicles: (A) trajectories; (B) Speed profiles.

Table 5 Specifications of scenarios S4 and S5

Characteristics	**Values**
Navigation area dimensions	8×5 m
Distance of trajectory	9 m
Safety distance	1 m
Obstacle diameter	0.5 m
Obstacle position in scenario S4	$X = 5$, $Y = 5.50$
Obstacle 1 position in scenario S5	$X = 3$, $Y = 5.55$
Obstacle 2 position in scenario S5	$X = 6.25$, $Y = 3.9$
Obstacle 3 position in scenario S5	$X = 6.25$, $Y = 6.75$
CDR	Changes of speed and changes of heading

CDR, conflict detection and resolution.

Obstacles are modeled as cylinders of 1-m height and 0.5-m radius. The distance between a UGV and an obstacle is calculated by the straight line between the position of vehicle and the position of obstacle. Scenario S4 has only one obstacle in the middle, while scenario S5 has three obstacles. The characteristics of scenarios S4 and S5 are given by Table 5.

The optimal criterion considered in this case is the minimum distance J_d. The goal is to find the conflict-free trajectories that minimize the optimal criterion and to meet the safety distance constraints of the problem.

4.3.4 Results of the S-Adaptive PS Method in Scenario S4

The trajectory planning results for scenario S4 are given by Fig. 22A. All UGVs moves to the center, as happens in scenario S3 but in this case they move round around the obstacle and keep moving to their goals. The safety distance between UGVs and obstacle clearance are maintained (see Fig. 22B and C). The speed profiles remain similar to those in scenario S3 (without obstacles) because all UGVs use the same strategy.

4.3.5 Results of the S-Adaptive PS Method in Scenario S5

Scenario S5 comprises four UGVs and three obstacles, increasing the number of problem constraints. The goal in this scenario is to analyze the behavior of the S-adaptive method in this context. Scenario S5 is much more difficult than scenario S4 because the free space for navigation is much smaller.

The trajectories obtained can be seen in Fig. 23A. Twenty collocation points are needed to solve the problem. Fig. 23B shows the distances

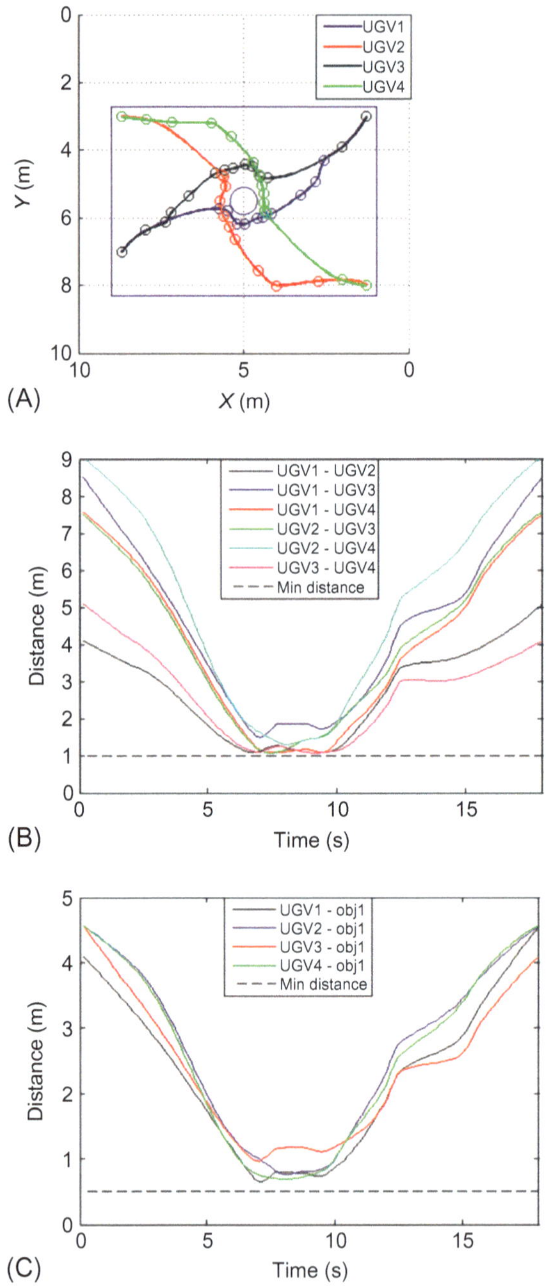

Fig. 22 Scenario S4: (A) trajectories; (B) safety distances between unmanned ground vehicles; (C) obstacle clearance.

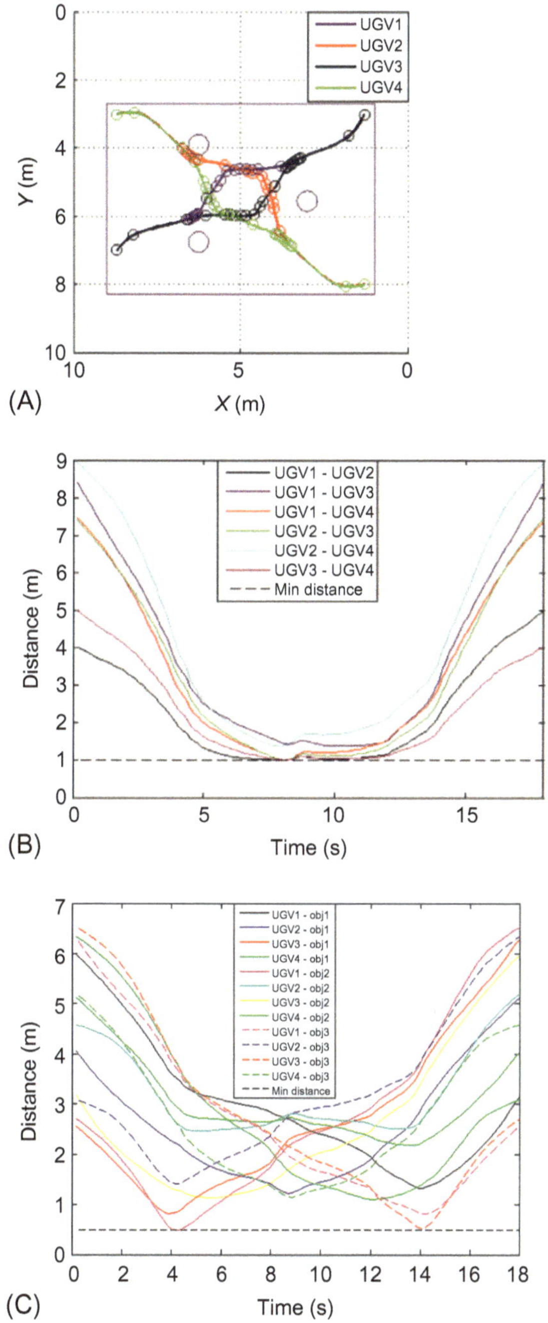

Fig. 23 Scenario S5: (a) trajectories; (b) safety distances between unmanned ground vehicles; (C) obstacle clearance.

between UGVs, while Fig. 23C shows the obstacle clearance for each UGV with respect to every obstacle. It can be seen that both the safety distances and obstacle clearance have been maintained throughout the experiment, meeting all safety constraints.

4.4 Experimental Validation on Unmanned Ground Vehicles

The experimental validation of the proposed algorithms with ground vehicles was performed in the Laboratory for Robotics and Intelligent Control Systems (LARICS) of the University of Zagreb, (Croatia). The UGVs use a global localization based on Robot Operating System (ROS) architecture. This system uses the odometry as well as laser scan values to localize on a two-dimensional map. The Sick LMS-500 laser range finder is used. The system implements an adaptive Monte Carlo localization approach, which uses a particle filter to track the position and orientation of a robot against a known map. Wireless communication between the master computer and mobile robots is performed by WiFi.

All software architecture is centralized in a master PC. This computer has ROS software which is responsible for receiving positioning data and the trajectory plan. It runs control software and sends control commands to the vehicles through a WiFi network. The navigation method used by the UGVs consists in the linear interpolation of the waypoints. These experiments were executed three times more slowly than in the simulation to ensure vehicle safety. By increasing the accuracy of the navigational system of the vehicles, we could execute the experiments without speed restrictions. Before the experiments with real robots, all scenarios were implemented in the ROS and the results were simulated by the Gazebo ROS simulator.

4.4.1 Collision Avoidance Without Fixed Obstacles

The test bed for the real scenario is based on a rectangular area of 10×5 m^2 and the trajectories are 8 m long. Four UGVs are considered. The conflict resolution is based on changes of speed and heading, and the safety distance considered between the vehicles is 1 m. The UGV speed limits are 0 and 0.5 m/s as the minimum and maximum speed, respectively.

Fig. 24A presents the results of experimental validation of scenario S3 (four UGVs without obstacles). It can be seen that all the UGVs move to their destinations through the center because this is the shortest trajectory. When they are centered, their positions interchange to maintain the safety

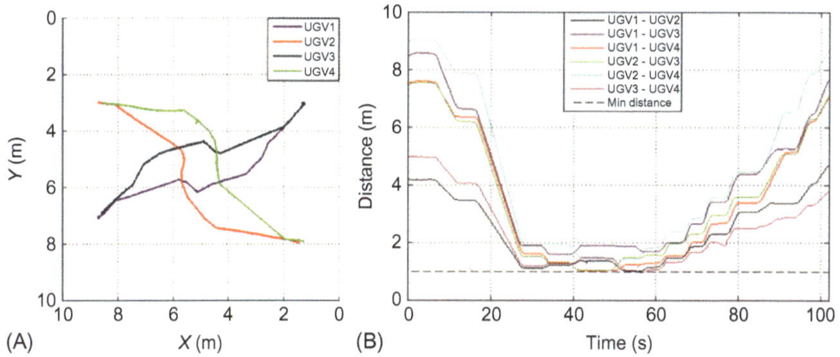

Fig. 24 Scenario S3 experimental results with optimal criterion J_d: (A) trajectories; (B) safety distances.

distance between them. In Fig. 24B, which shows the separation between every two UGVs, one can notice that the safety distance is maintained.

Moreover, it is seen that the distances are very close to the limits during the interval defined from instant 30 to instant 60. This period corresponds to the time in which all the UGVs are in the center of the scenario, turning around and interchanging their positions.

The curves of the distances are represented by jumps. This is due to the representation of interpolation in time of the waypoint of the navigation method of the vehicles. This method is based on two phases: the first in which the UGV turns around its base looking for the orientation to the next waypoint; in the second phase, when the UGV is oriented to the next waypoint, it starts to move to the waypoint with cruise speed and stops when it arrives.

4.4.2 Collision Avoidance With Fixed Obstacles

Two experiments based on scenario S4 (one obstacle) and scenario S5 (three obstacles) are presented in this section. The goal in this section is to demonstrate the performance of the S-adaptive method with UGVs when fixed obstacles are used. Obstacles are defined by small boxes of $0.5 \times 0.5 \times 0.5$ m. The safety distance considered is 1 m between UGVs as well as between the UGVs and the obstacles. The optimal criterion is J_d, the minimum distance of navigation.

Fig. 25A shows the trajectories when scenario S3 is considered. One object is used in this experiment, and it is shown in its real position by a

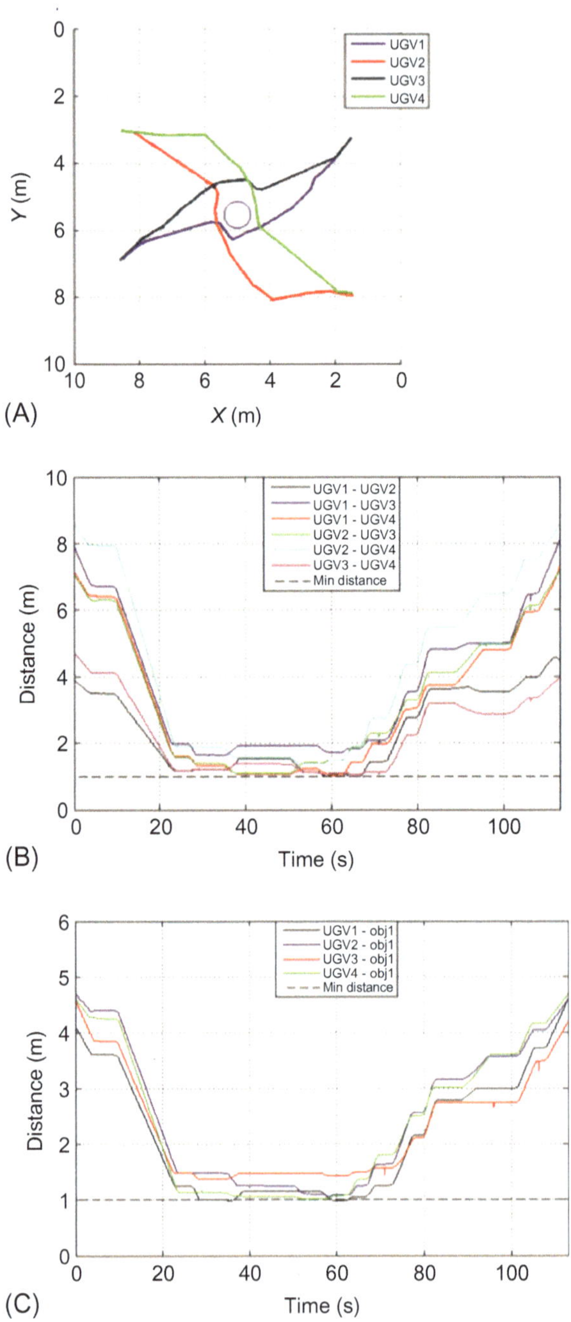

Fig. 25 Scenario S4 experimental results with criterion J_d: (A) trajectories; (B) safety distances; (C) obstacle clearance.

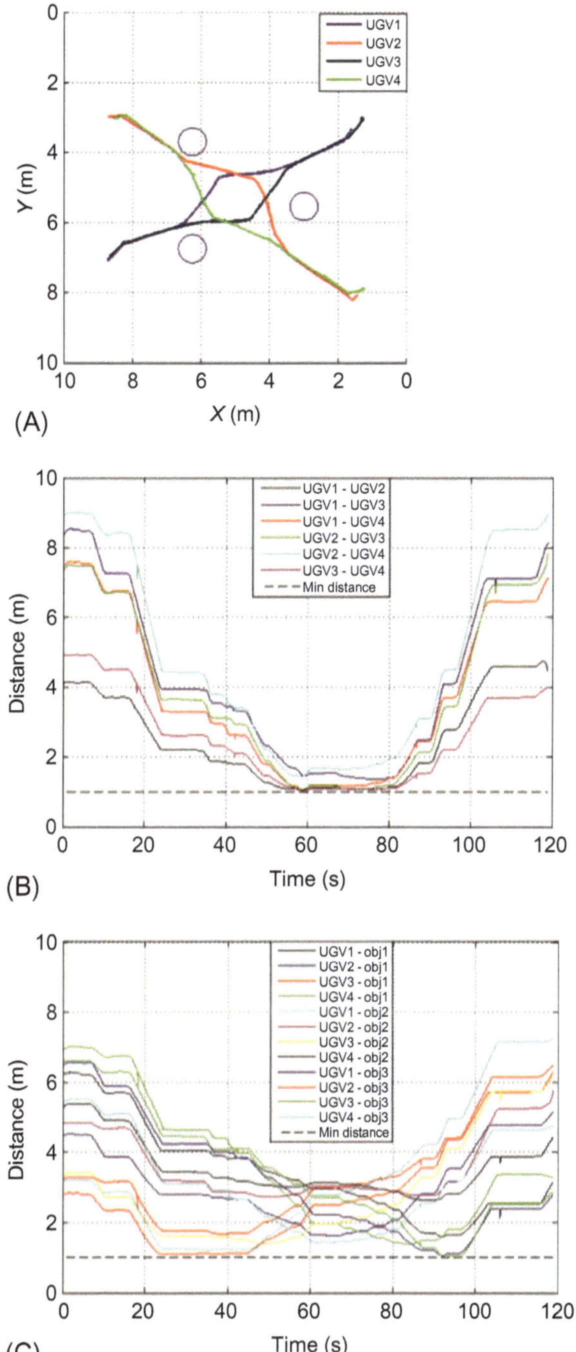

Fig. 26 Scenario S5 experimental results with criterion J_d: (A) trajectories; (B) safety distances; (C) obstacle clearance.

circle. The trajectories obtained are similar to the solution in the previous experiment.

The accuracy of the trajectories can be checked in Fig. 25B and C, which displays the distances between UGVs and obstacle clearance for each UGV. It can be seen that both the safety distance and obstacle clearance were maintained throughout the whole experiment.

4.4.3 Multiple Fixed Objects

As already stated, scenario S5 is more difficult than scenario S4 because there is less free space because of an increased number of obstacles. Thus the optimal control problem is more difficult to solve. The optimal criterion is J_d, the minimum distance of navigation.

Real trajectories are presented in Fig. 26A. The three objects are shown in their real positions. It can be seen that all UAVs move to the center of the scenario, avoiding all obstacles. In the center, each UGV moves around in order to interchange its position before finally moving to its destination.

Fig. 27 shows two pictures with different instances of time for the experiment at the multiple UGV test bed in the Laboratory for Robotics and Intelligent Control Systems.

The accuracy of the trajectories can be checked in Fig. 26B and C, which shows the separation between UGVs and the separation of UGVs from obstacles. It can be seen that each UGV maintains the minimum separation distance. As in the previous experiments, the distances between the UGVs are very close to the limit when all the UGVs are in the center interchanging their positions. Fig. 26C shows the distances between every UGV and all obstacles considered in this experiment. Two different phases can be seen: the first one between instants 20 and 40, and the second one

Fig. 27 Snapshots of the multiple unmanned ground vehicle test bed during the experiment for scenario S5. The initial state of the experiment is shown on the left. The exchange positions of four unmanned ground vehicles are shown on the right.

between instants 90 and 100. The first occurs when the UGVs move from the initial positions to the center of the scenario (avoiding the obstacles). The second one occurs then the UGVs move away from the center to their destinations (avoiding the obstacles again).

ACKNOWLEDGMENTS

This work was supported by the European Commission FP7-ICT-2011-7 project Estimation and Control for Safe Wireless High Mobility Cooperative Industrial Systems (EC-SAFEMOBIL, project no. 288082).

REFERENCES

[1] K.P. Bollino, L.R. Lewis, Collision-free multi-UAV optimal path planning and cooperative control for tactical applications, in: AIAA Guidance Navigation and Control Conference and Exhibit, Honolulu, Hawaii, 2008.
[2] K. Mohan, M.A. Patterson, A.V. Rao, Optimal trajectory and control generation for landing of multiple aircraft in the presence of obstacles, in: AIAA Guidance, Navigation, and Control Conference, Minneapolis, Minnesota, 2012.
[3] P. Gill, W. Murray, M. Saunders, M. Wright, SNOPT: an SQP algorithm for large-scale constrained optimization, SIAM J. Optim. 12 (2002) 979–1006.
[4] C. Goerzen, Z. Kong, B. Mettler, A survey of motion planning algorithms from the perspective of autonomous UAV guidance, J. Intell. Robot Syst. 57 (2010) 65–100.
[5] J.K. Kuchar, L.C. Yang, A review of conflict detection and resolution modeling methods, IEEE Trans. Intell. Transp. Syst. 1 (2000) 179–189.
[6] J.T. Betts, Practical Methods for Optimal Control Using Nonlinear Programming, SIAM, Philadelphia, PA, 2001.
[7] H.M. Prasanna, D. Ghosey, M.S. Bhat, C. Bhattacharyya, J. Umakant, Interpolation-aware trajectory optimization for a hypersonic vehicle using nonlinear programming, in: AIAA Guidance, Navigation, and Control Conference and Exhibit, San Francisco, CA, USA, 2005.
[8] A. Vela, S. Solak, W. Singhose, J.P. Clarke, A mixed integer program for flight-level assignment and speed control for conflict resolution, in: Proceedings of the 48th IEEE Conference on Decision and Control, 2009.
[9] L. Pallottino, E. Feron, A. Bicchi, Conflict resolution problems for air traffic management systems solved with mixed integer programming, IEEE Trans. Intell. Transp. Syst. 3 (2002) 3–11.
[10] D. Bauso, L. Giarre, R. Pesenti, Multiple UAV cooperative path planning via neuro-dynamic programming, in: 43rd IEEE Conference on Decision and Control, 2004.
[11] B.R. Geiger, J.F. Horn, G.L. Sinsley, J.A. Ross, L.N. Long, Flight testing a real time implementation of a UAV path planner using direct collocation, in: Proceeding of the AIAA Guidance, Navigation and Control Conference and Exhibit, Hilton Head, SC, USA, 2007.
[12] G. Basset, Y. Xua, O.A. Yakimenkob, Computing shorttime aircraft maneuvers using direct methods, J. Comput. Syst. Sci. Int. 49 (3) (2010) 481–513.
[13] J.C. Spall, Introduction to Stochastic Search and Optimization, Wiley-Interscience, Hoboken, NJ, 2003.
[14] Z.B. Zabinsky, Random Search Algorithms, in: Wiley Encyclopedia of Operations Research and Management, John Wiley & Sons, Chichester, UK, 2009.
[15] A. Chakrabarty, J.W. Langelaan, Flight path planning for UAV atmospheric energy harvesting using heuristic search, in: AIAA Guidance, Navigation, and Control Conference, Toronto, Canada, 2010.

[16] A. Stentz, I.C. Mellon, Int. J. Robot. Autom. 10 (1993) 89–100.
[17] S.M. Lavalle, J.J. Kuffner, Rapidly-exploring random trees: progress and prospects, in: Algorithmic and Computational Robotics: New Directions, 2000, pp. 293–308.
[18] M. Pontani, B.A. Conway, Particle swarm optimization applied to space trajectories, J. Guid. Control Dyn. 33 (2010) 1429–1441.
[19] R. Vivona, D. Karr, D. Roscoe, Pattern-based genetic algorithm for airborne conflict resolution, in: AIAA Guidance, Navigation, and Control Conference and Exhibit, Keystone, CO, USA, 2006.
[20] R. Conde, D. Alejo, J.A. Cobano, A. Viguria, A. Ollero, Conflict detection and resolution method for cooperating unmanned aerial vehicles, J. Intell. Robot. Syst. 65 (1–4) (2012) 495–505.
[21] A.J. Pohl, G.B. Lamont, Multi-objective UAV mission planning using evolutionary computation, in: Winter Simulation Conference, 2008.
[22] N. Durand, J.M. Alliot, Ant colony optimization for air traffic conflict resolution, in: Eight USA/Europe Air Traffic Management Research and Development Seminar, Napa, CA, USA, 2009.
[23] M. Xue, E.M. Atkins, Terminal area trajectory optimization using simulated annealing, in: 44th AIAA Aerospace Sciences Meeting and Exhibit, Reno, NV, USA, 2006.
[24] A. Bicchi, L. Pallottino, On optimal cooperative conflict resolution for Air Traffic Management Systems, IEEE Trans. Intell. Transp. Syst. 1 (2000) 221–231.
[25] A. Lecchini, W. Glover, J. Lygeros, J. Maciejowski, IEEE Trans. Intell. Transp. Syst. 7 (2006) 470–482.
[26] C. Tomlin, I. Mitchell, R. Ghosh, Safety verification of conflict resolution manoeuvres, IEEE Trans. Intell. Transp. Syst. 2 (2001) 110–120.
[27] M.A. Christodoulou, S.G. Kodaxakis, Automatic commercial aircraft-collision avoidance in free flight: the three-dimensional problem, IEEE Trans. Intell. Transp. Syst. 7 (2) (2006) 242–249.
[28] R. Ehrmanntraut, The potential of speed control, in: Proceedings of the 23rd Digital Avionics Systems Conference, 2004.
[29] D. Alejo, J.M. D.-B. nez, J.A. Cobano, P. Pérez-Lantero, A. Ollero, The velocity assignment problem for conflict resolution with multiple aerial vehicles sharing airspace, J. Intell. Robot. Syst. 69 (2013) 331–346.
[30] M. Ross, M. Karpenko, A review of pseudospectral optimal control: from theory to flight, Annu. Rev. Control 36 (2) (2012) 182–197.
[31] F. Fahroo, I.M. Ross, Advances in pseudospectral methods for optimal, in: AIAA Guidance, Navigation, and Control Conference, Honolulu, 2008.
[32] I.M. Ross, User's manual for DIDO: a MATLAB application package for solving optimal control problems, Monterey.
[33] A.V. Rao, D.A. Benson, C. Darby, M.A. Patterson, C. Francolin, I. Sanders, G.T. Huntington, GPOPS, a MATLAB software for solving multiple-phase optimal control problems using the Gauss pseudospectral method, ACM Trans. Math. Softw. 37 (2) (2010) 22:1–22:39.
[34] V.M. Becerra, PSOPT Optimal Control Solver User Manual, United Kingdom, 2009.
[35] S.W. Paris, C.R. Hargraves, OTIS 3.0 Manual, Seattle, 1996.
[36] J.T. Betts, Survey of numerical methods for trajectory optimization, J. Guid. Control Dyn. 21 (2) (1998) 193–207.
[37] O. von Stryk, User's Guide for DIRCOL (Version 2.1): A Direct Collocation Method for the Numerical Solution of Optimal Control Problems, Germany, 2000.
[38] C. Jansch, K.H. Well, K. Schnepper, GESOP—Eine Software Umgebung Zur Simulation Und Optimierung, in: Société Française de Biophysique (SFB), 1994.
[39] P. Falugi, E. Kerrigan, E. Van Wyk, Imperial College London Optimal Control Software User Guide (ICLOCS), 2010.

[40] B. Houska, H.J. Ferreau, M. Diehl, ACADO toolkit—an open-source framework for automatic control and dynamic optimization, Optim. Control Appl. Methods 32 (3) (2011) 298–312.
[41] I.M. Ross, A historical introduction to the covector mapping principle, in: Advances in the Astronautical Sciences: Astrodynamics, Paper AAS, vol. 122, 2005, pp. 05–332.
[42] C.R. Hargraves, S.W. Paris, Direct trajectory optimization using nonlinear programming and collocation, J. Guid. Control Dyn. 10 (4) (1987) 338–342.
[43] J.T. Betts, Practical Methods for Optimal Control and Estimation Using Nonlinear Programming, SIAM, Philadelphia, 2010.
[44] A.L. Dontchev, W.W. Hager, The Euler approximation in state constrained optimal control, Math. Comput. 70 (2001) 173–203.
[45] G. Elnagar, M.A. Kazemi, M. Razzaghi, The pseudospectral Legendre method for discretizing optimal control problems, IEEE Trans. Autom. Control 40 (10) (1995) 1793–1796.
[46] W.W. Hager, Runge-Kutta methods in optimal control and the transformed adjoint system, Numer. Math. 87 (2000) 247–282.
[47] P. Gill, W. Murray, M. Saunders, M. Wright, User's Guide for NPSOL 5.0: A Fortran Package for Nonlinear Programming, 2001.
[48] G. Elnagar, M. Razzaghi, A collocation-type method for linear quadratic optimal control problems, Optim. Control Appl. Methods 18 (3) (1998) 227–235.
[49] D.A. Benson, G.T. Huntington, T.P. Thorvalden, A.V. Rao, Direct trajectory optimization and costate estimation via an orthogonal collocation method, J. Guid. Control Dyn. 29 (6) (2006) 1435–1440.
[50] C. Canuto, M.Y. Hussaini, A. Quarteroni, T.A. Zang, Spectral Methods in Fluid Dynamics, Springer, Heidelberg, 1988.
[51] B. Fornberd, A Practical Guide to Pseudospectral Methods, Cambridge University Press, Cambridge, 1999.
[52] L.N. Trefethen, Spectral Methods Using MATLAB, SIAM, Philadelphia, 2000.
[53] D. Garg, W.W. Hager, A.V. Rao, Pseudospectral methods for solving infinite-horizon optimal control problems, Automatica 47 (4) (2011) 829–837.
[54] D.P.M. Garg, W.W. Hager, A.V. Rao, D.A. Benson, G.T. Huntington, A unified framework for the numerical solution of optimal control problems using pseudospectral methods, Automatica 46 (11) (2010) 1843–1851.
[55] S. Kameswaran, L.T. Biegler, Convergence rates for direct transcription of optimal control problems using collocation at Radau points, Comput. Optim. Appl. 41 (1) (2008) 81–126.
[56] Q. Gong, I.M. Ross, F. Fahroo, A Chebyshev pseudospectral method for nonlinear constrained optimal control problems, in: Joint 48th IEEE Conference on Decision and Control and 28th Chinese Control Conference, Shanghai, 2009.
[57] F. Fahroo, I.M. Ross, Costate estimation by a Legendre pseudospectral, J. Guid. Control Dyn. 24 (2) (2001) 270–277.
[58] Q. Gong, W. Kang, I.M. Ross, A pseudospectral method for the optimal control of constrained feedback linearizable systems, IEEE Trans. Autom. Control 51 (7) (2006) 1115–1129.
[59] Q. Gong, I.M. Ross, W. Kang, F. Fahroo, On the pseudospectral covector mapping theorem for nonlinear optimal control, in: 45th IEEE Conference on Decision and Control, San Diego, 2006.
[60] W. Kang, M. Ross, Q. Gong, Pseudospectral optimal control and its convergence theorems, in: Analysis and Design of Nonlinear Control Systems, 2008, pp. 109–124.
[61] C.L. Darby, W.W. Hager, A.V. Rao, An hp-adaptive pseudospectral method for solving optimal control problems, Optim. Control Appl. Methods 32 (4) (2010) 476–502.

[62] C.L. Darby, hp pseudospectral method for solving continuous-time nonlinear optimal control problems, Ph.D. thesis, Graduate School at the University of Florida, 2011.
[63] A.L. Herman, B.A. Conway, Direct optimization using collocation based on high-order Gauss-Lobatto quadrature rules, J. Guid. Control Dyn. 19 (3) (1996) 592–599.
[64] V. Coverstone-Carroll, J.E. Prussing, Optimal cooperative power-limited rendezvous with propellant constraints, J. Astronaut. Sci. 43 (3) (1995) 289–305.
[65] R.W. Beard, Quadrotor Dynamics and Control 2008.
[66] A. Ollero, Robótica: manipuladores y robots móviles, Marcombo, 2005.

CHAPTER TWENTY-TWO

Intelligent Control of a Prosthetic Ankle Joint Using Gait Recognition

A. Mai, S. Commuri
University of Oklahoma, Norman, OK, United States

1 INTRODUCTION

According to recent estimates [1], 1.6 million Americans, or roughly 1 in 190 persons, are currently living with the loss of a limb. Among them, 62% underwent an amputation procedure to the lower extremity. It is anticipated that the number of people living with limb loss will increase to 3.6 million by 2050, with more than 1.4 million of them being younger than 65 years. Clinical studies indicate that between 68 and 88% of amputees wear a prostheses at least 7 h a day [2, 3] and lead an active lifestyle [4]. These people need highly functional prostheses to maintain a healthy and good quality of life [1]. In the short term the use of a passive prosthetic foot with a fixed ankle can lead to asymmetric gait [5, 6], increased intact muscle contraction [7], and higher metabolic energy expenditure [8] in an individual. Long-term health complications such as osteoarthritis, osteoporosis, back pain, and other musculoskeletal problems can also be linked to the poor fit and improper alignment of the prosthesis and can lead to poor overall quality of life of individuals [9].

Research into amputee gait has shown that it is desirable that future prosthetic feet replicate the biomechanical characteristics of the biological ankle joint [10]. It is necessary for the prosthetic ankle joint to follow the typical joint displacement profile of a human ankle. Such a profile helps in absorbing shock due to the ground interaction and generates propulsion energy for the body to move forward, therefore guaranteeing stability and mobility during gait. A typical displacement profile of the ankle is shown in Fig. 1 [11] and is related to the phases and events of the gait [11–13]. This

Control of Complex Systems
http://dx.doi.org/10.1016/B978-0-12-805246-4.00022-7

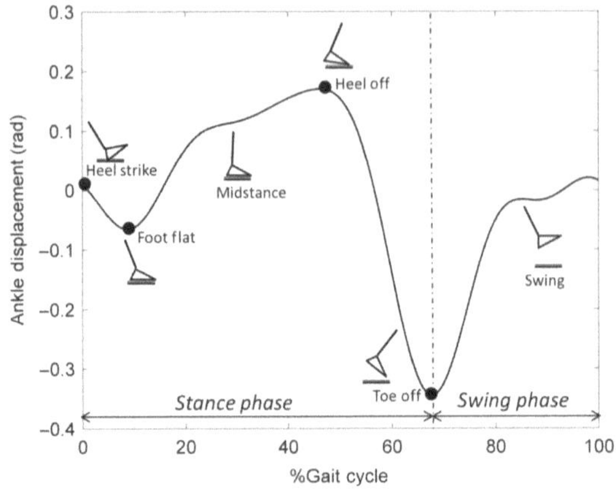

Fig. 1 Typical human ankle displacement profile.

profile is typical for all humans and is presumed to minimize the energy consumption during gait [14, 15].

Theoretical studies of human gait have also shown that there are optimization principles associated with the synthesis of the movement of the human body during locomotion [16–19]. These optimization principles can be parameterized [20] and defined by different metrics, such as dynamic effort (ie, integration of squares of all joint torques over time), deviations of the joint position from a desire trajectory, and metabolic energy [17, 21, 22]. By optimizing those performance metrics, one obtains optimal joint torques that can generate the desired motion. Several studies have shown that similar optimization principles can also be extended to not only generate a control torque and synthesize the gait for people with below-knee amputation [23, 24] but also to evaluate the performance of the prosthetic devices [25].

Next-generation prosthetic feet are likely to be equipped with controlled actuators to provide variable stiffness and movement of the artificial ankle joint and help achieve the desired gait performance [10]. However, control of the ankle prosthesis to replicate the movement of a healthy ankle by tracking a displacement profile during gait and provide an optimized energy metric is a difficult task for the following reasons:

1. The ideal ankle displacement profile depends on several factors, such as user gait (eg, stance time, swing time, step length, and stride length), selected walking speed, inclination of the terrain, and type of activity (eg, walking on level ground, ascending/descending stairs). Because of

changes in gait and terrain and unknown intent of the user, the ideal ankle displacement profile cannot be specified a priori.

2. During gait the movement of the prosthetic foot is influenced by the reaction force due to the interaction with the terrain. This ground reaction force plays a critical role in supporting the body weight, ensuring stability, and providing propulsion during gait [12]. The ground reaction force causes a reaction torque at the joint that has to be compensated for proper tracking behavior. The ground reaction force is usually measured with motion tracking systems and force plates in a laboratory setting [11]. Unfortunately, real-time measurement of the ground reaction force is not feasible during normal gait and is not available during the control of the prosthetic joint.
3. The dynamics of the foot are affected by the nonlinear coupling effects between the prosthetic ankle joint and the biological knee and hip joints of the amputee. These effects not only depend on anthropometric measurements of the human body but also vary with gait. Neglecting these interactions will lead to larger tracking errors for a specified ankle displacement profile.

Commercially available below-knee controlled prostheses such as Proprio Foot [26] and the BiOM Ankle system [27, 28] are capable of manipulating the movement of the ankle joint and adapting to different gaits and terrain [10, 29]. Improved gait performance obtained with these controlled prostheses has been reported in several case studies [30–32]. The performance of other active foot prototypes such as SPARKy [33] and Intelligent Prosthesis actuated by pleated Pneumatic Artificial Muscles (IPPAM) [34] in generating the required torque to manipulate the ankle movement is also promising. Generally, the control algorithms of these prostheses appear to neglect the dynamics of the ankle, its interaction with the remaining healthy joints of the residual limb, and the varying ground reaction torque. These devices are based on linearized dynamics and use proportional-derivative control with fixed control gains. While these controllers guarantee local stability, their performance might deteriorate quickly in the presence of unmodeled system dynamics and measurement noises. Advanced bionic feet reported in the recent literature can vary the ankle stiffness and generate power to support the gait. These designs are in the early development stage, and researchers are aware of the challenges in achieving satisfactory performance when gait conditions change [35]. It has been also shown that supervisory control schemes that implement different control strategies based on perceived gait help achieve the desired performance of the foot

[36]. However, the effects of gait and terrain changes, unmodeled dynamics, and noise on the performance of these supervisory controllers have not been systematically studied [29].

In this chapter we will address the problem of controlling the prosthetic ankle joint to follow a desired movement profile which is subjected to changing gait conditions and varying intent of the user. The goals for the control system of the prosthetic ankle joint are to:

- recognize the type of gait and detect the gait events in real time using actual gait data measured from amputees,
- determine an ankle joint displacement profile corresponding to the selected gait of amputees,
- compensate for the effect of ground reaction,
- implement a control algorithm that can adapt to the changing gait condition and generate an appropriate control torque that provides guaranteed tracking and optimized performance over time.

2 APPROACHES

The approaches described in the following sections will be used in the design of a control system that can achieve the aforementioned goals.

2.1 Modeling of the Prosthetic Ankle During Gait

The leg on the amputated side of a unilateral below-knee amputee is modeled by a link-segment diagram in the sagittal plane, where most of the ankle joint movements occur during gait. In this model the movement of the ankle joint will be controlled by an external actuator that generates an appropriate control torque for each gait condition. The prosthetic ankle will then be expected to follow a displacement profile similar to that of a natural ankle.

2.2 Recognition of the User's Intent

Recognition of gait and detection of gait events will be done with actual gait data, such as the residuum socket interface force (RSI) measured from the prosthetic socket of below-knee amputees. The RSI forces that the residual limb apply on the prosthetic socket during gait of an amputee will be captured by ultrathin FlexiForce sensors [37], which can be embedded inside the socket to provide nonintrusive force measurement in real time and during daily activities outside the laboratory setup. It has been shown that the recorded RSI forces relate to the gait [38] and can be used to distinguish between different gait types (normal self-paced walk, brisk walk, ascending/descending stairs, walking on a ramp, etc.) and detect gait parameters (heel strike, midstance, toe off, stance time, etc.) for each detected gait [39].

2.3 Generation of the Gait-Based Kinematic References

The gait-based kinematic references will be generated for each gait type recognized with use of actual gait data. Since the human gait and joint movements are cyclic in nature, the gait-based kinematic references will be generated with use of a Fourier series [16] representing a combination of periodic and bounded sine and cosine functions. The series is a function of the gait cycle, anthropometrics of the user, and parameters that represent different types of gait.

2.4 Compensation for the Ground Reaction Torque

Calculation of the ground reaction torque is usually attempted with use of kinematic tracking systems and force plates in a laboratory setting [11, 40], where the participants walk a short distance with a limited number of steps. Under normal gait conditions, direct measurement of the ground reaction force is not feasible for the purpose of controlling the prosthetic foot; therefore, the actual ground reaction torque is not available for the calculation of the ankle control torque. In this chapter the control system will generate a specific, gait-based ground reaction torque for the recognized gait using an empirical model describing the foot-ground contact as the compensation for the actual ground reaction torque. The empirical model will be a combination of a spring and a position-dependent damper [16, 41, 42], with the parameters obtained through experiments for each gait condition. The torque generated by such a model is shown to be not only similar to the actual quantities as observed in gait analysis [11, 40], but also related to the gait cycle. The resulted torque is also expected to be bounded because of its cyclic nature, finite human weight, and the passive mechanical components used.

2.5 Learning-Based Control of the Prosthetic Ankle Joint

In recent years several researchers have investigated the use of artificial neural networks for recognizing the user's intent during gait [36, 43–45]. However, the information gained was not used to directly control the prosthetic joint. In the proposed approach the intended gait of the amputee is estimated and a control strategy is implemented to optimize the energy consumption over time. The control algorithm will aim at learning and optimizing the performance metric which relates to the deviation of the ankle joint kinematics from its desired profiles for the gait. As the result, the tracking performance of the prosthetic ankle joint will be obtained and improved over time. For such an objective, an adaptive dynamic programming-based control structure, named *direct neural dynamic programming* (DNDP) [46] appears to be a good candidate. DNDP has been shown to be suitable for control of complex nonlinear systems with unknown dynamics and disturbances [47, 48]. Furthermore, this approach is also expected to minimize the long-term cost function in the sense of Bellman's principle of optimality.

In the DNDP framework a critic network is responsible for approximating the long-term cost function which satisfies Bellman's principle of optimality. The action network is responsible for generating a control signal which leads to the optimization of the approximated long-term cost (ie, output of the critic network). A learning-based control strategy including an DNDP-based controller and augmented learning rules will be implemented to generate an ankle torque which drives the prosthetic ankle joint along the designed kinematic patterns. Numerical simulations will be performed to evaluate the performance of the control approach.

3 FRAMEWORK FOR MODELING AND CONTROL OF THE PROSTHETIC ANKLE JOINT DURING GAIT

Mathematical models and experimental data can be effectively combined to study not only normal and pathological gaits [16, 20, 41, 49] but also the effect of prostheses on kinematic behaviors and other aspects of the gait [23, 50] without posing a risk of injury from prosthetic malfunctions when tests are being performed on humans [51]. In this section the leg on the amputated side of a person with a unilateral below-knee amputation is modeled by a link-segment diagram in the sagittal plane, where most of the ankle joint movements occur during gait. This representation includes a biological knee joint, a biological hip joint, and a prosthetic ankle joint on the prosthetic foot. The Euler-Lagrange approach is used to derive the dynamics of this representation. By assuming the total human control of the biological joints, we then concentrate on the dynamics and control of the ankle joint.

3.1 Link-Segment Diagram and Dynamics of the Ankle Joint

Fig. 2 shows the link-segment diagram of the leg on the amputated side of an individual in the sagittal plane. The dynamics in this diagram can be obtained with use the Euler-Lagrange approach [11, 52] and is as follows:

$$M(\theta)\ddot{\theta} + V\left(\theta, \dot{\theta}\right)\dot{\theta} + G(\theta) + \tau_{\mathrm{d}} = \tau, \tag{1}$$

with the time-dependent joint variable θ (rad), input torque τ (Nm), and disturbance torque τ_{d} (Nm). The joint variable $\theta(t) = \left[\ \theta_{\mathrm{a}}(t) \quad \theta_{\mathrm{k}}(t) \quad \theta_{\mathrm{h}}\ (t)\right]^{\mathrm{T}} \in \mathbb{R}^3$ represents the angular position of the prosthetic ankle joint, the biological knee joint, and the biological hip joint. The input torque to the model is $\tau(t) = \left[\ \tau_{\mathrm{a}} + \tau_{\mathrm{aG}} \quad \tau_{\mathrm{k}} \quad \tau_{\mathrm{h}}\ \right]^{\mathrm{T}}$, where $\tau_{\mathrm{a}}(t)$ is the external torque generated by the actuator at the prosthetic ankle joint;,$\tau_{\mathrm{aG}}(t)$ is the ground reaction torque and is caused by the interaction between the prosthetic foot and the ground during gait, and $\tau_{\mathrm{k}}(t)$ and $\tau_{\mathrm{h}}(t)$ describe the internal torques generated by the biological knee and hip joints, respectively. The additional component $\tau_{\mathrm{d}}(t) = \left[\ \tau_{\mathrm{ad}} \quad \tau_{\mathrm{kd}} \quad \tau_{\mathrm{hd}}\ \right]^{\mathrm{T}}$ represents the disturbance torque which is generated by the movement of the head, arm and trunk (HAT)—that is, the upper body—during gait. It is important to note that total human control is assumed at the biological knee joint and biological hip joint [11]. Therefore, knee torque $\tau_{\mathrm{k}}(t)$ and hip torque $\tau_{\mathrm{h}}(t)$ are purposely generated by the user to compensate for the

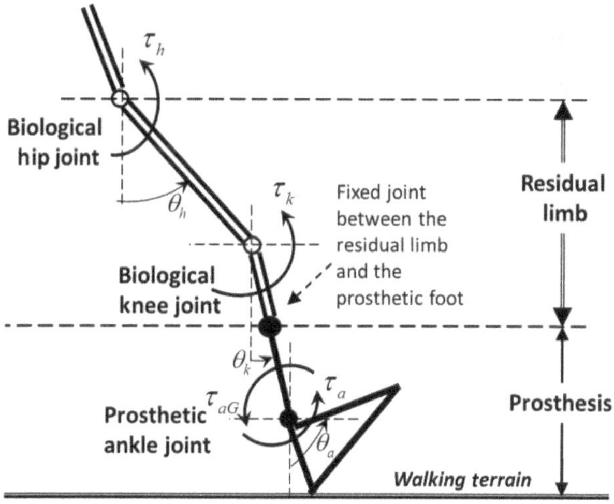

Fig. 2 Link-segment diagram of the residual limb and prosthetic foot.

coupling effects from the ankle joint, the effect of ground reaction, and disturbance torque from the HAT. Finally, the nonlinear terms in Eq. (1) include the inertia matrix $M(\theta)$, the vector of Coriolis and centripetal forces $V(\theta,\dot{\theta})\dot{\theta}$, and the vector representing gravitational forces $G(\theta)$.

Since total human control is assumed at the biological knee joint and the biological hip joint of the residual limb, we focus on the dynamics and control of the ankle prosthesis. The dynamics of the prosthetic ankle joint can be extracted from Eq. (1) as follows:

$$\underbrace{\underbrace{M_{aa}\ddot{\theta}_a + V_{aa}\dot{\theta}_a + G_a}_{\substack{\text{independent}\\ \text{ankle dynamics}}} + \underbrace{M_{ak}\ddot{\theta}_k + V_{ak}\dot{\theta}_k}_{\substack{\text{reflected dynamics}\\ \text{of knee joint}}} + \underbrace{M_{ah}\ddot{\theta}_h + V_{ah}\dot{\theta}_h}_{\substack{\text{reflected dynamics}\\ \text{of hip joint}}}}_{\text{unknown ankle dynamics}} + \underbrace{\tau_{ad}}_{\substack{\text{disturbance}\\ \text{from the HAT}}} = \underbrace{\tau_a}_{\substack{\text{control}\\ \text{torque}}} + \underbrace{\tau_{aG}}_{\substack{\text{ground reaction}\\ \text{torque}}}. \tag{2}$$

The unknown dynamics of the prosthetic ankle joint (2) include the dynamics that depend only on the ankle angular position and its derivatives $\left(M_{aa}\ddot{\theta}_a + V_{aa}\dot{\theta}_a + G_a\right)$, the coupling between the ankle joint and the biological knee joint $\left(M_{ak}\ddot{\theta}_k + V_{ak}\dot{\theta}_k\right)$, and the coupling between the ankle joint and the biological hip joint $\left(M_{ah}\ddot{\theta}_h + V_{ah}\dot{\theta}_h\right)$. The components M_{aa}, V_{aa}, G_a, M_{ak}, V_{ak}, M_{ah}, and V_{ah} are nonlinear functions of the joint angles. The ankle dynamics are also affected by the ground reaction torque τ_{aG} and the HAT disturbance torque τ_{ad}.

3.2 Framework for Control of the Prosthetic Ankle Joint

In this chapter the prosthetic ankle joint is controlled to replicate the movement of the healthy ankle by tracking a desired displacement profile during gait. Such a desired joint angular trajectory is generated for each type of gait which can be recognized by continuous measurement and interpretation of the gait data captured from users with below-knee amputation. This section first describes the structure of the control system of the prosthetic ankle joint and then discusses the approaches used to obtain the control action.

Fig. 3 shows the hierarchical control structure which aims at optimizing the performance metric which relates to the deviation of the ankle joint kinematics from its desired profiles for the gait.

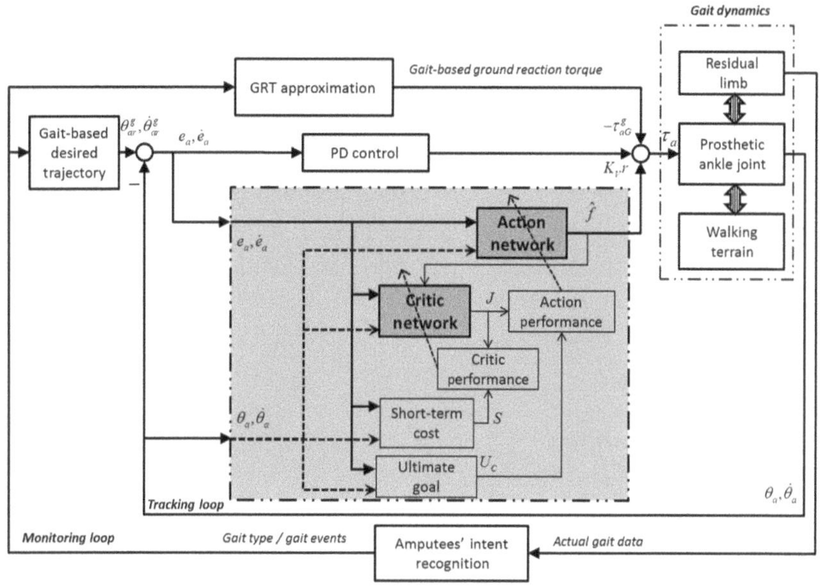

Fig. 3 The direct neural dynamic programming-based control of the prosthetic ankle joint.

The (pseudo) ankle joint tracking error is defined as

$$e_a = \theta_{ar}^{g} - \theta_a, \tag{3}$$

in which θ_a is the actual ankle angular position and θ_{ar}^{g} is the desired displacement profile which is generated for each particular type of gait.

We define the filtered tracking error as

$$r = \dot{e}_a + \lambda e_a, \tag{4}$$

with $\lambda > 0$. Then the closed loop error dynamics can be expressed as

$$M_{aa}\dot{r} = -V_{aa}r + f + \tau_{ad} - \tau_a - \tau_{aG}, \tag{5}$$

where

$$f = \left[M_{aa}\left(\ddot{\theta}_{ar}^{g} + \lambda\dot{e}_a\right) + V_{aa}\left(\dot{\theta}_{ar}^{g} + \lambda e_a\right) + G_a(\theta_a)\right] + \left[M_{ak}\ddot{\theta}_k + V_{ak}\dot{\theta}_k\right] + \left[M_{ah}\ddot{\theta}_h + V_{ah}\dot{\theta}_h\right] \tag{6}$$

represents the ankle dynamics and its interaction with the knee and hip joints.

In this structure the nonlinear function f will be approximated with the DNDP approach [46, 53], which has been shown to be suitable for control of complex nonlinear systems with unknown dynamics and disturbance while optimizing a gait-related performance index (see Section 3.2.1). Also in this control structure, the user's intent will be recognized in real time with use of actual gait data measured from the users (see Section 3.2.2). An appropriate desired displacement profile for the prosthetic ankle joint will be generated for the recognized gait type (see Section 3.2.3). The actual ground reaction torque, which is not available for the calculation of the control torque, will be compensated for by an empirical viscoelastic models (see Section 3.2.4).

3.2.1 Control Algorithm

As seen in the closed-loop error dynamics (5), the filtered tracking error r can be reduced only when the nonlinear function f and the actual ground reaction torque τ_{aG} are compensated for properly. Compensation for the foot-ground reaction by a viscoelastic contact model with passive mechanical components will be presented in Section 3.2.4. In this section the nonlinear function f will be approximated by a neural network, named the *action network*, which is responsible for generating the optimal output which leads to optimization of a user-defined cost function. Such a user-defined cost function describes the ultimate control goal for the prosthetic ankle and is learned and optimized by another neural network, named the *critic network*.

Critic Network

From the discussion of the optimization principles of human gait in Section 3.1 it can be hypothesized that there is a finite optimizable index which indicates the long-term performance of the prosthetic ankle joint in tracking of the desired gait-based trajectory. According to Bellman's principle of optimality, such a long-term performance index is expressed as the weighted sum of the short-term (instantaneous) cost at the subsequent iterations as follows:

$$\begin{aligned} L|_{[k-1]} &= S|_{[k]} + \alpha\, S|_{[k+1]} + \alpha^2\, S|_{[k+2]} + \cdots \\ &= S|_{[k]} + \alpha\, L|_{[k]}, \end{aligned} \tag{7}$$

in which $L|_{[k-1]}$ is the long-term performance index at iteration $k-1$, $S\big|_{[k]}$ is the instantaneous cost at iteration $[k]$, and so on, and $0 < \alpha < 1$ is the discount factor.

In the DNDP approach, that performance index is approximated by a multilayer neural network (critic network) with ideal weights $W_C \in \mathbb{R}^{N_{hC}\times 1}$ and $V_C \in \mathbb{R}^{N_{xC}\times N_{hC}}$ as

$$L(x_C)|_{[k]} = \left[W_C^T \sigma\left(V_C^T x_C\right) + \varepsilon_C\right]\Big|_{[k]}, \tag{8}$$

in which $x_C \in \mathbb{R}^{N_{xC}\times 1}$ is the vector of the neural network inputs, ε_C is the bounded approximation error (ie, $\|\varepsilon_C\| < \varepsilon_{BC} \in \mathbb{R}$), $\sigma(.)$ is a sigmoidal activation function, and N_{hC} is the number of nodes in the hidden layer. For simplifying purposes, the notation of the iteration $|_{[k]}$ will be presented only when needed. The ideal weights W_C and V_C in Eq. (8) are unknown but they can be approximated by adjustable weights $\hat{W}_C \in \mathbb{R}^{N_{hC}\times 1}$ and $\hat{V}_C \in \mathbb{R}^{N_{xC}\times N_{hC}}$ with the network weight errors

$$\begin{aligned} \tilde{W}_C &= W_C - \hat{W}_C, \\ \tilde{V}_C &= V_C - \hat{V}_C. \end{aligned} \tag{9}$$

As a result, the critic network will generate $J(x_C)$, which is the approximation of the cost function defined in Eq. (8):

$$J(x_C) = \hat{W}_C^T \hat{\sigma}_C\left(\hat{V}_C^T x_C\right) = \sum_{i=1}^{N_{hC}} \hat{W}_C^T(1,i)\,\hat{\sigma}_C\left(\sum_{j=1}^{N_{xC}} \hat{V}_C^T(i,j)\,x_C(j,1)\right), \tag{10}$$

where N_{hC} is the number of nodes in the hidden layer and N_{xC} is the number of inputs to the critic network. The backpropagation error of the critic network indicates how closely the approximated long-term cost function J follows Bellman's principle of optimality and is defined as follows:

$$e_C|_{[k]} = \underbrace{\left[J|_{[k-1]} - S|_{[k]}\right]}_{\text{target}} - \underbrace{\alpha J|_{[k]}}_{\text{current outcome}}. \tag{11}$$

Action Network

The action network is responsible for generating an approximation of f and leading to optimization of the approximated long-term cost function J (ie, the output of the critic network). During the gait of an individual with below-knee amputation, it is hypothesized that such an optimal control signal exists and is formulated as

$$\begin{aligned}\hat{f} &= W_{\mathrm{A}}^{\mathrm{T}}\sigma_{\mathrm{A}}\left(V_{\mathrm{A}}^{\mathrm{T}}x_{\mathrm{A}}\right) + \varepsilon_{\mathrm{A}} \\ &\approx \hat{W}_{\mathrm{A}}^{\mathrm{T}}\hat{\sigma}_{\mathrm{A}}\left(\hat{V}_{\mathrm{A}}^{\mathrm{T}}x_{\mathrm{A}}\right) = \sum_{i=1}^{N_{\mathrm{hA}}} \hat{W}_{\mathrm{A}}^{\mathrm{T}}(1,i)\,\hat{\sigma}_{\mathrm{A}}\left(\sum_{j=1}^{N_{\mathrm{xA}}} \hat{V}_{\mathrm{A}}^{\mathrm{T}}(i,j)\,x_{\mathrm{A}}(j,1)\right),\end{aligned} \tag{12}$$

in which $W_{\mathrm{A}} \in \mathbb{R}^{N_{\mathrm{hA}}\times 1}$ and $V_{\mathrm{A}} \in \mathbb{R}^{N_{\mathrm{xA}}\times N_{\mathrm{hA}}}$ are ideal network weights, $x_{\mathrm{A}} \in \mathbb{R}^{N_{\mathrm{xA}}\times 1}$ is the vector of the neural network inputs, ε_{A} is the bounded approximation error (ie, $\|\varepsilon_{\mathrm{A}}\| < \varepsilon_{\mathrm{BA}} \in \mathbb{R}$), $\sigma(.)$ is a sigmoidal activation function, N_{hA} is the number of nodes in the hidden layer, and N_{xA} is the number of inputs to the action network. The ideal weights W_{A} and V_{A} in Eq. (12) are unknown but they can be approximated by adjustable weights $\hat{W}_{\mathrm{A}} \in \mathbb{R}^{N_{\mathrm{hA}}\times 1}$ and $\hat{V}_{\mathrm{A}} \in \mathbb{R}^{N_{\mathrm{xA}}\times N_{\mathrm{hA}}}$ as in Eq. (12) with the network weight errors

$$\begin{aligned}\tilde{W}_{\mathrm{A}} &= W_{\mathrm{A}} - \hat{W}_{\mathrm{A}}, \\ \tilde{V}_{\mathrm{A}} &= V_{\mathrm{A}} - \hat{V}_{\mathrm{A}}.\end{aligned} \tag{13}$$

To quantify the performance of the action network in generating a control signal which leads to the optimization of the cost function J, the backpropagation error of the action network is calculated in terms of the ultimate control target as follows:

$$e_{\mathrm{A}}|_{[k]} = \underbrace{U_{\mathrm{C}}|_{[k]}}_{\text{target}} - \underbrace{J\big|_{[k]}}_{\text{current outcome}}, \tag{14}$$

where $U_{\mathrm{C}}|_{[k]}$ is an ultimate control goal, or the target for the long-term cost approximate $J\big|_{[k]}$.

Assumption

The ideal weights of the critic and action networks are constant and bounded—that is, $\|Z_{\mathrm{C}}\|_{\mathrm{F}} = \left\|\begin{bmatrix} W_{\mathrm{C}} & \Theta \\ \Theta & V_{\mathrm{C}} \end{bmatrix}\right\|_{\mathrm{F}} \le B_{\mathrm{CZ}} \in \mathbb{R}$ and $\|Z_{\mathrm{A}}\|_{\mathrm{F}} = \left\|\begin{bmatrix} W_{\mathrm{A}} & \Theta \\ \Theta & V_{\mathrm{A}} \end{bmatrix}\right\|_{\mathrm{F}} \le B_{\mathrm{AZ}} \in \mathbb{R}$, where $\|.\|_{\mathrm{F}}$ represents the Frobenius norm.

Theorem

Given the ankle dynamics in Eq. (2) and the control structure in Fig. 3, let the ankle torque be computed as

$$\tau_{\mathrm{a}} = \hat{f} - \tau_{\mathrm{aG}}^{\mathrm{g}} + K_{\mathrm{V}}r, \tag{15}$$

where $\hat{f}$ is control signal generated by the action network in Eq. (12), τ_{aG}^{g} is the compensation for the actual ground reaction torque τ_{aG}, r is the filtered tracking error ($r = \dot{e}_a + \lambda e_a$), and

$$K_V r = K_V(\dot{e}_a + \lambda e_a). \tag{16}$$

Further, let the critic and action neural network weights be updated according to

$$\Delta W_C = \alpha F_C e_C \hat{\sigma}_C - k_C F_C \|e_C\|_2 \hat{W}_C, \tag{17}$$

$$\Delta V_C = \alpha G_C e_C x_C \hat{W}_C^T \hat{\sigma}'_C - k_C G_C \|e_C\|_2 \hat{V}_C, \tag{18}$$

$$\Delta W_A = F_A e_A \hat{\sigma}_A \hat{V}_{CA} \hat{\sigma}'_C \hat{W}_C - F_A \hat{\sigma}'_A \hat{V}_A^T x_A r - k_A F_A \|e_A\|_2 \hat{W}_A, \tag{19}$$

$$\Delta V_A = G_A e_A x_A \hat{V}_{CA} \hat{\sigma}'_C \hat{W}_C \hat{W}_A^T \hat{\sigma}'_A - k_A G_A \|e_A\|_2 \hat{V}_A, \tag{20}$$

in which α is the discount factor, $\hat{V}_{CA}$ is contained in $\hat{V}_C$ to map $\hat{f}(x_A)$ to the hidden node outputs of the critic network, F_C, G_C, k_C, F_A, G_A, and k_A are real, positive design parameters, and $\hat{\sigma}'_C$ and $\hat{\sigma}'_A$ are Jacobian matrices defined as

$$\begin{aligned} \hat{\sigma}'_C &= \frac{\partial \hat{\sigma}_C\left(\hat{V}_C^T x_C\right)}{\partial\left(\hat{V}_C^T x_C\right)}, \\ \hat{\sigma}'_A &= \frac{\partial \hat{\sigma}_A\left(\hat{V}_A^T x_A\right)}{\partial\left(\hat{V}_A^T x_A\right)}. \end{aligned} \tag{21}$$

Then the actual tracking error e_a and the neural network weights errors $\tilde{W}_C$, $\tilde{V}_C$, $\tilde{W}_A$, and $\tilde{V}_A$ are uniformly ultimately bounded.

3.2.2 Recognition of the User's Intent

Recognition of gait and detection of gait events can be done with use of actual gait data such as the RSI force measured from the prosthetic socket of individuals with below-knee amputation. The RSI forces that the residual limb applies on the prosthetic socket during gait can be captured by ultrathin FlexiForce sensors [37] and a gait monitoring device as seen in Fig. 4. The force sensors can be embedded inside the socket to provide a nonintrusive force measurement in real time and during daily activities outside the laboratory setup.

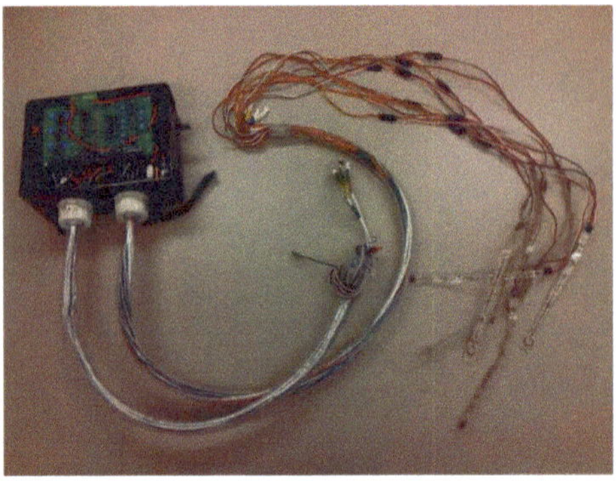

(A)

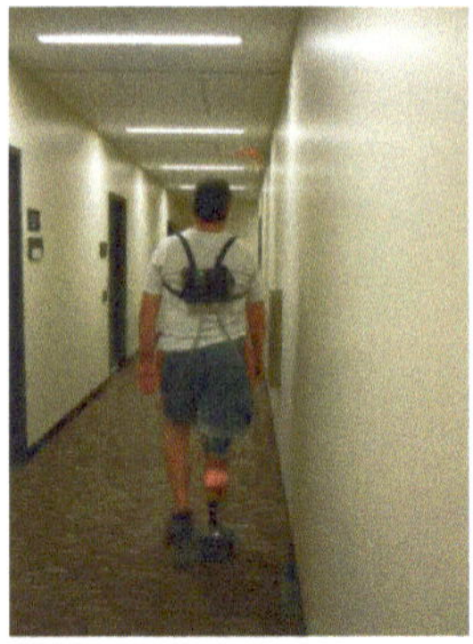

(B)

Fig. 4 (A) Data acquisition unit—OUPAM. (B) An individual wearing OUPAM during gait.

It has been shown that the recorded RSI forces relate to the gait [38] and can be used to distinguish between different gait types (eg, normal self-paced walk, brisk walk, ascending/descending stairs, walking on a ramp) and to detect gait parameters (heel strike, midstance, toe off, stance time, etc.) for each detected gait. The double-peak shape of the socket loading profile in Fig. 5 confirms that the mechanisms of the weight transferred between the prosthetic side and the intact side of people with unilateral below-knee amputation have characteristics in common with the mechanisms in people without amputation of the lower limb.

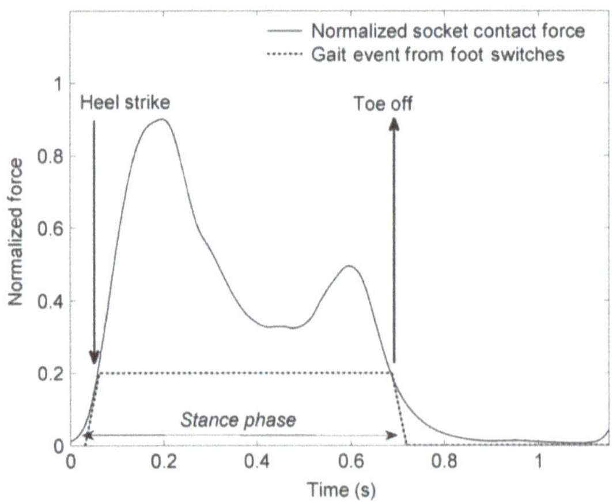

Fig. 5 Normalized socket contact force and gait events from foot switches.

On the basis of the recognized gait type and detected gait events, the updated gait-based kinematic references and approximated ground reaction torque are then specifically generated for the recognized gait type and are activated at the moment when ground contact (heel strike) is observed. Then the proposed control approach uses these gait-based quantities to calculate an appropriate control torque for the recognized gait.

3.2.3 Generation of the Gait-Based Kinematic References

In traditional control the tracking error is computed as the error between the ideal and actual displacement of the joint. In prosthetic ankle control the ideal joint profile is not available because of unknown user's intent and changes in walking terrain. Therefore, as the first step, an approximated ankle joint displacement profile is generated on the basis of the gait detected in real time.

Since the human gait and joint movements are cyclic in nature, the gait-based kinematic references can be generated with a Fourier series as follows [16]:

$$\theta_{\mathrm{r}}^{g}(t) = a_{\mathrm{g}}^{0} + \sum_{k_{\mathrm{g}}=1}^{5} a_{\mathrm{g}}^{k_g} \cos\left(k_{\mathrm{g}} w_{\mathrm{g}} t\right) + b_{\mathrm{g}}^{k_g} \sin\left(k_{\mathrm{g}} w_{\mathrm{g}} t\right), \tag{22}$$

where t denotes gait time and the parameters a_{g}^{0}, $a_{\mathrm{g}}^{k_g}$, and $b_{\mathrm{g}}^{k_g}$, w_{g} can be obtained for specific gait types. Since the sine and cosine functions in Eq. (22) are bounded, the gait-based kinematic patterns $\underline{\theta}_{\mathrm{r}}^{g}$ are also bounded. The ideal kinematic references $\underline{\theta}_{\mathrm{r}}$ are also bounded (ie, $\left\|\underline{\theta}_{\mathrm{r}}\right\| \leq B_{\theta_{\mathrm{r}}} \in \mathbb{R}$) because of the cyclic nature of gait and the human control of the movement of the residual limb to follow specific joint patterns which help minimize the metabolic energy consumption during gait [15]. Finally, the kinematic reference error $\underline{\tilde{\theta}}_{\mathrm{r}}$ is also bounded because $\underline{\theta}_{\mathrm{r}}$ and $\underline{\theta}_{\mathrm{r}}^{g}$ are both bounded and $\underline{\tilde{\theta}}_{\mathrm{r}} = \underline{\theta}_{\mathrm{r}} - \underline{\theta}_{\mathrm{r}}^{g}$.

3.2.4 Compensation of the Ground Reaction Torque

Calculation of the ground reaction torque τ_{aG} during gait requires kinematic tracking systems and force plates in the laboratory environments [11, 40]. Therefore under normal gait conditions, τ_{aG} is not available for the calculation of the control input τ_{a}. In the control scheme shown in Fig. 3, the gait-based ground reaction torque τ_{aG}^{g} is generated on the basis of the recognized gait and an empirical model describing the foot-ground contact as follows:

$$\tau_{\mathrm{aG}}^{g}(t) = d_{\mathrm{aZ}}(t)\,F_{\mathrm{X}}(t) + d_{\mathrm{aX}}(t)\,F_{\mathrm{Z}}(t)\,, \tag{23}$$

where t is the gait time, F_{X}(N) and F_{Z}(N) are the horizontal and vertical ground reaction forces, and d_{aX}(m) and d_{aZ}(m) are the horizontal and vertical distances between the ankle joint and the center of pressure (contact point) during gait. These quantities depend on the height and weight of the individual, as well as the terrain and type of gait.

During gait on level ground, ground reaction forces F_{X} and F_{Z} can be modeled by a combination of a spring and a position-dependent damper as follows [16, 41, 42]:

$$\begin{aligned} F_{\mathrm{Z}} &= \bar{k}\,(\bar{z}_{\mathrm{PEN}})^{\bar{e}} + \bar{c}_{\mathrm{d}}\dot{\bar{z}}_{\mathrm{PEN}}, \\ F_{\mathrm{X}} &= \bar{\mu}F_{\mathrm{Z}}\mathrm{sgn}\left(\dot{\bar{x}}_{\mathrm{P}}\right), \end{aligned} \tag{24}$$

in which $\bar{z}_{\mathrm{PEN}}(m)$ and $\dot{\bar{z}}_{\mathrm{PEN}}$ (m/s) are the penetration and penetration rate of the foot into the ground, $\bar{k}$ (N/m) and $\bar{e}$ are the spring coefficient and spring exponent, $\bar{c}_{\mathrm{d}}$ (N s/m) is the damping coefficient, $\bar{\mu}$ is the friction coefficient, $\dot{\bar{x}}_{\mathrm{P}}$ (m/s) is the horizontal velocity of the contact point with respect to the ground, and sgn (.) is the sign function. These parameters can be obtained from experiments similar to those reported in [16, 42]. The distances d_{aX} and d_{aZ} depend on the gait-based references $\underline{\theta}_{\mathrm{r}}^{g}$, temporal gait parameters, and the contact profiles assumed for the prosthetic feet.

It can be seen that τ_{aG}^{g} in Eq. (23) is bounded because F_{X} and F_{Z} in Eq. (24) are modeled by passive mechanical components and finite penetration, and d_{aX} and d_{aZ} are functions of the bounded kinematic references. The actual ground reaction torque is also bounded because of its cyclic nature and the finite body weight. Therefore the difference $\tilde{\tau}_{\mathrm{aG}} = \tau_{\mathrm{aG}} - \tau_{\mathrm{aG}}^{g}$ is bounded.

The dependence of the vertical ground reaction force generated by the empirical model on the gait cycle (24) is confirmed by the actual foot loading profile during a normal gait on level ground of an able-bodied person. The normalized foot loading profile in Fig. 6 is calculated as the summation of the forces measured by force sensors (FlexiForce A201 [37]) placed under the heel, the outer ball of the foot, and the toe of the foot. The typical double-peak shape of the modeled ground reaction force also agrees with the ground reaction force captured with force plates as found in the literature [11, 40].

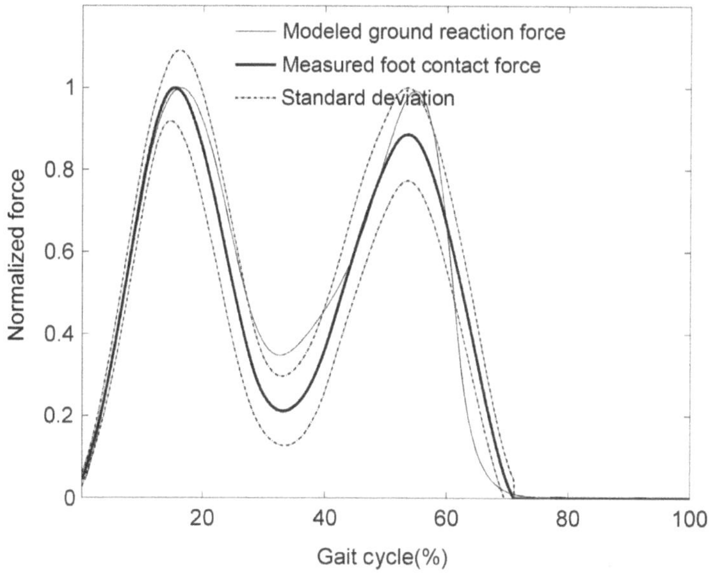

Fig. 6 Approximated ground reaction force and loading profile from actual foot pressure sensors.

4 DISCUSSION AND OUTLINE OF STABILITY AND PERFORMANCE ANALYSIS

4.1 Selection of the Performance Index

The short-term (ie, instantaneous) cost is calculated as follows:

$$S = -\frac{1}{2}\left(\frac{\theta_{\mathrm{ar}}^{\mathrm{g}} - \theta_{\mathrm{a}}}{\theta_{\mathrm{aM}}}\right)^2 - \frac{1}{2}\left(\frac{\dot{\theta}_{\mathrm{ar}}^{\mathrm{g}} - \dot{\theta}_{\mathrm{a}}}{\dot{\theta}_{\mathrm{aM}}}\right)^2, \tag{25}$$

where $\{\theta_{\mathrm{a}}, \dot{\theta}_{\mathrm{a}}\}$, $\{\theta_{\mathrm{ar}}^{\mathrm{g}}, \dot{\theta}_{\mathrm{ar}}^{\mathrm{g}}\}$, and $\{\theta_{\mathrm{aM}}, \dot{\theta}_{\mathrm{aM}}\}$ are the actual, desired, and maximal values of the ankle joint angular position and velocity. By our defining the cost in terms of the deviations of the prosthetic ankle joint position and velocity from their desired profiles, the performance measurement is related to the stability of the gait [17]. This selection also relates to the gait efficiency in the way that if the prosthetic ankle joint can perform as closely as possible to the biological ankle, then the hip and knee joints do not have to perform unnecessary work to compensate for gait degradation. As a result, people with below-knee amputation can walk with no or little unnecessary extra effort/energy consumption. Ultimately, the overall human-prosthetic system can perform a normal gait.

4.2 Selection of the Control Goal

The desire for optimization of the gait efficiency can be expressed as the optimization of the approximated long-term cost function J in the DNDP framework. Since the short-term cost function S is defined in terms of tracking errors as in Eq. (25), the ultimate control goal for J is selected as

$$U_C = 0. \tag{26}$$

4.3 Convergence of the Critic Network Output to the True Long-Term Performance Index

The learning algorithms for the critic and action networks that are presented in the next section are designed to make the backpropagation errors of the two networks go to zero (ie, $e_C \rightarrow 0$ and $e_A \rightarrow 0$). The relationship between the network errors going to zero and the selection of short-term cost function S and the ultimate control goal U_C can be explained as follows.

First, the critic network is trained so that its backpropagation error $e_C \rightarrow 0$. As $e_C \rightarrow 0$, it can easily be seen from Eq. (11) that $\alpha J\big|_{[k]} \rightarrow \left[J\big|_{[k-1]} - S|_{[k]}\right]$ or

$$J_{[k-1]} = S|_{[k]} + \alpha J\big|_{[k]}. \tag{27}$$

From a comparison of Eq. (27) with Eq. (7), it is clear that $J \rightarrow L$ (ie, the critic network generates the true approximation of the long-term performance index).

4.4 Convergence of the Tracking Error e_a

The action network is trained so that its output will make the backpropagation error $e_A \rightarrow 0$. Because $e_A = U_C - J$, the output of the critic network J will converge to the ultimate control goal U_C, which is 0. If the critic network is trained, then $J \rightarrow L$, and this implies that the long-term performance index L also goes to zero (ie, $L \rightarrow 0$). Because L is defined in terms of the short-term cost function S as in Eq. (7) and S is always a negative function as in Eq. (25), the instantaneous cost S also vanishes (ie, $S \rightarrow 0$). Since S is defined as the sum of the squares of the deviations of the ankle joint kinematics from its desired profiles, as $S \rightarrow 0$ the ankle joint tracking errors will go to zero (ie, $e_a \rightarrow 0$ and $\dot{e}_a \rightarrow 0$). This means the prosthetic ankle joint will follow the desired movement for a particular type of gait.

4.5 Optimization of the Long-Term Performance Index *L* and Its Approximation *J*

Finally, as the short-term cost S gets smaller, the smaller target for J in the subsequent iteration will be calculated. The critic network will repeat its training paradigm to generate an output that closely approximates the new target. As a result, the long-term performance index, which is approximated by the output of the critic network J, will be improved over time. That is the basic idea of adaptive dynamic programming-based controls which incorporate iterating and updating the approximated long-term performance index J over time (ie, reinforcement learning) and simultaneously satisfying Bellman's principle of optimality (ie, dynamic programming).

5 SIMULATION RESULTS

Both networks use sigmoid activation functions and are fully connected with randomly initialized weights in the range $[-1, 1]$. The critic network and action network weights are updated with use of Eqs. (17)–(20). The short-term cost at each time step is calculated by Eq. (25) Table 1 lists parameters used for the neural networks and the control algorithms.

The vector of inputs to the critic network is defined as

$$x_C = \begin{bmatrix} x_A^T & \hat{f} \end{bmatrix}^T, \tag{28}$$

in which x_A is the vector of inputs to the action network,

$$x_A = \begin{bmatrix} e_a & \dot{e}_a & \theta_a & \dot{\theta}_a \end{bmatrix}^T, \tag{29}$$

and $\hat{f}$ is the control signal generated by the action network as in Eq. (12).

Table 1 Direct neural dynamic programming-based control and neural network parameters

Parameters	Values
Control gain K_V	5
Design parameter λ	10
Discount factor α	0.95
Number of nodes in the input layer of the critic network N_{xC}	5
Number of nodes in the hidden layer of the critic network N_{hC}	10
Number of nodes in the output layer of the critic network	1
Number of nodes in the input layer of the action network N_{xA}	4
Number of nodes in the hidden layer of the action network N_{hA}	8
Number of nodes in the output layer of the action network	1

For the purpose of comparison, the simulation is repeated with other types of control at the ankle joint: proportional-derivative control and feedback linearization neural network-based control.

5.1 Ideal Condition

In the ideal condition the system is simulated with 20 steps of normal speed without any measurement and actuator noises.

Fig. 7 shows the tracking performance, short-term cost S, and accumulative cost $A = \Sigma S$ from one step during normal gait under the ideal condition (no noises, and no gait changes). All the controls can drive the ankle joint along the desired trajectory with a small tracking error. It can be seen that during the gait, although there are instances when the short-term cost S of the DNDP-based control exceeds the short-term cost of the feedback linearization neural network-based or proportional-derivative

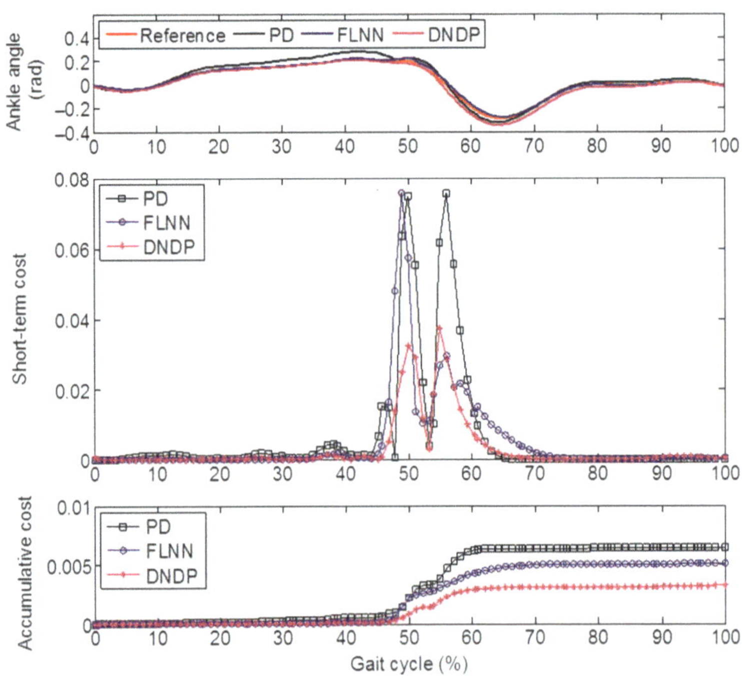

Fig. 7 Performance of the direct neural dynamic programming-based control during normal gait under ideal conditions.

control, the accumulative cost of the DNDP-based control during the entire gait cycle is lower than that of the other control algorithms.

5.2 Effect of Noise

Uniformly distributed measurement noises are added to the ankle position and angular velocity. Torque output generated for the ankle joint is also added with uniformly distributed actuator noise as follows:

$$\begin{aligned}\theta_a &= \theta_a + \rho\theta_a,\\ \dot{\theta}_a &= \dot{\theta}_a + \rho\dot{\theta}_a,\\ \tau_a &= \tau_a + \rho\tau_a,\end{aligned} \tag{30}$$

where ρ is in the range $[-2\%, 2\%]$ (or $[-5\%, 5\%]$). The system is simulated with 20 steps of normal walking speed and increasing measurement and/or actuator noises. Four simulation scenarios with different levels of measurement noises and actuator noises are considered: N1 (2% measurement noise), N2 (5% measurement noise), N3 (5% measurement noise and 2% actuator noise), and N4 (5% measurement noise and 5% actuator noise). In all scenarios (N1–N4), the measurement noises and actuator noises are added to the simulation model at the beginning of the sixth step.

The accumulative costs from one gait cycle of simulation scenarios N1–N4 with the presence of the measurement and actuator noises are shown in Fig. 8 for all control algorithms. It can be seen that as the measurement/actuator noises increase, the DNDP-based control outperforms other control methods by producing robust tracking performance with lower accumulative cost.

5.3 Conditions with Variations in Walking Speed

Control configurations similar to the previous simulation scenarios are repeated here to evaluate the performance of the DNDP-based control in the presence of variations in walking speed. The system is simulated with 5% measurement noise, 5% actuator noise, and four different walking setups: G1 (normal to fast), G2 (normal to slow), G3 (fast to Slow), and G4 (slow to fast). Three gaits labeled *slow*, *normal*, and *fast* indicate different walking speeds.

It can be seen from Fig. 9 that when the walking speed changes, the DNDP-based control is still able to provide lower accumulative performance cost compared with the other control strategies.

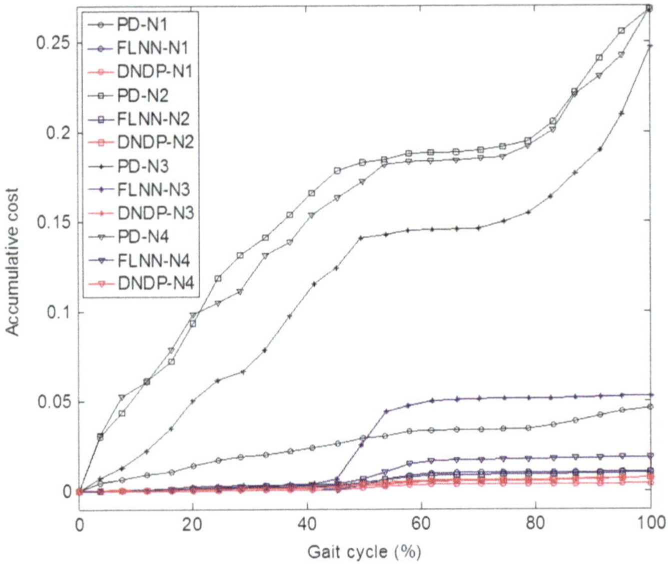

Fig. 8 Accumulative cost with the measurement and actuator noises.

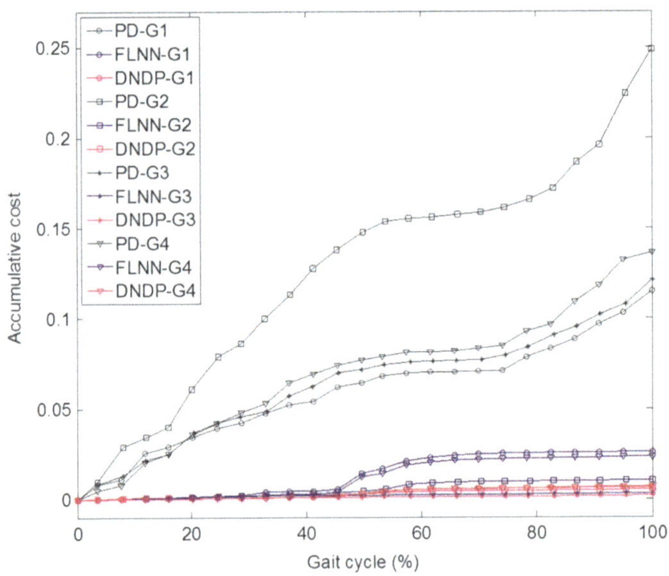

Fig. 9 Accumulative cost with the changing gaits.

6 CONCLUSIONS

We have evaluated the potential of a DNDP-based controller in improving the performance of a prosthetic ankle joint. We first addressed the formulation of the prosthetic ankle movement during gait, recognition of the user's intent using actual gait data, compensation for the actual ground reaction torque, and generation of the desired ankle displacement profile. The DNDP-based controller was then used to generate the control torque necessary for the ankle to track the generated displacement profile. By optimizing the performance metric, which is defined in terms of the deviation of the ankle kinematics from its desired profiles, the control approach described guarantees tracking performance as well as an increase in the efficiency of the gait. Simulation scenarios indicate that with the DNDP-based control, the prosthetic ankle joint was able to provide stable, robust tracking of the desired ankle displacement profiles at a reduced cost during gait. Future research will aim at evaluating the effectiveness of the proposed control structure with different performance indices, such as dynamic effort (ie, integration of the squares of all joint torques over time) and metabolic energy consumption. The implementation of the proposed approach on a prosthetic ankle and the evaluation of amputee gait is also being conducted.

REFERENCES

[1] K. Ziegler-Graham, E.J. MacKenzie, P.L. Ephraim, T.G. Travison, R. Brookmeyer, Estimating the prevalence of limb loss in the United States: 2005 to 2050, Arch. Phys. Med. Rehabil. 89 (2008) 422–429.
[2] T. Pohjolainen, H. Alaranta, M. Karkkainen, Prosthetic use and functional and social outcome following major lower limb amputation, Prosthet. Orthot. Int. 14 (1990) 75–79.
[3] C.R. Walker, R. Ingram, M. Hullin, S.W. McCreath, Lower limb amputation following injury: a survey of long-term functional outcome, Injury 25 (1994) 387–392.
[4] W.K. Brodzka, H. Thornhill, S. Zarapkar, J. Malloy, L. Weiss, Long-term function of persons with atherosclerotic bilateral below-knee amputation living in the inner city, Arch. Phys. Med. Rehabil. 71 (1990) 895–900.
[5] H. Bateni, S.J. Olney, Kinematic and kinetic variations of below-knee amputee gait, J. Prosthet. Orthot. 14 (2002) 2–13.
[6] I. Kovac, V. Medved, L. Ostojic, Spatial, temporal and kinematic characteristics of traumatic transtibial amputees' gait, Coll. Antropol. 34 (2010) 205–213.
[7] N.P. Fey, A.K. Silverman, R.R. Neptune, The influence of increasing steady-state walking speed on muscle activity in below-knee amputees, J. Electromyogr. Kinesiol. 20 (2010) 155–161.
[8] L. Torburn, C.M. Powers, R. Guiterrez, J. Perry, Energy expenditure during ambulation in dysvascular and traumatic below-knee amputees: a comparison of five prosthetic feet, J. Rehabil. Res. Dev. 32 (1995) 111–119.

[9] R. Gailey, K. Allen, J. Castles, J. Kucharik, M. Roeder, Review of secondary physical conditions associated with lower-limb amputation and long-term prosthesis use, J. Rehabil. Res. Dev. 45 (2008) 15–29.
[10] R. Versluys, P. Beyl, M.V. Damme, A. Desomer, R.V. Ham, D. Lefeber, Prosthetic feet: state-of-the-art review and the important of mimicking human ankle-foot biomechanics, Disabil. Rehabil. Assist. Technol. 4 (2009) 65–75.
[11] D.A. Winter, Biomechanics and Motor Control of Human Movement, Wiley, Hoboken, NJ, 2009.
[12] J. Perry, Gait Analysis: Normal and Pathological Function, SLACK Incorporated, Thorofare, NJ, 1992.
[13] H. Uustal, E. Baerga, Chapter 6: prosthetics and orthotics, in: S. Cuccurullo (Ed.), Physical Medicine and Rehabilitation Board Review, Demos Medical Publishing, New York, 2004.
[14] F.C. Anderson, M.G. Pandy, Dynamic optimization of human walking, J. Biomech. Eng. 123 (2001) 381–390.
[15] M. Ackermann, A.J. van den Bogert, Optimality principles for model-based prediction of human gait, J. Biomech. 43 (2010) 1055–1060.
[16] M. Peasgood, E. Kubica, J. McPhee, Stabilization of a dynamic walking gait simulation, ASME J. Comput. Nonlinear Dyn. 2 (2007) 65–72.
[17] Y. Xiang, J.S. Arora, K. Abdel-Malek, Physics-based modeling and simulation of human walking: a review of optimization-based and other approaches, Med. Bioeng. Appl. 42 (2010) 1–23.
[18] J.C. Ambrósio, A. Kecskeméthy, Multibody dynamics of biomechanical models for human motion via optimization, in: J.G. Orden, J. Goicolea, J. Cuadrado (Eds.), Multibody Dynamics, vol. 4, Springer, Dordrecht, 2007, pp. 245–272.
[19] R. Beckett, K. Chang, An evaluation of the kinematics of gait by minimum energy, J. Biomech. 1 (1968) 147–159.
[20] M.G. Pandy, Computer modeling and simulation of human movement, Annu. Rev. Biomed. Eng. 3 (2001) 245–273.
[21] M. Wojtyra, Multibody simulation model of human walking, Mech. Based Des. Struct. Mach. 31 (2003) 357–379.
[22] S. Srinivasan, I.A. Raptis, E.R. Westervelt, Low-dimensional sagittal plane model of normal human walking, J. Biomech. Eng. 130 (2008) 1–9.
[23] S. Pejhan, F. Farahmand, M. Parnianpour, Design optimization of an above-knee prosthesis based on the kinematics of gait, in: 30th Annual International IEEE EMBS Conference, Vancouver, BC, Canada, 2008, pp. 4274–4277.
[24] M.A. Shandiz, F. Farahmand, H. Zohour, Dynamic simulation of the biped normal and amputee human gait, in: 12th International Conference on Climbing and Walking Robots and the Support Technologies for Mobile Machines, Istanbul, Turkey, 2009, pp. 1113–1120.
[25] S. Srinivasan, E.R. Westervelt, A.H. Hansen, A low-dimensional sagittal-plane forward-dynamic model for asymmetric gait and its application to study the gait of transtibial prosthesis users, J. Biomech. Eng. 131 (2009) 1–9.
[26] Össur, Proprio Foot, 2015, Available: http://www.ossur.com/?PageID=12704.
[27] BIOM, BIOM—Personal Bionics, 2015, Available: http://www.biom.com/.
[28] S.K. Au, J. Weber, H. Herr, Powered ankle-foot prosthesis improves walking metabolic economy, IEEE Trans. Robot. 25 (2009) 51–66.
[29] R. Jiménez-Fabián, O. Verlinden, Review of control algorithms for robotic ankle systems in lower-limb orthoses, prostheses, and exoskeletons, Med. Eng. Phys. 34 (2012) 397–408.
[30] S.I. Wolf, M. Alimusaj, L. Fradet, J. Stegel, F. Braatz, Pressure characteristic at the stump/socket interface in transtibial amputees using an adaptive prosthetic foot, Clin. Biomech. 24 (2009) 860–865.

[31] A.E. Ferris, J.M. Aldridge, C.A. Rábago, J.M. Wilken, Evaluation of a powered ankle-foot prosthetic system during walking, Arch. Phys. Med. Rehabil. 93 (2012) 1911–1918.
[32] A.M. Grabowski, S. D'Andrea, Effects of a powered ankle-foot prosthesis on kinetic loading of the unaffected leg during level-ground walking, J. Neuroeng. Rehabil. 10 (2013) 1–12.
[33] J. Hitt, T. Sugar, M. Holgate, R. Bellman, K. Hollander, Robotic transtibial prosthesis with biomechanical energy regeneration, Int. J. Ind. Robot 36 (2009) 441–447.
[34] R. Versluys, A. Desomer, G. Lenaerts, M.V. Damme, P. Beyl, G.V. d. Perre, L. Peeraer, D. Lefeber, A pneumatically powered below-knee prosthesis: design specifications and first experiments with an amputee, in: 2nd Biennial IEEE/RAS-EMBS International Conference on Biomedical Robotics and Biomechanics, Scottsdale, AZ, USA, 2008, pp. 372–377.
[35] J. Markowitz, P. Krishnaswamy, M.F. Eilenberg, K. Endo, C. Barnhart, H. Herr, Speed adaptation in a powered transtibial prosthesis controlled with a neuromuscular model, Philos. Trans. R. Soc. Lond. B Biol. Sci. 366 (2011) 1621–1631.
[36] S. Au, M. Berniker, H. Herr, Powered ankle-foot prosthesis to assist level-ground and stair-descent gaits, Neural Netw. 21 (2008) 654–666.
[37] Tekscan, FlexiForce Sensors—User Manual 2015, Available: http://www.tekscan.com/pdf/A201-force-sensor.pdf.
[38] A. Mai, S. Commuri, C.P. Dionne, J. Day, W.J.J. Ertl, J.L. Regens, Effect of prosthetic foot on residuum-socket interface pressure and gait characteristics in an otherwise healthy man with transtibial osteomyoplastic amputation, J. Prosthet. Orthot. 24 (2012) 211–220.
[39] A. Mai, S. Commuri, Gait identification for an intelligent prosthetic foot, in: IEEE International Symposium on Intelligent Control (ISIC), Denver, CO, USA, 2011, pp. 1341–1346.
[40] C. Kirtley, Clinical Gait Analysis—Theory and Practice, Churchill Livingstone, Philadelphia, 2006.
[41] M. Millard, J. McPhee, E. Kubica, Multi-step forward dynamic gait simulation, Comput. Methods Appl. Sci. 12 (2008) 25–43.
[42] G.K. Klute, J.S. Berge, A.D. Segal, Heel-region properties of prosthetic feet and shoes, J. Rehabil. Res. Dev. 41 (2004) 535–546.
[43] D. Jin, R. Zhang, J. Zhang, R. Wang, W.A. Gruver, An intelligent above-knee prosthesis with EMG-based terrain identification, in: 2000 IEEE International Conference on Systems, Man, and Cybernetics, 2000, pp. 1859–1864.
[44] H.A. Varol, F. Sup, M. Goldfarb, Real-time gait model intent recognition of a powered knee and ankle prosthesis for standing and walking, in: 2nd Biennial IEEE/RAS-EMBS International Conference on Biomedical Robotics and Biomechatronics, Scottsdale, AZ, USA, 2008, pp. 66–72.
[45] R.R. Torrealba, C. Perez-D'Arpino, J. Cappelletto, G.F.-L. L. Fermın-Leon, J.C. Grieco, Through the development of a biomechatronic knee prosthesis for transfemoral amputees: mechanical design and manufacture, human gait characterization, intelligent control strategies and tests, in: IEEE International Conference on Robotics and Automation, Anchorage, AK, USA, 2010, pp. 2934–2939.
[46] J. Si, Y.-T. Wang, On-line learning control by association and reinforcement, IEEE Trans. Neural Netw. 12 (2001) 264–276.
[47] C. Lu, J. Si, X. Xie, Direct heuristic dynamic programming for damping oscillations in a large power system, IEEE Trans. Syst. Man Cybern. B: Cybern. 38 (2008) 1008–1013.
[48] R. Enns, J. Si, Helicopter trimming and tracking control using direct neural dynamic programming, IEEE Trans. Neural Netw. 14 (2003) 929–939.

[49] D.G. Thelen, F.C. Anderson, Using computed muscle control to generate forward dynamic simulations of human walking from experiment data, J. Biomech. 39 (2006) 1107–1115.
[50] P. Brugger, H.-B. Schemiedmayer, Simulating prosthetic gait—lessons to learn, Proc. Appl. Math. Mech. 3 (2003) 64–67.
[51] B. Dariush, V. Ng-Thow-Hing, D.G. Thelen, Simulation system, method and computer-readable medium for human augmentation devices, US Patent 7,251,593, 2007.
[52] F.M.L. Amirouche, Computational Methods in Multibody Dynamics, Prentice Hall, Englewood Cliffs, NJ, 1992.
[53] L. Yang, J. Si, K.S. Tsakalis, A.A. Rodriguez, Direct heuristic dynamic programming for nonlinear tracking control with filtered tracking error, IEEE Trans. Syst. Man Cybern. B: Cybern. 39 (2009) 1617–1622.

CHAPTER TWENTY-THREE

Novel Robust Adaptive Algorithms for Estimation and Control: Theory and Practical Examples

G. Herrmann*, **J. Na**†, **M. Nasiruddin Mahyuddin**‡

*University of Bristol, Bristol, United Kingdom
†Kunming University of Science and Technology, Kunming, China
‡University of Science, Malaysia, Pulau Pinang, Malaysia

1 INTRODUCTION

Adaptive control has been studied since the early 1950s. Early articles considered the control design for high-performance aircraft [1]. The basic principle of adaptive control is to update/estimate unknown system parameters online by use of appropriate errors, such that, satisfactory output tracking can be achieved even in the presence of unknown system parameters. Generally, most dynamic systems to be controlled have slowly varying uncertain parameters. Large parameter variations pose a great challenge in control. Many early appealing concepts have been proposed in which the notion of feedback systems was often labeled with the prefix of "self-tuning" and "adaptive" control [2].

Gain scheduling and several rudimentary model reference schemes were designed to overcome the sensitivity of a fixed-gain controller to large plant parameter variations. The goal in this scheme was to build a self-adjusting controller which yielded a closed-loop transfer function matching a prescribed reference model [3]. As a result of this early work, several schemes were proposed [4]—for example, the concept of self-adjusting control parameters using sensitivity functions and the famous MIT rule. The gradient descent approach and least squares (LS) methods were first developed for adaptive control of linear systems [3, 5]. The schemes were later advanced in the 1960s when Lyapunov theory was firmly established

Control of Complex Systems
http://dx.doi.org/10.1016/B978-0-12-805246-4.00023-9

as a very useful tool in proving convergence of adaptive control schemes. In the 1970s, Lyapunov proofs of model reference adaptive schemes were introduced by Narendra et al. [6] and Morse [7] using also a state-space approach. Input-output stability proof principles such as the Popov hyperstability criterion were suggested by Landau [8] and Egardt [9]. With these frameworks, asymptotic convergence of the tracking error and the boundedness of the estimated parameters can be rigorously shown when there are no modeling errors and external disturbances. A composite adaptive control was proposed in [1] to further improve the convergence performance of both the tracking error and parameter estimation. In this scheme, one designs the adaptive law by combining the tracking error and the prediction error; thus the advantages of self-tuning control and model reference adaptive control have been combined.

In the 1980s research evaluating the robustness of adaptive algorithms intensified: Rohrs et al. [10] highlighted the sensitivity of the adaptive methods to modeling uncertainties, which was previously ignored to simplify the controller design. This sparked waves of research into the robustness of adaptive algorithms [5]. In particular, σ-modification and *e*-modification were developed to retain the boundedness of the parameter estimation and thus to avoid the bursting phenomena. Moreover, the projection operator [5] depending on known upper/lower parameter bounds was also incorporated into the adaptive laws to retain the boundedness of the estimated parameters.

With regard to developments of nonlinear control schemes, the idea of adaptive control has also been explored for nonlinear systems: the feedback linearization method, adaptive backstepping, and adaptive neural network (NN) control have been developed in the past two decades [2, 11]. Many subsequent results have relaxed the required conditions (eg, *matching condition*) and remedied technical barriers for practical applications (eg, robot manipulators [12–14]).

Although nonlinear adaptive control theory can be used to prove stability and tracking performance for specific systems [15], much previous work did not provide a conclusion about parameter estimation convergence. It is well known that parameter convergence is a crucial performance trait as it can improve overall stability [1], which in turn is important for robustness of the designed adaptive system with disturbances and modeling uncertainties [16]. To address the parameter convergence problem, *persistent excitation* (PE) of the regressor vector is necessary for adaptive linear schemes. However, the online verification of such PE conditions is not trivial. This motivated Lin and Kanellakopoulos [15, 17] to develop an a priori verifiable condition for parameter convergence to obtain closed-loop systems that satisfy this condition and thus achieve global asymptotic stability. Their proposed analytic procedure allows one to determine if the estimated parameters can converge to their true values for a given nonlinear system with a "sufficiently rich" reference signal. It has also been shown that the system nonlinearities would introduce additional spectral lines, thus enhancing parameter convergence [15]. Other recent work [18] suggested a new concurrent learning method to relax the PE condition which records and processes specifically selected data concurrently with instantaneous data in the adaptation. Moreover, Adetola and Guay [19, 20] proposed a finite-time parameter estimation scheme which can reconstruct unknown parameters of the nonlinearity terms at the time instant when the PE condition is satisfied. This algorithm should test online the invertibility of the constructed regressor matrix and then compute its inverse.

Hence it has been the authors' wish to develop a robust adaptive estimation and control scheme with guaranteed convergence properties and to provide an intuitive method to verify the PE condition online. This chapter is a summary of work done by the authors on various aspects of adaptive control and estimation (eg, [21–31]). The original ideas were presented in [21, 22] and then further improved in [23] to achieve finite-time parameter estimation and control simultaneously. This estimation algorithm has also been practically evaluated to study parameter estimation in vehicle systems [26, 27, 31], adaptive control of robotics systems [30, 32, 33, 64], and adaptive optimal control [24, 25]. The extension of this new idea for estimation of fast time-varying parameters and for distributed control of multiagent systems was presented in [28–30, 34].

The basic idea of this new adaptive algorithm is to develop new leakage terms that contain the parameter estimation errors. With this new algorithm the estimated parameters converge to their true values and satisfactory robustness against bounded disturbances and modeling errors can be obtained. This can be achieved by application of stable filter operations on the system (eg, states, regressor, and the input). Then the parameter estimation error can be obtained explicitly on the basis of these auxiliary system variables. By constructing an adaptive law using the parameter estimation error, we can guarantee exponential error convergence when the filtered regressor matrix is positive definite. We have shown that this required condition holds provided that the conventional PE condition is satisfied. Moreover, the incorporation of a sliding-mode term, again employing the parameter estimation error, provides a novel scheme for finite-time reconstruction of the parameter estimates. The adoption of a special filter allows the scheme to avoid the measurement or computation of the state vector derivatives. Another salient feature is that these estimation schemes do not need to test online the invertibility of the regressor matrix and directly calculate the matrix inverse. Moreover, a novel forgetting mechanism in the dynamic structure of a filtered regressor matrix ensures that the matrix will not "blow up" in the absence of a PE signal [21–23]. In addition, a simplified and intuitive online validation of the PE condition is presented in which we prove that the PE condition is equivalent to the positive definiteness of an auxiliary (filtered regressor) matrix. This can be tested online. Evidently, the adaptive algorithm is easily used in conjunction with adaptive control and observer systems.

We will present this newly developed adaptive algorithm based on the parameter estimation error and discuss particular applications in the estimation of vehicle parameters, the control of robotic systems, and also the synthesis of online optimal control. We first present the basis of the gradient descent and LS schemes, showing our motivation for a new robust algorithm in Section 2, and then introduce the details of our new algorithm in Section 3. Then the application of this new method to practical problems of parameter estimation in vehicular systems and of an adaptive controller for a robotic arm and the development of a novel adaptive optimal control are given in Sections 4–6. Conclusions are given in Section 7.

2 REVIEW OF WELL-KNOWN ADAPTIVE ALGORITHMS AND MOTIVATION FOR INCREASED ROBUSTNESS

This section will briefly review several well-known adaptive parameter estimation schemes, and then show our motivation for investigating the new algorithm. For this purpose the following nonlinear system is studied

$$\dot{x} = \varphi(x, u) + \Phi(x, u)\Theta, \tag{1}$$

where $x \in \mathbb{R}^n$ is the measurable system state (ie, it is also an output), $u \in \mathbb{R}^m$ is the control input, $\Theta \in \mathbb{R}^N$ is the unknown parameter vector to be estimated, $\varphi(x, u) \in \mathbb{R}^n$ is a known function vector, and $\Phi(x, u) \in \mathbb{R}^{n\times N}$ is the known regressor matrix.

This nonlinear system structure will be adopted to discuss the formulation of our novel robust adaptive laws. Appropriate modifications can be made to fit the studies in the following sections. Without loss of generality, x and u are assumed to be bounded, and the functions $\varphi(x, u)$ and $\Phi(x, u)$

are continuous and thus bounded in a compact set. Moreover, for purpose of brevity, the following discussions will focus on the estimation of constant parameters only (ie, $\Theta \equiv$ const.).

It has been well recognized that the PE condition is essential for parameter estimation, which is presented here:

Definition 1. A vector/matrix function Φ is persistently excited when there exist $T > 0$ and $\varepsilon > 0$ and thus $\int_t^{t+T} \Phi(r)\Phi^T(r)dr \geq \varepsilon I$ for all $t \geq 0$.

For demonstration of well-known estimation schemes, system (1) is transformed into a generic parameterized system as

$$z = \phi\Theta, \tag{2}$$

where $z = \frac{s}{\Lambda(s)}[x] - \frac{1}{\Lambda(s)}[\varphi]$ and $\phi = \frac{1}{\Lambda(s)}[\Phi]$ are the filtered output and regressor, respectively, where $\Lambda(s) = s + \lambda$ for a positive constant $\lambda > 0$, s is the Laplace operator, and $\frac{1}{(s+\lambda)}$ is regarded as the first-order filter imposed on the corresponding variables. The above transformation from the differential equation into a generic parameterized system via the filter operation $\Lambda(s)$ can be extended to other high-order systems.

To accomplish the parameter estimation of Eq. (2), there are two well-known adaptive laws (ie, *gradient descent* and *LS methods*) which are widely used in the literature for control and estimation. These adaptive laws are fundamentally constructed on the basis of a given cost function of the parameter error, which should be minimized [5]. One can also view the basis of this cost function as a direct result of a Lyapunov stability analysis [3]. In general, there are two types of cost functions for the gradient descent methods: the *instantaneous* and *integral* cost functions. The most widely used gradient descent scheme is the one which is based on the instantaneous cost function [1, 3, 5]. On the other hand, the construct of the LS-based adaptive law is based on the principle of integrating the parameter estimation error over some time period. This endows the scheme with some degree of robustness against band-limited white noise and bounded disturbance [5]. However, one needs to compromise the robustness of the LS-based adaptive law against the convergence of the estimated parameters to their respective true values [1].

Next we present the construction of the aforementioned two adaptive laws. For brevity of notation, we assume that the system studied is a scalar system.

2.1 Gradient Descent Adaptive Law

The gradient method discussed in this section is the type of gradient descent method developed in the 1970s and 1980s. It has global stability properties [1, 3, 5]. The difference from the previous older scheme in the 1960s is the new formulations for the parameter estimation and the selection of different cost functions to be minimized. The gradient descent-based adaptive law based on an instantaneous cost function from [5] is presented. An algebraic estimation error equation is formed to motivate our use of an appropriate convex cost function $J(\hat{\Theta})$ of $\hat{\Theta}$. This cost function $J(\hat{\Theta})$ is minimized by application of the gradient descent method to update $\hat{\Theta}$ online.

The estimation $\hat{z}$ of z is given by

$$\hat{z} = \phi\hat{\Theta}, \tag{3}$$

where $\hat{\Theta}$ is the estimation of Θ. We define the prediction error ϵ as

$$\epsilon = z - \hat{z} = z - \phi\hat{\Theta}. \tag{4}$$

One can minimize the prediction error ϵ with respect to $\hat{\Theta}$ using the following quadratic instantaneous cost function:

$$J(\hat{\Theta}) = \frac{\epsilon^2}{2} = \frac{(z - \phi\hat{\Theta})^2}{2}. \tag{5}$$

By the use of the gradient descent method, the adaptive law for updating $\hat{\Theta}$ is given by the following differential equation:

$$\dot{\hat{\Theta}} = -\Gamma\nabla J(\hat{\Theta}), \tag{6}$$

where $\Gamma = \Gamma^T > 0$ is a diagonal matrix referred as the *adaptive learning gain*. On the basis of Eq. (5), one can obtain

$$\nabla J(\hat{\Theta}) = -(z - \phi\hat{\Theta})\phi = -\epsilon\phi, \tag{7}$$

and this will yield the adaptive law that generates $\hat{\Theta}$ as follows:

$$\dot{\hat{\Theta}} = \Gamma\epsilon\phi. \tag{8}$$

One can analyze the convergence property of the gradient descent adaptive law via a Lyapunov approach using $V(\tilde{\Theta}) = \frac{\tilde{\Theta}^T\Gamma^{-1}\tilde{\Theta}}{2}$, where $\tilde{\Theta} = \Theta - \hat{\Theta} \in \mathbb{R}$ is the parameter estimation error. Differentiating the Lyapunov function leads to $\dot{V} \leq -\tilde{\Theta}^T\phi\epsilon = -\epsilon^2 \leq 0$, implying that $V, \tilde{\Theta} \in \mathcal{L}_\infty$ and $\epsilon \in \mathcal{L}_2$. Then Barbalat's lemma implies that ϵ converges to zero. Under the condition that ϕ is persistently excited, $\hat{\Theta}$ converges to Θ *exponentially* as $t \to \infty$.

The adaptive law in Eq. (8) is essentially semistable, which may cause instability in the case of unknown disturbances. This phenomenon leads to the introduction of the σ-modification, the e-modification, or the projection operator [5]. Evidently, the robustness of the overall adaptive control system can be enhanced by use of such modifications. However, as a damping term is injected in the σ-modification and e-modification adaptations, the precise estimation of unknown parameters with these modified adaptive laws may be lost, which also affects the tracking performance of control systems using the adaptation scheme (ie, only ultimate uniform boundedness of the tracking error and estimation error can be claimed).

2.2 Least Squares Adaptive Law

The so-called *pure* LS-based adaptive law requires a strict assumption—that is, the regressor should be persistently excited or otherwise the adaptive gain will not be invertible). This will trigger the "blowing up" of the regressor matrix in the adaptation algorithm. Several variants of the LS-based adaptive law have evolved from the pure LS-based law—for example, the *recursive* LS-based adaptive law [1, 3, 5].

The RLS or nonrecursive LS method has been adopted to estimate unknown parameters [4, 5, 35, 36]. The basic idea of the LS methods is to minimize the following integral cost function:

$$J(\hat{\Theta}) = \frac{1}{2}\int_0^t e^{-\beta(t-\tau)}[z(\tau)-\phi(\tau)\hat{\Theta}]^2 d\tau + \frac{1}{2}e^{-\beta t}(\hat{\Theta}-\hat{\Theta}_0)^T Q_0(\hat{\Theta}-\hat{\Theta}_0), \tag{9}$$

where $Q_0 = Q_0^T > 0$ is a positive definite matrix, $\beta \geq 0$ is a scalar forgetting factor, and $\hat{\Theta}_0 = \hat{\Theta}(0)$ denotes the initial condition. The cost function (9) uses a discounting factor and a penalty term on the initial value $\hat{\Theta}_0$. Consider that $z, \phi \in \mathcal{L}_\infty$ and that $J(\hat{\Theta})$ is a convex function of $\hat{\Theta}$. The minimum of $J(\hat{\Theta})$ satisfies

$$\nabla J(\hat{\Theta}) = e^{-\beta t}Q_0(\hat{\Theta}(t)-\hat{\Theta}_0) - \int_0^t e^{-\beta(t-\tau)}\left(z(\tau)-\phi(\tau)\hat{\Theta}(t)\right)\phi(\tau)d\tau = 0. \tag{10}$$

Rearranging Eq. (10) for $\hat{\Theta}(t)$ yields the *nonrecursive LS* method:

$$\hat{\Theta} = \Gamma\left[e^{-\beta t}Q_0\hat{\Theta}_0 + \int_0^t e^{-\beta(t-\tau)}z(\tau)\phi(\tau)d\tau\right], \tag{11}$$

where

$$\Gamma = \left[e^{-\beta t}Q_0 + \int_0^t e^{-\beta(t-\tau)}\phi(\tau)\phi(\tau)^T d\tau\right]^{-1}. \tag{12}$$

We may wish to create Γ more efficiently by subjecting it to a differential equation. Using the matrix derivative identity $\frac{d}{dt}[\Gamma\Gamma^{-1}] = \dot{\Gamma}\Gamma^{-1} + \Gamma\frac{d}{dt}\left[\Gamma^{-1}\right] = 0$, we may express Eq. (12) as

$$\dot{\Gamma} = \beta\Gamma - \Gamma\phi\phi^T\Gamma, \quad \Gamma(0) = \Gamma_0 = Q_0^{-1}, \tag{13}$$

so the calculation of the inverse matrix in Eq. (12) is not needed. Instead, we can obtain Γ from Eq. (13). Differentiating $\hat{\Theta}(t)$ in Eq. (11) with respect to t and using Eq. (13) and $\epsilon = z - \hat{\Theta}^T\phi$, we obtain the following adaptive law:

$$\dot{\hat{\Theta}} = \Gamma\epsilon\phi, \tag{14}$$

where Γ is generated from Eq. (13). This adaptive law is also known as the *continuous-time RLS with a forgetting factor* β [1, 3, 5]. The *pure* version of the LS-based adaptive law (ie, without the forgetting factor) ensures only asymptotic convergence. The exponential convergence of the RLS-based adaptive law can be proven through Lyapunov analysis. The proof of convergence for the LS-based adaptive laws can be found in [3, 5].

2.3 Observer-Based Methods

For the purpose of estimating unknown parameters Θ for the linearly parameterized system (1) presented in a first-order differential equation, a straightforward method is to construct an adaptive observer or predictor as

$$\dot{\hat{x}} = \varphi(x, u) + \Phi(x, u)\hat{\Theta} + K(x - \hat{x}), \tag{15}$$

where $\hat{x}$ and $\hat{\Theta}$ are the estimation of x and Θ, and $K > 0$ is the observer gain.

The observer (15) is constructed to provide the output estimation error $x - \hat{x}$, which can then be used to estimate the unknown parameter Θ by the following adaptive law:

$$\dot{\hat{\Theta}} = \Gamma\Phi(x - \hat{x}), \tag{16}$$

with $\Gamma > 0$ being a constant diagonal gain matrix.

It was proven in [5] that asymptotic convergence of $\hat{\Theta}$ to Θ is achieved provided that the regressor matrix Φ fulfils the PE condition (see Chapters 4 and 5 in [5] for more details). However, the adaptive law (16) is driven by use of the output error $x - \hat{x}$, so the convergence performance of the estimation error $\Theta - \hat{\Theta}$ is a function of the observer error $x - \hat{x}$ although there are certain relationships between them. Thus the transient performance of $\hat{\Theta}$ depends heavily on the design of the observer (15), which could be unsatisfactory.

2.4 The Need for New Robust Adaptive Algorithms

In contrast to adaptive law (16), the error ϵ used in the adaptive gradient descent (8) and LS (14) laws can be expressed as a function of the parameter estimation error $\tilde{\Theta} = \Theta - \hat{\Theta}$. This follows after substitution of z from Eq. (2) into Eq. (4) to obtain $\epsilon = \phi\tilde{\Theta}$. In this sense, the adaptive laws (8) and (14) are driven by the parameter estimation as $\dot{\hat{\Theta}} = \Gamma\epsilon\phi = \Gamma\phi\phi^T\tilde{\Theta}$. It has been well recognized that the inclusion of the parameter estimation error in the adaptive laws can improve the estimation error convergence. However, the use of filter operation $1/\Lambda(s)$ on system (1) yielding Eq. (2) introduces a DC gain of $1/\lambda$ on the filtered regressor ϕ of Φ. This may reduce the amplitude of the regressor and thus affect the transient of the estimation performance [3]. It is thus evident that better estimation performance can be obtained as long as the idea of parameter estimation error-based adaptation (eg, gradient descent and LS methods can be further developed.

Moreover, although the LS method can provide some degree of robustness against bounded disturbances because of the integration of the parameter estimation error over some time period, the robustness of the adaptive law (16) and the gradient-descent scheme (8) cannot be guaranteed (ie, a modeling error or external disturbances may trigger unboundedness of estimated parameters). In contrast, our new adaptive law allows finite-time convergence using a sliding mode technique. Such a characteristic implies robustness to perturbation close to the origin and thus is more robust to model errors and exogenous disturbances. Finally, the online validation of the required PE condition for the parameter convergence is difficult, in particular for nonlinear systems, which has not been fully addressed in the adaptive laws discussed above.

Considering the above analysis, there are still obvious needs to develop new adaptive algorithms based on the parameter estimation error for system (1). This fact motivates our recent work to develop new robust finite-time adaptive algorithms with guaranteed convergence and robustness. We also suggest a simple and intuitive method to test online the standard PE condition.

3 NOVEL ADAPTIVE ALGORITHMS

In this section the basic idea of the new adaptive algorithm [21–23] based on the parameter estimation error is firstly shown. This method can estimate unknown parameters Θ of Eq. (1) by means of the estimation error information. The first step should derive such parameter estimation error.

For this purpose the filtered variables x_f, Φ_f, φ_f of x, Φ, and φ can be defined as

$$\begin{cases} k\dot{x}_f + x_f = x, \ x_f(0) = 0, \\ k\dot{\Phi}_f + \Phi_f = \Phi, \ \Phi_f(0) = 0, \\ k\dot{\varphi}_f + \varphi_f = \varphi, \ \varphi_f(0) = 0, \end{cases} \tag{17}$$

where $k > 0$ is a constant. We further define an auxiliary matrix P and vector Q as

$$\begin{cases} \dot{P} = -\ell P + \Phi_f^T \Phi_f, \ P(0) = 0, \\ \dot{Q} = -\ell Q + \Phi_f^T \left[\frac{(x-x_f)}{k} - \varphi_f\right], \ Q(0) = 0, \end{cases} \tag{18}$$

where $\ell > 0$ is another design parameter.

We finally define an auxiliary vector $M \in \mathbb{R}^N$ based on P and Q as

$$M = P\hat{\Theta} - Q, \tag{19}$$

where $\hat{\Theta}$ is the estimation of Θ, which is given by the following adaptive laws.

Lemma 1. *The vector M in Eq. (19) can be reformulated as $M = -P\tilde{\Theta}$, where $\tilde{\Theta} = \Theta - \hat{\Theta}$ defines the estimation error.*

Proof. The solution of matrix differential equation (18) can be obtained as

$$\begin{cases} P(t) = \int_0^t e^{-\ell(t-r)} \Phi_f^T(r)\Phi_f(r)dr, \\ Q(t) = \int_0^t e^{-\ell(t-r)} \Phi_f^T(r)\left[(x(r) - x_f(r))/k - \varphi_f(r)\right] dr. \end{cases} \tag{20}$$

On the other hand, by application of the linear filter operation (21) on both sides of Eq. (1), it can be obtained from Eq. (1) and the first equation in Eq. (17) that

$$\dot{x}_f = \frac{x - x_f}{k} = \varphi_f + \Phi_f \Theta. \tag{21}$$

From Eqs. (21) and (20) it can be verified that $Q = P\Theta$; thus

$$M = P\hat{\Theta} - Q = P\hat{\Theta} - P\Theta = -P\tilde{\Theta}, \tag{22}$$

with $\tilde{\Theta} = \Theta - \hat{\Theta}$ being the estimation error. □

Remark 1. It is shown in Eq. (22) that the variable M in Eq. (19) based on P and Q contains the information on the estimation error $\tilde{\Theta}$. Thus M can be used to drive parameter estimation without using an observer/predictor error. Moreover, the terms ℓP and ℓQ used in Eq. (18) serve as forgetting factors so as to retain the boundedness of P and Q. Finally, the estimation method can be extended to the case where the unknown parameter Θ is a matrix (see Section 6).

Before we present the estimation approach, we need to address the PE condition and introduce an intuitive method to test this condition online. Then we will show that the PE property of the original regressor vector $\Phi(x, u)$ in Eq. (1) is equal to the positive definiteness of matrix P—that is, $\lambda_{\min}(P(t)) > \sigma > 0$).[1]

Lemma 2. *If the regressor matrix $\Phi(x, u)$ in Eq. (1) is persistently excited, then the matrix P is positive definite—that is, $\lambda_{\min}(P(t)) > \sigma$ for $t > T$ and $\sigma > 0$, $T > 0$.*

Proof. We refer to our previous work [21, 22] for a detailed proof. □

Similar to general parameter estimation, the PE condition is needed in this chapter. On the basis of Lemma 2, the standard PE condition of the regressor vector Φ is sufficient to guarantee the positive definiteness of P. This observation provides an intuitive and numerically feasible approach to a *direct online* test for the required PE condition. This has been recognized as a well-known and difficult problem in the field. This can be achieved by testing for the positive definite property of P. It has also been shown in our work [21, 22] that the positive definiteness of P can be retained by application of a proper control u or a dither signal to system (1).

From the fact that $Q = P\Theta$ and the assumption that P is positive definite, one may have the following:

Corollary 1. *Assume Φ of system* (1) *defined in Eq. (20) is persistently excited. Then a direct estimation of the unknown parameter is given as $\Theta = P^{-1}Q$.*

Although Corollary 1 gives a direct method to obtain Θ, one needs to calculate the matrix inverse P^{-1}, which may be difficult because of potential numerical problems. Thus we will further develop three adaptive laws to obtain an estimated parameter $\hat{\Theta}$ online by using the parameter error information M in (22).

[1] Throughout this chapter, $\lambda_{\max}(\cdot)$ and $\lambda_{\min}(\cdot)$ denote the maximum and minimum eigenvalues of the corresponding matrices.

The first adaptive law to update $\hat{\Theta}$ is provided by

$$\dot{\hat{\Theta}} = -\Gamma M, \tag{23}$$

where $\Gamma > 0$ is a constant diagonal matrix.

Then we have the following theorem:

Theorem 1. *If the regressor Φ of system* (1) *is persistently excited, then the estimation error $\tilde{\Theta}$ of adaptive law* (23) *converges to zero exponentially, where the convergence rate is $\mu_1 = 2\sigma/\lambda_{\max}(\Gamma^{-1})$.*

Proof. It is shown that the matrix P satisfies $\lambda_{\min}(P) > \sigma > 0$ when Φ is persistently excited. For the Lyapunov function as $V_1 = \frac{1}{2}\tilde{\Theta}^T\Gamma^{-1}\tilde{\Theta}$, the derivative $\dot{V}_1$ along Eq. (23) is obtained as

$$\dot{V}_1 = \tilde{\Theta}^T\Gamma^{-1}\dot{\tilde{\Theta}} = \tilde{\Theta}^T M = -\tilde{\Theta}^T P\tilde{\Theta} \leq -\mu_1 V_1, \tag{24}$$

where $\mu_1 = 2\sigma/\lambda_{\max}(\Gamma^{-1})$ is positive for all $t > 0$. Then from Lyapunov's theorem and Eq. (24), we can conclude that $\tilde{\Theta}$ converges exponentially to zero. □

We can obtain the second adaptive law by modifying the adaptive law of Eq. (23) in terms of the sliding mode method to achieve finite-time convergence:

$$\dot{\hat{\Theta}} = -\Gamma\frac{P^T M}{\|M\|}, \tag{25}$$

where $\Gamma > 0$ is a diagonal constant matrix.

Theorem 2. *If the regressor vector Φ of system* (1) *is persistently excited, then the estimation error $\tilde{\Theta}$ of the adaptive law* (25) *converges to zero in finite time t_a satisfying $t_a \leq \left\|\tilde{\Theta}(0)\right\|\lambda_{\max}(\Gamma^{-1})/\sigma$.*

Proof. We select the Lyapunov function as $V_2 = \frac{1}{2}\tilde{\Theta}^T\Gamma^{-1}\tilde{\Theta}$. Then it can be verified with use of $\lambda_{\min}(P) > \sigma > 0$ and $M = -P\tilde{\Theta}$ that the derivative of V_2 can be computed as

$$\dot{V}_2 = \tilde{\Theta}^T\Gamma^{-1}\dot{\tilde{\Theta}} = -\tilde{\Theta}^T\frac{P^T M}{\left\|P\hat{\Theta} - Q\right\|} = -\frac{\tilde{\Theta}^T P^T P\tilde{\Theta}}{\left\|P\tilde{\Theta}\right\|} \leq -\left\|P\tilde{\Theta}\right\| \leq -\mu_2\sqrt{V_2}, \tag{26}$$

where $\mu_2 = \sigma\sqrt{2/\lambda_{\max}(\Gamma^{-1})}$ is positive. Then from Eq. (26), finite-time convergence of the parameter estimation error $\tilde{\Theta}$ can be guaranteed, and the time to achieve $\tilde{\Theta} = 0$ is estimated by $t_a \leq 2\sqrt{V_2(0)}/\mu_2$. Clearly, the convergence time t_a depends on the excitation level σ and the learning gain Γ (ie, larger σ and Γ will result in a smaller t_a). □

The third adaptive law is inspired by the fact shown in Corollary 1 (ie, $P^{-1}Q = \Theta$, where the calculation of the matrix inverse of P^{-1} is avoided). For this purpose, another auxiliary matrix K is introduced to approach the inverse of P as

$$\dot{K} = \ell K - K\Phi_f^T \Phi_f K, \quad K^{-1}(0) = K_0 > 0, \tag{27}$$

where $\ell > 0$ is the same constant as used in Eq. (18) and Φ_f is the filtered regressor defined in Eq. (17). According to the matrix equality $\frac{d}{dt}KK^{-1} = \dot{K}K^{-1} + K\frac{d}{dt}K^{-1} = 0$, one can obtain the solution of Eq. (27) as

$$K(t) = [\mathrm{e}^{-\ell t}K_0 + \int_0^t \mathrm{e}^{-\ell(t-r)}\Phi_f^T(r)\Phi_f(r)dr]^{-1} = [\mathrm{e}^{-\ell t}K_0 + P(t)]^{-1}, \tag{28}$$

where $K_0 = K_0^T = K^{-1}(0) > 0$ denotes the initial condition given by $K_0 = \eta I$, with $\eta > 0$ being a constant.

It is shown in Eq. (28) that K converges to P^{-1} subject to an exponential vanishing term $\mathrm{e}^{-\ell t}K_0$ of the initial condition. Then the third adaptive law is given as

$$\dot{\hat{\Theta}} = -\Gamma \frac{\hat{\Theta} - KQ}{\left\| \hat{\Theta} - KQ \right\|}. \tag{29}$$

Before discussing the convergence of the adaptive law (28), we first address the property of the term KQ in the following lemma (see [23] for a proof):

Lemma 3. *For the matrices K in Eq. (27) and Q in Eq. (21), the term KQ can be formulated as $K(t)Q(t) = [I - E(t)]\Theta$, where $E(t) \to 0$ for $t \to \infty$.*

Thus this property now permits us to state a result for convergence of the adaptive law (29) (see again [23] for a proof):

Theorem 3. *Consider the adaptive law $\hat{\Theta}$ from Eq. (29) with Q in Eq. (18) and K in Eq. (27) for $K^{-1}(0) = \eta I > 0$. If the regressor vector Φ of system* (1) *is persistently excited, then the estimation error $\tilde{\Theta}$ converges to a small bounded set $\left\| \tilde{\Theta} \right\| \le c$ in finite time, with $c > 0$ being a small positive constant.*

Following the analysis in Eq. (28) and Lemma 3, one can also provide an alternative estimation scheme as shown in the following corollary:

Corollary 2. *If the regressor Φ of system* (1) *is persistently excited, then $\lim_{t\to\infty} KQ = \Theta$ holds for matrices Q and K given in Eqs. (18) and* (27).

The proof can be obtained if one considers $P^{-1}Q = \Theta$ and $KP \to I$ as $t \to +\infty$.

Remark 2. K converges to P^{-1} because of $\lim_{t\to\infty} E(t) = 0$. In this case the estimation error of the adaptive law (29) converges to a small residual set in finite time, determined by the initial condition $K^{-1}(0) = \eta I \neq 0$. However, a salient feature for introducing matrix K is that the online test for the invertibility and the calculation of the inverse of P (when it is feasible) is avoided. Specifically, by introducing the auxiliary matrix K, one can conduct a practical and very intuitive test for the required excitation condition of adaptation (29) or of Corollary 2 *online* by checking $KP \approx I$, which implies nonsingularity of P.

This section has provided some new algorithms based on our motivation for deriving the estimation error information M, which is then used to drive the adaptive laws and also acts as a leakage term. The robustness of such algorithms can be found in [23]. Moreover, these new leakage terms can be used in the adaptive laws of observers and controllers, which will be addressed in the following sections.

4 EXAMPLE OF PARAMETER ESTIMATION IN VEHICULAR SYSTEMS

Real-time knowledge of a car's parameters such as mass and road gradient can improve active safety technologies and fuel efficiency [37–43]. It is important to note that active safety systems permit better vehicle control during aggressive maneuvers. This can minimize the likelihood of accidents due to rollover or skidding. Recent research shows that such rollover or skidding incidents can be reduced by as much as 33% or 59%, respectively [44]. Hence this section summarizes the application of the suggested new adaptive algorithm [26] for the online estimation of two vehicle parameters: vehicle mass and road gradient. In this case the novel algorithm is incorporated into an adaptive observer design. Practical experimental results on a small vehicle are provided, showing a significant improvement over previous work.

Recently there have been a large number of sensor-based approaches to obtain vehicle mass and road gradient [45–47]. Nevertheless, the nature of an inclinometer's construction may cause such methods to be susceptible to errors [48] as inclinometers are generally suited to measure only static values. Remote sensing techniques such as GPS [45] may help in road gradient measurement but at some cost. In addition, estimation of mass without information on the road gradient has proven to be ineffective.

The points above have been a strong motivation for us to obtain vehicle mass and road gradient simultaneously. The suggested approach for online estimation of the road gradient and vehicle mass uses only the velocity and driving torque of the vehicle. The estimation scheme is obtained on the basis of an adaptive nonlinear observer with our new adaptive law, which guarantees that the estimated parameters converge to the true values in finite time. Most importantly, the adaptation approach does not rely on possibly noisy acceleration measurements. We also validate the robustness of the adaptive method and verify the PE condition. This is achieved by computation of the condition number of the respective filtered regressor matrix. In contrast, the classical RLS algorithm cannot provide fast convergence guarantees, specifically exponential convergence, not to mention finite-time convergence [49].

4.1 Vehicle Longitudinal Dynamics

To further discuss the estimation of the road gradient and the vehicle's mass, the longitudinal vehicle dynamics that are used in the following development is presented first. Fig. 1 depicts a schematic model of the small model car used for experiments.

On the basis of Newton's second law, we can obtain the longitudinal model of a small vehicle as

$$m\ddot{x} = F_{\text{engine}} - mg\sin\theta - C_{vF}\dot{x} - C_{\mu}mg\cos\theta, \tag{30}$$

where m and θ are the vehicle mass and the track gradient which the vehicle is subject to. The velocity of the vehicular system is the scalar variable $\dot{x}$, C_{vF} defines the viscous damping coefficient, and C_{μ} is the rolling resistance coefficient. The parameters that will be estimated include the road gradient θ, the vehicle mass m, and the viscous damping coefficient C_{vf}.

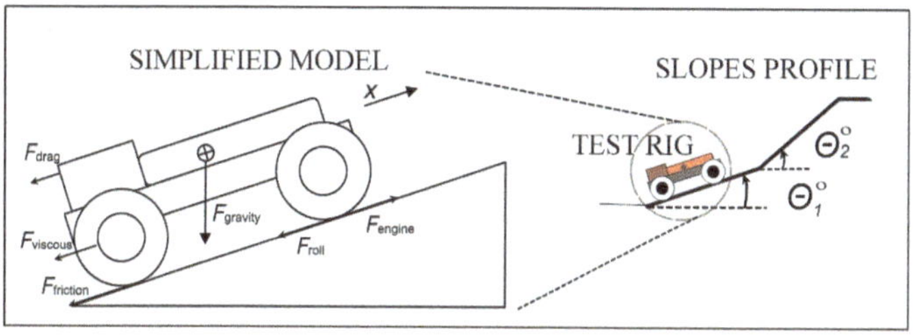

Fig. 1 Parameter estimation problem. *(Source: M. Mahyuddin, J. Na, G. Herrmann, X. Ren, P. Barber, Adaptive observer-based parameter estimation with application to road gradient and vehicle mass estimation, IEEE Trans. Ind. Electron. 61 (6) (2013) 2851–2863).*

Here the air drag F_{drag} is negligible at low speeds[2] below 0.2 m/s. The force F_{brake} due to braking is merged into the driving force, F_{engine}.

To facilitate the observer design and parameter estimation, we represent the system (30) in the following state-space form:

$$
\begin{aligned}
\begin{bmatrix} \dot{x} \\ \ddot{x} \end{bmatrix} &= \begin{bmatrix} 0 & 1 \\ 0 & 0 \end{bmatrix}\begin{bmatrix} x \\ \dot{x} \end{bmatrix} + \begin{bmatrix} 0 \\ \frac{1}{m} \end{bmatrix} F_{\text{engine}} - \begin{bmatrix} 0 \\ \sin\theta + C_\mu \cos\theta \end{bmatrix} g - \begin{bmatrix} 0 \\ \frac{C_{VF}}{m} \end{bmatrix}\dot{x} \\
&= \begin{bmatrix} 0 & 1 \\ 0 & 0 \end{bmatrix}\begin{bmatrix} x \\ \dot{x} \end{bmatrix} + \begin{bmatrix} 0 \\ 1 \end{bmatrix}\begin{bmatrix} -g & F_{\text{engine}} & -\dot{x} \end{bmatrix}\begin{bmatrix} s \\ b \\ f \end{bmatrix}, \\
y &= \begin{bmatrix} 1 & 0 \\ 0 & 1 \end{bmatrix}\begin{bmatrix} x \\ \dot{x} \end{bmatrix},
\end{aligned}
\tag{31}
$$

where s, b, and f denote $\sin\theta + C_\mu \cos\theta$, $\frac{1}{m}$, and $\frac{C_{VF}}{m}$, respectively. Without loss of generality, we assume that the driving force F_{engine} is bounded and the system states $\dot{x}$ are also bounded.

4.2 Observer Design

To facilitate the observer design, we consider the following nonlinear system:

$$
\begin{aligned}
\dot{x} &= Ax + B_1 u_1 + B_2 f(x, u_2) + \zeta, \\
y &= Cx,
\end{aligned}
\tag{32}
$$

where $A \in \mathbb{R}^{n\times n}$ is the known system matrix, $B_1 \in \mathbb{R}^{n\times m_1}$ and $B_2 \in \mathbb{R}^{n\times m_2}$ are known input matrices, $u_1 \in \mathbb{R}^{m_1}$ and $u_2 \in \mathbb{R}^{\bar{m}_2}$ are known inputs, $C \in \mathbb{R}^{p\times n}$ is the corresponding output matrix, and $\zeta \in L_\infty$ is a bounded disturbance. The function $f(x, u_2) : \mathbb{R}^n \times \mathbb{R}^{\bar{m}_2} \rightarrow \mathbb{R}^{m_2}$ is partially unknown, which can be represented in a linear parameterized form: $f(x, u_2) = \psi(x, u_2)\Theta$, where $\psi : \mathbb{R}^n \times \mathbb{R}^{m_2} \rightarrow \mathbb{R}^{m_2\times l}$ is a known Lipschitz continuous function and $\Theta = \text{const.}$; $\Theta \in \mathbb{R}^l$ is an unknown parameter vector to be estimated.

Nevertheless, it is generally assumed that the pair (A, B_1) is controllable, (C, A, B_2) is of minimum phase, and the matrix (CB_2) has full rank. We can specifically represent the vehicle system (31) studied using the structure of Eq. (32):

[2] In road vehicles the air drag cannot be easily omitted. In particular, at higher speeds the value of F_{drag} has to be estimated. Further discussions on this matter are avoided here.

$$A = \begin{bmatrix} 0 & 1 \\ 0 & 0 \end{bmatrix}, \quad C = \begin{bmatrix} 1 & 0 \\ 0 & 1 \end{bmatrix}, \quad B_2 = \begin{bmatrix} 0 \\ 1 \end{bmatrix}, \quad \Theta = \begin{bmatrix} s \\ b \\ f \end{bmatrix}, \quad \psi = \begin{bmatrix} g \\ F_{\text{engine}} \\ \dot{x} \end{bmatrix}^T. \tag{33}$$

An adaptive observer with a *novel* parameter estimation algorithm will be designed, and has the following form:

$$\dot{\hat{x}} = A\hat{x} + B_1 u_1 + B_2 \psi \hat{\Theta} + L(y - C\hat{x}), \tag{34}$$

where $\hat{x}$ is the observer state, $\hat{\Theta}$ is the estimation of the unknown parameter, and L is the observer feedback gain which makes $A_c = A - LC$ stable. Then from Proposition 6.3 in [50], there exist positive definite matrices P and Q such that

$$A_c^T P + PA_c = -Q \quad \text{and} \quad PB_2 = C^T F^T. \tag{35}$$

We can define the error dynamics from Eqs. (32) and (34) as $\tilde{x} = x - \hat{x}$, so

$$\dot{\tilde{x}} = (A - LC)\tilde{x} + B_2 \psi \tilde{\Theta} + \zeta = A_c \tilde{x} + B_2 \psi \tilde{\Theta} + \zeta, \tag{36}$$

where $\tilde{\Theta} = \Theta - \hat{\Theta}$ is the estimation error vector.

Now the problem is to design an adaptive law to obtain $\hat{\Theta}$. Thus Eq. (32) can be expressed as

$$\dot{x} = \varphi + \Phi\Theta + \zeta, \tag{37}$$

where $\varphi = Ax + B_1 u_1$ and $\Phi = B_2 \psi$. Then we denote the following filtered variables:

$$\begin{aligned} k\dot{x}_f + x_f &= x, & x_f(0) &= 0, \\ k\dot{\varphi}_f + \varphi_f &= \varphi, & \varphi_f(0) &= 0, \\ k\dot{\Phi}_f + \Phi_f &= \Phi, & \Phi_f(0) &= 0, \end{aligned} \tag{38}$$

where $k > 0$. Moreover, an auxiliary filter is applied for the bounded disturbance ζ as $k\dot{\zeta}_f + \zeta_f = \zeta, \zeta_f(0) = 0$ (this variable is used only for analysis), so $\zeta_f \in L_\infty$. Therefore one can obtain from Eqs. (37) and (38) that

$$\dot{x}_f = \frac{x - x_f}{k} \quad \text{and} \quad \frac{x - x_f}{k} - \varphi_f = \Phi_f \Theta + \zeta_f. \tag{39}$$

These filtered variables will be used in the construction of the filtered regressor matrix $P(t)$ and vector $Q(t)$ as

$$\dot{P}(t) = -\ell P(t) + \ell \Phi_f^T(t)\Phi_f(t), \quad P(0) = 0,$$
$$\dot{Q}(t) = -\ell Q(t) + \ell \Phi_f^T(t)\left(\frac{x - x_f}{k} - \varphi_f\right), \quad Q(0) = 0, \tag{40}$$

where $\ell > 0$ can be taken as a constant forgetting factor. Consequently, we can find from Eq. (40) that

$$Q(t) = P(t)\Theta + \int_0^t \mathrm{e}^{-\ell(t-r)} \ell \Phi_f^T(r)\zeta_f dr = P(t)\Theta + \zeta_N, \tag{41}$$

with $\zeta_N = \int_0^t \mathrm{e}^{-\ell(t-r)} \ell \Phi_f^T(r)\zeta_f dr$. We know that Φ is bounded because it is Lipschitz continuous and x and u_2 are bounded. This together with $\zeta_f \in L_\infty$ implies that $P(t)$, $Q(t)$, and ζ_N are bounded.

Then the adaptive law can be given as

$$\dot{\hat{\Theta}} = \Gamma[\Phi^T F(y - C\hat{x}) - R]. \tag{42}$$

The leakage term R is given as follows and has a sliding mode term to ensure fast parameter convergence:

$$R = \omega_1 P\Omega \frac{P\hat{\Theta} - Q}{\left\| P\hat{\Theta} - Q \right\|} + \omega_2 P\Omega(P\hat{\Theta} - Q), \tag{43}$$

where ω_1 and ω_2 are positive scalars. Γ and Ω are positive definite diagonal gain matrices given by

$$\Gamma = \mathrm{diag}(\gamma_1, \ldots, \gamma_l), \ \Omega = \mathrm{diag}(\omega_1, \ldots, \omega_l). \tag{44}$$

Compared with other adaptive laws driven by the observer error $\varphi^T F(y - C\hat{x})$, the extra leakage term R with parameter error information, $P\hat{\Theta} - Q$, is employed; this could enhance the parameter convergence performance [1]. In particular, we incorporate a sliding mode term into Eq. (43) such that finite-time convergence is guaranteed as stated in the following theorem:

Theorem 4. *Given a system (32) for which the PE condition of Φ is satisfied, an adaptive observer (34) is constructed and the adaptation law is provided in Eqs. (42)–(43) for which the unknown parameter vector Θ is estimated via $\hat{\Theta}$. The following can be shown:*

1. *For $\zeta \neq 0$, the adaptive observer and parameter estimation are ultimately uniformly bounded. Moreover, the observer error and parameter estimation error converge to a small bounded set in finite time.*
2. *For $\zeta = 0$, the adaptive observer and parameter estimation are exponentially stable. Moreover, finite-time convergence of the estimated parameters can be guaranteed.*

The proof of Theorem 4 is obtained by selecting the following Lyapunov function:

$$V = \frac{1}{2}\tilde{x}^T P \tilde{x} + \frac{1}{2} W^T P^{-1} \Gamma^{-1} P^{-1} W,$$

where $W = Q - P\hat{\Theta} = P\Theta + \zeta_N - P\hat{\Theta}$ so that $\tilde{\Theta} = P^{-1}[W - \zeta_N]$. The remainder of the proof can be found in [26].

4.3 Experimental Results

We will present the practical implementation of the proposed novel estimation algorithm on a previously built small model car [51] (see Fig. 2) to obtain the vehicle mass (nominal weight approximately 10kg) and the road gradient. Solely for reference, an inclinometer, SCA61T, was incorporated for measurement of the gradient. This has not been used in the algorithm. As a dedicated rapid prototyping system, a dSPACE MicroAutoBox is used to run the proposed estimation and control methods. The power train is enabled by a brushless DC motor (EC-i 40, Maxon) together with an EPOS 24/5 motor driver. The brushless motor is run in current-control mode using the MicroAutoBox. Considering that the current and motor torque are proportional, this subsequently produces the driving force F_{engine} for vehicle speed control. Agilent HEDL5540 encoders connected to the motor and the front wheel permit speed measurement. The test track is

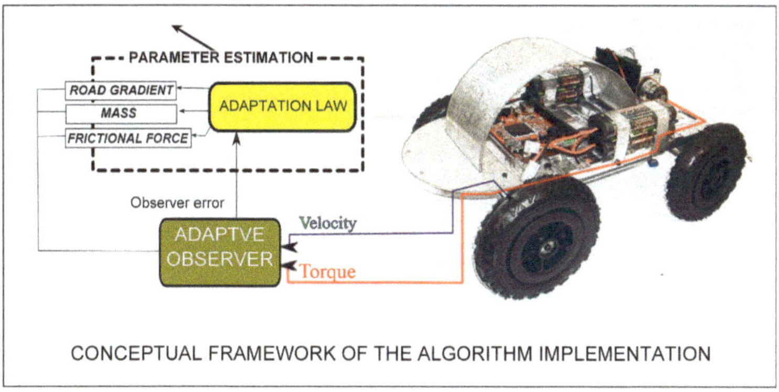

Fig. 2 Parameter estimation algorithm implementation concepts. *(Source: M. Mahyuddin, J. Na, G. Herrmann, X. Ren, P. Barber, Adaptive observer-based parameter estimation with application to road gradient and vehicle mass estimation, IEEE Trans. Ind. Electron. 61 (6) (2013) 2851–2863).*

built from two stiff wooden planks of 2 m length each. They are tilted and bolted together, allowing also variation of slope profiles (see Fig. 1).

The parameters used in the experiments are tuned as detailed in [26], and the performance of our proposed novel algorithm is compared against that of the conventional LS-based algorithm (equivalently setting the term $R(t) = 0$). The comparative results for parameter estimation along two slopes are presented in Figs. 3 and 4. It is evident that our proposed algorithm is significantly better than the LS algorithm in terms of accuracy. This is quantitatively assessed in [26].

A further test was conducted to evaluate the mass estimation performance of the proposed algorithm. Step changes of mass for the vehicle were introduced. Fig. 5 shows the graph for the mass estimate. Two extra preloaded masses, each weighing 1.85 kg, were applied, giving a total vehicle mass of 13.70 kg (the net mass is 10 kg). Consequently, the two load masses of 1.85 kg were removed within two time intervals. The original mass estimation signal (shown by the green line in the plot) was filtered by a Butterworth filter with a cutoff frequency of 3 rad/s for low-pass filtering. It is shown in Fig. 5 that the changes in the vehicle mass are sufficiently well reflected by the novel estimation algorithm. Differences in mass are picked up within 5–10 s, which is due to the finite-time convergence feature of the novel estimation algorithm.

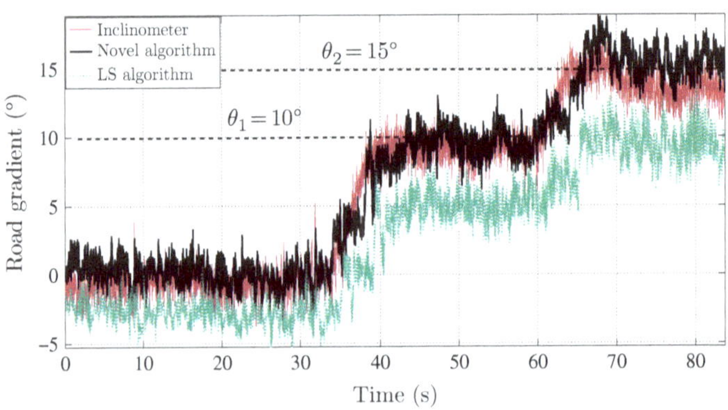

Fig. 3 Gradient estimate for slope profile 1 for a velocity of 0.14 m/s. *LS*, least squares. *(Source: M. Mahyuddin, J. Na, G. Herrmann, X. Ren, P. Barber, Adaptive observer-based parameter estimation with application to road gradient and vehicle mass estimation, IEEE Trans. Ind. Electron. 61 (6) (2013) 2851–2863).*

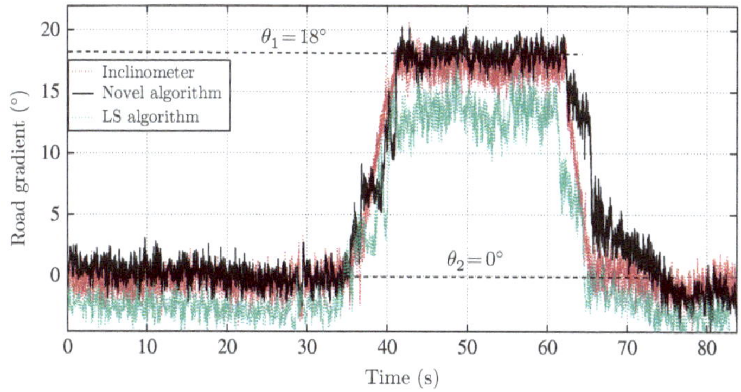

Fig. 4 Gradient estimate for slope profile 2 for a velocity of 0.14 m/s. *LS*, least squares. *(Source: M. Mahyuddin, J. Na, G. Herrmann, X. Ren, P. Barber, Adaptive observer-based parameter estimation with application to road gradient and vehicle mass estimation, IEEE Trans. Ind. Electron. 61 (6) (2013) 2851–2863).*

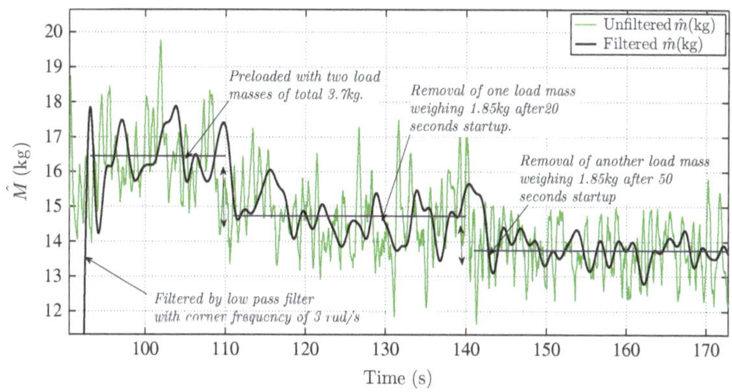

Fig. 5 Mass estimation, $\hat{M}$ (kg). *(Source: M. Mahyuddin, J. Na, G. Herrmann, X. Ren, P. Barber, Adaptive observer-based parameter estimation with application to road gradient and vehicle massestimation, IEEE Trans. Ind. Electron. 61 (6) (2013) 2851–2863).*

5 EXAMPLE IN CONTROL OF ROBOTIC SYSTEMS

This section will showcase the development of an adaptive control based on the proposed novel adaptive law for a humanoid robotic arm [33]. Adaptive robotic control research reached a turning point in the 1980s, when a linear parameterization scheme for nonlinear robotics was explicitly proposed [52]. Craig et al. [53] presented an adaptive computed-torque control and proved its convergence. The approach, however, has limited practical applicability because of the requirement of joint acceleration measurements and an inverse of the estimated inertia information. Measurements of acceleration can be inherently noisy. Moreover, although the adaptive control [13, 54] validated that the tracking error and prediction/estimation error can be combined to improve the performance, only exponential convergence can be theoretically claimed in [13, 52]. Inspired by these observations, we will extend the adaptive algorithm in [21, 33] to the control of a practical robotic arm. Consequently, we can guarantee robust adaptive control of the arm and finite-time convergence of parameter estimation. Moreover, inspired by Lewis et al. [14], we use strictly proper and monic stable filters on the torque dynamics such that the acceleration measurements in the regressor formulation are avoided. Finally, we also prove that the required PE condition can be satisfied by selecting the control demand signal so that it fulfills the sufficient richness condition.

5.1 Problem Formulation

The humanoid Bristol Elumotion Robotic Torso II (BERT II) robotic arm (see Fig. 6) is chosen as a plant. Its nonlinear and coupled system dynamics are presented as

$$M(q)\ddot{q} + V(q,\dot{q})\dot{q} + G(q) + T_d = \tau, \tag{45}$$

where $q, \dot{q}, \ddot{q} \in \mathbb{R}^n$ are the robot joint position, velocity, and acceleration, respectively, n is the number of degrees of freedom of the robot, $\tau \in \mathbb{R}^n$ is the input torque vector, T_d denotes the external disturbance, $M(q) \in \mathbb{R}^{n\times n}$, $M(q) > 0$, is the inertia matrix, $V(q,\dot{q})\dot{q}$ defines the Coriolis/centripetal torque including viscous and nonlinear damping, and $G(q) \in \mathbb{R}^n$ is the gravity torque.

To facilitate the control design, the following essential properties of Eq. (45) are presented [1, 14, 55]:

Property 1. *The matrix* $\dot{M}(q) - 2V(q,\dot{q})$ *is skew symmetric—that is,* $\xi^T\left(\dot{M}(q) - 2V(q,\dot{q})\right)\xi = 0$*—which implies the passive mapping from the input* τ *to the output* $\dot{q}$.

Property 2. *The dynamics in Eq. (45) can be linearly parameterized as*

$$M(q)\ddot{q} + V(q,\dot{q})\dot{q} + G(q) = \phi(q,\dot{q},\ddot{q})\Theta, \tag{46}$$

where $\Theta \in \mathbb{R}^l$ *is the unknown constant vector and* $\phi(q,\dot{q},\ddot{q}) \in \mathbb{R}^{n\times l}$ *is the known regressor matrix, which is a Lipschitz continuous function of* $(q,\dot{q},\ddot{q})$.

Property 3. *The demand* q_d *is continuously differentiable with respect time* t *and is sufficiently rich with respect to* $\phi(q_d,\dot{q}_d,\ddot{q}_d)$ *over a finite interval* $[t, t+T]$ *with length* $T > 0$ *that there are at least* l *time instances* t_i, $i = 1, 2, \ldots$ *so that for*

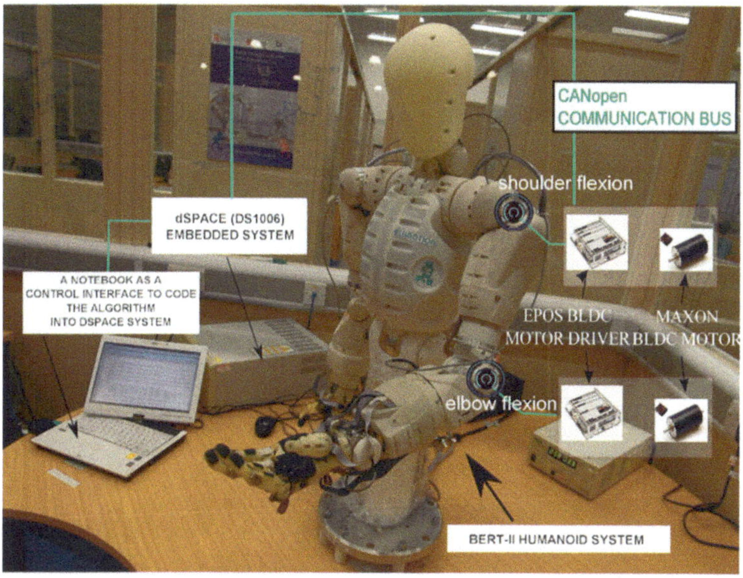

Fig. 6 Bristol Elumotion Robotic Torso II (BERT II) robotic arm system (BERT II). CANopen communication is established between dSPACE and the robot's actuator system. *BLDC*, brushless DC. *(Source: M. Mahyuddin, S. Khan, G. Herrmann, A novel robust adaptive control algorithm with finite-time online parameter estimation of a humanoid robot arm, Robot. Auton. Syst. 62 (3) (2014) 294–305).*

$t_1, t_2, \ldots, t_l \in [t, t+T]$

$$\Phi(q_d(\cdot)) = [(\phi(q_d(t_1)))^T, (\phi(q_d(t_2)))^T, \ldots, (\phi(q_d(t_l)))^T] \tag{47}$$

is of rank l. Thus there exists a constant $\delta > 0$ such that [3, 5]

$$\Phi^T(q_d(\cdot))\Phi(q_d(\cdot)) \geq \delta I, \quad \delta > 0. \tag{48}$$

Remark 3. According to Property 3, one may verify that

$$\int_t^{t+T} \phi^T(q_d(\nu), \dot{q}_d(\nu), \ddot{q}_d(\nu))\phi(q_d(\nu), \dot{q}_d(\nu), \ddot{q}_d(\nu))d\nu > \tilde{\delta} I, \tag{49}$$

where $\tilde{\delta} > 0$ is a constant. Property (49) implies the PE condition of the system studied, once the tracking controller follows this trajectory. This will be discussed later.

The original regressor matrix ϕ (46) has the acceleration as its argument. Thus in our following adaptive control design, the regressor matrix will be reformulated to eradicate the use of the joint acceleration. A similar approach was used to avoid acceleration measurements in [14], which motivates our present work.

5.2 Adaptive Control Design

To achieve tracking control, we define the error vectors as

$$u = \dot{q}_d - \Lambda e, \quad \dot{u} = \ddot{q}_d - \Lambda \dot{e}, \quad e = q_d - q, \quad \dot{e} = \dot{q}_d - \dot{q}, \tag{50}$$

where $e, \dot{e} \in \mathbb{R}^n$ denote the tracking errors between the desired joint position $q_d(t)$ and velocity $\dot{q}_d(t)$ and the system dynamics, and $\Lambda = \text{diag}(\lambda_1, \lambda_2, \ldots, \lambda_n)$ is a positive diagonal matrix. Moreover, a sliding mode-like error can be defined for each joint as

$$r_i = \dot{e}_i + \lambda_i e_i, \quad i = 1, \ldots, n, \quad r = [r_1, \ldots, r_n]^T, \tag{51}$$

where λ_i is the sliding mode gain for each respective joint that defines the sliding manifold r_i.

For output tracking control the control input is designed as

$$\tau = \hat{M}(q)\dot{u} + \hat{V}(q, \dot{q})u + \hat{G}(q) + K_{r_1} \frac{r}{\|r\|} + K_{r_2} r, \tag{52}$$

where K_{r_1} and K_{r_2} are scalar tracking gains. The matrices $\hat{M}(q)$ and $\hat{V}(q, \dot{q})$ are the estimation of $M(q)$ and $V(q, \dot{q})$.

Then substituting Eq. (52) into system (45) by using Eq. (50), we can compute the closed-loop computed torque equation as

$$\begin{aligned} \tilde{M}(q)\dot{u} + \tilde{V}(q, \dot{q})u + \tilde{G}(q) + T_d &= M(q)\dot{r} + V(q, \dot{q})r + K_{r_1} \frac{r}{\|r\|} \\ &+ K_{r_2} r = \phi(q, \dot{q}, u, \dot{u})\tilde{\Theta} + T_d, \end{aligned} \tag{53}$$

where $\tilde{M}(q) = M(q) - \hat{M}(q)$, $\tilde{V}(q, \dot{q}) = V(q, \dot{q}) - \hat{V}(q, \dot{q})$, and $\tilde{G}(q) = G(q) - \hat{G}(q)$ are the estimation errors and $\phi(q, \dot{q}, u, \dot{u}) \in \mathbb{R}^{n \times l}$ is the regressor defined in Property 2, with arguments u and $\dot{u}$ instead of $\ddot{q}$. It is clear that this new regressor is the function of joint position q, velocity $\dot{q}$, and the command vectors u and $\dot{u}$. The next section provides the basis to derive the adaptive law for $\hat{\Theta}$.

5.3 Design of the Adaptive Law

By filtering the torque measurement, one can avoid the acceleration measurements in the regressor $\phi(q, \dot{q}, \ddot{q})$. Hence the dynamics in Eq. (45) can be represented as

$$\tau = \dot{f} + T_d, \tag{54}$$

where

$$\dot{f} = \frac{d}{dt}\left[M(q)\dot{q}\right], \tag{55}$$

$$h = -\dot{M}(q)\dot{q} + V_m(q,\dot{q})\dot{q} + G(q) = h_1 + h_2, \tag{56}$$

with $h_1 = -\dot{M}(q)\dot{q}$ and $h_2 = V_m(q,\dot{q})\dot{q} + G(q)$. Moreover, on the basis of Property 2, these split terms are further parameterized as

$$f = M(q)\dot{q} = \varphi_{m1}(q,\dot{q})\Theta, \tag{57}$$

$$h_1 = -\dot{M}(q)\dot{q} = \varphi_{m2}(q,\dot{q})\Theta, \tag{58}$$

$$h_2 = V_m(q,\dot{q})\dot{q} + G(q) = \varphi_{vg}(q,\dot{q})\Theta. \tag{59}$$

Then we filter the terms $\varphi_{m1}, \varphi_{m2}$, and φ_{vg} and obtain

$$\kappa\dot{\varphi}_{m1f}(q,\dot{q}) + \varphi_{m1f}(q,\dot{q}) = \varphi_{m1}(q,\dot{q}), \quad \phi_{m1_f}(0) = 0, \tag{60}$$

$$\kappa\dot{\varphi}_{m2f}(q,\dot{q}) + \varphi_{m2f}(q,\dot{q}) = \varphi_{m2}(q,\dot{q}), \quad \phi_{m2_f}(0) = 0, \tag{61}$$

$$\kappa\dot{\varphi}_{vgf}(q,\dot{q}) + \varphi_{vgf}(q,\dot{q}) = \varphi_{vg}(q,\dot{q}), \quad \phi_{vg_f}(0) = 0. \tag{62}$$

To aid the analysis in the case of a disturbance, the additive bounded disturbance T_d is also filtered by

$$\kappa\dot{T}_{df} + T_{df} = T_d, \quad T_{df}(0) = 0.$$

Finally, the filtered torque can be given as

$$\kappa\dot{\tau}_f + \tau_f = \tau, \quad \tau_f(0) = 0. \tag{63}$$

Thus the filtered computed-torque equation by our considering Eqs. (45) and (46) can be rewritten as

$$\phi_f(q,\dot{q})\Theta + T_{df} = \left[\frac{\varphi_{m1}(q,\dot{q}) - \varphi_{m1f}(q,\dot{q})}{\kappa} + \varphi_{m2f}(q,\dot{q}) + \varphi_{vgf}(q,\dot{q})\right]\Theta + T_{df} = \tau_f, \tag{64}$$

where $\phi_f(q,\dot{q}) = [\frac{\varphi_{m1}(q,\dot{q}) - \varphi_{m1f}(q,\dot{q})}{\kappa} + \varphi_{m2f}(q,\dot{q}) + \varphi_{vgf}(q,\dot{q})] \in \mathbb{R}^{n\times l}$ is the filtered regressor, $\Theta \in \mathbb{R}^l$. In comparison with Eq. (46), the filtered regressor (64) clearly avoids the acceleration measurements, which are sometimes practically unavailable.

For the filtered system (64), the suggested algorithm in Section 3 can be further developed. For this purpose, we define a filtered matrix $P(t)$ and vector $Q(t)$ as

$$\dot{P}(t) = -\ell P(t) + \ell\phi_f^T(q,\dot{q})\phi_f(q,\dot{q}), \quad P(0) = \varrho I, \tag{65}$$

$$\dot{Q}(t) = -\ell Q(t) + \ell\phi_f^T(q,\dot{q})\tau_f, \quad Q(0) = 0, \tag{66}$$

where $\ell \in \mathbb{R}^+$ denotes the forgetting factor.

Then similar to what was done in Section 3, by solving the solution of $Q(t)$ and $P(t)$ in Eqs. (65) and (66), one can have

$$Q(t) = P(t)\Theta + T_{dN} - e^{-\ell t}\varrho\Theta, \tag{67}$$

where $T_{dN} = \int_0^t e^{-\ell(t-r)}\ell\phi_f^T(r)T_{df}(r)dr$.

The initial condition of $Q(t)$ is $Q(0) = 0$, whereas the initial condition of $P(t)$—that is, $P(0)$—is set to ϱI to avoid any potential singularity, where $\varrho > 0$ is some constant and $I \in \mathbb{R}^{l\times l}$ is the identity matrix. It is apparent that $P(t) \geq \varrho I e^{-\ell t}$ for $\varrho > 0$. Practically, ϱ should be chosen small. Further details about $P(t)$ can be found in [33].

The control law in Eq. (52) can be rewritten as

$$\tau = \phi(q, \dot{q}, u, \dot{u})\hat{\Theta} + K_{r_1}\frac{r}{\|r\|} + K_{r_2}r, \tag{68}$$

where $\hat{\Theta}$ is the estimation of the parameter vector Θ, which can be updated in terms of following adaptive law:

$$\dot{\hat{\Theta}} = \Gamma\phi^T(q, \dot{q}, u, \dot{u})r - \Gamma R, \tag{69}$$

where Γ is a diagonal, positive definite gain matrix—that is, $\Gamma = \mathrm{diag}(\gamma_1, \ldots, \gamma_l)$. The new leakage term $R(t)$ used to ensure fast parameter convergence is given by

$$R(t) = \omega_1\frac{P(t)\hat{\Theta} - Q(t)}{\left\|P(t)\hat{\Theta} - Q(t)\right\|} + \omega_2(P(t)\hat{\Theta} - Q(t)), \tag{70}$$

where ω_1 and ω_2 are positive scalars which can be chosen large enough to achieve robust stability. We can prove that finite-time convergence of the parameter error $\tilde{\Theta}$ to zero when $T_d = 0$. The following theorem provides the main result of this section:

Theorem 5. *Consider the adaptive control* (68) *and the adaptation law* (69) *for the robotic system with a suitably chosen bounded twice continuously differentiable, sufficiently rich reference q_d fulfilling Property 3. Then:*

1. *The trajectories $q(t)$ converge to the demand q_d in a semiglobal sense—that is, there exists a sliding mode gain matrix Λ and gains $K_{r_1}, K_{r_2}, \omega_1, \omega_2 \in \mathbb{R}^+$ large enough such that exponential convergence is guaranteed for bounded disturbance T_d.*
2. *The estimation error $\tilde{\Theta}$ will converge to zero in finite time when $T_d = 0$.*
3. *The estimation error $\tilde{\Theta}$ will converge to an ultimately bounded residual set in the case of $T_d \neq 0$.*

Theorem 5 can be proven with the Lyapunov function $V(t) = V_r + V_\Theta = \frac{1}{2}r^T M(q)r + \frac{1}{2}\tilde{Q}^T P^{-1}\Gamma^{-1}P^{-1}\tilde{Q}$, whereby $\tilde{Q}(t) = Q(t) - P(t)\hat{\Theta} = P(t)\tilde{\Theta} + T_{dN} - e^{-\ell t}\varrho I\Theta$. The first step is to show that the Lyapunov function $V_r(t)$ converges to values close to 0 within finite time. It follows that the estimation error $P^{-1}\tilde{Q}(t)$ and subsequently the tracking error r and converge to 0 in finite time. Clearly for the case of no disturbance, $\tilde{\Theta} \to 0$ in finite time. The details of the proof can be found in [33].

5.4 Results

The proposed novel robust adaptive control is implemented at a sampling frequency of 1 kHz for the Bristol Elumotion Robotic Torso II (BERT II) humanoid arm (shown in Fig. 6) by use of a dSPACE system. The seven degrees of freedom BERT II robotic arm is driven by Maxon brushless DC motors configured at each joint. In this experiment, the two joints corresponding to the shoulder flexion and the elbow flexion motion are to be controlled. The mass of each robot link, the *upper arm* M_1 and the *forearm* M_2, are the parameters in the mass matrix to be estimated—that is, $\hat{\Theta} = [\hat{M}_1, \hat{M}_2]^T$ defines the estimated parameter vector. The actual values of the approximated model of the BERT II arm are shown in Table 1.

Table 2 shows all the tuning parameters used in the proposed novel algorithm in the simulation and the experiment. The parameter estimation weight Γ are chosen to increase the sensitivity of the algorithm to the sliding error dynamics and also to increase the convergence rate. The forgetting factor ℓ is carefully selected so as to ensure that the inverse of $P(t)$ in Eq. (65) exists. For this, ℓ should not be chosen too large to permit sufficient data collection and to avoid too fast forgetting for $P(t)$ and $Q(t)$. However, a large forgetting factor ℓ within the regressor matrix $P(t)$ and vector $Q(t)$ is desirable to allow there to be faster estimation

Table 1 The model parameters of the Bristol Elumotion Robotic Torso II (BERT II) arm

Description	Link 1	Link 2
Mass (kg)	2.35	3
Link length (m)	0.2735	0.44
Radius of approximate cylindrical body (m)	0.05	0.05

Source: M. Mahyuddin, S. Khan, G. Herrmann, A novel robust adaptive control algorithm with finite-time online parameter estimation of a humanoid robot arm, Robot. Auton. Syst. 62 (3) (2014) 294–305.

Table 2 Adaptation mechanism parameters

Parameter description	Symbols	Values
Parameter estimation adaptive weights, Γ	γ_1	15
	γ_2	15
Forgetting factor (Eqs. 65 and 66)	ℓ	0.1
Filter constant (Eqs. 60–63)	κ	0.02
Sliding mode adaptive weights, Ω (Eq. 70)	ω_1	40
	ω_2	1
Control gains, K_{r_i} (Eq. 52)	K_{r_1}	10
	K_{r_2}	5
Sliding mode gain (Eqs. 50 and 51)	λ	0.5
	Λ	$0.5I$

Source: M. Mahyuddin, S. Khan, G. Herrmann, A novel robust adaptive control algorithm with finite-time online parameter estimation of a humanoid robot arm, Robot. Auton. Syst. 62 (3) (2014) 294–305.

error convergence. The control gains, K_{r_1} and K_{r_2}, should be chosen large enough to effectively reject disturbance.

5.4.1 Simulation Results

Fig. 7 provides the simulation results of the novel robust adaptive control and other two standard adaptive control laws—that is, a gradient descent-based and LS-based adaptive control [5] with the case of no disturbance, $T_d = 0$. Excellent tracking performances for both elbow flexion and shoulder flexion is obtained with the suggested novel adaptive law. Moreover, the estimates $\hat{M}_1$ and $\hat{M}_2$ obtained from the novel adaptive scheme converge faster to the true values (ie, 2.35 and 3 kg) when compared with the traditional gradient descent and LS algorithms (see Fig. 7C and D.).

Moreover, the robustness against external disturbances is verified. An additive bounded disturbance, T_d, is introduced in the robotic arm system. T_d comprises band-limited white noise with magnitude power of 0.1 and a sampling time $t_s = 0.02$ s. The inherent robustness property (both in the tracking performance and in the parameter estimation) of the proposed algorithm is apparent in contrast to that of the gradient descent-based and LS-based algorithms as seen in Fig. 8. It is interesting to see that the estimation resulting from the LS-based algorithm is quite sensitive to changes in the demand joint angle. Changes exceeding more than ± 15 degrees during the control transient have a significant effect on convergence of the Γ matrix (13). This is evident from Fig. 8 for the time interval between 0 and 10 s. For this, note that the design parameter, β, in the

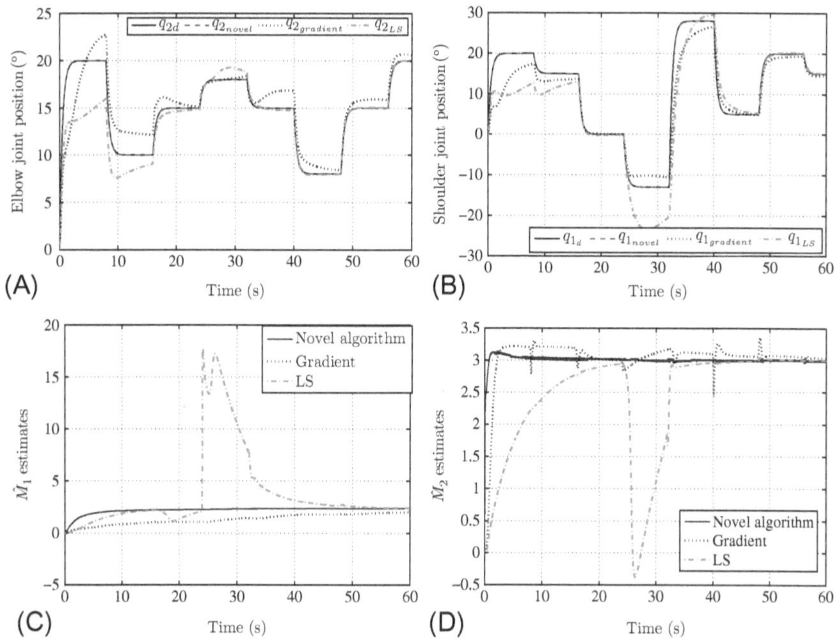

Fig. 7 Novel adaptive control of the manipulator in the case of no disturbance, $T_d = 0$. *LS*, least squares. (A) Control response of elbow joint. (B) Control response of shoulder joint. (C) Estimation of M_1. (D) Estimation of M_2. *(Source: M. Mahyuddin, S. Khan, G. Herrmann, A novel robust adaptive control algorithm with finite-time online parameter estimation of a humanoid robot arm, Robot. Auton. Syst. 62 (3) (2014) 294–305).*

LS-based algorithm (13) governs the speed of the parameter convergence (ie, it has to be carefully adjusted). A compromise should be made between robustness and fast convergence. In contrast, Fig. 8A and B shows that the tracking responses of both the elbow flexion motion and the shoulder flexion motion for the novel adaptation algorithm are excellent even in the presence of a disturbance.

The parameter estimation profile of the proposed algorithm (in contrast to the gradient descent-based and LS-based approaches) is evidently very good, exemplifying its ability to reject disturbances as seen in Fig. 8C and D. For the novel algorithm, the estimate for $\hat{M}_1$ converges in less than 20 s, while for $\hat{M}_2$, it achieves convergence in a mere 3 s. By contrast, the gradient descent-based algorithm's estimate for $\hat{M}_1$ converges at a much slower rate. The estimate of $\hat{M}_2$ by the LS-based algorithm shows a significant susceptibility to the disturbance. Although the LS-based algorithm achieves a better estimation performance in comparison with the

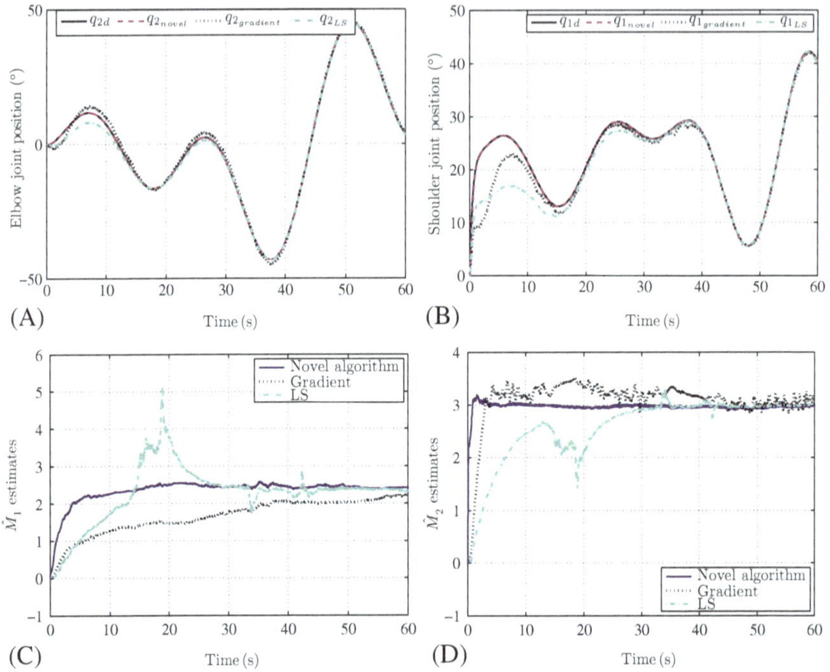

Fig. 8 Tracking and estimation performance of the novel adaptive law for the manipulator (*solid line* for estimation) in contrast to the gradient descent-based approach (*dashed line* for estimation) and least squares (LS)-based approach (*dash-dotted line* for estimation) for additive bounded disturbance. (A) Control response of elbow joint. (B) Control response of shoulder joint. (C) Estimation of M_1. (D) Estimation of M_2. *(Source: M. Mahyuddin, S. Khan, G. Herrmann, A novel robust adaptive control algorithm with finite-time online parameter estimation of a humanoid robot arm, Robot. Auton. Syst. 62 (3) (2014) 294–305).*

gradient descent-based approach for the mass estimation of the forearm mass $\hat{M}_1$, it suffers greatly when estimating the mass of the upper arm $\hat{M}_2$.

From the simulation results, it is evident that of the three algorithms, the LS-based algorithm is the worst in terms of tracking and estimation. Therefore we practically tested the performance of the two best algorithms. The practical results are illustrated in the following subsection.

5.4.2 Experimental Results

Fig. 9A provides the control results of the suggested novel adaptive control method for the BERT II humanoid system. The test results of the gradient descent-based approach are given in Fig. 9B. A sinusoidal command signal is used as the reference signal. The tracking response for the joint elbow

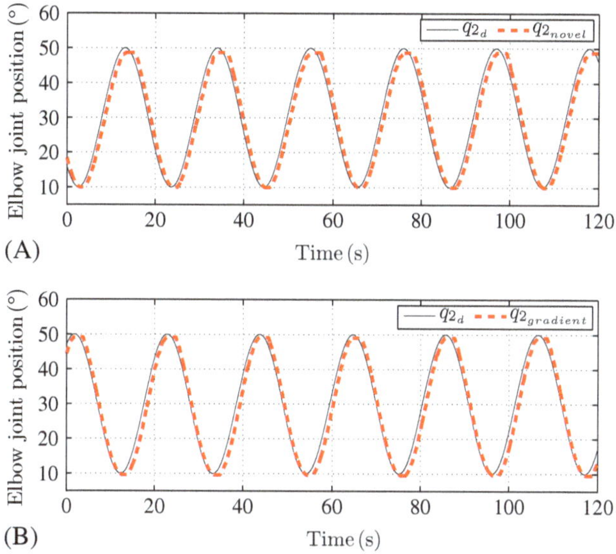

Fig. 9 Tracking response of the suggested adaptive control (A) and the gradient descent-based approach (B) for elbow flexion of Bristol Elumotion Robotic Torso II (BERT II). *(Source: M. Mahyuddin, S. Khan, G. Herrmann, A novel robust adaptive control algorithm with finite-time online parameter estimation of a humanoid robot arm, Robot. Auton. Syst. 62 (3) (2014) 294–305).*

flexion is comparable between the novel adaptive control algorithm and the gradient descent-based approach. The integral absolute error, $E_{I\bar{A}E} = \frac{\sum_T |q_i(t) - q_{di}(t)|}{T}$, computed for elbow tracking shows that the novel algorithm is slightly better (ie, its respective $E_{I\bar{A}E}$ is 2.0066 in comparison with that of the gradient descent-based approach, which is 2.1547). In addition, the link mass estimation of the BERT II robotic arm in Fig. 10A hovers around acceptable bounded values considering the fact that the proposed algorithm's regressor is formulated under the assumption that the robotic arm has a uniform cylindrical shape (which is not the real case). As in the gradient descent-based approach, the mass estimation settles at incorrect values below zero, as shown in Fig. 10B.

As in the case of successive step demands, Fig. 11 reveals the tracking profile for both joints (eg, elbow and shoulder flexion). The compensation from the fast parameter estimation approach is advantageous to the control effort. Moreover, the required torque on the motors validates that the control signals are bounded.

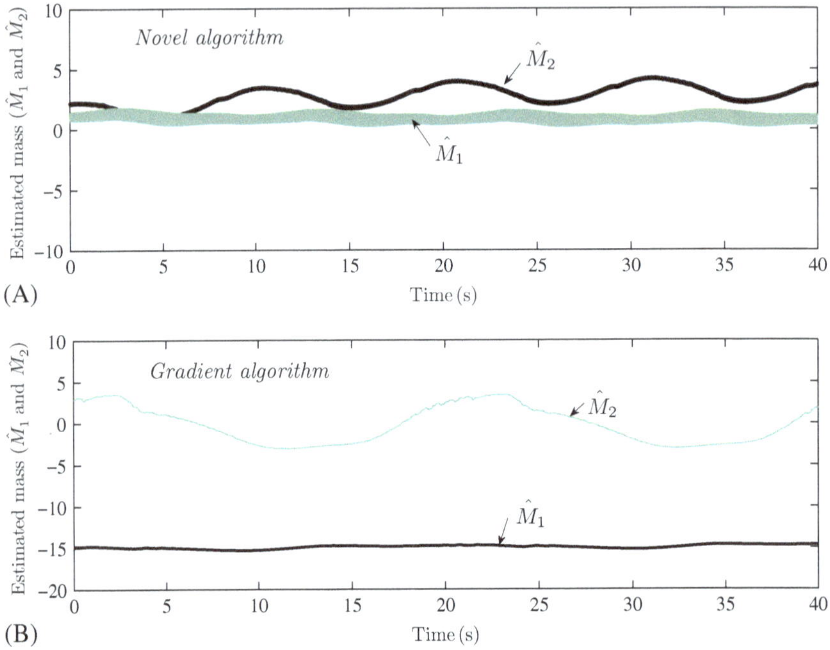

Fig. 10 The link mass estimation performance of the novel algorithm (A) in comparison with that of the gradient descent-based approach (B). *(Source: M. Mahyuddin, S. Khan, G. Herrmann, A novel robust adaptive control algorithm with finite-time online parameter estimation of a humanoid robot arm, Robot. Auton. Syst. 62 (3) (2014) 294–305).*

6 EXAMPLE IN ADAPTIVE OPTIMAL CONTROL

This section presents the application of the new adaptive algorithm as part of an adaptive optimal control design of continuous-time systems with uncertain dynamics [24]. The optimal control problem is usually studied by use of Pontryagin's minimum principle or by the solving of the Hamilton-Jacobi-Bellman (HJB) equation. In such a case, off-line operation is usually employed and the system dynamics are assumed to be completely known. On the other hand, adaptive control methods [56] have been used to address the problem of unknown system parameters in an online manner. However, classical adaptive control approaches are not optimal in general with respect to a specific optimization index.

As a consequence of the issues described above, there has been a desire to create adaptive optimal control approaches. Bioinspired reinforcement learning [57] or approximate dynamic programming (ADP) was proposed. Although ADP has been well developed for the design of discrete-time

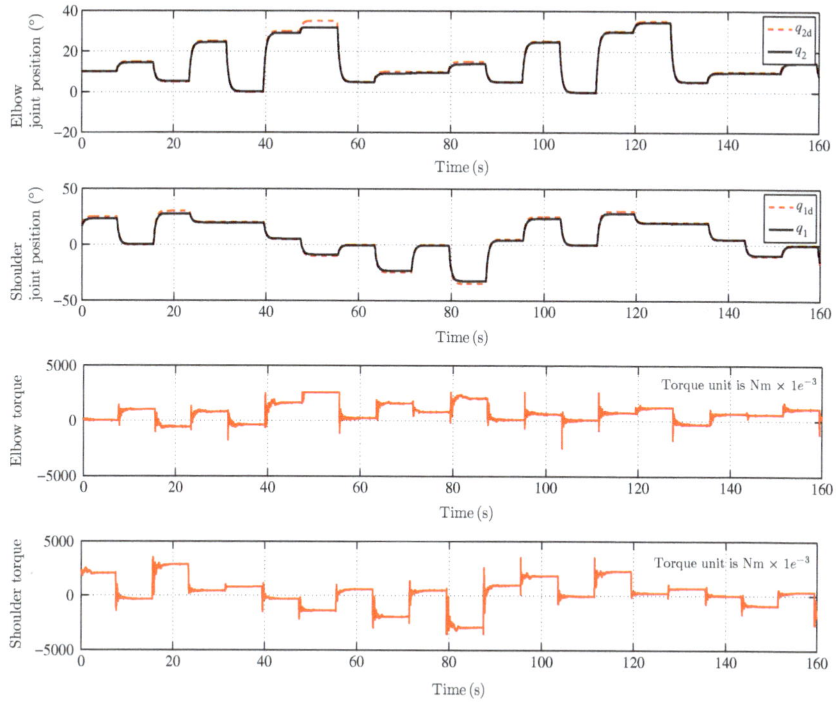

Fig. 11 Tracking control profiles and the applied torque of the novel adaptive control for the shoulder and elbow flexion joint of Bristol Elumotion Robotic Torso II (BERT II). *(Source: M. Mahyuddin, S. Khan, G. Herrmann, A novel robust adaptive control algorithm with finite-time online parameter estimation of a humanoid robot arm, Robot. Auton. Syst. 62 (3) (2014) 294–305).*

adaptive optimal control [58], the extension to continuous-time systems is challenging in proving stability and convergence of a model-free algorithm, which can be solved online. In particular, most ADP-based optimal control methods have been developed to address the stabilization or regulation problem, while only a few results have been reported for optimal tracking control [59].

Thus the main objective of this section is to propose a new ADP algorithm for solving the optimal tracking control problem of nonlinear continuous-time systems with uncertainties. The requirement of complete or at least partial knowledge of the system dynamics is eliminated by the construction of an adaptive NN identifier. The suggested tracking control consists of an adaptive steady-state control that can maintain the desired steady-state tracking response and an adaptive optimal control that can stabilize the tracking error in an optimal manner. To design such an optimal control, a critic NN is used to estimate *online* the solution of the HJB equation, which is then used to derive the optimal control action. Thus the actor NN widely used in the ADP literature is not necessary. This leads to a novel "identifier-critic"-based ADP framework. In particular, we will show how to incorporate the novel adaptive algorithm with the parameter estimation error into the online learning of the identifier and critic NN. Hence even when there are NN approximation errors, the estimated NN weights are guaranteed to converge to a small bounded set around their true values under a standard PE condition. Finally, we prove that the control obtained guarantees the tracking of a desired trajectory, while also asymptotically converging to a small bounded set around the optimal policy. Specifically, when there are no disturbances and approximation errors, convergence of the system parameters and the control policy to their optimal values can be achieved.

6.1 Problem Statement

Considering for Eq. (1) a special structure, we study the following continuous-time nonlinear system:

$$\dot{x} = Ax + Bu + \Theta^T \Phi(x) + \varepsilon, \tag{71}$$

where $x \in \mathbb{R}^n$ and $u \in \mathbb{R}^m$ are the measurable state and the input of the system studied, $A \in \mathbb{R}^{n\times n}$ and $B \in \mathbb{R}^{n\times m}$ are known matrices, $\Theta \in \mathbb{R}^{p\times n}$ is the unknown weight matrix, $\varepsilon \in \mathbb{R}^n$ is a bounded disturbance, and $\Phi(x) \in \mathbb{R}^p$ is a nonlinear regressor function vector, which is Lipschitz continuous such that $\left\|\Phi(x) - \Phi(y)\right\| \leq \kappa \left\|x - y\right\|$ holds for some positive constant $\kappa > 0$.

The purpose is to find an adaptive controller u which ensures that the system state x tracks a given command x_d and minimizes the following infinite-horizon cost function [60] (in a suboptimal sense):

$$V^*(e(t)) = \min_{u(\tau)\in\Psi(\Omega)} \int_t^\infty r(e(\tau), u(\tau))d\tau, \tag{72}$$

where $\Psi(\Omega)$ denotes the set of admissible control policies, $e = x - x_d$ defines the tracking error, and $r(\cdot,\cdot) : \mathbb{R}^n \times \mathbb{R}^m \to \mathbb{R}^n$, $r(e(\tau), u(\tau)) \geq 0$, is the utility function; the utility function will be defined later. The tracking error e rather than the system state x is used in the cost function (72) because the tracking control rather than the regulation problem is studied in this section.

Since the dynamics of Eq. (71) are mostly unknown, the optimal tracking control design can be divided into two steps [59]:

1. propose an adaptive identifier based on the input-output data to identify the unknown system parameters for Eq. (71);
2. design an adaptive optimal tracking controller using the identified dynamics and ADP methods.

6.2 Development of the Adaptive Identifier

In this section an adaptive identifier is suggested to reconstruct the unknown dynamics from available input-output measurements. To determine the unknown parameters Θ, we define the filtered version of variables x_f, u_f, and Φ_f as x, u, and f by

$$\begin{cases} k\dot{x}_f + x_f = x, & x_f(0) = 0, \\ k\dot{u}_f + u_f = u, & u_f(0) = 0, \\ k\dot{\Phi}_f + \Phi_f = \Phi, & \Phi_f(0) = 0, \end{cases} \tag{73}$$

where $k > 0$, $k \in \mathbb{R}$, is a constant filter parameter.

Then for any positive scalar $\ell > 0$, one can obtain the filtered and *integrated* regressor matrices $P_1 \in \mathbb{R}^{p\times p}$ and $Q_1 \in \mathbb{R}^{p\times n}$ as

$$\begin{cases} \dot{P}_1 = -\ell P_1 + \Phi_f \Phi_f^T, \quad P_1(0) = 0, \\ \dot{Q}_1 = -\ell Q_1 + \Phi_f \left[\frac{x-x_f}{k} - Ax_f - Bu_f\right]^T, \quad Q_1(0) = 0. \end{cases} \tag{74}$$

Finally, we define another auxiliary matrix $M_1 \in \mathbb{R}^{p\times n}$ based on P_1 and Q_1 as

$$M_1 = P_1\hat{\Theta} - Q_1, \tag{75}$$

where $\hat{\Theta}$ is the estimation of unknown parameter Θ.

An adaptive law for estimating $\hat{\Theta}$ is provided by

$$\dot{\hat{\Theta}} = -\Gamma_1 M_1, \tag{76}$$

with $\Gamma_1 > 0$ being a constant, positive definite gain matrix.

Similarly to the analysis shown in earlier sections, we can prove that the vector M_1 in Eq. (75) can be reformulated as $M_1 = -P_1\tilde{\Theta} + \psi_1$, where $\tilde{\Theta} = \Theta - \hat{\Theta}$ is the estimation error and $\psi_1(t) = -\int_0^t e^{-\ell(t-r)} f_f(r)\varepsilon_f^T(r)dr$ is a bounded variable—that is, $\|\psi_1\| \le \varepsilon_{1\psi}$ for a constant $\varepsilon_{1\psi} > 0$ because the NN regressor function $\Phi(x)$ and approximation error ε are all bounded for bounded x and u.

Moreover, we can also prove that if the regressor function vector $\Phi(x)$ defined in Eq. (71) is persistently excited, then the matrix P_1 in Eq. (74) is positive definite—that is, its minimum eigenvalue $\lambda_{\min}(P_1) > \sigma_1 > 0$ for positive constant σ_1.

Consequently, the following theorem holds:

Theorem 6. *If x and u in Eq. (71) are bounded and the regressor $\Phi(x)$ of system (71) is persistently excited, then for the adaptive law (76), we can show:*

1. *The estimation error $\tilde{\Theta}$ will converge to zero exponentially when there is no disturbance ($\varepsilon = 0$).*
2. *The estimation error $\tilde{\Theta}$ will converge to a residual set around zero when there is a bounded disturbance ($\varepsilon \neq 0$) only.*

The proof of Theorem 6 can be obtained by selecting a Lyapunov function candidate as $V_1 = \frac{1}{2}tr(\tilde{\Theta}^T\Gamma_1^{-1}\tilde{\Theta})$. Further details can be found in [24].

Remark 4. For the adaptive law (76), the variable M_1 in Eq. (75) represents the estimation error $P_1\tilde{\Theta}$, and the error ψ_1 can vanish for $\varepsilon \to 0$. Moreover, it is known that the NN approximation error $\varepsilon \to 0$ holds for a sufficient number of NN nodes in the identifier (71)—that is, $p \to +\infty$. In this sense the estimated parameter $\hat{\Theta}$ can be obtained directly without the use of an observer/predictor error in comparison with [61] (see Theorem 1 in [61]).

6.3 Tracking Control Design

We will now present the tracking control design. On the basis of the identifier shown earlier, system (71) can be represented as

$$\dot{x} = Ax + Bu + \hat{\Theta}^T\Phi(x) + \varepsilon + \varepsilon_N, \tag{77}$$

where $\varepsilon_N = \tilde{\Theta}\Phi(x)$ can be taken as an adaptation error, for which a later analysis will show the boundedness (ie, $\|\varepsilon_N\| \leq \phi_N$ for a constant $\phi_N > 0$ in the compact set Ω). Thus with the aim of solving the optimal control of Eq. (71), we consider the optimal control of Eq. (77) instead.

Consider the tracking error as $e = x - x_d$, so

$$\dot{e} = Ax + Bu + \hat{\Theta}^T\Phi(x) - \dot{x}_d + \varepsilon + \varepsilon_N. \tag{78}$$

To design an optimal tracking control, the overall control action u consists of two elements: $u = u_s + u_e$, where the first part u_s is adopted to retain the *steady-state* tracking error close to zero, and the second control u_e is used to stabilize the tracking error during the transient stage to minimize an optimal performance index.

Since the steady-state control u_s is designed to guarantee the steady-state tracking error is zero, it should be designed to retain for the steady-state dynamics the relationship $\dot{e} = \dot{x} - \dot{x}_d = 0$ in Eq. (78)—that is, $u = u_s$ needs to guarantee $x = x_d$ for $\varepsilon + \varepsilon_N = 0$. Thus the steady-state control signal u_s can be selected as

$$u_s = B^{\otimes}(\dot{x}_d - Ax_d - \hat{\Theta}^T \Phi(x_d)), \tag{79}$$

where $B^{\otimes}$ denotes the generalized inverse of B. The input gain B in Eq. (71) is assumed to be known but may not be invertible (ie, it may have rank lower than n).

Substituting Eq. (79) into Eq. (78), we can rewrite the tracking error dynamics as

$$\dot{e} = Ae + \hat{\Theta}^T [\Phi(x) - \Phi(x_d)] + Bu_e + \varepsilon + \varepsilon_N + \varepsilon_\varphi, \tag{80}$$

where $\varepsilon_\varphi = (BB^{\otimes} - I)(\dot{x}_d - Ax_d - \hat{\Theta}^T \Phi(x_d))$ denotes the residual error due to the generalized inverse of B, which is clearly bounded because x_d and $\dot{x}_d$ are bounded and ϕ is Lipschitz continuous—that is, $\|\varepsilon_\varphi\| \leq \phi_p$ for a constant $\phi_p > 0$. It is noted that ε_φ will be null under the so-called matching condition—for example, $(BB^{\otimes} - I)A = 0$ or $(BB^{\otimes} - I)\hat{\Theta}^T = 0$.

As shown above, the tracking control of original system (77) is reduced to a regulation problem of Eq. (80). Hence an optimal control u_e will be developed to stabilize the tracking error dynamics (80). In this case the optimal value function (72) for system (71) can be reformulated with use of u_e from system (80) to provide the value function

$$V(e(t)) = \int_t^{\infty} r(e(\tau), u_e(e(\tau)))d\tau, \tag{81}$$

where the utility function is chosen as $r(e(\tau), u_e(e(\tau))) = e^T Qe + u_e^T Ru_e$, and the matrices $Q \in \mathbb{R}^{n\times n}$ and $R \in \mathbb{R}^{m\times m}$ are symmetric and positive definite.

The remaining problem is to design an admissible control $u_e(e)$ which can minimize the cost (81) for system (80). Thus we further compute the Hamiltonian of system (80) as

$$\begin{aligned} H(e, u_e, V) = {} & V_e^T[Ae + \hat{\Theta}^T(\Phi(x) - \Phi(x_d)) + Bu_e + \varepsilon + \varepsilon_N + \varepsilon_\varphi] \\ & + e^T Qe + u_e^T Ru_e, \end{aligned} \tag{82}$$

where $V_e \triangleq \frac{\partial V}{\partial e}$ is the partial derivative of V with respect to e. Hence the optimal cost function $V^*(e)$ for Eq. (81) is

$$V^*(e) = \min_{u_e \in \Psi(\Omega)} \left(\int_t^{\infty} r(e(\tau), u_e(e(\tau)))d\tau, \right) \tag{83}$$

which satisfies the HJB equation

$$0 = \min_{u_e \in \Psi(\Omega)} \left[H(e, u_e^*, V^*) \right]. \tag{84}$$

Then we can obtain the optimal control u_e^* by solving $\partial H(e, u_e^*, V^*)/\partial u_e^* = 0$ as

$$u_e^* = -\frac{1}{2} R^{-1} B^T \frac{\partial V^*(e)}{\partial e}, \tag{85}$$

where V^* denotes the solution of the HJB equation (84). Thus, from the HJB equation (84), the optimal value function $V^*(e)$ is derived, which can be substituted into Eq. (85) to calculate the optimal control u_e^*. However, solving the HJB equation (84) is not trivial because it is a nonlinear partial differential equation. In the ADP literature the optimal control for Eq. (80) is addressed in terms of a critic-actor framework, where two NNs (critic NN and actor NN) are constructed to estimate the optimal value function and control policy, respectively. In contrast to those methods, a new online ADP algorithm with a dual approximation structure (ie, identifier and critic NN) is presented for the HJB equation (84), where the actor NN used in other ADP methods is not needed.

It is assumed that the optimal value function is continuous and defined on compact sets. Thus a single-layer NN can be used to estimate $V^*(e)$ and its derivative $\partial V^*(e)/\partial e$ with respect to e as

$$V^*(e) = W^{*T} \Phi(e) + \varepsilon_1 \tag{86}$$

and

$$\frac{\partial V^*(e)}{\partial e} = \nabla \Phi^T W^* + \nabla \varepsilon_1, \tag{87}$$

where $W^* \in \mathbb{R}^l$ are the unknown NN weights, $\Phi(e) = [\Phi_1, \ldots, \Phi_l]^T \in \mathbb{R}^l$ is the NN regressor function vector, ε_1 denotes the approximation error, l is the number of neurons, and $\nabla \Phi \overset{\Delta}{=} \partial \Phi / \partial e$ and $\nabla \varepsilon_1 \overset{\Delta}{=} \partial \varepsilon_1 / \partial e$ denote the partial derivatives of $\Phi(e)$ and ε_1 with respect to e, respectively.

Assumption 1 ([61, 62]). The unknown NN weights W^* are bounded—that is, $\|W^*\| \leq W_N$. The NN regressor vector $\Phi(\cdot)$ and the derivative $\nabla \Phi(\cdot)$ with respect to the argument e are bounded by $\|\nabla \Phi\| \leq \phi_M$. Moreover, the function approximation error ε_1 and its derivative $\nabla \varepsilon_1$ with respect to e are bounded by $\|\nabla \varepsilon_1\| \leq \phi_\varepsilon$.

In practical applications we can select the NN regressor functions $\{\Phi_i(e) : i = 1, \ldots, l\}$ such that $\Phi(e)$ gives a complete independent basis for $V^*(e)$ as long as $l \to +\infty$. Then according to Assumption 1 and the Weierstrass approximation theorem, we know that $V^*(e)$ and $\partial V^*(e)/\partial e$ can be approximated by NNs uniformly by Eqs. (86) and (87)—that is, as $l \to +\infty$, the approximation errors ε_1, $\nabla\varepsilon_1 \to 0$ as shown in [62, 63]. Then the critic NN $\hat{V}(e)$ that approximates the optimal value function $V^*(e)$ is given by

$$\hat{V}(e) = \hat{W}^T \Phi(e), \tag{88}$$

where $\hat{W}$ denotes the estimation of W^* in critic NN (86).

In this case, one may obtain the approximated optimal control as

$$\hat{u}_e = -\frac{1}{2} R^{-1} B^T \frac{\partial \hat{V}(e)}{\partial e} = -\frac{1}{2} R^{-1} B^T \nabla \Phi^T \hat{W} \tag{89}$$

such that the overall optimal tracking control for system (77) can be given as

$$u = u_s + \hat{u}_e, \tag{90}$$

with u_s being the steady-state control given in Eq. (79).

The ideal optimal control u_e^* can be determined from Eqs. (85) and (87) as

$$u_e^* = -\frac{1}{2} R^{-1} B^T \frac{\partial V^*(e)}{\partial e} = -\frac{1}{2} R^{-1} B^T (\nabla \Phi^T W^* + \nabla \varepsilon_1). \tag{91}$$

The estimated optimal control $\hat{u}_e$ of Eq. (89) substituted into the error dynamics (80) implies together with Eq. (91)

$$\begin{aligned}
\dot{e} &= Ae + \hat{\Theta}^T [\Phi(x) - \Phi(x_d)] + Bu_e + \varepsilon + \varepsilon_N + \varepsilon_\varphi \\
&= Ae + \hat{\Theta}^T [\Phi(x) - \Phi(x_d)] + B\Big[-\frac{1}{2} R^{-1} B^T \nabla \Phi^T \hat{W} + \frac{1}{2} R^{-1} B^T \\
&\quad (\nabla \Phi^T W^* + \nabla \varepsilon_1)\Big] + Bu_e^* + \varepsilon + \varepsilon_N + \varepsilon_\varphi \\
&= Ae + \hat{\Theta}^T [\Phi(x) - \Phi(x_d)] + \frac{1}{2} BR^{-1} B^T \nabla \Phi^T \tilde{W} + Bu_e^* \\
&\quad + \frac{1}{2} BR^{-1} B^T \nabla \varepsilon_1 + \varepsilon + \varepsilon_N + \varepsilon_\varphi.
\end{aligned} \tag{92}$$

The problem now is to estimate online the critic NN weights $\hat{W}$ such that $\hat{W}$ converges to a small bounded set around their ideal values W^*. To derive an adaptive law, we denote $\Phi_d = \Phi(x_d)$, substitute Eq. (87) into

Hamiltonian function (82), and thus rewrite the HJB equation (84) as

$$0 = H(e, u_e, V_e^*) = W^{*T}\nabla\Phi[Ae + \hat{\Theta}^T(\Phi - \Phi_d) + Bu_e] + e^T Qe + u_e^T Ru_e + \varepsilon_{\mathrm{HJB}}, \tag{93}$$

where $\varepsilon_{\mathrm{HJB}} = \nabla\varepsilon_1[Ae + \hat{\Theta}^T(\Phi - \Phi_d) + Bu_e + \varepsilon + \varepsilon_N + \varepsilon_\varphi] + W^{*T}\nabla \Phi(\varepsilon_N + \varepsilon_\varphi + \varepsilon)$ defines the effect of the residual NN approximation errors. Thus it can be made arbitrarily small by use of a sufficiently large number of NN nodes [62, 63]—that is, $\varepsilon \to 0$ for $p \to +\infty$ and $\nabla\varepsilon_1 \to 0$ for $l \to +\infty$. Equally, Theorem 6 implies that the estimation error, $\varepsilon_N = \tilde{\Theta}^T\Phi(x)$, converges to zero for $p \to +\infty$ and bounded control and states. Moreover, $\varepsilon_\varphi = 0$ when B is of rank n.

To design the adaptive law, one can define the known terms in Eq. (93) as $\Xi = \nabla\Phi\left[Ae + \hat{\Theta}^T(\Phi - \Phi_d) + Bu_e\right]$ and $\chi = e^T Qe + u_e^T Ru_e$, and then represent Eq. (93) as

$$\chi = -W^{*T}\Xi - \varepsilon_{\mathrm{HJB}}. \tag{94}$$

In Eq. (94) the critic NN weights W^* are in a linearly parameterized form. Thus they can be estimated "*directly*" by use of the parameter estimation error method proposed in Section 3.

Remark 5. It is shown in Eq. (93) that the residual error $\varepsilon_{\mathrm{HJB}}$ is created by the critic NN error $\nabla\varepsilon_1$ in Eq. (87), the identifier error $\varepsilon + \varepsilon_N$ in Eq. (77), and the matching error ε_φ. As claimed in [62, 63], the critic NN approximation error $\nabla\varepsilon_1$ converges to zero if the number of hidden-layer nodes increases—that is, $\nabla\varepsilon_1 \to 0$ as long as $l \to +\infty$. That is, for all $\mu > 0$, there exists $N(\mu) : \sup\|\nabla\varepsilon_1\| \le \mu$. Moreover, if there is no NN approximation error in (71)—that is, $\varepsilon = 0$—the effect of the identifier error ε_N in Eq. (77) will vanish (ie, $\varepsilon_N \to 0$ for $p \to +\infty$) because $\tilde{\Theta} \to 0$ holds for $\varepsilon = 0$ as proved in Theorem 6 for bounded state x and control input u. Finally, $\varepsilon_\varphi = 0$ is true under the matching condition in Eq. (80). Consequently, if there are no approximation errors ε in the identifier (77) and $\nabla\varepsilon_1$ in the critic NN (87) and the matching condition holds, the residual error in Eq. (94) is null (ie, $\varepsilon_{\mathrm{HJB}} = 0$).

To design the adaptive law, the auxiliary regressor matrix $P_2 \in \mathbb{R}^{l\times l}$ and vector $Q_2 \in \mathbb{R}^l$ are defined as

$$\begin{cases} \dot{P}_2 = -\ell P_2 + \Xi\Xi^T, & P_2(0) = 0, \\ \dot{Q}_2 = -\ell Q_2 + \Xi\chi, & Q_2(0) = 0, \end{cases} \tag{95}$$

and the auxiliary vector $M_2 \in \mathbb{R}^l$ can be calculated with P_2 and Q_2 as

$$M_2 = P_2\hat{W} + Q_2. \tag{96}$$

Then we can obtain $Q_2 = -P_2W^* + \psi_2$, with $\psi_2 = -\int_0^t e^{-\ell(t-r)}\varepsilon_{HJB}(r)\Xi(r)dr$ being a bounded variable for bounded state x and control u (eg, $\|\psi_2\| \le \varepsilon_{2\psi}$ for some $\varepsilon_{2\psi} > 0$). In this case, Eq. (96) can be rewritten as

$$M_2 = P_2\hat{W} + Q_2 = -P_2\tilde{W} + \psi_2, \tag{97}$$

where $\tilde{W} = W^* - \hat{W}$ is the NN weight error.

Then an adaptive law to update $\hat{W}$ is given by

$$\dot{\hat{W}} = -\Gamma_2 M_2, \tag{98}$$

with $\Gamma_2 > 0$ being a constant matrix.

Similarly, one may verify that the matrix P_2 is positive definite—that is, $\lambda_{\min}(P_2) > \sigma_2 > 0$—when the regressor function vector Ξ defined in Eq. (94) is persistently excited.

Then we have the following theorem:

Theorem 7. *Consider the adaptive law* (98) *used for the critic NN. Under the assumption that the regressor* Ξ *satisfies* $\lambda_{\min}(P_2) > \sigma_2 > 0$ *and the state* x *and control* u *are bounded:*

1. *The critic NN error* $\tilde{W}$ *exponentially converges to zero when there are no approximation errors,* $\varepsilon_{HJB} = 0$.
2. *The citric NN error* $\tilde{W}$ *will converge to a small bounded set around zero for bounded NN-approximation errors,* $\varepsilon_{HJB} \neq 0$.

On can obtain the proof by deriving the derivative of the Lyapunov function $V_2 = \frac{1}{2}\tilde{W}^T\Gamma_2^{-1}\tilde{W}$.

The main results concerning the closed-loop system stability and the convergence of the suggested adaptive control to its optimal value are summarized as follows:

Theorem 8. *System* (71) *with the control law of Eqs. (79) and* (89)*, and adaptive laws* (76) *and* (98) *are considered. If the regressors* Φ *and* Ξ *are persistently exciting, then the following semiglobal results hold:*

1. *Once there are no approximation errors, the tracking error* e *and the estimation errors* $\tilde{\Theta}$ *and* $\tilde{W}$ *converge to zero. Moreover, the adaptive optimal control* $\hat{u}_e$ *in Eq. (89) converges to its optimal solution* u_e^* *given in Eq. (85)—that is,* $\hat{u}_e \to u_e^*$ *if* $\nabla\varepsilon_1 = 0$.
2. *Once there are bounded approximation errors, the tracking error* e *and the estimation errors* $\tilde{\Theta}$ *and* $\tilde{W}$ *are uniformly ultimately bounded. Moreover, the*

adaptive optimal control $\hat{u}_e$ in Eq. (89) converges to a small bounded set around the optimal solution u_e^* given in Eq. (85). Thus $\left\|\hat{u}_e - u_e^*\right\| \le \varepsilon_u$ for a positive constant ε_u.

The proof uses the following Lyapunov functions:

$$V = \frac{1}{2}\text{tr}\,(\tilde{\Theta}^T \Gamma_1^{-1}\tilde{\Theta}) + \frac{1}{2}\tilde{W}^T \Gamma_2^{-1}\tilde{W} + \Gamma e^T e + KV^*(e) + \Sigma \psi_2^T \psi_2, \tag{99}$$

where the optimal value function $V^*(e)$ is given in Eq. (81), and $K > 0$, $\Gamma > 0$. and $\Sigma > 0$ are positive constants. For the case when there are bounded NN errors, it can be shown that the estimation errors $\tilde{\Theta}$, $\tilde{W}$, ψ_2, and e are uniformly ultimately bounded, and

$$\left\|\hat{u}_e - u_e^*\right\| \le \frac{1}{2}||R^{-1}B^T \nabla \Phi^T|| ||\tilde{W}|| + \frac{1}{2}||R^{-1}B^T|| \, \|\nabla\varepsilon_1\| \le \varepsilon_u, \tag{100}$$

where the bound ε_u depends on the critic NN errors $\tilde{W}$ and $\nabla\varepsilon_1$. If the critic NN approximation error is $\nabla\varepsilon_1 = 0$, we have

$$\lim_{t\to+\infty} \left\|\hat{u}_e - u_e^*\right\| \le \frac{1}{2}\phi_M ||R^{-1}B^T|| \lim_{t\to+\infty} ||\tilde{W}|| = 0. \tag{101}$$

Further details about the proof are available in [24].

6.4 Simulation Results

A numerical example is finally used to validate the effectiveness of the suggested approach. We consider a nonlinear system as

$$\begin{cases} \dot{x}_1 = -x_1 + x_2, \\ \dot{x}_2 = -0.5x_1 - 0.5x_2(1 - (\cos(2x_1) + 2)^2) + \cos(2x_1) + 2u. \end{cases} \tag{102}$$

Then the weight matrices Q and R of the cost function (72) are identity matrices. The objective is to find the control such that the system states x track a given trajectory $x_{1d} = \sin t$ and $x_{2d} = \cos t + \sin t$. Thus, we first adopt the identifier (71) to reconstruct the system dynamics, with $A = \begin{bmatrix} -1 & 1 \\ -0.5 & 0 \end{bmatrix}$ and $B = \begin{bmatrix} 0 \\ 2 \end{bmatrix}$ being known matrices and $\Theta = \begin{bmatrix} 0 & 0 & 0 \\ -0.5 & 0.5 & 1 \end{bmatrix}^T$ being the unknown identifier weight matrix which is to be estimated. The activation function is set as $\Phi(x) = \left[x_2, x_2(\cos(2x_1)+2)^2, \cos(2x_1)\right]^T$. The parameters used in the simulation are chosen as $k = 0.001$, $\ell = 1$, and $\Gamma_1 = 350$, and the initial estimated parameters are $\hat{\Theta}(0)=0$. Fig. 12 provides the response of the estimated identifier weights $\hat{\Theta}$ with adaptive law (76),

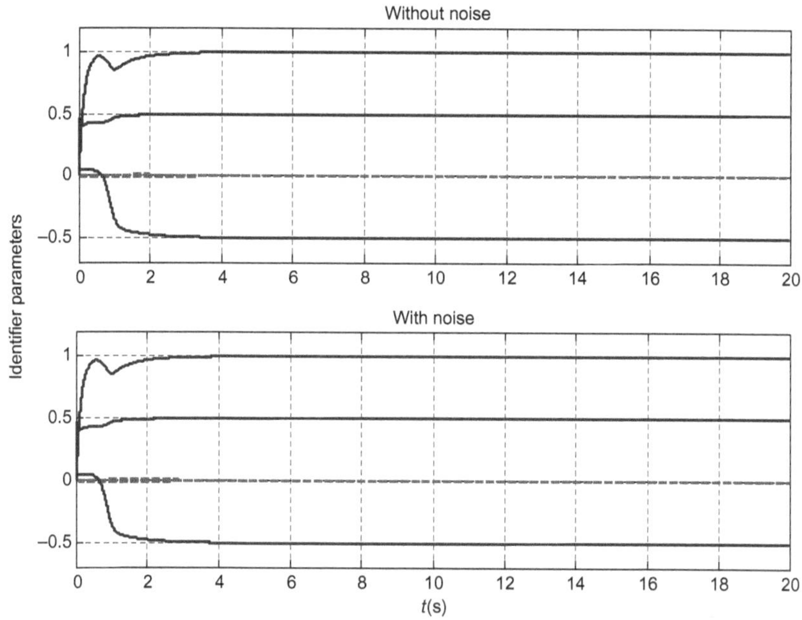

Fig. 12 Profiles of the identifier parameters $\hat{\Theta}$. *(Source: J. Na, G. Herrmann, Online adaptive approximate optimal tracking control with simplified dual approximation structure for continuous-time unknown nonlinearsystems, IEEE/CAA J. Autom. Sin. 1 (4) (2014) 412–422).*

where one sees that the weight estimates converge to their true values Θ after a 3.5-s transient. Moreover, the adaptive steady-state control (79) can be constructed for system (102) to maintain the steady-state performance. For the optimal control design the value function and the associated optimal control for system (102) are

$$V^*(e) = \frac{1}{2}e_1^2 + e_2^2 \quad \text{and} \quad u_e^* = -\frac{1}{2}R^{-1}B^T\frac{\partial V^*(e)}{\partial e} = -2e_2. \tag{103}$$

We select the regressor of the critic NN as $\Phi(e) = [e_1^2, e_1e_2, e_2^2]^T$. Thus the ideal critic NN weights are given by $W^* = [0.5, 0, 1]^T$. The time trace for the online-estimated critic NN weights $\hat{W}$ is shown in Fig. 13; this indicates that $\hat{W}$ converges in approximately 1.5 s. This implies that the proposed adaptive optimal control (89) converges to its optimal control action given in Eq. (103). An error in the weights is to be expected as $\varepsilon_\varphi \neq 0$. The novel identifier and critic NN weight update laws (76) and (98) achieve faster convergence of weights in comparison with results in [63] (see Fig. 4 in [63]). The controlled system states are given in Fig. 14, and converge to the given external command.

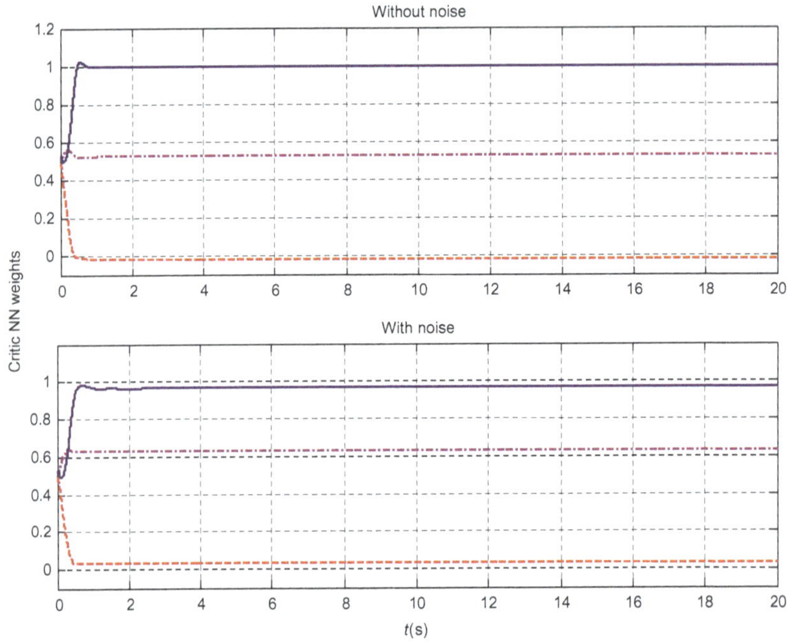

Fig. 13 Convergence of the critic neural network weights $\hat{W}$. *(Source: J. Na, G. Herrmann, Online adaptive approximate optimal tracking control with simplified dual approximation structure for continuous-time unknown nonlinear systems,IEEE/CAA J. Autom. Sin. 1 (4) (2014) 412–422).*

7 CONCLUSIONS AND FURTHER READING

We have provided a novel continuous-time adaptive parameter estimation algorithm which is robust to disturbances. Despite that strong feature of disturbance robustness, the algorithm guarantees also the estimation error convergence to the true values in the absence of disturbances. This has been shown in theoretical and also practical analysis. Basic robustness and stability features also show that the adaptive algorithm can act independently of a specific control or observer algorithm, although it is easily incorporated into an adaptive control or observer. Hence the novel algorithm can replace traditional adaptive estimation algorithms in continuous time. The novel algorithm has been proven to work as a parameter estimation algorithm in connection with an observer in a vehicular context and together with a demand tracking controller for a robotic arm. In both cases, the novel algorithm has provided superior results in terms of parameter estimation and control performance in comparison with the traditional LS method

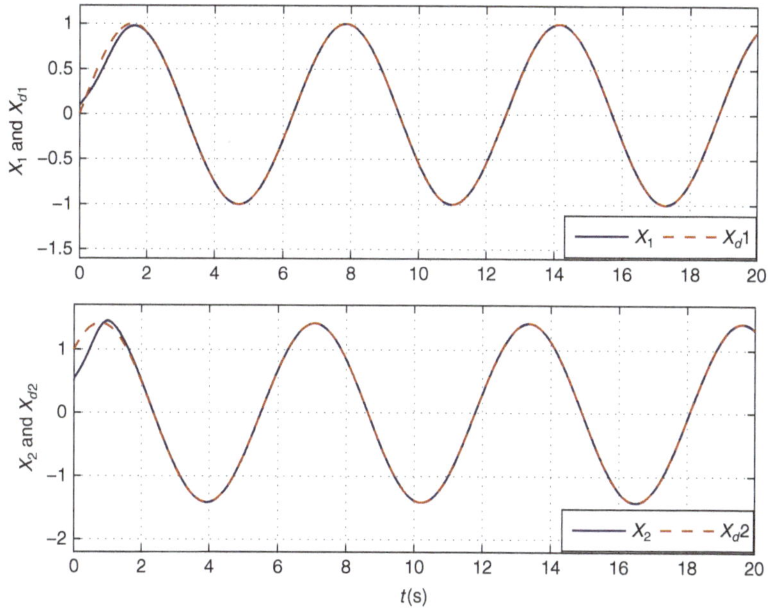

Fig. 14 Evaluation of tracking performance.

or the gradient descent algorithm. Moreover, it has been shown that the algorithm lends itself easily to adaptive optimal control. In fact, it has been shown that, in contrast to recent work, the true optimal controller can be obtained in the limit rather than in an ultimate bounded sense. The number of NNs needed for this adaptive optimal controller has been reduced to two, rather than three as reported in recent literature.

As mentioned earlier, we have summarized our recent work. In particular, we refer to [21–23], while other significant work has been published in [24–33]. The original ideas were presented in [21, 22], and then further improved in [23] to achieve finite-time parameter estimation and control simultaneously. Further details can be found in Section 1.

ACKNOWLEDGMENTS

The authors are highly grateful for years of inspiration, encouragement, and support from F.L. Lewis. This work was partially supported by a project (grant no. JP090823/61011130163) jointly funded by the Royal Society, United Kingdom, and the National Natural Science Foundation of China. Jing Na was supported by the National Natural Science Foundation of China (grant nos. 61203066 and 61573174) and the Marie Curie Intra-European Fellowships AECE Project (grant no. FP7-PEOPLE-2013-IEF-625531). Muhammad Nasiruddin Mahyuddin was supported by Universiti Sains Malaysia and the Ministry of Higher Education of Malaysia.

REFERENCES

[1] J. Slotine, W. Li, Applied Nonlinear Control, Prentice Hall, Englewood Cliffs, NJ, 1990.

[2] M. Krstić, I. Kanellakopoulos, P. Kokotović, Nonlinear and adaptive control design, in: Adaptive and Learning Systems for Signal Processing, Communications, and Control, Wiley, New York, 1995.

[3] S. Sastry, M. Bodson, Adaptive Control: Stability, Convergence and Robustness, Dover Publications, New York, 1989.

[4] K. Åström, B. Wittenmark, Adaptive Control, second ed., Addison-Wesley Longman Publishing Co., Inc., Boston, MA, USA, 1994.

[5] P. Ioannou, J. Sun, Robust Adaptive Control, first ed., Prentice Hall, Englewood Cliffs, NJ, 1996.

[6] K. Narendra, Y. Lin, L. Valavani, Stable adaptive controller design, Part II: Proof of stability, IEEE Trans. Autom. Control 25 (3) (1980) 440–448.

[7] A. Morse, Global stability of parameter-adaptive control systems, IEEE Trans. Autom. Control 25 (3) (1980) 433–439.

[8] Y. Landau, Adaptive Control: The Model Reference Approach, in: Control and Systems Theory Series, Dekker, New York, 1979.

[9] B. Egardt, Stability of adaptive controllers, in: Lecture Notes in Control and Information Sciences, Springer, Berlin, 1979.

[10] C. Rohrs, L. Valavani, M. Athans, G. Stein, Robustness of adaptive control algorithms in the presence of unmodeled dynamics, in: 21st IEEE Conference on Decision and Control, 1982, vol. 21, 1982, pp. 3–11.

[11] F. Lewis, S. Jagannathan, A. Yesildrak, Neural Network Control of Robots and Nonlinear Systems, Taylor and Francis, London, 1999.

[12] J. Craig, Adaptive Control of Mechanical Manipulators, Addison-Wesley, Reading, MA, 1988.

[13] W. Li, J. Slotine, An indirect adaptive robot controller, Syst. Control Lett. 12 (1989) 259–266.

[14] F. Lewis, D. Dawson, C. Abdallah, Robot Manipulator Control: Theory and Practice, second ed., Marcel Dekker, New York, 2004.

[15] J. Lin, I. Kanellakopoulos, Nonlinearities enhance parameter convergence in strict feedback systems, IEEE Trans. Autom. Control 44 (1) (1999) 89–94.

[16] Y. Zhang, P. Ioannou, C. Chien, Parameter convergence of a new class of adaptive controllers, IEEE Trans. Autom. Control 41 (10) (1996) 1489–1493.

[17] J. Lin, I. Kanellakopoulos, Nonlinearities enhance parameter convergence in output-feedback systems, IEEE Trans. Autom. Control 43 (2) (1998) 204–222.

[18] G. Chowdhary, T. Yucelen, M. Mühlegg, E.N. Johnson, Concurrent learning adaptive control of linear systems with exponentially convergent bounds, Int. J. Adapt. Control Signal Process. 27 (4) (2013) 280–301.

[19] V. Adetola, M. Guay, Finite-Time Parameter Estimation in Adaptive Control of Nonlinear Systems, American Control Conference, 2007. ACC '07, New York, 2007, pp. 1920–1924.

[20] V. Adetola, M. Guay, Finite-time parameter estimation in adaptive control of nonlinear systems, IEEE Trans. Autom. Control 53 (3) (2008) 807–811.

[21] J. Na, G. Herrmann, X. Ren, M. Mahyuddin, P. Barber, Robust adaptive finite-time parameter estimation and control of nonlinear systems, in: IEEE Multi-Conference on System and Control 2011, MSC2011, pp. 1014–1019.

[22] J. Na, M.N. Mahyuddin, G. Herrmann, X. Ren, Robust adaptive finite-time parameter estimation for linearly parameterized nonlinear systems, in: 32nd Chinese Control Conference (CCC), IEEE, 2013, pp. 1735–1741.

[23] J. Na, M.N. Mahyuddin, G. Herrmann, X. Ren, P. Barber, Robust adaptive finite-time parameter estimation and control for robotic systems, Int. J. Robust Nonlinear Control 25 (16) (2015) 3045–3071.
[24] J. Na, G. Herrmann, Online adaptive approximate optimal tracking control with simplified dual approximation structure for continuous-time unknown nonlinear systems, IEEE/CAA J. Autom. Sin. 1 (4) (2014) 412–422.
[25] Y. Lv, J. Na, Q. Yang, X. Wu, Y. Guo, Online adaptive optimal control for continuous-time nonlinear systems with completely unknown dynamics, Int. J. Control 89 (1) (2016) 99–112.
[26] M. Mahyuddin, J. Na, G. Herrmann, X. Ren, P. Barber, Adaptive observer-based parameter estimation with application to road gradient and vehicle mass estimation, IEEE Trans. Ind. Electron. 61 (6) (2014) 2851–2863.
[27] M. Mahyuddin, J. Na, G. Herrmann, X. Ren, P. Barber, An adaptive observer-based parameter estimation algorithm with application to road gradient and vehicle's mass estimation, in: 2012 UKACC International Conference on Control (CONTROL), 2012, pp. 102–107.
[28] M. Mahyuddin, G. Herrmann, J. Na, F. Lewis, Finite-time adaptive distributed control for double integrator leader-agent synchronisation, in: 2012 IEEE International Symposium on Intelligent Control (ISIC), 2012, pp. 714–720.
[29] M. Mahyuddin, G. Herrmann, F. Lewis, Distributed adaptive leader-following control for multi-agent multi-degree manipulators with finite-time guarantees, in: 2013 IEEE 52nd Annual Conference on Decision and Control (CDC), 2013, pp. 1496–1501.
[30] M. Mahyuddin, G. Herrmann, Cooperative Robot Manipulator Control with human 'pinning' for robot assistive task execution, in: Lecture Notes in Computer Science, vol. 8239, Springer International Publishing, Berlin, 2013 p. 521.
[31] J. Yang, J. Na, Y. Guo, X. Wu, Adaptive estimation of road gradient and vehicle parameters for vehicular systems, IET Control Theory Appl. 9 (6) (2015) 935–943.
[32] M. Mahyuddin, G. Herrmann, S. Khan, A novel adaptive control algorithm in application to a humanoid robot arm, in: G. Herrmann, M. Studley, M. Pearson, A. Conn, C. Melhuish, M. Witkowski, J.-H. Kim, P. Vadakkepat (Eds.), Advances in Autonomous Robotics, Lecture Notes in Computer Science, vol. 7429, Springer, Berlin/Heidelberg, 2012, pp. 25–36.
[33] M. Mahyuddin, S. Khan, G. Herrmann, A novel robust adaptive control algorithm with finite-time online parameter estimation of a humanoid robot arm, Robot. Auton. Syst. 62 (3) (2014) 294–305.
[34] J. Na, J. Yang, X. Ren, Y. Guo, Robust adaptive estimation of nonlinear system with time-varying parameters, Int. J. Adapt. Control Signal Process. 29 (8) (2015) 1055–1072.
[35] G. Goodwin, K. Sin, Adaptive Filtering Prediction and Control, Dover Publications, New York, 2009.
[36] L. Ljung, System Identification: Theory for the User, Pearson Education, Upper Saddle River, NJ, 1998.
[37] H. Bae, J. Ryu, J. Gerdes, Road grade and vehicle parameter estimation for longitudinal control using GPS, in: IEEE Conference on Intelligent Transportation Systems, Proceedings, ITSC, 2001, pp. 166–171.
[38] H. Jansson, E. Kozica, P. Sahlholm, K.-H. Johansson, Improved road grade estimation using sensor fusion, in: Proceedings of the 12th Reglermote in Stockholm, Sweden, 2006.
[39] P. Sahlholm, K. Johansson, Segmented road grade estimation for fuel efficient heavy duty vehicles, in: IEEE Conference on Decision and Control, 2010, pp. 1045–1050.
[40] M. McIntyre, T. Ghotikar, A. Vahidi, X. Song, D. Dawson, A two-stage Lyapunov-based estimator for estimation of vehicle mass and road grade, IEEE Trans. Veh. Technol. 7 (7) (2009) 3177–3185.

[41] S. Mangan, J. Wang, Development of a novel sensorless longitudinal road gradient estimation method based on vehicle CAN bus data, IEEE/ASME Trans. Mechatron. 12 (3) (2007) 375–386.

[42] V. Winstead, I. Kolmanovsky, Estimation of road grade and vehicle mass via model predictive control, in: Proceedings of 2005 IEEE Conference on Control Applications, 2005. CCA2005, 2005, pp. 1588–1593.

[43] A. Vahidi, M. Druzhinina, A. Stefanopoulou, H. Peng, Simultaneous mass and time-varying grade estimation for heavy-duty vehicles, in: Proceedings of the 2003 American Control Conference, vol. 6, IEEE, 2003, pp. 4951–4956.

[44] R. Frampton, P. Thomas, Effectiveness of Electronic Stability Control Systems in Great Britain, Technical Report, Department for Transport, Vehicle Safety Research Centre (VSRC), Loughborough, 2007, Report prepared for the Department for Transport.

[45] P. Sahlholm, K. Johansson, Road grade estimation for look-ahead vehicle control using multiple measurement runs, Control Eng. Pract. 18 (11) (2010) 1328–1341.

[46] P. Lingman, B. Schmidtbauer, Road slope and vehicle mass estimation using Kalman filtering, Veh. Syst. Dyn. 37 (2002) 12–23.

[47] J. Park, J. Kong, H. Jo, Y. Park, J. Lee, Measurement of road gradients for the development of driving modes including road gradients, Proc. Inst. Mech. Eng. Automob., vol. 215, 2001, pp. 997–986.

[48] S. Mangan, J. Wang, Q. Wu, Measurement of the road gradient using an inclinometer mounted on a moving vehicle, in: IEEE International Symposium on Computer Aided Control System and Design, 2002.

[49] S. Bittanti, P. Bolzern, M. Campi, Recursive least-square identification algorithms with incomplete excitation: convergence analysis and application to adaptive control, IEEE Trans. Autom. Control 35 (12) (1990) 1371–1373.

[50] C. Edwards, S. Spurgeon, Sliding Mode Control: Theory and Applications, CRC, Boca Raton, 1998.

[51] J. Foreman, M. Yazdouni, S. Russo, G. Chan, A. Spiers, G. Herrmann, P. Barber, Hardware in the loop validation of a gradient and weight estimation algorithm and longitudinal speed control using a laboratory model car, in: International Conference on Systems Engineering (ICSE), 2009.

[52] J. Slotine, W. Li, Composite adaptive control of robot manipulators, Automatica 25 (1989) 509–519.

[53] J. Craig, P. Hsu, S. Sastry, Adaptive control of mechanical manipulators, in: IEEE Int. Conf. on Robotics and Automation, San Francisco, California, 1986.

[54] J. Slotine, W. Li, Adaptive robot control. A case study, in: IEEE Int. Conf. on Robotics and Automation, Raleigh, North Carolina, 1987.

[55] M. Spong, M. Vidyasagar, Robot Dynamics and Control, Wiley, New York, 1989.

[56] F.L. Lewis, V.L. Syrmos, Optimal Control, John Wiley & Sons, New York, 1995.

[57] R.S. Sutton, A.G. Barto, Reinforcement Learning: An Introduction, vol. 1, MIT Press, Cambridge, 1998.

[58] A. Al-Tamimi, F.L. Lewis, M. Abu-Khalaf, Discrete-time nonlinear HJB solution using approximate dynamic programming: convergence proof, IEEE Trans. Syst. Man Cybern. B: Cybern. 38 (4) (2008) 943–949.

[59] H. Zhang, L. Cui, X. Zhang, Y. Luo, Data-driven robust approximate optimal tracking control for unknown general nonlinear systems using adaptive dynamic programming method, IEEE Trans. Neural Netw. 22 (12) (2011) 2226–2236.

[60] D. Vrabie, O. Pastravanu, M. Abu-Khalaf, F.L. Lewis, Adaptive optimal control for continuous-time linear systems based on policy iteration, Automatica 45 (2) (2009) 477–484.

[61] S. Bhasin, R. Kamalapurkar, M. Johnson, K.G. Vamvoudakis, F.L. Lewis, W.E. Dixon, A novel actor-critic-identifier architecture for approximate optimal control of uncertain nonlinear systems, Automatica 49 (1) (2013) 82–92.

[62] M. Abu-Khalaf, F.L. Lewis, Nearly optimal control laws for nonlinear systems with saturating actuators using a neural network HJB approach, Automatica 41 (5) (2005) 779–791.
[63] K.G. Vamvoudakis, F.L. Lewis, Online actor-critic algorithm to solve the continuous-time infinite horizon optimal control problem, Automatica 46 (5) (2010) 878–888.
[64] M. Mahyuddin, G. Herrmann, Distributed motion synchronisation control of humanoid arms, in: 5th International Conference on Advance Humanoid Robotic Research, FIRA RoboWorld Congress 2013, Springer, Berlin/Heidelberg, 2013.

AUTHOR INDEX

Note: Page numbers followed by *f* indicate figures.

C

D

E

F

G

H

I

J

K

L

M

N

V

W

SUBJECT INDEX

Note: Page numbers followed by *f* indicate figures and *t* indicate tables.

A

B

C

E

O

P

Kolibri 360 GmbH
Ferdinand-Jühlke-Straße 7
99095 Erfurt, Deutschland
produktsicherheit@kolibri360.de